ENGINEERING ECONOMY

McGraw-Hill Series in Industrial Engineering and Management Science

Consulting Editors

Kenneth E. Case, Department of Industrial Engineering and Management, Oklahoma State University

Philip M. Wolfe, Department of Industrial and Management Systems Engineering, Arizona State University

Blank and Tarquin: Engineering Economy
Chapra: Applied Numerical Methods with MATLAB for Engineers and Scientists
Grant and Leavenworth: Statistical Quality Control
Gryna: Quality Planning and Analysis: From Product Development through Use
Harrell, Ghosh, and Bowden: Simulation Using PROMODEL
Hillier and Lieberman: Introduction to Operations Research
Kelton, Sadowski, and Sturrock: Simulation with Arena
Law and Kelton: Simulation Modeling and Analysis
Navidi: Statistics for Engineers and Scientists
Niebel and Freivalds: Methods, Standards, and Work Design

Sixth Edition

ENGINEERING ECONOMY

Leland Blank, P. E.

American University of Sharjah, United Arab Emirates and
Texas A & M University

Anthony Tarquin, P. E.

University of Texas at El Paso

 Higher Education

Boston   Burr Ridge, IL   Dubuque, IA   Madison, WI   New York   San Francisco   St. Louis
Bangkok   Bogotá   Caracas   Kuala Lumpur   Lisbon   London   Madrid   Mexico City
Milan   Montreal   New Delhi   Santiago   Seoul   Singapore   Sydney   Taipei   Toronto

The McGraw·Hill Companies

ENGINEERING ECONOMY, SIXTH EDITION
International Edition 2008

Published by McGraw-Hill, a business unit of The McGraw-Hill Companies, Inc., 1221 Avenue of the Americas, New York, NY 10020. Copyright © 2005, 2002, 1998, 1989, 1983, 1976 by The McGraw-Hill Companies, Inc. All rights reserved. No part of this publication may be reproduced or distributed in any form or by any means, or stored in a database or retrieval system, without the prior written consent of The McGraw-Hill Companies, Inc., including, but not limited to, in any network or other electronic storage or transmission, or broadcast for distance learning.
Some ancillaries, including electronic and print components, may not be available to customers outside the United States.

10 09 08 07 06 05 04 03 02 01
20 09 08
CTF   SLP

When ordering this title, use ISBN: 978-007-127450-0 or MHID: 007-127450-2

Printed in Singapore

www.mhhe.com

Higher Education

This book is dedicated to our mothers for their ever-present encouragement to succeed in all aspects of life.

CONTENTS

Preface *xv*

LEVEL ONE	THIS IS HOW IT ALL STARTS	

Chapter 1 Foundations of Engineering Economy 4

1.1 Why Engineering Economy Is Important to Engineers (and Other Professionals) 6
1.2 Role of Engineering Economy in Decision Making 7
1.3 Performing an Engineering Economy Study 9
1.4 Interest Rate and Rate of Return 12
1.5 Equivalence 15
1.6 Simple and Compound Interest 17
1.7 Terminology and Symbols 23
1.8 Introduction to Solution by Computer 26
1.9 Minimum Attractive Rate of Return 28
1.10 Cash Flows: Their Estimation and Diagramming 30
1.11 Rule of 72: Estimating Doubling Time and Interest Rate 35
1.12 Spreadsheet Application—Simple and Compound Interest, and
Changing Cash Flow Estimates 36
Additional Examples 39
Chapter Summary 41
Problems 42
FE Review Problems 45
Extended Exercise—Effects of Compound Interest 45
Case Study—Describing Alternatives for Producing Refrigerator Shells 46

Chapter 2 Factors: How Time and Interest Affect Money 48

2.1 Single-Payment Factors (*F/P* and *P/F*) 50
2.2 Uniform-Series Present Worth Factor and Capital Recovery
Factor (*P/A* and *A/P*) 56
2.3 Sinking Fund Factor and Uniform-Series Compound Amount
Factor (*A/F* and *F/A*) 60
2.4 Interpolation in Interest Tables 63
2.5 Arithmetic Gradient Factors (*P/G* and *A/G*) 65
2.6 Geometric Gradient Series Factors 71
2.7 Determination of an Unknown Interest Rate 74
2.8 Determination of Unknown Number of Years 77
2.9 Spreadsheet Application—Basic Sensitivity Analysis 78
Additional Example 80
Chapter Summary 81
Problems 81
FE Review Problems 88
Case Study—What a Difference the Years and Compound Interest Can Make 90

Chapter 3   Combining Factors   **92**

3.1   Calculations for Uniform Series That Are Shifted   94
3.2   Calculations Involving Uniform-Series and Randomly Placed Single Amounts   98
3.3   Calculations for Shifted Gradients   103
3.4   Shifted Decreasing Arithmetic Gradients   108
3.5   Spreadsheet Application—Using Different Functions   110
        Additional Example   114
        Chapter Summary   115
        Problems   115
        FE Review Problems   121
        Extended Exercise—Preserving Land for Public Use   123

Chapter 4   Nominal and Effective Interest Rates   **124**

4.1   Nominal and Effective Interest Rate Statements   126
4.2   Effective Annual Interest Rates   130
4.3   Effective Interest Rates for Any Time Period   136
4.4   Equivalence Relations: Comparing Payment Period and Compounding
        Period Lengths (PP versus CP)   138
4.5   Equivalence Relations: Single Amounts with PP $\geq$ CP   139
4.6   Equivalence Relations: Series with PP $\geq$ CP   142
4.7   Equivalence Relations: Single Amounts and Series with PP $<$ CP   147
4.8   Effective Interest Rate for Continuous Compounding   149
4.9   Interest Rates That Vary over Time   151
        Chapter Summary   153
        Problems   154
        FE Review Problems   159
        Case Study—Financing a House   162

LEVEL TWO   TOOLS FOR EVALUATING ALTERNATIVES

Chapter 5   Present Worth Analysis   **168**

5.1   Formulating Mutually Exclusive Alternatives   170
5.2   Present Worth Analysis of Equal-Life Alternatives   172
5.3   Present Worth Analysis of Different-Life Alternatives   174
5.4   Future Worth Analysis   177
5.5   Capitalized Cost Calculation and Analysis   179
5.6   Payback Period Analysis   185
5.7   Life-Cycle Cost   190
5.8   Present Worth of Bonds   194
5.9   Spreadsheet Applications—PW Analysis and Payback Period   197
        Chapter Summary   202
        Problems   202
        FE Review Problems   210
        Extended Exercise—Evaluation of Social Security Retirement Estimates   212
        Case Study—Payback Evaluation of Ultralow-Flush Toilet Program   213

Chapter 6   Annual Worth Analysis   **216**

6.1   Advantages and Uses of Annual Worth Analysis   218
6.2   Calculation of Capital Recovery and AW Values   220
6.3   Evaluating Alternatives by Annual Worth Analysis   223
6.4   AW of a Permanent Investment   228
Chapter Summary   231
Problems   232
FE Review Problems   235
Case Study—The Changing Scene of an Annual Worth Analysis   236

Chapter 7   Rate of Return Analysis: Single Alternative   **238**

7.1   Interpretation of a Rate of Return Value   240
7.2   Rate of Return Calculation Using a PW or AW Equation   242
7.3   Cautions When Using the ROR Method   248
7.4   Multiple Rate of Return Values   249
7.5   Composite Rate of Return: Removing Multiple $i*$ Values   255
7.6   Rate of Return of a Bond Investment   261
Chapter Summary   263
Problems   264
FE Review Problems   270
Extended Exercise 1—The Cost of a Poor Credit Rating   272
Extended Exercise 2—When Is It Best to Sell a Business?   272
Case Study—Bob Learns About Multiple Rates of Return   273

Chapter 8   Rate of Return Analysis: Multiple Alternatives   **276**

8.1   Why Incremental Analysis is Necessary   278
8.2   Calculation of Incremental Cash Flows for ROR Analysis   279
8.3   Interpretation of Rate of Return on the Extra Investment   282
8.4   Rate of Return Evaluation Using PW: Incremental and Breakeven   283
8.5   Rate of Return Evaluation Using AW   291
8.6   Incremental ROR Analysis of Multiple, Mutually Exclusive Alternatives   292
8.7   Spreadsheet Application—PW, AW, and ROR Analyses All in One   297
Chapter Summary   300
Problems   300
FE Review Problems   306
Extended Exercise—Incremental ROR Analysis When Estimated
Alternative Lives are Uncertain   308
Case Study 1—So Many Options. Can a New Engineering Graduate
Help His Father?   309
Case Study 2—PW Analysis When Multiple Interest Rates Are Present   310

Chapter 9   Benefit/Cost Analysis and Public Sector Economics   **312**

9.1   Public Sector Projects   314
9.2   Benefit/Cost Analysis of a Single Project   319

9.3   Alternative Selection Using Incremental B/C Analysis   324

9.4   Incremental B/C Analysis of Multiple, Mutually Exclusive Alternatives   327

Chapter Summary   333

Problems   333

FE Review Problems   341

Extended Exercise—Costs to Provide Ladder Truck Service

for Fire Protectionn   342

Case Study—Freeway Lighting   343

Chapter 10   Making Choices: The Method, MARR, and Multiple Attributes   346

10.1   Comparing Mutually Exclusive Alternatives by Different Evaluation Methods   348

10.2   MARR Relative to the Cost of Capital   351

10.3   Debt-Equity Mix and Weighted Average Cost of Capital   354

10.4   Determination of the Cost of Debt Capital   357

10.5   Determination of the Cost of Equity Capital and the MARR   359

10.6   Effect of Debt-Equity Mix on Investment Risk   362

10.7   Multiple Attribute Analysis: Identification and Importance of Each Attribute   364

10.8   Evaluation Measure for Multiple Attributes   369

Chapter Summary   371

Problems   372

Extended Exercise—Emphasizing the Right Things   381

Case Study—Which Way to Go—Debt or Equity Financing?   382

LEVEL THREE   MAKING DECISIONS ON REAL-WORLD PROJECTS

Chapter 11   Replacement and Retention Decisions   386

11.1   Basics of the Replacement Study   388

11.2   Economic Service Life   391

11.3   Performing a Replacement Study   397

11.4   Additional Considerations in a Replacement Study   403

11.5   Replacement Study over a Specified Study Period   404

Chapter Summary   410

Problems   410

FE Review Problems   418

Extended Exercise—Economic Service Life Under Varying Conditions   419

Case Study—Replacement Analysis for Quarry Equipment   420

Chapter 12   Selection from Independent Projects Under Budget Limitation   422

12.1   An Overview of Capital Rationing Among Projects   424

12.2   Capital Rationing Using PW Analysis of Equal-Life Projects   426

12.3   Capital Rationing Using PW Analysis of Unequal-Life Projects   428

12.4   Capital Budgeting Problem Formulation Using Linear Programming   432

     Chapter Summary   436

     Problems   437

     Case Study—Lifelong Engineering Education in a Web Environment   440

Chapter 13   Breakeven Analysis   442

13.1   Breakeven Analysis for a Single Project   444

13.2   Breakeven Analysis Between Two Alternatives   451

13.3   Spreadsheet Application—Using Excel's SOLVER for Breakeven Analysis   455

     Chapter Summary   459

     Problems   459

     Case Study—Water Treatment Plant Process Costs   464

LEVEL FOUR   ROUNDING OUT THE STUDY

Chapter 14   Effects of Inflation   470

14.1   Understanding the Impact of Inflation   472

14.2   Present Worth Calculations Adjusted for Inflation   474

14.3   Future Worth Calculations Adjusted for Inflation   480

14.4   Capital Recovery Calculations Adjusted for Inflation   485

     Chapter Summary   486

     Problems   487

     FE Review Problems   491

     Extended Exercise—Fixed-Income Investments versus the Forces of Inflationn   492

Chapter 15   Cost Estimation and Indirect Cost Allocation   494

15.1   Understanding How Cost Estimation Is Accomplished   496

15.2   Cost Indexes   499

15.3   Cost Estimating Relationships: Cost-Capacity Equations   503

15.4   Cost Estimating Relationships: Factor Method   505

15.5   Traditional Indirect Cost Rates and Allocation   508

15.6   Activity-Based Costing (ABC) for Indirect Costs   512

     Chapter Summary   516

     Problems   517

     FE Review Problems   525

     Case Study—Total Cost Estimates for Optimizing Coagulant Dosage   525

     Case Study—Indirect Cost Comparison of Medical Equipment Sterilization Unit   528

Chapter 16   Depreciation Methods   **530**

    16.1   Depreciation Terminology   532
    16.2   Straight Line (SL) Depreciation   535
    16.3   Declining Balance (DB) and Double Declining Balance (DDB) Depreciation 536
    16.4   Modified Accelerated Cost Recovery System (MACRS)   541
    16.5   Determining the MACRS Recovery Period   545
    16.6   Depletion Methods   545
            Chapter Summary   548
            Problems   550
            FE Review Problems   554
    16A.1   Sum-of-Year Digits (SYD) Depreciation   555
    16A.2   Switching Between Depreciation Methods   557
    16A.3   Determination of MACRS Rates   562
            Appendix Problems   566

Chapter 17   After-Tax Economic Analysis   **568**

    17.1   Income Tax Terminology and Relations for Corporations (and Individuals) 570
    17.2   Before-Tax and After-Tax Cash Flow   574
    17.3   Effect on Taxes of Different Depreciation Methods and Recovery Periods 578
    17.4   Depreciation Recapture and Capital Gains (Losses): for Corporations   581
    17.5   After-Tax PW, AW, and ROR Evaluation   586
    17.6   Spreadsheet Applications—After-Tax Incremental ROR Analysis   592
    17.7   After-Tax Replacement Study   595
    17.8   After-Tax Value-Added Analysis   599
    17.9   After-Tax Analysis for International Projects   603
            Chapter Summary   605
            Problems   606
            Case Study—After-Tax Evaluation of Debt and Equity Financing   617

Chapter 18   Formalized Sensitivity Analysis and Expected Value Decisions   **620**

    18.1   Determining Sensitivity to Parameter Variation   622
    18.2   Formalized Sensitivity Analysis Using Three Estimates   629
    18.3   Economic Variability and the Expected Value   632
    18.4   Expected Value Computations for Alternatives   633
    18.5   Staged Evaluation of Alternatives Using a Decision Tree   635
            Chapter Summary   640
            Problems   641
            Extended Exercise—Looking at Alternatives from Different Angles   649
            Case Study—Sensitivity Analysis of Public Sector
            Projects—Water Supply Plans   649

Chapter 19   More on Variation and Decision Making Under Risk   **654**

    19.1   Interpretation of Certainty, Risk, and Uncertainty   656
    19.2   Elements Important to Decision Making Under Risk   660

19.3	Random Samples	666
19.4	Expected Value and Standard Deviation	671
19.5	Monte Carlo Sampling and Simulation Analysis	677
	Additional Examples	686
	Chapter Summary	691
	Problems	691
	Extended Exercise—Using Simulation and the Excel RNG for Sensitivity Analysis	696

Appendix A   Using Spreadsheets and Microsoft Excel©   **697**

A.1	Introduction to Using Excel	697
A.2	Organization (Layout) of the Spreadsheet	701
A.3	Excel Functions Important to Engineering Economy (alphabetical order)	703
A.4	SOLVER—An Excel Tool for Breakeven and "What If?" Analysis	712
A.5	List of Excel Financial Functions	713
A.6	Error Messages	715

Appendix B   Basics of Accounting Reports and Business Ratios   **716**

B.1	The Balance Sheet	716
B.2	Income Statement and Cost of Goods Sold Statement	718
B.3	Business Ratios	719
	Problems	723

Reference Materials	*725*
Factor Tables	*727*
Index	*757*

PREFACE

The primary purpose of this text is to present the principles and applications of economic analysis in a clearly written fashion, supported by a large number and wide range of engineering-oriented examples, end-of-chapter exercises, and electronic-based learning options. Through all editions of the book, our objective has been to present the material in the clearest, most concise fashion possible without sacrificing coverage or true understanding on the part of the learner. The sequence of topics and flexibility of chapter selection used to accommodate different course objectives are described later in the preface.

EDUCATION LEVEL AND USE OF TEXT

This text is best used in learning and teaching at the university level, and as a reference book for the basic computations of engineering economic analysis. It is well suited for a one-semester or one-quarter undergraduate course in engineering economic analysis, project analysis, or engineering cost analysis. Additionally, because of its behavioral-based structure, it is perfect for individuals who wish to learn the material for the first time completely on their own, and for individuals who simply want to review. Students should be at least at the sophomore level, and preferably of junior standing, so that they can better appreciate the engineering context of the problems. A background in calculus is not necessary to understand the calculations, but a basic familiarization with engineering terminology makes the material more meaningful and therefore easier and more enjoyable to learn. Nevertheless, the building-block approach used in the text's design allows a practitioner unacquainted with economics and engineering principles to use the text to learn, understand, and correctly apply the principles and techniques for effective decision making.

NEW TO THIS EDITION

The basic design and structure of previous editions have been retained for the sixth edition. However, considerable changes have been made. The most significant changes include:

- More than 80% of the end-of-chapter problems are new or revised for this edition.
- Time-based materials such as tax rates and cost indexes have been updated.
- The international dimension of the book is more apparent.
- Many of the Fundamentals of Engineering (FE) Review Problems are new to this edition.

STRUCTURE OF TEXT AND OPTIONS FOR PROGRESSION THROUGH THE CHAPTERS

The text is written in a modular form, providing for topic integration in a variety of ways that serve different course purposes, structures, and time limitations.

There are a total of 19 chapters in four levels. As indicated in the flowchart on the next page, some of the chapters have to be covered in sequential order; however, the modular design allows for great flexibility in the selection and sequencing of topics. The chapter progression graphic (which follows the flowchart) shows some of the options for introducing chapters earlier than their numerical order. For example, if the course is designed to emphasize after-tax analysis early in the semester or quarter, Chapter 16 and the initial sections of Chapter 17 may be introduced at any point after Chapter 6 without loss of foundation preparation. There are clear primary and alternate entry points for the major categories of inflation, estimation, taxes, and risk. Alternative entries are indicated by a dashed arrow on the graphic.

The material in Level One emphasizes basic computational skills, so these chapters are prerequisites for all the others in the book. The chapters in Level Two are primarily devoted to the most common analytical techniques for comparing alternatives. While it is advisable to cover all the chapters in this level, only the first two (Chapters 5 and 6) are widely used throughout the remainder of the text. The three chapters of Level Three show how any of the techniques in Level Two can be used to evaluate presently owned assets or independent alternatives, while the chapters in Level Four emphasize the tax consequences of decision making and some additional concepts in cost estimation, activity-based costing, sensitivity analysis, and risk, as treated using Monte Carlo simulation.

Organization of Chapters and End-of-Chapter Exercises   Each chapter contains a purpose and a series of progressive learning objectives, followed by the study material. Section headings correspond to each learning objective; for example, Section 5.1 contains the material pertaining to the first objective of the chapter. Each section contains one or more illustrative examples solved by hand, or by both hand and computer methods. Examples are separated from the textual material and include comments about the solution and pertinent connections to other topics in the book. The crisp end-of-chapter summaries neatly tie together the concepts and major topics covered to reinforce the learner's understanding prior to engaging in the end-of chapter exercises.

The end-of-chapter unsolved problems are grouped and labeled in the same general order as the sections in the chapter. This approach provides an opportunity to apply material on a section-by-section basis or to schedule problem solving when the chapter is completed.

Appendices A and B contain supplementary information: a basic introduction to the use of spreadsheets (Microsoft Excel) for readers unfamiliar with them and the basics of accounting and business reports. Interest factor tables are located at the end of the text for easy access. Finally, the inside front covers offer a quick reference to factor notation, formulas, and cash flow diagrams, plus a guide to the format for commonly used spreadsheet functions. A glossary of common terms and symbols used in engineering economy appears inside the back cover.

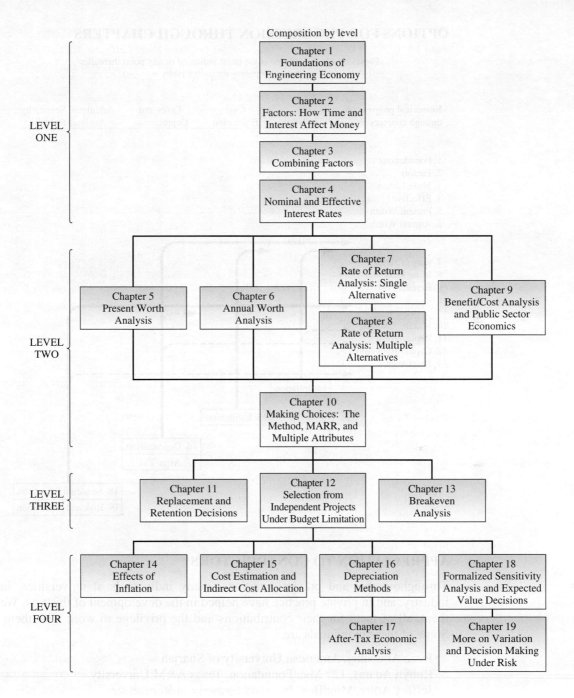

Composition by level

LEVEL
ONE

Chapter 1
Foundations of
Engineering Economy

Chapter 2
Factors: How Time and
Interest Affect Money

Chapter 3
Combining Factors

Chapter 4
Nominal and Effective
Interest Rates

LEVEL
TWO

Chapter 5
Present Worth
Analysis

Chapter 6
Annual Worth
Analysis

Chapter 7
Rate of Return
Analysis: Single
Alternative

Chapter 8
Rate of Return
Analysis: Multiple
Alternatives

Chapter 9
Benefit/Cost Analysis
and Public Sector
Economics

Chapter 10
Making Choices: The
Method, MARR, and
Multiple Attributes

LEVEL
THREE

Chapter 11
Replacement and
Retention Decisions

Chapter 12
Selection from
Independent Projects
Under Budget Limitation

Chapter 13
Breakeven
Analysis

LEVEL
FOUR

Chapter 14
Effects of
Inflation

Chapter 15
Cost Estimation and
Indirect Cost Allocation

Chapter 16
Depreciation
Methods

Chapter 18
Formalized Sensitivity
Analysis and Expected
Value Decisions

Chapter 17
After-Tax Economic
Analysis

Chapter 19
More on Variation
and Decision Making
Under Risk

OPTIONS FOR PROGRESSION THROUGH CHAPTERS

Topics may be introduced at the point indicated or any point thereafter
(Alternative entry points are indicated by ◄ – – –)

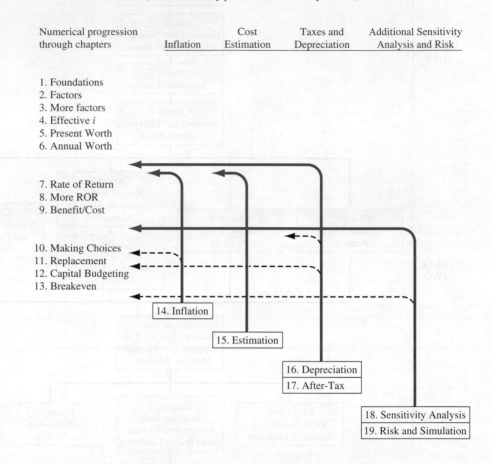

Numerical progression through chapters	Inflation	Cost Estimation	Taxes and Depreciation	Additional Sensitivity Analysis and Risk

1. Foundations
2. Factors
3. More factors
4. Effective i
5. Present Worth
6. Annual Worth

7. Rate of Return
8. More ROR
9. Benefit/Cost

10. Making Choices
11. Replacement
12. Capital Budgeting
13. Breakeven

14. Inflation

15. Estimation

16. Depreciation
17. After-Tax

18. Sensitivity Analysis
19. Risk and Simulation

APPRECIATION TO CONTRIBUTORS

Throughout this and previous editions, many individuals at universities, in industry, and in private practice have helped in the development of this text. We thank all of them for their contributions and the privilege to work with them. Some of these individuals are

Roza Abubaker, American University of Sharjah
Robyn Adams, 12th Man Foundation, Texas A&M University
Jeffrey Adler, MindBox, Inc., and formerly of Rensselaer
    Polytechnic Institute
Richard H. Bernhard, North Carolina State University
Stanley F. Bullington, Mississippi State University
Peter Chan, CSA Engineering, Inc.

Ronald T. Cutwright, Florida A&M University
John F. Dacquisto, Gonzaga University
John Yancey Easley, Mississippi State University
Nader D. Ebrahimi, University of New Mexico
Charles Edmonson, University of Dayton, Ohio
Sebastian Fixson, University of Michigan
Louis Gennaro, Rochester Institute of Technology
Joseph Hartman, Lehigh University
John Hunsucker, University of Houston
Cengiz Kahraman, Istanbul Technical University, Turkey
Walter E. LeFevre, University of Arkansas
Kim LaScola Needy, University of Pittsburgh
Robert Lundquist, Ohio State University
Gerald T. Mackulak, Arizona State University
Mike Momot, University of Wisconsin, Platteville
James S. Noble, University of Missouri–Columbia
Richard Patterson, University of Florida
Antonio Pertence Jr., Faculdade de Sabara, Minas Gerais, Brazil
William R. Peterson, Old Dominion University
Stephen M. Robinson, University of Wisconsin–Madison
David Salladay, San Jose State University
Mathew Sanders, Kettering University
Tep Sastri, formerly of Texas A&M University
Michael J. Schwandt, Tennessee Technological University
Frank Sheppard, III, The Trust for Public Land
Sallie Sheppard, American University of Sharjah
Don Smith, Texas A&M University
Alan Stewart, Accenture LLP
Mathias Sutton, Purdue University
Ghassan Tarakji, San Francisco State University
Ciriaco Valdez-Flores, Sielken and Associates Consulting
Richard West, CPA, Sanders and West

We would also like to thank Jack Beltran for his accuracy check of this and previous editions. His work will help make this text a success.

Finally, we welcome any comments or suggestions you may have to help improve the textbook or the Online Learning Center. You can reach us at lblank@ausharjah.edu or lblank@tamu.edu and atarquin@utep.edu. We look forward to hearing from you.

Lee Blank
Tony Tarquin

GUIDED TOUR

CHAPTER EXAMPLES AND EXERCISES   Users of this book have numerous ways to reinforce the concepts they've learned. The end-of-chapter *problems, in-chapter examples, extended exercises, case studies, and FE (Fundamentals of Engineering) review problems* offer students the opportunity to learn economic analysis in a variety of ways. The various exercises range from working relatively simple, one-step review problems to answering a series of comprehensive, in-depth questions based on real-world cases. In-chapter examples are also helpful in reinforcing concepts learned.

END-OF-CHAPTER PROBLEMS

As in previous editions, each chapter contains many homework problems representative of real-world situations. 80% of the end-of-chapter problems have been revised or are new to this edition.

PROBLEMS

Types of Projects

5.1 What is meant by *service alternative*?

5.2 When you are evaluating projects by the present worth method, how do you know which one(s) to select if the projects

are (*a*) independent and (*b*) mutually exclusive?

5.3 Read the statement in the following problems and determine if the cash flows define

EXTENDED EXERCISES

The extended exercises are designed to require spreadsheet analysis with a general emphasis on sensitivity analysis.

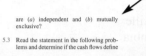

FUNDAMENTALS OF ENGINEERING (FE) EXAM REVIEW PROBLEMS

The FE exam review problems cover the same topics as the FE exam and are written in the same multiple-choice format as the exam. *All of these problems are new to this edition.*

FE REVIEW PROBLEMS

Note: The sign convention on the FE exam may be opposite of that used here. That is, on the FE exam, costs may be positive and receipts negative.

6.23 In comparing alternatives that have different lives by the annual worth method,
(*a*) The annual worth value of both alternatives must be calculated over a time period equal to the life of the shorter-lived one.
(*b*) The annual worth value of both alternatives must be calculated over a time period equal to the life of the longer-lived asset.
(*c*) The annual worth values must be calculated over a time period equal to the least common multiple of the lives.

(*d*) The annual worth values calculated over one life cycle of each alternative can be compared.

Problems 6.24 through 6.26 are based on the following estimates.

Alternative	A	B
First cost, $	−50,000	−80,000
Annual cost, $/year	−20,000	−10,000
Salvage value, $	10,000	25,000
Life, years	3	6

Use an interest rate of 10% per year.

6.24 The equivalent annual worth of alternative A is closest to:
(*a*) $−25,130
(*b*) $−37,100

CASE STUDIES

All the case studies present real-world, in-depth treatments and exercises that cover the wide spectrum of economic analysis in the engineering profession.

IN-CHAPTER EXAMPLES

Examples within the chapters are relevant to all engineering disciplines that use this text, including industrial, civil, environmental, mechanical, petroleum, and electrical engineering as well as engineering management and engineering technology programs.

USE OF SPREADSHEETS

The text integrates spreadsheets and shows how easy they are to use in solving virtually any type of engineering economic analysis problem and how powerful they can be for altering estimates to achieve a better understanding of sensitivity and economic consequences of the uncertainties inherent in all forecasts. Beginning in Chapter 1, Blank and Tarquin illustrate their spreadsheet discussions with screenshots from Microsoft Excel™*.

When a single-cell, built-in Excel function may be used to solve a problem, a checkered flag icon labeled *Q-Solv* (for *quick solution*) appears in the margin.

The thunderbolt *E-Solve* icon indicates that a more complex, sophisticated spreadsheet is developed to solve the problem. The spreadsheet will contain data and several functions and possibly an Excel chart or graph to illustrate the answer and sensitivity analysis of the solution to changing data.

For both Q-Solv and E-Solve examples, the authors have included cells that show the exact Excel function needed to obtain the value in a specific cell. The E-Solve icon is also used throughout chapters to point out descriptions of how to best use the computer to address the engineering economy topic under discussion.

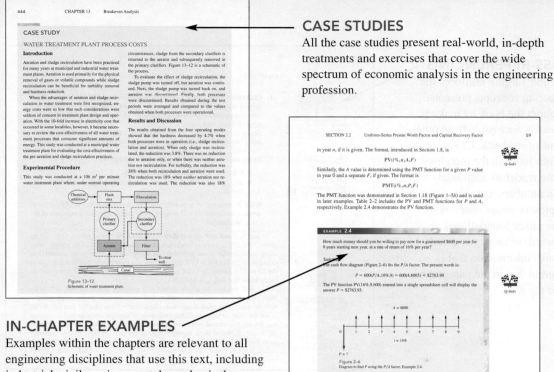

CROSS-REFERENCING

Blank and Tarquin reinforce the engineering concepts presented throughout the book by making them easily accessible from other sections of the book. Cross-reference icons in the margins refer the reader to additional section numbers, specific examples, or entire chapters that contain either foundational (backward) or more advanced (forward) information that is relevant to that in the paragraph next to the icon.

Sec. 7.4

Multiple-root tests

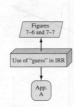

Incremental
ROR
Must Justify
Incremental
Investment

Figures
7–6 and 7–7

Use of "guess" in IRR

App.
A

4. Count the number of sign changes in the incremental cash flow series to determine if multiple rates of return may be present. If necessary, use Norstrom's criterion on the cumulative incremental cash flow series to determine if a single positive root exists.

5. Set up the PW equation for the incremental cash flows in the form of Equation [7.1], and determine Δi^*_{B-A} using trial and error by hand or spreadsheet functions.

6. Select the economically better alternative as follows:

> **If Δi^*_{B-A} < MARR, select alternative A.**
> **If Δi^*_{B-A} ≥ MARR, the extra investment is justified; select alternative B.**

If the incremental i^* is exactly equal to or very near the MARR, noneconomic considerations will most likely be used to help in the selection of the "best" alternative.

In step 5, if trial and error is used to calculate the rate of return, time may be saved if the Δi^*_{B-A} value is bracketed, rather than approximated by a point value using linear interpolation, provided that a single ROR value is not needed. For example, if the MARR is 15% per year and you have established that Δi^*_{B-A} is in the 15 to 20% range, an exact value is not necessary to accept B since you already know that Δi^*_{B-A} ≥ MARR.

The IRR function on a spreadsheet will normally determine one Δi^* value. Multiple guess values can be input to find multiple roots in the range −100% to ∞ for a nonconventional series, as illustrated in Examples 7.4 and 7.5. If this is not the case, to be correct, the indication of multiple roots in step 4 requires that the net-investment procedure, Equation [7.6], be applied in step 5 to make $\Delta i' = \Delta i^*$. If one of these multiple roots is the same as the expected reinvestment rate c, this root can be used as the ROR value, and the net-investment procedure is not necessary. In this case only $\Delta i' = \Delta i^*$, as concluded at the end of Section 7.5.

INTERNATIONAL APPEAL

The international dimensions of this book are more apparent throughout the sixth edition. Examples and new sections on international corporate depreciation and taxation considerations and international forms of contracts, such as the BOT method of subcontracting, are included. The impact of hyperinflation and deflationary cycles are discussed from an international perspective.

17.9   AFTER-TAX ANALYSIS FOR INTERNATIONAL PROJECTS

Primary questions to be answered prior to performing a corporate-based after-tax analysis for international settings revolve on tax-deductible allowances—depreciation, business expenses, capital asset evaluation—and the effective tax rate needed for Equation [17.6], taxes = $TI(T_e)$. As discussed in Chapter 16, most governments of the world recognize and use the straight line (SL) and declining balance (DB) methods of depreciation with some variations to determine the annual tax-deductible allowance. Expense deductions vary widely from country to country. By way of example, some of these are summarized here.

Canada

Depreciation: This is deductible and is normally based on DB calculations, although SL may be used. An equivalent of the half-year convention is applied in the first year of ownership. The annual tax-deductible allowance is termed *capital cost allowance (CCA)*. As in the U.S. system,

ADDITIONAL RESOURCES

The sixth edition of Blank and Tarquin features an Online Learning Center (OLC) available to students and professors who use the text. The URL for the site is http://www.mhhe.com/blank6.

The OLC will house the solutions to end-of-chapter problems, FE (Fundamentals of Engineering) exam prep quiz, spreadsheet exercises, matching and true/false quizzes, links to important websites, chapter objectives, lecture slides, end-of chapter summaries and more!

ENGINEERING ECONOMY sixth edition
Blank and Tarquin

Help   Feedback

Information Center

- Sample Chapters
- Table of Contents
- About the Authors
- Book Preface
- What's New
- PageOut
- Guided Tour

Online Learning Center

Student Edition
Instructor Edition

Engineering Economy, 6/e

Leland Blank, American Univ. of Sharjah and Texas A&M Univ.
Anthony Tarquin, University of Texas - El Paso

ISBN: 0072918632
Copyright year: 2005

Engineering Economy, 6th edition , provides undergraduate students and practicing professionals with a solid preparation in the financial understanding of engineering problems and projects, as well as the techniques needed for evaluating and making sound economic decisions. Information on cost estimation, depreciation, and taxes has been updated to conform to new tax laws and a majority of the end-of-chapter problems are revised or new to this edition.

Distinguishing pedagogical characteristics of this market-leading text include its easy-to-read writing style, chapter objectives, worked examples, integrated spreadsheets, case studies, Fundamentals of Engineering (FE) exam questions, and numerous end-of-chapter problems. Graphical cross-referencing is indicated so users are able to locate additional material on any one subject in the text. Quick-solve (Q-Solv) and Excel-solve (E-Solve) icons found in the text indicate the difficulty of a problem, example, or spreadsheet.

While the chapters are progressive, over three-quarters can stand alone, allowing instructors flexibility for meeting course needs. A complete Online Learning Center (OLC) offers supplemental practice problems, spreadsheet exercises, review questions for the Fundamentals of Engineering (FE) exam, and more!

ENGINEERING ECONOMY sixth edition

Leland Blank • Anthony Tarquin

To obtain an instructor login for this Online Learning Center, ask your local sales representative. If you're an instructor thinking about adopting this textbook, request a free copy for review.

New to this edition is McGraw-Hill's new database management tool. Complete Online Solutions Manual Organization System (C.O.S.M.O.S.). C.O.S.M.O.S. is delivered via CD-ROM and helps instructors to organize solutions and distribute and track problem sets as they are assigned to students in the course. This helps instructors to quickly find solutions and keep a record of problems assigned, to avoid duplication of tests and quizzes in subsequent semesters. The ISBN for the *Engineering Economy* C.O.S.M.O.S. CD-ROM is 0–07–298450–3. Contact your McGraw-Hill representative to get a copy.

ENGINEERING ECONOMY

LEVEL ONE

THIS IS HOW IT ALL STARTS

LEVEL ONE This Is How It All Starts	LEVEL TWO Tools for Evaluating Alternatives	LEVEL THREE Making Decisions on Real-World Projects	LEVEL FOUR Rounding Out the Study
Chapter 1 Foundations of Engineering Economy	**Chapter 5** Present Worth Analysis	**Chapter 11** Replacement and Retention Decisions	**Chapter 14** Effects of Inflation
Chapter 2 Factors: How Time and Interest Affect Money	**Chapter 6** Annual Worth Analysis	**Chapter 12** Selection from Independent Projects under Budget Limitation	**Chapter 15** Cost Estimation and Indirect Cost Allocation
Chapter 3 Combining Factors	**Chapter 7** Rate of Return Analysis: Single Alternative	**Chapter 13** Breakeven Analysis	**Chapter 16** Depreciation Methods
Chapter 4 Nominal and Effective Interest Rates	**Chapter 8** Rate of Return Analysis: Multiple Alternatives		**Chapter 17** After-Tax Economic Analysis
	Chapter 9 Benefit/Cost Analysis and Public Sector Economics		**Chapter 18** Formalized Sensitivity Analysis and Expected Value Decisions
	Chapter 10 Making Choices: The Method, MARR, and Multiple Attributes		**Chapter 19** More on Variation and Decision Making under Risk

The foundations of engineering economy are introduced in these four chapters. When you have completed level one, you will be able to understand and work problems that account for the *time value of money*, *cash flows* occurring at different times with different amounts, and *equivalence* at different interest rates. The techniques you master here form the basis of how an engineer in any discipline can take *economic value* into account in virtually any project environment.

The eight factors commonly used in all engineering economy computations are introduced and applied in this level. Combinations of these factors assist in moving monetary values forward and backward through time and at different interest rates. Also, after these four chapters, you should be comfortable with using many of the Excel spreadsheet functions to solve problems.

Foundations of Engineering Economy

The need for engineering economy is primarily motivated by the work that engineers do in performing analysis, synthesizing, and coming to a conclusion as they work on projects of all sizes. In other words, engineering economy is at the heart of *making decisions*. These decisions involve the fundamental elements of *cash flows of money*, *time*, and *interest rates*. This chapter introduces the basic concepts and terminology necessary for an engineer to combine these three essential elements in organized, mathematically correct ways to solve problems that will lead to better decisions. Many of the terms common to economic decision making are introduced here and used in later chapters of the text. Icons in the margins serve as back and forward cross-references to more fundamental and additional material throughout the book.

The case studies included after the end-of-chapter problems focus on the development of engineering economy alternatives.

LEARNING OBJECTIVES

Purpose: Understand the fundamental concepts of engineering economy.

This chapter will help you:

Questions	1. Understand the types of questions engineering economy can answer.
Decision making	2. Determine the role of engineering economy in the decision-making process.
Study approach	3. Identify what is needed to successfully perform an engineering economy study.
Interest rate	4. Perform calculations about interest rates and rate of return.
Equivalence	5. Understand what equivalence means in economic terms.
Simple and compound interest	6. Calculate simple interest and compound interest for one or more interest periods.
Symbols	7. Identify and use engineering economy terminology and symbols.
Spreadsheet functions	8. Identify the Excel© spreadsheet functions commonly used to solve engineering economy problems.
Minimum Attractive Rate of Return	9. Understand the meaning and use of Minimum Attractive Rate of Return (MARR).
Cash flows	10. Understand cash flows, their estimation, and how to graphically represent them.
Doubling time	11. Use the rule of 72 to estimate a compound interest rate or number of years for a present worth amount to double.
Spreadsheets	12. Develop a spreadsheet that involves simple and compound interest, incorporating sensitivity analysis.

1.1 WHY ENGINEERING ECONOMY IS IMPORTANT TO ENGINEERS (and other professionals)

Decisions made by engineers, managers, corporation presidents, and individuals are commonly the result of choosing one alternative over another. Decisions often reflect a person's educated choice of how to best invest funds, also called *capital*. The amount of capital is usually restricted, just as the cash available to an individual is usually limited. The decision of how to invest capital will invariably change the future, hopefully for the better; that is, it will be *value adding*. Engineers play a major role in capital investment decisions based on their analysis, synthesis, and design efforts. The factors considered in making the decision are a combination of economic and noneconomic factors. Additional factors may be intangible, such as convenience, goodwill, friendship, and others.

Fundamentally, engineering economy involves formulating, estimating, and evaluating the economic outcomes when alternatives to accomplish a defined purpose are available. Another way to define engineering economy is as a collection of mathematical techniques that simplify economic comparison.

For many corporations, especially larger ones, many of the projects and services are international in scope. They may be developed in one country for application in another. People and plants located in sites around the world routinely separate product design and manufacturing from each other, and from the customers who utilize the product. The approaches presented here are easily implemented in multinational settings or within a single country or location. Correct use of the techniques of engineering economy is especially important, since virtually any project—local, national, or international—will affect costs and/or revenues.

Some of the typical questions that can be addressed using the material in this book are posed below.

For Engineering Activities

- Should a new bonding technique be incorporated into the manufacture of automobile brake pads?
- If a computer-vision system replaces the human inspector in performing quality tests on an automobile welding line, will operating costs decrease over a time horizon of 5 years?
- Is it an economically wise decision to upgrade the composite material production center of an airplane factory in order to reduce costs by 20%?
- Should a highway bypass be constructed around a city of 25,000 people, or should the current roadway through the city be expanded?
- Will we make the required rate of return if we install the newly offered technology onto our medical laser manufacturing line?

For Public Sector Projects and Government Agencies

- How much new tax revenue does the city need to generate to pay for an upgrade to the electric distribution system?
- Do the benefits outweigh the costs of a bridge over the intracoastal waterway at this point?

- Is it cost-effective for the state to cost-share with a contractor to construct a new toll road?
- Should the state university contract with a local community college to teach foundation-level undergraduate courses or have university faculty teach them?

For Individuals

- Should I pay off my credit card balance with borrowed money?
- What are graduate studies worth financially over my professional career?
- Are federal income tax deductions for my home mortgage a good deal, or should I accelerate my mortgage payments?
- Exactly what rate of return did we make on our stock investments?
- Should I buy or lease my next car, or keep the one I have now and pay off the loan?

EXAMPLE **1.1**

Two lead engineers with a mechanical design company and a structural analysis firm work together often. They have decided that, due to their joint and frequent commercial airline travel around the region, they should evaluate the purchase of a plane co-owned by the two companies. What are some of the economics-based questions the engineers should answer as they evaluate the alternatives to (1) co-own a plane or (2) continue to fly commercially?

Solution

Some questions (and what is needed to respond) for each alternative are as follows:

- How much will it cost each year? (Cost estimates are needed.)
- How do we pay for it? (A financing plan is needed.)
- Are there tax advantages? (Tax law and tax rates are needed.)
- What is the basis for selecting an alternative? (A selection criterion is needed.)
- What is the expected rate of return? (Equations are needed.)
- What happens if we fly more or less than we estimate now? (Sensitivity analysis is needed.)

1.2 ROLE OF ENGINEERING ECONOMY IN DECISION MAKING

People make decisions; computers, mathematics, and other tools do not. The techniques and models of engineering economy *assist people in making decisions.* Since decisions affect what will be done, the time frame of engineering economy is primarily *the future.* Therefore, numbers used in an engineering economic analysis are *best estimates of what is expected to occur.* These estimates often involve the three essential elements mentioned earlier: cash flows, time of occurrence, and interest rates. These estimates are about the future, and will be somewhat different than what actually occurs, primarily because of changing circumstances and unplanned-for events. In other words, the *stochastic nature* of estimates will likely make the observed value in the future differ from the estimate made now.

Commonly, *sensitivity analysis* is performed during the engineering economic study to determine how the decision might change based on varying estimates,

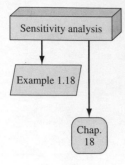

Sensitivity analysis

Example 1.18

Chap. 18

especially those that may vary widely. For example, an engineer who expects initial software development costs to vary as much as ±20% from an estimated $250,000 should perform the economic analysis for first-cost estimates of $200,000, $250,000, and $300,000. Other uncertain estimates about the project can be "tweaked" using sensitivity analysis. (Sensitivity analysis is quite easy to perform using electronic spreadsheets. Tabular and graphical displays make analysis possible by simply changing the estimated values. The power of spreadsheets is used to advantage throughout this text and on the supporting website.)

Engineering economy can be used equally well to analyze outcomes of *the past.* Observed data are evaluated to determine if the outcomes have met or not met a specified criterion, such as a rate of return requirement. For example, suppose that 5 years ago, a United States–based engineering design company initiated a detailed-design service in Asia for automobile chassis. Now, the company president wants to know if the actual return on the investment has exceeded 15% per year.

There is an important procedure used to address the development and selection of alternatives. Commonly referred to as the *problem-solving approach* or the *decision-making process,* the steps in the approach follow.

1. **Understand the problem and define the objective.**
2. **Collect relevant information.**
3. **Define the feasible alternative solutions and make realistic estimates.**
4. **Identify the criteria for decision making using one or more attributes.**
5. **Evaluate each alternative, using sensitivity analysis to enhance the evaluation.**
6. **Select the best alternative.**
7. **Implement the solution.**
8. **Monitor the results.**

Engineering economy has a major role in all steps and is primary to steps 2 through 6. Steps 2 and 3 establish the alternatives and make the estimates for each one. Step 4 requires the analyst to identify attributes for alternative selection. This sets the stage for the technique to apply. Step 5 utilizes engineering economy models to complete the evaluation and perform any sensitivity analysis upon which a decision is based (step 6).

EXAMPLE 1.2

Reconsider the questions presented for the engineers in the previous example about co-owning an airplane. State some ways in which engineering economy contributes to decision making between the two alternatives.

Solution
Assume that the objective is the same for each engineer—available, reliable transportation that minimizes total cost. Use the steps above.

Steps 2 and 3: The framework for an engineering economy study assists in identifying what should be estimated or collected. For alternative 1 (buy the plane), estimate the purchase cost, financing method and interest rate, annual operating costs, possible increase in annual sales revenue, and income tax deductions. For alternative 2 (fly commercial) estimate commercial transportation costs, number of trips, annual sales revenue, and other relevant data.

Step 4: The selection criterion is a numerically valued attribute called a *measure of worth.* Some measures of worth are

Present worth (PW)	Future worth (FW)	Payback period
Annual worth (AW)	Rate of return (ROR)	Economic value added
Benefit/cost ratio (B/C)	Capitalized cost (CC)	

When determining a measure of worth, the fact that money today is worth a different amount in the future is considered; that is, the *time value of money* is accounted for.

There are many noneconomic attributes—social, environmental, legal, political, personal, to name a few. This multiple-attribute environment may result in less reliance placed on the economic results in step 6. But this is exactly why the decision maker must have adequate information for all factors—economic and noneconomic—to make an informed selection. In our case, the economic analysis may favor the co-owned plane (alternative 1), but because of noneconomic factors, one or both engineers may select alternative 2.

Steps 5 and 6: The actual computations, sensitivity analysis, and alternative selection are accomplished here.

> Multiple attributes
>
> Secs.
> 10.7 and 10.8

The concept of the *time value of money* was mentioned above. It is often said that money makes money. The statement is indeed true, for if we elect to invest money today, we inherently expect to have more money in the future. If a person or company borrows money today, by tomorrow more than the original loan principal will be owed. This fact is also explained by the time value of money.

The change in the amount of money over a given time period is called the *time value of money;* it is the most important concept in engineering economy.

1.3   PERFORMING AN ENGINEERING ECONOMY STUDY

Consider the terms *engineering economy, engineering economic analysis, economic decision making, capital allocation study, economic analysis,* and similar terms to be synonymous throughout this book. There is a general approach, called the *Engineering Economy Study Approach,* that provides an overview of the engineering economic study. It is outlined in Figure 1–1 for two alternatives. The decision-making process steps are keyed to the blocks in Figure 1–1.

Alternative Description   The result of decision-making process step 1 is a basic understanding of what the problem requires for solution. There may initially be many alternatives, but only a few will be feasible and actually evaluated. If alternatives *A*, *B*, and *C* have been identified for analysis, when method *D*, though not recognized as an alternative, is the most attractive, the wrong decision is certain to be made.

Alternatives are stand-alone options that involve a word description and best estimates of parameters, such as *first cost* (including purchase price, development, installation), *useful life, estimated annual incomes and expenses, salvage value* (resale or trade-in value), an *interest rate* (rate of return), and possibly *inflation* and *income tax effects*. Estimates of annual expenses are usually lumped together and called annual operating costs (AOC) or maintenance and operation (M&O) costs.

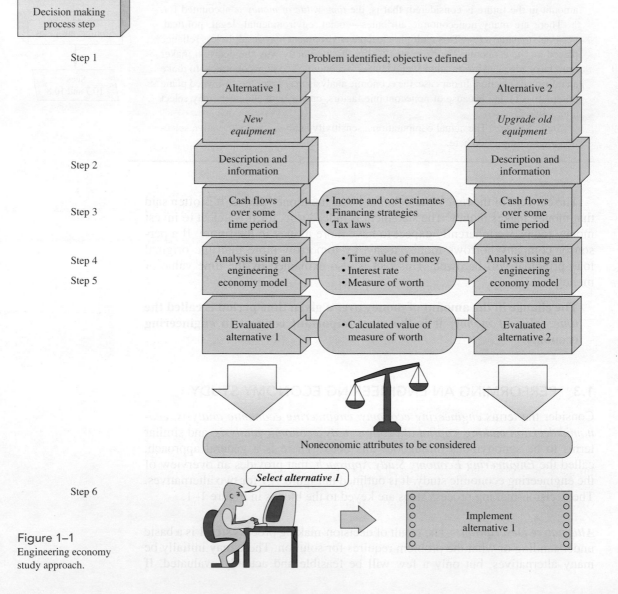

Decision making process step

Step 1 — Problem identified; objective defined

Alternative 1 — *New equipment* | Alternative 2 — *Upgrade old equipment*

Step 2 — Description and information | Description and information

Step 3 — Cash flows over some time period • Income and cost estimates • Financing strategies • Tax laws — Cash flows over some time period

Step 4 / Step 5 — Analysis using an engineering economy model • Time value of money • Interest rate • Measure of worth — Analysis using an engineering economy model

Evaluated alternative 1 • Calculated value of measure of worth — Evaluated alternative 2

Noneconomic attributes to be considered

Select alternative 1

Step 6 — Implement alternative 1

Figure 1–1
Engineering economy study approach.

Cash Flows   The estimated inflows (revenues) and outflows (costs) of money are called cash flows. These estimates are made for each alternative (step 3). Without cash flow estimates over a stated time period, no engineering economy study can be conducted. Expected variation in cash flows indicates a real need for sensitivity analysis in step 5.

Analysis Using Engineering Economy   Computations that consider the time value of money are performed on the cash flows of each alternative to obtain the measure of worth.

Alternative Selection   The measure-of-worth values are compared, and an alternative is selected. This is the result of the engineering economy analysis. For example, the result of a rate-of-return analysis may be: Select alternative 1, where the rate of return is estimated at 18.4% per year, over alternative 2 with an expected 10% per year return. Some combination of economic criteria using the measure of worth, and the noneconomic and intangible factors, may be applied to help select one alternative.

If only one feasible alternative is defined, a second is often present in the form of the *do-nothing alternative*. This is the *as-is* or *status quo* alternative. Do nothing can be selected if no alternative has a favorable measure of worth.

Whether we are aware of it or not, we use criteria every day to choose between alternatives. For example, when you drive to campus, you decide to take the "best" route. But how did you define *best?* Was the best route the safest, shortest, fastest, cheapest, most scenic, or what? Obviously, depending upon which criterion or combination of criteria is used to identify the best, a different route might be selected each time. In economic analysis, *financial units* (dollars or other currency) are generally used as the tangible basis for evaluation. Thus, when there are several ways of accomplishing a stated objective, the alternative with the lowest overall cost or highest overall net income is selected.

An *after-tax analysis* is performed during project evaluation, usually with only significant effects for asset depreciation and income taxes accounted for. Taxes imposed by local, state, federal, and international governments usually take the form of an income tax on revenues, value-added tax (VAT), import taxes, sales taxes, real estate taxes, and others. Taxes affect alternative estimates for cash flows; they tend to *improve* cash flow estimates for expenses, cost savings, and asset depreciation, while they *reduce* cash flow estimates for revenue and after-tax net income. This text delays the details of after-tax analysis until the fundamental tools and techniques of engineering economy are covered. Until then, it is assumed that all alternatives are taxed equally by prevailing tax laws. (If the effects of taxes must be considered earlier, it is recommended that Chapters 16 and 17 be covered after Chapter 6, 8 or 11.)

Now, we turn to some fundamentals of engineering economy that are applicable in the everyday life of engineering practice, as well as personal decision making.

1.4 INTEREST RATE AND RATE OF RETURN

Interest is the manifestation of the time value of money. Computationally, interest is the difference between an ending amount of money and the beginning amount. If the difference is zero or negative, there is no interest. There are always two perspectives to an amount of interest—interest paid and interest earned. Interest is *paid* when a person or organization borrowed money (obtained a loan) and repays a larger amount. Interest is *earned* when a person or organization saved, invested, or lent money and obtains a return of a larger amount. It is shown below that the computations and numerical values are essentially the same for both perspectives, but there are different interpretations.

Interest paid on borrowed funds (a loan) is determined by using the relation

$$\text{Interest} = \text{amount owed now} - \text{original amount} \qquad [1.1]$$

When interest paid over a *specific time unit* is expressed as a percentage of the original amount (principal), the result is called the *interest rate*.

$$\text{Interest rate (\%)} = \frac{\text{interest accrued per time unit}}{\text{original amount}} \times 100\% \qquad [1.2]$$

The time unit of the rate is called the *interest period*. By far the most common interest period used to state an interest rate is 1 year. Shorter time periods can be used, such as, 1% per month. Thus, the interest period of the interest rate should always be included. If only the rate is stated, for example, 8.5%, a 1-year interest period is assumed.

Loan interest rate

Borrower   Bank

EXAMPLE 1.3

An employee at LaserKinetics.com borrows $10,000 on May 1 and must repay a total of $10,700 exactly 1 year later. Determine the interest amount and the interest rate paid.

Solution
The perspective here is that of the borrower since $10,700 repays a loan. Apply Equation [1.1] to determine the interest paid.

$$\text{Interest earned} = \$10,700 - 10,000 = \$700$$

Equation [1.2] determines the interest rate paid for 1 year.

$$\text{Percent interest rate} = \frac{\$700}{\$10,000} \times 100\% = 7\% \text{ per year}$$

EXAMPLE 1.4

Stereophonics, Inc., plans to borrow $20,000 from a bank for 1 year at 9% interest for new recording equipment. (*a*) Compute the interest and the total amount due after 1 year. (*b*) Construct a column graph that shows the original loan amount and total amount due after 1 year used to compute the loan interest rate of 9% per year.

Solution

(*a*) Compute the total interest accrued by solving Equation [1.2] for interest accrued.

$$\text{Interest} = \$20{,}000(0.09) = \$1800$$

The total amount due is the sum of principal and interest.

$$\text{Total due} = \$20{,}000 + 1800 = \$21{,}800$$

(*b*) Figure 1–2 shows the values used in Equation [1.2]: $1800 interest, $20,000 original loan principal, 1-year interest period.

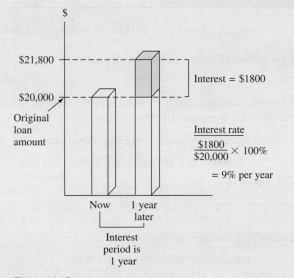

Figure 1–2
Values used to compute an interest rate of 9% per year, Example 1.4.

Comment
Note that in part (*a*), the total amount due may also be computed as

$$\text{Total due} = \text{principal}(1 + \text{interest rate}) = \$20{,}000(1.09) = \$21{,}800$$

Later we will use this method to determine future amounts for times longer than one interest period.

From the perspective of a saver, a lender, or an investor, interest earned is the final amount minus the initial amount, or principal.

$$\text{Interest earned} = \text{total amount now} - \text{original amount} \qquad [1.3]$$

Interest earned over a specific period of time is expressed as a percentage of the original amount and is called *rate of return (ROR)*.

$$\textbf{Rate of return (\%)} = \frac{\textbf{interest accrued per time unit}}{\textbf{original amount}} \times \textbf{100\%} \qquad [1.4]$$

Investment rate of return

Saver   Bank

The time unit for rate of return is called the *interest period,* just as for the borrower's perspective. Again, the most common period is 1 year.

The term *return on investment (ROI)* is used equivalently with ROR in different industries and settings, especially where large capital funds are committed to engineering-oriented programs.

The numerical values in Equation [1.2] and Equation [1.4] are the same, but the term *interest rate paid* is more appropriate for the borrower's perspective, while the *rate of return earned* is better for the investor's perspective.

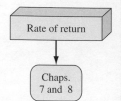

Rate of return

Chaps. 7 and 8

EXAMPLE 1.5

(*a*) Calculate the amount deposited 1 year ago to have $1000 now at an interest rate of 5% per year.
(*b*) Calculate the amount of interest earned during this time period.

Solution

(*a*) The total amount accrued ($1000) is the sum of the original deposit and the earned interest. If X is the original deposit,

$$\text{Total accrued} = \text{original} + \text{original(interest rate)}$$

$$\$1000 = X + X(0.05) = X(1 + 0.05) = 1.05X$$

The original deposit is

$$X = \frac{1000}{1.05} = \$952.38$$

(*b*) Apply Equation [1.3] to determine interest earned.

$$\text{Interest} = \$1000 - 952.38 = \$47.62$$

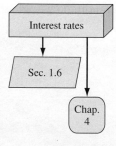

Interest rates

Sec. 1.6

Chap. 4

In Examples 1.3 to 1.5 the interest period was 1 year, and the interest amount was calculated at the end of one period. When more than one interest period is involved (e.g., if we wanted the amount of interest owed after 3 years in Example 1.4), it is necessary to state whether the interest is accrued on a *simple* or *compound* basis from one period to the next.

An additional economic consideration for any engineering economy study is *inflation.* Several comments about the fundamentals of inflation are warranted at this early stage. First, inflation represents a decrease in the value of a given currency. That is, $1 now will not purchase the same number of apples (or most other things) as $1 did 20 years ago. The changing value of the currency affects market interest rates. In simple terms, bank interest rates reflect two things: a so-called real rate of return *plus* the expected inflation rate. The real rate of return allows the investor to purchase more than he or she could have purchased before the investment. The safest investments (such as U.S. government bonds) typically have a 3% to 4% real rate of return built into their overall interest rates. Thus, an interest rate of, say, 9% per year on a U.S. government bond means that investors expect the inflation rate to be in the range of 5% to 6% per year. Clearly, then, inflation causes interest rates to rise.

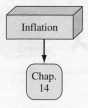

Inflation

Chap. 14

From the borrower's perspective, the rate of inflation is simply another interest rate *tacked on to the real interest rate*. And, from the vantage point of the saver or investor in a fixed-interest account, inflation *reduces the real rate of return* on the investment. Inflation means that cost and revenue cash flow estimates increase over time. This increase is due to the changing value of money that is forced upon a country's currency by inflation, thus making a unit of currency (one dollar) worth less relative to its value at a previous time. We see the effect of inflation in that money purchases less now than it did at a previous time. Inflation contributes to

- A reduction in purchasing power of the currency.
- An increase in the CPI (consumer price index).
- An increase in the cost of equipment and its maintenance.
- An increase in the cost of salaried professionals and hourly employees.
- A reduction in the real rate of return on personal savings and certain corporate investments.

In other words, inflation can materially contribute to changes in corporate and personal economic analysis.

Commonly, engineering economy studies assume that inflation affects all estimated values equally. Accordingly, an interest rate or rate of return, such as 8% per year, is applied throughout the analysis without accounting for an additional inflation rate. However, if inflation were explicitly taken into account, and it was reducing the value of money at, say, an average of 4% per year, it would be necessary to perform the economic analysis using an inflated interest rate of 12.32% per year. (The relevant relations are derived in Chapter 14.) On the other hand, if the stated ROR on an investment is 8% with inflation included, the same inflation rate of 4% per year results in a real rate of return of only 3.85% per year!

1.5   EQUIVALENCE

Equivalent terms are used very often in the transfer from one scale to another. Some common equivalencies or conversions are as follows:

Length: 100 centimeters = 1 meter   1000 meters = 1 kilometer
            12 inches = 1 foot   3 feet = 1 yard   39.370 inches = 1 meter

Pressure: 1 atmosphere = 1 newton/meter^2
           1 atmosphere = 10^3 pascal = 1 kilopascal

Many equivalent measures are a combination of two or more scales. For example, 110 kilometers per hour (kph) is equivalent to 68 miles per hour (mph) or 1.133 miles per minute, based on the equivalence that 1 mile = 1.6093 kilometers and 1 hour = 60 minutes. We can further conclude that driving at approximately 68 mph for 2 hours is equivalent to traveling a total of about 220 kilometers, or 136 miles. Three scales—time in hours, length in miles, and length in kilometers—are combined to develop equivalent statements. An additional use of these equivalencies is to estimate driving time in hours between two cities using two maps, one indicating distance in miles, a second showing kilometers. Note that throughout these statements the fundamental relation 1 mile = 1.6093 kilometers is used. If this relation changes, then the other equivalencies would be in error.

Figure 1–3
Equivalence of three amounts at a 6% per year interest rate.

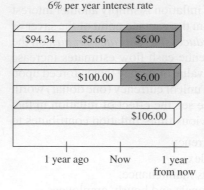

6% per year interest rate

$94.34 | $5.66 | $6.00

$100.00 | $6.00

$106.00

1 year ago   Now   1 year from now

When considered together, the time value of money and the interest rate help develop the concept of *economic equivalence,* which means that different sums of money at different times would be equal in economic value. For example, if the interest rate is 6% per year, $100 today (present time) is equivalent to $106 one year from today.

$$\text{Amount accrued} = 100 + 100(0.06) = 100(1 + 0.06) = \$106$$

So, if someone offered you a gift of $100 today or $106 one year from today, it would make no difference which offer you accepted from an economic perspective. In either case you have $106 one year from today. However, the two sums of money are equivalent to each other *only* when the interest rate is 6% per year. At a higher or lower interest rate, $100 today is not equivalent to $106 one year from today.

In addition to future equivalence, we can apply the same logic to determine equivalence for previous years. A total of $100 now is equivalent to $100/1.06 = $94.34 one year ago at an interest rate of 6% per year. From these illustrations, we can state the following: $94.34 last year, $100 now, and $106 one year from now are equivalent at an interest rate of 6% per year. The fact that these sums are equivalent can be verified by computing the two interest rates for 1-year interest periods.

$$\frac{\$6}{\$100} \times 100\% = 6\% \text{ per year}$$

and

$$\frac{\$5.66}{\$94.34} \times 100\% = 6\% \text{ per year}$$

Figure 1–3 indicates the amount of interest each year necessary to make these three different amounts equivalent at 6% per year.

EXAMPLE **1.6**

AC-Delco makes auto batteries available to General Motors dealers through privately owned distributorships. In general, batteries are stored throughout the year, and a 5% cost increase is added each year to cover the inventory carrying charge for the distributorship owner. Assume you own the City Center Delco facility. Make the calculations necessary to show which of the following statements are true and which are false about battery costs.

(a) The amount of $98 now is equivalent to a cost of $105.60 one year from now.
(b) A truck battery cost of $200 one year ago is equivalent to $205 now.
(c) A $38 cost now is equivalent to $39.90 one year from now.
(d) A $3000 cost now is equivalent to $2887.14 one year ago.
(e) The carrying charge accumulated in 1 year on an investment of $2000 worth of batteries is $100.

Solution
(a) Total amount accrued = 98(1.05) = $102.90 ≠ $105.60; therefore, it is false. Another way to solve this is as follows: Required original cost is 105.60/1.05 = $100.57 ≠ $98.
(b) Required old cost is 205.00/1.05 = $195.24 ≠ $200; therefore, it is false.
(c) The cost 1 year from now is $38(1.05) = $39.90; true.
(d) Cost now is 2887.14(1.05) = $3031.50 ≠ $3000; false.
(e) The charge is 5% per year interest, or 2000(0.05) = $100; true.

1.6   SIMPLE AND COMPOUND INTEREST

The terms *interest, interest period,* and *interest rate* (introduced in Section 1.4) are useful in calculating equivalent sums of money for one interest period in the past and one period in the future. However, for more than one interest period, the terms *simple interest* and *compound interest* become important.

Simple interest is calculated using the principal only, ignoring any interest accrued in preceding interest periods. The total simple interest over several periods is computed as

$$\textbf{Interest} = \textbf{(principal)(number of periods)(interest rate)} \qquad [1.5]$$

where the interest rate is expressed in decimal form.

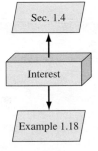

Sec. 1.4

Interest

Example 1.18

EXAMPLE 1.7

Pacific Telephone Credit Union loaned money to an engineering staff member for a radio-controlled model airplane. The loan is for $1000 for 3 years at 5% per year simple interest. How much money will the engineer repay at the end of 3 years? Tabulate the results.

Solution
The interest for each of the 3 years is

$$\text{Interest per year} = 1000(0.05) = \$50$$

Total interest for 3 years from Equation [1.5] is

$$\text{Total interest} = 1000(3)(0.05) = \$150$$

The amount due after 3 years is

$$\text{Total due} = \$1000 + 150 = \$1150$$

The $50 interest accrued in the first year and the $50 accrued in the second year do not earn interest. The interest due each year is calculated only on the $1000 principal.

The details of this loan repayment are tabulated in Table 1–1 from the perspective of the borrower. The year zero represents the present, that is, when the money is borrowed. No payment is made until the end of year 3. The amount owed each year increases uniformly by $50, since simple interest is figured on only the loan principal.

TABLE 1–1   Simple Interest Computations

(1) End of Year	(2) Amount Borrowed	(3) Interest	(4) Amount Owed	(5) Amount Paid
0	$1000			
1	—	$50	$1050	$ 0
2	—	50	1100	0
3	—	50	1150	1150

For *compound interest,* the interest accrued for each interest period is calculated on the *principal plus the total amount of interest accumulated in all previous periods.* Thus, compound interest means interest on top of interest. Compound interest reflects the effect of the time value of money on the interest also. Now the interest for one period is calculated as

$$\textbf{Interest} = \textbf{(principal + all accrued interest)(interest rate)} \qquad \textbf{[1.6]}$$

EXAMPLE 1.8

If an engineer borrows $1000 from the company credit union at 5% per year compound interest, compute the total amount due after 3 years. Graph and compare the results of this and the previous example.

Solution
The interest and total amount due each year are computed separately using Equation [1.6].

$$\text{Year 1 interest:} \qquad \$1000(0.05) = \$50.00$$
$$\text{Total amount due after year 1:} \qquad \$1000 + 50.00 = \$1050.00$$
$$\text{Year 2 interest:} \qquad \$1050(0.05) = \$52.50$$
$$\text{Total amount due after year 2:} \qquad \$1050 + 52.50 = \$1102.50$$
$$\text{Year 3 interest:} \qquad \$1102.50(0.05) = \$55.13$$
$$\text{Total amount due after year 3:} \qquad \$1102.50 + 55.13 = \$1157.63$$

TABLE 1–2 Compound Interest Computations, Example 1.8				
(1) End of Year	(2) Amount Borrowed	(3) Interest	(4) Amount Owed	(5) Amount Paid
0	$1000			
1	—	$50.00	$1050.00	$ 0
2	—	52.50	1102.50	0
3	—	55.13	1157.63	1157.63

The details are shown in Table 1–2. The repayment plan is the same as that for the simple interest example—no payment until the principal plus accrued interest is due at the end of year 3.

Figure 1–4 shows the amount owed at the end of each year for 3 years. The difference due to the time value of money is recognized for the compound interest case. An extra $1157.63 − $1150 = $7.63 of interest is paid compared with simple interest over the 3-year period.

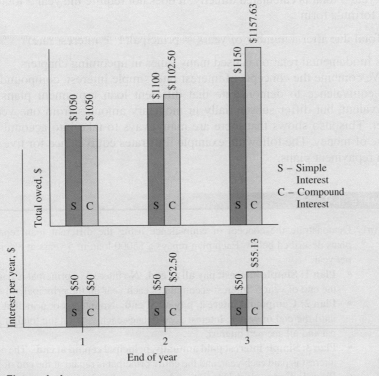

S – Simple
Interest
C – Compound
Interest

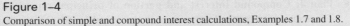

Figure 1–4
Comparison of simple and compound interest calculations, Examples 1.7 and 1.8.

Comment

The difference between simple and compound interest grows each year. If the computations are continued for more years, for example, 10 years, the difference is $128.90; after 20 years compound interest is $653.30 more than simple interest.

If $7.63 does not seem like a significant difference in only 3 years, remember that the beginning amount here is $1000. If we make these same calculations for an initial amount of $100,000 or $1 million, multiply the difference by 100 or 1000, and we are talking real money. This indicates that the power of compounding is vitally important in all economics-based analyses.

Another and shorter way to calculate the total amount due after 3 years in Example 1.8 is to combine calculations rather than perform them on a year-by-year basis. The total due each year is as follows:

$$\text{Year 1:} \qquad \$1000(1.05)^1 = \$1050.00$$
$$\text{Year 2:} \qquad \$1000(1.05)^2 = \$1102.50$$
$$\text{Year 3:} \qquad \$1000(1.05)^3 = \$1157.63$$

The year 3 total is calculated directly; it does not require the year 2 total. In general formula form

$$\text{Total due after a number of years} = \text{principal}(1 + \text{interest rate})^{\text{number of years}}$$

This fundamental relation is used many times in upcoming chapters.

We combine the concepts of interest rate, simple interest, compound interest, and equivalence to demonstrate that different loan repayment plans may be equivalent, but differ substantially in monetary amounts from one year to another. This also shows that there are many ways to take into account the time value of money. The following example illustrates equivalence for five different loan repayment plans.

EXAMPLE 1.9

(*a*) Demonstrate the concept of equivalence using the different loan repayment plans described below. Each plan repays a $5000 loan in 5 years at 8% interest per year.

- **Plan 1: Simple interest, pay all at end.** No interest or principal is paid until the end of year 5. Interest accumulates each year on the principal only.
- **Plan 2: Compound interest, pay all at end.** No interest or principal is paid until the end of year 5. Interest accumulates each year on the total of principal and all accrued interest.
- **Plan 3: Simple interest paid annually, principal repaid at end.** The accrued interest is paid each year, and the entire principal is repaid at the end of year 5.
- **Plan 4: Compound interest and portion of principal repaid annually.** The accrued interest and one-fifth of the principal (or $1000) is repaid each

year. The outstanding loan balance decreases each year, so the interest for each year decreases.

- **Plan 5: Equal payments of compound interest and principal made annually.** Equal payments are made each year with a portion going toward principal repayment and the remainder covering the accrued interest. Since the loan balance decreases at a rate slower than that in plan 4 due to the equal end-of-year payments, the interest decreases, but at a slower rate.

(b) Make a statement about the equivalence of each plan at 8% simple or compound interest, as appropriate.

Solution

(a) Table 1–3 presents the interest, payment amount, total owed at the end of each year, and total amount paid over the 5-year period (column 4 totals).

TABLE 1–3 Different Repayment Schedules Over 5 Years for $5000 at 8% Per Year Interest

(1) End of Year	(2) Interest Owed for Year	(3) Total Owed at End of Year	(4) End-of-Year Payment	(5) Total Owed after Payment
Plan 1: Simple Interest, Pay All at End				
0				$5000.00
1	$400.00	$5400.00	—	5400.00
2	400.00	5800.00	—	5800.00
3	400.00	6200.00	—	6200.00
4	400.00	6600.00	—	6600.00
5	400.00	7000.00	$7000.00	
Totals			$7000.00	
Plan 2: Compound Interest, Pay All at End				
0				$5000.00
1	$400.00	$5400.00	—	5400.00
2	432.00	5832.00	—	5832.00
3	466.56	6298.56	—	6298.56
4	503.88	6802.44	—	6802.44
5	544.20	7346.64	$7346.64	
Totals			$7346.64	
Plan 3: Simple Interest Paid Annually; Principal Repaid at End				
0				$5000.00
1	$400.00	$5400.00	$400.00	5000.00
2	400.00	5400.00	400.00	5000.00
3	400.00	5400.00	400.00	5000.00
4	400.00	5400.00	400.00	5000.00
5	400.00	5400.00	5400.00	
Totals			$7000.00	

TABLE 1–3 (Continued)

(1) End of Year	(2) Interest Owed for Year	(3) Total Owed at End of Year	(4) End-of-Year Payment	(5) Total Owed after Payment
Plan 4: Compound Interest and Portion of Principal Repaid Annually				
0				$5000.00
1	$400.00	$5400.00	$1400.00	4000.00
2	320.00	4320.00	1320.00	3000.00
3	240.00	3240.00	1240.00	2000.00
4	160.00	2160.00	1160.00	1000.00
5	80.00	1080.00	1080.00	
Totals			$6200.00	
Plan 5: Equal Annual Payments of Compound Interest and Principal				
0				$5000.00
1	$400.00	$5400.00	$1252.28	4147.72
2	331.82	4479.54	1252.28	3227.25
3	258.18	3485.43	1252.28	2233.15
4	178.65	2411.80	1252.28	1159.52
5	92.76	1252.28	1252.28	
Totals			$6261.41	

The amounts of interest (column 2) are determined as follows:

Plan 1 Simple interest = (original principal)(0.08)
Plan 2 Compound interest = (total owed previous year)(0.08)
Plan 3 Simple interest = (original principal)(0.08)
Plan 4 Compound interest = (total owed previous year)(0.08)
Plan 5 Compound interest = (total owed previous year)(0.08)

Note that the amounts of the annual payments are different for each repayment schedule and that the total amounts repaid for most plans are different, even though each repayment plan requires exactly 5 years. The difference in the total amounts repaid can be explained (1) by the time value of money, (2) by simple or compound interest, and (3) by the partial repayment of principal prior to year 5.

(*b*) Table 1–3 shows that $5000 at time 0 is equivalent to each of the following:

Plan 1 $7000 at the end of year 5 at 8% simple interest.
Plan 2 $7346.64 at the end of year 5 at 8% compound interest.
Plan 3 $400 per year for 4 years and $5400 at the end of year 5 at 8% simple interest.
Plan 4 Decreasing payments of interest and partial principal in years 1 ($1400) through 5 ($1080) at 8% compound interest.
Plan 5 $1252.28 per year for 5 years at 8% compound interest.

An engineering economy study uses plan 5; interest is compounded, and a constant amount is paid each period. This amount covers accrued interest and a partial amount of principal repayment.

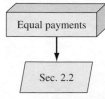

Equal payments

Sec. 2.2

1.7 TERMINOLOGY AND SYMBOLS

The equations and procedures of engineering economy utilize the following terms and symbols. Sample units are indicated.

P = value or amount of money at a time designated as the present or time 0. Also P is referred to as present worth (PW), present value (PV), net present value (NPV), discounted cash flow (DCF), and capitalized cost (CC); dollars

F = value or amount of money at some future time. Also F is called future worth (FW) and future value (FV); dollars

A = series of consecutive, equal, end-of-period amounts of money. Also A is called the annual worth (AW) and equivalent uniform annual worth (EUAW); dollars per year, dollars per month

n = number of interest periods; years, months, days

i = interest rate or rate of return per time period; percent per year, percent per month; percent per day

t = time, stated in periods; years, months, days

The symbols P and F represent one-time occurrences: A occurs with the same value once each interest period for a specified number of periods. It should be clear that a present value P represents a single sum of money at some time prior to a future value F or prior to the first occurrence of an equivalent series amount A.

It is important to note that the symbol A always represents a uniform amount (i.e., the same amount each period) that extends through *consecutive* interest periods. Both conditions must exist before the series can be represented by A.

The interest rate i is assumed to be a compound rate, unless specifically stated as simple interest. The rate i is expressed in percent per interest period, for example, 12% per year. Unless stated otherwise, assume that the rate applies throughout the entire n years or interest periods. The decimal equivalent for i is always used in engineering economy computations.

All engineering economy problems involve the element of time n and interest rate i. In general, every problem will involve at least four of the symbols P, F, A, n, and i, with at least three of them estimated or known.

EXAMPLE 1.10

A new college graduate has a job with Boeing Aerospace. She plans to borrow $10,000 now to help in buying a car. She has arranged to repay the entire principal plus 8% per year interest after 5 years. Identify the engineering economy symbols involved and their values for the total owed after 5 years.

Solution
In this case, P and F are involved, since all amounts are single payments, as well as n and i. Time is expressed in years.

$$P = \$10,000 \qquad i = 8\% \text{ per year} \qquad n = 5 \text{ years} \qquad F = ?$$

The future amount F is unknown.

EXAMPLE 1.11

Assume you borrow $2000 now at 7% per year for 10 years and must repay the loan in equal yearly payments. Determine the symbols involved and their values.

Solution
Time is in years.

$$P = \$2000$$

$$A = ? \text{ per year for 5 years}$$

$$i = 7\% \text{ per year}$$

$$n = 10 \text{ years}$$

In Examples 1.10 and 1.11, the P value is a receipt to the borrower, and F or A is a disbursement from the borrower. It is equally correct to use these symbols in the reverse roles.

EXAMPLE 1.12

On July 1, 2002, your new employer Ford Motor Company deposits $5000 into your money market account, as part of your employment bonus. The account pays interest at 5% per year. You expect to withdraw an equal annual amount for the following 10 years. Identify the symbols and their values.

Solution
Time is in years.

$$P = \$5000$$

$$A = ? \text{ per year}$$

$$i = 5\% \text{ per year}$$

$$n = 10 \text{ years}$$

EXAMPLE 1.13

You plan to make a lump-sum deposit of $5000 now into an investment account that pays 6% per year, and you plan to withdraw an equal end-of-year amount of $1000 for 5 years, starting next year. At the end of the sixth year, you plan to close your account by withdrawing the remaining money. Define the engineering economy symbols involved.

Solution
Time is expressed in years.

$$P = \$5000$$
$$A = \$1000 \text{ per year for 5 years}$$
$$F = ? \text{ at end of year 6}$$
$$i = 6\% \text{ per year}$$
$$n = 5 \text{ years for the } A \text{ series and 6 for the } F \text{ value}$$

EXAMPLE 1.14

Last year Jane's grandmother offered to put enough money into a savings account to generate $1000 this year to help pay Jane's expenses at college. (a) Identify the symbols, and (b) calculate the amount that had to be deposited exactly 1 year ago to earn $1000 in interest now, if the rate of return is 6% per year.

Solution
(a) Time is in years.

$$P = ?$$
$$i = 6\% \text{ per year}$$
$$n = 1 \text{ year}$$
$$F = P + \text{interest}$$
$$= ? + \$1000$$

(b) Refer to Equations [1.3] and [1.4]. Let F = total amount now and P = original amount. We know that $F - P = \$1000$ is the accrued interest. Now we can determine P for Jane and her grandmother.

$$F = P + P(\text{interest rate})$$

The $1000 interest can be expressed as

$$\text{Interest} = F - P = [P + P(\text{interest rate})] - P$$
$$= P(\text{interest rate})$$
$$\$1000 = P(0.06)$$
$$P = \frac{1000}{0.06} = \$16,666.67$$

1.8   INTRODUCTION TO SOLUTION BY COMPUTER

The functions on a computer spreadsheet can greatly reduce the amount of hand and calculator work for equivalency computations involving *compound interest* and the terms P, F, A, i, and n. The power of the electronic spreadsheet often makes it possible to enter a predefined spreadsheet function into one cell and obtain the final answer immediately. Any spreadsheet system can be used—one off the shelf, such as Microsoft Excel©, or one specially developed with built-in financial functions and operators. Excel is used throughout this book because it is readily available and easy to use.

Appendix A is a primer on using spreadsheets and Excel. The functions used in engineering economy are described there in detail, with explanations of all the parameters (also called arguments) placed between parentheses after the function identifier. The Excel online help function provides similar information. Appendix A also includes a section on spreadsheet layout that is useful when the economic analysis is presented to someone else—a coworker, a boss, or a professor.

A total of six Excel functions can perform most of the fundamental engineering economy calculations. However, these functions are no substitute for knowing how the time value of money and compound interest work. The functions are great supplemental tools, but they do not replace the understanding of engineering economy relations, assumptions, and techniques.

Using the symbols P, F, A, i, and n exactly as defined in the previous section, the Excel functions most used in engineering economic analysis are formulated as follows.

To find the present value P: PV($i\%$,n,A,F)

To find the future value F: FV($i\%$,n,A,P)

To find the equal, periodic value A: PMT($i\%$,n,P,F)

To find the number of periods n: NPER($i\%$,A,P,F)

To find the compound interest rate i: RATE(n,A,P,F)

To find the compound interest rate i: IRR(first_cell:last_cell)

To find the present value P of any series: NPV($i\%$, second_cell:last_cell) + first_cell

If some of the parameters don't apply to a particular problem, they can be omitted and zero is assumed. If the parameter omitted is an interior one, the comma must be entered. The last two functions require that a series of numbers be entered into contiguous spreadsheet cells, but the first five can be used with no supporting data. In all cases, the function must be preceded by an equals sign ($=$) in the cell where the answer is to be displayed.

Each of these functions will be introduced and illustrated at the point in this text where they are most useful. However, to get an idea of how they work, look back at Examples 1.10 and 1.11. In Example 1.10, the future amount F is unknown, as indicated by $F = ?$ in the solution. In the next chapter, we will learn how the time value of money is used to find F, given P, i, and n. To find F in this example using a spreadsheet, simply enter the FV function preceded by an equals

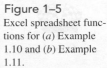

Figure 1–5
Excel spreadsheet functions for (*a*) Example 1.10 and (*b*) Example 1.11.

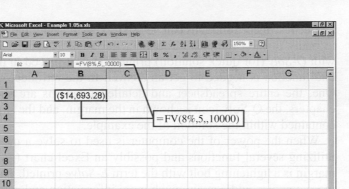

(*a*)

(*b*)

sign into any cell. The format is =FV(*i*%,*n*,,*P*) or =FV(8%,5,,10000). The comma is entered because there is no *A* involved. Figure 1–5*a* is a screen image of the Excel spreadsheet with the FV function entered into cell B2. The answer of $−14,693.28 is displayed. The answer is in red on the actual Excel screen to indicate a negative amount from the borrower's perspective to repay the loan after 5 years. The FV function is shown in the formula bar above the worksheet itself. Also, we have added a cell tag to show the format of the FV function.

In Example 1.11, the uniform annual amount *A* is sought, and *P*, *i*, and *n* are known. Find *A*, using the function PMT(*i*%,*n*,*P*) or, in this example, PMT(7%,10,2000). Figure 1–5*b* shows the result in cell C4. The format of the FV function is shown in the formula bar and the cell tag.

Q-Solv

Because these functions can be used so easily and rapidly, we will detail them in many of the examples throughout the book. A special checkered-flag icon, with *Q-Solv* (for quick solution) printed on it, is placed in the margin when just one function is needed to get an answer. In the introductory chapters of Level One, the entire spreadsheet and detailed functions are shown. In succeeding chapters, the Q-Solv icon is shown in the margin, and the spreadsheet function is contained within the solution of the example.

When the power of the computer is used to solve a more complex problem utilizing several functions and possibly an Excel chart (graph), the icon in the margin is a lightening bolt with the term *E-Solve* printed. These spreadsheets are more complex and contain much more information and computation, especially when sensitivity analysis is performed. The Solution by Computer answer to an example is always presented after the Solution by Hand. As mentioned earlier, the spreadsheet function is not a replacement for the correct understanding and application of the engineering economy relations. Therefore, the hand and computer solutions complement each other.

1.9   MINIMUM ATTRACTIVE RATE OF RETURN

For any investment to be profitable, the investor (corporate or individual) expects to receive more money than the amount invested. In other words, a fair *rate of return,* or *return on investment,* must be realizable. The definition of ROR in Equation [1.4] is used in this discussion, that is, amount earned divided by the original amount.

Engineering alternatives are evaluated upon the prognosis that a reasonable ROR can be expected. Therefore, some reasonable rate must be established for the selection criteria phase of the engineering economy study (Figure 1–1). The reasonable rate is called the *Minimum Attractive Rate of Return (MARR)* and is higher than the rate expected from a bank or some safe investment that involves minimal investment risk. Figure 1–6 indicates the relations between different rate of return values. In the United States, the current U.S. Treasury bill return is sometimes used as the benchmark safe rate.

The MARR is also referred to as the *hurdle rate* for projects; that is, to be considered financially viable the expected ROR must meet or exceed the MARR or hurdle rate. Note that the MARR is not a rate that is calculated like a ROR. The MARR is established by (financial) managers and is used as a criterion against which an alternative's ROR is measured, when making the accept/reject decision.

To develop a foundation-level understanding of how a MARR value is established and used, we must return to the term *capital* introduced in Section 1.1. Capital is also referred to as *capital funds* and *capital investment money.* It always costs money in the form of interest to raise capital. The interest, stated as a percentage rate, is called the *cost of capital.* For example, if you want to purchase a new music system, but don't have sufficient money (capital), you could obtain a credit union loan at some interest rate, say, 9% per year and use that cash to pay the merchant now. Or, you could use your (newly acquired) credit card and

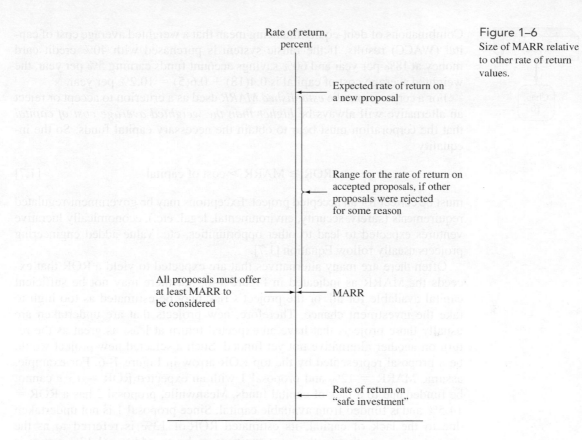

Figure 1–6
Size of MARR relative to other rate of return values.

Rate of return, percent

Expected rate of return on a new proposal

Range for the rate of return on accepted proposals, if other proposals were rejected for some reason

All proposals must offer at least MARR to be considered

MARR

Rate of return on "safe investment"

pay off the balance on a monthly basis. This approach will probably cost you at least 18% per year. Or, you could use funds from your savings account that earns 5% per year and pay cash. The 9%, 18%, and 5% rates are your cost of capital estimates to raise the capital for the system by different methods of capital financing. In analogous ways, corporations estimate the cost of capital from different sources to raise funds for engineering projects and other types of projects.

In general, capital is developed in two ways—equity financing and debt financing. A combination of these two is very common for most projects. Chapter 10 covers these in greater detail, but a snapshot description follows.

Equity financing. The corporation uses its own funds from cash on hand, stock sales, or retained earnings. Individuals can use their own cash, savings, or investments. In the example above, using money from the 5% savings account is equity financing.

Debt financing. The corporation borrows from outside sources and repays the principal and interest according to some schedule, much like the plans in Table 1–3. Sources of debt capital may be bonds, loans, mortgages, venture capital pools, and many others. Individuals, too, can utilize debt sources, such as the credit card and credit union options described in the music system example.

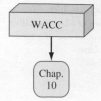

Combinations of debt-equity financing mean that a weighted average cost of capital (WACC) results. If the music system is purchased with 40% credit card money at 18% per year and 60% savings account funds earning 5% per year, the weighted average cost of capital is 0.4(18) + 0.6(5) = 10.2% per year.

For a corporation, the *established MARR* used as a criterion to accept or reject an alternative will always be *higher than the weighted average cost of capital* that the corporation must bear to obtain the necessary capital funds. So the inequality

$$\text{ROR} \geq \text{MARR} > \text{cost of capital} \qquad [1.7]$$

must be correct for an accepted project. Exceptions may be government-regulated requirements (safety, security, environmental, legal, etc.), economically lucrative ventures expected to lead to other opportunities, etc. Value-added engineering projects usually follow Equation [1.7].

Often there are many alternatives that are expected to yield a ROR that exceeds the MARR as indicated in Figure 1–6, but there may not be sufficient capital available for all, or the project's risk may be estimated as too high to take the investment chance. Therefore, new projects that are undertaken are usually those projects that have an expected return at least as great as the return on another alternative not yet funded. Such a selected new project would be a proposal represented by the top ROR arrow in Figure 1–6. For example, assume MARR = 12% and proposal 1 with an expected ROR = 13% cannot be funded due to a lack of capital funds. Meanwhile, proposal 2 has a ROR = 14.5% and is funded from available capital. Since proposal 1 is not undertaken due to the lack of capital, its estimated ROR of 13% is referred to as the *opportunity cost;* that is, the opportunity to make an additional 13% return is forgone.

1.10 CASH FLOWS: THEIR ESTIMATION AND DIAGRAMMING

In Section 1.3 cash flows are described as the inflows and outflows of money. These cash flows may be estimates or observed values. Every person or company has cash receipts—revenue and income (inflows); and cash disbursements—expenses, and costs (outflows). These receipts and disbursements are the cash flows, with a plus sign representing cash inflows and a minus sign representing cash outflows. Cash flows occur during specified periods of time, such as 1 month or 1 year.

Of all the elements of the engineering economy study approach (Figure 1–1), cash flow estimation is likely the most difficult and inexact. Cash flow estimates are just that—estimates about an uncertain future. Once estimated, the techniques of this book guide the decision making process. But the time-proven accuracy of an alternative's estimated cash inflows and outflows clearly dictates the quality of the economic analysis and conclusion.

Cash inflows, or receipts, may be comprised of the following, depending upon the nature of the proposed activity and the type of business involved.

Samples of Cash Inflow Estimates

Revenues (usually *incremental* resulting from an alternative).

Operating cost reductions (resulting from an alternative).

Asset salvage value.

Receipt of loan principal.

Income tax savings.

Receipts from stock and bond sales.

Construction and facility cost savings.

Saving or return of corporate capital funds.

Cash outflows, or disbursements, may be comprised of the following, again depending upon the nature of the activity and type of business.

Samples of Cash Outflow Estimates

First cost of assets.

Engineering design costs.

Operating costs (annual and incremental).

Periodic maintenance and rebuild costs.

Loan interest and principal payments.

Major expected/unexpected upgrade costs.

Income taxes.

Expenditure of corporate capital funds.

Background information for estimates may be available in departments such as accounting, finance, marketing, sales, engineering, design, manufacturing, production, field services, and computer services. The accuracy of estimates is largely dependent upon the experiences of the person making the estimate with similar situations. Usually *point estimates* are made; that is, a single-value estimate is developed for each economic element of an alternative. If a statistical approach to the engineering economy study is undertaken, a *range estimate* or *distribution estimate* may be developed. Though more involved computationally, a statistical study provides more complete results when key estimates are expected to vary widely. We will use point estimates throughout most of this book. Final chapters discuss decision making under risk.

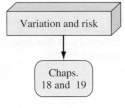

Once the cash inflow and outflow estimates are developed, the net cash flow can be determined.

$$\text{Net cash flow} = \text{receipts} - \text{disbursements}$$

$$= \text{cash inflows} - \text{cash outflows} \qquad [1.8]$$

Since cash flows normally take place at varying times within an interest period, a simplifying assumption is made.

> The *end-of-period convention* means that all cash flows are assumed to occur at the end of an interest period. When several receipts and disbursements occur within a given interest period, the *net* cash flow is assumed to occur at the *end* of the interest period.

However, it should be understood that, although *F* or *A* amounts are located at the end of the interest period by convention, the end of the period is not necessarily December 31. In Example 1.12 the deposit took place on July 1, 2002, and the withdrawals will take place on July 1 of each succeeding year for 10 years. *Thus, end of the period means end of interest period, not end of calendar year.*

The *cash flow diagram* is a very important tool in an economic analysis, especially when the cash flow series is complex. It is a graphical representation of cash flows drawn on a time scale. The diagram includes what is known, what is estimated, and what is needed. That is, once the cash flow diagram is complete, another person should be able to work the problem by looking at the diagram.

Cash flow diagram time $t = 0$ is the present, and $t = 1$ is the end of time period 1. We assume that the periods are in years for now. The time scale of Figure 1–7 is set up for 5 years. Since the end-of-year convention places cash flows at the end of years, the "1" marks the end of year 1.

While it is not necessary to use an exact scale on the cash flow diagram, you will probably avoid errors if you make a neat diagram to approximate scale for both time and relative cash flow magnitudes.

The direction of the arrows on the cash flow diagram is important. A vertical arrow pointing up indicates a positive cash flow. Conversely, an arrow pointing down indicates a negative cash flow. Figure 1–8 illustrates a receipt (cash inflow) at the end of year 1 and equal disbursements (cash outflows) at the end of years 2 and 3.

The perspective or vantage point must be determined prior to placing a sign on each cash flow and diagramming it. As an illustration, if you borrow $2500 to buy a $2000 used Harley-Davidson for cash, and you use the remaining $500 for a new paint job, there may be several different perspectives taken. Possible

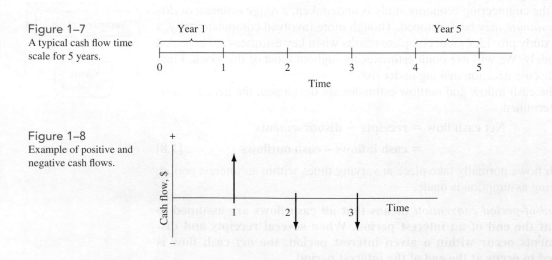

Figure 1–7
A typical cash flow time scale for 5 years.

Figure 1–8
Example of positive and negative cash flows.

perspectives, cash flow signs, and amounts are as follows.

Perspective	Cash Flow, $
Credit union	−2500
You as borrower	+2500
You as purchaser,	−2000
and as paint customer	−500
Used cycle dealer	+2000
Paint shop owner	+500

EXAMPLE 1.15

Reread Example 1.10, where $P = \$10,000$ is borrowed at 8% per year and F is sought after 5 years. Construct the cash flow diagram.

Solution
Figure 1–9 presents the cash flow diagram from the vantage point of the borrower. The present sum P is a cash inflow of the loan principal at year 0, and the future sum F is the cash outflow of the repayment at the end of year 5. The interest rate should be indicated on the diagram.

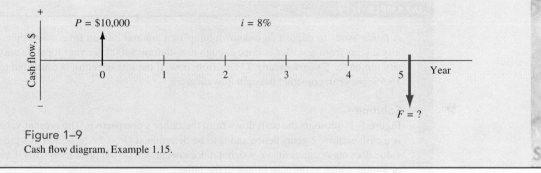

Figure 1–9
Cash flow diagram, Example 1.15.

EXAMPLE 1.16

Each year Exxon-Mobil expends large amounts of funds for mechanical safety features throughout its worldwide operations. Carla Ramos, a lead engineer for Mexico and Central American operations, plans expenditures of $1 million now and each of the next 4 years just for the improvement of field-based pressure-release valves. Construct the cash flow diagram to find the equivalent value of these expenditures at the end of year 4, using a cost of capital estimate for safety-related funds of 12% per year.

Solution
Figure 1–10 indicates the uniform and negative cash flow series (expenditures) for five periods, and the unknown F value (positive cash flow equivalent) at exactly the same

time as the fifth expenditure. Since the expenditures start immediately, the first $1 million is shown at time 0, not time 1. Therefore, the last negative cash flow occurs at the end of the fourth year, when F also occurs. To make this diagram appear similar to that of Figure 1–9 with a full 5 years on the time scale, the addition of the year -1 prior to year 0 completes the diagram for a full 5 years. This addition demonstrates that year 0 is the end-of-period point for the year -1.

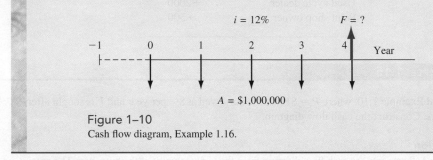

Figure 1–10
Cash flow diagram, Example 1.16.

EXAMPLE 1.17

A father wants to deposit an unknown lump-sum amount into an investment opportunity 2 years from now that is large enough to withdraw $4000 per year for state university tuition for 5 years starting 3 years from now. If the rate of return is estimated to be 15.5% per year, construct the cash flow diagram.

Solution

Figure 1–11 presents the cash flows from the father's perspective. The present value P is a cash outflow 2 years hence and is to be determined ($P = ?$). Note that this present value does not occur at time $t = 0$, but it does occur one period prior to the first A value of $4000, which is the cash inflow to the father.

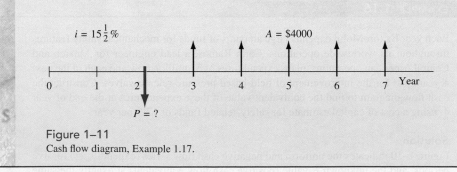

Figure 1–11
Cash flow diagram, Example 1.17.

Additional Examples 1.19 and 1.20.

1.11 RULE OF 72: ESTIMATING DOUBLING TIME AND INTEREST RATE

Sometimes it is helpful to estimate the number of years n or the rate of return i required for a single cash flow amount to double in size. The *rule of 72* for compound interest rates can be used to estimate i or n, given the other value. The estimation is simple; the time required for an initial single amount to double in size with compound interest is approximately equal to 72 divided by the rate of return in percent.

$$\text{Estimated } n = \frac{72}{i} \qquad [1.9]$$

For example, at a rate of 5% per year, it would take approximately $72/5 =$ 14.4 years for a current amount to double. (The actual time required is 14.3 years, as will be shown in Chapter 2.) Table 1–4 compares the times estimated from the rule of 72 to the actual times required for doubling at several compounded rates. As you can see, very good estimates are obtained.

Alternatively, the compound rate i in percent required for money to double in a specified period of time n can be estimated by dividing 72 by the specified n value.

$$\text{Estimated } i = \frac{72}{n} \qquad [1.10]$$

In order for money to double in a time period of 12 years, for example, a compound rate of return of approximately $72/12 = 6\%$ per year would be required. The exact answer is 5.946% per year.

If the interest is simple, a rule of 100 may be used in the same way. In this case the answers obtained will always be exactly correct. As illustrations, money doubles in exactly 12 years at $100/12 = 8.33\%$ simple interest. Or, at 5% simple interest it takes exactly $100/5 = 20$ years to double.

TABLE **1–4**	Doubling Time Estimates Using the Rule of 72 and the Actual Time Using Compound Interest Calculations	
	Doubling Time, Years	
Rate of Return, % per Year	Rule-of-72 Estimate	Actual Years
1	72	70
2	36	35.3
5	14.4	14.3
10	7.2	7.5
20	3.6	3.9
40	1.8	2.0

E-Solve

1.12   SPREADSHEET APPLICATION—SIMPLE AND COMPOUND INTEREST, AND CHANGING CASH FLOW ESTIMATES

The example below demonstrates how an Excel spreadsheet can be used to obtain equivalent future values. A key feature is the use of mathematical relations developed in the cells to perform sensitivity analysis for changing cash flow estimates and the interest rate. To answer these basic questions using hand solution can be time-consuming; the spreadsheet makes it much easier.

EXAMPLE **1.18**

A Japan-based architectural firm has asked a United States–based software engineering group to infuse GIS (geographical information system) sensing capability via satellite into monitoring software for high-rise structures in order to detect greater-than-expected horizontal movements. This software could be very beneficial as an advance warning of serious tremors in earthquake-prone areas in Japan and the United States. The inclusion of accurate GIS data is estimated to increase annual revenue over that for the current software system by $200,000 for each of the next 2 years, and by $300,000 for each of years 3 and 4. The planning horizon is only 4 years due to the rapid advances made internationally in building-monitoring software. Develop spreadsheets to answer the questions below.

(*a*)   Determine the equivalent future value in year 4 of the increased cash flows, using an 8% per year rate of return. Obtain answers for both simple and compound interest.

(*b*)   Rework part (*a*) if the cash flow estimates in years 3 and 4 increase from $300,000 to $600,000.

(*c*)   The financial manager of the U.S. company wants to consider the effects of 4% per year inflation in the analysis of part (*a*). As mentioned in Section 1.4, inflation reduces the real rate of return. For the 8% rate of return, an inflation rate of 4% per year compounded each year reduces the return to 3.85% per year.

Solution by Computer

Refer to Figure 1–12*a* to *c* for the solutions. All three spreadsheets contain the same information, but the cell values are altered as required by the question. (Actually, all the questions posed here can be answered on one spreadsheet by simply changing the numbers. Three spreadsheets are shown here for explanation purposes only.)

The Excel functions are constructed with reference to the cells, not the values themselves, so that sensitivity analysis can be performed without function changes. This approach treats the value in a cell as a *global variable* for the spreadsheet. For example, the 8% (simple or compound interest) rate in cell B4 will be referenced in all functions as B4, not 8%. Thus, a change in the rate requires only one alteration in the cell B4 entry, not in every spreadsheet relation and function where 8% is used. Key Excel relations are detailed in the cell tags.

(*a*)   *8% simple interest.* Refer to Figure 1–12*a*, columns C and D, for the answers. Simple interest earned each year (column C) incorporates Equation [1.5] one year

Microsoft Excel - Example 1.18(a-c)

File Edit View Insert Format Tools Data Window Help

A22

	A	B	C	D	E	F	G	H
1	Example 1.18 (contains 3 worksheets)							
2	Part (a) - Find F in year 4							
3								
4	Rate of return	8.00%						
5								
6								
7			Simple interest		Compound interest			
8				Cumulative		Cumulative		
9	End of year		Interest earned	equivalent	Interest earned	equivalent		
10	(EOY)	Cash flow	during year	EOY cash flow	during year	EOY cash flow		
11	0	$ -						
12	1	$ 200,000	$ -	$ 200,000	$ -	$ 200,000		
13	2	$ 200,000	$ 16,000	$ 416,000	$ 16,000	$ 416,000		
14	3	$ 300,000	$ 32,000	$ 748,000	$ 33,280	$ 749,280		
15	4	$ 300,000	$ 56,000	$ 1,104,000	$ 59,942	$ 1,109,222		
16	Total	$1,000,000	$ 104,000		$ 109,222			

=B15+E15+F14

=B12*B4

=C14+B14*B4

=SUM(B15:C15)+D14

=F14*B4

Ex 1.18(a) / Ex 1.18(b) / Ex 1.18(c) / Sheet4 / Sheet5 / Sheet6 / Sheet7

Ready

(a)

Microsoft Excel - Example 1.18(a-c)

File Edit View Insert Format Tools Data Window Help

A22

	A	B	C	D	E	F	G
1							
2			(b) Find F with increased cash flow estimates				
3							
4	Rate of return	8.00%					
5							
6							
7			Simple interest		Compound interest		
8				Cumulative		Cumulative	
9	End of year		Interest earned	equivalent	Interest earned	equivalent	
10	(EOY)	Cash flow	during year	EOY cash flow	during year	EOY cash flow	
11	0	$ -					
12	1	$ 200,000	$ -	$ 200,000	$ -	$ 200,000	
13	2	$ 200,000	$ 16,000	$ 416,000	$ 16,000	$ 416,000	
14	3	$ 600,000	$ 32,000	$ 1,048,000	$ 33,280	$ 1,049,280	
15	4	$ 600,000	$ 80,000	$ 1,728,000	$ 83,942	$ 1,733,222	
16	Total	$1,600,000	$ 128,000		$ 133,222		

Ex 1.18(a) / Ex 1.18(b) / Ex 1.18(c) / Sheet4 / Sheet5 / Sheet6 / Sheet7

Ready   NUM

(b)

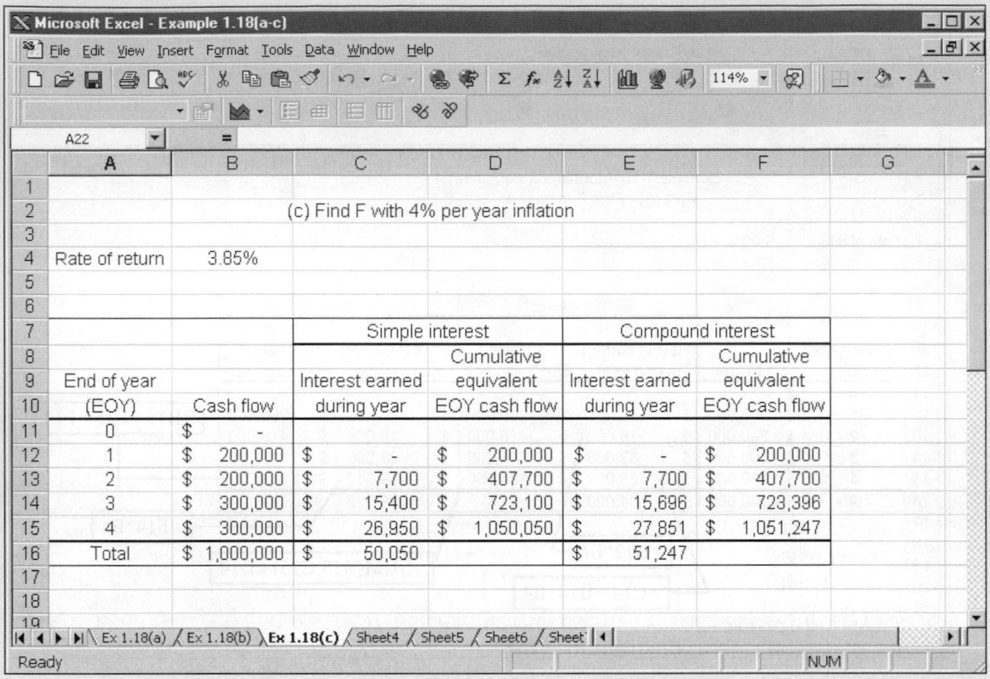

(c)

Figure 1–12
Spreadsheet solution including sensitivity analysis, Example 1.18(a)–(c).

at a time into the interest relation by using only the end-of-year (EOY) cash flow amounts ($200,000 or $300,000) to determine interest for the next year. This interest is added to the interest from all previous years. In $1000 units,

Year 2:   C13 = B12*B4 = $200(0.08) = $16   (see the cell tag)

Year 3:   C14 = C13 + B13*B4 = $16 + 200(0.08) = $32

Year 4:   C15 = C14 + B14*B4 = $32 + 300(0.08) = $56   (see the cell tag)

Remember, an = sign must precede each relation in the spreadsheet. Cell C16 contains the function SUM(C12:C15) to display the total simple interest of $104,000 over the 4 years. The future value is in D15. It is F = $1,104,000 which includes the cumulative amount of all cash flows and all simple interest. In $1000 units, example functions are

Year 2:   D13 = SUM(B13:C13) + D12 = ($200 +16) + 200 = $416

Year 4:   D15 = SUM(B15:C15) + D14 = ($300 + 56) + 748 = $1104

8% compound interest. See Figure 1–12a, columns E and F. The spreadsheet structure is the same, except that Equation [1.6] is incorporated into the compound interest values in column E, thus adding interest on top of earned interest.

Interest at 8% is based on the accumulated cash flow at the end of the previous year. In $1000 units,

Year 2 interest: E13 = F12*B4 = $200(0.08) = $16
Cumulative cash flow: F13 = B13 + E13 + F12 = $200 + 16 + 200 = $416

Year 4 interest: E15 = F14*B4 = $749.28(0.08) = $59.942
                                (see the cell tag)
Cumulative cash flow: F15 = B15 + E15 + F14
                                = $300 + 59.942 + 749.280 = $1109.222

The equivalent future value is in cell F15, where $F = \$1,109,222$ is shown.

   The cash flows are equivalent to $1,104,000 at a simple 8% interest rate, and $1,109,222 at a compound 8% interest rate. Using a compound interest rate increases the F value by $5222.

   Note that it is not possible to use the FV function in this case because the A values are not the same for all 4 years. We will learn how to use all the basic functions in more versatile ways in the next few chapters.

(b) Refer to Figure 1–12b. In order to initialize the spreadsheet with the two increased cash flow estimates, replace the $300,000 values in B14 and B15 with $600,000. All spreadsheet relations are identical, and the new interest and accumulated cash flow values are shown immediately. The equivalent fourth-year F values have increased for both the 8% simple and compound interest rates (D15 and F15, respectively).

(c) Figure 1–12c is identical to the spreadsheet in Figure 1–12a, except the cell B4 now contains the rate of 3.85%. The corresponding F value for compound interest in F15 has decreased to $1,051,247 from the $1,109,222 at 8%. This represents an effect of inflation of $57,975 in only four years. It is no surprise that governments, corporations, engineers, and all individuals are concerned when inflation rises and the currency is worth less over time.

Comment

When working with an Excel spreadsheet, it is possible to display all of the entries and functions on the screen by simultaneously touching the <Ctrl> and <`> keys, which may be in the upper left of the keyboard on the key with <~>. Additionally, it may be necessary to widen some columns in order to display the entire function statement.

ADDITIONAL EXAMPLES

EXAMPLE **1.19**

CASH FLOW DIAGRAMS

A rental company spent $2500 on a new air compressor 7 years ago. The annual rental income from the compressor has been $750. Additionally, the $100 spent on maintenance during the first year has increased each year by $25. The company plans to sell the compressor at the end of next year for $150. Construct the cash flow diagram from the company's perspective.

Solution
Use now as time $t = 0$. The incomes and costs for years -7 through 1 (next year) are tabulated below with net cash flow computed using Equation [1.8]. The net cash flows (one negative, eight positive) are diagrammed in Figure 1–13.

End of year	Income	Cost	Net Cash Flow
-7	$ 0	$2500	$-2500
-6	750	100	650
-5	750	125	625
-4	750	150	600
-3	750	175	575
-2	750	200	550
-1	750	225	525
0	750	250	500
1	750 + 150	275	625

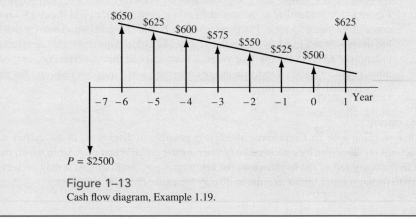

Figure 1–13
Cash flow diagram, Example 1.19.

EXAMPLE 1.20

CASH FLOW DIAGRAMS

An electrical engineer wants to deposit an amount P now such that she can withdraw an equal annual amount of $A_1 = \$2000$ per year for the first 5 years starting 1 year after the deposit, and a different annual withdrawal of $A_2 = \$3000$ per year for the following 3 years. How would the cash flow diagram appear if $i = 8.5\%$ per year?

Solution

The cash flows are shown in Figure 1–14. The negative cash outflow P occurs now. The first withdrawal (positive cash inflow) for the A_1 series occurs at the end of year 1, and A_2 occurs in years 6 through 8.

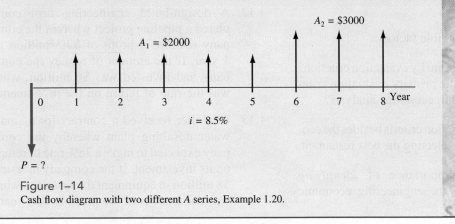

Figure 1–14
Cash flow diagram with two different A series, Example 1.20.

CHAPTER SUMMARY

Engineering economy is the application of economic factors and criteria to evaluate alternatives, considering the time value of money. The engineering economy study involves computing a specific economic measure of worth for estimated cash flows over a specific period of time.

The concept of *equivalence* helps in understanding how different sums of money at different times are equal in economic terms. The differences between simple interest (based on principal only) and compound interest (based on principal and interest upon interest) have been described in formulas, tables, and graphs. This power of compounding is very noticeable, especially over long periods of time, as is the effect of inflation, introduced here.

The MARR is a reasonable rate of return established as a hurdle rate to determine if an alternative is economically viable. The MARR is always higher than the return from a safe investment.

Also, we learned about cash flows:

Difficulties with their estimation.
Difference between estimated and actual value.
End-of-year convention for cash flow location.
Net cash flow computation.
Different perspectives in determining the cash flow sign.
Construction of a cash flow diagram.

PROBLEMS

Basic Concepts

1.1 What is meant by the term *time value of money?*

1.2 List three intangible factors.

1.3 (*a*) What is meant by evaluation criterion?
(*b*) What is the primary evaluation criterion used in economic analysis?

1.4 List three evaluation criteria besides the economic one for selecting the best restaurant.

1.5 Discuss the importance of *identifying* alternatives in the engineering economic process.

1.6 What is the difference between simple and compound interest?

1.7 What is meant by minimum attractive rate of return?

1.8 What is the difference between debt and equity financing? Give an example of each.

Interest Rate and Rate of Return

1.9 Trucking giant Yellow Corp agreed to purchase rival Roadway for $966 million in order to reduce so-called back-office costs (e.g., payroll and insurance) by $45 million per year. If the savings were realized as planned, what would be the rate of return on the investment?

1.10 If Ford Motor Company's profits increased from 22 cents per share to 29 cents per share in the April–June quarter compared to the previous quarter, what was the rate of increase in profits for that quarter?

1.11 A broadband service company borrowed $2 million for new equipment and repaid the principal of the loan plus $275,000 interest after 1 year. What was the interest rate on the loan?

1.12 A design-build engineering firm completed a pipeline project wherein the company realized a profit of $2.3 million in 1 year. If the amount of money the company had invested was $6 million, what was the rate of return on the investment?

1.13 US Filter received a contract for a small water desalting plant whereby the company expected to make a 28% rate of return on its investment. If the company invested $8 million in equipment the first year, what was the amount of the profit in that year?

1.14 A publicly traded construction company reported that it just paid off a loan that it received 1 year earlier. If the total amount of money the company paid was $1.6 million and the interest rate on the loan was 10% per year, how much money did the company borrow 1 year ago?

1.15 A start-up chemical company has established a goal of making at least a 35% per year rate of return on its investment. If the company acquired $50 million in venture capital, how much did it have to earn in the first year?

Equivalence

1.16 At an interest rate of 8% per year, $10,000 today is equivalent to how much (*a*) 1 year from now and (*b*) 1 year ago?

1.17 A medium-size consulting engineering firm is trying to decide whether it should replace its office furniture now or wait and do it 1 year from now. If it waits 1 year, the cost is expected to be $16,000. At an interest rate of 10% per year, what would be the equivalent cost now?

1.18 An investment of $40,000 one year ago and $50,000 now are equivalent at what interest rate?

1.19 At what interest rate would $100,000 now be equivalent to $80,000 one year ago?

Simple and Compound Interest

1.20 Certain certificates of deposit accumulate interest at 10% per year simple interest. If a company invests $240,000 now in these certificates for the purchase of a new machine 3 years from now, how much will the company have at the end of the 3-year period?

1.21 A local bank is offering to pay compound interest of 7% per year on new savings accounts. An e-bank is offering 7.5% per year simple interest on a 5-year certificate of deposit. Which offer is more attractive to a company that wants to set aside $1,000,000 now for a plant expansion 5 years from now?

1.22 Badger Pump Company invested $500,000 five years ago in a new product line that is now worth $1,000,000. What rate of return did the company earn (*a*) on a simple interest basis and (*b*) on a compound interest basis?

1.23 How long will it take for an investment to double at 5% per year (*a*) simple interest and (*b*) compound interest?

1.24 A company that manufactures regenerative thermal oxidizers made an investment 10 years ago that is now worth $1,300,000. How much was the initial investment at an interest rate of 15% per year (*a*) simple interest and (*b*) compound interest?

1.25 Companies frequently borrow money under an arrangement that requires them to make periodic payments of only interest and then pay the principal of the loan all at once. A company that manufactures odor control chemicals borrowed $400,000 for 3 years at 10% per year compound interest under such an arrangement. What is the difference in the *total amount paid* between this arrangement (identified as plan 1) and plan 2, in which the company makes no interest payments until the loan is due and then pays it off in one lump sum?

1.26 A company that manufactures in-line mixers for bulk manufacturing is considering borrowing $1.75 million to update a production line. If it borrows the money now, it can do so at an interest rate of 7.5% per year simple interest for 5 years. If it borrows next year, the interest rate will be 8% per year compound interest, but it will be for only 4 years. (*a*) How much interest (total) will be paid under each scenario, and (*b*) should the company borrow now or 1 year from now? Assume the total amount due will be paid when the loan is due in either case.

Symbols and Spreadsheets

1.27 Define the symbols involved when a construction company wants to know how much money it can spend 3 years from now in lieu of spending $50,000 now to purchase a new truck, when the compound interest rate is 15% per year.

1.28 State the purpose for each of the following built-in Excel functions:
  (*a*) FV(*i*%,*n*,*A*,*P*)
  (*b*) IRR(first_cell:last_cell)
  (*c*) PMT(*i*%,*n*,*P*,*F*)
  (*d*) PV(*i*%,*n*,*A*,*F*)

1.29 What are the values of the engineering economy symbols *P*, *F*, *A*, *i*, and *n* in the following Excel functions? Use a ? for the symbol that is to be determined.
  (*a*) FV(7%,10,2000,9000)
  (*b*) PMT(11%,20,14000)
  (*c*) PV(8%,15,1000,800)

1.30 Write the engineering economy symbol that corresponds to each of the following Excel functions.
  (a) PV
  (b) PMT
  (c) NPER
  (d) IRR
  (e) FV

1.31 In a built-in Excel function, if a certain parameter does not apply, under what circumstances can it be left blank? When must a comma be entered in its place?

MARR and Cost of Capital

1.32 Identify each of the following as either a safe investment or a risky one.
  (a) New restaurant business
  (b) Savings account in a bank
  (c) Certificate of deposit
  (d) Government bond
  (e) Relative's "get-rich-quick" idea

1.33 Identify each of the following as either equity or debt financing.
  (a) Money from savings
  (b) Money from a certificate of deposit
  (c) Money from a relative who is a partner in the business
  (d) Bank loan
  (e) Credit card

1.34 Rank the following from highest to lowest rate of return or interest rate: government bond, corporate bond, credit card, bank loan to new business, interest on checking account.

1.35 Rank the following from highest to lowest interest rate: cost of capital, acceptable rate of return on a risky investment, minimum attractive rate of return, rate of return on a safe investment, interest on checking account, interest on savings account.

1.36 Five separate projects have calculated rates of return of 8, 11, 12.4, 14, and 19% per year. An engineer wants to know which projects to accept on the basis of rate of return. She learns from the finance department that company funds, which have a cost of capital of 18% per year, are commonly used to fund 25% of all capital projects. Later, she is told that borrowed money is currently costing 10% per year. If the MARR is established at exactly the weighted average cost of capital, which projects should she accept?

Cash Flows

1.37 What is meant by the end-of-period convention?

1.38 Identify the following as cash inflows or outflows to Daimler-Chrysler: income taxes, loan interest, salvage value, rebates to dealers, sales revenues, accounting services, cost reductions.

1.39 Construct a cash flow diagram for the following cash flows: $10,000 outflow at time zero, $3000 per year outflow in years 1 through 3 and $9000 inflow in years 4 through 8 at an interest rate of 10% per year, and an unknown future amount in year 8.

1.40 Construct a cash flow diagram to find the present worth of a future outflow of $40,000 in year 5 at an interest rate of 15% per year.

Doubling the Value

1.41 Use the rule of 72 to estimate the time it would take for an initial investment of $10,000 to accumulate to $20,000 at a compound rate of 8% per year.

1.42 Estimate the time it would take (according to the rule of 72) for money to quadruple in value at a compound interest rate of 9% per year.

1.43 Use the rule of 72 to estimate the interest rate that would be required for $5000 to accumulate to $10,000 in 4 years.

1.44 If you now have $62,500 in your retirement account and you want to retire when the account is worth $2 million, estimate the rate of return that the account must earn if you want to retire in 20 years without adding any more money to the account.

FE REVIEW PROBLEMS

1.45 An example of an intangible factor is
   (a) Taxes
   (b) Cost of materials
   (c) Morale
   (d) Rent

1.46 The time it would take for money to double at a simple interest rate of 5% per year is closest to
   (a) 10 years
   (b) 12 years
   (c) 15 years
   (d) 20 years

1.47 At a compound interest rate of 10% per year, $10,000 one year ago is now equivalent to
   (a) $8264
   (b) $9091
   (c) $11,000
   (d) $12,100

1.48 An investment of $10,000 nine years ago has accumulated to $20,000 now. The compound rate of return earned on the investment is closest to
   (a) 6%
   (b) 8%
   (c) 10%
   (d) 12%

1.49 In most engineering economy studies, the best alternative is the one that
   (a) Will last the longest time
   (b) Is easiest to implement
   (c) Costs the least
   (d) Is most politically correct

1.50 The cost of tuition at a certain public university was $160 per credit-hour 5 years ago. The cost today (exactly 5 years later) is $235. The annual rate of increase is closest to
   (a) 4%
   (b) 6%
   (c) 8%
   (d) 10%

EXTENDED EXERCISE

EFFECTS OF COMPOUND INTEREST

In an effort to maintain compliance with noise emission standards on the processing floor, National Semiconductors requires the use of noise-measuring instruments. The company plans to purchase new portable systems at the end of next year at a cost of $9000 each. National estimates the maintenance cost to be $500 per year for 3 years, after which they will be salvaged for $2000 each.

Questions

1. Construct the cash flow diagram. For a compound interest rate of 8% per year, find the equivalent F value after 4 years, using calculations by hand.
2. Find the F value in question 1, using a spreadsheet.
3. Find the F value if the maintenance costs are $300, $500, and $1000 for each of the 3 years. By how much has the F value changed?
4. Find the F value in question 1 in terms of dollars needed in the future with an adjustment for inflation of 4% per year. This increases the interest rate from 8% to 12.32% per year.

CASE STUDY

DESCRIBING ALTERNATIVES FOR PRODUCING REFRIGERATOR SHELLS

Background

Large refrigerator manufacturers like Whirlpool, General Electric, Frigidaire, and others may subcontract the molding of their plastic liners and door panels. One prime national subcontractor is Innovations Plastics. It is expected that in about 2 years improvements in mechanical properties will allow the molded plastic to sustain increased vertical and horizontal loading, thus significantly reducing the need for attached metal anchors for some shelving. However, improved molding equipment will be needed to enter this market. The company president wants a recommendation on whether Innovations should plan on offering the new technology to the major manufacturers and an estimate of the necessary capital investment to enter this market early.

You work as an engineer for Innovations. At this stage, you are not expected to perform a complete engineering economic analysis, for not enough information is available. You are asked to formulate reasonable alternatives, determine what data and estimates are needed for each one, and ascertain what criteria (economic and noneconomic) should be utilized to make the final decision.

Information

Some information useful at this time is as follows:

- The technology and equipment are expected to last about 10 years before new methods are developed.
- Inflation and income taxes will not be considered in the analysis.
- The expected returns on capital investment used for the last three new technology projects were compound rates of 15, 5, and 18%. The 5% rate was the criterion for enhancing an employee-safety system on an existing chemical-mixing process.
- Equity capital financing beyond $5 million is not possible. The amount of debt financing and its cost are unknown.
- Annual operating costs have been averaging 8% of first cost for major equipment.
- Increased annual training costs and salary requirements for handling the new plastics and operating new equipment can range from $800,000 to $1.2 million.

There are two manufacturers working on the new-generation equipment. You label these options as alternatives A and B.

Case Study Exercises

1. Use the first four steps of the decision-making process to generally describe the alternatives and identify what economic-related estimates you will need to complete an engineering economy analysis for the president.

2. Identify any noneconomic factors and criteria to be considered in making the alternative selection.

3. During your inquiries about alternative B from its manufacturer, you learn that this company has already produced a prototype molding machine and has sold it to a company in Germany for $3 million (U.S. dollars). Upon inquiry, you further discover that the German company already has unused capacity on the equipment for manufacturing plastic shells. The company is willing to sell time on the equipment to Innovations immediately to produce its own shells for U.S. delivery. This could allow an earlier market entry into the United States. Consider this as alternative C, and develop the estimates necessary to evaluate C at the same time as alternatives A and B.

Factors: How Time and Interest Affect Money

In the previous chapter we learned the basic concepts of engineering economy and their role in decision making. The cash flow is fundamental to every economic study. Cash flows occur in many configurations and amounts— isolated single values, series that are uniform, and series that increase or decrease by constant amounts or constant percentages. This chapter develops derivations for all the commonly used engineering economy factors that take the time value of money into account.

The application of factors is illustrated using their mathematical forms and a standard notation format. Spreadsheet functions are introduced in order to rapidly work with cash flow series and to perform sensitivity analysis.

The case study focuses on the significant impacts that compound interest and time make on the value and amount of money.

LEARNING OBJECTIVES

Purpose: Derive and use the engineering economy factors to account for the time value of money.

This chapter will help you:

F/P and P/F factors	1.	Derive and use the compound amount factor and present worth factor for single payments.
P/A and A/P factors	2.	Derive and use the uniform series present worth and capital recovery factors.
F/A and A/F factors	3.	Derive and use the uniform series compound amount and sinking fund factors.
Interpolate factor values	4.	Linearly interpolate to determine a factor value.
P/G and A/G factors	5.	Derive and use the arithmetic gradient present worth and uniform series factors.
Geometric gradient	6.	Derive and use the geometric gradient series formulas.
Calculate i	7.	Determine the interest rate (rate of return) for a sequence of cash flows.
Calculate n	8.	Determine the number of years required for equivalence in a cash flow series.
Spreadsheets	9.	Develop a spreadsheet to perform basic sensitivity analysis using spreadsheet functions.

2.1 SINGLE-PAYMENT FACTORS (F/P AND P/F)

The most fundamental factor in engineering economy is the one that determines the amount of money F accumulated after n years (or periods) from a *single* present worth P, with interest compounded one time per year (or period). Recall that compound interest refers to interest paid on top of interest. Therefore, if an amount P is invested at time $t = 0$, the amount F_1 accumulated 1 year hence at an interest rate of i percent per year will be

$$F_1 = P + Pi$$
$$= P(1 + i)$$

where the interest rate is expressed in decimal form. At the end of the second year, the amount accumulated F_2 is the amount after year 1 plus the interest from the end of year 1 to the end of year 2 on the entire F_1.

$$F_2 = F_1 + F_1 i$$
$$= P(1 + i) + P(1 + i)i \quad\quad\quad [2.1]$$

This is the logic used in Chapter 1 for compound interest, specifically in Examples 1.8 and 1.18. The amount F_2 can be expressed as

$$F_2 = P(1 + i + i + i^2)$$
$$= P(1 + 2i + i^2)$$
$$= P(1 + i)^2$$

Similarly, the amount of money accumulated at the end of year 3, using Equation [2.1], will be

$$F_3 = F_2 + F_2 i$$

Substituting $P(1 + i)^2$ for F_2 and simplifying, we get

$$F_3 = P(1 + i)^3$$

From the preceding values, it is evident by mathematical induction that the formula can be generalized for n years to

$$F = P(1 + i)^n \quad\quad\quad [2.2]$$

The factor $(1 + i)^n$ is called the *single-payment compound amount factor* (SPCAF), but it is usually referred to as the F/P *factor*. This is the conversion factor that, when multiplied by P, yields the future amount F of an initial amount P after n years at interest rate i. The cash flow diagram is seen in Figure 2–1a.

Reverse the situation to determine the P value for a stated amount F that occurs n periods in the future. Simply solve Equation [2.2] for P.

$$P = F\left[\frac{1}{(1 + i)^n}\right] \qu\quad\quad [2.3]$$

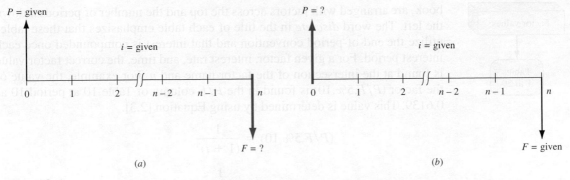

Figure 2–1
Cash flow diagrams for single-payment factors: (*a*) find F and (*b*) find P.

The expression is brackets is known as the *single-payment present worth factor* (SPPWF), or the *P/F factor*. This expression determines the present worth P of a given future amount F after n years at interest rate i. The cash flow diagram is shown in Figure 2–1*b*.

Note that the two factors derived here are for *single payments; that is,* they are used to find the present or future amount when only one payment or receipt is involved.

A standard notation has been adopted for all factors. The notation includes two cash flow symbols, the interest rate, and the number of periods. It is always in the general form $(X/Y,i,n)$. The letter X represents what is sought, while the letter Y represents what is given. For example, F/P means *find F when given P.* The i is the interest rate in percent, and n represents the number of periods involved. Thus, $(F/P,6\%,20)$ represents the factor that is used to calculate the future amount F accumulated in 20 periods if the interest rate is 6% per period. The P is given. The standard notation, simpler to use than formulas and factor names, will be used hereafter.

Table 2–1 summarizes the standard notation and equations for the F/P and P/F factors. This information is also included inside the front cover.

To simplify routine engineering economy calculations, tables of factor values have been prepared for interest rates from 0.25 to 50% and time periods from 1 to large n values, depending on the i value. These tables, found at the rear of the

TABLE 2–1	F/P and P/F Factors: Notation and Equations				
	Factor		**Standard Notation**	**Equation**	**Excel**
Notation	**Name**	**Find/Given**	**Equation**	**with Factor Formula**	**Functions**
$(F/P,i,n)$	Single-payment compound amount	F/P	$F = P(F/P,i,n)$	$F = P(1 + i)^n$	FV($i\%,n,,P$)
$(P/F,i,n)$	Single-payment present worth	P/F	$P = F(P/F,i,n)$	$P = F[1/(1 + i)^n]$	PV($i\%,n,,F$)

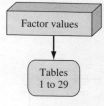

Factor values

↓

Tables
1 to 29

book, are arranged with factors across the top and the number of periods *n* down the left. The word *discrete* in the title of each table emphasizes that these tables utilize the end-of-period convention and that interest is compounded once each interest period. For a given factor, interest rate, and time, the correct factor value is found at the intersection of the factor name and *n*. For example, the value of the factor (*P/F*,5%,10) is found in the *P/F* column of Table 10 at period 10 as 0.6139. This value is determined by using Equation [2.3].

$$(P/F,5\%,10) = \frac{1}{(1 + i)^n}$$

$$= \frac{1}{(1.05)^{10}}$$

$$= \frac{1}{1.6289} = 0.6139$$

Q-Solv

For *solution by computer,* the *F* value is calculated by the FV function using the format

$$\textbf{FV}(i\%,n,,P)$$

An = sign must precede the function when it is entered. The amount *P* is determined using the PV function with the format

$$\textbf{PV}(i\%,n,,F)$$

These functions are included in Table 2–1. Refer to Appendix A or Excel online help for more information on the FV and PV functions. Examples 2.1 and 2.2 illustrate solutions by computer using both of these functions.

EXAMPLE 2.1

An industrial engineer received a bonus of $12,000 that he will invest now. He wants to calculate the equivalent value after 24 years, when he plans to use all the resulting money as the down payment on an island vacation home. Assume a rate of return of 8% per year for each of the 24 years. (*a*) Find the amount he can pay down, using both the standard notation and the factor formula. (*b*) Use a computer to find the amount he can pay down.

(a) Solution by Hand
The symbols and their values are

$$P = \$12,000 \qquad F = ? \qquad i = 8\% \text{ per year} \qquad n = 24 \text{ years}$$

The cash flow diagram is the same as that in Figure 2–1*a*.

Standard notation: Determine *F*, using the *F/P* factor for 8% and 24 years. Table 13 provides the factor value.

$$F = P(F/P,i,n) = 12,000(F/P,8\%,24)$$

$$= 12,000(6.3412)$$

$$= \$76,094.40$$

Factor formula: Apply Equation [2.2] to calculate the future worth F.

$$F = P(1 + i)^n - 12,000(1 + 0.08)^{24}$$

$$= 12,000(6.341181)$$

$$= \$76,094.17$$

The slight difference in answers is due to the round-off error introduced by the tabulated factor values. An equivalence interpretation of this result is that $12,000 today is worth $76,094 after 24 years of growth at 8% per year, compounded annually.

(b) Solution by Computer

To find the future value use the FV function that has the format FV($i\%,n,A,P$). The spreadsheet will look like the one in Figure 1–5a, except the cell entry is FV(8%,24,,12000). The F value displayed by Excel is ($76,094.17) in red to indicate a cash outflow. The FV function has performed the computation $F = P(1 + i)^n = 12,000(1 + 0.08)^{24}$ and presented the answer on the screen.

Q-Solv

EXAMPLE 2.2

Hewlett-Packard has completed a study indicating that $50,000 in reduced maintenance this year (i.e., year zero) on one processing line resulted from improved integrated circuit (IC) fabrication technology based on rapidly changing designs.

(a) If Hewlett-Packard considers these types of savings worth 20% per year, find the equivalent value of this result after 5 years.

(b) If the $50,000 maintenance savings occurs now, find its equivalent value 3 years earlier with interest at 20% per year.

(c) Develop a spreadsheet to answer the two parts above at compound rates of 20% and 5% per year. Additionally develop an Excel column chart indicating the equivalent values at the three different times for both rate-of-return values.

Solution

(a) The cash flow diagram appears as in Figure 2–1a. The symbols and their values are

$$P = \$50,000 \qquad F = ? \qquad i = 20\% \text{ per year} \qquad n = 5 \text{ years}$$

Use the F/P factor to determine F after 5 years.

$$F = P(F/P,i,n) = \$50,000(F/P,20\%,5)$$

$$= 50,000(2.4883)$$

$$= \$124,415.00$$

The function FV(20%,5,,50000) provides the same answer. See Figure 2–2a, cell C4.

Q-Solv

(b) In this case, the cash flow diagram appears as in Figure 2–1b with F placed at time $t = 0$ and the P value placed 3 years earlier at $t = -3$. The symbols and their values are

$$P = ? \qquad F = \$50,000 \qquad i = 20\% \text{ per year} \qquad n = 3 \text{ years}$$

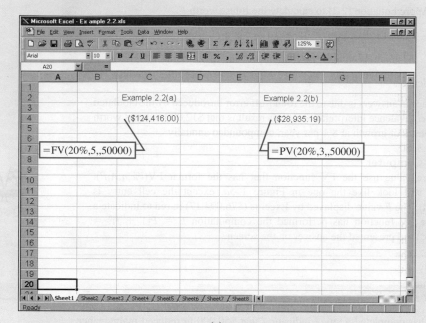

(a)

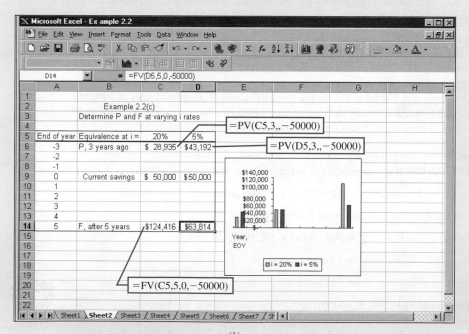

(b)

Figure 2–2
(a) Q-solv spreadsheet for Example 2.2(a) and (b); (b) complete spreadsheet with graphic, Example 2.2.

Use the P/F factor to determine P three years earlier.

$$P = F(P/F,i,n) = \$50,000(P/F,20\%,3)$$
$$= 50,000(0.5787) = \$28,935.00$$

An equivalence statement is that $28,935 three years ago is the same as $50,000 today, which will grow to $124,415 five years from now, provided a 20% per year compound interest rate is realized each year.

Use the PV function PV($i\%,n,A,F$) and omit the A value. Figure 2–2a shows the result of entering PV(20%,3,,50000) in cell F4 to be the same as using the P/F factor.

Q-Solv

Solution by Computer

E-Solve

(c)   Figure 2–2b is a complete spreadsheet solution on one worksheet with the chart. Two columns are used for 20% and 5% computations primarily so the graph can be developed to compare the F and P values. Row 14 shows the F values using the FV function with the format FV($i\%,5,0,-50000$) where the i values are taken from cells C5 and D5. The future worth $F = \$124,416$ in cell C14 is the same (round-off considered) as that calculated above. The minus sign on 50,000 makes the result a positive number for the chart.

   The PV function is used to find the P values in row 6. For example, the present worth at 20% in year -3 is determined in cell C6 using the PV function. The result $P = \$28,935$ is the same as that obtained by using the P/F factor previously. The chart graphically shows the noticeable difference that 20% versus 5% makes over the 8-year span.

EXAMPLE 2.3

An independent engineering consultant reviewed records and found that the cost of office supplies varied as shown in the pie chart of Figure 2–3. If the engineer wants to know the equivalent value in year 10 of only the three largest amounts, what is it at an interest rate of 5% per year?

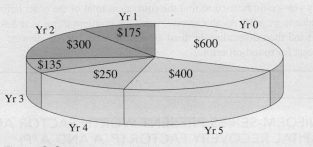

Figure 2–3
Pie chart of costs, Example 2.3.

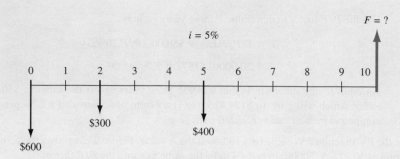

Figure 2–4
Diagram for a future worth in year 10, Example 2.3.

Solution

Draw the cash flow diagram for the values $600, $300, and $400 from the engineer's perspective (Figure 2–4). Use F/P factors to find F in year 10.

$$F = 600(F/P,5\%,10) + 300(F/P,5\%,8) + 400(F/P,5\%,5)$$
$$= 600(1.6289) + 300(1.4775) + 400(1.2763)$$
$$= \$1931.11$$

The problem could also be solved by finding the present worth in year 0 of the $300 and $400 costs using the P/F factors and then finding the future worth of the total in year 10.

$$P = 600 + 300(P/F,5\%,2) + 400(P/F,5\%,5)$$
$$= 600 + 300(0.9070) + 400(0.7835)$$
$$= \$1185.50$$
$$F = 1185.50(F/P,5\%,10) = 1185.50(1.6289)$$
$$= \$1931.06$$

Comment

It should be obvious that there are a number of ways the problem could be worked, since any year could be used to find the equivalent total of the costs before finding the future value in year 10. As an exercise, work the problem using year 5 for the equivalent total and then determine the final amount in year 10. All answers should be the same except for round-off error.

2.2   UNIFORM-SERIES PRESENT WORTH FACTOR AND CAPITAL RECOVERY FACTOR (P/A AND A/P)

The equivalent present worth P of a uniform series A of end-of-period cash flows is shown in Figure 2–5a. An expression for the present worth can be determined by considering each A value as a future worth F, calculating its present worth

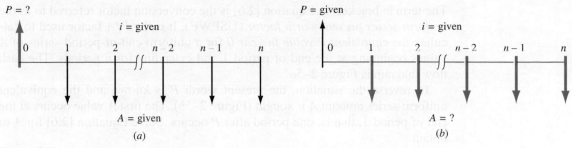

Figure 2-5
Cash flow diagrams used to determine (a) P of a uniform series and (b) A for a present worth.

with the P/F factor, and summing the results. The equation is

$$P = A\left[\frac{1}{(1+i)^1}\right] + A\left[\frac{1}{(1+i)^2}\right] + A\left[\frac{1}{(1+i)^3}\right] + \cdots$$

$$+ A\left[\frac{1}{(1+i)^{n-1}}\right] + A\left[\frac{1}{(1+i)^n}\right]$$

The terms in brackets are the P/F factors for years 1 through n, respectively. Factor out A.

$$P = A\left[\frac{1}{(1+i)^1} + \frac{1}{(1+i)^2} + \frac{1}{(1+i)^3} + \cdots + \frac{1}{(1+i)^{n-1}} + \frac{1}{(1+i)^n}\right] \quad [2.4]$$

To simplify Equation [2.4] and obtain the P/A factor, multiply the n-term geometric progression in brackets by the $(P/F,i\%,1)$ factor, which is $1/(1+i)$. This results in Equation [2.5] below. Then subtract the two equations, [2.4] from [2.5], and simplify to obtain the expression for P when $i \neq 0$ (Equation [2.6]). This progression follows.

$$\frac{P}{1+i} = A\left[\frac{1}{(1+i)^2} + \frac{1}{(1+i)^3} + \frac{1}{(1+i)^4} + \cdots + \frac{1}{(1+i)^n} + \frac{1}{(1+i)^{n+1}}\right]$$

$$[2.5]$$

$$\frac{1}{1+i}P = A\left[\frac{1}{(1+i)^2} + \frac{1}{(1+i)^3} + \cdots + \frac{1}{(1+i)^n} + \frac{1}{(1+i)^{n+1}}\right]$$

$$-P = A\left[\frac{1}{(1+i)^1} + \frac{1}{(1+i)^2} + \cdots + \frac{1}{(1+i)^{n-1}} + \frac{1}{(1+i)^n}\right]$$

$$\frac{-i}{1+i}P = A\left[\frac{1}{(1+i)^{n+1}} - \frac{1}{(1+i)^1}\right]$$

$$P = \frac{A}{-i}\left[\frac{1}{(1+i)^n} - 1\right]$$

$$P = A\left[\frac{(1+i)^n - 1}{i(1+i)^n}\right] \qquad i \neq 0 \qquad\qquad [2.6]$$

The term in brackets in Equation [2.6] is the conversion factor referred to as the *uniform-series present worth factor* (USPWF). It is the P/A factor used to calculate the *equivalent P value in year 0* for a uniform end-of-period series of A values beginning at the end of period 1 and extending for n periods. The cash flow diagram is Figure 2–5a.

To reverse the situation, the present worth P is known and the equivalent uniform-series amount A is sought (Figure 2–5b). The first A value occurs at the end of period 1, that is, one period after P occurs. Solve Equation [2.6] for A to obtain

$$A = P\left[\frac{i(1 + i)^n}{(1 + i)^n - 1}\right] \qquad [2.7]$$

The term in brackets is called the *capital recovery factor* (CRF), or A/P *factor*. It calculates the equivalent uniform annual worth A over n years for a given P in year 0, when the interest rate is i.

These formulas are derived with the present worth P and the first uniform annual amount A one year (period) apart. That is, the present worth P must always be located one period prior to the first A.

The factors and their use to find P and A are summarized in Table 2–2, and inside the front cover. The standard notations for these two factors are $(P/A,i\%,n)$ and $(A/P,i\%,n)$. Tables 1 through 29 at the end of the text include the factor values. As an example, if $i = 15\%$ and $n = 25$ years, the P/A factor value from Table 19 is $(P/A,15\%,25) = 6.4641$. This will find the equivalent present worth at 15% per year for any amount A that occurs uniformly from years 1 through 25. When the bracketed relation in Equation [2.6] is used to calculate the P/A factor, the result is the same except for round-off errors.

$$(P/A,15\%,25) = \frac{(1 + i)^n - 1}{i(1 + i)^n} = \frac{(1.15)^{25} - 1}{0.15(1.15)^{25}} = \frac{31.91895}{4.93784} = 6.46415$$

Spreadsheet functions are capable of determining both P and A values in lieu of applying the P/A and A/P factors. The PV function that we used in the last section also calculates the P value for a given A over n years, and a separate F value

TABLE 2–2 P/A and A/P Factors: Notation and Equations					
Factor			**Factor**	**Standard**	**Excel**
Notation	**Name**	**Find/Given**	**Formula**	**Notation Equation**	**Function**
$(P/A,i,n)$	Uniform-series present worth	P/A	$\dfrac{(1 + i)^n - 1}{i(1 + i)^n}$	$P = A(P/A,i,n)$	$PV(i\%,n,A)$
$(A/P,i,n)$	Capital recovery	A/P	$\dfrac{i(1 + i)^n}{(1 + i)^n - 1}$	$A = P(A/P,i,n)$	$PMT(i\%,n,P)$

in year n, if it is given. The format, introduced in Section 1.8, is

$$PV(i\%,n,A,F)$$

Similarly, the A value is determined using the PMT function for a given P value in year 0 and a separate F, if given. The format is

$$PMT(i\%,n,P,F)$$

The PMT function was demonstrated in Section 1.18 (Figure 1–5b) and is used in later examples. Table 2–2 includes the PV and PMT functions for P and A, respectively. Example 2.4 demonstrates the PV function.

EXAMPLE 2.4

How much money should you be willing to pay now for a guaranteed $600 per year for 9 years starting next year, at a rate of return of 16% per year?

Solution
The cash flow diagram (Figure 2–6) fits the P/A factor. The present worth is:

$$P = 600(P/A,16\%,9) = 600(4.6065) = \$2763.90$$

The PV function PV(16%,9,600) entered into a single spreadsheet cell will display the answer $P = \$2763.93$.

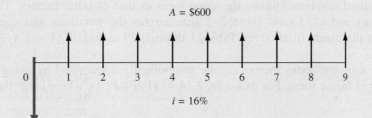

Figure 2–6
Diagram to find P using the P/A factor, Example 2.4.

Comment
Another solution approach is to use P/F factors for each of the nine receipts and add the resulting present worths to get the correct answer. Another way is to find the future worth F of the $600 payments and then find the present worth of the F value. There are many ways to solve an engineering economy problem. Only the most direct method is presented here.

2.3 SINKING FUND FACTOR AND UNIFORM-SERIES COMPOUND AMOUNT FACTOR (A/F AND F/A)

The simplest way to derive the A/F factor is to substitute into factors already developed. If P from Equation [2.3] is substituted into Equation [2.7], the following formula results.

$$A = F\left[\frac{1}{(1+i)^n}\right]\left[\frac{i(1+i)^n}{(1+i)^n - 1}\right]$$

$$A = F\left[\frac{i}{(1+i)^n - 1}\right] \qquad [2.8]$$

The expression in brackets in Equation [2.8] is the A/F or sinking fund factor. It determines the uniform annual series that is equivalent to a given future worth F. This is shown graphically in Figure 2–7a.

The uniform series A begins at the end of period 1 and continues *through the period of the given F*.

Equation [2.8] can be rearranged to find F for a stated A series in periods 1 through n (Figure 2–7b).

$$F = A\left[\frac{(1+i)^n - 1}{i}\right] \qquad [2.9]$$

The term in brackets is called the *uniform-series compound amount factor* (USCAF), or F/A factor. When multiplied by the given uniform annual amount A, it yields the future worth of the uniform series. It is important to remember that the future amount F occurs in the same period as the last A.

Standard notation follows the same form as that of other factors. They are $(F/A,i,n)$ and $(A/F,i,n)$. Table 2–3 summarizes the notations and equations, as does the inside front cover. Tables 1 through 29 include F/A and A/F factor values.

The uniform-series factors can be symbolically determined by using an abbreviated factor form. For example, $F/A = (F/P)(P/A)$, where cancellation of

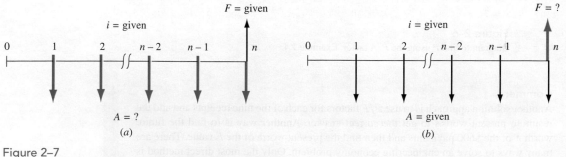

Figure 2–7
Cash flow diagrams to (a) find A, given F, and (b) find F, given A.

	Factor		**Factor**	**Standard**	**Excel**
Notation	**Name**	**Find/Given**	**Formula**	**Notation Equation**	**Functions**
$(F/A,i,n)$	Uniform-series compound amount	F/A	$\dfrac{(1 + i)^n - 1}{i}$	$F = A(F/A,i,n)$	$FV(i\%,n,A)$
$(A/F,i,n)$	Sinking fund	A/F	$\dfrac{i}{(1 + i)^n - 1}$	$A = F(A/F,i,n)$	$PMT(i\%,n,,F)$

TABLE 2–3 F/A and A/F Factors: Notation and Equations

the P is correct. Using the factor formulas, we have

$$(F/A,i,n) = [(1 + i)^n]\left[\frac{(1 + i)^n - 1}{i(1 + i)^n}\right] = \frac{(1 + i)^n - 1}{i}$$

Also the A/F factor in Equation [2.8] may be derived from the A/P factor by subtracting i.

$$(A/F,i,n) = (A/P,i,n) - i$$

This relation can be verified empirically in any interest factor table in the rear of the text, or mathematically by simplifying the equation to derive the A/F factor formula. This relation is used later to compare alternatives by the annual worth method.

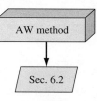

AW method

Sec. 6.2

For solution by computer, the FV spreadsheet function calculates F for a stated A series over n years. The format is

$$\mathbf{FV}(i\%,n,A,P)$$

Q-Solv

The P may be omitted when no separate present worth value is given. The PMT function determines the A value for n years, given F in year n, and possibly a separate P value in year 0. The format is

$$\mathbf{PMT}(i\%,n,P,F)$$

If P is omitted, the comma must be entered so the computer knows the last entry is an F value. These functions are included in Table 2–3. The next two examples include the FV and PMT functions.

EXAMPLE 2.5

Formasa Plastics has major fabrication plants in Texas and Hong Kong. The president wants to know the equivalent future worth of a $1 million capital investment each year for 8 years, starting 1 year from now. Formasa capital earns at a rate of 14% per year.

Solution

The cash flow diagram (Figure 2–8) shows the annual payments starting at the end of year 1 and ending in the year the future worth is desired. Cash flows are indicated in $1000 units. The F value in 8 years is

$$F = 1000(F/A,14\%,8) = 1000(13.2328) = \$13,232.80$$

The actual future worth is $13,232,800. The FV function is FV(14%,8,1000000).

Q-Solv

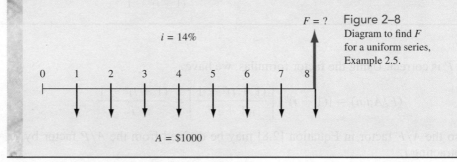

Figure 2–8
Diagram to find F for a uniform series, Example 2.5.

EXAMPLE 2.6

How much money must Carol deposit every year starting 1 year from now at 5½% per year in order to accumulate $6000 seven years from now?

Solution

The cash flow diagram from Carol's perspective (Figure 2–9a) fits the A/F factor.

$$A = \$6000(A/F,5.5\%,7) = 6000(0.12096) = \$725.76 \text{ per year}$$

The A/F factor value of 0.12096 was computed using the factor formula in Equation [2.8]. Alternatively, use the PMT function as shown in Figure 2–9b to obtain $A = \$725.79$ per year.

Q-Solv

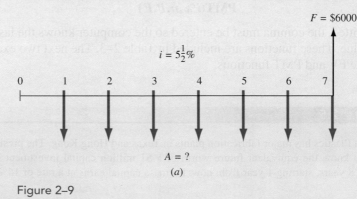

Figure 2–9
(a) Cash flow diagram and (b) PMT function to determine A, Example 2.6.

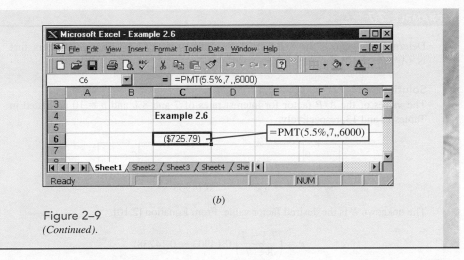

(b)

Figure 2–9
(Continued).

2.4 INTERPOLATION IN INTEREST TABLES

When it is necessary to locate a factor value for an i or n not in the interest tables, the desired value can be obtained in one of two ways: (1) by using the formulas derived in Sections 2.1 to 2.3 or (2) by linearly interpolating between the tabulated values. It is generally easier and faster to use the formulas from a calculator or spreadsheet that has them preprogrammed. Furthermore, the value obtained through linear interpolation is not exactly correct, since the equations are nonlinear. Nevertheless, interpolation is sufficient in most cases as long as the values of i or n are not too distant from one another.

The first step in linear interpolation is to set up the known (values 1 and 2) and unknown factors, as shown in Table 2–4. A ratio equation is then set up and solved for c, as follows:

$$\frac{a}{b} = \frac{c}{d} \qquad \text{or} \qquad c = \frac{a}{b}d \qquad\qquad [2.10]$$

where a, b, c, and d represent the differences between the numbers shown in the interest tables. The value of c from Equation [2.10] is added to or subtracted from value 1, depending on whether the factor is increasing or decreasing in value, respectively. The following examples illustrate the procedure just described.

TABLE 2–4 Linear Interpolation Setup	
i or n	**Factor**
tabulated	value 1
desired	unlisted
tabulated	value 2

EXAMPLE 2.7

Determine the value of the A/P factor for an interest rate of 7.3% and n of 10 years, that is, $(A/P,7.3\%,10)$.

Solution

The values of the A/P factor for interest rates of 7 and 8% and $n = 10$ are listed in Tables 12 and 13, respectively.

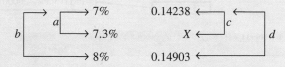

The unknown X is the desired factor value. From Equation [2.10],

$$c = \left(\frac{7.3 - 7}{8 - 7}\right)(0.14903 - 0.14238)$$

$$= \frac{0.3}{1}(0.00665) = 0.00199$$

Since the factor is increasing in value as the interest rate increases from 7 to 8%, the value of c must be *added* to the value of the 7% factor. Thus,

$$X = 0.14238 + 0.00199 = 0.14437$$

Comment

It is good practice to check the reasonableness of the final answer by verifying that X lies *between* the values of the known factors in approximately the correct proportions. In this case, since 0.14437 is less than 0.5 of the distance between 0.14238 and 0.14903, the answer seems reasonable. If Equation [2.7] is applied, the exact factor value is 0.144358.

EXAMPLE 2.8

Find the value of the $(P/F,4\%,48)$ factor.

Solution

From Table 9 for 4% interest, the values of the P/F factor for 45 and 50 years are found.

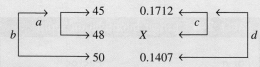

From Equation [2.10],

$$c = \frac{a}{b}(d) = \frac{48 - 45}{50 - 45}(0.1712 - 0.1407) = 0.0183$$

Since the value of the factor decreases as *n* increases, *c* is subtracted from the factor value for *n* = 45.

$$X = 0.1712 - 0.0183 = 0.1529$$

Comment

Though it is possible to perform two-way linear interpolation, it is much easier and more accurate to use the factor formula or a spreadsheet function.

2.5   ARITHMETIC GRADIENT FACTORS (*P/G* AND *A/G*)

An *arithmetic gradient* is a *cash flow series* that either increases or decreases by a constant amount. The cash flow, whether income or disbursement, changes by the same arithmetic amount each period. The *amount* of the increase or decrease is the *gradient*. For example, if a manufacturering engineer predicts that the cost of maintaining a robot will increase by $500 per year until the machine is retired, a gradient series is involved and the amount of the gradient is $500.

Formulas previously developed for an *A* series have year-end amounts of equal value. In the case of a gradient, each year-end cash flow is different, so new formulas must be derived. First, assume that the cash flow at the end of year 1 is not part of the gradient series, but is rather a *base amount*. This is convenient because in actual applications, the base amount is usually larger or smaller than the gradient increase or decrease. For example, if you purchase a used car with a 1-year warranty, you might expect to pay the gasoline and insurance costs during the first year of operation. Assume these cost $1500; that is, $1500 is the base amount. After the first year, you absorb the cost of repairs, which could reasonably be expected to increase each year. If you estimate that total costs will increase by $50 each year, the amount the second year is $1550, the third $1600, and so on to year *n*, when the total cost is $1500 + (n - 1)50$. The cash flow diagram is shown in Figure 2–10. Note that the gradient ($50) is first observed between year 1 and year 2, and the base amount ($1500 in year 1) is not equal to the gradient.

Define the symbol *G* for gradients as

G = constant arithmetic change in the magnitude of receipts or disbursements from one time period to the next; *G* may be positive or negative.

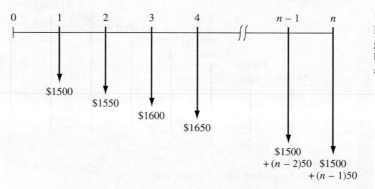

Figure 2–10
Diagram of an arithmetic gradient series with a base amount of $1500 and a gradient of $50.

Figure 2–11
Conventional arithmetic
gradient series without
the base amount.

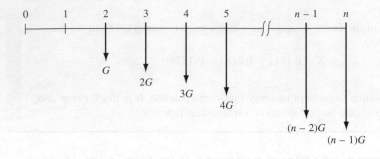

The cash flow in year n (CF_n) may be calculated as

$$CF_n = \text{base amount} + (n - 1)G$$

If the base amount is ignored, a generalized arithmetic (increasing) gradient cash flow diagram is as shown in Figure 2–11. Note that the gradient begins between years 1 and 2. This is called a *conventional gradient*.

EXAMPLE 2.9

A sports apparel company has initiated a logo-licensing program. It expects to realize a revenue of $80,000 in fees next year from the sale of its logo. Fees are expected to increase uniformly to a level of $200,000 in 9 years. Determine the arithmetic gradient and construct the cash flow diagram.

Solution
The base amount is $80,000 and the total revenue increase is

$$\text{Increase in 9 years} = 200,000 - 80,000 = 120,000$$

$$\text{Gradient} = \frac{\text{increase}}{n - 1}$$

$$= \frac{120,000}{9 - 1} = \$15,000 \text{ per year}$$

The cash flow diagram is shown in Figure 2–12.

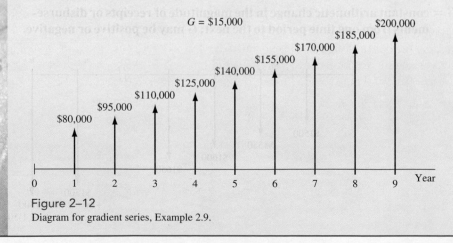

Figure 2–12
Diagram for gradient series, Example 2.9.

In this text, three factors are derived for arithmetic gradients: the P/G factor for present worth, the A/G factor for annual series and the F/G factor for future worth. There are several ways to derive them. We use the single-payment present worth factor $(P/F,i,n)$, but the same result can be obtained using the F/P, F/A, or P/A factor.

In Figure 2–11, the present worth at year 0 of only the gradient is equal to the sum of the present worths of the individual values, where each value is considered a future amount.

$$P = G(P/F,i,2) + 2G(P/F,i,3) + 3G(P/F,i,4) + \cdots$$
$$+ [(n-2)G](P/F,i,n-1) + [(n-1)G](P/F,i,n)$$

Factor out G and use the P/F formula.

$$P = G\left[\frac{1}{(1+i)^2} + \frac{2}{(1+i)^3} + \frac{3}{(1+i)^4} + \cdots + \frac{n-2}{(1+i)^{n-1}} + \frac{n-1}{(1+i)^n}\right] \quad [2.11]$$

Multiplying both sides of Equation [2.11] by $(1+i)^1$ yields

$$P(1+i)^1 = G\left[\frac{1}{(1+i)^1} + \frac{2}{(1+i)^2} + \frac{3}{(1+i)^3} + \cdots + \frac{n-2}{(1+i)^{n-2}} + \frac{n-1}{(1+i)^{n-1}}\right]$$
$$[2.12]$$

Subtract Equation [2.11] from Equation [2.12] and simplify.

$$iP = G\left[\frac{1}{(1+i)^1} + \frac{1}{(1+i)^2} + \cdots + \frac{1}{(1+i)^{n-1}} + \frac{1}{(1+i)^n}\right] - G\left[\frac{n}{(1+i)^n}\right]$$
$$[2.13]$$

The left bracketed expression is the same as that contained in Equation [2.4], where the P/A factor was derived. Substitute the closed-end form of the P/A factor from Equation [2.6] into Equation [2.13] and solve for P to obtain a simplified relation.

$$P = \frac{G}{i}\left[\frac{(1+i)^n - 1}{i(1+i)^n} - \frac{n}{(1+i)^n}\right] \quad [2.14]$$

Equation [2.14] is the general relation to convert an arithmetic gradient G (not including the base amount) for n years into a present worth at year 0. Figure 2–13*a* is converted into the equivalent cash flow in Figure 2–13*b*. The *arithmetic-gradient present worth factor,* or *P/G factor,* may be expressed in two forms:

$$(P/G,i,n) = \frac{1}{i}\left[\frac{(1+i)^n - 1}{i(1+i)^n} - \frac{n}{(1+i)^n}\right]$$

or

$$(P/G,i,n) = \frac{(1+i)^n - in - 1}{i^2(1+i)^n} \quad [2.15]$$

Remember: The gradient starts in year 2 and P is located in year 0. Equation [2.14] expressed as an engineering economy relation is

$$P = G(P/G,i,n) \quad [2.16]$$

Figure 2–13
Conversion diagram from
an arithmetic gradient to a
present worth.

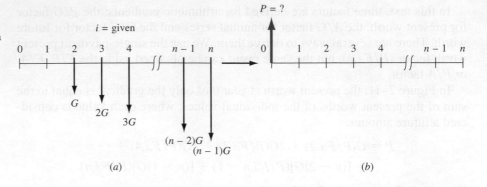

The equivalent uniform annual series (A value) for an arithmetic gradient G is found by multiplying the present worth in Equation [2.16] by the $(A/P,i,n)$ factor expression. In standard notation form, the equivalent of algebraic cancellation of P can be used to obtain the $(A/G,i,n)$ factor.

$$A = G(P/G,i,n)(A/P,i,n)$$
$$= G(A/G,i,n)$$

In equation form,

$$A = \frac{G}{i}\left[\frac{(1+i)^n - 1}{i(1+i)^n} - \frac{n}{(1+i)^n}\right]\left[\frac{i(1+i)^n}{(1+i)^n - 1}\right]$$

$$= G\left[\frac{1}{i} - \frac{n}{(1+i)^n - 1}\right] \qquad [2.17]$$

The expression in brackets in Equation [2.17] is called the *arithmetic-gradient uniform-series factor* and is identified by $(A/G,i,n)$. This factor converts Figure 2–14a into Figure 2–14b.

The P/G and A/G factors and relations are summarized inside the front cover. Factor values are tabulated in the two rightmost columns of Tables 1 through 29 at the rear of this text.

Figure 2–14
Conversion diagram of an
arithmetic gradient series
to an equivalent uniform
annual series.

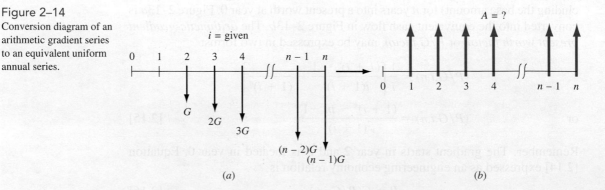

There is no direct, single-cell spreadsheet function to calculate P or A for an arithmetic gradient. Use the NPV function for P, and the PMT function for A, after all cash flows are entered into cells. (The use of NPV and PMT functions for this type of cash flow series is illustrated in Chapter 3.)

An F/G factor (*arithmetic-gradient future worth factor*) can be derived by multiplying the P/G and F/P factors. The resulting factor, $(F/G,i,n)$, in brackets, and engineering economy relation is

$$F = G\left[\left(\frac{1}{i}\right)\left(\frac{(1 + i)^n - 1}{i} - n\right)\right]$$

The total present worth P_T for a gradient series must consider the base and the gradient separately. Thus, for cash flow series involving conventional gradients:

- The *base amount* is the uniform-series amount A that begins in year 1 and extends through year n. Its present worth is represented by P_A.
- For an increasing gradient, the *gradient amount* must be added to the uniform-series amount. The present worth is P_G.
- For a decreasing gradient, the gradient amount must be subtracted from the uniform-series amount. The present worth is $-P_G$.

The general equations for calculating total present worth P_T of conventional arithmetic gradients are

$$P_T = P_A + P_G \quad \text{and} \quad P_T = P_A - P_G \qquad [2.18]$$

Similarly, the equivalent total annual series are

$$A_T = A_A + A_G \quad \text{and} \quad A_T = A_A - A_G \qquad [2.19]$$

where A_A is the annual base amount and A_G is the equivalent annual amount of the gradient series.

EXAMPLE 2.10

Three contiguous counties in Florida have agreed to pool tax resources already designated for county-maintained bridge refurbishment. At a recent meeting, the county engineers estimated that a total of \$500,000 will be deposited at the end of next year into an account for the repair of old and safety-questionable bridges throughout the three-county area. Further, they estimate that the deposits will increase by \$100,000 per year for only 9 years thereafter, then cease. Determine the equivalent (*a*) present worth and (*b*) annual series amounts if county funds earn interest at a rate of 5% per year.

Solution
(*a*)   The cash flow diagram from the county's perspective is shown in Figure 2–15. Two computations must be made and added: the first for the present worth of the base amount P_A and a second for the present worth of the gradient P_G. The total present worth P_T occurs in year 0. This is illustrated by the partitioned cash flow diagram in Figure 2–16. In \$1000 units, the present worth, from

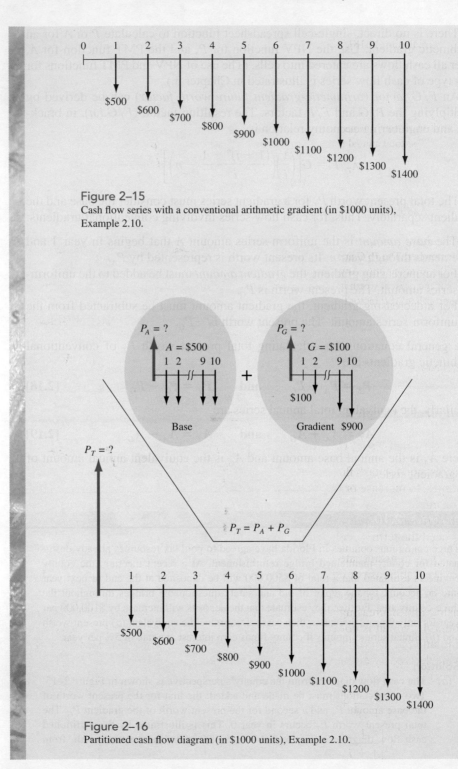

Figure 2–15
Cash flow series with a conventional arithmetic gradient (in $1000 units), Example 2.10.

Figure 2–16
Partitioned cash flow diagram (in $1000 units), Example 2.10.

Equation [2.18], is

$$P_T = 500(P/A,5\%,10) + 100(P/G,5\%,10)$$

$$= 500(7.7217) + 100(31.652)$$

$$= \$7026.05 \quad (\$7,026,050)$$

(b) Here, too, it is necessary to consider the gradient and the base amount separately. The total annual series A_T is found by using Equation [2.19].

$$A_T = 500 + 100(A/G,5\%,10) = 500 + 100(4.0991)$$

$$= \$909.91 \text{ per year} \quad (\$909,910)$$

And A_T occurs from year 1 through year 10.

Comment

Remember: The P/G and A/G factors determine the present worth and annual series of the *gradient only*. Any other cash flow must be considered separately.

   If the present worth is already calculated [as in part (a)], P_T can be multiplied by the appropriate A/P factor to get A_T.

$$A_T = P_T(A/P,5\%,10) = 7026.05(0.12950)$$

$$= \$909.87 \quad (\$909,870)$$

Round-off accounts for the \$40 difference.

Additional Example 2.16.

2.6 GEOMETRIC GRADIENT SERIES FACTORS

It is common for cash flow series, such as operating costs, construction costs, and revenues, to increase or decrease from period to period by a *constant percentage*, for example, 5% per year. This uniform rate of change defines a *geometric gradient series* of cash flows. In addition to the symbols i and n used thus far, we now need the term

> g = constant rate of change, in decimal form, by which amounts
>   increase or decrease from one period to the next

Figure 2–17 presents cash flow diagrams for geometric gradient series with increasing and decreasing uniform rates. The series starts in year 1 at an initial amount A_1, which is *not* considered a base amount as in the arithmetic gradient. The relation to determine the total present worth P_g *for the entire cash flow series* may be derived by multiplying each cash flow in Figure 2–17a by the P/F factor $1/(1+i)^n$.

$$P_g = \frac{A_1}{(1+i)^1} + \frac{A_1(1+g)}{(1+i)^2} + \frac{A_1(1+g)^2}{(1+i)^3} + \cdots + \frac{A_1(1+g)^{n-1}}{(1+i)^n}$$

$$= A_1\left[\frac{1}{1+i} + \frac{1+g}{(1+i)^2} + \frac{(1+g)^2}{(1+i)^3} + \cdots + \frac{(1+g)^{n-1}}{(1+i)^n}\right] \qquad [2.20]$$

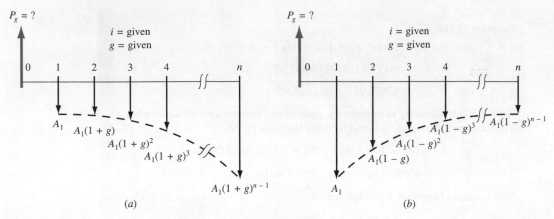

Figure 2–17
Cash flow diagram of (a) increasing and (b) decreasing geometric gradient series and present worth P_g.

Multiply both sides by $(1 + g)/(1 + i)$, subtract Equation [2.20] from the result, factor out P_g, and obtain

$$P_g\left(\frac{1 + g}{1 + i} - 1\right) = A_1\left[\frac{(1 + g)^n}{(1 + i)^{n+1}} - \frac{1}{1 + i}\right]$$

Solve for P_g and simplify.

$$P_g = A_1\left[\frac{1 - \left(\frac{1 + g}{1 + i}\right)^n}{i - g}\right] \qquad g \neq i \qquad [2.21]$$

The term in brackets in Equation [2.21] is the geometric-gradient-series present worth factor for values of g not equal to the interest rate i. The standard notation used is $(P/A,g,i,n)$. When $g = i$, substitute i for g in Equation [2.20] to obtain

$$P_g = A_1\left(\frac{1}{(1 + i)} + \frac{1}{(1 + i)} + \frac{1}{(1 + i)} + \cdots + \frac{1}{(1 + i)}\right)$$

The term $1/(1 + i)$ appears n times, so

$$P_g = \frac{nA_1}{(1 + i)} \qquad\qquad [2.22]$$

In summary, the engineering economy relation and factor formulas to calculate P_g in period $t = 0$ for a geometric gradient series starting in period 1 in the amount A_1 and increasing by a constant rate of g each period are

$$P_g = A_1(P/A,g,i,n) \qquad\qquad [2.23]$$

$$(P/A,g,i,n) = \begin{cases} \dfrac{1 - \left(\dfrac{1 + g}{1 + i}\right)^n}{i - g} & g \neq i \\[2ex] \dfrac{n}{1 + i} & g = i \end{cases} \qquad [2.24]$$

It is possible to derive factors for the equivalent A and F values; however, it is easier to determine the P_g amount and then multiply by the A/P or F/P factor.

As with the arithmetic gradient series, there are no direct spreadsheet functions for geometric gradients series. Once the cash flows are entered, P and A are determined by using the NPV and PMT functions, respectively. However, it is always an option to develop on the spreadsheet a function that uses the factor equation to determine a P, F, or A value. Example 2.11 demonstrates this approach to find the present worth of a geometric gradient series using Equations [2.24].

EXAMPLE 2.11

Engineers at SeaWorld, a division of Busch Gardens, Inc., have completed an innovation on an existing water sports ride to make it more exciting. The modification costs only $8000 and is expected to last 6 years with a $1300 salvage value for the solenoid mechanisms. The maintenance cost is expected to be high at $1700 the first year, increasing by 11% per year thereafter. Determine the equivalent present worth of the modification and maintenance cost by hand and by computer. The interest rate is 8% per year.

Solution by Hand

The cash flow diagram (Figure 2–18) shows the salvage value as a positive cash flow and all costs as negative. Use Equation [2.24] for $g \neq i$ to calculate P_g. The total P_T is

$$P_T = -8000 - P_g + 1300(P/F,8\%,6)$$

$$= -8000 - 1700\left[\frac{1 - (1.11/1.08)^6}{0.08 - 0.11}\right] + 1300(P/F,8\%,6)$$

$$= -8000 - 1700(5.9559) + 819.26 = \$-17{,}305.85 \qquad\qquad [2.25]$$

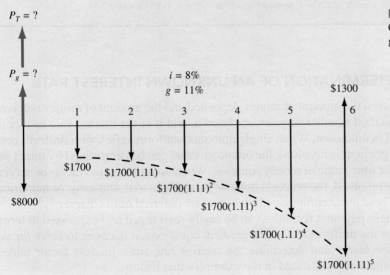

$P_T = ?$

$P_g = ?$

$i = 8\%$
$g = 11\%$

$1300

$1700 $1700(1.11)

$1700(1.11)^2

$8000 $1700(1.11)^3

$1700(1.11)^4

$1700(1.11)^5

Figure 2–18
Cash flow diagram of a geometric gradient, Example 2.11.

Solution by Computer

Figure 2–19 presents a spreadsheet with the total present worth in cell B13. The function used to determine $P_T = \$-17{,}305.89$ is detailed in the cell tag. It is a rewrite of Equation [2.25]. Since it is complex, column C and D cells also contain the three elements of P_T, which are summed in D13 to obtain the same result.

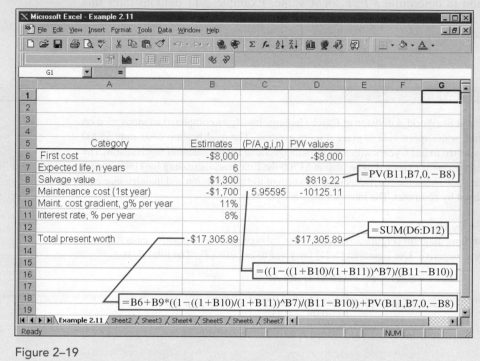

Figure 2–19

Spreadsheet used to determine present worth of a geometric gradient with $g = 11\%$, Example 2.11.

2.7   DETERMINATION OF AN UNKNOWN INTEREST RATE

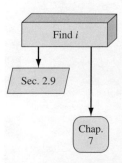

In some cases, the amount of money deposited and the amount of money received after a specified number of years are known, and it is the interest rate or rate of return that is unknown. When single amounts, uniform series, or a uniform conventional gradient is involved, the unknown rate i can be determined by direct solution of the time value of money equation. When nonuniform payments or several factors are involved, the problem must be solved by a trial-and-error or numerical method. These more complicated problems are deferred until Chapter 7.

The single-payment formulas can be easily rearranged and expressed in terms of i, but for the uniform series and gradient equations, it is easier to *solve for the value of the factor* and determine the interest rate from interest factor tables. Both situations are illustrated in the examples that follow.

EXAMPLE 2.12

If Laurel can make an investment in a friend's business of $3000 now in order to receive $5000 five years from now, determine the rate of return. If Laurel can receive 7% per year interest on a certificate of deposit, which investment should be made?

Solution

Since only single payment amounts are involved, i can be determined directly from the P/F factor.

$$P = F(P/F,i,n) = F\frac{1}{(1+i)^n}$$

$$3000 = 5000 \frac{1}{(1+i)^5}$$

$$0.600 = \frac{1}{(1+i)^5}$$

$$i = \left(\frac{1}{0.6}\right)^{0.2} - 1 = 0.1076 \ (10.76\%)$$

Alternatively, the interest rate can be found by setting up the standard notation P/F relation, solving for the factor value, and interpolating in the tables.

$$P = F(P/F,i,n)$$

$$\$3000 = 5000(P/F,i,5)$$

$$(P/F,i,5) = \frac{3000}{5000} = 0.60$$

From the interest tables, a P/F factor of 0.6000 for $n = 5$ lies between 10 and 11%. Interpolate between these two values to obtain $i = 10.76\%$.

Since 10.76% is greater than the 7% available from a certificate of deposit, Laurel should make the business investment. Since the higher rate of return would be received on the business investment, Laurel would probably select this option instead of the certificate of deposit. However, the degree of risk associated with the business investment was not specified. Obviously, risk is an important parameter that may cause selection of the lower rate of return investment. Unless specified to the contrary, equal risk for all alternatives is assumed in this text.

The IRR spreadsheet function is one of the most useful of all those available. IRR means internal rate of return, which is a topic unto itself, discussed in detail in Chapter 7. However, even at this early stage of engineering economic analysis, the IRR function can be used beneficially to find the interest rate (or rate of return) for any cash flow series that is entered into a series of contiguous spreadsheet cells, vertical or horizontal. It is very important that any years (periods) with a zero cash flow have an entry of '0' in the cell. A cell left blank is not sufficient, because an incorrect value of i will be displayed by the IRR function. The basic format is

E-Solve

IRR(first_cell:last_cell)

The first_cell and last_cell are the cell references for the start and end of the cash flow series. Example 2.13 illustrates the IRR function.

The RATE function, also very useful, is an alternative to IRR. RATE is a one-cell function that displays the compound interest rate (or rate of return) only when the annual cash flows, that is, A values, are the same. Present and future values different from the A value can be entered. The format is

$$\textbf{RATE(number\_years},A,P,F\textbf{)}$$

The F value does not include the amount A that occurs in year n. No entry into spreadsheet cells of each cash flow is necessary to use RATE, so it should be used whenever there is a uniform series over n years with associated P and/or F values stated. Example 2.13 illustrates the RATE function.

EXAMPLE 2.13

Professional Engineers, Inc., requires that $500 per year be placed into a sinking fund account to cover any unexpected major rework on field equipment. In one case, $500 was deposited for 15 years and covered a rework costing $10,000 in year 15. What rate of return did this practice provide to the company? Solve by hand and by computer.

Solution by Hand
The cash flow diagram is shown in Figure 2–20. Either the A/F or F/A factor can be used. Using A/F,

$$A = F(A/F,i,n)$$

$$500 = 10,000(A/F,i,15)$$

$$(A/F,i,15) = 0.0500$$

From interest Tables 8 and 9 under the A/F column for 15 years, the value 0.0500 lies between 3 and 4%. By interpolation, $i = 3.98\%$. (This is considered a low return for an engineering project.)

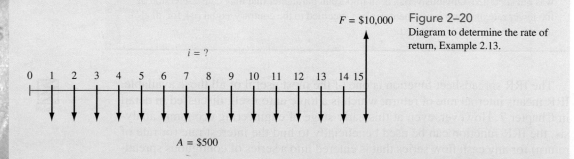

Figure 2–20
Diagram to determine the rate of return, Example 2.13.

Solution by Computer
Refer to the cash flow diagram (Figure 2–20) while completing the spreadsheet (Figure 2–21). A single-cell solution using the RATE function can be applied since $A = \$-500$ occurs each year and the $F = \$10,000$ value takes place in the last year of the series. Cell

A3 contains the function RATE(15,−500,,10000), and the answer displayed is 3.98%. The minus sign on 500 indicates the annual deposit. The extra comma is necessary to indicate that no *P* value is present. This function is fast, but it allows only limited sensitivity analysis; all the *A* values have to change by the same amount. The IRR function is much better for answering "what if?" questions.

To apply the IRR function and obtain the same answer, enter the value 0 in a cell (for year 0), followed by −500 for 14 years and 9500 (from 10,000 − 500) in year 15. Figure 2–21 contains these numbers in cells D2 through D17. In any cell on the spreadsheet, enter the IRR function IRR(D2:D17). The answer *i* = 3.98% is displayed in cell E3. It is advisable to enter the year numbers 0 through *n* (15 in this example) in the column immediately to the left of the cash flow entries. The IRR function does not need these numbers, but it makes the cash flow entry activity easier and more accurate. Now any cash flow can be changed, and a new rate will be displayed immediately via IRR.

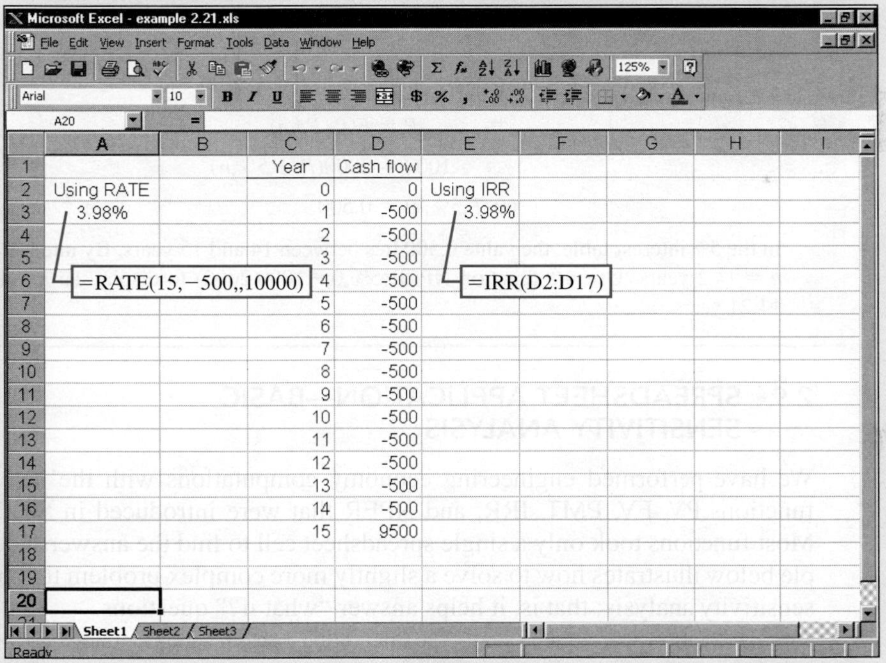

Figure 2–21
Spreadsheet solution for rate of return using the RATE and IRR functions, Example 2.13.

2.8 DETERMINATION OF UNKNOWN NUMBER OF YEARS

It is sometimes necessary to determine the number of years (periods) required for a cash flow series to provide a stated rate of return. Other times it is desirable to determine when specified amounts will be available from an investment. In both cases, the unknown value is *n*. Techniques similar to those of the preceding section are used to find *n*. Some problems can be solved directly for *n* by manipulation of the single-payment and uniform-series formulas. In other cases, n is found through interpolation in the interest tables, as illustrated below.

Q-Solv

The spreadsheet function NPER can be used to quickly find the number of years (periods) n for given A, P, and/or F values. The format is

$$\text{NPER}(i\%,A,P,F)$$

If the future value F is not involved, F is omitted; however, a present worth P and uniform amount A must be entered. The A entry can be zero when only single amounts P and F are known, as in the next example. At least one of the entries must have a sign opposite the others to obtain an answer from NPER.

EXAMPLE 2.14

How long will it take for $1000 to double if the interest rate is 5% per year?

Solution
The n value can be determined using either the F/P or P/F factor. Using the P/F factor,

$$P = F(P/F,i,n)$$

$$1000 = 2000(P/F,5\%,n)$$

$$(P/F,5\%,n) = 0.500$$

Q-Solv

In the 5% interest table, the value 0.500 lies between 14 and 15 years. By interpolation, $n = 14.2$ years. Use the function NPER(5%,0,−1000,2000) to display an n value of 14.21 years.

2.9 SPREADSHEET APPLICATION—BASIC SENSITIVITY ANALYSIS

We have performed engineering economy computations with the spreadsheet functions PV, FV, PMT, IRR, and NPER that were introduced in Section 1.8. Most functions took only a single spreadsheet cell to find the answer. The example below illustrates how to solve a slightly more complex problem that involves sensitivity analysis; that is, it helps answer "what if?" questions.

EXAMPLE 2.15

An engineer and a medical doctor have teamed up to develop a major improvement in laparoscopic surgery for gallbladder operations. They formed a small business corporation to handle the financial aspects of their partnership. The company has invested $500,000 in the project already this year ($t = 0$), and it expects to spend $500,000 annually for the next 4 years, and possibly for more years. Develop a spreadsheet that helps answer the following questions.

E-Solve

(*a*) Assume the $500,000 is expended for only 4 additional years. If the company sells the rights to use the new technology at the end of year 5 for $5 million, what is the anticipated rate of return?

(b) The engineer and doctor estimate that they will need $500,000 per year for more than 4 additional years. How many years from now do they have to finish their development work and receive the $5 million license fee to make at least 10% per year? Assume the $500,000 per year is expended through the year immediately *prior* to the receipt for the $5 million.

Solution by Computer

Figure 2–22 presents the spreadsheet, with all financial values in $1000 units. The IRR function is used throughout.

(a) The function IRR(B6:B11) in cell B15 displays $i = 24.07\%$. Note there is a cash flow of -500 in year 0. The equivalence statement is: Spending $500,000 now and $500,000 each year for 4 more years is equivalent to receiving $5 million at the end of year 5, when the interest rate is 24.07% per year.

(b) Find the rate of return for an increasing number of years that the $500 is expended. Columns C and D in Figure 2–22 present the results of IRR functions with the $5000 cash flow in different years. Cells C15 and D15 show returns on opposite sides of 10%. Therefore, the $5 million must be received some time prior to the end of year 7 to make more than the 8.93% shown in cell D15. The engineer and doctor have less than 6 years to complete their development work.

	A	B	C	D	E
2					
3		Part (a)	Part (b)		
4		Find i	Fnd n such that i > 10%		
5	Year	Get $5 million in yr 5	Get $5 million in yr 6	Get $5 million in yr 7	
6	0	$ (500)	$ (500)	$ (500)	
7	1	$ (500)	$ (500)	$ (500)	
8	2	$ (500)	$ (500)	$ (500)	
9	3	$ (500)	$ (500)	$ (500)	
10	4	$ (500)	$ (500)	$ (500)	
11	5	$ 5,000	$ (500)	$ (500)	
12	6		$ 5,000	$ (500)	
13	7			$ 5,000	
14					
15	Rate of returns	24.07%	14.80%	8.93%	
16					
17					
18		=IRR(B6:B11)	=IRR(C6:C12)	=IRR(D6:D13)	
19					

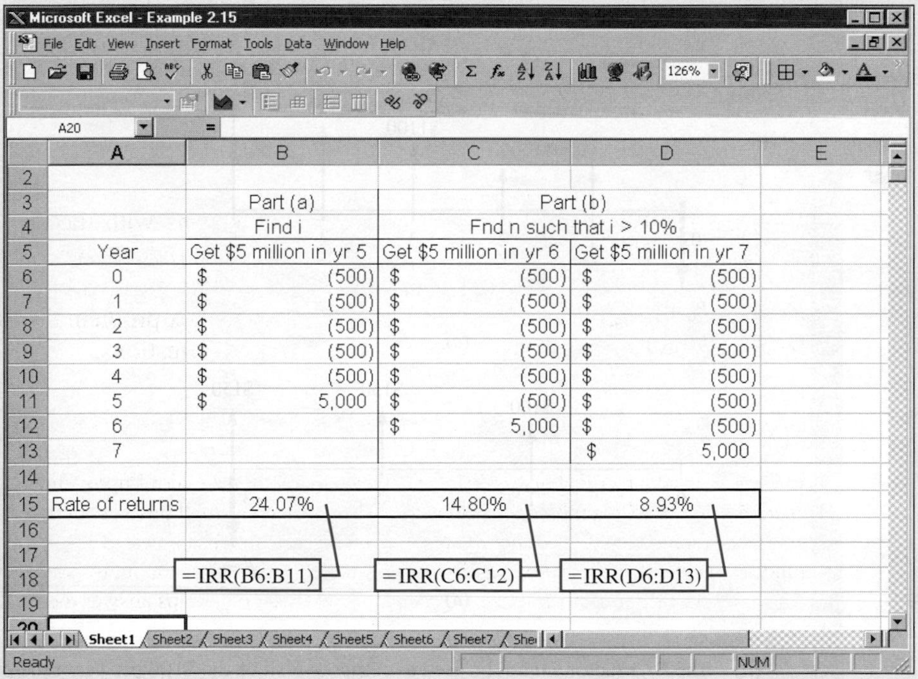

Figure 2–22
Spreadsheet solution including sensitivity analysis, Example 2.15.

ADDITIONAL EXAMPLE

EXAMPLE 2.16

P, F, AND *A* CALCULATIONS

Explain why the uniform series factors cannot be used to compute *P* or *F* directly for any of the cash flows shown in Figure 2–23.

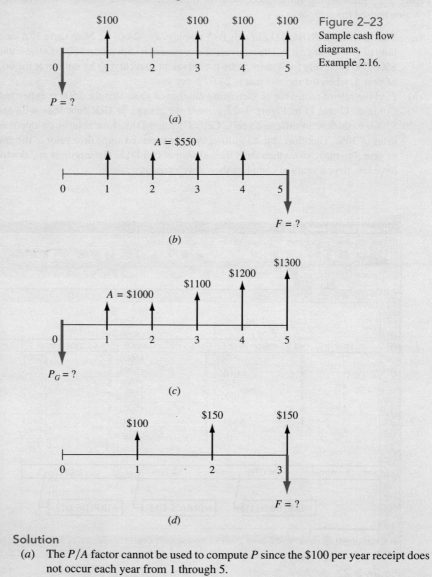

Figure 2–23
Sample cash flow diagrams, Example 2.16.

Solution

(a)   The *P/A* factor cannot be used to compute *P* since the $100 per year receipt does not occur each year from 1 through 5.

(b) Since there is no $A = \$550$ in year 5, the F/A factor cannot be used. The relation $F = 550(F/A,i,4)$ would furnish the future worth in year 4, not year 5.

(c) The first gradient amount $G = \$100$ occurs in year 3. Use of the relation $P_G = 100(P/G,i\%,4)$ will compute P_G in year 1, not year 0. (The present worth of the base amount of $1000 is not included here.)

(d) The receipt values are unequal; thus the relation $F = A(F/A,i,3)$ cannot be used to compute F.

CHAPTER SUMMARY

Formulas and factors derived and applied in this chapter perform equivalence calculations for present, future, annual, and gradient cash flows. Capability in using these formulas and their standard notation manually and with spreadsheets is critical to complete an engineering economy study. Using these formulas and spreadsheet functions, you can convert single cash flows into uniform cash flows, gradients into present worths, and much more. You can solve for rate of return i or time n. A thorough understanding of how to manipulate cash flows using the material in this chapter will help you address financial questions in professional practice as well as in everyday living.

PROBLEMS

Use of Interest Tables

2.1 Find the correct numerical value for the following factors from the interest tables.
1. $(F/P,8\%,25)$
2. $(P/A,3\%,8)$
3. $(P/G,9\%,20)$
4. $(F/A,15\%,18)$
5. $(A/P,30\%,15)$

Determination of F, P, and A

2.2 The U.S. Border Patrol is considering the purchase of a new helicopter for aerial surveillance of the New Mexico–Texas border with Mexico. A similar helicopter was purchased 4 years ago at a cost of $140,000. At an interest rate of 7% per year, what would be the equivalent value today of that $140,000 expenditure?

2.3 Pressure Systems, Inc., manufactures high-accuracy liquid-level transducers. It is investigating whether it should update certain equipment now or wait to do it later. If the cost now is $200,000, what will the equivalent amount be 3 years from now at an interest rate of 10% per year?

2.4 Petroleum Products, Inc., is a pipeline company that provides petroleum products to wholesalers in the northern United States and Canada. The company is considering purchasing insertion turbine flowmeters to allow for better monitoring of pipeline integrity. If these meters would prevent one major disruption (through early detection of product loss) valued at $600,000 four years from now, how much could the company afford to spend now at an interest rate of 12% per year?

2.5 Sensotech Inc., a maker of microelectromechanical systems, believes it can reduce product recalls by 10% if it purchases new software for detecting faulty parts. The cost of the new software is $225,000. (*a*) How much would the company have to save each year for 4 years to recover its investment if it uses a minimum attractive rate of return of 15% per year? (*b*) What was the cost of recalls per year before the software was purchased if the company did exactly recover its investment in 4 years from the 10% reduction?

2.6 Thompson Mechanical Products is planning to set aside $150,000 now for possibly replacing its large synchronous refiner motors whenever it becomes necessary. If the replacement isn't needed for 7 years, how much will the company have in its investment set-aside account if it achieves a rate of return of 18% per year?

2.7 French car maker Renault signed a $75 million contract with ABB of Zurich, Switzerland, for automated underbody assembly lines, body assembly workshops, and line control systems. If ABB will be paid in 2 years (when the systems are ready), what is the present worth of the contract at 18% per year interest?

2.8 Atlas Long-Haul Transportation is considering installing Valutemp temperature loggers in all of its refrigerated trucks for monitoring temperatures during transit. If the systems will reduce insurance claims by $100,000 two years from now, how much should the company be willing to spend now if it uses an interest rate of 12% per year?

2.9 GE Marine Systems is planning to supply a Japanese shipbuilder with aero-derivative gas turbines to power 11 DD-class destroyers for the Japanese Self-Defense Force. The buyer can pay the total contract price of $1,700,000 now or an equivalent amount 1 year from now (when the turbines will be needed). At an interest rate of 18% per year, what is the equivalent future amount?

2.10 What is the present worth of a future cost of $162,000 to Corning, Inc., 6 years from now at an interest rate of 12% per year?

2.11 How much could Cryogenics Inc., a maker of superconducting magnetic energy storage systems, afford to spend now on new equipment in lieu of spending $125,000 five years from now if the company's rate of return is 14% per year?

2.12 V-Tek Systems is a manufacturer of vertical compactors, and it is examining its cash flow requirements for the next 5 years. The company expects to replace office machines and computer equipment at various times over the 5-year planning period. Specifically, the company expects to spend $9000 two years from now, $8000 three years from now, and $5000 five years from now. What is the present worth of the planned expenditures at an interest rate of 10% per year?

2.13 A proximity sensor attached to the tip of an endoscope could reduce risks during eye surgery by alerting surgeons to the location of critical retinal tissue. If a certain eye surgeon expects that by using this technology, he will avoid lawsuits of $1.25 and $0.5 million 2 and 5 years from now, respectively, how much could he afford to spend now if his out-of-pocket costs for the lawsuits would be only 10% of the total amount of each suit? Use an interest rate of 8% per year.

2.14 The current cost of liability insurance for a certain consulting firm is $65,000. If the insurance cost is expected to increase by 4% each year, what will be the cost 5 years from now?

2.15 American Gas Products manufactures a device called a Can-Emitor that empties the contents of old aerosol cans in 2 to 3 seconds. This eliminates having to dispose of the cans as hazardous wastes. If a certain paint company can save $75,000 per year in waste disposal costs, how much could the company afford to spend now on the Can-Emitor if it wants to recover its investment in 3 years at an interest rate of 20% per year?

2.16 Atlantic Metals and Plastic uses austenitic nickel-chromium alloys to manufacture resistance heating wire. The company is considering a new annealing-drawing process to reduce costs. If the new process will cost $1.8 million now, how much must be saved each year to recover the investment in 6 years at an interest rate of 12% per year?

2.17 A green algae, *Chlamydomonas reinhardtii,* can produce hydrogen when temporarily deprived of sulfur for up to 2 days at a time. A small company needs to purchase equipment costing $3.4 million to commercialize the process. If the company wants to earn a rate of return of 20% per year and recover its investments in 8 years, what must be the net value of the hydrogen produced each year?

2.18 How much money could RTT Environmental Services borrow to finance a site reclamation project if it expects revenues of $280,000 per year over a 5-year cleanup period? Expenses associated with the project are expected to be $90,000 per year. Assume the interest rate is 10% per year.

2.19 Western Playland and Aquatics Park spends $75,000 each year in consulting services for ride inspection. New actuator element technology enables engineers to simulate complex computer-controlled movements in any direction. How much could the amusement park afford to spend now on the new technology if the annual consulting services will no longer be needed? Assume the park uses an interest rate of 15% per year and it wants to recover its investment in 5 years.

2.20 Under an agreement with the Internet Service Providers (ISPs) Association, SBC Communications reduced the price it charges ISPs to resell its high-speed digital subscriber line (DSL) service from $458 to $360 per year per customer line. A particular ISP, which has 20,000 customers, plans to pass 90% of the savings along to its customers. What is the total future worth of these savings over a 5-year horizon at an interest rate of 8% per year?

2.21 To improve crack detection in aircraft, the U.S. Air Force combined ultrasonic inspection procedures with laser heating to identify fatigue cracks. Early detection of cracks may reduce repair costs by as much as $200,000 per year. What is the present worth of these savings over a 5-year period at an interest rate of 10% per year?

2.22 A recent engineering graduate passed the FE exam and was given a raise (beginning in year 1) of $2000. At an interest rate of 8% per year, what is the present value of the $2000 per year over her expected 35-year career?

2.23 Southwestern Moving and Storage wants to have enough money to purchase a new tractor-trailer in 3 years. If the unit will cost $250,000, how much should the company set aside each year if the account earns 9% per year?

2.24 Vision Technologies, Inc., is a small company that uses ultra-wideband technology to develop devices that can detect objects (including people) inside buildings, behind walls, or below ground. The company expects to spend $100,000 per year for labor and $125,000 per year for supplies before a product can be marketed. At an interest rate of 15% per year, what is the total equivalent future amount of the company's expenses at the end of 3 years?

Factor Values

2.25 Find the numerical value of the following factors by (a) interpolation and (b) using the appropriate formula.
1. $(P/F,18\%,33)$
2. $(A/G,12\%,54)$

2.26 Find the numerical value of the following factors by (a) interpolation and (b) using the appropriate formula.
1. $(F/A,19\%,20)$
2. $(P/A,26\%,15)$

Arithmetic Gradient

2.27 A cash flow sequence starts in year 1 at $3000 and decreases by $200 each year through year 10. (a) Determine the value of the gradient G; (b) determine the amount of cash flow in year 8; and (c) determine the value of n for the gradient.

2.28 Cisco Systems expects sales to be described by the cash flow sequence $(6000 + 5k)$, where k is in years and cash flow is in millions. Determine (a) the value of the gradient G; (b) the amount of cash flow in

year 6; and (c) the value of n for the gradient if the cash flow ends in year 12.

2.29 For the cash flow sequence that starts in year 1 and is described by $900 - 100k$, where k represents years 1 through 5, (a) determine the value of the gradient G and (b) determine the cash flow in year 5.

2.30 Omega Instruments has budgeted $300,000 per year to pay for certain ceramic parts over the next 5 years. If the company expects the cost of the parts to increase uniformly according to an arithmetic gradient of $10,000 per year, what is it expecting the cost to be in year 1, if the interest rate is 10% per year?

2.31 Chevron-Texaco expects receipts from a group of stripper wells (wells that produce less than 10 barrels per day) to decline according to an arithmetic gradient of $50,000 per year. This year's receipts are expected to be $280,000 (i.e., end of year 1), and the company expects the useful life of the wells to be 5 years. (a) What is the amount of the cash flow in year 3, and (b) what is the equivalent uniform annual worth in years 1 through 5 of the income from the wells at an interest rate of 12% per year?

2.32 Income from cardboard recycling at Fort Bliss has been increasing at a constant rate of $1000 in each of the last 3 years. If this year's income (i.e., end of year 1) is expected to be $4000 and the increased income trend continues through year 5, (a) what will the income be 3 years from now (i.e., end of year 3) and (b) what is the present worth of the income over that 5-year period at an interest rate of 10% per year?

2.33 Amazon is considering purchasing a sophisticated computer system to "cube" a book's dimensions—measure its height, length, and width so that the proper box size will be used for shipment. This will save packing material, cardboard, and labor. If the savings will be $150,000 the first year, $160,000 the second year, and amounts increasing by $10,000 each year for 8 years, what is the present worth of the system at an interest rate of 15% per year?

2.34 West Coast Marine and RV is considering replacing its wired pendant controllers on its heavy-duty cranes with new portable infrared keypad controllers. The company expects to achieve cost savings of $14,000 the first year and amounts increasing by $1500 each year thereafter for the next 4 years. At an interest rate of 12% per year, what is the equivalent annual worth of the savings?

2.35 Ford Motor Company was able to reduce by 80% the cost required for installing data acquisition instrumentation on test vehicles by using MTS-developed spinning wheel force transducers. (a) If this year's cost (i.e., end of year 1) is expected to be $2000, what was the cost the year before installation of the transducers? (b) If the costs are expected to increase by $250 each year for the next 4 years (i.e., through year 5), what is the equivalent annual worth of the costs (years 1 through 5) at an interest rate of 18% per year?

2.36 For the cash flow shown below, determine the value of G that will make the future worth in year 4 equal to $6000 at an interest rate of 15% per year.

Year	0	1	2	3	4
Cash Flow	0	$2000	2000−G	2000−2G	2000−3G

2.37 A major drug company anticipates that in future years it could be involved in litigation regarding perceived side effects of one of its antidepressant drugs. To prepare a "war chest," the company wants to have money available 6 years from now that has a present worth today of $50 million. The company expects to set aside $6 million the first year and uniformly increasing amounts in each of the next 5 years. If the company can earn 12% per year on the money it sets aside, by how much must it increase the amount set aside each year to achieve its goal?

2.38 A start-up direct marketer of car parts expects to spend $1 million the first year for advertising, with amounts decreasing by $100,000 each year. Income is expected to be $4 million the first year, increasing by $500,000 each year. Determine the equivalent annual worth in years 1 through 5 of the company's *net cash flow* at an interest rate of 16% per year.

Geometric Gradient

2.39 Assume you were told to prepare a table of factor values (like those in the back of this book) for calculating the present worth of a geometric gradient series. Determine the first three values (i.e., for $n = 1, 2,$ and 3) for an interest rate of 10% per year and a rate of change g of 4% per year.

2.40 A chemical engineer planning for her retirement will deposit 10% of her salary each year into a high-technology stock fund. If her salary this year is $60,000 (i.e., end of year 1) and she expects her salary to increase by 4% each year, what will be the present worth of the fund after 15 years if it earns 4% per year?

2.41 The effort required to maintain a scanning electron microscope is known to increase by a fixed percentage each year. A high-tech equipment maintenance company has offered it services for a fee of $25,000 for the first year (i.e., end of year 1) with increases of 6% per year thereafter. If a biotechnology company wants to pay for a 3-year contract up front to take advantage of a temporary tax loophole, how much should it be willing to pay if it uses an interest rate of 15% per year?

2.42 Hughes Cable Systems plans to offer its employees a salary enhancement package that has revenue sharing as its main component. Specifically, the company will set aside 1% of total sales for year-end bonuses for all its employees. The sales are expected to be $5 million the first year, $6 million the second year, and amounts increasing by 20% each year for the next 5 years. At an interest rate of 10% per year, what is the equivalent annual worth in years 1 through 5 of the bonus package?

2.43 Determine how much money would be in a savings account that started with a deposit of $2000 in year 1 with each succeeding amount increasing by 10% per year. Use an interest rate of 15% per year and a 7-year period.

2.44 The future worth in year 10 of a geometric gradient series of cash flows was found to be $80,000. If the interest rate was 15% per year and the annual rate of increase was 9% per year, what was the cash flow amount in year 1?

2.45 Thomasville Furniture Industries offers several types of high-performance fabrics that are capable of withstanding chemicals as harsh as chlorine. A certain midwestern manufacturing company that uses fabric in several products has a report showing that the present worth of fabric purchases over a certain 5-year period was $900,000. If the costs were known to geometrically

increase by 5% per year during that time and the company used an interest rate of 15% per year for investments, what was the cost of the fabric in year 2?

2.46 Find the present worth of a series of investments that starts at $1000 in year 1 and increases by 10% per year for 20 years. Assume the interest rate is 10% per year.

2.47 A northern California consulting firm wants to start saving money for replacement of network servers. If the company invests $3000 at the end of year 1 and increases the amount invested by 5% each year, how much will be in the account 4 years from now if it earns interest at a rate of 8% per year?

2.48 A company that manufactures purgable hydrogen sulfide monitors is planning to make deposits such that each one is 5% larger than the preceding one. How large must the first deposit be (at the end of year 1) if the deposits extend through year 10 and the fourth deposit is $1250? Use an interest rate of 10% per year.

Interest Rate and Rate of Return

2.49 What compound interest rate per year is equivalent to a 12% per year simple interest rate over a 15-year period?

2.50 A publicly traded consulting engineering firm pays a bonus to each engineer at the end of the year based on the company's profit for that year. If the company's initial investment was $1.2 million, what rate of return has it made on its investment if each engineer's bonus has been $3000 per year for the past 10 years? Assume the company has six engineers and that the bonus money represents 5% of the company's profit.

2.51 Danson Iron Works, Inc., manufactures angular contact ball bearings for pumps that operate in harsh environments. If the company invested $2.4 million in a process that resulted in profits of $760,000 per year for 5 years, what rate of return did the company make on its investment?

2.52 An investment of $600,000 increased to $1,000,000 over a 5-year period. What was the rate of return on the investment?

2.53 A small company that specializes in powder coating expanded its building and purchased a new oven that is large enough to handle automobile frames. The building and oven cost $125,000, but new business from hot-rodders has increased annual income by $520,000. If operating expenses for gas, materials, labor, etc., amount to $470,000 per year, what rate of return will be made on the investment if only the cash flows that occur over the next 4 years are included in the calculation?

2.54 The business plan for a start-up company that manufactures multigas portable detectors showed equivalent annual cash flows of $400,000 for the first 5 years. If the cash flow in year 1 was $320,000 and the increase thereafter was $50,000 per year, what interest rate was used in the calculation?

2.55 A new company that makes medium-voltage soft starters spent $85,000 to build a new website. Net income was $60,000 the first year, increasing by $15,000 each year. What rate of return did the company make in its first 5 years?

Number of Years

2.56 A company that manufactures plastic control valves has a fund for equipment replacement that contains $500,000. If the company spends $75,000 per year on new equipment, how many years will it take to reduce the fund to less than $75,000 at an interest rate of 10% per year?

2.57 An A&E firm is considering purchasing the building it currently occupies under a long-term lease because the owner of the building suddenly put it up for sale. The building is being offered at a price of $170,000. Since the lease is already paid for this year, the next annual lease payment of $30,000 isn't due until the end of this year. Because the A&E firm has been a good tenant, the owner has offered to sell to them for $160,000. If the firm purchases the building with no down payment, how long will it be before the company recovers its investment at an interest rate of 12% per year?

2.58 An engineer who invested very well plans to retire now because she has $2,000,000 in her ORP account. How long will she be able to withdraw $100,000 per year (beginning 1 year from now) if her account earns interest at a rate of 4% per year?

2.59 A company that manufactures ultrasonic wind sensors invested $1.5 million 2 years ago to acquire part ownership in an innovative chip-making company. How long would it take (from the date of the initial investment) for its share of the chip company to be worth $3 million if that company is growing at a rate of 20% per year?

2.60 A certain mechanical engineer plans to retire when he has $1.6 million in his brokerage account. If he started with $100,000 in the account, how long will it be (from the time he started) before he can retire if the account makes a rate of return of 18% per year?

2.61 How many years will it take for a uniform annual deposit of size A to accumulate to

10 times the size of a single deposit if the rate of return is 10% per year?

2.62 How many years would it take for an investment of $10,000 in year 1 with increases of 10% per year to have a present worth of $1,000,000 at an interest rate of 7% per year?

FE REVIEW PROBLEMS

2.64 A construction company has an option to purchase a certain bulldozer for $61,000 at any time between now and 4 years from now. If the company plans to purchase the dozer 4 years from now, the equivalent present amount that the company is paying for the dozer at 6% per year interest is closest to
 (a) $41,230
 (b) $46,710
 (c) $48,320
 (d) Over $49,000

2.65 The cost of tuition at a certain public university was $160 per credit-hour 5 years ago. The cost today (exactly 5 years later) is $235. The annual rate of increase is closest to
 (a) 4%
 (b) 6%
 (c) 8%
 (d) 10%

2.66 The present worth of an increasing geometric gradient is $23,632. The interest rate is 6% per year, and the rate of change is 4% per year. If the cash flow amount in year 1 is $3000, the year in which the gradient ends is year
 (a) 7
 (b) 9
 (c) 11
 (d) 12

2.63 You were told that a certain cash flow sequence started at $3000 in year 1 and increased by $2000 each year. How many years were required for the equivalent annual worth of the sequence to be $12,000 at an interest rate of 10% per year?

2.67 The winner of a multistate megamillions lottery jackpot worth $175 million was given the option of taking payments of $7 million per year for 25 years, beginning 1 year now, or taking $109.355 million now. At what interest rate are the two options equivalent to each other?
 (a) 4%
 (b) 5%
 (c) 6%
 (d) 7%

2.68 A manufacturer of toilet flush valves wants to have $2,800,000 available 10 years from now so that a new product line can be initiated. If the company plans to deposit money each year, starting 1 year from now, how much will it have to deposit each year at 6% per year interest in order to have the $2,800,000 available immediately after the last deposit is made?
 (a) Less than $182,000
 (b) $182,500
 (c) $191,300
 (d) Over $210,000

2.69 Rubbermaid Plastics Corp. invested $10,000,000 in manufacturing equipment for producing small wastebaskets. If the company uses an interest rate of 15% per year, how much money would it have to earn each year if it wanted to recover its investment in 7 years?

(a) $2,403,600
(b) $3,530,800
(c) $3,941,800
(d) Over $4,000,000

2.70 An engineer deposits $8000 in year 1, $8500 in year 2, and amounts increasing by $500 per year through year 10. At an interest rate of 10% per year, the present worth in year 0 is closest to

(a) $60,600
(b) $98,300
(c) $157,200
(d) $173,400

2.71 The amount of money that could be spent 7 years from now in lieu of spending $50,000 now at an interest rate 18% per year is closest to

(a) $15,700
(b) $159,300
(c) $199,300
(d) $259,100

2.72 A deposit of $10,000 twenty years from now at an interest rate of 10% per year will have a present value closest to

(a) $1720
(b) $1680
(c) $1590
(d) $1490

2.73 Income from sales of an injector-cleaning gasoline additive has been averaging $100,000 per year. At an interest rate of 18% per year, the future worth of the income in years 1 through 5 is closest to

(a) $496,100
(b) $652,200
(c) $715,420
(d) Over $720,000

2.74 Chemical costs associated with a packed-bed flue gas incinerator (for odor control) have been decreasing uniformly for 5 years because of increases in efficiency. If

the cost in year 1 was $100,000 and it decreased by $5000 per year through year 5, the present worth of the costs at 10% per year is closest to

(a) Less than $350,000
(b) $402,200
(c) $515,400
(d) Over $520,000

2.75 The future worth in year 10 of a present investment of $20,000 at an interest rate of 12% per year is closest to

(a) $62,120
(b) $67,560
(c) $71,900
(d) $81,030

2.76 A manufacturing company borrows $100,000 with a promise to repay the loan with equal annual payments over a 5-year period. At an interest rate of 12% per year, the annual payment will be closest to

(a) $23,620
(b) $27,740
(c) $29,700
(d) $31,800

2.77 Simpson Electronics wants to have $100,000 available in 3 years to replace a production line. The amount of money that would have to be deposited each year at an interest rate of 12% per year would be closest to

(a) $22,580
(b) $23,380
(c) $29,640
(d) Over $30,000

2.78 A civil engineer deposits $10,000 per year into a retirement account that achieves a rate of return of 12% per year. The amount of money in the account at the end of 25 years is closest to

(a) $670,500
(b) $902,800
(c) $1,180,900
(d) $1,333,300

2.79 The future worth (in year 8) of $10,000 in year 3, $10,000 in year 5, and $10,000 in year 8 at an interest rate of 12% per year is closest to
(a) $32,100
(b) $39,300
(c) $41,670
(d) $46,200

2.80 Maintenance costs for a regenerative thermal oxidizer have been increasing uniformly for 5 years. If the cost in year 1 was $8000 and it increased by $900 per year through year 5, the present worth of the costs at an interest rate of 10% per year is closest to
(a) $31,670
(b) $33,520
(c) $34,140
(d) Over $36,000

2.81 An investment of $100,000 resulted in income of $20,000 per year for 10 years. The rate of return on the investment was closest to
(a) 15%
(b) 18%
(c) 21%
(d) 25%

2.82 A construction company invested $60,000 in a new bulldozer. If the income from temporary leasing of the bulldozer is expected to be $15,000 per year, the length of time required to recover the investment at an interest rate of 18% per year is closest to
(a) 5 years
(b) 8 years
(c) 11 years
(d) 13 years

CASE STUDY

WHAT A DIFFERENCE THE YEARS AND COMPOUND INTEREST CAN MAKE

Two Real-World Situations

1. Manhattan Island purchase. History reports that Manhattan Island in New York was purchased for the equivalent of $24 in the year 1626. In the year 2001, the 375th anniversary of the purchase of Manhattan was recognized.

2. Stock-option program purchases. A young BS-graduate from a California university engineering college went to work for a company at the age of 22 and placed $50 per month into the stock purchase option. He left the company after a full 60 months of employment at age 27, and he did not sell the stock. The engineer did not inquire about the value of the stock until age 57, some 30 years later.

Case Study Exercises

About the Manhattan Island purchase:

1. Public sector investments are evaluated at 6% per year. Assume New York had invested the $24 at the conservative rate of 6%. Determine the worth of the Manhattan Island purchase in 2001 at (a) 6% per year simple interest and (b) 6% per year compound interest. Observe the significant difference that compounding has made at 6% over a long period of time—375 years, in this case.

2. What is the equivalent amount that New York would have had to commit in 1626 and *each year* thereafter to exactly equal the amount in (1) above at 6% per year *compounded* annually?

About the stock purchase program:

1. Construct the cash flow diagram for ages 22 through 57.
2. The engineer has learned that over the 35 intervening years, the stock earned at a rate of 1.25% per month. Determine the value of the $50 per month when the engineer left the company after a total of 60 purchases.
3. Determine the value of the engineer's company stock at age 57. Again, observe the significant difference that 30 years have made at a 15% per year compound rate.
4. Assume the engineer did not leave the funds invested in the stock at age 27. Now determine the amount he would have to deposit each year, starting at age 50, to be equivalent to the value at age 57 you calculated in (3) above. Assume the 7 years of deposits make a return of 15% per year.
5. Finally, compare the total amount of money deposited during the 5 years when the engineer was in his twenties with the total amount he would have to deposit during the 7 years in his fifties to have the equal and equivalent amount at age 57, as determined in (3) above.

Combining Factors

Most estimated cash flow series do not fit exactly the series for which the factors and equations in Chapter 2 were developed. Therefore, it is necessary to combine the equations. For a given sequence of cash flows, there are usually several correct ways to determine the equivalent present worth *P*, future worth *F*, or annual worth *A*. This chapter explains how to combine the engineering economy factors to address more complex situations involving shifted uniform series and gradient series. Spreadsheet functions are used to speedup the computations.

LEARNING OBJECTIVES

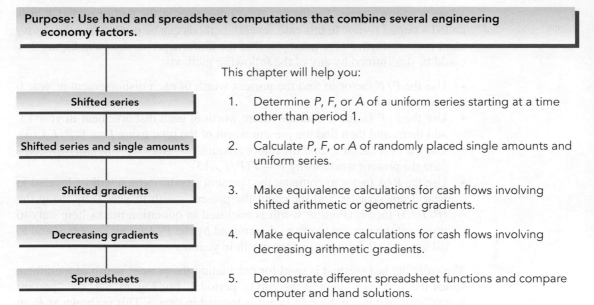

Purpose: Use hand and spreadsheet computations that combine several engineering economy factors.

This chapter will help you:

Shifted series	1.	Determine *P*, *F*, or *A* of a uniform series starting at a time other than period 1.
Shifted series and single amounts	2.	Calculate *P*, *F*, or *A* of randomly placed single amounts and uniform series.
Shifted gradients	3.	Make equivalence calculations for cash flows involving shifted arithmetic or geometric gradients.
Decreasing gradients	4.	Make equivalence calculations for cash flows involving decreasing arithmetic gradients.
Spreadsheets	5.	Demonstrate different spreadsheet functions and compare computer and hand solutions.

3.1 CALCULATIONS FOR UNIFORM SERIES THAT ARE SHIFTED

When a uniform series begins at a time other than at the end of period 1, it is called a *shifted series*. In this case several methods can be used to find the equivalent present worth P. For example, P of the uniform series shown in Figure 3–1 could be determined by any of the following methods:

- Use the P/F factor to find the present worth of each disbursement at year 0 and add them.
- Use the F/P factor to find the future worth of each disbursement in year 13, add them, and then find the present worth of the total using $P = F(P/F,i,13)$.
- Use the F/A factor to find the future amount $F = A(F/A,i,10)$, and then compute the present worth using $P = F(P/F,i,13)$.
- Use the P/A factor to compute the "present worth" (which will be located in year 3 not year 0), and then find the present worth in year 0 by using the $(P/F,i,3)$ factor. (Present worth is enclosed in quotation marks here only to represent the present worth as determined by the P/A factor in year 3, and to differentiate it from the present worth in year 0.)

Typically the last method is used for calculating the present worth of a uniform series that does not begin at the end of period 1. For Figure 3–1, the "present worth" obtained using the P/A factor is located in year 3. This is shown as P_3 in Figure 3–2. Note that a P value is always located *1 year or period prior* to the beginning of the first series amount. Why? Because the P/A factor was derived with P in time period 0 and A beginning at the end of period 1. The most common mistake made in working problems of this type is improper placement of P. Therefore, it is extremely important to remember:

> **The present worth is always located one period prior to the first uniform-series amount when using the P/A factor.**

To determine a future worth or F value, recall that the F/A factor derived in Section 2.3 had the F located in the *same* period as the last uniform-series amount. Figure 3–3 shows the location of the future worth when F/A is used for Figure 3–1 cash flows.

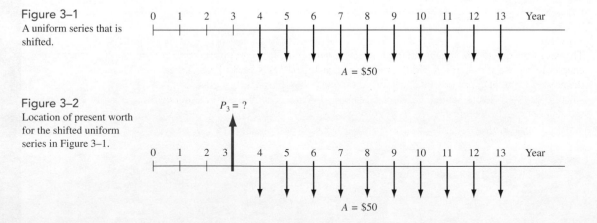

Figure 3–1
A uniform series that is shifted.

Figure 3–2
Location of present worth for the shifted uniform series in Figure 3–1.

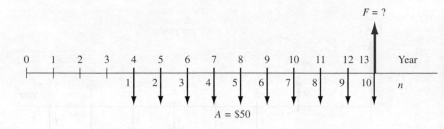

Figure 3–3
Placement of F and
renumbering for n for the
shifted uniform series of
Figure 3–1.

The future worth is always located in the same period as the last uniform-series amount when using the F/A factor.

It is also important to remember that the number of periods n in the P/A or F/A factor is equal to the number of uniform-series values. It may be helpful to *renumber* the cash flow diagram to avoid errors in counting. Figure 3–3 shows Figure 3–1 renumbered to determine $n = 10$.

As stated above, there are several methods that can be used to solve problems containing a uniform series that is shifted. However, it is generally more convenient to use the uniform-series factors than the single-amount factors. There are specific steps that should be followed in order to avoid errors:

1. Draw a diagram of the positive and negative cash flows.
2. Locate the present worth or future worth of each series on the cash flow diagram.
3. Determine n for each series by renumbering the cash flow diagram.
4. Draw another cash flow diagram representing the desired equivalent cash flow.
5. Set up and solve the equations.

These steps are illustrated below.

EXAMPLE **3.1**

An engineering technology group just purchased new CAD software for $5000 now and annual payments of $500 per year for 6 years starting 3 years from now for annual upgrades. What is the present worth of the payments if the interest rate is 8% per year?

Solution

The cash flow diagram is shown in Figure 3–4. The symbol P_A is used throughout this chapter to represent the present worth of a uniform annual series A, and P'_A represents the present worth at a time other than period 0. Similarly, P_T represents the total present worth at time 0. The correct placement of P'_A and the diagram renumbering to obtain n are also indicated. Note that P'_A is located in actual year 2, not year 3. Also, $n = 6$, not 8, for the P/A factor. First find the value of P'_A of the shifted series.

$$P'_A = \$500(P/A,8\%,6)$$

Since P'_A is located in year 2, now find P_A in year 0.

$$P_A = P'_A(P/F,8\%,2)$$

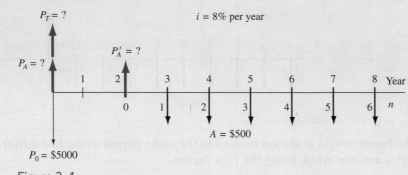

Figure 3–4
Cash flow diagram with placement of P values, Example 3.1.

The total present worth is determined by adding P_A and the initial payment P_0 in year 0.

$$P_T = P_0 + P_A$$
$$= 5000 + 500(P/A,8\%,6)(P/F,8\%,2)$$
$$= 5000 + 500(4.6229)(0.8573)$$
$$= \$6981.60$$

The more complex that cash flow series become, the more useful are the spreadsheet functions. When the uniform series A is shifted, the NPV function is used to determine P, and the PMT function finds the equivalent A value. The NPV function, like the PV function, determines the P values, but NPV can handle any combination of cash flows directly from the cells in the same way as the IRR function. Enter the net cash flows in contiguous cells (column or row), making sure to enter '0' for all zero cash flows. Use the format

$$\textbf{NPV}(i\%,\textbf{second\_cell:last\_cell}) + \textbf{first\_cell}$$

First_cell contains the cash flow for year 0 and must be listed separately for NPV to correctly account for the time value of money. The cash flow in year 0 may be 0.

The easiest way to find an equivalent A over n years for a shifted series is with the PMT function, where the P value is from the NPV function above. The format is the same as we learned earlier, but the entry for P is a cell reference, not a number.

$$\textbf{PMT}(i\%,n,\textbf{cell\_with\_P},F)$$

Alternatively, the same technique can be used when an F value was obtained using the FV function. Now the last entry in PMT is "cell_with_F."

It is very fortunate that any parameter in a spreadsheet function can itself be a function. Thus, it is possible to write the PMT function in a single cell by embedding the NPV function (and FV function, if needed). The format is

$$\text{PMT}(i\%,n,\text{NPV}(i\%,\text{second\_cell:last\_cell})+\text{first\_cell},F)$$

Of course, the answer for A is the same for the two-cell operation or a single-cell, embedded function. All three of these functions are illustrated in the next example.

Q-Solv

EXAMPLE 3.2

Recalibration of sensitive measuring devices costs $8000 per year. If the machine will be recalibrated for each of 6 years starting 3 years after purchase, calculate the 8-year equivalent uniform series at 16% per year. Show hand and computer solutions.

Solution by Hand
Figure 3–5a and b shows the original cash flows and the desired equivalent diagram. To convert the $8000 shifted series into an equivalent uniform series over all periods, first

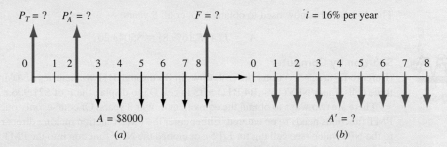

Figure 3–5
(a) Original and (b) equivalent cash flow diagrams; and (c) spreadsheet functions to determine P and A, Example 3.2.

convert the uniform series into a present worth or future worth amount. then either the A/P factor or the A/F factor can be used. Both methods are illustrated here.

Present worth method. (Refer to Figure 3–5a.) Calculate P'_A for the shifted series in year 2 and P_T in year 0.

$$P'_A = 8000(P/A,16\%,6)$$

$$P_T = P'_A(P/F,16\%,2) = 8000(P/A,16\%,6)(P/F,16\%,2)$$

$$= 8000(3.6847)(0.7432) = \$21,907.75$$

The equivalent series A' *for 8 years* can now be determined via the A/P factor.

$$A' = P_T(A/P,16\%,8) = \$5043.60$$

Future worth method. (Refer to Figure 3–5a.) First calculate the future worth F in year 8.

$$F = 8000(F/A,16\%,6) = \$71,820$$

The A/F factor is now used to obtain A' over all 8 years.

$$A' = F(A/F,16\%,8) = \$5043.20$$

Q-Solv

Solution by Computer

(Refer to Figure 3–5c.) Enter the cash flows in B3 through B11 with entries of '0' in the first three cells. Enter NPV(16%,B4:B11)+B3 in cell D5 to display the P of $21,906.87.

There are two ways to obtain the equivalent A over 8 years. Of course, only one of these PMT functions needs to be entered. Either enter the PMT function making direct reference to the NPV value (see cell tag for E/F5), or embed the NPV function into the PMT function (see cell tag for E/F8 or the formula bar).

3.2 CALCULATIONS INVOLVING UNIFORM-SERIES AND RANDOMLY PLACED SINGLE AMOUNTS

When a cash flow includes both a uniform series and randomly placed single amounts, the procedures of Section 3.1 are applied to the uniform series and the single-amount formulas are applied to the one-time cash flows. This approach, illustrated in Examples 3.3 and 3.4, is merely a combination of previous ones. For spreadsheet solutions, it is necessary to enter the net cash flows before using the NPV and other functions.

EXAMPLE 3.3

An engineering company in Wyoming that owns 50 hectares of valuable land has decided to lease the mineral rights to a mining company. The primary objective is to obtain long-term income to finance ongoing projects 6 and 16 years from the present time. The engineering company makes a proposal to the mining company that it pay $20,000 per year for 20 years beginning 1 year from now, plus $10,000 six years from now and $15,000 sixteen years from now. If the mining company wants to pay off its lease immediately, how much should it pay now if the investment should make 16% per year?

Solution

The cash flow diagram is shown in Figure 3–6 from the owner's perspective. Find the present worth of the 20-year uniform series and add it to the present worth of the two one-time amounts.

$$P = 20{,}000(P/A,16\%,20) + 10{,}000(P/F,16\%,6) + 15{,}000(P/F,16\%,16)$$

$$= \$124{,}075$$

Note that the $20,000 uniform series starts at the end of year 1, so the P/A factor determines the present worth at year 0.

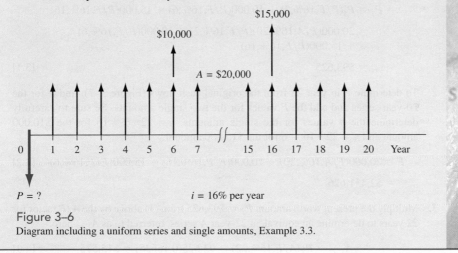

Figure 3–6
Diagram including a uniform series and single amounts, Example 3.3.

When you calculate the A value for a cash flow series that includes randomly placed single amounts and uniform series, *first convert everything to a present worth or a future worth.* Then the A value is obtained by multiplying P or F by the appropriate A/P or A/F factor. Example 3.4 illustrates this procedure.

EXAMPLE 3.4

Assume similar cash flow estimates to those projected in the previous example (Example 3.3) for the engineering company planning to lease its mineral rights. However, move the beginning year for the $20,000 per year series forward 2 years to start in year 3. It will now continue through year 22. Utilize engineering economy relations by hand and by computer to determine the five *equivalent values* listed below at 16% per year.

1. Total present worth P_T in year 0
2. Future worth F in year 22
3. Annual series over all 22 years

4. Annual series over the first 10 years
5. Annual series over the last 12 years

Solution by Hand

Figure 3–7 presents the cash flows with equivalent P and F values indicated in the correct years for the P/A, P/F, and F/A factors.

1. First determine the present worth of the series in year 2. Then the total present worth P_T is the sum of three P values: the series present worth value moved back to $t = 0$ with the P/F factor, and the P values at $t = 0$ for the two single amounts in years 6 and 16.

$$P'_A = 20{,}000(P/A,16\%,20)$$

$$P_T = P'_A(P/F,16\%,2) + 10{,}000(P/F,16\%,6) + 15{,}000(P/F,16\%,16)$$

$$= 20{,}000(P/A,16\%,20)(P/F,16\%,2) + 10{,}000(P/F,16\%,6)$$
$$+ 15{,}000(P/F,16\%,16)$$

$$= \$93{,}625 \tag{3.1}$$

2. To determine F in year 22 from the original cash flows (Figure 3–7), find F for the 20-year series and add the F values for the two single amounts. Be sure to carefully determine the n values for the single amounts: $n = 22-6 = 16$ for the $10,000 amount and $n = 22-16 = 6$ for the $15,000 amount.

$$F = 20{,}000(F/A,16\%,20) + 10{,}000(F/P,16\%,16) + 15{,}000(F/P,16\%,6) \tag{3.2}$$

$$= \$2{,}451{,}626$$

3. Multiply the present worth amount $P_T = \$93{,}625$ from (1) above by the A/P factor for 22 years to determine an equivalent 22-year A series, referred to as A_{1-22} here.

$$A_{1-22} = P_T(A/P,16\%,22) = 93{,}625(0.16635) = \$15{,}575 \tag{3.3}$$

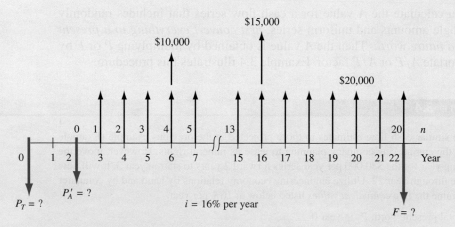

Figure 3–7
Diagram from Figure 3–6 with the A series shifted 2 years forward, Example 3.4.

An alternate way to determine the 22-year series uses the F value from (2) above. In this case, the computation is $A_{1-22} = F(A/F,16\%,22) = \$15,575$. Note that in both methods, the equivalent total P or F value is determined first, then the A/P or A/F factor for 22 years is applied.

4. This and (5), which follows, are special cases that often occur in engineering economy studies. The equivalent A series is calculated for a number of years different from that covered by the original cash flows. This occurs when a defined *study period* or *planning horizon* is preset for the analysis. (More is mentioned about study periods later.) To determine the equivalent A series for years 1 through 10 only (call it A_{1-10}) the P_T value *must be used* with the A/P factor for $n = 10$. This computation will transform the original cash flows in Figure 3–7 into the equivalent series A_{1-10} in Figure 3–8a.

$$A_{1-10} = P_T(A/P,16\%,10) = 93,625(0.20690) = \$19,371 \qquad [3.4]$$

5. For the equivalent 12-year series for years 11 through 22 (call it A_{11-22}), the F value *must be used* with the A/F factor for 12 years. This transforms Figure 3–7 into the 12-year series A_{11-22} in Figure 3–8b.

$$A_{11-22} = F(A/F,16\%,12) = 2,451,626(0.03241) = \$79,457 \qquad [3.5]$$

Notice the huge difference of more than \$60,000 in equivalent annual amounts that occurs when the present worth of \$93,625 is allowed to compound at 16% per year for the first 10 years. This is another demonstration of the time value of money.

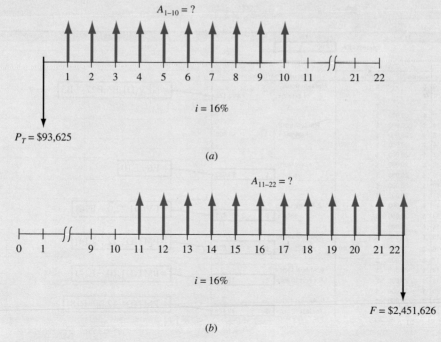

Figure 3–8
Cash flows of Figure 3–7 converted to equivalent uniform series for (a) years 1 to 10 and (b) years 11 to 22.

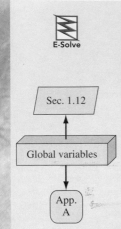

Solution by Computer

Figure 3–9 is a spreadsheet image with answers for all five questions. The $20,000 series and the two single amounts have been entered into separate columns, B and C. The zero cash flow values are all entered so that the functions will work correctly. This is an excellent example demonstrating the versatility of the NPV, FV, and PMT functions. In order to prepare for sensitivity analysis, the functions are developed using cell reference format or global variables, as indicated in the cell tags. This means that virtually any number—the interest rate, any cash flow estimate in the series or the single amounts, and the timing within the 22-year time frame—can be changed and the new answers will be immediately displayed. This is the general spreadsheet structure utilized for performing an engineering economy analysis with sensitivity analysis on the estimates.

1. Present worth values for the series and single amounts are determined in cells E6 and E10, respectively, using the NPV function. The sum of these in E14 is P_T = $93,622, which corresponds to the value in Equation [3.1].

Figure 3–9
Spreadsheet using cell reference format, Example 3.4.

2. The FV function in cell E18 uses the P value in E14 (preceded by a minus sign) to determine F twenty-two years later. This is significantly easier than Equation [3.2], which determines the three separate F values and adds them to obtain $F = \$2,451,626$. Of course, either method is correct.

3. To find the 22-year A series of \$15,574 starting in year 1, the PMT function in E21 references the P value in cell E14. This is effectively the same procedure used in Equation [3.3] to obtain A_{1-22}.

    For the spreadsheet enthusiast, it is possible to find the 22-year A series value in E21 directly by using the PMT function with embedded NPV functions. The cell reference format would be PMT(D1,22,−(NPV(D1,B6:B27)+B5 + NPV(D1,C6:C27)+C5)).

4. and 5.   It is quite simple to determine an equivalent uniform series over any number of periods using a spreadsheet, provided the series starts one period after the P value is located or ends in the same period that the F value is located. These are both true for the series requested here—the first 10-year series can reference P in cell E14, and the last 12-year series can anchor on F in cell E18. The results in E24 and E27 are the same as A_{1-10} and A_{11-22} in Equations [3.4] and [3.5], respectively.

Comment

Remember that some round-off error will always be present when comparing hand and computer results. The spreadsheet functions carry more decimal places than the tables during calculations. Also, be very careful when constructing spreadsheet functions. It is easy to miss a value, such as the P or F in PMT and FV functions, or a minus sign between entries. Always check your function entries carefully before touching <Enter>.

Additional Example 3.10.

3.3   CALCULATIONS FOR SHIFTED GRADIENTS

In Section 2.5, we derived the relation $P = G(P/G,i,n)$ to determine the present worth of the arithmetic gradient series. The P/G factor, Equation [2.15], was derived for a present worth in year 0 with the gradient starting between periods 1 and 2.

The present worth of an arithmetic gradient will always be located *two periods before the gradient starts*.

Refer to Figure 2–13 as a refresher for the cash flow diagrams.

The relation $A = G(A/G,i,n)$ was also derived in Section 2.5. The A/G factor in Equation [2.17] performs the equivalence transformation of a gradient only into an A series from years 1 through n, as indicated in Figure 2–14. Recall that when there is a base amount, it and the arithmetic gradient must be treated separately. Then the equivalent P or A values can be summed to obtain the equivalent total present worth P_T and total annual series A_T, according to Equations [2.18] and [2.19].

A conventional gradient series starts between periods 1 and 2 of the cash flow sequence. A gradient starting at any other time is called a *shifted gradient*. The n value in the P/G and A/G factors for a shifted gradient is determined by renumbering the time scale. The period in which the *gradient first appears is labeled period 2*. The *n* value for the factor is determined by the renumbered period where the last gradient increase occurs.

Partitioning a cash flow series into the arithmetic gradient series and the remainder of the cash flows can make very clear what the gradient *n* value should be. Example 3.5 illustrates this partitioning.

EXAMPLE 3.5

Gerri, an engineer at Fujitsu, Inc., has tracked the average inspection cost on a robotics manufacturing line for 8 years. Cost averages were steady at $100 per completed unit for the first 4 years, but have increased consistently by $50 per unit for each of the last 4 years. Gerri plans to analyze the gradient increase using the P/G factor. Where is the present worth located for the gradient? What is the general relation used to calculate total present worth in year 0?

Solution
Gerri constructed the cash flow diagram in Figure 3–10a. It shows the base amount $A = \$100$ and the arithmetic gradient $G = \$50$ starting between periods 4 and 5.

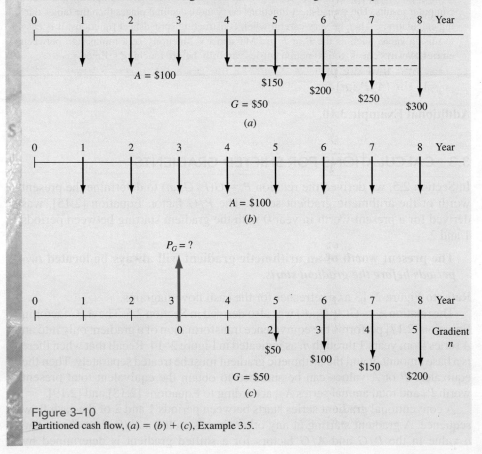

Figure 3–10
Partitioned cash flow, (*a*) = (*b*) + (*c*), Example 3.5.

Figure 3–10b and c partitions these two series. Gradient year 2 is placed in year 5 of the entire sequence in Figure 3–10c. It is clear that $n = 5$ for the P/G factor. The $P_G = ?$ arrow is correctly placed in gradient year 0, which is year 3 in the cash flow series.

The general relation for P_T is taken from Equation [2.18]. The uniform series $A = \$100$ occurs for all 8 years, and the $G = \$50$ gradient present worth appears in year 3.

$$P_T = P_A + P_G = 100(P/A,i,8) + 50(P/G,i,5)(P/F,i,3)$$

The P/G and A/G factor values for the shifted gradients in Figure 3–11 are shown below each diagram. Determine the factors and compare the answers with these values.

It is important to note that the A/G factor *cannot* be used to find an equivalent A value in periods 1 through n for cash flows involving a shifted gradient. Consider the cash flow diagram of Figure 3–11b. To find the equivalent annual series in years 1 through 10 for the gradient series only, first find the present worth of the gradient in year 5, take this present worth back to year 0, and then annualize the present worth for 10 years with the A/P factor. If you apply the annual series gradient factor $(A/G,i,5)$ directly, the gradient is converted into an equivalent annual series over years 6 through 10 only. Remember:

To find the equivalent A series of a shifted gradient through all the periods, first find the present worth of the gradient at actual time 0, then apply the $(A/P,i,n)$ factor.

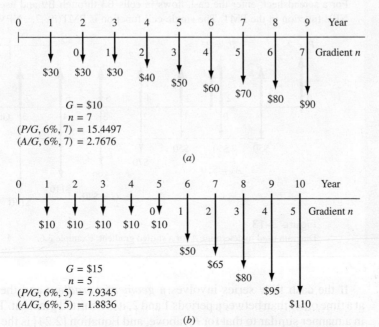

$G = \$10$
$n = 7$
$(P/G, 6\%, 7) = 15.4497$
$(A/G, 6\%, 7) = 2.7676$

(a)

$G = \$15$
$n = 5$
$(P/G, 6\%, 5) = 7.9345$
$(A/G, 6\%, 5) = 1.8836$

(b)

Figure 3–11
Determination of G and n values used in factors for shifted gradients.

EXAMPLE 3.6

Set up the engineering economy relations to compute the equivalent annual series in years 1 through 7 for the cash flow estimates in Figure 3–12.

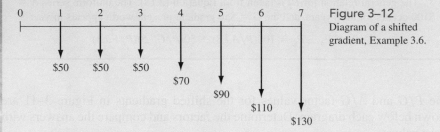

Figure 3–12
Diagram of a shifted gradient, Example 3.6.

Solution
The base amount annual series is $A_B = \$50$ for all 7 years (Figure 3–13). Find the present worth P_G in year 2 of the $20 gradient that starts in actual year 4. The gradient year is $n = 5$.

$$P_G = 20(P/G,i,5)$$

Bring the gradient present worth back to actual year 0.

$$P_0 = P_G(P/F,i,2) = 20(P/G,i,5)(P/F,i,2)$$

Annualize the gradient present worth from year 0 through year 7 to obtain A_G.

$$A_G = P_0(A/P,i,7)$$

Finally, add the base amount to the gradient annual series.

$$A = 20(P/G,i,5)(P/F,i,2)(A/P,i,7) + 50$$

For a spreadsheet, enter the cash flows in cells B3 through B9 and use an embedded NPV function in the PMT. The single-cell function is PMT($i\%$,7,$-$NPV($i\%$, B3:B9)).

Q-Solv

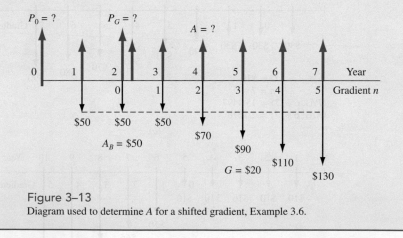

Figure 3–13
Diagram used to determine A for a shifted gradient, Example 3.6.

If the cash flow series involves a *geometric gradient* and the gradient starts at a time other than between periods 1 and 2, it is a shifted gradient. The P_g is located in a manner similar to that for P_G above, and Equation [2.24] is the factor formula.

Sec. 2.6

$(P/A,g,i,n)$ factor

EXAMPLE 3.7

Chemical engineers at a Coleman Industries plant in the Midwest have determined that a small amount of a newly available chemical additive will increase the water repellency of Coleman's tent fabric by 20%. The plant superintendent has arranged to purchase the additive through a 5-year contract at $7000 per year, starting 1 year from now. He expects the annual price to increase by 12% per year thereafter for the next 8 years. Additionally, an initial investment of $35,000 was made now to prepare a site suitable for the contractor to deliver the additive. Use $i = 15\%$ to determine the equivalent total present worth for all these cash flows.

Solution

Figure 3–14 presents the cash flows. The total present worth P_T is found using $g = 0.12$ and $i = 0.15$. Equation [2.24] is used to determine the present worth P_g for the entire geometric series at actual year 4, which is moved to year 0 using $(P/F,15\%,4)$.

$$P_T = 35{,}000 + A(P/A,15\%,4) + A_1(P/A,12\%,15\%,9)(P/F,15\%,4)$$

$$= 35{,}000 + 7000(2.8550) + \left[7000 \, \frac{1 - (1.12/1.15)^9}{0.15 - 0.12}\right](0.5718)$$

$$= 35{,}000 + 19{,}985 + 28{,}247$$

$$= \$83{,}232$$

Note that $n = 4$ in the $(P/A,15\%,4)$ factor because the $7000 in year 5 is the initial amount A_1 in Equation [2.23].

For solution by computer, enter the cash flows of Figure 3–14. If cells B1 through B14 are used, the function to find $P = \$83{,}230$ is

$$NPV(15\%,B2:B14)+B1$$

The fastest way to enter the geometric series is to enter $7840 for year 6 (into cell B7) and set up each succeeding cell multiplied by 1.12 for the 12% increase.

Q-Solv

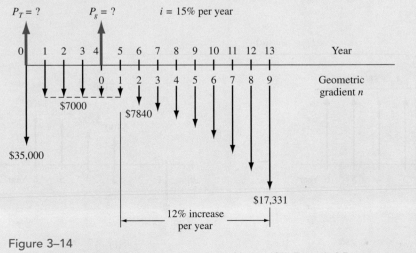

Figure 3–14
Cash flow diagram including a geometric gradient with $g = 12\%$, Example 3.7.

3.4 SHIFTED DECREASING ARITHMETIC GRADIENTS

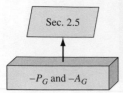

Sec. 2.5

$-P_G$ and $-A_G$

The use of the arithmetic gradient factors is the same for increasing and decreasing gradients, except that in the case of decreasing gradients the following are true:

1. The base amount is equal to the *largest* amount in the gradient series, that is, the amount in period 1 of the series.
2. The gradient amount is *subtracted* from the base amount instead of added to it.
3. The term $-G(P/G,i,n)$ or $-G(A/G,i,n)$ is used in the computations and in Equations [2.18] and [2.19] for P_T and A_T, respectively.

The present worth of the gradient will still take place two periods before the gradient starts, and the equivalent A value will start at period 1 of the gradient series and continue through period n.

Figure 3–15 partitions a decreasing gradient series with $G = \$-100$ that is shifted 1 year forward. P_G occurs in actual year 1, and P_T is the sum of three components.

$$P_T = \$800(P/F,i,1) + 800(P/A,i,5)(P/F,i,1) - 100(P/G,i,5)(P/F,i,1)$$

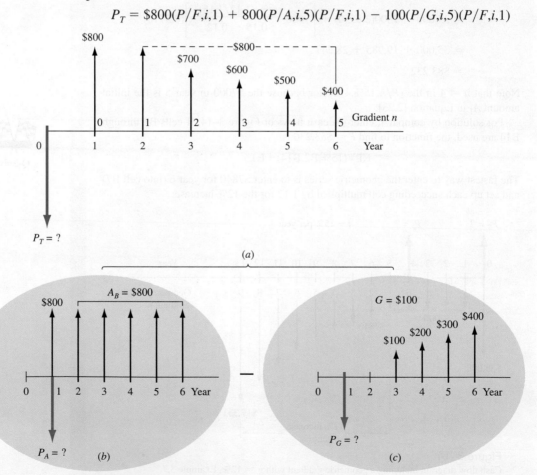

Figure 3–15
Partitioned cash flow of a shifted arithmetic gradient, $(a) = (b) - (c)$.

EXAMPLE **3.8**

Assume that you are planning to invest money at 7% per year as shown by the increasing gradient of Figure 3–16. Further, you expect to withdraw according to the decreasing gradient shown. Find the net present worth and equivalent annual series for the entire cash flow sequence and interpret the results.

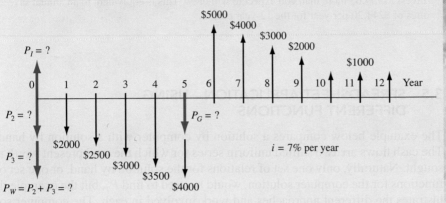

Figure 3–16
Investment and withdrawal series, Example 3.8.

Solution

For the investment sequence, G is $500, the base amount is $2000, and $n = 5$. For the withdrawal sequence through year 10, G is $-1000, the base amount is $5000, and $n = 5$. There is a 2-year annual series with $A = $1000 in years 11 and 12. For the investment series,

$$P_I = \text{present worth of deposits}$$

$$= 2000(P/A,7\%,5) + 500(P/G,7\%,5)$$

$$= 2000(4.1002) + 500(7.6467)$$

$$= \$12,023.75$$

For the withdrawal series, let P_W represent the present worth of the withdrawal base amount and gradient series in years 6 through 10 (P_2), plus the present worth of withdrawals in years 11 and 12 (P_3). Then

$$P_W = P_2 + P_3$$

$$= P_G(P/F,7\%,5) + P_3$$

$$= [5000(P/A,7\%,5) - 1000(P/G,7\%,5)](P/F,7\%,5)$$

$$\quad + 1000(P/A,7\%,2)(P/F,7\%,10)$$

$$= [5000(4.1002) - 1000(7.6467)](0.7130) + 1000(1.8080)(0.5083)$$

$$= \$9165.12 + 919.00 = \$10,084.12$$

Since P_I is actually a negative cash flow and P_W is positive, the net present worth is

$$P = P_W - P_I = 10,084.12 - 12,023.75 = \$-1939.63$$

The A value may be computed using the $(A/P,7\%,12)$ factor.

$$A = P(A/P,7\%,12)$$
$$= \$-244.20$$

The interpretation of these results is as follows: In present-worth equivalence, you will invest $1939.63 more than you expect to withdraw. This is equivalent to an annual savings of $244.20 per year for the 12-year period.

3.5 SPREADSHEET APPLICATION—USING DIFFERENT FUNCTIONS

E-Solve

The example below compares a solution by computer with a solution by hand. The cash flows are two shifted uniform series for which the total present worth is sought. Naturally, only one set of relations for the solution by hand, or one set of functions for the computer solution, would be used to find P_T, but the example illustrates the different approaches and work involved in each. The computer solution is faster, but the solution by hand helps in the understanding of how the time value of money is accounted for by engineering economy factors.

EXAMPLE 3.9

Determine the total present worth P_T in period 0 at 15% per year for the two shifted uniform series in Figure 3–17. Use two approaches: by computer with different functions and by hand using three different factors.

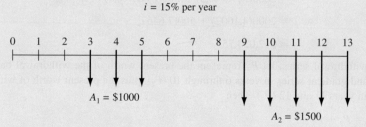

Figure 3–17
Uniform series used to compute present worth by several methods, Example 3.9.

Solution by Computer
Figure 3–18 finds P_T using the NPV and PV functions.

NPV function: This is by far the easiest way to determine $P_T = \$3370$. The cash flows are entered into cells, and the NPV function is developed using the format

Q-Solv

NPV($i\%$,second_cell:last_cell)+first_cell   or   NPV(B1,B6:B18)+B5

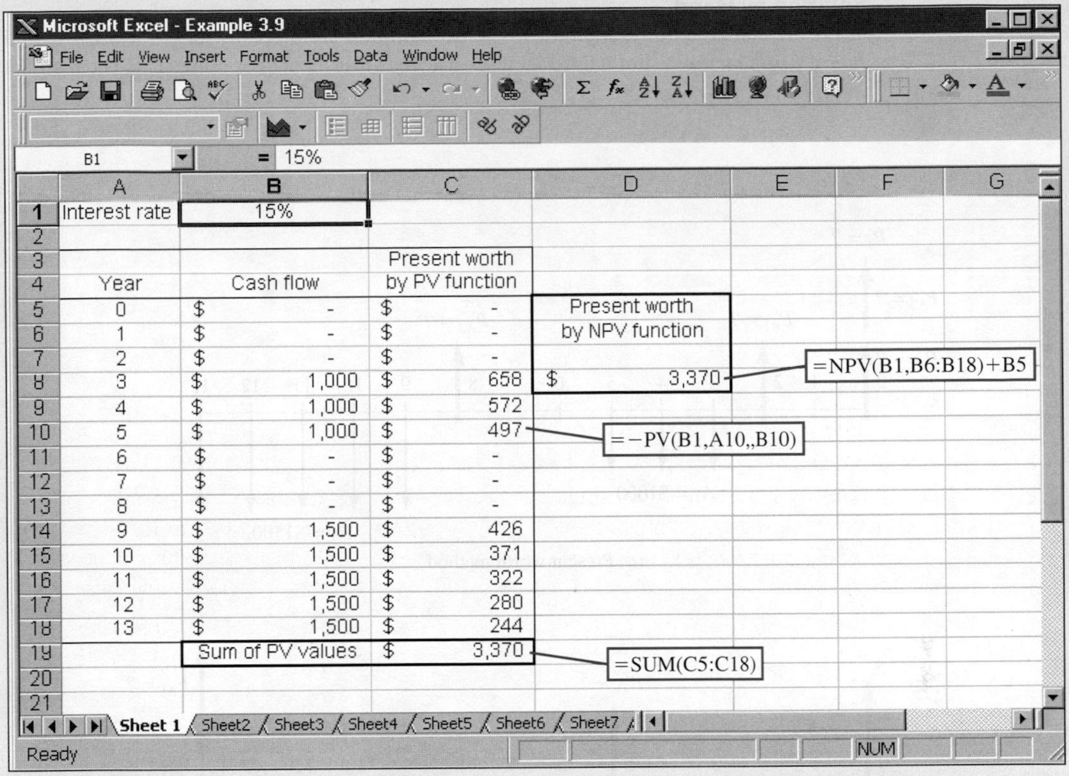

Figure 3–18
Spreadsheet determination of total present worth using NPV and PV functions, Example 3.9.

The value $i = 15\%$ is in cell B1. With the NPV parameters in cell-reference form, any value can be changed, and the new P_T value is displayed immediately. Additionally, if more than 13 years are needed, simply add the cash flows at the end of column B and increase the B18 entry accordingly. *Remember: The NPV function requires that all spreadsheet cells representing a cash flow have an entry, including those periods that have a zero cash flow value.* The wrong answer is generated if cells are left blank.

PV function: Column C entries in Figure 3–18 include PV functions that determine P in period 0 for each cash flow separately. They are added in C19 using the SUM function. This approach takes more keyboard time, but it does provide the P value for each cash flow, if these amounts are needed. Also, the PV function does not require that each zero cash flow entry be made.

FV function: It is not efficient to determine P_T using the FV function, because the FV format does not allow direct entry of cell references like the NPV function. Each cash flow must first be taken to the last period using the general format FV(15%,years_remaining,,cash_flow), where they are summed using the SUM function. This SUM is then moved back to period 0 via the PV(15%,13,,SUM) function. In this case, both the NPV and PV functions, especially NPV, provide a much more efficient use of spreadsheet capabilities than FV.

Solution by Hand

There are numerous ways to find P_T. The two simplest are probably the *present worth* and *future worth methods*. For a third method, use year 7 as an anchor point. This is called the *intermediate year method*.

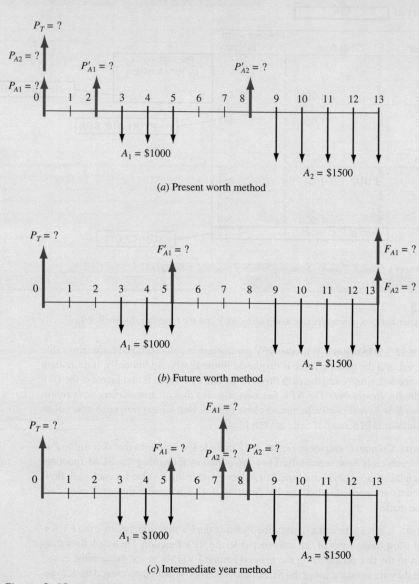

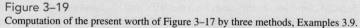

Figure 3–19
Computation of the present worth of Figure 3–17 by three methods, Examples 3.9.

Present worth method: See Figure 3–19a. The use of P/A factors for the uniform series, followed by the use of P/F factors to obtain the present worth in year 0, finds P_T.

$$P_T = P_{A1} + P_{A2}$$

$$P_{A1} = P'_{A1}(P/F,15\%,2) = A_1(P/A,15\%,3)(P/F,15\%,2)$$

$$= 1000(2.2832)(0.7561)$$

$$= \$1726$$

$$P_{A2} = P'_{A2}(P/F,15\%,8) = A_2(P/A,15\%,5)(P/F,15\%,8)$$

$$= 1500(3.3522)(0.3269)$$

$$= \$1644$$

$$P_T = 1726 + 1644 = \$3370$$

Future worth method: See Figure 3–19b. Use the F/A, F/P, and P/F factors.

$$P_T = (F_{A1} + F_{A2})(P/F,15\%,13)$$

$$F_{A1} = F'_{A1}(F/P,15\%,8) = A_1(F/A,15\%,3)(F/P,15\%,8)$$

$$= 1000(3.4725)(3.0590) = \$10,622$$

$$F_{A2} = A_2(F/A,15\%,5) = 1500(6.7424) = \$10,113$$

$$P_T = (F_{A1} + F_{A2})(P/F,15\%,13) = 20,735(0.1625) = \$3369$$

Intermediate year method: See Figure 3–19c. Find the equivalent worth of both series at year 7 and then use the P/F factor.

$$P_T = (F_{A1} + P_{A2})(P/F,15\%,7)$$

The P_{A2} value is computed as a present worth; but to find the total value P_T at year 0, it must be treated as an F value. Thus,

$$F_{A1} = F'_{A1}(F/P,15\%,2) = A_1(F/A,15\%,3)(F/P,15\%,2)$$

$$= 1000(3.4725)(1.3225) = \$4592$$

$$P_{A2} = P'_{A2}(P/F,15\%,1) = A_2(P/A,15\%,5)(P/F,15\%,1)$$

$$= 1500(3.3522)(0.8696) = \$4373$$

$$P_T = (F_{A1} + P_{A2})(P/F,15\%,7)$$

$$= 8965(0.3759) = \$3370$$

ADDITIONAL EXAMPLE

EXAMPLE 3.10

PRESENT WORTH BY COMBINING FACTORS

Calculate the total present worth of the following series of cash flows at $i = 18\%$ per year.

Year	0	1	2	3	4	5	6	7
Cash Flow, $	+460	+460	+460	+460	+460	+460	+460	−5000

Solution
The cash flow diagram is shown in Figure 3–20. Since the receipt in year 0 is equal to the A series in years 1 through 6, the P/A factor can be used for either 6 or 7 years. The problem is worked both ways.

Using P/A and $n = 6$: The receipt P_0 in year 0 is added to the present worth of the remaining amounts, since the P/A factor for $n = 6$ will place P_A in year 0.

$$P_T = P_0 + P_A - P_F$$

$$= 460 + 460(P/A,18\%,6) - 5000(P/F,18\%,7)$$

$$= \$499.40$$

Using P/A and $n = 7$: By using the P/A factor for $n = 7$, the "present worth" is located in year -1, not year 0, because the P is one period prior to the first A. It is necessary to move the P_A value 1 year forward with the F/P factor.

$$P = 460(P/A,18\%,7)(F/P,18\%,1) - 5000(P/F,18\%,7)$$

$$= \$499.38$$

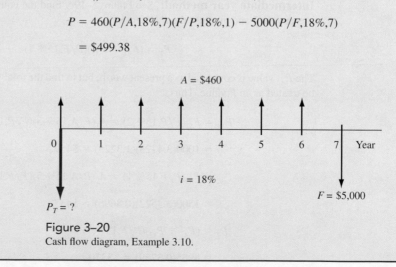

Figure 3–20
Cash flow diagram, Example 3.10.

CHAPTER SUMMARY

In Chapter 2, we derived the equations to calculate the present, future, or annual worth of specific cash flow series. In this chapter, we have shown that these equations apply to cash flow series different from those for which the basic relations are derived. For example, when a uniform series does not begin in period 1, we still use the P/A factor to find the "present worth" of the series, except the P value is located one period ahead of the first A value, not at time 0. For arithmetic and geometric gradients, the P value is two periods ahead of where the gradient starts. With this information, it is possible to solve for any symbol—P, A, or F— for any conceivable cash flow series.

We have experienced some of the power of spreadsheet functions in determining P, F, and A values once the cash flow estimates are entered into spreadsheet cells.

PROBLEMS

Present Worth Calculations

3.1 Because unintended lane changes by distracted drivers are responsible for 43% of all highway fatalities, Ford Motor Co. and Volvo launched a program to develop technologies to prevent accidents by sleepy drivers. A device costing $260 tracks lane markings and sounds an alert during lane changes. If these devices are included in 100,000 new cars per year beginning 3 years from now, what would be the present worth of their cost over a 10-year period at an interest rate of 10% per year?

3.2 One plan to raise money for Texas schools involves an "enrichment tax" that could collect $56 for every student in a certain school district. If there are 50,000 students in the district and the cash flow begins 2 years from now, what is the present worth of the enrichment plan over a 5-year planning period at an interest rate of 8% per year?

3.3 Amalgamated Iron and Steel purchased a new machine for ram cambering large I-beams. The company expects to bend 80 beams at $2000 per beam in each of the first 3 years, after which the company expects to bend 100 beams per year at $2500 per beam through year 8. If the company's minimum attractive rate of return is 18% per year, what is the present worth of the expected income?

3.4 Rubbermaid Plastics plans to purchase a rectilinear robot for pulling parts from an injection molding machine. Because of the robot's speed, the company expects production costs to decrease by $100,000 per year in each of the first 3 years and by $200,000 per year in the next 2 years. What is the present worth of the cost savings if the company uses an interest rate of 15% per year on such investments?

3.5 Toyco Watercraft has a contract with a parts supplier that involves purchases amounting to $150,000 per year, with the first purchase to be made now, followed by similar purchases over the next 5 years. Determine the present worth of the contract at an interest rate of 10% per year.

3.6 Calculate the present worth in year 0 of the following series of disbursements. Assume that $i = 10\%$ per year.

Year	Disbursement, $	Year	Disbursement, $
0	0	6	5000
1	3500	7	5000
2	3500	8	5000
3	3500	9	5000
4	5000	10	5000
5	5000		

Year	Income, $/Year	Expense, $/Year
0	10,000	2000
1–6	800	200
7–10	900	300

Annual Worth Calculations

3.7 Cisco's *gross revenue* (the percentage of revenue left after subtracting the cost of goods sold) was 70.1% of total revenue over a certain 4-year period. If the *total revenue* was $5.4 billion for the first 2 years and $6.1 billion for the last 2 years, what was the equivalent annual worth of the *gross revenue* over that 4-year period at an interest rate of 20% per year?

3.8 BKM Systems sales revenues are shown below. Calculate the equivalent annual worth (years 1 through 7), using an interest rate of 10% per year.

Year	Disbursement, $	Year	Disbursement, $
0	0	4	5000
1	4000	5	5000
2	4000	6	5000
3	4000	7	5000

3.9 A metallurgical engineer decides to set aside money for his newborn daughter's college education. He estimates that her needs will be $20,000 on her 17th, 18th, 19th, and 20th birthdays. If he plans to make uniform deposits starting 3 years from now and continue through year 16, what should be the size of each deposit, if the account earns interest at a rate of 8% per year?

3.10 Calculate the annual worth in years 1 through 10 of the following series of incomes and expenses, if the interest rate is 10% per year.

3.11 How much money would you have to pay each year in 8 equal payments, starting 2 years from today, to repay a $20,000 loan received from a relative today, if the interest rate is 8% per year?

3.12 An industrial engineer is planning for his early retirement 25 years from now. He believes he can comfortably set aside $10,000 each year starting *now*. If he plans to start withdrawing money 1 year after he makes his last deposit (i.e., year 26), what uniform amount could he withdraw each year for 30 years, if the account earns interest at a rate of 8% per year?

3.13 A rural utility company provides standby power to pumping stations using diesel-powered generators. An alternative has arisen whereby the utility could use natural gas to power the generators, but it will be a few years before the gas is available at remote sites. The utility estimates that by switching to gas, it will save $15,000 per year, starting 2 years from now. At an interest rate of 8% per year, determine the equivalent annual worth (years 1 through 10) of the projected savings.

3.14 The operating cost of a pulverized coal cyclone furnace is expected to be $80,000 per year. If the steam produced will be needed only for 5 years beginning now (i.e., years 0 through 5), what is the equivalent annual worth in years 1 through 5 of the operating cost at an interest rate of 10% per year?

3.15 An entrepreneurial electrical engineer has approached a large water utility with a proposal that promises to reduce the utility's power bill by at least 15% per year for the next 5 years through installation of patented surge protectors. The proposal states that the engineer will get $5000 now and annual payments that are equivalent to 75% of the power savings achieved from the devices. Assuming the savings are the same every year (i.e., 15%) and that the utility's power bill is $1 million per year, what would be the equivalent uniform amount (years 1 through 5) of the payments to the engineer? Assume the utility uses an interest rate of 6% per year.

3.16 A large water utility is planning to upgrade its SCADA system for controlling well pumps, booster pumps, and disinfection equipment, so that everything can be controlled from one site. The first phase will reduce labor and travel costs by $28,000 per year. The second phase will reduce costs by an additional $20,000 per year. If phase I savings occur in years 0, 1, 2, and 3 and phase II occurs in years 4 through 10, what is the equivalent annual worth of the upgraded system in years 1 through 10 at an interest rate of 8% per year?

3.17 A mechanical engineer who recently graduated with a master's degree is contemplating starting his own commercial heating and cooling company. He can purchase a Web page design package aimed at delivering information only for $600 per year. If his business is successful, he will purchase a more elaborate e-commerce package costing $4000 per year. If the engineer purchases the less expensive page now (beginning-of-year payments) and he purchases the e-commerce package 1 year from now (also beginning-of-year payments), what is the equivalent annual worth of costs for the website for a 5-year period (years 1 through 5) at an interest rate of 12% per year?

Future Worth Calculations

3.18 Lifetime savings accounts, known as LSAs, would allow people to invest after-tax

money without being taxed on any of the gains. If an engineer invests $10,000 now and $10,000 each year for the next 20 years, how much will be in the account immediately after the last deposit if the account grows by 15% per year?

3.19 How much money was deposited each year for 5 years if the account is now worth $100,000 and the last deposit was made 10 years ago? Assume the account earned interest at 7% per year.

3.20 Calculate the future worth (in year 11) of the following income and expenses, if the interest rate is 8% per year.

Year	Income, $	Expense, $
0	12,000	3000
1–6	800	200
7–11	900	200

Random Placement and Uniform Series

3.21 What is the equivalent worth in year 5 of the following series of income and disbursements, if the interest rate is 12% per year?

Year	Income, $	Expense, $
0	0	9000
1–5	6000	6000
6–8	6000	3000
9–14	8000	5000

3.22 Use the cash flow diagram below to calculate the amount of money in year 5 that is equivalent to all the cash flows shown, if the interest rate is 12% per year.

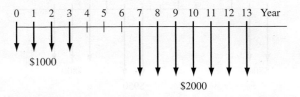

3.23 By spending $10,000 now and $25,000 three years from now, a plating company can increase its income in years 4 through 10. At an interest rate of 12% per year, how much extra income per year would be needed in years 4 through 10 to recover the investment?

3.24 Sierra Electric Company is considering the purchase of a hillside ranch for possible use as a windmill farm sometime in the future. The owner of the 500-acre ranch will sell for $3000 per acre if the company will pay her in two payments—one payment now and another that is twice as large 3 years from now. If the transaction interest rate is 8% per year, what is the amount of the first payment?

3.25 Two equal deposits made 20 and 21 years ago, respectively, will allow a retiree to withdraw $10,000 now and $10,000 per year for 14 more years. If the account earned interest at 10% per year, how large was each deposit?

3.26 A concrete and building materials company is trying to bring the company-funded portion of its employee retirement fund into compliance with HB-301. The company has already deposited $20,000 in each of the last 5 years. How much must be deposited now in order for the fund to have $350,000 three years from now, if the fund grows at a rate of 15% per year?

3.27 Find the value of x below such that the positive cash flows will be exactly equivalent to the negative cash flows, if the interest rate is 14% per year.

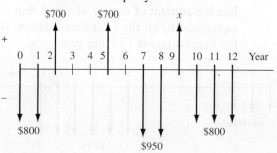

3.28 In attempting to obtain a swing loan from a local bank, a general contractor was asked to provide an estimate of annual expenses. One component of the expenses is shown in the cash flow diagram below. Convert the amounts shown into an equivalent uniform annual amount in years 1 through 8, using an interest rate of 12% per year.

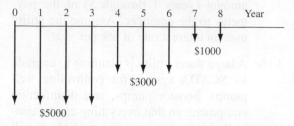

3.29 Determine the value in year 8 that is equivalent to the cash flows below. Use an interest rate of 12% per year.

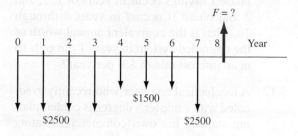

3.30 Find the value of x in the diagram below that will make the equivalent present worth of the cash flow equal to $15,000, if the interest rate is 15% per year.

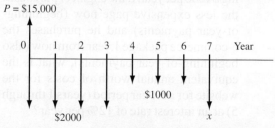

3.31 Calculate the amount of money in year 3 that is equivalent to the following cash flows, if the interest rate is 16% per year.

Year	Amount, $	Year	Amount, $
0	900	5	3000
1	900	6	−1500
2	900	7	500
3	900	8	500
4	3000		

3.32 Calculate the annual worth (years 1 through 7) of the following series of disbursements. Assume that $i = 12\%$ per year.

Year	Disbursement, $	Year	Disbursement, $
0	5000	4	5000
1	3500	5	5000
2	3500	6	5000
3	3500	7	5000

3.33 Calculate the value of x for the cash flows below such that the equivalent total value in year 8 is $20,000, using an interest rate of 15% per year.

Year	Cash Flow, $	Year	Cash Flow, $
0	2000	6	x
1	2000	7	x
2	x	8	x
3	x	9	1000
4	x	10	1000
5	x	11	1000

Shifted Arithmetic Gradients

3.34 San Antonio is considering various options for providing water in its 50-year plan, including desalting. One brackish aquifer is expected to yield desalted water that will generate revenue of $4.1 million per year for the first 4 years, after which less production will decrease revenue each year by $50,000 per year. If the aquifer will be totally depleted in 25 years, what is the present worth of the desalting option at an interest rate of 6% per year?

3.35 Exxon-Mobil is planning to sell a number of producing oil wells. The wells are expected to produce 100,000 barrels of oil per year for 8 more years at a selling price of $28 per barrel for the next 2 years, increasing by $1 per barrel through year 8. How much should an independent refiner be willing to pay for the wells now, if the interest rate is 12% per year?

3.36 Burlington Northern is considering the elimination of a railroad grade crossing by constructing a dual-track overpass. The railroad subcontracts for maintenance of its crossing gates at $11,500 per year. Beginning 4 years from now, however, the costs are expected to increase by $1000 per year into the foreseeable future (that is, $12,500 4 years from now, $13,500 five years from now, etc.). The overpass will cost $1.4 million (now) to build, but it will eliminate 100% of the auto-train collisions that have cost an average of $250,000 per year. If the railroad uses a 10-year study period and an interest rate of 10% per year, determine whether the railroad should build the overpass.

3.37 Levi Strauss has some of its jeans stone-washed under a contract with independent U.S. Garment Corp. If U.S. Garment's operating cost per machine is $22,000 per year for years 1 and 2 and then it increases by $1000 per year through year 5, what is the equivalent uniform annual cost per machine (years 1 through 5) at an interest rate of 12% per year?

3.38 Herman Trucking Company's receipts and disbursements (in $1000) are shown below. Calculate the future worth in year 7 at an interest rate of 10% per year.

Year	Cash Flow, $	Year	Cash Flow, $
0	−10,000	4	5,000
1	4,000	5	−1,000
2	3,000	6	7,000
3	4,000	7	8,000

3.39 Peyton Packing has a ham cooker that has the cost stream below. If the interest rate is

15% per year, determine the annual worth (in years 1 through 7) of the costs.

Year	Cost, $	Year	Cost, $
0	4,000	4	6,000
1	4,000	5	8,000
2	3,000	6	10,000
3	2,000	7	12,000

3.40 A start-up company selling color-keyed carnuba car wax borrows $40,000 at an interest rate of 10% per year and wishes to repay the loan over a 5-year period with annual payments such that the third through fifth payments are $2000 greater than the first two. Determine the size of the first two payments.

3.41 For the cash flows below, find the value of x that makes the present worth in year 0 equal to $11,000 at an interest rate of 12% per year.

Year	Cash Flow, $	Year	Cash Flow, $
0	200	5	700
1	300	6	800
2	400	7	900
3	x	8	1000
4	600	9	1100

Shifted Geometric Gradients

3.42 In an effort to compensate for shrinking land-line customers, SBC and Bell South (owners of Cingular Wireless LLC) got into a bidding war with Vodaphone to acquire AT&T Wireless. The initial $11 per share offer escalated to $13 for the 2.73 billion shares of AT&T Wireless. If the buyout took exactly 1 year to close (i.e., end of year 1), what would be the present worth today (time 0) of the acquisition for profits of $5.3 billion in year 2, increasing by 9% per year through year 11? Assume SBC and Bell South use a rate of return of 15% per year.

3.43 A successful alumnus is planning to make a contribution to the community college from which he graduated. The donation is to be made over a 5-year period beginning now, a total of six payments. It will support five preengineering students per year for 20 years, with the first scholarship to be awarded immediately (a total of 21 scholarships). The cost of tuition at the school is $4000 per year and is expected to stay at that amount for 3 more years. After that time (i.e., year 4), the tuition is expected to increase by 8% per year. If the college can invest the money and earn interest at a rate of 10% per year, what size must the donations be?

3.44 Calculate the present worth (year 0) of a lease that requires a payment of $20,000 now and amounts increasing by 5% per year through year 10. Use an interest rate of 14% per year.

3.45 Calculate the present worth for a machine that has an initial cost of $29,000, a life of 10 years, and an annual operating cost of $13,000 for the first 4 years, increasing by 10% per year thereafter. Use an interest rate of 10% per year.

3.46 A-1 Box Company is planning to lease a computer system that will cost (with service) $15,000 in year 1, $16,500 in year 2, and amounts increasing by 10% each year thereafter. Assume the lease payments must be made at the beginning of the year and that a 5-year lease is planned. What is the present worth (year 0) if the company uses a minimum attractive rate of return of 16% per year?

3.47 Dakota Hi-C Steel signed a contract that will generate revenue of $210,000 now, $222,600 in year 1, and amounts increasing by 8% per year through year 5. Calculate the future worth of the contract at an interest rate of 8% per year.

Shifted Decreasing Gradients

3.48 Find the present worth (at time 0) of the chrome plating costs in the cash flow diagram. Assume $i = 12\%$ per year.

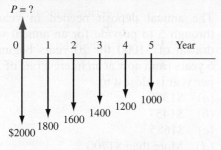

3.49 Compute the present worth (year 0) of the following cash flows at $i = 12\%$ per year.

Year	Amount, $	Year	Amount, $
0	5000	8	700
1–5	1000	9	600
6	900	10	500
7	800	11	400

3.50 For the cash flow tabulation, calculate the equivalent uniform annual worth in periods 1 through 10, if the interest rate is 10% per year.

Year	Amount, $	Year	Amount, $
0	2000	6	2400
1	2000	7	2300
2	2000	8	2200
3	2000	9	2100
4	2000	10	2000
5	2500		

FE REVIEW PROBLEMS

3.54 The annual worth in years 4 through 8 of an amount of money x that will be received 2 years from now is $4000. At an interest rate of 10% per year, the value of x is closest to
(*a*) Less than $12,000
(*b*) $12,531
(*c*) $12,885
(*d*) More than $13,000

3.51 Prudential Realty has an escrow account for one of its property management clients that currently contains $20,000. How long will it take to deplete the account if the client withdraws $5000 now, $4500 one year from now, and amounts decreasing by $500 each year thereafter, if the account earns interest at a rate of 8% per year?

3.52 The cost of spacers used around fuel rods in liquid-metal fast breeder reactors has been decreasing because of the availability of improved temperature-resistant ceramic materials. Determine the present worth (in year 0) of the costs shown in the diagram below, using an interest rate of 15% per year.

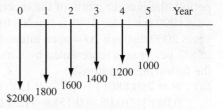

3.53 Compute the future worth in year 10 at $i = 10\%$ per year for the cash flow shown below.

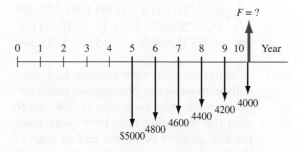

3.55 The multistate Powerball Lottery, worth $182 million, was won by a single individual who had purchased five tickets at $1 each. The individual was given two choices: Receive 26 payments of $7 million each, with the first payment to be made *now* and the rest to be made at the end of each of the next 25 years; or receive

a single lump-sum payment *now* that would be equivalent to the 26 payments of $7 million each. If the state uses an interest rate of 4% per year, the amount of the lump-sum payment is closest to
(*a*) Less than $109 million
(*b*) $109,355,000
(*c*) $116,355,000
(*d*) Over $117 million

3.56 The attendance at the annual El Paso Livestock Show and Rodeo has been declining for the past 5 years. The attendance was 25,880 in 2000 and 13,500 in 2004 (a 15% per year decrease). If the average ticket price was $10 per person over that time period, the present worth of the income in year 1999 (i.e., year 1999 is time 0) for the years 2000 through 2004 at an interest rate of 8% per year is represented by which of the following equations?
(*a*) $P = 250,880\{1 - [(1 + 0.15)^5/(1 + 0.08)^5]\}/(0.08 - 0.15) = \$1,322,123$
(*b*) $P = 250,880\{1 - [(1 - 0.15)^4/(1 + 0.08)^4]\}/(0.08 + 0.15) = \$672,260$
(*c*) $P = 250,880\{1 - [(1 + 0.15)^4/(1 + 0.08)^4]\}/(0.08 - 0.15) = \$1,023,489$
(*d*) $P = 250,880\{1 - [(1 - 0.15)^5/(1 + 0.08)^5]\}/(0.08 + 0.15) = \$761,390$

3.57 A company that manufactures hydrogen sulfide monitors is planning to make deposits such that each one is 5% larger than the preceding one. How large must the first deposit be (at the end of year 1) if the deposits extend through year 10 and the fourth deposit is $1250? Use an interest rate of 10% per year.
(*a*) $1312.50
(*b*) $1190.48
(*c*) $1133.79
(*d*) $1079.80

3.58 If $10,000 is borrowed now at 10% per year interest, the balance at the end of year 2 after payments of $3000 in year 1 and $3000 in year 2 will be closest to

(*a*) Less than $5000
(*b*) $5800
(*c*) $6100
(*d*) More than $7000

3.59 The annual deposit needed in years 1 through 5 to provide for an annual withdrawal of $1000 for 20 years beginning 6 years from now at an interest rate of 10% per year is closest to
(*a*) $1395
(*b*) $1457
(*c*) $1685
(*d*) More than $1700

3.60 The maintenance cost for a certain machine is $1000 per year for the first 5 years and $2000 per year for the next 5 years. At an interest rate of 10% per year, the annual worth in years 1 through 10 of the maintenance cost is closest to
(*a*) $1255
(*b*) $1302
(*c*) $1383
(*d*) $1426

3.61 If a company wants to have $100,000 in a contingency fund 10 years from now, the amount the company must deposit each year in years 6 through 9, at an interest rate of 10% per year, is closest to
(*a*) $19,588
(*b*) $20,614
(*c*) $21,547
(*d*) $22,389

3.62 If a person begins saving money by depositing $1000 now and then increases the deposit by $500 each year through year 10, the amount that will be in the account in year 10 at an interest rate of 10% per year is closest to
(*a*) $21,662
(*b*) $35,687
(*c*) $43,872
(*d*) $56,186

3.63   If a sum of $5000 is deposited now, $7000
two years from now, and $2000 per year in
years 6 through 10, the amount in year 10
at an interest rate of 10% per year will be
closest to

(a)   Less than $40,000
(b) $40,185
(c) $42,200
(d) $43,565

EXTENDED EXERCISE

PRESERVING LAND FOR PUBLIC USE

The Trust for Public Land is a national organization that purchases and oversees the
improvement of large land sites for government agencies at all levels. Its mission
is to ensure the preservation of the natural resources, while providing necessary,
but minimal, development for recreational use by the public. All Trust projects are
evaluated at 7% per year, and Trust reserve funds earn 7% per year.

A southern U.S. state, which has long-term groundwater problems, has asked
the Trust to manage the purchase of 10,000 acres of aquifer recharge land and the
development of three parks of different use types on the land. The 10,000 acres
will be acquired in increments over the next 5 years with $4 million expended im-
mediately on purchases. Total annual purchase amounts are expected to decrease
25% each year through the fifth year and then cease for this particular project.

A city with 1.5 million citizens immediately to the southeast of this acreage
relies heavily on the aquifer's water. Its citizens passed a bond issue last year,
and the city government now has available $3 million for the purchase of land.
The bond interest rate is an effective 7% per year.

The civil engineers working on the park plan intend to complete all the development over a 3-year period starting in year 4, when the amount budgeted is
$550,000. Increases in construction costs are expected to be $100,000 each year
through year 6.

At a recent meeting, the following agreements were made:

- Purchase the initial land increment now. Use the bond issue funds to assist
  with this purchase. Take the remaining amount from Trust reserves.
- Raise the remaining project funds over the next 2 years in equal annual amounts.
- Evaluate a financing alternative (suggested informally by one individual at
  the meeting) in which the Trust provides all funds, except the $3 million
  available now, until the parks development is initiated in year 4.

Questions

Use hand or computer computations to find the following.

1.   For each of the 2 years, what is the equivalent annual amount necessary to
     supply the remaining project funds?
2.   If the Trust did agree to fund all costs except the $3 million bond proceeds now
     available, determine the equivalent annual amount that must be raised in years
     4 through 6 to supply all remaining project funds. Assume the Trust will not
     charge any extra interest over the 7% to the state or city on the borrowed funds.

4

Nominal and Effective Interest Rates

In all engineering economy relations developed thus far, the interest rate has been a constant, annual value. For a substantial percentage of the projects evaluated by professional engineers in practice, the interest rate is compounded more frequently than once a year; frequencies such as semiannual, quarterly, and monthly are common. In fact, weekly, daily, and even continuous compounding may be experienced in some project evaluations. Also, in our own personal lives, many of the financial considerations we make—loans of all types (home mortgages, credit cards, automobiles, boats), checking and savings accounts, investments, stock option plans, etc.—have interest rates compounded for a time period shorter than 1 year. This requires the introduction of two new terms—nominal and effective interest rates.

This chapter explains how to understand and use nominal and effective interest rates in engineering practice and in daily life situations. The flowchart on calculating an effective interest rate in the appendix to this chapter serves as a reference throughout the sections on nominal and effective rates, as well as continuous compounding of interest. This chapter also develops equivalence calculations for any compounding frequency in combination with any cash flow frequency.

The case study includes an evaluation of several financing plans for the purchase of a house.

LEARNING OBJECTIVES

Purpose: Make economic calculations for interest rates and cash flows that occur on a time basis other than 1 year.

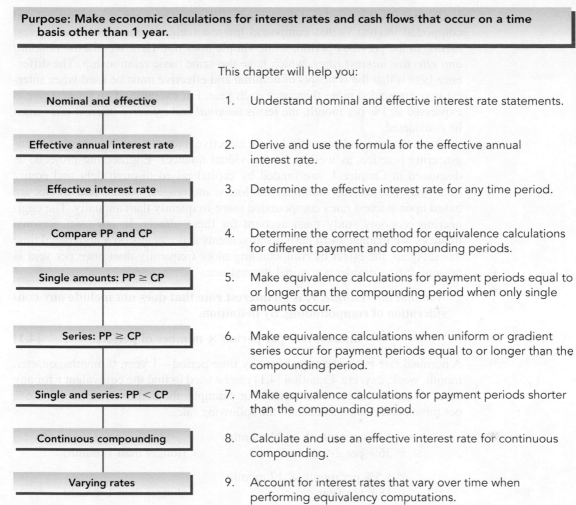

This chapter will help you:

Nominal and effective	1. Understand nominal and effective interest rate statements.
Effective annual interest rate	2. Derive and use the formula for the effective annual interest rate.
Effective interest rate	3. Determine the effective interest rate for any time period.
Compare PP and CP	4. Determine the correct method for equivalence calculations for different payment and compounding periods.
Single amounts: PP ≥ CP	5. Make equivalence calculations for payment periods equal to or longer than the compounding period when only single amounts occur.
Series: PP ≥ CP	6. Make equivalence calculations when uniform or gradient series occur for payment periods equal to or longer than the compounding period.
Single and series: PP < CP	7. Make equivalence calculations for payment periods shorter than the compounding period.
Continuous compounding	8. Calculate and use an effective interest rate for continuous compounding.
Varying rates	9. Account for interest rates that vary over time when performing equivalency computations.

4.1 NOMINAL AND EFFECTIVE INTEREST RATE STATEMENTS

In Chapter 1, we learned that the primary difference between simple interest and compound interest is that compound interest includes interest on the interest earned in the previous period, while simple does not. Here we discuss *nominal and effective interest rates*, which have the same basic relationship. The difference here is that the concepts of nominal and effective must be used when interest is compounded more than once each year. For example, if an interest rate is expressed as 1% per month, the terms *nominal* and *effective* interest rates must be considered.

To understand and correctly handle effective interest rates is important in engineering practice, as well as for individual finances. Engineering projects, as discussed in Chapter 1, are funded by capital raised through debt and equity financing. The interest amounts for loans, mortgages, bonds, and stocks are based upon interest rates compounded more frequently than annually. The engineering economy study must account for these effects. In our own personal finances, we manage most cash disbursements and receipts on a nonannual time basis. Again, the effect of compounding more frequently than once per year is present. First, consider a nominal interest rate.

> **Nominal interest rate, *r*, is an interest rate that does not include any consideration of compounding. By definition,**

$$r = \text{interest rate per period} \times \text{number of periods} \qquad [4.1]$$

A nominal rate *r* may be stated for any time period—1 year, 6 months, quarter, month, week, day, etc. Equation [4.1] can be used to find the equivalent *r* for any other shorter or longer time period. For example, the nominal rate of $r = 1.5\%$ per month is the same as each of the following rates.

$$
\begin{aligned}
r &= 1.5\% \text{ per month} \times 24 \text{ months} \\
&= 36\% \text{ per 2-year period} \qquad \text{(longer than 1 month)}
\end{aligned}
$$

$$
\begin{aligned}
&= 1.5\% \text{ per month} \times 12 \text{ months} \\
&= 18\% \text{ per year} \qquad \text{(longer than 1 month)}
\end{aligned}
$$

$$
\begin{aligned}
&= 1.5\% \text{ per month} \times 6 \text{ months} \\
&= 9\% \text{ per semiannual period} \qquad \text{(longer than 1 month)}
\end{aligned}
$$

$$
\begin{aligned}
&= 1.5\% \text{ per month} \times 3 \text{ months} \\
&= 4.5\% \text{ per quarter} \qquad \text{(longer than 1 month)}
\end{aligned}
$$

$$
\begin{aligned}
&= 1.5\% \text{ per month} \times 1 \text{ month} \\
&= 1.5\% \text{ per month} \qquad \text{(equal to 1 month)}
\end{aligned}
$$

$$
\begin{aligned}
&= 1.5\% \text{ per month} \times 0.231 \text{ month} \\
&= 0.346\% \text{ per week} \qquad \text{(shorter than 1 month)}
\end{aligned}
$$

Note that none of these nominal rates make mention of the compounding frequency. They all have the format "*r*% per time period *t*."

Now, consider an effective interest rate.

***Effective interest rate* is the actual rate that applies for a stated period of time. The compounding of interest during the time period of the corresponding nominal rate is accounted for by the effective interest rate. It is commonly expressed on an annual basis as the effective annual rate i_a, but any time basis can be used.**

An effective rate has the compounding frequency attached to the nominal rate statement. If the compounding frequency is not stated, it is assumed to be the same as the time period of r, in which case the nominal and effective rates have the same value. All the following are nominal rate statements; however, they will not have the same effective interest rate value over all time periods, due to the different compounding frequencies.

4% per year, compounded monthly	(compounding more often than time period)
12% per year, compounded quarterly	(compounding more often than time period)
9% per year, compounded daily	(compounding more often than time period)
3% per quarter, compounded monthly	(compounding more often than time period)
6% per 6 months, compounded weekly	(compounding more often than time period)
3% per quarter, compounded daily	(compounding more often than time period)

Note that all these rates make mention of the compounding frequency. They all have the format "r% per time period t, compounded m-ly." The m is a month, a quarter, a week, or some other time unit. The formula to calculate the effective interest rate value for any nominal or effective rate statement is discussed in the next section.

All the interest formulas, factors, tabulated values, and spreadsheet relations must have the effective interest rate to properly account for the time value of money.

Therefore, it is very important to determine the effective interest rate before performing time value of money calculations in the engineering economy study. This is especially true when the cash flows occur at other than annual intervals.

The terms *APR* and *APY* are used in many individual financial situations instead of nominal and effective interest rates. The Annual Percentage Rate (APR) is the same as the nominal interest rate, and Annual Percentage Yield (APY) is used in lieu of effective interest rate. All definitions and interpretations are the same as those developed in this chapter.

Based on these descriptions, there are always three time-based units associated with an interest rate statement.

Time period—the period over which the interest is expressed. This is the *t* in the statement of *r*% per time period *t*, for example, 1% *per month*. The time unit of 1 year is by far the most common. It is assumed when not stated otherwise.

Compounding period (CP)—the shortest time unit over which interest is charged or earned. This is defined by the compounding term in the interest rate statement, for example, 8% per year *compounded monthly*. If not stated, it is assumed to be 1 year.

Compounding frequency—the number of times that *m* compounding occurs within the time period *t*. If the compounding period CP and the time period *t* are the same, the compounding frequency is 1, for example, 1% *per month compounded monthly*.

Consider the rate 8% per year, compounded monthly. It has a time period *t* of 1 year, a compounding period CP of 1 month, and a compounding frequency *m* of 12 times per year. A rate of 6% per year, compounded weekly, has *t* = 1 year, CP = 1 week, and *m* = 52, based on the standard of 52 weeks per year.

In previous chapters, all interest rates have *t* and *m* values of 1 year. This means the rates are both the effective and the nominal rate because the same time unit of 1 year is used. It is common to express the effective rate on the same time basis as the compounding period. The corresponding effective rate per CP is determined by using the relation

$$\text{Effective rate per CP} = \frac{r\% \text{ per time period } t}{m \text{ compounding periods per } t} = \frac{r}{m} \quad [4.2]$$

As an illustration, assume *r* = 9% per year, compounded monthly; then *m* = 12. Equation [4.2] is used to obtain the effective rate of 9%/12 = 0.75% per month, compounded monthly. It is important to note that changing the basic time period *t* does not alter the compounding period, which is 1 month in this illustration.

EXAMPLE **4.1**

The different bank loan rates for three separate electric generation equipment projects are listed below. Determine the effective rate on the basis of the compounding period for each quote.

   (*a*)   9% per year, compounded quarterly.
   (*b*)   9% per year, compounded monthly.
   (*c*)   4.5% per 6-months, compounded weekly.

Solution
Apply Equation [4.2] to determine the effective rate per CP for different compounding frequencies. The accompanying graphic indicates how the interest rate is distributed over time.

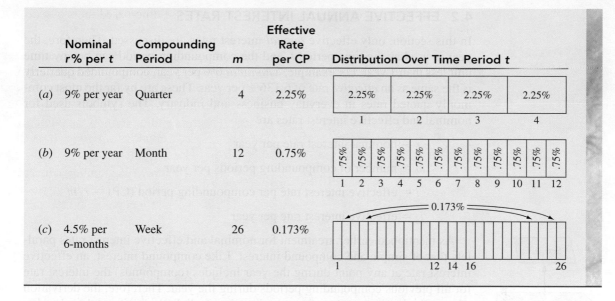

Sometimes it is not obvious whether a stated rate is a nominal or an effective rate. Basically there are three ways to express interest rates, as detailed in Table 4–1. The right column includes a statement about the effective rate. For the first format, no statement of nominal or effective is given, but the compounding frequency is stated. The effective rate must be calculated (discussed in the next sections). In the second format, the stated rate is identified as effective (or APY could also be used), so the rate is used directly in computations.

In the third format, no compounding frequency is identified, for example, 8% per year. This rate is effective only over the time (compounding) period of a year, in this case. The effective rate for any other time period must be calculated.

TABLE 4–1   Various Ways to Express Nominal and Effective Interest Rates

Format of Rate Statement	Examples of Statement	What about the Effective Rate?
Nominal rate stated, compounding period stated	8% per year, compounded quarterly	Find effective rate
Effective rate stated	Effective 8.243% per year, compounded quarterly	Use effective rate directly
Interest rate stated, no compounding period stated	8% per year or 2% per quarter	Rate is effective only for time period stated; find effective rate for all other time periods

4.2 EFFECTIVE ANNUAL INTEREST RATES

In this section, only effective *annual* interest rates are discussed. Therefore, the year is used as the time period t, and the compounding period can be any time unit less than 1 year. For example, a *nominal* 6% per year, compounded quarterly is the same as an *effective* rate of 6.136% per year. These are by far the most commonly quoted rates in everyday business and industry. The symbols used for nominal and effective interest rates are

$$r = \text{nominal interest rate per year}$$

$$m = \text{number of compounding periods per year}$$

$$i = \text{effective interest rate per compounding period (CP)} = r/m$$

$$i_a = \text{effective interest rate per year}$$

Sec. 2.1

Future worth

As mentioned earlier, treatment for nominal and effective interest rates parallels that of simple and compound interest. Like compound interest, an effective interest rate at any point during the year includes (compounds) the interest rate for all previous compounding periods during the year. Therefore, the derivation of an effective interest rate formula directly parallels the logic used to develop the future worth relation $F = P(1 + i)^n$.

The future worth F at the end of 1 year is the principal P plus the interest $P(i)$ through the year. Since interest may be compounded several times during the year, replace i with the effective annual rate i_a. Now write the relation for F at the end of 1 year.

$$F = P + Pi_a = P(1 + i_a) \qquad [4.3]$$

As indicated in Figure 4–1, the rate i per CP must be compounded through all m periods to obtain the total effect of compounding by the end of the year. This

Figure 4–1

Future worth calculation at a rate i, compounded m times in a year.

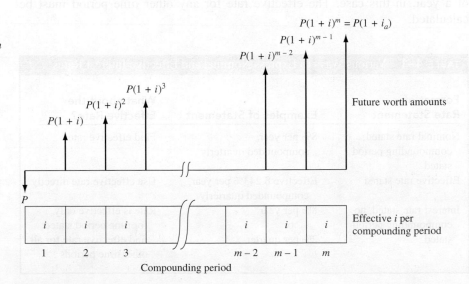

means that F can also be written as

$$F = P(1 + i)^m \qquad [4.4]$$

Consider the F value for a present worth P of \$1. By equating the two expressions for F and substituting \$1 for P, the *effective annual interest rate* formula for i_a is derived.

$$1 + i_a = (1 + i)^m$$

$$i_a = (1 + i)^m - 1 \qquad [4.5]$$

So Equation [4.5] calculates the effective annual interest rate for any number of compounding periods when i is the rate for one compounding period.

If the effective annual rate i_a and compounding frequency m are known, Equation [4.5] can be solved for i to determine the *effective interest rate per compounding period.*

$$i = (1 + i_a)^{1/m} - 1 \qquad [4.6]$$

Further, it is possible to determine the *nominal annual rate r* using the definition of i stated above, namely, $i = r/m$.

$$r\% \text{ per year} = (i\% \text{ per CP})(\text{no. of CPs per year}) = (i)(m) \qquad [4.7]$$

This is the same as Equation [4.1] where CP is the period of time.

EXAMPLE 4.2

Jacki obtained a new credit card from a national bank, MBNA, with a stated rate of 18% per year, compounded monthly. For a \$1000 balance at the beginning of the year, find the effective annual rate and the total amount owed to MBNA after 1 year, provided no payments are made during the year.

Solution
There are 12 compounding periods per year. Thus, $m = 12$ and $i = 18\%/12 = 1.5\%$ per month. For a \$1000 balance that is not reduced during the year, apply Equation [4.5], then [4.3] to provide Jacki with the information.

$$i_a = (1 + 0.015)^{12} - 1 = 1.19562 - 1 = 0.19562$$

$$F = \$1000(1.19562) = \$1195.62$$

Jacki will pay 19.562%, or \$195.62 plus the \$1000 balance, for the use of the bank's money during the year.

Table 4–2 utilizes the rate of 18% per year, compounded over different times (yearly to weekly) to determine the effective annual interest rates over these various compounding periods. In each case, the compound period rate i is applied

TABLE 4–2 Effective Annual Interest Rates Using Equation [4.5]

r = 18% per year, compounded m-ly

Compounding Period	Times Compounded per Year, m	Rate per Compound Period, i	Distribution of i over the Year of Compounding Periods	Effective Annual Rate, i_a
Year	1	18%	18% (period 1)	$(1.18)^1 - 1 = 18\%$
6 months	2	9%	9% (1), 9% (2)	$(1.09)^2 - 1 = 18.81\%$
Quarter	4	4.5%	4.5% (1), 4.5% (2), 4.5% (3), 4.5% (4)	$(1.045)^4 - 1 = 19.252\%$
Month	12	1.5%	1.5% in each (1–12)	$(1.015)^{12} - 1 = 19.562\%$
Week	52	0.34615%	0.34615% in each (1–52)	$(1.0034615)^{52} - 1 = 19.684\%$

Nominal Rate r%	Semiannually (m = 2)	Quarterly (m = 4)	Monthly (m = 12)	Weekly (m = 52)	Daily (m = 365)	Continuously (m = ∞; e^r − 1)
0.25	0.250	0.250	0.250	0.250	0.250	0.250
0.50	0.501	0.501	0.501	0.501	0.501	0.501
1.00	1.003	1.004	1.005	1.005	1.005	1.005
1.50	1.506	1.508	1.510	1.511	1.511	1.511
2	2.010	2.015	2.018	2.020	2.020	2.020
3	3.023	3.034	3.042	3.044	3.045	3.046
4	4.040	4.060	4.074	4.079	4.081	4.081
5	5.063	5.095	5.116	5.124	5.126	5.127
6	6.090	6.136	6.168	6.180	6.180	6.184
7	7.123	7.186	7.229	7.246	7.247	7.251
8	8.160	8.243	8.300	8.322	8.328	8.329
9	9.203	9.308	9.381	9.409	9.417	9.417
10	10.250	10.381	10.471	10.506	10.516	10.517
12	12.360	12.551	12.683	12.734	12.745	12.750
15	15.563	15.865	16.076	16.158	16.177	16.183
18	18.810	19.252	19.562	19.684	19.714	19.722
20	21.000	21.551	21.939	22.093	22.132	22.140
25	26.563	27.443	28.073	28.325	28.390	28.403
30	32.250	33.547	34.489	34.869	34.968	34.986
40	44.000	46.410	48.213	48.954	49.150	49.182
50	56.250	60.181	63.209	64.479	64.816	64.872

TABLE 4–3 Effective Annual Interest Rates For Selected Nominal Rates

m times during the year. Table 4–3 summarizes the effective annual rate for frequently quoted nominal rates using Equation [4.5]. A standard of 52 weeks and 365 days per year is used throughout. The values in the continuous-compounding column are discussed in Section 4.8.

When Equation [4.5] is applied, the result is usually not an integer. Therefore, the engineering economy factor cannot be obtained directly from the interest factor tables. There are three alternatives to find the factor value.

- Linearly interpolate between two tabulated rates (as discussed in Section 2.4).
- Use the factor formula with the i_a rate substituted for i.
- Develop a spreadsheet using i_a or $i = r/m$ in the functions, as required by the spreadsheet function.

We use the second method in examples that are solved by hand and the last one in solutions by computer.

EXAMPLE **4.3**

Joshua works for Watson Bio, a genetics engineering R&D company. He has just received a $10,000 bonus. He wants to invest the money for the next 5 years.

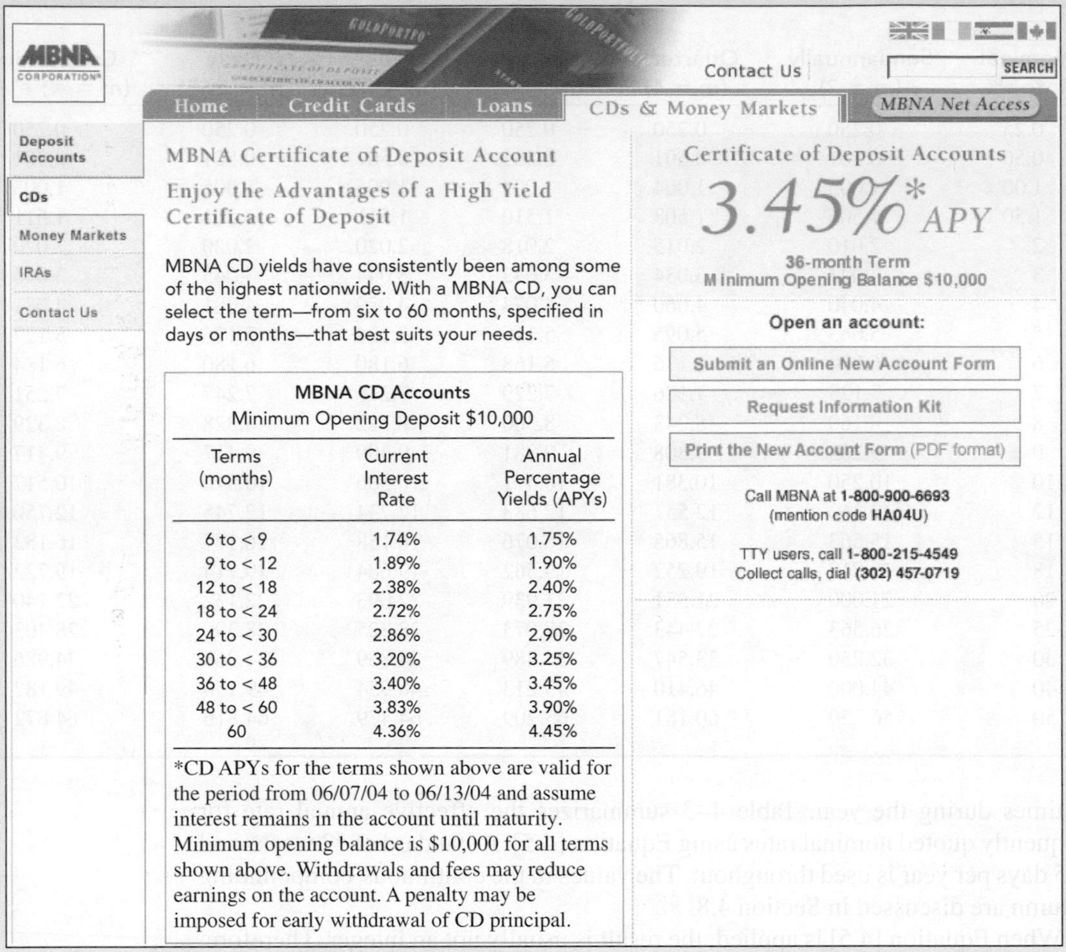

Figure 4–2
Internet ad showing interest rates on Certificates of Deposit. The ad depicted is a sample, similar to one that appeared on
June 11, 2004, on the MBNA America Bank website www.mbna.com. The rates shown are not current.

Joshua saw an MBNA America bank website ad for Certificate of Deposit (CD) interest
rates (Figure 4–2). He is thinking of putting the $10,000 into a CD for 5 years for capital
preservation. Alternatively, he is thinking of investing it all for the next 2 years in stocks,
where he estimates he can make an effective annual rate of 10%. Once he has made this
larger up-front return, he will then be more conservative and place the entire amount in a
CD for the final 3 years. Help Joshua with the following:

(a) Determine the compounding period for the 3- and 5-year CDs, since they are not
    included in the website ad. Get as close as possible to the quoted APY rounded to
    two decimals.

(b) Determine the total amount he will have after 5 years for the two options he has
    outlined.

Solution

(a) The annual interest rate is stated, but the compounding period or frequency is not. Substitute different m values into Equation [4.5] to obtain the corresponding i_a value. (Use Equation [4.12] for continuous compounding.) Compare it with the listed APY rate in the Web ad (Figure 4–2). From the results below and rounding to two decimal places for the estimated APY rates, it appears that the bank applies monthly compounding to its stated current interest rates.

For an Investment Term of:	Stated Interest Rate is:	If the Compounding Frequency m Is Assumed to be:	The Effective Annual Rate i_a, or Estimated APY, is:	The Most Likely Compounding Period Used by the Bank is:
3 years	3.40%	4 quarters	3.444	
		12 months	3.453	Monthly
		52 weeks	3.457	
5 years	4.36%	4 quarters	4.432	
		12 months	4.448	Monthly
		52 weeks	4.455	
		Continuous	4.456	

(b) *Option 1: 5-year CD.* Use the APY rate of 4.45% (from Figure 4–2) in the F/P factor or the Excel FV function.

$$F = 10,000(F/P,4.45\%,5) = 10,000(1.2432) = \$12,432$$

Option 2: 2 years in stocks, then a 3-year CD. This is a higher-risk option since the return on stocks is not certain. Use 10% per year for the stocks, which is the estimated effective annual rate, followed by 3 years at the 36-month CD effective annual rate of 3.45%. (It is unlikely that the CD rate will remain at this level for 2 more years, but this is the best estimate available now.)

$$F = 10,000(F/P,10\%,2)(F/P,3.45\%,3)$$
$$= 10,000(1.21)(1.1071) = \$13,396$$

The second option is estimated to earn $964 more over the 5 years.

Comment

The interest rates and compounding periods used in this example are only representative; they change frequently and vary from one institution to another. Check the website of any financial institution offering Internet banking services to learn about the current rates.

All the economic situations discussed in this section involve annual nominal and effective rates and annual cash flows. When cash flows are non-annual, it is necessary to remove the year assumption in the interest rate statement "$r\%$ per year, compounded m-ly." This is the topic of the next section.

Figure 4–3
One-year cash flow diagram for a monthly payment period (PP) and semiannual compounding period (CP).

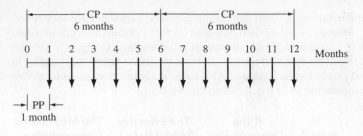

r = nominal 14% per year, compounded semiannually

4.3 EFFECTIVE INTEREST RATES FOR ANY TIME PERIOD

The concepts of nominal and effective annual interest rates have been introduced. Now, in addition to the compounding period (CP), it is necessary to consider the frequency of the payments or receipts, that is, the cash flow transaction period. For simplicity, this is called the *payment period (PP)*. It is important to distinguish between the compounding period and the payment period because in many instances the two do not coincide. For example, if a company deposits money each month into an account that pays a nominal interest rate of 14% per year, compounded semiannually, the payment period is 1 month while the compounding period is 6 months (Figure 4–3). Similarly, if a person deposits money each year into a savings account which compounds interest quarterly, the payment period is 1 year, while the compounding period is 3 months.

To evaluate cash flows that occur more frequently than annually, that is, PP < 1 year, the effective interest rate over the PP must be used in the engineering economy relations. The effective annual interest rate formula is easily generalized to any nominal rate by substituting r/m for the period interest rate in Equation [4.5].

$$\text{Effective } i = (1 + r/m)^m - 1 \qquad [4.8]$$

where:

r = nominal interest rate per payment period (PP)

m = number of compounding periods per payment period (CP per PP)

Instead of i_a, this general expression uses i as the symbol for effective interest. This conforms to all other uses of i for the remainder of the text. With Equation [4.8], it is possible to take a nominal rate ($r\%$ per year or any other time period) and convert it to an effective rate i for any time basis, the most common of which will be the PP time period. The next two examples illustrate how to do this.

EXAMPLE 4.4

Visteon, a spin-off company of Ford Motor Company, supplies major automobile components to auto manufacturers worldwide and is Ford's largest supplier. An engineer is on a Visteon committee to evaluate bids for new-generation coordinate-measuring machinery to be directly linked to the automated manufacturing of high-precision components. Three vendor bids include the interest rates on the next page. Visteon will make payments on a semiannual basis only. The engineer is confused about the effective interest rates—what they are annually and over the payment period of 6-months.

Bid #1: 9% per year, compounded quarterly
Bid #2: 3% per quarter, compounded quarterly
Bid #3: 8.8% per year, compounded monthly

(a) Determine the effective rate for each bid on the basis of semiannual payments, and construct cash flow diagrams similar to Figure 4–3 for each bid rate.
(b) What are the effective annual rates? These are to be a part of the final bid selection.
(c) Which bid has the lowest effective annual rate?

Solution
(a) Set the payment period (PP) at 6-months, convert the nominal rate $r\%$ to a semiannual basis, then determine m. Finally, use Equation [4.8] to calculate the effective semiannual interest rate i. For Bid #1, the following are correct:

$$PP = 6 \text{ months}$$

$$r = 9\% \text{ per year} = 4.5\% \text{ per 6-months}$$

$$m = 2 \text{ quarters per 6-months}$$

$$\text{Effective } i\% \text{ per 6-months} = \left(1 + \frac{0.045}{2}\right)^2 - 1 = 1.0455 - 1 = 4.55\%$$

Table 4–4 (left section) summarizes the effective semiannual rates for all three bids. Figure 4–4a is the cash flow diagram for bids #1 and #2, semiannual payments (PP = 6 months) and quarterly compounding (CP − 1 quarter). Figure 4–4b is the same for monthly compounding (bid #3).

(b) For the effective annual rate, the time basis in Equation [4.8] is 1 year. This is the same as PP = 1 year. For bid #1,

$$r = 9\% \text{ per year} \qquad m = 4 \text{ quarters per year}$$

$$\text{Effective } i\% \text{ per year} = \left(1 + \frac{0.09}{4}\right)^4 - 1 = 1.0931 - 1 = 9.31\%$$

The right section of Table 4–4 includes a summary of the effective annual rates.

(c) Bid #3 includes the lowest effective annual rate of 9.16%, which is equivalent to an effective semiannual rate of 4.48%.

TABLE 4–4 Effective Semiannual and Annual Interest Rates for Three Bid Rates, Example 4.4

	Semiannual Rates			Annual Rates		
Bid	Nominal per 6 Months, r	CP per PP, m	Equation [4.8], Effective i	Nominal per Year, r	CP per Year, m	Equation [4.8], Effective i
#1	4.5%	2	4.55%	9%	4	9.31%
#2	6.0%	2	6.09%	12%	4	12.55%
#3	4.4%	6	4.48%	8.8%	12	9.16%

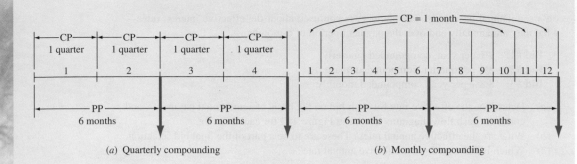

Figure 4–4
Cash flow diagram showing CP and PP for (*a*) bids 1 and 2 and (*b*) bid 3, Example 4.4.

Comment

The effective rates for bid #2 only may be found directly in Table 4–3. For the effective semiannual rate, look at the nominal 6% line under $m = 2$, which is the number of quarters per 6-months. The effective semiannual rate is 6.09%. Similarly, for the nominal 12% rate, there are $m = 4$ quarters per year, so effective annual $i = 12.551\%$. Although Table 4–3 was originally designed for nominal annual rates, it is correct for other nominal rate periods, provided the appropriate m value is included in the column headings.

EXAMPLE 4.5

A dot-com company plans to place money in a new venture capital fund that currently returns 18% per year, compounded daily. What effective rate is this (*a*) yearly and (*b*) semiannually?

Solution

(*a*) Use Equation [4.8], with $r = 0.18$ and $m = 365$.

$$\text{Effective } i\% \text{ per year} = \left(1 + \frac{0.18}{365}\right)^{365} - 1 = 19.716\%$$

(*b*) Here $r = 0.09$ per 6-months and $m = 182$ days.

$$\text{Effective } i\% \text{ per 6-months} = \left(1 + \frac{0.09}{182}\right)^{182} - 1 = 9.415\%$$

4.4 EQUIVALENCE RELATIONS: COMPARING PAYMENT PERIOD AND COMPOUNDING PERIOD LENGTHS (PP VERSUS CP)

In a large percentage of equivalency computations, the frequency of cash flows does not equal the frequency of interest compounding. For example, cash flows may occur monthly, and compounding occurs annually, quarterly, or more often.

TABLE **4–5** Section References for Equivalence Calculations Based on Payment Period and Compounding Period Comparison

Length of Time	Involves Single Amounts (P and F Only)	Involves Uniform Series or Gradient Series (A, G, or g)
PP = CP	Section 4.5	Section 4.6
PP > CP	Section 4.5	Section 4.6
PP < CP	Section 4.7	Section 4.7

Consider deposits made to a savings account each month, while the earning rate is compounded quarterly. The length of the CP is a quarter, while the PP is a month. To correctly perform any equivalence computation, it is essential that the compounding period and payment period be placed on the same time basis, and that the interest rate be adjusted accordingly.

The next three sections describe procedures to determine correct i and n values for engineering economy factors and spreadsheet solutions. First, compare the length of PP and CP, then identify the cash flow series as only single amounts (P and F) or as a series (A, G, or g). Table 4–5 provides the section reference. When only single amounts are involved, there is no payment period PP per se defined by the cash flows. The length of PP is, therefore, defined by the time period t of the interest rate statement. If the rate is 8% per 6-months, compounded quarterly, the PP is 6-months, the CP is 3 months, and PP > CP.

Note that the section references in Table 4–5 are the same when PP = CP and PP > CP. The equations to determine i and n are the same. Additionally, the technique to account for the time value of money is the same because it is only when cash flows occur that the effect of the interest rate is determined. For example, assume that cash flows occur every 6 months (PP is semiannual), and that interest is compounded each 3 months (CP is a quarter). After 3 months there is no cash flow and no need to determine the effect of quarterly compounding. However, at the 6-month time point, it is necessary to consider the interest accrued during the previous two quarterly compounding periods.

4.5 EQUIVALENCE RELATIONS: SINGLE AMOUNTS WITH PP ≥ CP

When only single-amount cash flows are involved, there are two equally correct ways to determine i and n for P/F and F/P factors. Method 1 is easier to apply, because the interest tables in the back of the text can usually provide the factor value. Method 2 likely requires a factor formula calculation, because the resulting effective interest rate is not an integer. For spreadsheets, either method is acceptable; however, method 1 is usually easier.

Method 1: Determine the effective interest rate over the *compounding period CP*, and set n equal to the number of compounding periods between P and F. The relations to calculate P and F are

$$P = F(P/F, \text{effective } i\% \text{ per CP, total number of periods } n) \qquad [4.9]$$

$$F = P(F/P, \text{effective } i\% \text{ per CP, total number of periods } n) \qquad [4.10]$$

For example, assume that a nominal 15% per year, compounded monthly, is the stated credit card rate. Here CP is a month. To find P or F over a 2-year span, calculate the effective monthly rate of $15\%/12 = 1.25\%$ and the total months of $2(12) = 24$. Then 1.25% and 24 are used in the P/F and F/P factors.

Any time period can be used to determine the effective interest rate; however, CP is the best basis. The CP is best because only over the CP can the effective rate have the same numerical value as the nominal rate over the same time period as the CP. This was discussed in Section 4.1 and Table 4–1. This means that the effective rate over CP is usually a whole number. Therefore, the factor tables in the back of the text can be used.

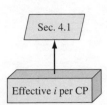

Method 2: Determine the effective interest rate for the *time period t of the nominal rate,* and set n equal to the total number of periods using this same time period. The P and F relations are the same as in Equations [4.9] and [4.10] with the term *effective i% per t* substituted for the interest rate.

For a credit card rate of 15% per year, compounded monthly, the time period t is 1 year. The effective rate over 1 year and n values are

$$\text{Effective } i\% \text{ per year} = \left(1 + \frac{0.15}{12}\right)^{12} - 1 = 16.076\%$$

$$n = 2 \text{ years}$$

The P/F factor is the same by both methods: $(P/F,1.25\%,24) = 0.7422$ using Table 5; and $(P/F,16.076\%,2) = 0.7422$ using the P/F factor formula.

EXAMPLE 4.6

An engineer working as a private consultant made deposits into a special account to cover unreimbursed travel expenses. Figure 4–5 is the cash flow diagram. Find the amount in the account after 10 years at an interest rate of 12% per year, compounded semiannually.

Solution
Only P and F values are involved. Both methods are illustrated to find F in year 10.

Method 1: Use the semiannual CP to express the effective semiannual rate of 6% per 6-month period. There are $n = (2)$(number of years) semiannual periods for each cash flow. Using the factor values from Table 11, we see that the future worth by Equation [4.10] is

$$F = 1000(F/P,6\%,20) + 3000(F/P,6\%,12) + 1500(F/P,6\%,8)$$

$$= 1000(3.2071) + 3000(2.0122) + 1500(1.5938)$$

$$= \$11{,}634$$

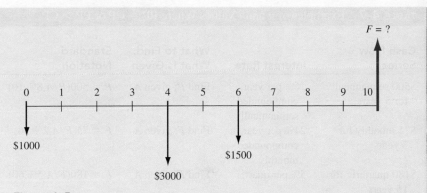

Figure 4–5
Cash flow diagram, Example 4.6.

Method 2: Express the effective annual rate, based on semiannual compounding.

$$\text{Effective } i\% \text{ per year} = \left(1 + \frac{0.12}{2}\right)^2 - 1 = 12.36\%$$

The n value is the actual number of years. Use the factor formula $(F/P,i,n) = (1.1236)^n$ and Equation [4.10] to obtain the same answer as with method 1.

$$F = 1000(F/P,12.36\%,10) + 3000(F/P,12.36\%,6) + 1500(F/P,12.36\%,4)$$

$$= 1000(3.2071) + 3000(2.0122) + 1500(1.5938)$$

$$= \$11,634$$

Comment

For single-amount cash flows, any combination of i and n derived from the stated nominal rate can be used in the factors, provided they are on the same time basis. Using 12% per year, *compounded monthly,* Table 4–6 presents various acceptable combinations of i and n. Other combinations are correct, such as the effective weekly rate for i and weeks for n.

TABLE 4–6	Various i and n Values for Single-Amount Equations Using $r = 12\%$ per Year, Compounded Monthly
Effective Rate i	**Units for n**
1% per month	Months
3.03% per quarter	Quarters
6.15% per 6 months	Semiannual periods
12.68% per year	Years
26.97% per 2 years	2-year periods

TABLE 4–7   Examples of n and i Values Where PP = CP or PP > CP

Cash Flow Series	Interest Rate	What to Find; What Is Given	Standard Notation
$500 semiannually for 5 years	16% per year, compounded semiannually	Find P; given A	$P = 500(P/A,8\%,10)$
$75 monthly for 3 years	24% per year, compounded monthly	Find F; given A	$F = 75(F/A,2\%,36)$
$180 quarterly for 15 years	5% per quarter	Find F; given A	$F = 180(F/A,5\%,60)$
$25 per month increase for 4 years	1% per month	Find P; given G	$P = 25(P/G,1\%,48)$
$5000 per quarter for 6 years	1% per month	Find A; given P	$A = 5000(A/P,3.03\%,24)$

4.6   EQUIVALENCE RELATIONS: SERIES WITH PP ≥ CP

When uniform or gradient series are included in the cash flow sequence, the procedure is basically the same as method 2 above, except that PP is now defined by
the frequency of the cash flows. This also establishes the time unit of the effective interest rate. For example, if cash flows occur on a *quarterly* basis, PP is a
quarter and the effective *quarterly* rate is necessary. The n value is the total number of *quarters*. If PP is a quarter, 5 years translates to an n value of 20 quarters.
This is a direct application of the following general guideline:

> When cash flows involve a series (i.e., *A, G, g*) and the payment period
> equals or exceeds the compounding period in length,
>
> - Find the effective i per payment period.
> - Determine n as the total number of payment periods.

In performing equivalence computations for series, *only* these values of i and n
can be used in interest tables, factor formulas, and spreadsheet functions. In other
words, there are no other combinations that give the correct answers, as there are
for single amount cash flows.

Table 4–7 shows the correct formulation for several cash flow series and
interest rates. Note that n is always equal to the total number of payment periods
and i is an effective rate expressed over the same time period as n.

EXAMPLE 4.7

For the past 7 years, a quality manager has paid $500 every 6 months for the software main-
tenance contract of a LAN. What is the equivalent amount after the last payment, if these
funds are taken from a pool that has been returning 20% per year, compounded quarterly?

Solution

The cash flow diagram is shown in Figure 4–6. The payment period (6 months) is longer than the compounding period (quarter); that is, PP > CP. Applying the guideline, we need to determine an effective semiannual interest rate. Use Equation [4.8] with $r = 0.10$ per 6-month period and $m = 2$ quarters per semiannual period.

$$\text{Effective } i\% \text{ per 6-months} = \left(1 + \frac{0.10}{2}\right)^2 - 1 = 10.25\%$$

The effective semiannual interest rate can also be obtained from Table 4–3 by using the r value of 10% and $m = 2$ to get $i = 10.25\%$.

The value $i = 10.25\%$ seems reasonable, since we expect the effective rate to be slightly higher than the nominal rate of 10% per 6-month period. The total number of semiannual payment periods is $n = 2(7) = 14$. The relation for F is

$$F = A(F/A,10.25\%,14)$$

$$= 500(28.4891)$$

$$= \$14,244.50$$

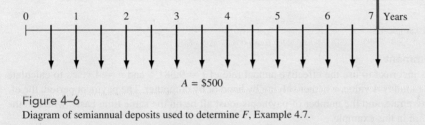

$i = 20\%$ per year, compounded quarterly

$F = ?$

$A = \$500$

Figure 4–6
Diagram of semiannual deposits used to determine F, Example 4.7.

EXAMPLE **4.8**

Suppose you plan to purchase a car and carry a loan of $12,500 at 9% per year, compounded monthly. Payments will be made monthly for 4 years. Determine the monthly payment. Compare the computer and hand solutions.

Solution

A monthly series A is sought; the PP and CP are both a month. Use the steps for PP = CP when a uniform series is present. The effective interest per month is $9\%/12 = 0.75\%$, and the number of payments is (4 years)(12 months per year) = 48.

Enter PMT(9%/12,48,−12500) into any cell to display $311.06.

Figure 4–7 shows a complete spreadsheet with the PMT function in cell B5 using cell reference format. This monthly payment of $311.06 is equivalent to the following solution by hand, using standard notation and the factor tables.

Q-Solv

$$A = \$12,500(A/P,0.75\%,48) = 12,500(0.02489) = \$311.13$$

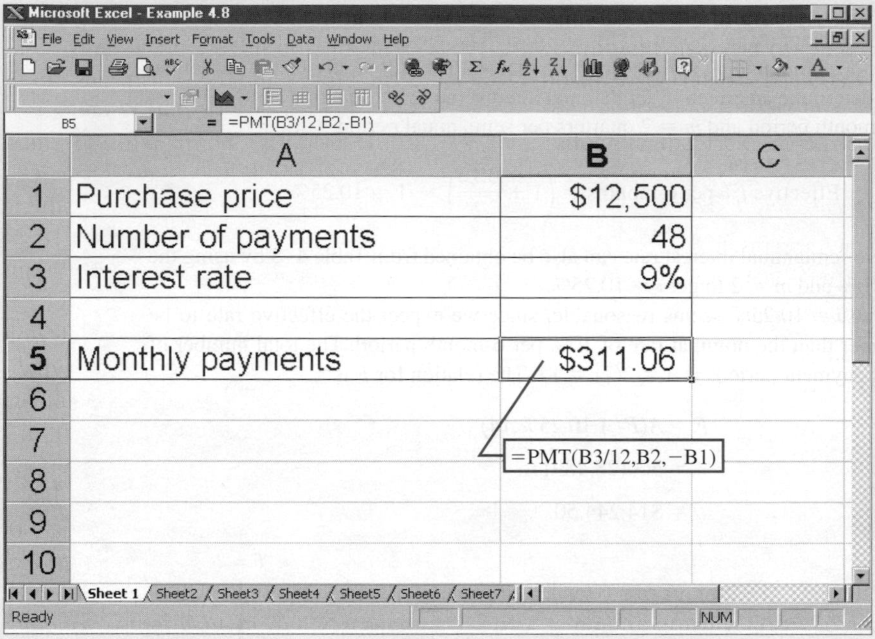

Figure 4–7
Spreadsheet for Example 4.8.

Comment

It is incorrect to use the effective annual rate of $i = 9.381\%$ and $n = 4$ years to calculate the monthly A value, whether solving by hand or by computer. The payment period, the effective rate, and the number of payments must all be on the same time basis, which is the *month* in this example.

EXAMPLE 4.9

The Scott and White Health Plan (SWHP) has purchased a robotized prescription fulfillment system for faster and more accurate delivery to patients with stable, pill-form medication for chronic health problems, such as diabetes, thyroid, and high blood pressure. Assume this high-volume system costs $3 million to install and an estimated $200,000 per year for all materials, operating, personnel, and maintenance costs. The expected life is 10 years. An SWHP biomedical engineer wants to estimate the total revenue requirement for each 6-month period that is necessary to recover the investment, interest, and annual costs. Find this semiannual A value both by hand and by computer, if capital funds are evaluated at 8% per year using two different compounding periods:

1. 8% per year, compounded *semiannually*.
2. 8% per year, compounded *monthly*.

Solution

Figure 4–8 shows the cash flow diagram. Throughout the 20 semiannual periods, the annual cost occurs every other period, and the capital recovery series is sought for every 6-month period. This pattern makes the solution by hand quite involved if the P/F factor, not the P/A factor, is used to find P for the 10 annual \$200,000 costs. The computer solution is recommended in cases such as this.

Solution by hand—rate 1: Steps to find the semiannual A value are summarized below:

PP = CP at 6-months; find the effective rate per semiannual period.

Effective semiannual $i = 8\%/2 = 4\%$ per 6-months, compounded semiannually.

Number of semiannual periods $n = 2(10) = 20$.

Calculate P, using the P/F factor for $n = 2, 4, \ldots, 20$ periods since the costs are annual, not semiannual. Then use the A/P factor over 20 periods to find the semiannual A.

$$P = 3,000,000 + 200,000\left[\sum_{k=2,4}^{20}(P/F,4\%,k)\right]$$

$$= 3,000,000 + 200,000(6.6620) = \$4,332,400$$

$$A = \$4,332,400(A/P,4\%,20) = \$318,778$$

Conclusion: Revenue of \$318,778 is necessary every 6 months to cover all costs and interest at 8% per year, compounded semiannually.

Solution by hand—rate 2: The PP is semiannual, but the CP is now monthly; therefore, PP > CP. To find the effective semiannual rate, the effective interest rate, Equation [4.8], is applied with $r = 4\%$ and $m = 6$ months per semiannual period.

$$\text{Effective semiannual } i = \left(1 + \frac{0.04}{6}\right)^6 - 1 = 4.067\%$$

$$P = 3,000,000 + 200,000\left[\sum_{k=2,4}^{20}(P/F,4.067\%,k)\right]$$

$$= 3,000,000 + 200,000(6.6204) = \$4,324,080$$

$$A = \$4,324,080(A/P,4.067\%,20) = \$320,064$$

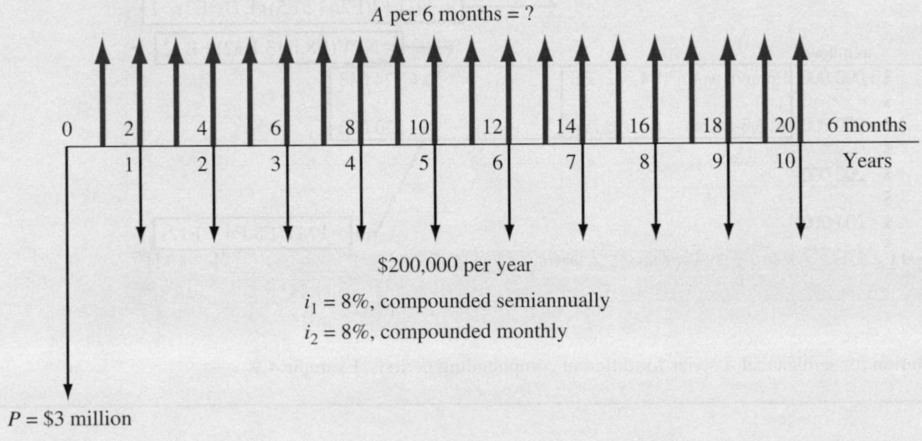

A per 6 months = ?

$200,000 per year

$i_1 = 8\%$, compounded semiannually

$i_2 = 8\%$, compounded monthly

P = \$3 million

Figure 4–8
Cash flow diagram with two different compounding periods, Example 4.9.

Now, $320,064, or $1286 more semiannually, is required to cover the more frequent compounding of the 8% per year interest. Note that all P/F and A/P factors must be calculated with factor formulas at 4.067%. This method is usually more calculation-intensive and error-prone than with a spreadsheet solution.

Solution by computer—rates 1 and 2: Figure 4–9 presents a general solution for the problem at both rates. (Several rows at the bottom of the spreadsheet are not printed. They continue the cash flow pattern of $200,000 every other 6 months through cell B32.) The functions in C8 and E8 are general expressions for the effective rate per PP, expressed in months. This allows some sensitivity analysis to be performed for different PP and CP values. Note the function in C7 and E7 to determine *m* for the effective rate relations. This technique works well for spreadsheets once PP and CP are entered in the time unit of the CP.

Each 6-month period is included in the cash flows, including the $0 entries, so the NPV and PMT functions work correctly. The final *A* values in D14 ($318,784) and F14 ($320,069) are the same (except for rounding) as those above.

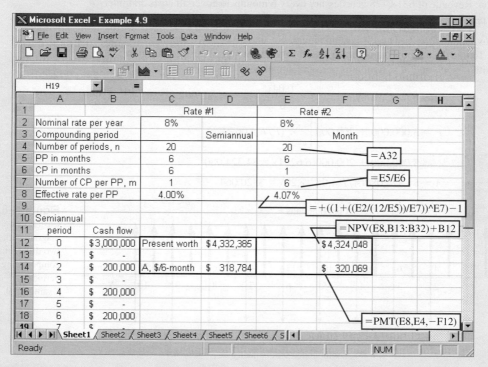

Figure 4–9
Spreadsheet solution for semiannual *A* series for different compounding periods, Example 4.9.

4.7 EQUIVALENCE RELATIONS: SINGLE AMOUNTS AND SERIES WITH PP < CP

If a person deposits money each month into a savings account where interest is compounded quarterly, do all the monthly deposits earn interest before the next quarterly compounding time? If a person's credit card payment is due with interest on the 15th of the month, and if the full payment is made on the 1st, does the financial institution reduce the interest owed, based on early payment? The usual answers are 'No'. However, if a monthly payment on a $10 million, quarterly-compounded, bank loan were made early by a large corporation, the corporate financial officer would likely insist that the bank reduce the amount of interest due, based on early payment. These are examples of PP < CP. The timing of cash flow transactions between compounding points introduces the question of how *interperiod compounding* is handled. Fundamentally, there are two policies: interperiod cash flows earn *no interest,* or they earn *compound interest.*

For a no-interperiod-interest policy, deposits (negative cash flows) are all regarded as *deposited at the end of the compounding period,* and withdrawals are all regarded as *withdrawn at the beginning.* As an illustration, when interest is compounded quarterly, all monthly deposits are moved to the end of the quarter (no interperiod interest is earned), and all withdrawals are moved to the beginning (no interest is paid for the entire quarter). This procedure can significantly alter the distribution of cash flows before the effective quarterly rate is applied to find P, F, or A. This effectively forces the cash flows into a PP = CP situation, as discussed in Sections 4.5 and 4.6. Example 4.10 illustrates this procedure and the economic fact that, within a one-compounding-period time frame, there is no interest advantage to making payments early. Of course, non-economic factors may be present.

EXAMPLE 4.10

Rob is the on-site coordinating engineer for Alcoa Aluminum, where an under-renovation mine has new ore refining equipment being installed by a local contractor. Rob developed the cash flow diagram in Figure 4–10a in $1000 units from the project perspective. Included are payments to the contractor he has authorized for the current year and approved advances from Alcoa's home office. He knows that the interest rate on equipment "field projects" such as this is 12% per year, compounded quarterly, and that Alcoa does not bother with interperiod compounding of interest. Will Rob's project finances be in the "red" or the "black" at the end of the year? By how much?

Solution

With no interperiod interest considered, Figure 4–10b reflects the moved cash flows. The future worth after four quarters requires an F at an effective rate per quarter of $12\%/4 = 3\%$. Figure 4–10b shows all negative cash flows (payments to contractor) moved to the end of the respective quarter, and all positive cash flows (receipts from home office)

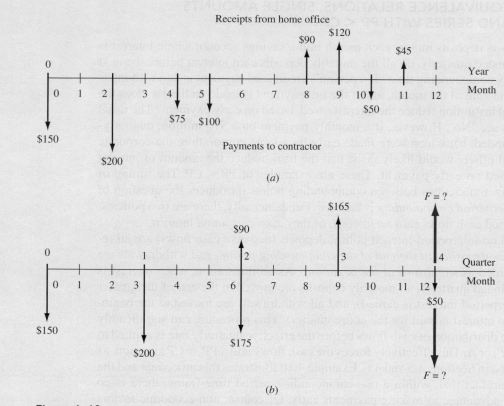

Figure 4–10
(*a*) Actual and (*b*) moved cash flows (in $1000) for quarterly compounding periods using no interperiod interest, Example 4.10.

moved to the beginning of the respective quarter. Calculate the *F* value at 3%.

$$F = 1000[-150(F/P,3\%,4) - 200(F/P,3\%,3) + (-175 + 90)(F/P,3\%,2)$$
$$+ 165(F/P,3\%,1) - 50]$$
$$= \$-357,592$$

Rob can conclude that the on-site project finances will be in the red about $357,600 by the end of the year.

If PP < CP and interperiod compounding is earned, the cash flows are not moved, and the equivalent *P*, *F*, or *A* values are determined using the effective interest rate per payment period. The engineering economy relations are determined in the same way as in the previous two sections for PP ≥ CP. The effective interest rate formula will have an *m* value less than 1, because there is only a fractional part of the CP within one PP. For example, weekly cash flows and quarterly compounding require that $m = 1/13$ of a quarter. When the nominal rate is 12% per year, compounded quarterly (the same as 3% per quarter, compounded quarterly), the effective rate per PP is

$$\text{Effective weekly } i\% = (1.03)^{1/13} - 1 = 0.228\% \text{ per week}$$

4.8 EFFECTIVE INTEREST RATE FOR CONTINUOUS COMPOUNDING

If we allow compounding to occur more and more frequently, the compounding period becomes shorter and shorter. Then m, the number of compounding periods per payment period, increases. This situation occurs in businesses that have a very large number of cash flows every day, so it is correct to consider interest as compounded continuously. As m approaches infinity, the effective interest rate, Equation [4.8], must be written in a new form. First, recall the definition of the natural logarithm base.

$$\lim_{h \to \infty}\left(1 + \frac{1}{h}\right)^h = e = 2.71828+ \qquad [4.11]$$

The limit of Equation [4.8] as m approaches infinity is found by using $r/m = 1/h$, which makes $m = hr$.

$$\lim_{m \to \infty} i = \lim_{m \to \infty}\left(1 + \frac{r}{m}\right)^m - 1$$

$$= \lim_{h \to \infty}\left(1 + \frac{1}{h}\right)^{hr} - 1 = \lim_{h \to \infty}\left[\left(1 + \frac{1}{h}\right)^h\right]^r - 1$$

$$i = e^r - 1 \qquad [4.12]$$

Equation [4.12] is used to compute the *effective continuous interest rate,* when the time periods on i and r are the same. As an illustration, if the nominal annual $r = 15\%$ per year, the effective continuous rate per year is

$$i\% = e^{0.15} - 1 = 16.183\%$$

For convenience, Table 4–3 includes effective continuous rates for the nominal rates listed.

EXAMPLE **4.11**

(a) For an interest rate of 18% per year, compounded continuously, calculate the effective monthly and annual interest rates.

(b) An investor requires an effective return of at least 15%. What is the minimum annual nominal rate that is acceptable for continuous compounding?

Solution

(*a*) The nominal monthly rate is $r = 18\%/12 = 1.5\%$, or 0.015 per month. By Equation [4.12], the effective monthly rate is

$$i\% \text{ per month} = e^r - 1 = e^{0.015} - 1 = 1.511\%$$

Similarly, the effective annual rate using $r = 0.18$ per year is

$$i\% \text{ per year} = e^r - 1 = e^{0.18} - 1 = 19.72\%$$

(*b*) Solve Equation [4.12] for *r* by taking the natural logarithm.

$$e^r - 1 = 0.15$$
$$e^r = 1.15$$
$$\ln e^r = \ln 1.15$$
$$r\% = 13.976\%$$

Therefore, a rate of 13.976% per year, compounded continuously, will generate an effective 15% per year return.

Comment

The general formula to find the nominal rate, given the effective continuous rate *i*, is $r = \ln(1 + i)$

EXAMPLE 4.12

Engineers Marci and Suzanne both invest $5000 for 10 years at 10% per year. Compute the future worth for both individuals if Marci receives annual compounding and Suzanne receives continuous compounding.

Solution

Marci: For annual compounding the future worth is

$$F = P(F/P,10\%,10) = 5000(2.5937) = \$12,969$$

Suzanne: Using Equation [4.12], first find the effective *i* per year for use in the F/P factor.

$$\text{Effective } i\% = e^{0.10} - 1 = 10.517\%$$
$$F = P(F/P,10.517\%,10) = 5000(2.7183) = \$13,591$$

Continuous compounding causes a $622 increase in earnings. For comparison, daily compounding yields an effective rate of 10.516% ($F = \$13,590$), only slightly less than the 10.517% for continuous compounding.

For some business activities, cash flows occur throughout the day. Examples of costs are energy and water costs, inventory costs, and labor costs. A realistic model for these activities is to increase the frequency of the cash flows to become continuous. In these cases, the economic analysis can be performed for continuous cash flow (also called continuous funds flow) and the continuous compounding of interest as discussed above. Different expressions must be derived for the factors for these cases. In fact, the monetary differences for continuous cash flows relative to the discrete cash flow and discrete compounding assumptions are usually not large. Accordingly, most engineering economy studies do not require the analyst to utilize these mathematical forms to make a sound economic project evaluation and decision.

4.9 INTEREST RATES THAT VARY OVER TIME

Real-world interest rates for a corporation vary from year to year, depending upon the financial health of the corporation, its market sector, the national and international economies, forces of inflation, and many other elements. Loan rates may increase from one year to another. Home mortgages financed using ARM (adjustable rate mortgage) interest is a good example. The mortgage rate is slightly adjusted annually to reflect the age of the loan, the current cost of mortgage money, etc. An example of interest rates that rise over time is inflation-protected bonds that are issued by the U.S. government and other agencies. The dividend rate that the bond pays remains constant over its stated life, but the lump-sum amount due to the owner when the bond reaches maturity is adjusted upward with the inflation index of the Consumer Price Index (CPI). This means the annual rate of return will increase annually in accordance with observed inflation. (Bonds and inflation are visited again in Chapters 5 and 14, respectively.)

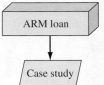

When P, F, and A values are calculated using a constant or average interest rate over the life of a project, rises and falls in i are neglected. If the variation in i is large, the equivalent values will vary considerably from those calculated using the constant rate. Although an engineering economy study can accommodate varying i values mathematically, it is more involved computationally to do so.

To determine the P value for future cash flow values (F_t) at different i values (i_t) for each year t, we will assume *annual compounding*. Define

i_t = effective annual interest rate for year t (t = years 1 to n)

To determine the present worth, calculate the P of each F_t value, using the applicable i_t, and sum the results. Using standard notation and the P/F factor,

$$P = F_1(P/F,i_1,1) + F_2(P/F,i_1,1)(P/F,i_2,1) + \cdots$$
$$+ F_n(P/F,i_1,1)(P/F,i_2,1) \cdots (P/F,i_n,1) \qquad [4.13]$$

When only single amounts are involved, that is, one P and one F in the final year n, the last term in Equation [4.13] is the expression for the present worth of the future cash flow.

$$P = F_n(P/F,i_1,1)(P/F,i_2,1) \cdots (P/F,i_n,1) \qquad [4.14]$$

If the equivalent uniform series A over all n years is needed, first find P using either of the last two equations, then substitute the symbol A for each F_t symbol. Since the equivalent P has been determined numerically using the varying rates, this new equation will have only one unknown, namely, A. The following example illustrates this procedure.

EXAMPLE 4.13

CE, Inc., leases large earth tunneling equipment. The net profit from the equipment for each of the last 4 years has been decreasing, as shown below. Also shown are the annual rates of return on invested capital. The return has been increasing. Determine the present worth P and equivalent uniform series A of the net profit series. Take the annual variation of rates of return into account.

Year	1	2	3	4
Net Profit	$70,000	$70,000	$35,000	$25,000
Annual Rate	7%	7%	9%	10%

Solution

Figure 4–11 shows the cash flows, rates for each year, and the equivalent P and A. Equation [4.13] is used to calculate P. Since for both years 1 and 2 the net profit is $70,000 and the annual rate is 7%, the P/A factor can be used for these 2 years only.

$$P = [70(P/A,7\%,2) + 35(P/F,7\%,2)(P/F,9\%,1)$$
$$+ 25(P/F,7\%,2)(P/F,9\%,1)(P/F,10\%,1)](1000)$$
$$= [70(1.8080) + 35(0.8013) + 25(0.7284)](1000)$$
$$= \$172,816 \qquad [4.15]$$

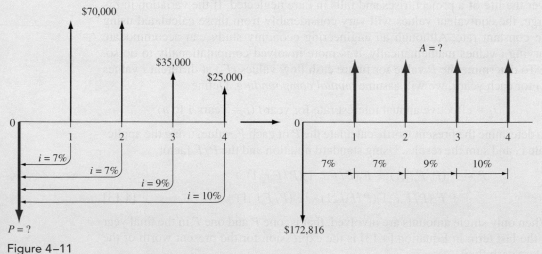

Figure 4–11
Equivalent P and A values for varying interest rates, Example 4.13.

To determine an equivalent annual series, substitute the symbol A for all net profit values on the right side of Equation [4.15], set it equal to $P = \$172,816$ and solve for A. This equation accounts for the varying i values each year. See Figure 4–11 for the cash flow diagram transformation.

$$\$172,816 = A[(1.8080) + (0.8013) + (0.7284)] = A[3.3377]$$

$$A = \$51,777 \text{ per year}$$

Comment
If the average of the four annual rates, that is, 8.25%, is used, the result is $A = \$52,467$. This is a $690 per year overestimate of the required equivalent amount.

When there is a cash flow in year 0 and interest rates vary annually, this cash flow must be included when one is determining P. In the computation for the equivalent uniform series A over all years, including year 0, it is important to include this initial cash flow at $t = 0$. This is accomplished by inserting the factor value for $(P/F,i_0,0)$ into the relation for A. This factor value is always 1.00. It is equally correct to find the A value using a future worth relation for F in year n. In this case, the A value is determined using the F/P factor, and the cash flow in year n is accounted for by including the factor $(F/P,i_n,0) = 1.00$.

CHAPTER SUMMARY

Since many real-world situations involve cash flow frequencies and compounding periods other than 1 year, it is necessary to use nominal and effective interest rates. When a nominal rate r is stated, the effective interest rate per payment period is determined by using the effective interest rate equation.

$$\textbf{Effective } i = \left(1 + \frac{r}{m}\right)^m - 1$$

The m is the number of compounding periods (CP) per payment period (PP). If interest compounding becomes more and more frequent, the length of a CP approaches zero, continuous compounding results, and the effective i is $e^r - 1$.

All engineering economy factors require the use of an effective interest rate. The i and n values placed in a factor depend upon the type of cash flow series. If only single amounts (P and F) are present, there are several ways to perform equivalence calculations using the factors. However, when series cash flows (A, G, and g) are present, only one combination of the effective rate i and number of periods n is correct for the factors. This requires that the relative lengths of PP and CP be considered as i and n are determined. *The interest rate and payment periods must have the same time unit* for the factors to correctly account for the time value of money.

From one year (or interest period) to the next, interest rates will vary. To accurately perform equivalence calculations for P and A when rates vary significantly, the applicable interest rate should be used, not an average or constant rate. Whether performed by hand or by computer, the procedures and factors are the same as those for constant interest rates; however, the number of calculations increases.

PROBLEMS

Nominal and Effective Rates

4.1 Identify the compounding period for the following interest statements: (a) 1% per month; (b) 2.5% per quarter; and (c) 9.3% per year compounded semiannually.

4.2 Identify the compounding period for the following interest statements: (a) Nominal 7% per year compounded quarterly; (b) effective 6.8% per year compounded monthly; and (c) effective 3.4% per quarter compounded weekly.

4.3 Determine the number of times interest would be compounded in 1 year for the following interest statements: (a) 1% per month; (b) 2% per quarter; and (c) 8% per year compounded semiannually.

4.4 For an interest rate of 10% per year compounded quarterly, determine the number of times interest would be compounded (a) per quarter, (b) per year, and (c) per 3 years.

4.5 For an interest rate of 0.50% per quarter, determine the nominal interest rate per (a) semiannual period, (b) year, and (c) 2 years.

4.6 For an interest rate of 12% per year compounded every 2 months, determine the nominal interest rate per (a) 4 months, (b) 6 months, and (c) 2 years.

4.7 For an interest rate of 10% per year, compounded quarterly, determine the nominal rate per (a) 6 months and (b) 2 years.

4.8 Identify the following interest rate statements as either nominal or effective: (a) 1.3% per month; (b) 1% per week, compounded weekly; (c) nominal 15% per year, compounded monthly; (d) effective 1.5% per month, compounded daily; and (e) 15% per year, compounded semiannually.

4.9 What effective interest rate per 6 months is equivalent to 14% per year, compounded semiannually?

4.10 An interest rate of 16% per year, compounded quarterly, is equivalent to what effective interest rate per year?

4.11 What nominal interest rate per year is equivalent to an effective 16% per year, compounded semiannually?

4.12 What effective interest rate per year is equivalent to an effective 18% per year, compounded semiannually?

4.13 What compounding period is associated with nominal and effective rates of 18% and 18.81% per year, respectively?

4.14 An interest rate of 1% per month is equivalent to what effective rate per 2 months?

4.15 An interest rate of 12% per year, compounded monthly, is equivalent to what nominal and effective interest rates per 6 months?

4.16 (*a*) An interest rate of 6.8% per semiannual period, compounded weekly, is equivalent to what weekly interest rate?

(*b*) Is the weekly rate a nominal or effective rate? Assume 26 weeks per 6 months.

Payment and Compounding Periods

4.17 Deposits of $100 per week are made into a savings account that pays interest of 6% per year, compounded quarterly. Identify the payment and compounding periods.

4.18 A certain national bank advertises quarterly compounding for business checking accounts. What payment and compounding periods are associated with deposits of daily receipts?

4.19 Determine the F/P factor for 3 years at an interest rate of 8% per year, compounded quarterly.

4.20 Determine the P/G factor for 5 years at an effective interest rate of 6% per year, compounded semiannually.

Equivalence for Single Amounts and Series

4.21 A company that specializes in online security software development wants to have $85 million available in 3 years to pay stock dividends. How much money must the company set aside *now* in an account that earns interest at a rate of 8% per year, compounded quarterly?

4.22 Because testing of nuclear bombs was halted in 1992, the U.S. Department of Energy has been developing a laser project that will allow engineers to simulate (in a laboratory) conditions in a thermonuclear reaction. Due to soaring cost overruns, a congressional committee undertook an investigation and discovered that the estimated development cost of the project increased at an average rate of 3% per month over a 5-year period. If the original cost was estimated to be $2.7 billion 5 years ago, what is the expected cost today?

4.23 A present sum of $5000 at an interest rate of 8% per year, compounded semiannually, is equivalent to how much money 8 years ago?

4.24 In an effort to ensure the safety of cell phone users, the Federal Communications Commission (FCC) requires cell phones to have a specific absorbed radiation (SAR) number of 1.6 watts per kilogram (W/kg) of tissue or less. A new cell phone company estimates that by advertising its favorable 1.2 SAR number, it will increase sales by $1.2 million 3 months from now when its phones go on sale. At an interest rate of 20% per year, compounded quarterly, what is the maximum amount the company can afford to spend *now* for advertising in order to break even?

4.25 Radio Frequency Identification (RFID) is technology that is used by drivers with speed passes at toll booths and ranchers who track livestock from farm to fork. Wal-Mart expects to begin using the technology to track products within its stores. If RFID-tagged products will result in better inventory control that will save the company $1.3 million per month beginning 3 months from now, how much could the company afford to spend now to implement the technology at an interest rate of 12% per year, compounded monthly, if it wants to recover its investment in 2½ years?

4.26 The patriot missile, developed by Lock-heed Martin for the U.S. Army, is designed to shoot down aircraft and other missiles. The Patriot Advanced Capability-3 was originally promised to cost $3.9 billion, but due to extra time needed to write computer code and scrapped tests (due to high winds) at White Sands Missile Range, the actual cost was much higher. If the total project development time was 10 years and costs increased at a rate of 0.5% per month, what was the final cost of the project?

4.27 Video cards based on Nvidia's highly praised GeForce2 GTS processor typically cost $250. Nvidia released a light version of the chip that costs $150. If a certain video game maker was purchasing 3000 chips per quarter, what was the present worth of the savings associated with the cheaper chip over a 2-year period at an interest rate of 16% per year, compounded quarterly?

4.28 A 40-day strike at Boeing resulted in 50 fewer deliveries of commercial jetliners at the end of the first quarter of 2000. At a cost of $20 million per plane, what was the equivalent end-of-year cost of the strike (i.e., end of fourth quarter) at an interest rate of 18% per year, compounded monthly?

4.29 The optical products division of Panasonic is planning a $3.5 million building expansion for manufacturing its powerful Lumix DMC digital zoom camera. If the company uses an interest rate of 20% per year, compounded quarterly, for all new investments, what is the uniform amount per quarter the company must make to recover its investment in 3 years?

4.30 Thermal Systems, a company that specializes in odor control, made deposits of $10,000 now, $25,000 at the end of month 6, and $30,000 at the end of month 9. Determine the future worth (end of year 1) of the deposits at an interest rate of 16% per year, compounded quarterly.

4.31 Lotus Development has a software rental plan called SmartSuite that is available on the World Wide Web. A number of programs are available at $2.99 for 48 hours. If a construction company uses the service an average of 48 hours per week, what is the present worth of the rental costs for 10 months at an interest rate of 1% per month, compounded weekly? (Assume 4 weeks per month.)

4.32 Northwest Iron and Steel is considering getting involved in electronic commerce. A modest e-commerce package is available for $20,000. If the company wants to recover the cost in 2 years, what is the equivalent amount of new income that must be realized every 6 months, if the interest rate is 3% per quarter?

4.33 Metropolitan Water Utilities purchases surface water from Elephant Butte Irrigation District at a cost of $100,000 per month in the months of February through September. Instead of paying monthly, the utility makes a single payment of $800,000 at the end of the year (i.e., end of December) for the water it used. The delayed payment essentially represents a subsidy by the irrigation district to the water utility. At an interest rate of 0.25% per month, what is the amount of the subsidy?

4.34 Scott Specialty Manufacturing is considering consolidating all its electronic services with one company. By purchasing a digital phone from AT&T Wireless, the company can buy wireless e-mail and fax services for $6.99 per month. For $14.99

per month, the company will get unlimited Web access and personal organization functions. For a 2-year contract period, what is the present worth of the *difference* between the services at an interest rate of 12% per year, compounded monthly?

4.35 Magnetek Instrument and Controls, a manufacturer of liquid-level sensors, expects sales for one of its models to increase by 20% every 6 months into the foreseeable future. If the sales 6 months from now are expected to be $150,000, determine the equivalent semiannual worth of sales for a 5-year period at an interest rate of 14% per year, compounded semiannually.

4.36 Metalfab Pump and Filter projects that the cost of steel bodies for certain valves will increase by $2 every 3 months. If the cost for the first quarter is expected to be $80, what is the present worth of the costs for a 3-year period at an interest rate of 3% per quarter?

4.37 Fieldsaver Technologies, a manufacturer of precision laboratory equipment, borrowed $2 million to renovate one of its testing labs. The loan was repaid in 2 years through quarterly payments that increased by $50,000 each time. At an interest rate of 3% per quarter, what was the size of the first quarterly payment?

4.38 For the cash flows shown below, determine the present worth (time 0), using an interest rate of 18% per year, compounded monthly.

Month	Cash Flow, $/Month
0	1000
1–12	2000
13–28	3000

4.39 The cash flows (in thousands) associated with Fisher-Price's Touch learning system are shown below. Determine the uniform quarterly series in quarters 0 through 8 that would be equivalent to the cash flows shown at an interest rate of 16% per year, compounded quarterly.

Quarter	Cash Flow, $/Quarter
1	1000
2–3	2000
5–8	3000

Equivalence When PP < CP

4.40 An engineer deposits $300 per month into a savings account that pays interest at a rate of 6% per year, compounded semiannually. How much will be in the account at the end of 15 years? Assume no interperiod compounding.

4.41 At time $t = 0$, an engineer deposited $10,000 into an account that pays interest at 8% per year, compounded semiannually. If she withdrew $1000 in months 2, 11, and 23, what was the total value of the account at the end of 3 years? Assume no interperiod compounding.

4.42 For the transactions shown below, determine the amount of money in the account at the end of year 3 if the interest rate is 8% per year, compounded semiannually. Assume no interperiod compounding.

End of Quarter	Amount of Deposit, $/Quarter	Amount of Withdrawal, $/Quarter
1	900	
2–4	700	
7	1000	2600
11	—	1000

4.43 The New Mexico State Police and Public Safety Department owns a helicopter that it uses to provide transportation and logistical support for high-level state officials. The $495 hourly rate covers operating expenses and the pilot's salary. If the governor uses the helicopter an average of 2 days per month for 6 hours each day, what is the equivalent future worth of the costs for 1 year at an interest rate of 6% per year, compounded quarterly (treat the costs as deposits)?

Continuous Compounding

4.44 What effective interest rate per year, compounded continuously, is equivalent to a nominal rate of 13% per year?

4.45 What effective interest rate per 6 months is equal to a nominal 2% per month, compounded continuously?

4.46 What nominal rate per quarter is equivalent to an effective rate of 12.7% per year, compounded continuously?

4.47 Corrosion problems and manufacturing defects rendered a gasoline pipeline between El Paso and Phoenix subject to longitudinal weld seam failures. Therefore, pressure was reduced to 80% of the design value. If the reduced pressure results in delivery of $100,000 per month less product, what will be the value of the lost revenue after a 2-year period at an interest rate of 15% per year, compounded continuously?

4.48 Because of a chronic water shortage in Santa Fe, new athletic fields must use artificial turf or xeriscape landscaping. If the value of the water saved each month is $6000, how much can a private developer afford to spend on artifical turf if he wants

to recover his investment in 5 years at an interest rate of 18% per year, compounded continuously?

4.49 A Taiwan-based chemical company had to file for bankruptcy because of a nationwide phase-out of methyl tertiary butyl ether (MTBE). If the company reorganizes and invests $50 million in a new ethanol production facility, how much money must it make each month if it wants to recover its investment in 3 years at an interest rate of 2% per month, compounded continuously?

4.50 In order to have $85,000 four years from now for equipment replacement, a construction company plans to set aside money today in government-insured bonds. If the bonds earn interest at a rate of 6% per year, compounded continuously, how much money must the company invest?

4.51 How long would it take for a lump-sum investment to double in value at an interest rate of 1.5% per month, compounded continuously?

4.52 What effective interest rate per month, compounded continuously, would be required for a single deposit to triple in value in 5 years?

Varying Interest Rates

4.53 How much money could the maker of fluidized-bed scrubbers afford to spend now instead of spending $150,000 in year 5 if the interest rate is 10% in years 1 through 3 and 12% in years 4 and 5?

4.54 What is the future worth in year 8 of a present sum of $50,000 if the interest rate is 10% per year in years 1 through 4 and 1% per month in years 5 through 8?

4.55 For the cash flows shown below, determine (*a*) the future worth in year 5 and (*b*) the equivalent *A* value for years 0 through 5.

Year	Cash Flow, $/Year	Interest Rate per Year, %
0	5000	12
1–4	6000	12
5	9000	20

4.56 For the cash flow series shown below, find the equivalent *A* value in years 1 through 5.

Year	Cash Flow, $/Year	Interest Rate per Year, %
0	0	
1–3	5000	10
4–5	7000	12

FE REVIEW PROBLEMS

4.57 An interest rate of effective 14% per month, compounded weekly, is
   (*a*) An effective rate per year
   (*b*) An effective rate per month
   (*c*) A nominal rate per year
   (*d*) A nominal rate per month

4.58 An interest rate of 2% per month is the same as
   (*a*) 24% per year, compounded monthly
   (*b*) A nominal 24% per year, compounded monthly
   (*c*) An effective 24% per year, compounded monthly
   (*d*) Both (*a*) and (*b*)

4.59 An interest rate of 12% per year, compounded monthly, is nearest to
   (*a*) 12.08% per year
   (*b*) 12.28% per year
   (*c*) 12.48% per year
   (*d*) 12.68% per year

4.60 An interest rate of 1.5% per month, compounded continuously, is closest to an effective
   (*a*) 1.51% per quarter
   (*b*) 4.5% per quarter

   (*c*) 4.6% per quarter
   (*d*) 9% per 6 months

4.61 An interest rate of 2% per quarter is the same as
   (*a*) Nominal 2% per quarter
   (*b*) Nominal 6% per year, compounded quarterly
   (*c*) Effective 2% every 4 months
   (*d*) Effective 2% every 3 months

4.62 An interest rate expressed as an effective 12% per year, compounded monthly, is the same as
   (*a*) 12% per year
   (*b*) 1% per month
   (*c*) 12.68% per year
   (*d*) Any of the above

4.63 A 1-*year* interest rate of 20% per year, compounded continuously, is closest to the following interest rate:
   (*a*) Simple 22% per year
   (*b*) 21% per year, compounded quarterly
   (*c*) 21% per year, compounded monthly
   (*d*) 22% per year, compounded semi-annually

4.64 For an interest rate of 1% per quarter, compounded continuously, the effective semiannual interest rate is closest to
(a) Less than 2.0%
(b) 2.02%
(c) 2.20%
(d) Over 2.25%

4.65 The only time you change the amount and timing of the original cash flows in problems involving a uniform series is when
(a) Payment period is longer than the compounding period.
(b) Payment period is equal to the compounding period.
(c) Payment period is shorter than the compounding period.
(d) It may be done in any of the cases above, depending upon how the effective interest rate is calculated.

4.66 Exotic Faucets and Sinks, Ltd., guarantees that its new infrared sensor faucet will save any household that has two or more children at least $30 per month in water costs beginning 1 month after the faucet is installed. If the faucet is under full warranty for 5 years, the minimum amount a family could afford to spend now on such a faucet at an interest rate of 6% per year, compounded monthly, is closest to
(a) $149
(b) $1552
(c) $1787
(d) $1890

4.67 The multistate Powerball Lottery, worth $182 million, was won by a single individual who had purchased five tickets at $1 each. The individual was given two choices: Receive 26 payments of $7 million each, with the first payment to be made *now* and the rest to be made at the end of each of the next 25 years; or receive a single lump-sum payment *now* that

would be equivalent to the 26 payments of $7 million each. If the state uses an interest rate of 4% per year, the amount of the lump-sum payment would be closest to
(a) Less than $109 million
(b) $109,355,000
(c) $116,355,000
(d) Over $117 million

4.68 Royalties paid to holders of mineral rights tend to decrease with time as resources become depleted. In one particular case, the right holder received a royalty check of $18,000 six months after the lease was signed. She continued to receive checks at 6-month intervals, but the amount decreased by $2000 each time. At an interest rate of 6% per year, compounded semiannually, the equivalent uniform semiannual worth of the royalty payments through the first 4 years is represented by
(a) A = $18,000 - 2000(A/G, 3\%, 8)$
(b) A = $18,000 - 2000(A/G, 6\%, 4)$
(c) A = $18,000(A/P, 3\%, 8) - 2000$
(d) A = $18,000 + 2000(A/G, 3\%, 8)$

4.69 The cost for increasing the production capacity in a certain manufacturing facility is expected to increase by 7% per year over the next 5-year period. If the cost at the end of year 1 is $39,000 and the interest rate is 10% per year, the present worth of the costs through the end of the 5-year period is represented by
(a) $P = 39,000\{1 - [(1 + 0.07)^6/(1 + 0.10)^6]\}/(0.10 - 0.07)$
(b) $P = 39,000\{1 - [(1 + 0.07)^5/(1 + 0.10)^5]\}/(0.10 + 0.07)$
(c) $P = 39,000\{1 - [(1 + 0.07)^4/(1 + 0.10)^4]\}/(0.10 - 0.07)$
(d) $P = 39,000\{1 - [(1 + 0.07)^5/(1 + 0.10)^5]\}/(0.10 - 0.07)$

4.70 A plant manager wants to know the present worth of the maintenance costs for a certain assembly line. An industrial engineer, who designed the system, estimates that the maintenance costs are expected to be zero for the first 3 years, $2000 in year 4, $2500 in year 5, and amounts increasing by $500 each year through year 10. At an interest rate of 8% per year, compounded semiannually, the value of n to use in the P/G equation for this problem is

(a) 7
(b) 8
(c) 10
(d) 14

4.71 A public relations company hired by the city of El Paso to increase tourism to the Sun City proposed that the city build the world's only roller coaster that travels through two different countries. The idea is to build the ride along the Rio Grande River and have part of the tracks in the United States and part in Mexico. The ride would be built such that coaster cars could be launched from either side of the border, but riders would get off at the same point where they got on. After the ride becomes operational, tourism revenue is projected to be $1 million initially (i.e., at time 0), $1.05 million after the first month, $1.1025 million after the second month, and amounts increasing by 5% each month through the first year. At an interest rate of 12% per year, compounded monthly, the present worth (time 0) of the tourism revenue generated by the ride is closest to

(a) $15.59 million
(b) $16.59 million
(c) $17.59 million
(d) Over $18 million

4.72 In problems that involve an arithmetic gradient G wherein the payment period is longer than the interest period, the interest rate to use in the equations

(a) Can be any effective rate, as long as the time units on i and n are the same
(b) Must be the interest rate that is exactly as it is stated in the problem
(c) Must be an effective interest rate that is expressed over a 1-year period
(d) Must be the effective interest rate that is expressed over the time period equal to the time where the first change equal to G occurs

4.73 An engineer analyzing cost data discovered that the information for the first three years was missing. However, he knew that the cost in year 4 was $1250 and that it increased by 5% per year thereafter. If the same trend applied to the first three years, the cost in year 1 was closest to

(a) $1235.70
(b) $1191.66
(c) $1133.79
(d) $1079.80

4.74 Encon Environmental Testing needs to purchase $40,000 worth of equipment 2 years from now. At an interest rate of 20% per year, compounded quarterly, the uniform quarterly worth of the equipment (quarters 1 through 8) is closest to

(a) $3958
(b) $4041
(c) $4189
(d) Over $4200

4.75 Border Steel invested $800,000 in a new shearing unit. At an interest rate of 12% per year, compounded quarterly, the quarterly income required to recover the investment in 3 years is closest to

(a) $69,610
(b) $75,880
(c) $80,370
(d) $83,550

CASE STUDY

FINANCING A HOUSE

Introduction

When a person or a couple decide to purchase a house, one of the most important considerations is the financing. There are many methods of financing the purchase of residential property, each having advantages which make it the method of choice under a given set of circumstances. The selection of one method from several for a given set of conditions is the topic of this case study. Three methods of financing are described in detail. Plans A and B are evaluated; you are asked to evaluate plan C and perform some additional analyses.

The criterion used here is: Select the financing plan which has the largest amount of money remaining at the end of a 10-year period. Therefore, calculate the future worth of each plan, and select the one with the largest future worth value.

Plan	Description
A	30-year fixed rate of 10% per year interest, 5% down payment
B	30-year adjustable-rate mortgage (ARM), 9% first 3 years, 9½% in year 4, 10¼% in years 5 through 10 (assumed), 5% down payment
C	15-year fixed rate of 9½% per year interest, 5% down payment

Other information:

- Price of house is $150,000.
- House will be sold in 10 years for $170,000 (net proceeds after selling expenses).
- Taxes and insurance (T&I) are $300 per month.
- Amount available: maximum of $40,000 for down payment, $1600 per month, including T&I.
- New loan expenses: origination fee of 1%, appraisal fee $300, survey fee $200, attorney's fee $200, processing fee $350, escrow fees $150, other costs $300.

- Any money not spent on the down payment or monthly payments will earn tax-free interest at ¼% per month.

Analysis of Financing Plans

Plan A: 30-Year Fixed Rate
The amount of money required up front is

(a)	Down payment (5% of $150,000)	$7,500
(b)	Origination fee (1% of $142,500)	1,425
(c)	Appraisal	300
(d)	Survey	200
(e)	Attorney's fee	200
(f)	Processing	350
(g)	Escrow	150
(h)	Other (recording, credit report, etc.)	300
	Total	$10,425

The amount of the loan is $142,500. The equivalent monthly principal and interest (P&I) payment is determined at $10\%/12$ per month for $30(12) = 360$ months.

$$A = 142{,}500(A/P,10\%/12,360)$$

$$= \$1250.56$$

When T&I are added to P&I, the total monthly payment PMT_A is

$$PMT_A = 1250.56 + 300$$

$$= \$1550.56$$

We can now determine the future worth of plan A by summing three future worth amounts: the remaining funds not used for the down payment and up-front fees (F_{1A}) and for monthly payments (F_{2A}), and the increase in the value of the house (F_{3A}). Since non-expended money earns interest at ¼% per month, in 10 years the first future worth is

$$F_{1A} = (40{,}000 - 10{,}425)(F/P,0.25\%,120)$$

$$= \$39{,}907.13$$

The available money not spent on monthly payments is $49.44 = $1600 − 1550.56. Its future worth after 10 years is

$$F_{2A} = 49.44(F/A,0.25\%,120) = \$6908.81$$

Net money available from the sale of the house is the difference between the net selling price and the balance of the loan. The balance of the loan is

$$\text{Loan balance} = 142,500(F/P,10\%/12,120)$$
$$- 1250.56(F/A,10\%/12,120)$$
$$= 385,753.40 - 256,170.92$$
$$= \$129,582.48$$

Since the net proceeds from the sale of the house are $170,000,

$$F_{3A} = 170,000 - 129,582.48 = \$40,417.52$$

The total future worth of plan A is

$$F_A = F_{1A} + F_{2A} + F_{3A}$$
$$= 39,907.13 + 6908.81 + 40,417.52$$
$$= \$87,233.46$$

Plan B: 30-Year Adjustable Rate Mortgage

Adjustable rate mortgages are tied to some index such as U.S. Treasury bonds. For this example, we have assumed that the rate is 9% for the first 3 years, 9½% in year 4, and 10¼% in years 5 through 10. Since this option also requires 5% down, the up-front money required will be the same as for plan A, that is, $10,425.

The monthly P&I amount for the first 3 years is based on 9% per year for 30 years.

$$A = 142,500(A/P,9\%/12,360) = \$1146.58$$

The total monthly payment for the first 3 years is

$$\text{PMT}_B = \$1146.58 + 300 = \$1446.58$$

At the end of year 3, the interest rate changes to 9½% per year. This new rate applies to the balance of the loan at that time:

Loan balance
at end of year 3 $= 142,500(F/P,0.75\%,36)$
$\qquad\qquad - 1146.58(F/A,0.75\%,36)$
$\qquad\qquad = \$139,297.08$

The P&I monthly payment for year 4 is now

$$A = 139,297.08(A/P,9.5\%/12,324) = \$1195.67$$

The total payment for year 4 is

$$\text{PMT}_B = 1195.67 + 300 = \$1495.67$$

At the end of year 4, the interest rate changes again, this time to 10¼% per year, and it stays at this rate for the remainder of the 10-year period. The loan balance at the end of year 4 is

Loan balance
at end of year 4 $= 139,297.08(F/P,9.5\%/12,12)$
$\qquad\qquad - 1195.67(F/A,9.5\%/12,12)$
$\qquad\qquad = \$138,132.42$

The new P&I amount is

$$A = 138,132.42(A/P,10.25\%/12,312) = \$1269.22$$

The new total payment for years 5 through 10 is

$$\text{PMT}_B = 1269.22 + 300 = \$1569.22$$

The loan balance at the end of 10 years is

Loan balance
after 10 years $= 138,132.42(F/P,10.25\%/12,72)$
$\qquad\qquad - 1269.22(F/A,10.25\%/12,72)$
$\qquad\qquad = \$129,296.16$

The future worth of plan B can now be determined using the same three future worth values. The future worth of the money not spent on a down payment is the same as for plan A.

$$F_{1B} = (40,000 - 10,425)(F/P,0.25\%,120)$$
$$= \$39,907.13$$

The future worth of the money not spent on monthly payments is more complex than in plan A.

$$F_{2B} = (1600 - 1446.58)(F/A,0.25\%,36)$$
$$\times (F/P,0.25\%,84) + (1600 - 1495.67)$$
$$\times (F/A,0.25\%,12)(F/P,0.25\%,72)$$
$$+ (1600 - 1569.22)(F/A,0.25\%,72)$$
$$= 7118.61 + 1519.31 + 2424.83$$
$$= \$11,062.75$$

The amount of money left from the sale of the house is

$$F_{3B} = 170,000 - 129,296.16 = \$40,703.84$$

The total future worth of plan B is

$$F_B = F_{1B} + F_{2B} + F_{3B} = \$91,673.72$$

Case Study Exercises

1. Evaluate plan C and select the best financing method.
2. What is the total amount of interest paid in plan A through the 10-year period?

3. What is the total amount of interest paid in plan B through year 4?
4. What is the maximum amount of money available for a down payment under plan A, if $40,000 is the total amount available?
5. By how much does the payment increase in plan A for each 1% increase in interest rate?
6. If you wanted to "buy down" the interest rate from 10% to 9% in plan A, how much extra down payment would you have to make?

CHAPTER 4 APPENDIX: CALCULATION OF AN EFFECTIVE INTEREST RATE

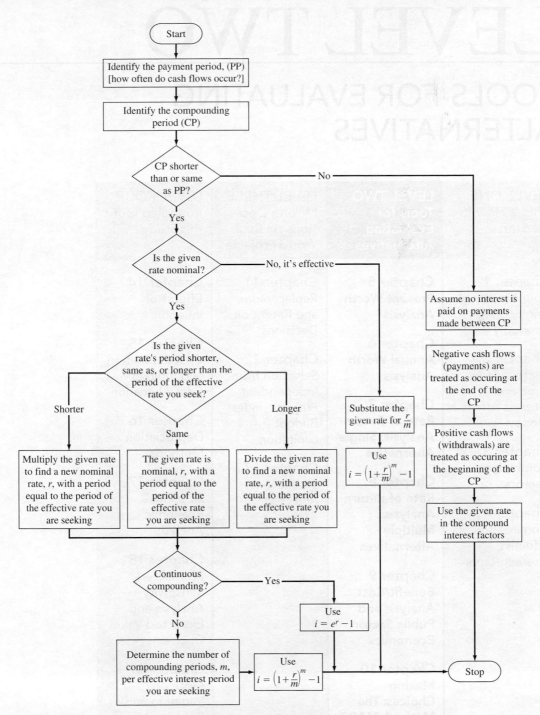

Contributed by Dr. Mathias Sutton, Purdue University

LEVEL TWO

TOOLS FOR EVALUATING ALTERNATIVES

LEVEL ONE
This Is How It All Starts

Chapter 1
Foundations of Engineering Economy

Chapter 2
Factors: How Time and Interest Affect Money

Chapter 3
Combining Factors

Chapter 4
Nominal and Effective Interest Rates

LEVEL TWO
Tools for Evaluating Alternatives

Chapter 5
Present Worth Analysis

Chapter 6
Annual Worth Analysis

Chapter 7
Rate of Return Analysis: Single Alternative

Chapter 8
Rate of Return Analysis: Multiple Alternatives

Chapter 9
Benefit/Cost Analysis and Public Sector Economics

Chapter 10
Making Choices: The Method, MARR, and Multiple Attributes

LEVEL THREE
Making Decisions on Real-World Projects

Chapter 11
Replacement and Retention Decisions

Chapter 12
Selection from Independent Projects under Budget Limitation

Chapter 13
Breakeven Analysis

LEVEL FOUR
Rounding Out the Study

Chapter 14
Effects of Inflation

Chapter 15
Cost Estimation and Indirect Cost Allocation

Chapter 16
Depreciation Methods

Chapter 17
After-Tax Economic Analysis

Chapter 18
Formalized Sensitivity Analysis and Expected Value Decisions

Chapter 19
More on Variation and Decision Making under Risk

One or more engineering alternatives are formulated to solve a problem or provide specified results. In engineering economics, each alternative has cash flow estimates for the initial investment, periodic (usually annual) incomes and/or costs, and possibly a salvage value at the end of its estimated life. The chapters in this level develop the four different methods by which one or more alternatives can be evaluated economically using the factors and formulas learned in the previous Level One.

In professional practice, it is typical that the evaluation method and parameter estimates necessary for the economic study are not specified. The last chapter in this level begins with a focus on selecting the best evaluation method for the study. It continues by treating the fundamental question of what MARR to use and the historic dilemma of how to consider noneconomic factors when selecting an alternative.

Important note: If depreciation and/or after tax analysis is to be considered along with the evaluation methods in Chapters 5 through 9, Chapter 16 and/or Chapter 17 should be covered, preferably after Chapter 6.

CHAPTER 5

Present Worth Analysis

A future amount of money converted to its equivalent value now has a present worth (PW) that is always less than that of the actual cash flow, because for any interest rate greater than zero, all P/F factors have a value less than 1.0. For this reason, present worth values are often referred to as *discounted cash flows* (DCF). Similarly, the interest rate is referred to as the *discount rate*. Besides PW, two other terms frequently used are present value (PV) and net present value (NPV). Up to this point, present worth computations have been made for one project or alternative. In this chapter, techniques for comparing two or more mutually exclusive alternatives by the present worth method are treated.

Several extensions to PW analysis are covered here—future worth, capitalized cost, payback period, life-cycle costing, and bond analysis—these all use present worth relations to analyze alternatives.

In order to understand how to organize an economic analysis, this chapter begins with a description of independent and mutually exclusive projects, as well as revenue and service alternatives.

The case study examines the payback period and sensitivity for a public sector project.

LEARNING OBJECTIVES

Purpose: Compare mutually exclusive alternatives on a present worth basis, and apply extensions of the present worth method.

This chapter will help you:

Formulating alternatives	1.	Identify mutually exclusive and independent projects, and define a service and a revenue alternative.
PW of equal-life alternatives	2.	Select the best of equal-life alternatives using present worth analysis.
PW of different-life alternatives	3.	Select the best of different-life alternatives using present worth analysis.
FW analysis	4.	Select the best alternative using future worth analysis.
Capitalized cost (CC)	5.	Select the best alternative using capitalized cost calculations.
Payback period	6.	Determine the payback period at $i = 0\%$ and $i > 0\%$, and state the shortcomings of payback analysis.
Life-cycle cost (LCC)	7.	Perform a life-cycle cost analysis for the acquisition and operations phases of a (system) alternative.
PW of bonds	8.	Calculate the present worth of a bond investment.
Spreadsheets	9.	Develop spreadsheets that use PW analysis and its extensions, including payback period.

5.1 FORMULATING MUTUALLY EXCLUSIVE ALTERNATIVES

Section 1.3 explains that the economic evaluation of an alternative requires cash flow estimates over a stated time period and a criterion for selecting the best alternative. The alternatives are developed from project proposals to accomplish a stated purpose. This progression is depicted in Figure 5–1. Some projects are economically and technologically viable, and others are not. Once the viable projects are defined, it is possible to formulate the alternatives. For example, assume Med-supply.com, an internet-based medical supply provider, wants to challenge its storefront competitors by significantly shortening the time between order placement and delivery to the hospital or clinic. Three projects have been proposed: closer networking with UPS and FedEx for shortened delivery time; partnering with local medical supply houses in major cities to provide same-day delivery; and developing a 3-d fax-like machine to ship items not physically larger than the machine. Economically (and technologically) only the first two project proposals can be pursued at this time; they are the two alternatives to evaluate.

The description above correctly treats project proposals as precursors to economic alternatives. To help formulate alternatives, *categorize each project* as one of the following:

- **Mutually exclusive.** *Only one of the viable projects can be selected* by the economic analysis. Each viable project *is* an alternative.
- **Independent.** *More than one viable project may be selected* by the economic analysis. (There may be dependent projects requiring a particular project to be selected before another, and contingent projects where one project may be substituted for another.)

The *do-nothing* (*DN*) option is usually understood to be an alternative when the evaluation is performed. If it is absolutely required that one of the defined alternatives be selected, do nothing is not considered an option. (This may occur when a mandated function must be installed for safety, legal, or other purposes.) Selection of the DN alternative means that the current approach is maintained; nothing new is initiated. No new costs, revenues, or savings are generated by the DN alternative.

A mutually exclusive alternative selection takes place, for example, when an engineer must select the one best diesel-powered engine from several competing models. Mutually exclusive alternatives are, therefore, the same as the viable projects; each one is evaluated, and the one best alternative is chosen. Mutually exclusive alternatives *compete with one another* in the evaluation. All the analysis techniques through Chapter 9 are developed to compare mutually exclusive alternatives. Present worth is discussed in the remainder of this chapter. If no mutually exclusive alternative is considered economically acceptable, it is possible to reject all alternatives and (by default) accept the DN alternative. (This option is indicated in Figure 5–1 by colored shading on the DN mutually exclusive alternative.)

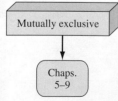

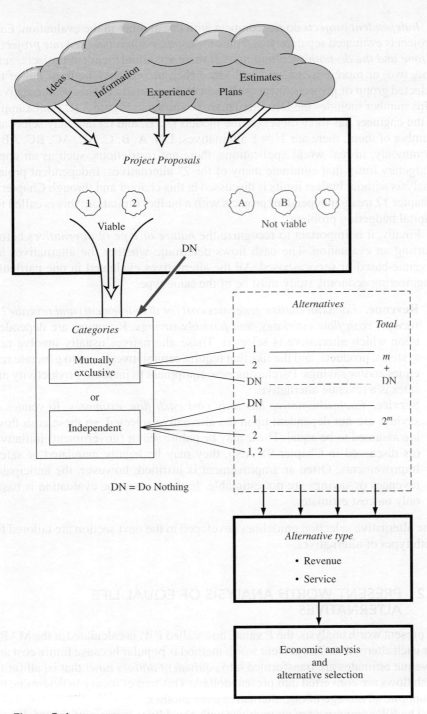

Figure 5–1
Progression from projects to alternatives to economic analysis.

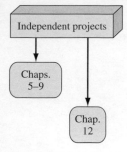

Independent projects do not compete with one another in the evaluation. Each project is evaluated separately, and thus the *comparison is between one project at a time and the do-nothing alternative.* If there are *m* independent projects, zero, one, two, or more may be selected. Since each project may be in or out of the selected group of projects, there are a total of 2^m mutually exclusive alternatives. This number includes the DN alternative, as shown in Figure 5–1. For example, if the engineer has three diesel engine models (A, B, and C) and may select any number of them, there are $2^3 = 8$ alternatives: DN, A, B, C, AB, AC, BC, ABC. Commonly, in real-world applications, there are restrictions, such as an upper budgetary limit, that eliminate many of the 2^m alternatives. Independent project analysis without budget limits is discussed in this chapter and through Chapter 9. Chapter 12 treats independent projects with a budget limitation; this is called the capital budgeting problem.

Finally, it is important to recognize the *nature or type of alternatives* before starting an evaluation. The cash flows determine whether the alternatives are revenue-based or service-based. All the alternatives evaluated in one particular engineering economy study must be of the same type.

- **Revenue.** *Each alternative generates cost (or disbursement) and revenue (or receipt) cash flow estimates, and possibly savings.* Revenues are dependent upon which alternative is selected. These alternatives usually involve new systems, products, and the like that require capital investment to generate revenues and/or savings. Purchasing new equipment to increase productivity and sales is a revenue alternative.
- **Service.** *Each alternative has only cost cash flow estimates.* Revenues or savings are not dependent upon the alternative selected, so these cash flows are assumed to be equal. These may be public sector (government) initiatives (as discussed in Chapter 9). Also, they may be legally mandated or safety improvements. Often an improvement is justified; however, the anticipated revenues or savings are not estimable. In these cases the evaluation is based only on cost estimates.

The alternative selection guidelines developed in the next section are tailored for both types of alternatives.

5.2 PRESENT WORTH ANALYSIS OF EQUAL-LIFE ALTERNATIVES

In present worth analysis, the *P* value, now called *PW*, is calculated at the MARR for each alternative. The present worth method is popular because future cost and revenue estimates are transformed into *equivalent dollars now;* that is, all future cash flows are converted into present dollars. This makes it easy to determine the economic advantage of one alternative over another.

The PW comparison of alternatives with equal lives is straightforward. If both alternatives are used in identical capacities for the same time period, they are termed *equal-service* alternatives.

Whether mutually exclusive alternatives involve disbursements only (service) or receipts and disbursements (revenue), the following guidelines are applied to select one alternative.

One alternative.   Calculate PW at the MARR. If PW ≥ 0, the requested MARR is met or exceeded and the alternative is financially viable.

Two or more alternatives.   Calculate the PW of each alternative at the MARR. *Select the alternative with the PW value that is numerically largest,* **that is, less negative or more positive, indicating a lower PW of cost cash flows or larger PW of net cash flows of receipts minus disbursements.**

Note that the guideline to select one alternative with the lowest cost or the highest income uses the criterion of *numerically largest*. This is *not the absolute value* of the PW amount, because the sign matters. The selections below correctly apply the guideline for the listed PW values.

PW₁	PW₂	Selected Alternative
$-1500	$-500	2
-500	+1000	2
+2500	-500	1
+2500	+1500	1

If the projects are *independent,* the selection guideline is as follows:

For one or more independent projects, select all projects with PW ≥ 0 at the MARR.

This compares each project with the do-nothing alternative. The projects must have positive and negative cash flows to obtain a PW value that exceeds zero; that is, they must be revenue projects.

A PW analysis requires a MARR for use as the *i* value in all PW relations. The bases used to establish a realistic MARR were summarized in Chapter 1 and are discussed in detail in Chapter 10.

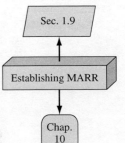

Sec. 1.9

Establishing MARR

Chap. 10

EXAMPLE 5.1

Perform a present worth analysis of equal-service machines with the costs shown below, if the MARR is 10% per year. Revenues for all three alternatives are expected to be the same.

	Electric-Powered	Gas-Powered	Solar-Powered
First cost, $	-2500	-3500	-6000
Annual operating cost (AOC), $	-900	-700	-50
Salvage value S, $	200	350	100
Life, years	5	5	5

Solution

These are service alternatives. The salvage values are considered a "negative" cost, so a + sign precedes them. (If it costs money to dispose of an asset, the estimated disposal cost has a − sign.) The PW of each machine is calculated at $i = 10\%$ for $n = 5$ years. Use subscripts E, G, and S.

$$PW_E = -2500 - 900(P/A,10\%,5) + 200(P/F,10\%,5) = \$-5788$$

$$PW_G = -3500 - 700(P/A,10\%,5) + 350(P/F,10\%,5) = \$-5936$$

$$PW_S = -6000 - 50(P/A,10\%,5) + 100(P/F,10\%,5) = \$-6127$$

The electric-powered machine is selected since the PW of its costs is the lowest; it has the numerically largest PW value.

5.3   PRESENT WORTH ANALYSIS OF DIFFERENT-LIFE ALTERNATIVES

When the present worth method is used to compare mutually exclusive alternatives that have different lives, the procedure of the previous section is followed with one exception:

The PW of the alternatives must be compared over the same number of years and end at the same time.

This is necessary, since a present worth comparison involves calculating the equivalent present value of all future cash flows for each alternative. A fair comparison can be made only when the PW values represent costs (and receipts) associated with equal service. Failure to compare equal service will always favor a shorter-lived alternative (for costs), even if it is not the most economical one, because fewer periods of costs are involved. The equal-service requirement can be satisfied by either of two approaches:

- Compare the alternatives over a period of time equal to the *least common multiple (LCM)* of their lives.
- Compare the alternatives using a *study period of length n years,* which does not necessarily take into consideration the useful lives of the alternatives. This is also called the *planning horizon* approach.

In either case, the PW of each alternative is calculated at the MARR, and the selection guideline is the same as that for equal-life alternatives. The LCM approach automatically makes the cash flows for all alternatives extend to the same time period. For example, alternatives with expected lives of 2 and 3 years are compared over a 6-year time period. Such a procedure requires that some assumptions be made about subsequent life cycles of the alternatives.

The assumptions of a PW analysis of different-life alternatives for the LCM method are as follows:

1. **The service provided by the alternatives will be needed for the LCM of years or more.**
2. **The selected alternative will be repeated over each life cycle of the LCM in exactly the same manner.**
3. **The cash flow estimates will be the same in every life cycle.**

As will be shown in Chapter 14, the third assumption is valid only when the cash flows are expected to change by exactly the inflation (or deflation) rate that is applicable through the LCM time period. If the cash flows are expected to change by any other rate, then the PW analysis must be conducted using constant-value dollars, which considers inflation (Chapter 14). A study period analysis is necessary if the first assumption about the length of time the alternatives are needed cannot be made. A present worth analysis over the LCM requires that the estimated salvage values be included in each life cycle.

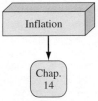

For the study period approach, a time horizon is chosen over which the economic analysis is conducted, and only those cash flows which occur during that time period are considered relevant to the analysis. All cash flows occurring beyond the study period are ignored. An estimated market value at the end of the study period must be made. The time horizon chosen might be relatively short, especially when short-term business goals are very important. The study period approach is often used in replacement analysis. It is also useful when the LCM of alternatives yields an unrealistic evaluation period, for example, 5 and 9 years.

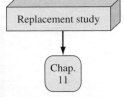

Example 5.2 includes evaluations based on the LCM and study period approaches. Also, Example 5.12 in Section 5.9 illustrates the use of spreadsheets in PW analysis for both different lives and a study period.

EXAMPLE **5.2**

A project engineer with EnvironCare is assigned to start up a new office in a city where a 6-year contract has been finalized to take and to analyze ozone-level readings. Two lease options are available, each with a first cost, annual lease cost, and deposit-return estimates shown below.

	Location A	Location B
First cost, $	−15,000	−18,000
Annual lease cost, $ per year	−3,500	−3,100
Deposit return, $	1,000	2,000
Lease term, years	6	9

(a) Determine which lease option should be selected on the basis of a present worth comparison, if the MARR is 15% per year.

(b) EnvironCare has a standard practice of evaluating all projects over a 5-year period. If a study period of 5 years is used and the deposit returns are not expected to change, which location should be selected?

(c) Which location should be selected over a 6-year study period if the deposit return at location B is estimated to be $6000 after 6 years?

Solution

(a) Since the leases have different terms (lives), compare them over the LCM of 18 years. For life cycles after the first, the first cost is repeated in year 0 of each new cycle, which is the last year of the previous cycle. These are years 6 and 12 for location A and year 9 for B. The cash flow diagram is in Figure 5–2. Calculate PW at 15% over 18 years.

$$PW_A = -15,000 - 15,000(P/F,15\%,6) + 1000(P/F,15\%,6)$$
$$- 15,000(P/F,15\%,12) + 1000(P/F,15\%,12) + 1000(P/F,15\%,18)$$
$$- 3500(P/A,15\%,18)$$

$$= \$-45,036$$

$$PW_B = -18,000 - 18,000(P/F,15\%,9) + 2000(P/F,15\%,9)$$
$$+ 2000(P/F,15\%,18) - 3100(P/A,15\%,18)$$

$$= \$-41,384$$

Location B is selected, since it costs less in PW terms; that is, the PW_B value is numerically larger than PW_A.

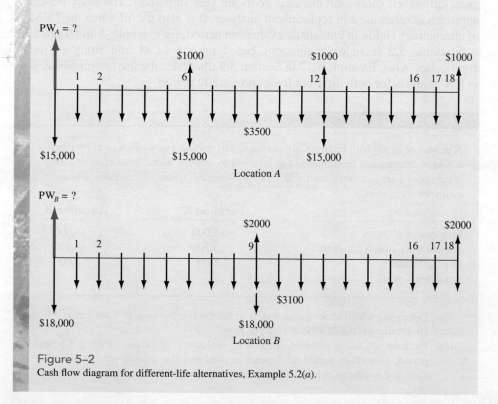

Figure 5–2
Cash flow diagram for different-life alternatives, Example 5.2(a).

(b) For a 5-year study period no cycle repeats are necessary. The PW analysis is

$$PW_A = -15,000 - 3500(P/A,15\%,5) + 1000(P/F,15\%,5)$$

$$= \$-26,236$$

$$PW_B = -18,000 - 3100(P/A,15\%,5) + 2000(P/F,15\%,5)$$

$$= \$-27,397$$

Location A is now the better choice.

(c) For a 6-year study period, the deposit return for B is $6000 in year 6.

$$PW_A = -15,000 - 3500(P/A,15\%,6) + 1000(P/F,15\%,6) = \$-27,813$$

$$PW_B = -18,000 - 3100(P/A,15\%,6) + 6000(P/F,15\%,6) = \$-27,138$$

Location B now has a small economic advantage. Noneconomic factors are likely to enter into the final decision.

Comments

In part (a) and Figure 5–2, the deposit return for each lease is recovered *after each life cycle,* that is, in years 6, 12, and 18 for A and in years 9 and 18 for B. In part (c), the increase of the deposit return from $2000 to $6000 (one year later), switches the selected location from A to B. The project engineer should reexamine these estimates before making a final decision.

5.4 FUTURE WORTH ANALYSIS

The future worth (FW) of an alternative may be determined directly from the cash flows by determining the future worth value, or by multiplying the PW value by the F/P factor, at the established MARR. Therefore, it is an extension of present worth analysis. The n value in the F/P factor depends upon which time period has been used to determine PW—the LCM value or a specified study period. Analysis of one alternative, or the comparison of two or more alternatives, using FW values is especially applicable to large capital investment decisions when a prime goal is to maximize the *future wealth* of a corporation's stockholders.

Future worth analysis is often utilized if the asset (equipment, a corporation, a building, etc.) might be sold or traded at some time after its start-up or acquisition, but before the expected life is reached. An FW value at an intermediate year estimates the alternative's worth at the time of sale or disposal. Suppose an entrepreneur is planning to buy a company and expects to trade it within 3 years. FW analysis is the best method to help with the decision to sell or keep it 3 years hence. Example 5.3 illustrates this use of FW analysis. Another excellent application of FW analysis is for projects that will not come online until the end of the investment period. Alternatives such as electric generation facilities, toll roads, hotels, and the like can be analyzed using the FW value of investment commitments made during construction.

Once the FW value is determined, the selection guidelines are the same as with PW analysis; FW ≥ 0 means the MARR is met or exceeded (one alternative). For two (or more) mutually exclusive alternatives, select the one with the numerically larger (largest) FW value.

EXAMPLE 5.3

A British food distribution conglomerate purchased a Canadian food store chain for $75 million (U.S.) three years ago. There was a net loss of $10 million at the end of year 1 of ownership. Net cash flow is increasing with an arithmetic gradient of $+5 million per year starting the second year, and this pattern is expected to continue for the foreseeable future. This means that breakeven net cash flow was achieved this year. Because of the heavy debt financing used to purchase the Canadian chain, the international board of directors expects a MARR of 25% per year from any sale.

(a) The British conglomerate has just been offered $159.5 million (U.S.) by a French company wishing to get a foothold in Canada. Use FW analysis to determine if the MARR will be realized at this selling price.

(b) If the British conglomerate continues to own the chain, what selling price must be obtained at the end of 5 years of ownership to make the MARR?

Solution

(a) Set up the future worth relation in year 3 (FW_3) at $i = 25\%$ per year and an offer price of $159.5 million. Figure 5–3a presents the cash flow diagram in million $ units.

$$FW_3 = -75(F/P,25\%,3) - 10(F/P,25\%,2) - 5(F/P,25\%,1) + 159.5$$

$$= -168.36 + 159.5 = \$-8.86 \text{ million}$$

No, the MARR of 25% will not be realized if the $159.5 million offer is accepted.

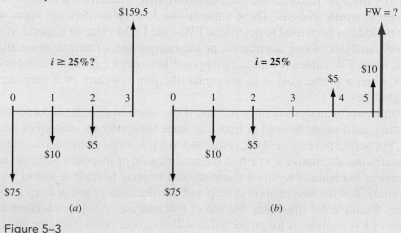

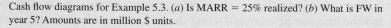

Figure 5–3
Cash flow diagrams for Example 5.3. (a) Is MARR = 25% realized? (b) What is FW in year 5? Amounts are in million $ units.

(b)   Determine the future worth 5 years from now at 25% per year. Figure 5–3b presents the cash flow diagram. The A/G and F/A factors are applied to the arithmetic gradient.

$$FW_5 = -75(F/P,25\%,5) - 10(F/A,25\%,5) + 5(A/G,25\%,5)(F/A,25\%,5)$$

$$= \$-246.81 \text{ million}$$

The offer must be for at least \$246.81 million to make the MARR. This is approximately 3.3 times the purchase price only 5 years earlier, in large part based on the required MARR of 25%.

Comment

If the 'rule of 72' in Equation [1.9] is applied at 25% per year, the sales price must double approximately every $72/25\% = 2.9$ years. This does not consider any annual net positive or negative cash flows during the years of ownership.

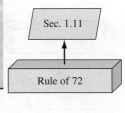

5.5   CAPITALIZED COST CALCULATION AND ANALYSIS

Capitalized cost (CC) is the present worth of an alternative that will last "forever." Public sector projects such as bridges, dams, irrigation systems, and railroads fall into this category. In addition, permanent and charitable organization endowments are evaluated using the capitalized cost methods.

The formula to calculate CC is derived from the relation $P = A(P/A,i,n)$, where $n = \infty$. The equation for P using the P/A factor formula is

$$P = A\left[\frac{(1 + i)^n - 1}{i(1 + i)^n}\right]$$

Divide the numerator and denominator by $(1 + i)^n$.

$$P = A\left[\frac{1 - \dfrac{1}{(1 + i)^n}}{i}\right]$$

As n approaches ∞, the bracketed term becomes $1/i$, and the symbol CC replaces PW and P.

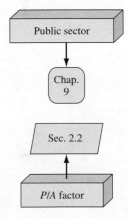

$$CC = \frac{A}{i} \qquad\qquad [5.1]$$

If the A value is an annual worth (AW) determined through equivalence calculations of cash flows over n years, the CC value is

$$CC = \frac{AW}{i} \qquad\qquad [5.2]$$

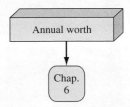

The validity of Equation [5.1] can be illustrated by considering the time value of money. If \$10,000 earns 20% per year, compounded annually, the maximum amount of money that can be withdrawn at the end of every year for *eternity* is $2000, or the interest accumulated each year. This leaves the original $10,000 to earn interest so that another $2000 will be accumulated the next year. Mathematically, the amount A of new money generated each consecutive interest period for an infinite number of periods is

$$A = Pi = CC(i) \tag{5.3}$$

The capitalized cost calculation in Equation [5.1] is Equation [5.3] solved for P and renamed CC.

For a public sector alternative with an infinite or very long life, the A value determined by Equation [5.3] is used when the benefit/cost (B/C) ratio is the comparison basis for public projects. This method is covered in Chapter 9.

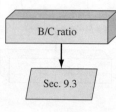

The cash flows (costs or receipts) in a capitalized cost calculation are usually of two types: *recurring,* also called periodic, and *nonrecurring.* An annual operating cost of $50,000 and a rework cost estimated at $40,000 every 12 years are examples of recurring cash flows. Examples of nonrecurring cash flows are the initial investment amount in year 0 and one-time cash flow estimates at future times, for example, $500,000 in royalty fees 2 years hence. The following procedure assists in calculating the CC for an infinite sequence of cash flows.

1. Draw a cash flow diagram showing all nonrecurring (one-time) cash flows and at least two cycles of all recurring (periodic) cash flows.
2. Find the present worth of all nonrecurring amounts. This is their CC value.
3. Find the equivalent uniform annual worth (A value) through *one life cycle* of all recurring amounts. This is the same value in all succeeding life cycles, as explained in Chapter 6. Add this to all other uniform amounts occurring in years 1 through infinity and the result is the total equivalent uniform annual worth (AW).
4. Divide the AW obtained in step 3 by the interest rate i to obtain a CC value. This is an application of Equation [5.2].
5. Add the CC values obtained in steps 2 and 4.

Drawing the cash flow diagram (step 1) is more important in CC calculations than elsewhere, because it helps separate nonrecurring and recurring amounts. In step 5 the present worths of all component cash flows have been obtained; the total capitalized cost is simply their sum.

EXAMPLE **5.4**

The property appraisal district for Marin County has just installed new software to track residential market values for property tax computations. The manager wants to know the total equivalent cost of all future costs incurred when the three county judges agreed to purchase the software system. If the new system will be used for the indefinite future, find the equivalent value (*a*) now and (*b*) for each year hereafter.

The system has an installed cost of $150,000 and an additional cost of $50,000 after 10 years. The annual software maintenance contract cost is $5000 for the first 4 years and

$8000 thereafter. In addition, there is expected to be a recurring major upgrade cost of $15,000 every 13 years. Assume that $i = 5\%$ per year for county funds.

Solution

(a) The five-step procedure is applied.

1. Draw a cash flow diagram for two cycles (Figure 5–4).
2. Find the present worth of the nonrecurring costs of $150,000 now and $50,000 in year 10 at $i = 5\%$. Label this CC_1.

$$CC_1 = -150,000 - 50,000(P/F,5\%,10) = \$-180,695$$

3. Convert the recurring cost of $15,000 every 13 years into an annual worth A_1 for the first 13 years.

$$A_1 = -15,000(A/F,5\%,13) = \$-847$$

The same value, $A_1 = \$-847$, applies to all the other 13-year periods as well.

4. The capitalized cost for the two annual maintenance cost series may be determined in either of two ways: (1) consider a series of $-5000 from now to infinity and find the present worth of $-\$8000 - (\$-5000) = \$-3000$ from year 5 on; or (2) find the CC of $-5000 for 4 years and the present worth of $-8000 from year 5 to infinity. Using the first method, the annual cost (A_2) is $-5000 forever. The capitalized cost CC_2 of $-3000 from year 5 to infinity is found using Equation [5.1] times the P/F factor.

$$CC_2 = \frac{-3000}{0.05}(P/F,5\%,4) = \$-49,362$$

The two annual cost series are converted into a capitalized cost CC_3.

$$CC_3 = \frac{A_1 + A_2}{i} = \frac{-847 + (-5000)}{0.05} = \$-116,940$$

5. The total capitalized cost CC_T is obtained by adding the three CC values.

$$CC_T = -180,695 - 49,362 - 116,940 = \$-346,997$$

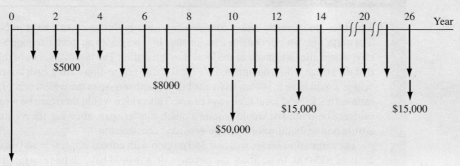

Figure 5–4
Cash flows for two cycles of recurring costs and all nonrecurring amounts, Example 5.4.

(b)   Equation [5.3] determines the A value forever.

$$A = Pi = CC_T(i) = \$346,997(0.05) = \$17,350$$

Correctly interpreted, this means Marin County officials have committed the equivalent of $17,350 forever to operate and maintain the property appraisal software.

Comment

The CC_2 value is calculated using $n = 4$ in the P/F factor because the present worth of the annual $3000 cost is located in year 4, since P is always one period ahead of the first A. Rework the problem using the second method suggested for calculating CC_2.

For the comparison of two or more alternatives on the basis of capitalized cost, use the procedure above to find CC_T for each alternative. Since the capitalized cost represents the total present worth of financing and maintaining a given alternative forever, the alternatives will automatically be compared for the same number of years (i.e., infinity). The alternative with the smaller capitalized cost will represent the more economical one. This evaluation is illustrated in Example 5.5.

As in present worth analysis, it is only the differences in cash flow between the alternatives that must be considered for comparative purposes. Therefore, whenever possible, the calculations should be simplified by eliminating the elements of cash flow which are common to both alternatives. On the other hand, if true capitalized cost values are needed to reflect actual financial obligations, actual cash flows should be used.

EXAMPLE 5.5

Two sites are currently under consideration for a bridge to cross a river in New York. The north site, which connects a major state highway with an interstate loop around the city, would alleviate much of the local through traffic. The disadvantages of this site are that the bridge would do little to ease local traffic congestion during rush hours, and the bridge would have to stretch from one hill to another to span the widest part of the river, railroad tracks, and local highways below. This bridge would therefore be a suspension bridge. The south site would require a much shorter span, allowing for construction of a truss bridge, but it would require new road construction.

The suspension bridge will cost $50 million with annual inspection and maintenance costs of $35,000. In addition, the concrete deck would have to be resurfaced every 10 years at a cost of $100,000. The truss bridge and approach roads are expected to cost $25 million and have annual maintenance costs of $20,000. The bridge would have to be painted every 3 years at a cost of $40,000. In addition, the bridge would have to be sandblasted every 10 years at a cost of $190,000. The cost of purchasing right-of-way is expected to be $2 million for the suspension bridge and $15 million for the truss bridge. Compare the alternatives on the basis of their capitalized cost if the interest rate is 6% per year.

Solution

Construct the cash flow diagrams over two cycles (20 years).

Capitalized cost of suspension bridge (CC_S):

CC_1 = capitalized cost of initial cost

$= -50.0 - 2.0 = \$-52.0$ million

The recurring operating cost is $A_1 = \$-35,000$, and the annual equivalent of the resurface cost is

$$A_2 = -100,000(A/F,6\%,10) = \$-7587$$

$$CC_2 = \text{capitalized cost of recurring costs} = \frac{A_1 + A_2}{i}$$

$$= \frac{-35,000 + (-7587)}{0.06} = \$-709,783$$

The total capitalized cost is

$$CC_S = CC_1 + CC_2 = \$-52.71 \text{ million}$$

Capitalized cost of truss bridge (CC_T):

$CC_1 = -25.0 + (-15.0) = \$-40.0$ million

$A_1 = \$-20,000$

$A_2 = \text{annual cost of painting} = -40,000(A/F,6\%,3) = \$-12,564$

$A_3 = \text{annual cost of sandblasting} = -190,000(A/F,6\%,10) = \$-14,415$

$$CC_2 = \frac{A_1 + A_2 + A_3}{i} = \frac{\$-46,979}{0.06} = \$-782,983$$

$CC_T = CC_1 + CC_2 = \$-40.78$ million

Conclusion: Build the truss bridge, since its capitalized cost is lower.

If a finite-life alternative (e.g., 5 years) is compared to one with an indefinite or very long life, capitalized costs can be used for the evaluation. To determine capitalized cost for the alternative with a finite life, calculate the equivalent A value for one life cycle and divide by the interest rate (Equation [5.1]). This procedure is illustrated in the next example.

EXAMPLE **5.6**

APSco, a large electronics subcontractor for the Air Force, needs to immediately acquire 10 soldering machines with specially prepared jigs for assembling components onto printed circuit boards. More machines may be needed in the future. The lead production engineer has outlined below two simplified, but viable, alternatives. The company's MARR is 15% per year.

> *Alternative LT (long-term).* For $8 million now, a contractor will provide the necessary number of machines (up to a maximum of 20), now and in the future, for as long as APSco needs them. The annual contract fee is a total of $25,000 with no additional per-machine annual cost. There is no time limit placed on the contract, and the costs do not escalate.

> *Alternative ST (short-term).* APSco buys its own machines for $275,000 each and expends an estimated $12,000 per machine in annual operating cost (AOC). The useful life of a soldering system is 5 years.

Perform a capitalized cost evaluation by hand and by computer. Once the evaluation is complete, use the spreadsheet for sensitivity analysis to determine the maximum number of soldering machines that can be purchased now and still have a capitalized cost less than that of the long-term alternative.

Solution by Hand

For the LT alternative, find the CC of the AOC using Equation [5.1], $CC = A/i$. Add this amount to the initial contract fee, which is already a capitalized cost (present worth) amount.

$$CC_{LT} = CC \text{ of contract fee} + CC \text{ of AOC}$$

$$= -8 \text{ million} - 25,000/0.15 = \$-8,166,667$$

For the ST alternative, first calculate the equivalent annual amount for the purchase cost over the 5-year life, and add the AOC values for all 10 machines. Then determine the total CC using Equation [5.2].

$$AW_{ST} = AW \text{ for purchase} + AOC$$

$$= -2.75 \text{ million}(A/P,15\%,5) - 120,000 = \$-940,380$$

$$CC_{ST} = -940,380/0.15 = \$-6,269,200$$

The ST alternative has a lower capitalized cost by approximately $1.9 million present value dollars.

Solution by Computer

Figure 5–5 contains the solution for 10 machines in column B. Cell B8 uses the same relation as in the solution by hand. Cell B15 uses the PMT function to determine the equivalent annual amount A for the purchase of 10 machines, to which the AOC is added. Cell B16 uses Equation [5.2] to find the total CC for the ST alternative. As expected, alternative ST is selected. (Compare CC_{ST} for the hand and computer solutions to note that the round-off error using the tabulated interest factors gets larger for large P values.)

The type of sensitivity analysis requested here is easy to perform once a spreadsheet is developed. The PMT function in B15 is expressed generally in terms of cell B12, the

X	Microsoft Excel - Example 5.6							_ □ ×

File   Edit   View   Insert   Format   Tools   Data   Window   Help   _ ⊟ ×

C12   = 13

	A	B	C	D	E	F	G
1							
2	Interest rate	15%	15%	15%			
3							
4	**Alternative LT (long term)**						
5	Initial contract cost	$ (8,000,000)	$ (8,000,000)	$ (8,000,000)			
6	Annual cost	$ (25,000)	$ (25,000)	$ (25,000)			
7							
8	**Capitalized cost for LT**	$ (8,166,667)	$ (8,166,667)	$ (8,166,667)		=D5+D6/D2	
9							
10	**Alternative ST (short term)**						
11	Initial cost per machine	$ (275,000)	$ (275,000)	$ (275,000)			
12	Number of machines	10	13	14			
13	Expected life, years	5	5	5			
14	AOC per machine	$ (12,000)	$ (12,000)	$ (12,000)			
15	Equivalent annual value (AW)	$ (940,368)	$ (1,222,478)	$ (1,316,515)			
16	**Capitalized cost for ST**	$ (6,269,118)	$ (8,149,854)	$ (8,776,766)			
17							
18			=+B15/B2				
19							
20	=PMT(B2,B13,−B11*B12)+B14*B12						
21							

Sheet 1 / Sheet2 / Sheet3 / Sheet4 / Sheet5 / Sheet6 / Sheet7 / Sh

Ready   NUM

Figure 5–5
Spreadsheet solution for capitalized cost comparison, Example 5.6.

number of machines purchased. Columns C and D replicate the evaluation for 13 and 14 machines. Thirteen is the maximum number of machines that can be purchased and have a CC for the ST alternative that is less than that of the LT contract. This conclusion is easily reached by comparing total CC values in rows 8 and 16. (*Note:* It is not necessary to duplicate column B into C and D to perform this sensitivity analysis. Changing the entry in cell B12 upward from 10 will provide the same information. Duplication is shown here in order to view all the results on one spreadsheet.)

5.6   PAYBACK PERIOD ANALYSIS

Payback analysis (also called payout analysis) is another extension of the present worth method. Payback can take two forms: one for $i > 0\%$ (also called *discounted payback analysis*) and another for $i = 0\%$. There is a logical linkage between payback and breakeven analysis, which is used in several chapters and discussed in detail in Chapter 13.

The payback period n_p is the estimated time, usually in years, it will take for the estimated revenues and other economic benefits *to recover the initial investment*

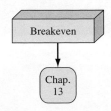

and a stated rate of return. The n_p value is generally not an integer. It is important to remember the following:

> **The payback period n_p should never be used as the primary measure of worth to select an alternative. Rather, it should be determined in order to provide initial screening or supplemental information in conjunction with an analysis performed using present worth or another method.**

The payback period should be calculated using a required return that is greater than 0%. However, in practice the payback period is often determined with a no-return requirement ($i = 0\%$) to initially screen a project and determine whether it warrants further consideration.

To find the discounted payback period at a stated rate $i > 0\%$, calculate the years n_p that make the following expression correct.

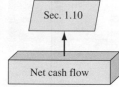

$$0 = -P + \sum_{t=1}^{t=n_p} \text{NCF}_t(P/F,i,t) \qquad [5.4]$$

The amount P is the initial investment or first cost, and NCF is the estimated net cash flow for each year t as determined by Equation [1.8], NCF = receipts − disbursements. If the NCF values are expected to be equal each year, the P/A factor may be used, in which case the relation is

$$0 = -P + \text{NCF}(P/A,i,n_p) \qquad [5.5]$$

After n_p years, the cash flows will recover the investment and a return of $i\%$. If, in reality, the asset or alternative is used for more than n_p years a larger return may result; but if the useful life is less than n_p years, there is not enough time to recover the initial investment and the $i\%$ return. It is very important to realize that in payback analysis *all net cash flows occurring after n_p years are neglected.* Since this is significantly different from the approach of PW (or annual worth, or rate of return, as discussed later), where all cash flows for the entire useful life are included in the economic analysis, payback analysis can unfairly bias alternative selection. So use payback analysis only as a screening or supplemental technique.

When $i > 0\%$ is used, the n_p value does provide a sense of the risk involved if the alternative is undertaken. For example, if a company plans to produce a product under contract for only 3 years and the payback period for the equipment is estimated to be 6 years, the company should not undertake the contract. Even in this situation, the 3-year payback period is only supplemental information, not a good substitute for a complete economic analysis.

No-return payback (or simple payback) analysis determines n_p at $i = 0\%$. This n_p value serves merely as an initial indicator that a proposal is a viable alternative worthy of a full economic evaluation. Use $i = 0\%$ in Equation [5.4] and find n_p.

$$0 = -P + \sum_{t=1}^{t=n_p} \text{NCF}_t \qquad [5.6]$$

For a uniform net cash flow series, Equation [5.6] is solved for n_p directly.

$$n_p = \frac{P}{\text{NCF}}$$ [5.7]

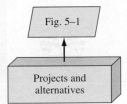

Fig. 5–1

Projects and alternatives

An example use of n_p as an initial screening of proposed projects is a corporation president who absolutely insists that every project must return the investment in 3 years or less. Therefore, no proposed project with $n_p > 3$ should become an alternative.

It is incorrect to use the no-return payback period to make final alternative selections because it:

1. **Neglects any required return, since the time value of money is omitted.**
2. **Neglects all net cash flows after time n_p, including positive cash flows that may contribute to the return on the investment.**

As a result, the selected alternative may be different from that selected by an economic analysis based on PW (or AW) computations. This fact is demonstrated later in Example 5.8.

EXAMPLE 5.7

The board of directors of Halliburton International has just approved an $18 million worldwide engineering construction design contract. The services are expected to generate new annual net cash flows of $3 million. The contract has a potentially lucrative repayment clause to Halliburton of $3 million at any time that the contract is canceled by either party during the 10 years of the contract period. (*a*) If $i = 15\%$, compute the payback period. (*b*) Determine the no-return payback period and compare it with the answer for $i = 15\%$. This is an initial check to determine if the board made a good economic decision.

Solution

(*a*) The net cash flow each year is $3 million. The single $3 million payment (call it CV for cancellation value) could be received at any time within the 10-year contract period. Equation [5.5] is altered to include CV.

$$0 = -P + \text{NCF}(P/A,i,n) + \text{CV}(P/F,i,n)$$

In $1,000,000 units,

$$0 = -18 + 3(P/A,15\%,n) + 3(P/F,15\%,n)$$

The 15% payback period is $n_p = 15.3$ years. During the period of 10 years, the contract will not deliver the required return.

(*b*) If Halliburton requires absolutely no return on its $18 million investment, Equation [5.6] results in $n_p = 5$ years, as follows (in million $):

$$0 = -18 + 5(3) + 3$$

There is a very significant difference in n_p for 15% and 0%. At 15% this contract would have to be in force for 15.3 years, while the no-return payback period requires only 5 years. A longer time is always required for $i > 0\%$ for the obvious reason that the time value of money is considered.

Use NPER(15%,3,−18,3) to display 15.3 years. Change the rate from 15% to 0% to display the no-return payback period of 5 years.

Comment

The payback calculation provides the number of years required to recover the invested dollars. But from the points of view of engineering economic analysis and the time value of money, no-return payback analysis is not a reliable method for alternative selection.

If two or more alternatives are evaluated using payback periods to indicate that one may be better than the other(s), the second shortcoming of payback analysis (neglect of cash flows after n_p) may lead to an economically incorrect decision. When cash flows that occur after n_p are neglected, it is possible to favor short-lived assets even when longer-lived assets produce a higher return. In these cases, PW (or AW) analysis should always be the primary selection method. Comparison of short- and long-lived assets in Example 5.8 illustrates this incorrect use of payback analysis.

EXAMPLE 5.8

Two equivalent pieces of quality inspection equipment are being considered for purchase by Square D Electric. Machine 2 is expected to be versatile and technologically advanced enough to provide net income longer than machine 1.

	Machine 1	Machine 2
First cost, $	12,000	8,000
Annual NCF, $	3,000	1,000 (years 1–5), 3,000 (years 6–14)
Maximum life, years	7	14

The quality manager used a return of 15% per year and a PC-based economic analysis package. The software utilized Equations [5.4] and [5.5] to recommend machine 1 because it has a shorter payback period of 6.57 years at $i = 15\%$. The computations are summarized here.

Machine 1: $n_p = 6.57$ years, which is less than the 7-year life.

Equation used: $$0 = -12,000 + 3000(P/A,15\%,n_p)$$

Machine 2: $n_p = 9.52$ years, which is less than the 14-year life.

Equation used: $$0 = -8000 + 1000(P/A,15\%,5)$$
$$+ 3000(P/A,15\%,n_p-5)(P/F,15\%,5)$$

Recommendation: Select machine 1.

Now, use a 15% PW analysis to compare the machines and comment on any difference in the recommendation.

Solution

For each machine, consider the net cash flows for all years during the estimated (maximum) life. Compare them over the LCM of 14 years.

$$PW_1 = -12{,}000 - 12{,}000(P/F,15\%,7) + 3000(P/A,15\%,14) = \$663$$

$$PW_2 = -8000 + 1000(P/A,15\%,5) + 3000(P/A,15\%,9)(P/F,15\%,5)$$
$$= \$2470$$

Machine 2 is selected since its PW value is numerically larger than that of machine 1 at 15%. This result is the opposite of the payback period decision. The PW analysis accounts for the increased cash flows for machine 2 in the later years. As illustrated in Figure 5–6 (for one life cycle for each machine), payback analysis neglects all cash flow amounts that may occur after the payback time has been reached.

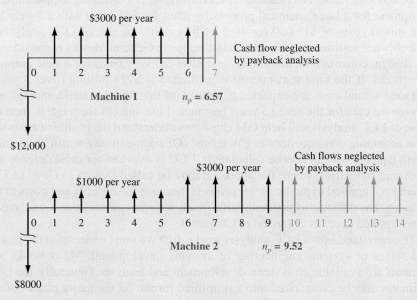

Figure 5–6
Illustration of payback periods and neglected net cash flows, Example 5.8.

Comment

This is a good example of why payback analysis is best used for initial screening and supplemental risk assessment. Often a shorter-lived alternative evaluated by payback analysis may appear to be more attractive, when the longer-lived alternative has cash flows estimated later in its life that make it more economically attractive.

5.7 LIFE-CYCLE COST

Life-cycle cost (LCC) is another extension of present worth analysis. The PW value at a stated MARR is utilized to evaluate one or more alternatives. The LCC method, as its name implies, is commonly applied to alternatives with cost estimates over the entire *system life span.* This means that costs from the very early stage of the project (needs assessment) through the final stage (phaseout and disposal) are estimated. Typical applications for LCC are buildings (new construction or purchases), new product lines, manufacturing plants, commercial aircraft, new automobile models, defense systems, and the like.

A PW analysis with all definable costs (and possibly incomes) estimated may be considered a LCC analysis. However, the broad definition of the LCC term *system life span* requires cost estimates not usually made for a regular PW analysis. Also, for large long-life projects, the longer-term estimates are less accurate. This implies that life-cycle cost analysis is not necessary in most alternative analysis. *LCC is most effectively applied when a substantial percentage of the total costs over the system life span, relative to the initial investment, will be operating and maintenance costs* (postpurchase costs such as labor, energy, upkeep, and materials). For example, if Exxon-Mobil is evaluating the purchase of equipment for a large chemical processing plant for $150,000 with a 5-year life and annual costs of $15,000 (or 10% of first cost), the use of LCC analysis is probably not justified. On the other hand, suppose General Motors is considering the design, construction, marketing, and after-delivery costs for a new automobile model. If the total start-up cost is estimated at $125 million (over 3 years) and total annual costs are expected to be 20% of this figure to build, market, and service the cars for the next 15 years (estimated life span of the model), then the logic of LCC analysis will help GM engineers understand the profile of costs and their economic consequences in PW terms. (Of course, future worth and annual worth equivalents can also be calculated). LCC is required for most defense and aerospace industries, where the approach may be called Design to Cost. LCC is usually not applied to public sector projects, because the benefits and costs to the citizenry are difficult to estimate with much accuracy. Benefit/cost analysis is better applied here, as discussed in Chapter 9.

To understand how a LCC analysis works, first we must understand the phases and stages of systems engineering or systems development. Many books and manuals are available on systems development and analysis. Generally, the LCC estimates may be categorized into a simplified format for the major phases of *acquisition* and *operation,* and their respective stages.

Acquisition phase: all activities prior to the delivery of products and services.
- Requirements definition stage—Includes determination of user/customer needs, assessing them relative to the anticipated system, and preparation of the system requirements documentation.
- Preliminary design stage—Includes feasibility study, conceptual, and early-stage plans; final go–no go decision is probably made here.
- Detailed design stage—Includes detailed plans for resources—capital, human, facilities, information systems, marketing, etc.; there is some acquisition of assets, if economically justifiable.

Operations phase: all activities are functioning, products and services are available.

- Construction and implementation stage—Includes purchases, construction, and implementation of system components; testing; preparation, etc.
- Usage stage—Uses the system to generate products and services.
- Phaseout and disposal stage—Covers time of clear transition to new system; removal/recycling of old system.

EXAMPLE **5.9**

In the 1860s General Mills Inc. and Pillsbury Inc. both started in the flour business in the Twin Cities of Minneapolis–St. Paul, Minnesota. In the 2000–2001 time frame, General Mills purchased Pillsbury for a combination cash and stock deal worth more than $10 billion. The General Mills promise was to develop Pillsbury's robust food line to meet consumer needs, especially in the "one hand free" prepared-food markets in order to appeal to the rapidly changing eating habits and nutrition needs of people at work and play who have no time for or interest in preparing meals. Food engineers, food designers, and food safety experts made many cost estimates as they determined the needs of consumers and the combined company's ability to technologically and safely produce and market new food products. At this point only cost estimates have been addressed—no revenues or profits.

Assume that the major cost estimates below have been made based on a 6-month study about two new products that could have a 10-year life span for the company. Some cost elements were not estimated (e.g., raw food stuffs, product distribution, and phaseout). Use LCC analysis at the industry MARR of 18% to determine the size of the commitment in PW dollars. (Time is indicated in product-years. Since all estimates are for costs, they are not preceded by a minus sign.)

Consumer habits study (year 0)	$0.5 million
Preliminary food product design (year 1)	0.9 million
Preliminary equipment/plant design (year 1)	0.5 million
Detail product designs and test marketing (years 1, 2)	1.5 million each year
Detail equipment/plant design (year 2)	1.0 million
Equipment acquisition (years 1 and 2)	$2.0 million each year
Current equipment upgrades (year 2)	1.75 million
New equipment purchases (years 4 and 8)	2.0 million (year 4) + 10% per purchase thereafter
Annual equipment operating cost (AOC) (years 3–10)	200,000 (year 3) + 4% per year thereafter
Marketing, year 2	$8.0 million
years 3–10	5.0 million (year 3) and −0.2 million per year thereafter
year 5 only	3.0 million extra
Human resources, 100 new employees for 2000 hours per year (years 3–10)	$20 per hour (year 3) + 5% per year

Solution

LCC analysis can get complicated rapidly due to the number of elements involved. Calculate the PW by phase and stage, then add all PW values. Values are in million $ units.

Acquisition phase:

Requirements definition: consumer study

$$PW = \$0.5$$

Preliminary design: product and equipment

$$PW = 1.4(P/F,18\%,1) = \$1.187$$

Detailed design: product and test marketing, and equipment

$$PW = 1.5(P/A,18\%,2) + 1.0(P/F,18\%,2) = \$3.067$$

Operations phase:

Construction and implementation: equipment and AOC

$$PW = 2.0(P/A,18\%,2) + 1.75(P/F,18\%,2) + 2.0(P/F,18\%,4) + 2.2(P/F,18\%,8)$$

$$+ \, 0.2 \left[\frac{1 - \left(\dfrac{1.04}{1.18} \right)^8}{0.14} \right] (P/F,18\%,2) = \$6.512$$

Use: marketing

$$PW = 8.0(P/F,18\%,2) + [5.0(P/A,18\%,8) - 0.2(P/G,18\%,8)](P/F,18\%,2)$$

$$+ \, 3.0(P/F,18\%,5)$$

$$= \$20.144$$

Use: human resources: (100 employees)(2000 h/yr)($20/h) = $4.0 million in year 3

$$PW = 4.0 \left[\frac{1 - \left(\dfrac{1.05}{1.18} \right)^8}{0.13} \right] (P/F,18\%,2) = \$13.412$$

The total LCC commitment at this time is the sum of all PW values.

$$PW = \$44.822 \text{ (effectively \$45 million)}$$

As a point of interest, over 10 years at 18% per year, the future worth of the General Mills commitment, thus far, is FW = PW$(F/P,18\%,10)$ = $234.6 million.

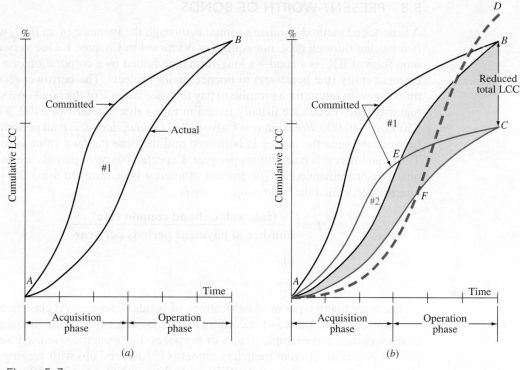

Figure 5–7
LCC envelopes for committed and actual costs: (*a*) design 1, (*b*) improved design 2.

The total LCC for a system is established or locked in early. It is not unusual to have 75 to 85% of the entire life span LCC committed during the preliminary and detail design stages. As shown in Figure 5–7*a*, the actual or observed LCC (bottom curve *AB*) will trail the committed LCC throughout the life span (unless some major design flaw increases the total LCC of design #1 above point *B*). *The potential for significantly reducing total LCC occurs primarily during the early stages.* A more effective design and more efficient equipment can reposition the envelope to design #2 in Figure 5–7*b*. Now the committed LCC curve *AEC* is below *AB* at all points, as is the actual LCC curve *AFC*. It is this lower envelope #2 we seek. The shaded area represents the reduction in actual LCC.

Even though an effective LCC envelope may be established early in the acquisition phase, it is not uncommon that unplanned cost-saving measures are introduced during the acquisition phase and early operation phase. These apparent "savings" may actually increase the total LCC, as shown by curve *AFD*. This style of ad hoc cost savings, often imposed by management early in the design stage and/or construction stage, can substantially increase costs later, especially in the after-sale portion of the use stage. For example, the use of inferior-strength concrete and steel has been the cause of structural failures many times, thus increasing the overall life span LCC.

5.8   PRESENT WORTH OF BONDS

A time-tested method of raising capital is through the issuance of an IOU, which is financing through debt, not equity, as discussed in Chapter 1. One very common form of IOU is a bond—a long-term note issued by a corporation or a government entity (the borrower) to finance major projects. The borrower receives money now in return for a promise to pay the *face value V* of the bond on a stated maturity date. Bonds are usually issued in face value amounts of $100, $1000, $5000, or $10,000. *Bond interest I,* also called *bond dividend,* is paid periodically between the time the money is borrowed and the time the face value is repaid. The bond interest is paid *c* times per year. Expected payment periods are usually quarterly or semiannually. The amount of interest is determined using the stated interest rate, called the *bond coupon rate b.*

$$I = \frac{\text{(face value)(bond coupon rate)}}{\text{number of payment periods per year}}$$

$$I = \frac{Vb}{c} \tag{5.8}$$

There are many types or classifications of bonds. Four general classifications are summarized in Table 5–1 according to their issuing entity, some fundamental characteristics, and example names or purposes. For example, *Treasury securities* are issued in different monetary amounts ($1000 and up) with varying periods of time to the maturity date (Bills up to 1 year; Notes for 2 to 10 years). In

TABLE 5–1   Classification and Characteristics of Bonds

Classification	Issued by	Characteristics	Examples
Treasury securities	Federal government	Backed by U.S. government	Bills (≤ 1 year) Notes (2–10 years) Bonds (10–30 years)
Municipal	Local governments	Federal tax-exempt Issued against taxes received	General obligation Revenue Zero coupon Put
Mortgage	Corporation	Backed by specified assets or mortgage Low rate/low risk on first mortgage Foreclosure, if not repaid	First mortgage Second mortgage Equipment trust
Debenture	Corporation	Not backed by collateral, but by reputation of corporation Bond rate may 'float' Higher interest rates and higher risks	Convertible Subordinated Junk or high yield

the United States, Treasury securities are considered a very safe bond purchase because they are backed with the "full faith and credit of the U.S. government." The safe investment rate indicated in Figure 1–6 as the lowest level for establishing a MARR is the coupon rate on a U.S. Treasury security. As another illustration, *debenture bonds* are issued by corporations in order to raise capital, but they are not backed by any particular form of collateral. The corporation's reputation attracts bond purchasers, and the corporation may make the bond interest rate 'float' to further attract buyers. Often debenture bonds are *convertible* to common stock of the corporation at a fixed rate prior to their maturity date.

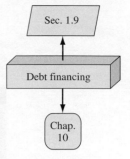

EXAMPLE 5.10

Procter and Gamble Inc. has issued $5,000,000 worth of $5000 ten-year debenture bonds. Each bond pays interest quarterly at 6%. (*a*) Determine the amount a purchaser will receive each 3 months and after 10 years. (*b*) Suppose a bond is purchased at a time when it is discounted by 2% to $4900. What are the quarterly interest amounts and the final payment amount at the maturity date?

Solution
(*a*)   Use Equation [5.8] for the quarterly interest amount.

$$I = \frac{(5000)(0.06)}{4} = \$75$$

The face value of $5000 is repaid after 10 years.
(*b*)   Purchasing the bond at a discount from face value does not change the interest or final repayment amounts. Therefore, $75 per quarter and $5000 after 10 years remain the amounts.

Finding the PW value of a bond is another extension of present worth analysis. When a corporation or government agency offers bonds, potential purchasers can determine how much they should be willing to pay in PW terms for a bond of a stated denomination. The amount paid at purchase time establishes the rate of return for the remainder of the bond life. The steps to calculate the PW of a bond are as follows:

1. Determine *I*, the interest per payment period, using Equation [5.8].
2. Construct the cash flow diagram of interest payments and face value repayment.
3. Establish the required MARR or rate of return.
4. Calculate the PW value of the bond interest payments and the face value at *i* = MARR. (If the bond interest payment period is not equal to the MARR compounding period, that is, PP ≠ CP, first use Equation [4.8] to determine the effective rate per payment period. Use this rate and the logic of Section 4.6 for PP ≥ CP to complete the PW calculations.)

Use the following logic:

> PW ≥ bond purchase price; MARR is met or exceeded, buy the bond.
> PW < bond purchase price; MARR is not met, do not buy the bond.

EXAMPLE 5.11

Determine the purchase price you should be willing to pay now for a 4.5% $5000 10-year bond with interest paid semiannually. Assume your MARR is 8% per year, compounded quarterly.

Solution
First, determine the semiannual interest.

$$I = 5000(0.045)/2 = \$112.50 \text{ every 6 months}$$

The present worth of all bond payments to you (Figure 5–8) is determined in either of two ways.

1. *Effective semiannual rate.* Use the approach of Section 4.6. The cash flow period is PP = 6 months, and the compounding period is CP = 3 months; PP > CP. Find the effective semiannual rate, then apply P/A and P/F factors to the interest payments and $5000 receipt in year 10. The nominal semiannual MARR is $r = 8\%/2 = 4\%$. For $m = 2$ quarters per 6-months, Equation [4.8] yields

$$\text{Effective } i = \left(1 + \frac{0.04}{2}\right)^2 - 1 = 4.04\% \text{ per 6-months}$$

    The PW of the bond is determined for $n = 2(10) = 20$ semiannual periods.

$$PW = \$112.50(P/A,4.04\%,20) + 5000(P/F,4.04\%,20) = \$3788$$

2. *Nominal quarterly rate.* Find the PW of each $112.50 semiannual bond interest receipt in year 0 separately with a P/F factor, and add the PW of the $5000 in year 10. The nominal quarterly MARR is $8\%/4 = 2\%$. The total number of

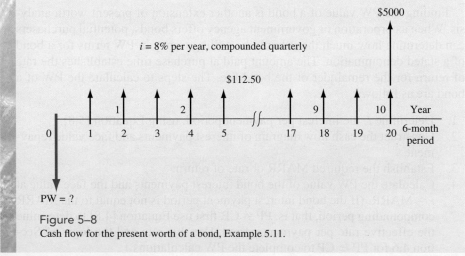

Figure 5–8
Cash flow for the present worth of a bond, Example 5.11.

periods is $n = 4(10) = 40$ quarters, double those shown in Figure 5–8, since the payments are made semiannually while the MARR is compounded quarterly.

$$PW = 112.50(P/F,2\%,2) + 112.50(P/F,2\%,4) + \cdots + 112.50(P/F,2\%,40)$$
$$+ 5000(P/F,2\%,40)$$

$$= \$3788$$

If the asking price is more than $3788 for the bond, which is a discount of more than 24%, you will not make the MARR.

The spreadsheet function PV(4.04%,20,112.50,5000) displays the PW value of $3788.

Q-Solv

5.9   SPREADSHEET APPLICATIONS—PW ANALYSIS AND PAYBACK PERIOD

Example 5.12 illustrates how to set up a spreadsheet for PW analysis for different-life alternatives and for a specified study period. Example 5.13 demonstrates the technique and shortcomings of payback period analysis for $i > 0\%$. Both hand and computer solutions are presented for this second example.

Some general guidelines help organize spreadsheets for any PW analysis. The LCM of the alternatives dictates the number of row entries for initial investment and salvage/market values, based on the repurchase assumption that PW analysis requires. Some alternatives will be service-based (cost cash flows only); others are revenue-based (cost and income cash flows). Place the annual cash flows in separate columns from the investment and salvage amounts. This reduces the amount of number processing you have to do before entering a cash flow value. Determine the PW values for all columns pertinent to an alternative, and add them to obtain the final PW value.

Spreadsheets can become crowded very rapidly. However, placing the NPV functions at the head of each cash flow column and inserting a separate summary table make the component and total PW values stand out. Finally, place the MARR value in a separate cell, so sensitivity analysis on the required return can be easily accomplished. Example 5.12 illustrates these guidelines.

EXAMPLE 5.12

Southeastern Cement plans to open a new rock pit. Two plans have been devised for movement of raw material from the quarry to the plant. Plan A requires the purchase of two earthmovers and construction of an unloading pad at the plant. Plan B calls for construction of a conveyor system from the quarry to the plant. The costs for each plan are detailed in Table 5–2. (*a*) Using spreadsheet-based PW analysis, determine which plan should be selected if money is worth 15% per year. (*b*) After only 6 years of operation a major environmental problem made Southeastern stop all operations at the rock pit. Use a 6-year study period to determine if plan A or B was economically better. The market value of each mover after 6 years is $20,000, and the trade-in value of the conveyor after 6 years is only $25,000. The pad can be salvaged for $2000.

TABLE 5–2	Estimates for Plans to Move Rock from Quarry to Cement Plant		
	Plan A		Plan B
	Mover	Pad	Conveyor
Initial cost, $	−45,000	−28,000	−175,000
Annual operating cost, $	−6,000	−300	−2,500
Salvage value, $	5,000	2,000	10,000
Life, years	8	12	24

Solution

(a) Evaluation must take place over the LCM of 24 years. Reinvestment in the two movers will occur in years 8 and 16, and the unloading pad must be rebuilt in year 12. No reinvestment is necessary for plan B. First, construct the cash flow diagrams for plans A and B over 24 years to better understand the spreadsheet analysis in Figure 5–9. Columns B, D, and F include all investments, reinvestments, and salvage values. (*Remember to enter zeros in all cells with no cash flows, or the NPV function will give an incorrect PW value.*) These are service-based alternatives, so columns C, E, and G display the AOC estimates, labeled "Annual CF". NPV functions provide the PW amounts in row 8 cells. These are added by alternative in cells H19 and H22.

Conclusion: Select plan B because the PW of costs is smaller.

(b) Both alternatives are abruptly terminated after 6 years, and current market or trade-in values are estimated. To perform the PW analysis for a severely truncated study period, Figure 5–10 uses the same format as that for the 24-year analysis, except for two major alterations. Cells in row 16 now include the market and trade-in amounts, and all rows after 16 are deleted. See the cell tags in row 9 for the new NPV functions for the 6 years of cash flows. Cells D20 and D21 are the PW values found by summing the appropriate PW values in row 9.

Conclusion: Plan A should have been selected, had the termination after 6 years been known at the design stage of the rock pit.

Comment

The spreadsheet solution for part (*b*) was developed by initially copying the entire worksheet in part (*a*) to sheet 2 of the Excel workbook. Then the changes outlined above were made to the copy. Another method uses the same worksheet to build the new NPV functions as shown in Figure 5–10 cell tags, but on the Figure 5–9 worksheet after inserting a new row 16 for year 6 cash flows. This approach is faster and less formal than the method demonstrated here. There is one real danger in using the one-worksheet approach to solving this (or any sensitivity analysis) problem. The altered worksheet now solves a different problem, so the functions display new answers. For example, when the cash flows are truncated to a 6-year study period, the old NPV functions in row 8 must be changed, or the new NPV functions must be added in row 9. But now the NPV functions of the old 24-year PW analysis display incorrect answers, or possibly an Excel error message. This introduces error possibilities into the decision making. For accurate, correct results, take the time to copy the first sheet to a new worksheet and make the changes on the copy. Store both solutions after documenting what each sheet is designed to analyze. This provides a historical record of what was altered during the sensitivity analysis.

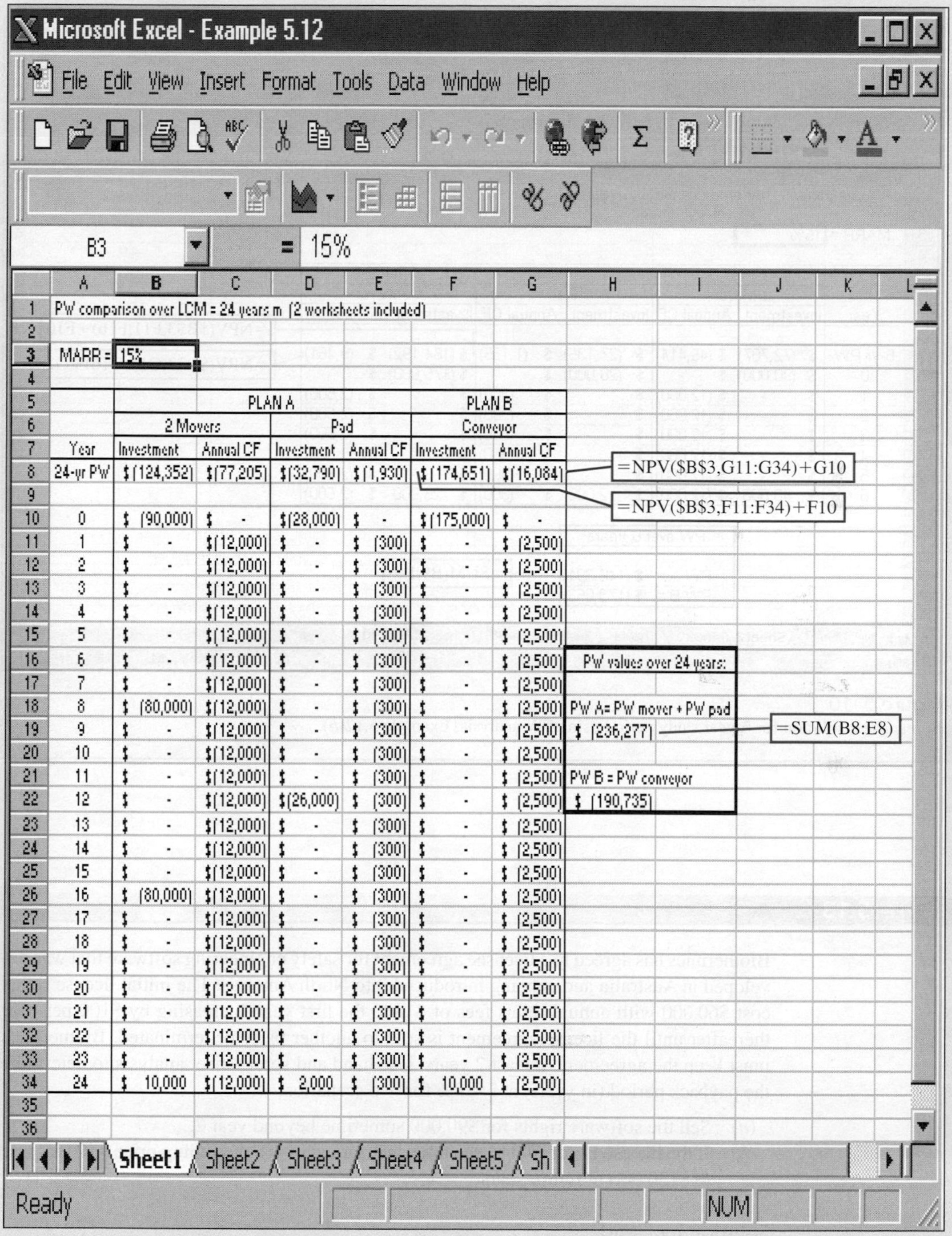

	A	B	C	D	E	F	G	H
1	PW comparison over LCM = 24 years m (2 worksheets included)							
2								
3	MARR =	15%						
4								
5		PLAN A				PLAN B		
6		2 Movers		Pad		Conveyor		
7	Year	Investment	Annual CF	Investment	Annual CF	Investment	Annual CF	
8	24-yr PW	$(124,352)	$(77,205)	$(32,790)	$(1,930)	$(174,651)	$(16,084)	
9								
10	0	$ (90,000)	$ -	$(28,000)	$ -	$(175,000)	$ -	
11	1	$ -	$(12,000)	$ -	$ (300)	$ -	$ (2,500)	
12	2	$ -	$(12,000)	$ -	$ (300)	$ -	$ (2,500)	
13	3	$ -	$(12,000)	$ -	$ (300)	$ -	$ (2,500)	
14	4	$ -	$(12,000)	$ -	$ (300)	$ -	$ (2,500)	
15	5	$ -	$(12,000)	$ -	$ (300)	$ -	$ (2,500)	
16	6	$ -	$(12,000)	$ -	$ (300)	$ -	$ (2,500)	PW values over 24 years:
17	7	$ -	$(12,000)	$ -	$ (300)	$ -	$ (2,500)	
18	8	$ (80,000)	$(12,000)	$ -	$ (300)	$ -	$ (2,500)	PW A= PW mover + PW pad
19	9	$ -	$(12,000)	$ -	$ (300)	$ -	$ (2,500)	$ (236,277)
20	10	$ -	$(12,000)	$ -	$ (300)	$ -	$ (2,500)	
21	11	$ -	$(12,000)	$ -	$ (300)	$ -	$ (2,500)	PW B = PW conveyor
22	12	$ -	$(12,000)	$(26,000)	$ (300)	$ -	$ (2,500)	$ (190,735)
23	13	$ -	$(12,000)	$ -	$ (300)	$ -	$ (2,500)	
24	14	$ -	$(12,000)	$ -	$ (300)	$ -	$ (2,500)	
25	15	$ -	$(12,000)	$ -	$ (300)	$ -	$ (2,500)	
26	16	$ (80,000)	$(12,000)	$ -	$ (300)	$ -	$ (2,500)	
27	17	$ -	$(12,000)	$ -	$ (300)	$ -	$ (2,500)	
28	18	$ -	$(12,000)	$ -	$ (300)	$ -	$ (2,500)	
29	19	$ -	$(12,000)	$ -	$ (300)	$ -	$ (2,500)	
30	20	$ -	$(12,000)	$ -	$ (300)	$ -	$ (2,500)	
31	21	$ -	$(12,000)	$ -	$ (300)	$ -	$ (2,500)	
32	22	$ -	$(12,000)	$ -	$ (300)	$ -	$ (2,500)	
33	23	$ -	$(12,000)	$ -	$ (300)	$ -	$ (2,500)	
34	24	$ 10,000	$(12,000)	$ 2,000	$ (300)	$ 10,000	$ (2,500)	

Figure 5–9
Spreadsheet solution using PW analysis of different-life alternatives, Example 5.12(*a*).

Microsoft Excel - Example 5.12

PW comparison over study period of 6 years

MARR = 15%

	Plan A				Plan B	
	2 Movers		Pad		Conveyor	
Year	Investment	Annual CF	Investment	Annual CF	Investment	Annual CF
6-yr PW	$ (72,707)	$ (45,414)	$ (27,135)	$ (1,135)	$ (164,192)	$ (9,461)
0	$ (90,000)	$ -	$ (28,000)	$ -	$ (175,000)	$ -
1	$ -	$ (12,000)	$ -	$ (300)	$ -	$ (2,500)
2	$ -	$ (12,000)	$ -	$ (300)	$ -	$ (2,500)
3	$ -	$ (12,000)	$ -	$ (300)	$ -	$ (2,500)
4	$ -	$ (12,000)	$ -	$ (300)	$ -	$ (2,500)
5	$ -	$ (12,000)	$ -	$ (300)	$ -	$ (2,500)
6	$ 40,000	$ (12,000)	$ 2,000	$ (300)	$ 25,000	$ (2,500)

=NPV($B$3,F11:F16)+F10

=NPV($B$3,G11:G16)+G10

PW over 6 years	
PW A =	$ (146,391)
PW B =	$ (173,653)

=SUM(B9:E9)

Figure 5–10
Spreadsheet solution for 6-year study period using PW analysis, Example 5.12(*b*).

EXAMPLE 5.13

Biothermics has agreed to a licensee agreement for safety engineering software that was developed in Australia and is being introduced into North America. The initial license rights cost \$60,000 with annual rights fees of \$1800 the first year, increasing by \$100 per year thereafter until the license agreement is sold to another party or terminated. Biothermics must keep the agreement at least 2 years. Use hand and spreadsheet analysis to determine the payback period (in years) at $i = 8\%$ for two scenarios:

(*a*) Sell the software rights for \$90,000 sometime beyond year 2.

(*b*) If the license is not sold by the time determined in (*a*), the selling price will increase to \$120,000 in future years.

Solution by Hand

(*a*) From Equation [5.4], it is necessary that PW = 0 at the 8% payback period n_p. Set up the PW relation for $n \geq 3$ years, and determine the number of years at which PW crosses the zero value.

$$0 = -60{,}000 - 1800(P/A,8\%,n) - 100(P/G,8\%,n) + 90{,}000(P/F,8\%,n)$$

n, Years	3	4	5
PW Value	\$6562	\$$-$274	\$$-$6672

The 8% payback is between 3 and 4 years. By linear interpolation, $n_p = 3.96$ years.

(b) If the license is not sold prior to 4 years, the price goes up to \$120,000. The PW relation for 4 or more years and the PW values for n are

$$0 = -60{,}000 - 1800(P/A,8\%,n) - 100(P/G,8\%,n) + 120{,}000(P/F,8\%,n)$$

n, Years	5	6	7
PW Value	\$13,748	\$6247	\$$-$755

The 8% payback is now between 6 and 7 years. By interpolation, $n_p = 6.90$ years.

Solution by Computer

(a and b) Figure 5–11 presents a spreadsheet that lists the software rights costs (column B) and expected selling price (columns C and E). The NPV functions in column D (selling

E-Solve

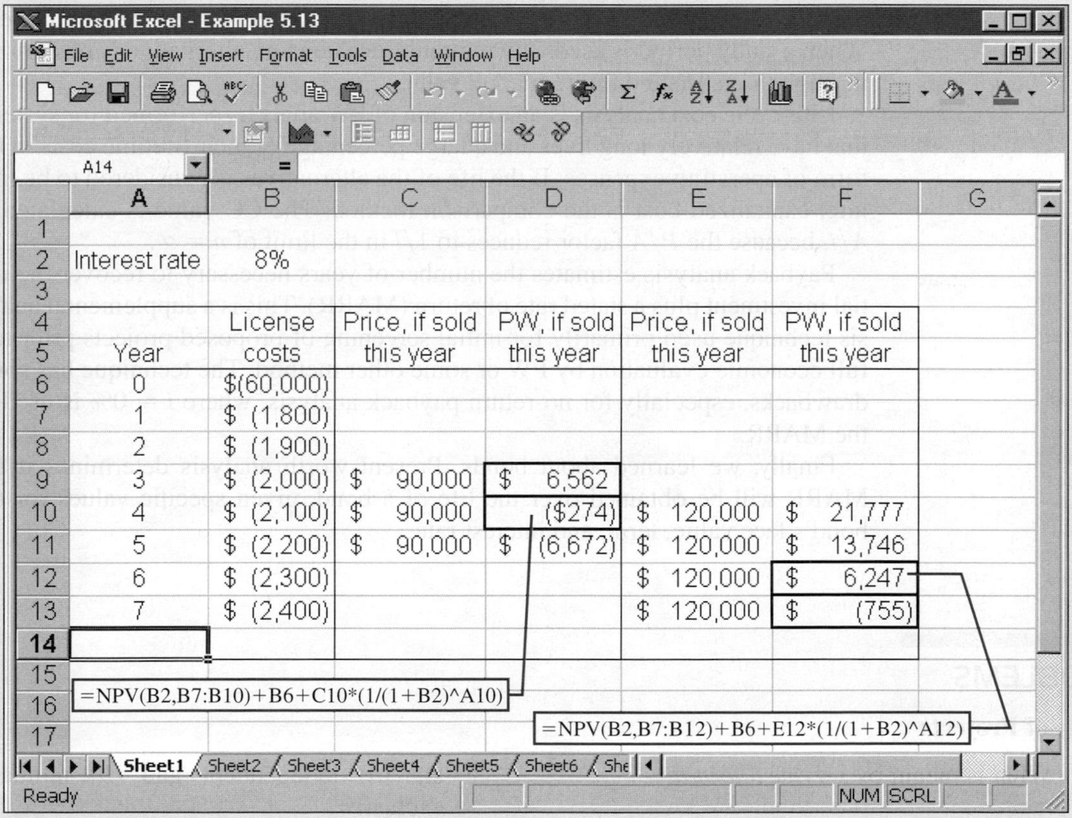

X Microsoft Excel - Example 5.13

File Edit View Insert Format Tools Data Window Help

A14

	A	B	C	D	E	F	G
1							
2	Interest rate	8%					
3							
4		License	Price, if sold	PW, if sold	Price, if sold	PW, if sold	
5	Year	costs	this year	this year	this year	this year	
6	0	\$(60,000)					
7	1	\$ (1,800)					
8	2	\$ (1,900)					
9	3	\$ (2,000)	\$ 90,000	\$ 6,562			
10	4	\$ (2,100)	\$ 90,000	(\$274)	\$ 120,000	\$ 21,777	
11	5	\$ (2,200)	\$ 90,000	\$ (6,672)	\$ 120,000	\$ 13,746	
12	6	\$ (2,300)			\$ 120,000	\$ 6,247	
13	7	\$ (2,400)			\$ 120,000	\$ (755)	
14							
15							
16	=NPV(B2,B7:B10)+B6+C10*(1/(1+B2)^A10)						
17			=NPV(B2,B7:B12)+B6+E12*(1/(1+B2)^A12)				

Sheet1 / Sheet2 / Sheet3 / Sheet4 / Sheet5 / Sheet6 / She

Ready NUM SCRL

Figure 5–11

Determination of payback period using a spreadsheet, Example 5.13(a) and (b).

price $90,000) show the payback period to be between 3 and 4 years, while the NPV results in column F (selling price $120,000) indicate PW switching from positive to negative between 6 and 7 years. The NPV functions reflect the relations presented in the hand solution, except the cost gradient of $100 has been incorporated into the costs in column B.

If more exact payback values are needed, interpolate between the PW results on the spreadsheet. The values will be the same as in the solution by hand, namely, 3.96 and 6.90 years.

CHAPTER SUMMARY

The present worth method of comparing alternatives involves converting all cash flows to present dollars at the MARR. The alternative with the numerically larger (or largest) PW value is selected. When the alternatives have different lives, the comparison must be made for equal-service periods. This is done by performing the comparison over either the LCM of lives or a specific study period. Both approaches compare alternatives in accordance with the equal-service requirement. When a study period is used, any remaining value in an alternative is recognized through the estimated future market value.

Life-cycle cost analysis is an extension of PW analysis performed for systems that have relatively long lives and a large percentage of their lifetime costs in the form of operating expenses. If the life of the alternatives is considered to be infinite, capitalized cost is the comparison method. The CC value is calculated as A/i, because the P/A factor reduces to $1/i$ in the limit of $n = \infty$.

Payback analysis estimates the number of years necessary to recover the initial investment plus a stated rate of return (MARR). This is a supplemental analysis technique used primarily for initial screening of proposed projects prior to a full economic evaluation by PW or some other method. The technique has some drawbacks, especially for no-return payback analysis, where $i = 0\%$ is used as the MARR.

Finally, we learned about bonds. Present worth analysis determines if the MARR will be obtained over the life of a bond, given specific values for the bond's face value, term, and interest rate.

PROBLEMS

Types of Projects

5.1 What is meant by *service alternative?*

5.2 When you are evaluating projects by the present worth method, how do you know which one(s) to select if the projects

are (*a*) independent and (*b*) mutually exclusive?

5.3 Read the statement in the following problems and determine if the cash flows define a revenue or a service project: (*a*) Problem 2.12, (*b*) Problem 2.31, (*c*) Problem 2.51, (*d*) Problem 3.6, (*e*) Problem 3.10, and (*f*) Problem 3.14.

5.4 A rapidly growing city is dedicated to neighborhood integrity. However, increasing traffic and speed on a through street are of concern to residents. The city manager has proposed five independent options to slow traffic:

1. Stop sign at corner A.
2. Stop sign at corner B.
3. Low-profile speed bump at point C.
4. Low-profile speed bump at point D.
5. Speed dip at point E.

There cannot be any of the following combinations in the final alternatives:

- No combination of dip and one or two bumps
- Not two bumps
- Not two stop signs

Use the five independent options and the restrictions to determine (*a*) the *total number* of mutually exclusive alternatives possible and (*b*) the *acceptable* mutually exclusive alternatives.

5.5 What is meant by the term *equal service?*

5.6 What two approaches can be used to satisfy the equal service requirement?

5.7 Define the term *capitalized cost* and give a real-world example of something that might be analyzed using that technique.

Alternative Comparison—Equal Lives

5.8 Lennon Hearth Products manufactures glass-door fireplace screens that have two types of mounting brackets for the frame. An L-shaped bracket is used for relatively small fireplace openings, and a U-shaped bracket is used for all others. The company includes both types of brackets in the box with the product, and the purchaser discards the one not needed. The cost of these two brackets with screws and other parts is $3.50. If the frame of the fireplace screen is redesigned, a single universal bracket can be used that will cost $1.20 to make. However, retooling will cost $6000. In addition, inventory write-downs will amount to another $8000. If the company sells 1200 fireplace units per year, should the company keep the old brackets or go with the new ones, assuming the company uses an interest rate of 15% per year and it wants to recover its investment in 5 years? Use the present worth method.

5.9 Two methods can be used for producing expansion anchors. Method A costs $80,000 initially and will have a $15,000 salvage value after 3 years. The operating cost with this method will be $30,000 per year. Method B will have a first cost of $120,000, an operating cost of $8000 per year, and a $40,000 salvage value after its 3-year life. At an interest rate of 12% per year, which method should be used on the basis of a present worth analysis?

5.10 Sales of bottled water in the United States totaled 16.3 gallons per person in 2004. Evian Natural Spring Water costs 40¢ per bottle. A municipal water utility provides tap water for $2.10 per 1000 gallons. If the average person drinks 2 bottles of water per day or uses 5 gallons per day in getting that amount of water from the tap, what are the present worth values of drinking bottled water or tap water per person for 1 year? Use an interest rate of 6% per year, compounded monthly, and 30 days per month.

5.11 A software package created by Navarro & Associates can be used for analyzing and designing three-sided guyed towers and three- and four-sided self-supporting towers. A single-user license will cost

$4000 per year. A site license has a one-time cost of $15,000. A structural engineering consulting company is trying to decide between two alternatives: first, to buy one single-user license *now* and one each year for the next 4 years (which will provide 5 years of service); or second, to buy a site license now. Determine which strategy should be adopted at an interest rate of 12% per year for a 5-year planning period, using the present worth method of evaluation.

5.12 A company that manufactures amplified pressure transducers is trying to decide between the machines shown below. Compare them on the basis of their present worth values, using an interest rate of 15% per year.

	Variable Speed	Dual Speed
First cost, $	−250,000	−224,000
Annual operating cost, $/year	−231,000	−235,000
Overhaul in year 3, $	—	−26,000
Overhaul in year 4, $	−140,000	—
Salvage value, $	50,000	10,000
Life, years	6	6

Alternative Comparison over Different Time Periods

5.13 NASA is considering two materials for use in a space vehicle. The costs are shown below. Which should be selected on the basis of a present worth comparison at an interest rate of 10% per year?

	Material JX	Material KZ
First cost, $	−205,000	−235,000
Maintenance cost, $/year	−29,000	−27,000
Salvage value, $	2,000	20,000
Life, years	2	4

5.14 Two processes can be used for producing a polymer that reduces friction loss in engines. Process K will have a first cost of $160,000, an operating cost of $7000 per quarter, and a salvage value of $40,000 after its 2-year life. Process L will have a first cost of $210,000, an operating cost of $5000 per quarter, and a $26,000 salvage value after its 4-year life. Which process should be selected on the basis of a present worth analysis at an interest rate of 8% per year, compounded quarterly?

5.15 Two methods are under consideration for producing the case for a portable hazardous material photoionization monitor. A plastic case will require an initial investment of $75,000 and will have an annual operating cost of $27,000 with no salvage after 2 years. An aluminum case will require an investment of $125,000 and will have annual costs of $12,000. Some of the equipment can be sold for $30,000 after its 3-year life. At an interest rate of 10% per year, which case should be used on the basis of a present worth analysis?

5.16 Three different plans were presented to the GAO by a high-technology facilities manager for operating a small weapons production facility. Plan A would involve renewable 1-year contracts with payments of $1 million at the beginning of each year. Plan B would be a 2-year contract, and it would require four payments of $600,000 each, with the first one to be made now and the other three at 6-month intervals. Plan C would be a 3-year contract, and it would entail a payment of $1.5 million now and another payment of $0.5 million 2 years from now. Assuming that the GAO could renew any of the plans under the same conditions if it wants to do so, which plan is better on the basis of a present worth analysis at an interest rate of 6% per year, compounded semiannually?

Future Worth Comparison

5.17 A remotely located air sampling station can be powered by solar cells or by running an electric line to the site and using conventional power. Solar cells will cost $12,600 to install and will have a useful life of 4 years with no salvage value. Annual costs for inspection, cleaning, etc., are expected to be $1400. A new power line will cost $11,000 to install, with power costs expected to be $800 per year. Since the air sampling project will end in 4 years, the salvage value of the line is considered to be zero. At an interest rate of 10% per year, which alternative should be selected on the basis of a future worth analysis?

5.18 The Department of Energy is proposing new rules mandating a 20% increase in clothes washer efficiency by 2005 and a 35% increase by 2008. The 20% increase is expected to add $100 to the current price of a washer, while the 35% increase will add $240 to the price. If the cost for energy is $80 per year with the 20% increase in efficiency and $65 per year with the 35% increase, which one of the two proposed standards is more economical on the basis of a future worth analysis at an interest rate of 10% per year? Assume a 15-year life for all washer models.

5.19 A small strip-mining coal company is trying to decide whether it should purchase or lease a new clamshell. If purchased, the shell will cost $150,000 and is expected to have a $65,000 salvage value in 6 years. Alternatively, the company can lease a clamshell for $30,000 per year, but the lease payment will have to be made at the *beginning* of each year. If the clamshell is purchased, it will be leased to other strip-mining companies whenever possible, an activity that is expected to yield revenues

of $12,000 per year. If the company's minimum attractive rate of return is 15% per year, should the clamshell be purchased or leased on the basis of a future worth analysis?

5.20 Three types of drill bits can be used in a certain manufacturing operation. A bright high-speed steel (HSS) bit is the least expensive to buy, but it has a shorter life than either gold oxide or titanium nitride bits. The HSS bits will cost $3500 to buy and will last for 3 months under the conditions in which they will be used. The operating cost for these bits will be $2000 per month. The gold oxide bits will cost $6500 to buy and will last for 6 months with an operating cost of $1500 per month. The titanium nitride bits will cost $7000 to buy and will last 6 months with an operating cost of $1200 per month. At an interest rate of 12% per year, compounded monthly, which type of drill bit should be used on the basis of a future worth analysis?

5.21 El Paso Electric is considering two alternatives for satisfying state regulations regarding pollution control for one of its generating stations. This particular station is located at the outskirts of the city and a short distance from Juarez, Mexico. The station is currently producing excess VOCs and oxides of nitrogen. Two plans have been proposed for satisfying the regulators. Plan A involves replacing the burners and switching from fuel oil to natural gas. The cost of the option will be $300,000 initially and an extra $900,000 per year in fuel costs. Plan B involves going to Mexico and running gas lines to many of the "backyard" brickmaking sites that now use wood, tires, and other combustible waste materials for firing the bricks. The idea behind plan B is that by reducing the particulate pollution responsible for smog in El Paso, there would be greater benefit to U.S. citizens than would be achieved through plan A. The initial cost of plan B will be $1.2 million for installation of the lines. Additionally, the electric company would subsidize the cost of gas for the brickmakers to the extent of $200,000 per year. Extra air monitoring associated with this plan will cost an additional $150,000 per year. For a 10-year project period and no salvage value for either plan, which one should be selected on the basis of a future worth analysis at an interest rate of 12% per year?

Capitalized Costs

5.22   The cost of painting the Golden Gate Bridge is $400,000. If the bridge is painted now and every 2 years hereafter, what is the capitalized cost of painting at an interest rate of 6% per year?

5.23   The cost of extending a certain road at Yellowstone National Park is $1.7 million. Resurfacing and other maintenance are expected to cost $350,000 every 3 years. What is the capitalized cost of the road at an interest rate of 6% per year?

5.24   Determine the capitalized cost of an expenditure of $200,000 at time 0, $25,000 in years 2 through 5, and $40,000 per year from year 6 on. Use an interest rate of 12% per year.

5.25   A city that is attempting to attract a professional football team is planning to build a new stadium costing $250 million. Annual upkeep is expected to amount to $800,000 per year. The artificial turf will have to be replaced every 10 years at a cost of $950,000. Painting every 5 years will cost $75,000. If the city expects to maintain the facility indefinitely, what will be its capitalized cost at an interest rate of 8% per year?

5.26   A certain manufacturing alternative has a first cost of $82,000, an annual maintenance cost of $9000, and a salvage value of $15,000 after its 4-year life. What is its capitalized cost at an interest rate of 12% per year?

5.27   If you want to be able to withdraw $80,000 per year forever beginning 30 years from now, how much will you have to have in your retirement account (that earns 8% per year interest) in (a) year 29 and (b) year 0?

5.28   What is the capitalized cost (absolute value) of the *difference* between the following two plans at an interest rate of 10% per year? Plan A will require an expenditure of $50,000 every 5 years forever (beginning in year 5). Plan B will require an expenditure of $100,000 every 10 years forever (beginning in year 10).

5.29   What is the capitalized cost of expenditures of $3,000,000 now, $50,000 in months 1 through 12, $100,000 in months 13 through 25, and $50,000 in months 26 through infinity if the interest rate is 12% per year, compounded monthly?

5.30   Compare the following alternatives on the basis of their capitalized cost at an interest rate of 10% per year.

	Petroleum-Based Feedstock	Inorganic-Based Feedstock
First cost, $	−250,000	−110,000
Annual operating cost, $/year	−130,000	−65,000
Annual revenues, $/year	400,000	270,000
Salvage value, $	50,000	20,000
Life, years	6	4

5.31 An alumna of Ohio State University wanted to set up an endowment fund that would award scholarships to female engineering students totaling $100,000 per year forever. The first scholarships are to be granted *now* and continue each year forever. How much must the alumna donate now, if the endowment fund is expected to earn interest at a rate of 8% per year?

5.32 Two large-scale conduits are under consideration by a large municipal utility district (MUD). The first involves construction of a steel pipeline at a cost of $225 million. Portions of the pipeline will have to be replaced every 40 years at a cost of $50 million. The pumping and other operating costs are expected to be $10 million per year. Alternatively, a gravity flow canal can be constructed at a cost of $350 million. The M&O costs for the canal are expected to be $0.5 million per year. If both conduits are expected to last forever, which should be built at an interest rate of 10% per year?

5.33 Compare the alternatives shown below on the basis of their capitalized costs, using an interest rate 12% per year, compounded quarterly.

	Alternative E	Alternative F	Alternative G
First cost, $	−200,000	−300,000	−900,000
Quarterly income, $/quarter	30,000	10,000	40,000
Salvage value, $	50,000	70,000	100,000
Life, years	2	4	∞

Payback Analysis

5.34 What is meant by no-return payback or simple payback?

5.35 Explain why the alternative that recovers its initial investment at a specified rate of return in the shortest time is *not necessarily* the most economically attractive one.

5.36 Determine the payback period for an asset that has a first cost of $40,000, a salvage value of $8000 anytime within 10 years of its purchase, and generates income of $6000 per year. The required return is 8% per year.

5.37 Accusoft Systems is offering business owners a software package that keeps track of many accounting functions from the company's bank transactions sales invoices. The site license will cost $22,000 to install and will involve a quarterly fee of $2000. If a certain small company can save $3500 every quarter and have the security of managing its books in-house, how long will it take for the company to recover its investment at an interest rate of 4% per quarter?

5.38 Darnell Enterprises constructed an addition to its building at a cost of $70,000. Extra annual expenses are expected to be $1850, but extra income will be $14,000 per year. How long will it take for the company to recover its investment at an interest rate of 10% per year?

5.39 A new process for manufacturing laser levels will have a first cost of $35,000 with annual costs of $17,000. Extra income associated with the new process is expected to be $22,000 per year. What is the payback period at (*a*) *i* = 0% and (*b*) *i* = 10% per year?

5.40 A multinational engineering consulting firm that wants to provide resort accommodations to certain clients is considering the purchase of a three-bedroom lodge in upper Montana that will cost $250,000.

The property in that area is rapidly appreciating in value because people anxious to get away from urban developments are bidding up the prices. If the company spends an average of $500 per month for utilities and the investment increases at a rate of 2% per month, how long would it be before the company could sell the property for $100,000 more than it has invested in it?

5.41 A window frame manufacturer is searching for ways to improve revenue from its triple-insulated sliding windows, sold primarily in the far northern areas of the United States. Alternative A is an increase in TV and radio marketing. A total of $300,000 spent now is expected to increase revenue by $60,000 per year. Alternative B requires the same investment for enhancements to the in-plant manufacturing process that will improve the temperature retention properties of the seals around each glass pane. New revenues start slowly for this alternative at an estimated $10,000 the first year, with growth of $15,000 per year as the improved product gains reputation among builders. The MARR is 8% per year, and the maximum evaluation period is 10 years for either alternative. Use both payback analysis and present worth analysis at 8% (for 10 years) to select the more economical alternative. State the reason(s) for any difference in the alternative chosen between the two analyses.

Life-Cycle Costs

5.42 A high-technology defense contractor has been asked by the Pentagon to estimate the life-cycle cost (LCC) for a proposed light-duty support vehicle. Its list of items included the following general categories: R&D costs (R&D), nonrecurring investment (NRI) costs, recurring investment (RI) costs, scheduled and unscheduled maintenance costs (Maint), equipment usage costs (Equip), and disposal costs (Disp). The costs (in millions) for the 20-year life cycle are as indicated. Calculate the LCC at an interest rate of 7% per year.

Year	R&D	NRI	RI	Maint	Equip	Disp
0	5.5	1.1				
1	3.5					
2	2.5					
3	0.5	5.2	1.3	0.6	1.5	
4		10.5	3.1	1.4	3.6	
5		10.5	4.2	1.6	5.3	
6–10			6.5	2.7	7.8	
11 on			2.2	3.5	8.5	
18–20						2.7

5.43 A manufacturing software engineer at a major aerospace corporation has been assigned the management responsibility of a project to design, build, test, and implement AREMSS, a new-generation automated scheduling system for routine and expedited maintenance. Reports on the disposition of each service will also be entered by field personnel, then filed and archived by the system. The initial application will be on existing Air Force in-flight refueling aircraft. The system is expected to be widely used over time for other aircraft maintenance scheduling. Once it is fully implemented, enhancements will have to be made, but the system is expected to serve as a worldwide scheduler for up to 15,000 separate aircraft. The engineer, who must make a presentation next week of the best estimates of costs over a 20-year life period, has decided to use the life-cycle cost approach of cost estimations. Use the following information to determine the current LCC at 6% per year for the AREMSS scheduling system.

Cost Category	Cost in Year ($ Millions)							
	1	2	3	4	5	6 on	10	18
Field study	0.5							
Design of system	2.1	1.2	0.5					
Software design		0.6	0.9					
Hardware purchases			5.1					
Beta testing			0.1	0.2				
User's manual development			0.1	0.1	0.2	0.2	0.06	
System implementation				1.3	0.7			
Field hardware				0.4	6.0	2.9		
Training trainers				0.3	2.5	2.5	0.7	
Software upgrades						0.6	3.0	3.7

5.44 The U.S. Army received two proposals for a turnkey design-build project for barracks for infantry unit soldiers in training. Proposal A involves an off-the-shelf barebones design and standard-grade construction of walls, windows, doors, and other features. With this option, heating and cooling costs will be greater, maintenance costs will be higher, and replacement will be sooner than for proposal B. The initial cost for A will be $750,000. Heating and cooling costs will average $6000 per month, with maintenance costs averaging $2000 per month. Minor remodeling will be required in years 5, 10, and 15 at a cost of $150,000 each time in order to render the units usable for 20 years. They will have no salvage value.

Proposal B will include tailored design and construction costs of $1.1 million initially, with estimated heating and cooling costs of $3000 per month and maintenance costs of $1000 per month. There will be no salvage value at the end of the 20-year life.

Which proposal should be accepted on the basis of a life-cycle cost analysis, if the interest rate is 0.5% per month?

5.45 A medium-size municipality plans to develop a software system to assist in project selection during the next 10 years. A life-cycle cost approach has been used to categorize costs into development, programming, operating, and support costs for each alternative. There are three alternatives under consideration, identified as A (tailored system), B (adapted system), and C (current system). The costs are summarized below. Use a life-cycle cost approach to identify the best alternative at 8% per year.

Alternative	Cost Component	Cost
A	Development	$250,000 now, $150,000 years 1 through 4
	Programming	$45,000 now, $35,000 years 1, 2
	Operation	$50,000 years 1 through 10
	Support	$30,000 years 1 through 5
B	Development	$10,000 now
	Programming	$45,000 year 0, $30,000 years 1 through 3
	Operation	$80,000 years 1 through 10
	Support	$40,000 years 1 through 10
C	Operation	$175,000 years 1 through 10

Bonds

5.46 A mortgage bond with a face value of $10,000 has a bond interest rate of 6% per year payable quarterly. What are the amount and frequency of the interest payments?

5.47 What is the face value of a municipal bond that has a bond interest rate of 4% per year with semiannual interest payments of $800?

5.48   What is the bond interest rate on a $20,000 bond that has semiannual interest payments of $1500 and a 20-year maturity date?

5.49   What is the present worth of a $50,000 bond that has interest of 10% per year, payable quarterly? The bond matures in 20 years. The interest rate in the marketplace is 10% per year, compounded quarterly.

5.50   What is the present worth of a $50,000 municipal bond that has an interest rate of 4% per year, payable quarterly? The bond matures in 15 years, and the market interest rate is 8% per year, compounded quarterly.

5.51   General Electric issued 1000 debenture bonds 3 years ago with a face value of $5000 each and a bond interest rate of 8% per year payable semiannually. The bonds have a maturity date of 20 years *from the date they were issued*. If the interest rate in the market place is 10% per year, compounded semiannually, what is the present worth of one bond to an investor who wishes to purchase it today?

5.52   Charleston Independent School District needs to raise $200 million to refurbish

its existing schools and build new ones. The bonds will pay interest semiannually at a rate of 7% per year, and they will mature in 30 years. Brokerage fees associated with the sale of the bonds will be $1 million. If the interest rate in the marketplace rises to 8% per year, compounded semiannually, before the bonds are issued, what will the face value of the bonds have to be for the school district to net $200 million?

5.53   An engineer planning for his retirement thinks that the interest rates in the marketplace will decrease before he retires. Therefore, he plans to invest in corporate bonds. He plans to buy a $50,000 bond that has a bond interest rate of 12% per year, payable quarterly with a maturity date 20 years from now.
   (*a*)   How much should he be able to sell the bond for in 5 years if the market interest rate is 8% per year, compounded quarterly?
   (*b*)   If he invested the interest he received at an interest rate of 12% per year, compounded quarterly, how much will he have (total) immediately after he sells the bond 5 years from now?

FE REVIEW PROBLEMS

5.54   For the mutually exclusive alternatives shown below, determine which one(s) should be selected.

Alternative	Present Worth, $
A	−25,000
B	−12,000
C	10,000
D	15,000

   (*a*)   Only A
   (*b*)   Only D
   (*c*)   Only A and B
   (*d*)   Only C and D

5.55   The present worth of $50,000 now, $10,000 per year in years 1 through 15, and $20,000 per year in years 16 through infinity at 10% per year is closest to

(a) Less than $169,000
(b) $169,580
(c) $173,940
(d) $195,730

5.56 A certain donor wishes to start an endowment at her alma mater that will provide scholarship money of $40,000 per year beginning in year 5 and continuing indefinitely. If the university earns 10% per year on the endowment, the amount she must donate now is closest to
(a) $225,470
(b) $248,360
(c) $273,200
(d) $293,820

5.57 At an interest rate of 10% per year, the amount you must deposit in your retirement account each year in years 0 through 9 (i.e., 10 deposits) if you want to withdraw $50,000 per year forever beginning 30 years from now is closest to
(a) $4239
(b) $4662
(c) $4974
(d) $5471

Problems 5.58 through 5.60 are based on the following estimates. The cost of money is 10% per year.

	Machine X	Machine Y
Initial cost, $	−66,000	−46,000
Annual cost, $/year	−10,000	−15,000
Salvage value, $	10,000	24,000
Life, years	6	3

5.58 The present worth of machine X is closest to
(a) $−65,270
(b) $−87,840
(c) $−103,910
(d) $−114,310

5.59 In comparing the machines on a present worth basis, the present worth of machine Y is closest to
(a) $−65,270
(b) $−97,840
(c) $−103,910
(d) $−114,310

5.60 The capitalized cost of machine X is closest to
(a) $−103,910
(b) $−114,310
(c) $−235,990
(d) $−238,580

5.61 The cost of maintaining a public monument in Washington, D.C., occurs as periodic outlays of $10,000 every 5 years. If the first outlay is now, the capitalized cost of the maintenance at an interest rate of 10% per year is closest to
(a) $−16,380
(b) $−26,380
(c) $−29,360
(d) $−41,050

5.62 The alternatives shown below are to be compared on the basis of their capitalized costs. At an interest rate of 10% per year, compounded continuously, the equation that represents the capitalized cost of alternative A is

	Alternative A	Alternative B
First cost, $	−50,000	−90,000
Annual cost, $/year	−10,000	−4,000
Salvage value, $	13,000	15,000
Life, years	3	6

(a) $PW_A = -50,000 - 10,000(P/A, 10.52\%,6) - 37,000(P/F,10.52\%,3) + 13,000(P/F,10.52\%,6)$
(b) $PW_A = -50,000 - 10,000(P/A, 10.52\%,3) + 13,000(P/F, 10.52\%,3)$

(c) $PW_A = [-50,000(A/P,10.52\%,3) -$
      $10,000 + 13,000(A/F,10.52\%,3)]/$
      0.1052
(d) $PW_A = [-50,000(A/P,10\%,3) -$
      $10,000 + 13,000(A/F,10\%,3)]/0.10$

5.63 A corporate bond has a face value of
      $10,000, a bond interest rate of 6% per
      year payable semiannually, and a maturity
      date of 20 years from now. If a person pur-
      chases the bond for $9000 when the inter-
      est rate in the marketplace is 8% per year,
      compounded semiannually, the size and
      frequency of the interest payments the
      person will receive are closest to
      (a) $270 every 6 months
      (b) $300 every 6 months
      (c) $360 every 6 months
      (d) $400 every 6 months

5.64 A municipal bond that was issued 3 years
      ago has a face value of $5000 and a bond
      interest rate of 4% per year payable semi-
      annually. The bond has a maturity date of

20 years *from the date it was issued.* If the
interest rate in the marketplace is 8% per
year, compounded quarterly, the value of n
that must be used in the P/A equation to
calculate the present worth of the bond is
(a)   34
(b)   40
(c)   68
(d)   80

5.65 A $10,000 bond has an interest rate of 6%
      per year payable quarterly. The bond ma-
      tures 15 years from now. At an interest rate
      of 8% per year, compounded quarterly, the
      present worth of the bond is represented
      by which of the equations below
      (a) $PW = 150(P/A,1.5\%,60) + 10,000$
            $(P/F,1.5\%,60)$
      (b) $PW = 150(P/A,2\%,60) + 10,000$
            $(P/F,2\%,60)$
      (c) $PW = 600(P/A,8\%,15) + 10,000$
            $(P/F,8\%,15)$
      (d) $PW = 600(P/A,2\%,60) + 10,000$
            $(P/F,2\%,60)$

EXTENDED EXERCISE

EVALUATION OF SOCIAL SECURITY RETIREMENT ESTIMATES

Charles is a senior engineer who has worked for 18 years since he graduated
from college. Yesterday in the mail, he received a report from the U.S. Social
Security Administration. In short, it stated that if he continues to earn at the same
rate, social security will provide him with the following estimated monthly retirement benefits:

• Normal retirement at age 66; full benefit of $1500 per month starting at
  age 66.
• Early retirement at age 62; benefit reduced by 25% starting at age 62.
• Extended retirement at age 70; benefit increased by 30% starting at age 70.

Charles never thought much about social security; he usually thought of it as a
monthly deduction from his paycheck that helped pay for his parents' retirement

benefits from social security. But this time he decided an analysis should be performed. Charles decided to neglect the effect of the following over time: income taxes, cost-of-living increases, and inflation. Also, he assumed the retirement benefits are all received at the end of each year; that is, no compounding effect occurs during the year. Using an expected rate of return on investments of 8% per year and an anticipated death just after his 85th birthday, use a spreadsheet to do the following for Charles:

1. Calculate the total future worth of each benefit scenario through the age of 85.
2. Plot the annual accumulated future worth for each benefit scenario through the age of 85.

The report also mentioned that if Charles dies this year, his spouse is eligible at full retirement age for a benefit of $1600 per month for the remainder of her life. If Charles and his wife are both 40 years old today, determine the following about his wife's survivor benefits, if she starts at age 66 and lives through her 85th birthday:

3. Present worth now.
4. Future worth for his wife after her 85th birthday.

CASE STUDY

PAYBACK EVALUATION OF ULTRALOW-FLUSH TOILET PROGRAM

Introduction

In many cities in the southwestern part of the United States, water is being withdrawn from subsurface aquifers faster than it is being replaced. The attendant depletion of groundwater supplies has forced some of these cities to take actions ranging from restrictive pricing policies to mandatory conservation measures in residential, commercial, and industrial establishments. Beginning in the mid-1990s, a city undertook a project to encourage installation of ultralow-flush toilets in existing houses. To evaluate the cost-effectiveness of the program, an economic analysis was conducted.

Background

The heart of the toilet replacement program involved a rebate of 75% of the cost of the fixture (up to $100 per

unit), providing the toilet used no more than 1.6 gallons of water per flush. There was no limit on the number of toilets any individual or business could have replaced.

Procedure

To evaluate the water savings achieved (if any) through the program, monthly water use records were searched for 325 of the household participants, representing a sample size of approximately 13%. Water consumption data were obtained for 12 months before and 12 months after installation of the ultralow-flush toilets. If the house changed ownership during the evaluation period, that account was not included in the evaluation. Since water consumption increases dramatically during the hot summer months for lawn watering, evaporative cooling, car washing, etc., only the winter months of December, January, and February were used to evaluate water consumption before and after installation of the toilet. Before any calculations were made, high-volume water users (usually businesses) were screened out by eliminating all records whose average monthly consumption exceeded 50 CCF (1 CCF = 100 cubic feet = 748 gallons). Additionally, accounts which had monthly averages of 2 CCF or less (either before or after installation) were also eliminated because it was believed that such low consumption rates probably represented an abnormal condition, such as a house for sale which was vacant during part of the study period. The 268 records that remained after the screening procedures were then used to quantify the effectiveness of the program.

Results

Water Consumption

Monthly consumption before and after installation of the ultralow-flush toilets was found to be 11.2 and 9.1 CCF, respectively, for an average reduction of 18.8%. When only the months of January and February were used in the before and after calculations, the respective values were 11.0 and 8.7 CCF, resulting in a water savings rate of 20.9%.

Economic Analysis

The following table shows some of the program totals through the first 1¾ years of the program.

Program Summary

Number of households participating	2466
Number of toilets replaced	4096
Number of persons	7981
Average cost of toilet	$115.83
Average rebate	$76.12

The results in the previous section indicated monthly water savings of 2.1 CCF. For the average program participant, the payback period n_p in years with *no interest considered* is calculated using Equation [5.7].

$$n_p = \frac{\text{net cost of toilets} + \text{installation cost}}{\text{net annual savings for water and sewer charges}}$$

The lowest rate block for water charges is $0.76 per CCF. The sewer surcharge is $0.62 per CCF. Using these values and a $50 cost for installation, the payback period is

$$n_p = \frac{(115.83 - 76.12) + 50}{(2.1 \text{ CCF/month} \times 12 \text{ months}) \times (0.76 + 0.62)/\text{CCF}}$$

$$= 2.6 \text{ years}$$

Less expensive toilets or lower installation costs would reduce the payback period accordingly, while consideration of the time value of money would lengthen it.

From the standpoint of the utility which supplies water, the cost of the program must be compared against the marginal cost of water delivery and wastewater treatment. The marginal cost c may be represented as

$$c = \frac{\text{cost of rebates}}{\begin{array}{c}\text{volume of water not delivered}\\ + \text{volume of wastewater not treated}\end{array}}$$

Theoretically, the reduction in water consumption would go on for an infinite period of time, since replacement will never be with a less efficient model. But for a worst-case condition, it is assumed the toilet would have a "productive" life of only 5 years, after which it would leak and not be repaired. The cost to the city for the water not delivered or wastewater not treated would be

$$c = \frac{\$76.12}{(2.1 + 2.1 \text{ CCF/month})(12 \text{ months})(5 \text{ years})}$$

$$= \frac{\$0.302}{\text{CCF}} \text{ or } \frac{\$0.40}{1000 \text{ gallons}}$$

Thus, unless the city can deliver water and treat the resulting wastewater for less than $0.40 per 1000 gallons, the toilet replacement program would be considered economically attractive. For the city, the operating costs alone, that is, without the capital expense, for water and wastewater services that were not expended were about $1.10 per 1000 gallons, which far exceeds $0.40 per 1000 gallons. Therefore, the toilet replacement program was clearly very cost-effective.

Case Study Exercises

1. For an interest rate of 8% and a toilet life of 5 years, what would the participant's payback period be?

2. Is the participant's payback period more sensitive to the interest rate used or to the life of the toilet?

3. What would the cost to the city be if an interest rate of 6% per year were used with a toilet life of 5 years? Compare the cost in $/CCF and $/1000 gallons to those determined at 0% interest.

4. From the city's standpoint, is the success of the program sensitive to (*a*) the percentage of toilet cost rebated, (*b*) the interest rate, if rates of 4% to 15% are used, or (*c*) the toilet life, if lives of 2 to 20 years are used?

5. What other factors might be important to (*a*) the participants and (*b*) the city in evaluating whether the program is a success?

CHAPTER

Annual Worth Analysis

In this chapter, we add to our repertoire of alternative comparison tools. In the last chapter we learned the PW method. Here we learn the equivalent annual worth, or AW, method. AW analysis is commonly considered the more desirable of the two methods because the AW value is easy to calculate; the measure of worth—AW in dollars per year—is understood by most individuals; and its assumptions are essentially identical to those of the PW method.

Annual worth is also known by other titles. Some are equivalent annual worth (EAW), equivalent annual cost (EAC), annual equivalent (AE), and EUAC (equivalent uniform annual cost). The resulting equivalent annual worth amount is the same for all name variations. The alternative selected by the AW method will always be the same as that selected by the PW method, and all other alternative evaluation methods, provided they are performed correctly.

In the case study, the estimates made when an AW analysis was performed are found to be substantially different after the equipment is installed. Spreadsheets, sensitivity analysis, and annual worth analysis work together to evaluate the situation.

LEARNING OBJECTIVES

Purpose: Make annual worth calculations and compare alternatives using the annual worth method.

This chapter will help you:

One life cycle	1. Demonstrate that AW needs to be calculated over only one life cycle.
AW calculation	2. Calculate capital recovery (CR) and AW using two methods.
Alternative selection by AW	3. Select the best alternative on the basis of an AW analysis.
Permanent investment AW	4. Calculate the AW of a permanent investment.

6.1 ADVANTAGES AND USES OF ANNUAL WORTH ANALYSIS

For many engineering economic studies, the AW method is the best to use, when compared to PW, FW, and rate of return (next two chapters). Since the AW value is the equivalent uniform annual worth of all estimated receipts and disbursements during the life cycle of the project or alternative, AW is easy to understand by any individual acquainted with annual amounts, that is, dollars per year. The AW value, which has the same economic interpretation as A used thus far, is equivalent to the PW and FW values at the MARR for n years. All three can be easily determined from each other by the relation

$$AW = PW(A/P,i,n) = FW(A/F,i,n) \qquad [6.1]$$

The n in the factors is the number of years for equal-service comparison. This is the LCM or the stated study period of the PW or FW analysis.

When all cash flow estimates are converted to an AW value, this value applies for every year of the life cycle, and for *each additional life cycle*. In fact, a prime computational and interpretation advantage is that

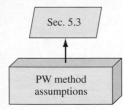

> **The AW value has to be calculated for *only one life cycle*. Therefore, it is not necessary to use the LCM of lives, as it is for PW and FW analyses.**

Therefore, determining the AW over one life cycle of an alternative determines the AW for all future life cycles. As with the PW method, there are three fundamental assumptions of the AW method that should be understood.

> **When alternatives being compared have different lives, the AW method makes the assumptions that**
>
> 1. **The services provided are needed for at least the LCM of the lives of the alternatives.**
> 2. **The selected alternative will be repeated for succeeding life cycles in exactly the same manner as for the first life cycle.**
> 3. **All cash flows will have the same estimated values in every life cycle.**

In practice, no assumption is precisely correct. If, in a particular evaluation, the first two assumptions are not reasonable, a study period must be established for the analysis. Note that for assumption 1, the length of time may be the indefinite future (forever). In the third assumption, all cash flows are expected to change exactly with the inflation (or deflation) rate. If this is not a reasonable assumption, new cash flow estimates must be made for each life cycle, and, again, a study period must be used. AW analysis for a stated study period is discussed in Section 6.3.

EXAMPLE 6.1

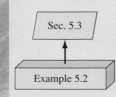

In Example 5.2 about office lease options, a PW analysis was performed over 18 years, the LCM of 6 and 9 years. Consider only location A, which has a 6-year life cycle. The diagram in Figure 6–1 shows the cash flows for all three life cycles (first cost $15,000; annual costs $3500; deposit return $1000). Demonstrate the equivalence at $i = 15\%$ of PW over three life cycles and AW over one cycle. In the previous example, present worth for location A was calculated as PW = $-45,036.

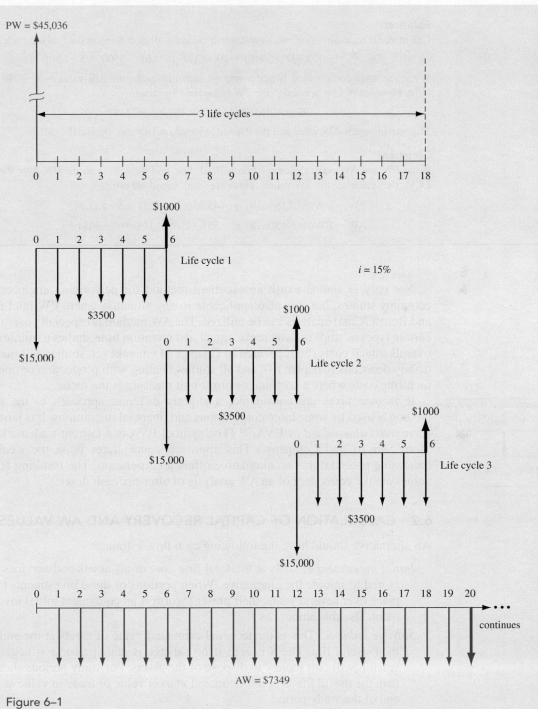

Figure 6–1
PW and AW values for three life cycles, Example 6.1.

Solution

Calculate the equivalent uniform annual worth value for all cash flows in the first life cycle.

$$AW = -15,000(A/P,15\%,6) + 1000(A/F,15\%,6) - 3500 = \$-7349$$

When the same computation is performed on each life cycle, the AW value is $\$-7349$. Now, Equation [6.1] is applied to the PW value for 18 years.

$$AW = -45,036(A/P,15\%,18) = \$-7349$$

The one-life-cycle AW value and the PW value based on 18 years are equal.

Comment

If the FW and AW equivalence relation is used, first find the FW from the PW over the LCM, then calculate the AW value. (There are small round-off errors.)

$$FW = PW(F/P,15\%,18) = -45,036(12.3755) = \$-557,343$$

$$AW = FW(A/F,15\%,18) = -557,343(0.01319) = \$-7351$$

Not only is annual worth an excellent method for performing engineering economy studies, but it is also applicable in any situation where PW (and FW and Benefit/Cost) analysis can be utilized. The AW method is especially useful in certain types of studies: asset replacement and retention time studies to minimize overall annual costs (both covered in Chapter 11), breakeven studies and make-or-buy decisions (Chapter 13), and all studies dealing with production or manufacturing costs where a cost/unit or profit/unit measure is the focus.

If income taxes are considered, a slightly different approach to the AW method is used by some large corporations and financial institutions. It is termed *economic value added* or EVA.™ (The symbol EVA is a current trademark of Stern Stewart and Company.) This approach concentrates upon the wealth-increasing potential that an alternative offers a corporation. The resulting EVA values are the equivalent of an AW analysis of after-tax cash flows.

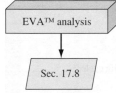

6.2   CALCULATION OF CAPITAL RECOVERY AND AW VALUES

An alternative should have the following cash flow estimates:

Initial investment P. This is the total first cost of all assets and services required to initiate the alternative. When portions of these investments take place over several years, their present worth is an equivalent initial investment. Use this amount as *P.*

Salvage value S. This is the terminal estimated value of assets at the end of their useful life. The *S* is zero if no salvage is anticipated; *S* is negative when it will cost money to dispose of the assets. For study periods shorter than the useful life, *S* is the estimated market value or trade-in value at the end of the study period.

Annual amount A. This is the equivalent annual amount (costs only for service alternatives; costs and receipts for revenue alternatives). Often this is the annual operating cost (AOC), so the estimate is already an equivalent *A* value.

The annual worth (AW) value for an alternative is comprised of two components: capital recovery for the initial investment P at a stated interest rate (usually the MARR) and the equivalent annual amount A. The symbol CR is used for the capital recovery component. In equation form,

$$AW = -CR - A \qquad [6.2]$$

Both CR and A have minus signs because they represent costs. The total annual amount A is determined from uniform recurring costs (and possibly receipts) and nonrecurring amounts. The P/A and P/F factors may be necessary to first obtain a present worth amount, then the A/P factor converts this amount to the A value in Equation [6.2]. (If the alternative is a revenue project, there will be positive cash flow estimates present in the calculation of the A value.)

The recovery of an amount of capital P committed to an asset, plus the time value of the capital at a particular interest rate, is a very fundamental principle of economic analysis. *Capital recovery is the equivalent annual cost of owning the asset plus the return on the initial investment.* The A/P factor is used to convert P to an equivalent annual cost. If there is some anticipated positive salvage value S at the end of the asset's useful life, its equivalent annual value is removed using the A/F factor. This action reduces the equivalent annual cost of owning the asset. Accordingly, CR is

$$CR = -[P(A/P,i,n) - S(A/F,i,n)] \qquad [6.3]$$

The computation of CR and AW is illustrated in Example 6.2.

EXAMPLE 6.2

Lockheed Martin is increasing its booster thrust power in order to win more satellite launch contracts from European companies interested in opening up new global communications markets. A piece of earth-based tracking equipment is expected to require an investment of $13 million, with $8 million committed now and the remaining $5 million expended at the end of year 1 of the project. Annual operating costs for the system are expected to start the first year and continue at $0.9 million per year. The useful life of the tracker is 8 years with a salvage value of $0.5 million. Calculate the AW value for the system, if the corporate MARR is currently 12% per year.

Solution
The cash flows (Figure 6–2a) for the tracker system must be converted to an equivalent AW cash flow sequence over 8 years (Figure 6–2b). (All amounts are expressed in $1 million units.) The AOC is $A = \$-0.9$ per year, and the capital recovery is calculated by using Equation [6.3]. The present worth P in year 0 of the two separate investment amounts of $8 and $5 is determined *before* multiplying by the A/P factor.

$$CR = -\{[8.0 + 5.0(P/F,12\%,1)](A/P,12\%,8) - 0.5(A/F,12\%,8)\}$$

$$= -\{[12.46](0.2013) - 0.040\}$$

$$= \$-2.47$$

The correct interpretation of this result is very important to Lockheed Martin. It means that each and every year for 8 years, the equivalent total revenue from the tracker must

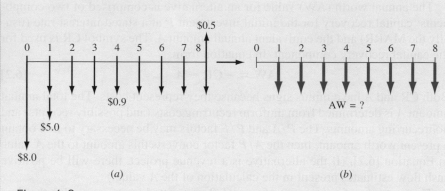

Figure 6–2

(*a*) Cash flow diagram for satellite tracker costs, and (*b*) conversion to an equivalent AW (in $1 million), Example 6.2.

be at least $2,470,000 *just to recover the initial present worth investment plus the required return of 12% per year.* This does not include the AOC of $0.9 million each year.

Since this amount, CR = $−2.47 million, is an *equivalent annual cost,* as indicated by the minus sign, total AW is found by Equation [6.2].

$$AW = -2.47 - 0.9 = \$-3.37 \text{ million per year}$$

This is the AW for all future life cycles of 8 years, provided the costs rise at the same rate as inflation, and the same costs and services are expected to apply for each succeeding life cycle.

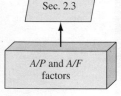

Sec. 2.3

A/P and A/F factors

There is a second, equally correct way to determine CR. Either method results in the same value. In Section 2.3, a relation between the A/P and A/F factors was stated as

$$(A/F,i,n) = (A/P,i,n) - i$$

Both factors are present in the CR Equation [6.3]. Substitute for the A/F factor to obtain

$$CR = -\{P(A/P,i,n) - S[(A/P,i,n) - i]\}$$

$$= -[(P - S)(A/P,i,n) + S(i)] \qquad [6.4]$$

There is a basic logic to this formula. Subtracting S from the initial investment P before applying the A/P factor recognizes that the salvage value will be recovered. This reduces CR, the annual cost of asset ownership. However, the fact that S is not recovered until year n of ownership is compensated for by charging the annual interest $S(i)$ against the CR.

In Example 6.2, the use of this second way to calculate CR results in the same value.

$$CR = -\{[8.0 + 5.0(P/F,12\%,1) - 0.5](A/P,12\%,8) + 0.5(0.12)\}$$

$$= -\{[12.46 - 0.5](0.2013) + 0.06\} = \$-2.47$$

Although either CR relation results in the same amount, it is better to consistently use the same method. The first method, Equation [6.3], will be used in this text.

For solution by computer, use the PMT function to determine CR only in a single spreadsheet cell. The general function $PMT(i\%,n,P,F)$ is rewritten using the initial investment as P and $-S$ for the salvage value. The format is

$$PMT(i\%,n,P,-S)$$

Q-Solv

As an illustration, determine the CR only in Example 6.2 above. Since the initial investment is distributed over 2 years—$8 million in year 0 and $5 million in year 1—embed the PV function into PMT to find the equivalent P in year 0. The complete function for only the CR amount (in $1 million units) is $PMT(12\%,8,8+PV(12\%,1,-5),-0.5)$, where the embedded PV function is in italic. The answer of -2.47 (million) will be displayed in the spreadsheet cell.

6.3 EVALUATING ALTERNATIVES BY ANNUAL WORTH ANALYSIS

The annual worth method is typically the easiest of the evaluation techniques to perform, when the MARR is specified. The alternative selected has the lowest equivalent annual cost (service alternatives), or highest equivalent income (revenue alternatives). This means that the selection guidelines are the same as for the PW method, but using the AW value.

For mutually exclusive alternatives, calculate AW at the MARR.

One alternative: AW ≥ 0, MARR is met or exceeded.

Two or more alternatives: Choose the lowest cost or highest income (numerically largest) AW value.

If an assumption in Section 6.1 is not acceptable for an alternative, a study period analysis must be used. Then the cash flow estimates over the study period are converted to AW amounts. This is illustrated later in Example 6.4.

EXAMPLE 6.3

PizzaRush, which is located in the general Los Angeles area, fares very well with its competition in offering fast delivery. Many students at the area universities and community colleges work part-time delivering orders made via the web at PizzaRush.com. The owner, a software engineering graduate of USC, plans to purchase and install five portable, in-car systems to increase delivery speed and accuracy. The systems provide a link between the web order-placement software and the On-Star© system for satellite-generated directions to any address in the Los Angeles area. The expected result is faster, friendlier service to customers, and more income for PizzaRush.

Each system costs $4600, has a 5-year useful life, and may be salvaged for an estimated $300. Total operating cost for all systems is $650 for the first year, increasing by $50 per year thereafter. The MARR is 10%. Perform an annual worth evaluation for the owner that answers the following questions. Perform the solution by hand and by computer, as requested below.

(a) How much new annual income is necessary to recover the investment at the MARR of 10% per year? Generate this value by hand and by computer.

(b) The owner conservatively estimates increased income of $1200 per year for all five systems. Is this project financially viable at the MARR? Solve by hand and by computer.

(c) Based on the answer in part (b), use the computer to determine how much new income PizzaRush must have to economically justify the project. Operating costs remain as estimated.

Solution by Hand

(a and b) The CR and AW values will answer these two questions. Cash flow is presented in Figure 6–3 for all five systems. Use Equation [6.3] for the capital recovery at 10%.

$$CR = 5(4600)(A/P,10\%,5) - 5(300)(A/F,10\%,5)$$
$$= \$5822$$

The financial viability can be determined without calculating the AW value. The $1200 in new income is substantially lower than the CR of $5822, which does not yet include the annual costs. The purchase is clearly not economically justified. However, to complete the analysis, determine AW. The annual operating costs and incomes form an arithmetic gradient series with a base of $550 in year 1, decreasing by $50 per year for 5 years. The AW relation is

$$AW = -\text{capital recovery} + \text{equivalent net income}$$
$$= -5822 + 550 - 50(A/G,10\%,5)$$
$$= \$-5362$$

This is the equivalent 5-year net amount needed to return the investment and recover the estimated operating costs at a 10% per year return. This shows, once again, that the alternative is clearly not financially viable at MARR = 10%. Note that the estimated extra $1200 per year income, offset by the operating costs, has reduced the required annual amount from $5822 to $5362.

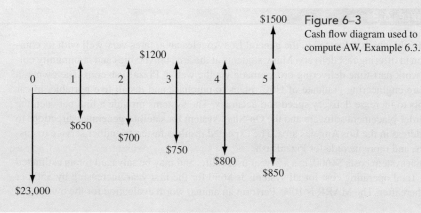

Figure 6–3
Cash flow diagram used to compute AW, Example 6.3.

Solution by Computer

The spreadsheet layout (Figure 6–4) shows the cash flows for the investment, the operating costs, and annual income in separate columns. The functions use global variable format for faster sensitivity analysis.

E-Solve

(*a* and *b*) The capital recovery value of $5822 is displayed in cell B7, which is determined by a PMT function with an embedded NPV function. Cells C7 and D7 also use the PMT function to find the annual equivalent for costs and incomes, again with an embedded NPV function.

Cell F11 displays the final answer of AW = $-5362, which is the sum of all three AW components in row 7.

(*c*) To find the income (column D) necessary to justify the project, a value of AW = $0 must be displayed in cell F11. Other estimates remain the same. Because all the annual incomes in column D receive their value from cell B4, change the entry in B4 until F11 displays '$0'. This occurs at $6562. (These amounts are not shown in B4 and F11 of Figure 6–4.) The owner of PizzaRush would have to increase the estimate of extra income for the new system from $1200 to $6562 per year to make a 10% return. This is a substantial increase.

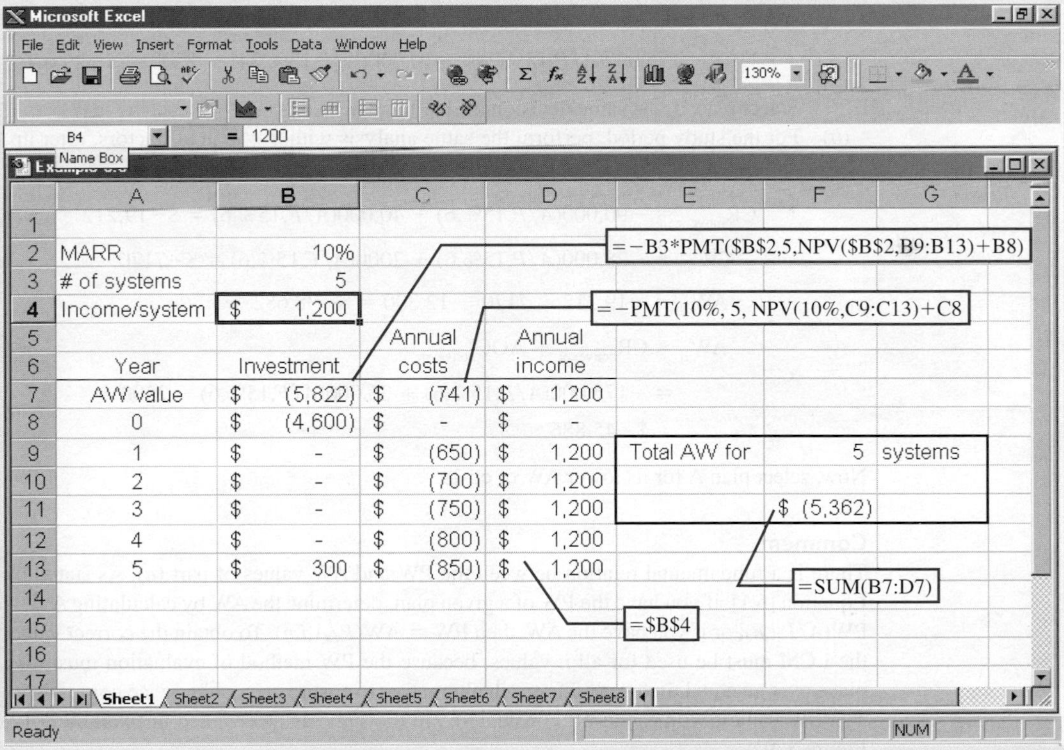

	A	B	C	D	E	F	G
1							
2	MARR	10%					
3	# of systems	5					
4	Income/system	$ 1,200					
5			Annual	Annual			
6	Year	Investment	costs	income			
7	AW value	$ (5,822)	$ (741)	$ 1,200			
8	0	$ (4,600)	$ -	$ -			
9	1	$ -	$ (650)	$ 1,200	Total AW for	5	systems
10	2	$ -	$ (700)	$ 1,200			
11	3	$ -	$ (750)	$ 1,200		$ (5,362)	
12	4	$ -	$ (800)	$ 1,200			
13	5	$ 300	$ (850)	$ 1,200			

Figure 6–4

Spreadsheet solution, Example 6.3(a) and (b).

EXAMPLE 6.4

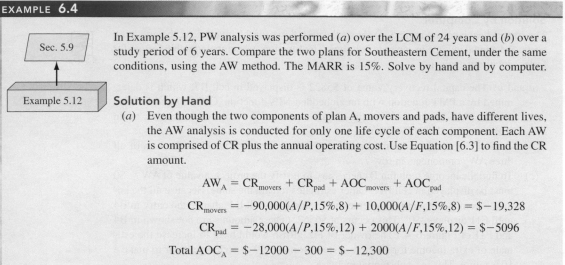

Sec. 5.9

Example 5.12

In Example 5.12, PW analysis was performed (a) over the LCM of 24 years and (b) over a study period of 6 years. Compare the two plans for Southeastern Cement, under the same conditions, using the AW method. The MARR is 15%. Solve by hand and by computer.

Solution by Hand

(a) Even though the two components of plan A, movers and pads, have different lives, the AW analysis is conducted for only one life cycle of each component. Each AW is comprised of CR plus the annual operating cost. Use Equation [6.3] to find the CR amount.

$$AW_A = CR_{movers} + CR_{pad} + AOC_{movers} + AOC_{pad}$$

$$CR_{movers} = -90,000(A/P,15\%,8) + 10,000(A/F,15\%,8) = \$-19,328$$

$$CR_{pad} = -28,000(A/P,15\%,12) + 2000(A/F,15\%,12) = \$-5096$$

$$\text{Total AOC}_A = \$-12000 - 300 = \$-12,300$$

The total AW for each plan is

$$AW_A = -19,328 - 5096 - 12,300 = \$-36,724$$

$$AW_B = CR_{conveyor} + AOC_{conveyor}$$

$$= -175,000(A/P,15\%,24) + 10,000(A/F,15\%,24) - 2500 = \$-29,646$$

Select plan B, the same decision as that for PW analysis.

(b) For the study period, perform the same analysis with $n = 6$ in all factors, after updating the salvage values to the residual values.

$$CR_{movers} = -90,000(A/P,15\%,6) + 40,000(A/F,15\%,6) = \$-19,212$$

$$CR_{pad} = -28,000(A/P,15\%,6) + 2000(A/F,15\%,6) = \$-7170$$

$$AW_A = -19,212 - 7170 - 12,300 = \$-38,682$$

$$AW_B = CR_{conveyor} + AOC_{conveyor}$$

$$= -175,000(A/P,15\%,6) + 25,000(A/F,15\%,6) - 2500$$

$$= \$-45,886$$

Now, select plan A for its lower AW of costs.

Comment

There is a fundamental relation between the PW and AW values of part (a). As stated by Equation [6.1], if you have the PW of a given plan, determine the AW by calculating $AW = PW(A/P,i,n)$; or if you have the AW, then $PW = AW(P/A,i,n)$. To obtain the correct value, the LCM must be used for all n values, because the PW method of evaluation must take place over an equal time period for each alternative to ensure an equal-service comparison. The PW values, with round-off considered, are the same as determined in Example 5.12, Figure 5–9.

$$PW_A = AW_A(P/A,15\%,24) = \$-236,275$$

$$PW_B = AW_B(P/A,15\%,24) = \$-190,736$$

Solution by Computer

(a) See Figure 6–5a. This is exactly the same format as that used for the PW evaluation over the LCM of 24 years (Figure 5–9), except only the *cash flows for one life cycle* are shown here, and the NPV functions at the head of each column are now *PMT functions with the NPV function embedded*. The cell tags detail two of the PMT functions, where the initial minus sign ensures the result is a cost amount in the total AW for each plan (cells H19 and H22). (The bottom portion of the spreadsheet is not shown. Plan B continues through its entire life with the $10,000 salvage value in year 24, and the annual cost of $2500 continues through year 24.)

   The resulting CR and AW values obtained here are the same as those for the solution by hand. Plan B is selected.

Plan A: $CR_{movers} = \$-19{,}328$ (in B8) $CR_{pad} = \$-5097$ (in D8)

   $AW_A = \$-36{,}725$ (in H19)

Plan B: $CR_{conveyor} = \$-27{,}146$ (in F8) $AW_B = \$-29{,}646$ (in H22)

(b) In Figure 6–5b, the lives are shortened to the study period of 6 years. The estimated residual values in year 6 are entered (row 16 cells), and all AOC amounts

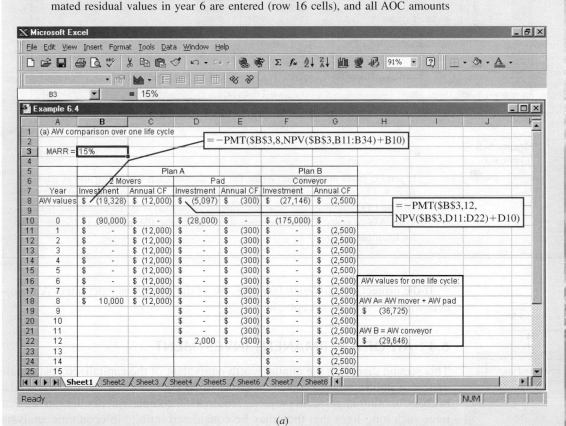

(a)

Figure 6–5
Spreadsheet solution using AW comparison of two alternatives: (*a*) one life cycle, (*b*) study period of 6 years, Example 6.4.

Microsoft Excel

File Edit View Insert Format Tools Data Window Help

100%

B3 = 15%

Example 6.4

	A	B	C	D	E	F	G	H	I	J
1	AW comparison over study period of 6 years									
2										
3	MARR =	15%				= −PMT($B$3,6,NPV($B$3,D11:D16)+D10)				
4										
5				Plan A			Plan B			
6			2 Movers		Pad		Conveyor			
7	Year	Investment	Annual CF	Investment	Annual CF	Investment	Annual CF			
8										
9	6-yr AW	$ (19,212)	$ (12,000)	$ (7,170)	$ (300)	$ (43,386)	$ (2,500)			
10	0	$ (90,000)	$ -	$ (28,000)	$ -	$ (175,000)	$ -			
11	1	$ -	$ (12,000)	$ -	$ (300)	$ -	$ (2,500)			
12	2	$ -	$ (12,000)	$ -	$ (300)	$ -	$ (2,500)			
13	3	$ -	$ (12,000)	$ -	$ (300)	$ -	$ (2,500)			
14	4	$ -	$ (12,000)	$ -	$ (300)	$ -	$ (2,500)			
15	5	$ -	$ (12,000)	$ -	$ (300)	$ -	$ (2,500)			
16	6	$ 40,000	$ (12,000)	$ 2,000	$ (300)	$ 25,000	$ (2,500)			
17										
18			AW over 6 years		=SUM(B9:E9)					
19										
20			AW A =	$ (38,682)						
21			AW B =	$ (45,886)						
22										

Sheet1 / Sheet2 / Sheet3 / Sheet4 / Sheet5 / Sheet6 / Sheet7 / Sheet8

Ready NUM

(b)

Figure 6–5
(*Continued*).

beyond 6 years are deleted. When the *n* value in each PMT function is adjusted from 8, 12, or 24 years to 6 in every case, new CR values are displayed, and cells D20 and D21 display the new AW values. Now plan A is selected since it has a lower AW of costs. This is the same result as for the PW analysis in Figure 5–10 for Example 5.12*b*.

If the projects are *independent,* the AW at the MARR is calculated. All projects with AW ≥ 0 are acceptable.

6.4 AW OF A PERMANENT INVESTMENT

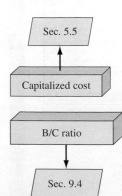

Sec. 5.5

Capitalized cost

B/C ratio

Sec. 9.4

This section discusses the annual worth equivalent of the capitalized cost. Evaluation of public sector projects, such as flood control dams, irrigation canals, bridges, or other large-scale projects, requires the comparison of alternatives that have such long lives that they may be considered infinite in economic analysis terms. For this type of analysis, the annual worth of the initial investment is the perpetual annual interest earned on the initial investment, that is, $A = Pi$. This is Equation [5.3]; however, the *A* value is also the capital recovery amount. (This same relation will be used again when benefit/cost ratios are discussed.)

Cash flows recurring at regular or irregular intervals are handled exactly as in conventional AW computations; they are converted to equivalent uniform annual amounts A for one cycle. This automatically annualizes them for each succeeding life cycle, as discussed in Section 6.1. Add all the A values to the CR amount to find total AW, as in Equation [6.2].

EXAMPLE 6.5

The U.S. Bureau of Reclamation is considering three proposals for increasing the capacity of the main drainage canal in an agricultural region of Nebraska. Proposal A requires dredging the canal in order to remove sediment and weeds which have accumulated during previous years' operation. The capacity of the canal will have to be maintained in the future near its design peak flow because of increased water demand. The Bureau is planning to purchase the dredging equipment and accessories for $650,000. The equipment is expected to have a 10-year life with a $17,000 salvage value. The annual operating costs are estimated to total $50,000. To control weeds in the canal itself and along the banks, environmentally safe herbicides will be sprayed during the irrigation season. The yearly cost of the weed control program is expected to be $120,000.

Proposal B is to line the canal with concrete at an initial cost of $4 million. The lining is assumed to be permanent, but minor maintenance will be required every year at a cost of $5000. In addition, lining repairs will have to be made every 5 years at a cost of $30,000.

Proposal C is to construct a new pipeline along a different route. Estimates are: an initial cost of $6 million, annual maintenance of $3000 for right-of-way, and a life of 50 years.

Compare the alternatives on the basis of annual worth, using an interest rate of 5% per year.

Solution

Since this is an investment for a permanent project, compute the AW for one cycle of all recurring costs. For proposals A and C, the CR values are found using Equation [6.3], with $n_A = 10$ and $n_C = 50$, respectively. For proposal B, the CR is simply $P(i)$.

Proposal A	
CR of dredging equipment:	
$-650,000(A/P,5\%,10) + 17,000(A/F,5\%,10)$	$\$ -82,824$
Annual cost of dredging	$-50,000$
Annual cost of weed control	$-120,000$
	$\$-252,824$
Proposal B	
CR of initial investment: $-4,000,000(0.05)$	$\$-200,000$
Annual maintenance cost	$-5,000$
Lining repair cost: $-30,000(A/F,5\%,5)$	$-5,429$
	$-210,429$
Proposal C	
CR of pipeline: $-6,000,000(A/P,5\%,50)$	$\$-328,680$
Annual maintenance cost	$-3,000$
	$\$-331,680$

Proposal B is selected due to its lowest AW of costs.

Comment

Note the use of the A/F factor for the lining repair cost in proposal B. The A/F factor is used instead of A/P, because the lining repair cost begins in year 5, not year 0, and continues indefinitely at 5-year intervals.

If the 50-year life of proposal C is considered infinite, CR $= P(i) = \$-300,000$, instead of $\$-328,680$ for $n = 50$. This is a small economic difference. How long lives of 40 or more years are treated economically is a matter of "local" practice.

EXAMPLE 6.6

An engineer with Becker Consulting has just received a bonus of $10,000. If she deposits it now at an interest rate of 8% per year, how many years must the money accumulate before she can withdraw $2000 per year forever? Use a computer to find the answer.

Solution by Computer

Figure 6–6 presents the cash flow diagram. The first step is to find the total amount of money, call it P_n, that must be accumulated in year n, just 1 year prior to the first withdrawal of the perpetual $A = \$2000$ per year series. That is,

$$P_n = \frac{A}{i} = \frac{2000}{0.08} = \$25,000$$

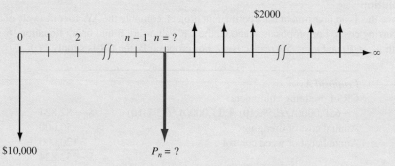

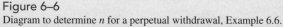

Figure 6–6
Diagram to determine n for a perpetual withdrawal, Example 6.6.

Q-Solv

E-Solve

Use the NPER function in one cell to determine when the initial $10,000 deposit will accumulate to $25,000 (Figure 6–7, cell B4). The answer is 11.91 years. If the engineer leaves the money in for 12 years, and if 8% is earned every year, forever, the $2000 per year is ensured.

Figure 6–7 also presents a more general spreadsheet solution in cells B7 through B11. Cell B10 determines the amount to accumulate in order to receive any amount (cell B9) forever at 8% (cell B7), and B11 includes the NPER function developed in cell reference format for any interest rate, deposit, and accumulated amount.

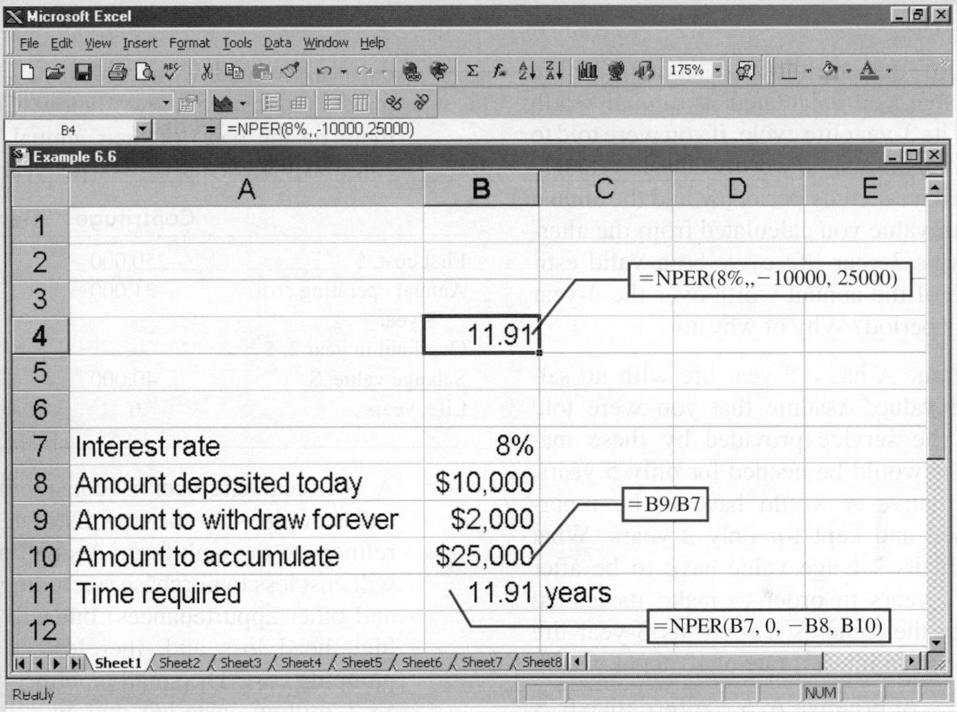

Figure 6–7
Two spreadsheet solutions to find an *n* value using the NPER function, Example 6.6.

CHAPTER SUMMARY

The annual worth method of comparing alternatives is often preferred to the present worth method, because the AW comparison is performed for only one life cycle. This is a distinct advantage when comparing different-life alternatives. AW for the first life cycle is the AW for the second, third, and all succeeding life cycles, under certain assumptions. When a study period is specified, the AW calculation is determined for that time period, regardless of the lives of the alternatives. As in the present worth method, the remaining value of an alternative at the end of the study period is recognized by estimating a market value.

For infinite-life alternatives, the initial cost is annualized simply by multiplying *P* by *i*. For finite-life alternatives, the AW through one life cycle is equal to the perpetual equivalent annual worth.

PROBLEMS

6.1   Assume that an alternative has a 3-year life and that you calculated its annual worth over its 3-year life cycle. If you were told to provide the annual worth of that alternative for a 4-year study period, would the annual worth value you calculated from the alternative's 3-year life cycle be a valid estimate of the annual worth over the 4-year study period? Why or why not?

6.2   Machine A has a 3-year life with no salvage value. Assume that you were told that the service provided by these machines would be needed for only 5 years. Alternative A would have to be repurchased and kept for only 2 years. What would its salvage value have to be after the 2 years in order to make its annual worth the same as it is for its 3-year life cycle at an interest rate of 10% per year?

Year	Alternative A, $	Alternative B, $
0	−10,000	−20,000
1	−7,000	−5,000
2	−7,000	−5,000
3	−7,000	−5,000
4		−5,000
5		−5,000

Alternatives Comparison

6.3   A consulting engineering firm is considering two models of SUVs for the company principals. A GM model will have a first cost of $26,000, an operating cost of $2000, and a salvage value of $12,000 after 3 years. A Ford model will have a first cost of $29,000, an operating cost of $1200, and a $15,000 resale value after 3 years. At an interest rate of 15% per year, which model should the consulting firm buy? Conduct an annual worth analysis.

6.4   A large textile company is trying to decide which sludge dewatering process it should use ahead of its sludge drying operation. The costs associated with centrifuge and belt press systems are shown below. Compare them on the basis of their annual worths, using an interest rate of 10% per year.

	Centrifuge	Belt Press
First cost, $	−250,000	−170,000
Annual operating cost, $/year	−31,000	−35,000
Overhaul in year 2, $	—	−26,000
Salvage value, $	40,000	10,000
Life, years	6	4

6.5   A chemical engineer is considering two styles of pipes for moving distillate from a refinery to the tank farm. A small pipeline will cost less to purchase (including valves and other appurtenances) but will have a high head loss and, therefore, a higher pumping cost. The small pipeline will cost $1.7 million installed and will have an operating cost of $12,000 per month. A larger-diameter pipeline will cost $2.1 million installed, but its operating cost will be only $8000 per month. Which pipe size is more economical at an interest rate of 1% per month on the basis of an annual worth analysis? Assume the salvage value is 10% of the first cost for each pipeline at the end of the 10-year project period.

6.6   Polymer Molding, Inc., is considering two processes for manufacturing storm drains. Plan A involves conventional injection molding that will require making a steel mold at a cost of $2 million. The cost for inspecting, maintaining, and cleaning the molds is expected to be $5000 per month. Since the cost of materials for this plan is expected to be the same as for the other plan, this cost will not be included in the comparison. The salvage value for plan A is expected to be 10% of the first cost. Plan B involves using an innovative process known as virtual engineered composites wherein a floating mold uses an operating system that constantly adjusts the water pressure around the mold and the chemicals entering the process. The first cost to tool the floating mold is only $25,000, but because of the newness of the process, personnel and product-reject costs are expected to be higher than those for a conventional process. The company expects the operating costs to be $45,000 per month for the first 8 months and then to decrease to $10,000 per month thereafter. There will be no salvage value with this plan. At an interest rate of 12% per year, compounded monthly, which process should the company select on the basis of an annual worth analysis over a 3-year study period?

6.7 An industrial engineer is considering two robots for purchase by a fiber-optic manufacturing company. Robot X will have a first cost of $85,000, an annual maintenance and operation (M&O) cost of $30,000, and a $40,000 salvage value. Robot Y will have a first cost of $97,000, an annual M&O cost of $27,000, and a $48,000 salvage value. Which should be selected on the basis of an annual worth comparison at an interest rate of 12% per year? Use a 3-year study period.

6.8 Accurate airflow measurement requires straight unobstructed pipe for a minimum of 10 diameters upstream and 5 diameters downstream of the measuring device. In one particular application, physical constraints compromised the pipe layout, so the engineer was considering installing the airflow probes in an elbow, knowing that flow measurement would be less accurate but good enough for process control. This was plan A, which would be acceptable for only 2 years, after which a more accurate flow measurement system with the same costs as plan A will be available. This plan would have a first cost of $25,000 with annual maintenance estimated at $4000. Plan B involved installation of a recently

designed submersible airflow probe. The stainless steel probe could be installed in a drop pipe with the transmitter located in a waterproof enclosure on the handrail. The cost of this system would be $88,000, but because it is accurate, it would not have to be replaced for at least 6 years. Its maintenance cost is estimated to be $1400 per year. Neither system will have a salvage value. At an interest rate of 12% per year, which one should be selected on the basis of an annual worth comparison?

6.9 A mechanical engineer is considering two types of pressure sensors for a low-pressure steam line. The costs are shown below. Which should be selected based on an annual worth comparison at an interest rate of 12% per year?

	Type X	Type Y
First cost, $	−7,650	−12,900
Maintenance cost, $/year	−1,200	−900
Salvage value, $	0	2,000
Life, years	2	4

6.10 The machines shown below are under consideration for an improvement to an automated candy bar wrapping process. Determine which should be selected on the basis of an annual worth analysis using an interest rate of 15% per year.

	Machine C	Machine D
First cost, $	−40,000	−65,000
Annual cost, $/year	−10,000	−12,000
Salvage value, $	12,000	25,000
Life, years	3	6

6.11 Two processes can be used for producing a polymer that reduces friction loss in engines. Process K will have a first cost of $160,000, an operating cost of $7000 per month, and a salvage value of $40,000 after its 2-year life. Process L will have a first cost of $210,000, an operating cost of $5000 per month, and a $26,000 salvage value after its 4-year life. Which process should be selected on the basis of an annual worth analysis at an interest rate of 12% per year, compounded monthly?

6.12 Two mutually exclusive projects have the estimated cash flows shown below. Use an annual worth analysis to determine which should be selected at an interest rate of 10% per year.

	Project Q	Project R
First cost, $	−42,000	−80,000
Annual cost, $/year	−6,000	−7,000 year 1, increasing by $1000 per year
Salvage value, $	0	4,000
Life, years	2	4

6.13 An environmental engineer is considering three methods for disposing of a nonhazardous chemical sludge: land application, fluidized-bed incineration, and private disposal contract. The details of each method are shown below. Determine which has the least cost on the basis of an annual worth comparison at 12% per year.

	Land Application	Inciner-ation	Contract
First cost, $	−110,000	−800,000	0
Annual cost, $/year	−95,000	−60,000	−190,000
Salvage value, $	15,000	250,000	0
Life, years	3	6	2

6.14 A state highway department is trying to decide whether it should "hot-patch" a short section of an existing county road or resurface it. If the hot-patch method is used, approximately 300 cubic meters of material would be required at a cost of $700 per cubic meter (in place). Additionally, the shoulders will have to be improved at the same time at a cost of $24,000. These improvements will last 2 years, at which time they will have to be redone. The annual cost of routine maintenance on the patched up road would be $5000. Alternatively, the state can resurface the road at a cost of $850,000. This surface will last 10 years if

the road is maintained at a cost of $2000 per year beginning 3 years from now. No matter which alternative is selected, the road will be completely rebuilt in 10 years. At an interest rate of 8% per year, which alternative should the state select on the basis of an annual worth analysis?

Permanent Investments and Projects

6.15 How much must you deposit in your retirement account starting *now* and continuing each year through year 9 (i.e., 10 deposits) if you want to be able to withdraw $80,000 per year forever beginning 30 years from now? Assume the account earns interest at 10% per year.

6.16 What is the *difference* in annual worth between an investment of $100,000 per year for 100 years and an investment of $100,000 per year forever at an interest rate of 10% per year?

6.17 A stockbroker claims she can consistently earn 15% per year on an investor's money. If she invests $20,000 now, $40,000 two years from now, and $10,000 per year through year 11 starting 4 years from now, how much money can the client withdraw every year forever, beginning 12 years from now, if the stockbroker delivers what she said and the account earns 6% per year from year 12 forward? Disregard taxes.

6.18 Determine the perpetual equivalent annual worth (in years 1 through infinity) of an investment of $50,000 at time 0 and $50,000 per year thereafter (forever) at an interest rate of 10% per year.

6.19 The cash flow associated with landscaping and maintaining a certain monument in Washington, D.C., is $100,000 now and $50,000 every 5 years forever. Determine its perpetual equivalent annual worth (in years 1 through infinity) at an interest rate of 8% per year.

6.20 The cost associated with maintaining rural highways follows a predictable pattern.

There are usually no costs for the first 3 years, but thereafter maintenance is required for restriping, weed control, light replacement, shoulder repairs, etc. For one section of a particular highway, these costs are projected to be $6000 in year 3, $7000 in year 4, and amounts increasing by $1000 per year through the highway's expected 30-year life. Assuming it is replaced with a similar roadway, what is its perpetual equivalent annual worth (in years 1 through infinity) at an interest rate of 8% per year?

6.21 A philanthropist working to set up a permanent endowment wants to deposit money each year, starting *now* and making 10 more (i.e., 11) deposits, so that money will be available for research related to planetary colonization. If the size of the first deposit is $1 million and each succeeding one is $100,000 larger than the previous one, how much will be available forever beginning in year 11, if the fund earns interest at a rate of 10% per year?

FE REVIEW PROBLEMS

Note: The sign convention on the FE exam may be opposite of that used here. That is, on the FE exam, costs may be positive and receipts negative.

6.24 For the mutually exclusive alternatives shown below, determine which one(s) should be selected.

Alternative	Annual Worth, $/yr
A	−25,000
B	−12,000
C	10,000
D	15,000

(a) Only A
(b) Only D
(c) Only A and B
(d) Only C and D

6.22 For the cash flow sequence shown below (in thousands of dollars), determine the amount of money that can be withdrawn annually for an infinite period of time, if the first withdrawal is to be made in year 10 and the interest rate is 12% per year.

Year	0	1	2	3	4	5	6
Deposit amount, $	100	90	80	70	60	50	40

6.23 A company that manufactures magnetic membrane switches is investigating three production options that have the estimated cash flows below. (a) Determine which option is preferable at an interest rate of 15% per year. (b) If the options are independent, determine which are economically acceptable. (All dollar values are in millions.)

	In-house	License	Contract
First cost, $	−30	−2	0
Annual cost, $/year	−5	−0.2	−2
Annual income, $/year	14	1.5	2.5
Salvage value, $	7	—	—
Life, years	10	∞	5

6.25 The annual worth (in years 1 through infinity) of $50,000 now, $10,000 per year in years 1 through 15, and $20,000 per year in years 16 through infinity at 10% per year is closest to
(a) Less than $16,900
(b) $16,958
(c) $17,394
(d) $19,573

6.26 An alumnus of West Virginia University wishes to start an endowment that will provide scholarship money of $40,000 per year beginning in year 5 and continuing indefinitely. The donor plans to give money *now* and for each of the next 2 years. If the size of each donation is exactly the same, the amount that must be donated each year at $i = 8\%$ per year is closest to

(a) $190,820
(b) $122,280
(c) $127,460
(d) $132,040

6.27. How much must you deposit in your retirement account each year for 10 years starting *now* (i.e., years 0 through 9) if you want to be able to withdraw $50,000 per year forever beginning 30 years from now? Assume the account earns interest at 10% per year.
(a) $4239
(b) $4662
(c) $4974
(d) $5471

6.28 Assume that a grateful engineering economy graduate starts an endowment at UTEP by donating $100,000 now. The conditions of the donation are that scholarships totaling $10,000 per year are to be given to engineering economy students beginning *now* and continuing through year 5. After that (i.e., year 6), scholarships are to be given in an amount equal to the interest that is generated on the investment. If the investment earns an effective rate of 10% per year, compounded continuously, how much money will be available for scholarships from year 6 on?
(a) $7380
(b) $8389
(c) $10,000
(d) $11,611

Problems 6.29 through 6.31 are based on the following cash flows and an interest rate of 10% per year, compounded semiannually.

	Alternative X	Alternative Y
First cost, $	−200,000	−800,000
Annual cost, $/year	−60,000	−10,000
Salvage value, $	20,000	150,000
Life, years	5	∞

6.29 In comparing the alternatives by the *annual worth method,* the annual worth of alternative X is represented by
(a) −200,000(0.1025) − 60,000 + 20,000(0.1025)
(b) −200,000(A/P,10%,5) − 60,000 + 20,000(A/F,10%,5)
(c) −200,000(A/P,5%,10) − 60,000 + 20,000(A/F,5%,10)
(d) −200,000(A/P,10.25%,5) − 60,000 + 20,000(A/F,10.25%,5)

6.30 The annual worth of *perpetual service* for alternative X is represented by
(a) −200,000(0.1025) − 60,000 + 20,000(0.1025)
(b) −200,000(A/P,10%,5) − 60,000 + 20,000(A/F,10%,5)
(c) −200,000(0.10) − 60,000 + 20,000(0.10)
(d) −200,000(A/P,10.25%,5) − 60,000 + 20,000(A/F,10.25%,5)

6.31 The annual worth of alternative Y is closest to
(a) $−50,000
(b) $−76,625
(c) $−90,000
(d) $−92,000

CASE STUDY

THE CHANGING SCENE OF AN ANNUAL WORTH ANALYSIS

Harry, owner of an automobile battery distributorship in Atlanta, Georgia, performed an economic analysis 3 years ago when he decided to place surge protectors in-line for all his major pieces of testing equipment.

The estimates used and the annual worth analysis at MAAR = 15% are summarized here. Two different manufacturers' protectors were compared.

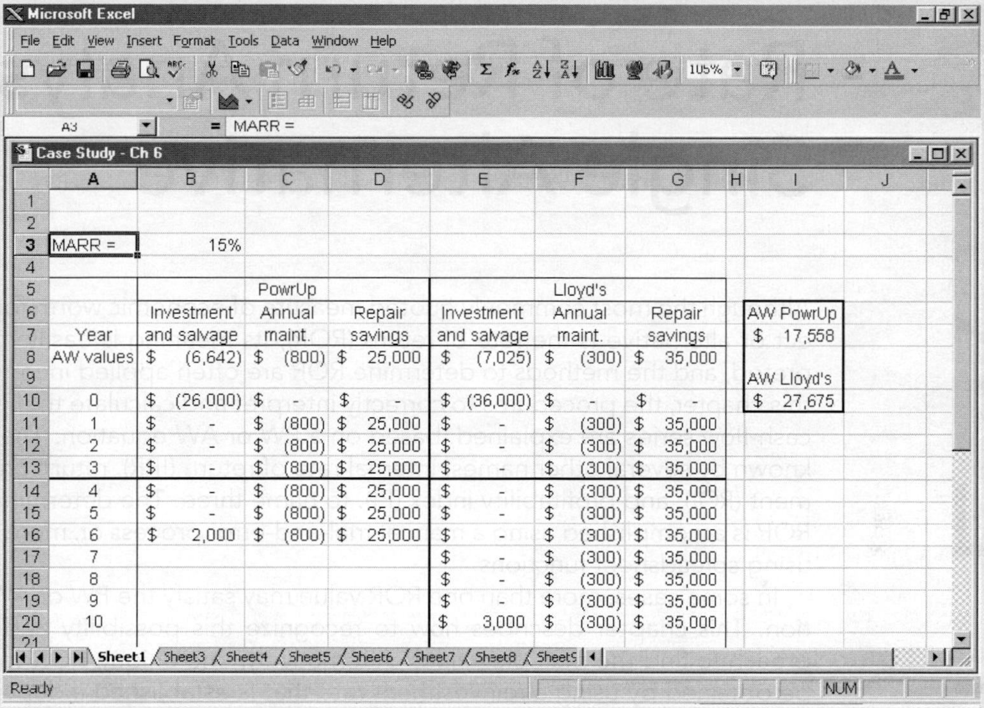

	A	B	C	D	E	F	G	H	I
1									
2									
3	MARR =	15%							
4									
5			PowrUp			Lloyd's			
6		Investment	Annual	Repair	Investment	Annual	Repair		AW PowrUp
7	Year	and salvage	maint.	savings	and salvage	maint.	savings		$ 17,558
8	AW values	$ (6,642)	$ (800)	$ 25,000	$ (7,025)	$ (300)	$ 35,000		
9									AW Lloyd's
10	0	$ (26,000)	$ -	$ -	$ (36,000)	$ -	$ -		$ 27,675
11	1	$ -	$ (800)	$ 25,000	$ -	$ (300)	$ 35,000		
12	2	$ -	$ (800)	$ 25,000	$ -	$ (300)	$ 35,000		
13	3	$ -	$ (800)	$ 25,000	$ -	$ (300)	$ 35,000		
14	4	$ -	$ (800)	$ 25,000	$ -	$ (300)	$ 35,000		
15	5	$ -	$ (800)	$ 25,000	$ -	$ (300)	$ 35,000		
16	6	$ 2,000	$ (800)	$ 25,000	$ -	$ (300)	$ 35,000		
17	7				$ -	$ (300)	$ 35,000		
18	8				$ -	$ (300)	$ 35,000		
19	9				$ -	$ (300)	$ 35,000		
20	10				$ 3,000	$ (300)	$ 35,000		

Figure 6–8
AW analysis of two surge protector proposals, case study, Chapter 6.

	PowrUp	Lloyd's
Cost and installation	$−26,000	$−36,000
Annual maintenance cost	−800	−300
Salvage value	2000	3000
Equipment repair savings	25,000	35,000
Useful life, years	6	10

The spreadsheet in Figure 6–8 is the one Harry used to make the decision. Lloyd's was the clear choice due to its substantially large AW value. The Lloyd's protectors were installed.

During a quick review this last year (year 3 of operation), it was obvious the maintenance costs and repair savings have not followed (and will not follow) the estimates made 3 years ago. In fact, the maintenance contract cost (which includes quarterly inspection) is going from $300 to $1200 per year next year and will then increase 10% per year for the next 10 years. Also, the repair savings for the last 3 years were $35,000, $32,000, and $28,000, as best as Harry can determine. He believes savings will decrease by $2000 per year hereafter. Finally, these 3-year-old protectors are worth nothing on the market now, so the salvage in 7 years is zero, not $3000.

Case Study Exercises

1. Plot a graph of the newly estimated maintenance costs and repair savings projections, assuming the protectors last for 7 more years.
2. With these new estimates, what is the recalculated AW for the Lloyd's protectors? Use the old first cost and maintenance cost estimates for the first 3 years. If these estimates had been made 3 years ago, would Lloyd's still have been the economic choice?
3. How has the capital recovery amount changed for the Lloyd's protectors with these new estimates?

CHAPTER 7

Rate of Return Analysis: Single Alternative

Although the most commonly quoted measure of economic worth for a project or alternative is the rate of return (ROR), its meaning is easily misinterpreted, and the methods to determine ROR are often applied incorrectly. In this chapter, the procedures to correctly interpret and calculate the ROR of a cash flow series are explained, based on a PW or AW equation. The ROR is known by several other names: internal rate of return (IRR), return on investment (ROI), and profitability index (PI), to name three. The determination of ROR is accomplished using a manual trial-and-error process or, more rapidly, using spreadsheet functions.

In some cases, more than one ROR value may satisfy the PW or AW equation. This chapter describes how to recognize this possibility and an approach to find the multiple values. Alternatively, one unique ROR value can be obtained by using a reinvestment rate that is established independently of the project cash flows.

Only one alternative is considered here; the next chapter applies these same principles to multiple alternatives. Finally, the rate of return for a bond investment is discussed here.

The case study focuses on a cash flow series that has multiple rates of return.

LEARNING OBJECTIVES

Purpose: Understand the meaning of rate of return (ROR) and perform ROR calculations for one alternative.

This chapter will help you:

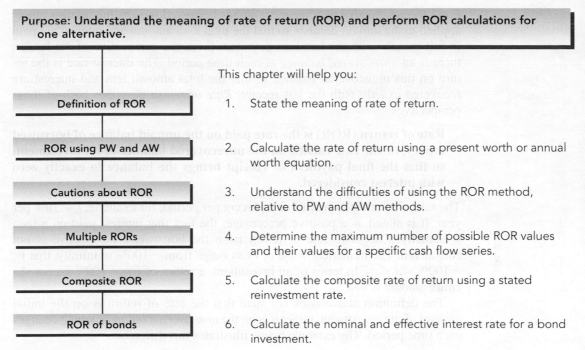

Definition of ROR	1. State the meaning of rate of return.
ROR using PW and AW	2. Calculate the rate of return using a present worth or annual worth equation.
Cautions about ROR	3. Understand the difficulties of using the ROR method, relative to PW and AW methods.
Multiple RORs	4. Determine the maximum number of possible ROR values and their values for a specific cash flow series.
Composite ROR	5. Calculate the composite rate of return using a stated reinvestment rate.
ROR of bonds	6. Calculate the nominal and effective interest rate for a bond investment.

7.1 INTERPRETATION OF A RATE OF RETURN VALUE

From the perspective of someone who has borrowed money, the interest rate is applied to the *unpaid balance* so that the total loan amount and interest are paid in full exactly with the last loan payment. From the perspective of the lender, there is an *unrecovered balance* at each time period. The interest rate is the return on this unrecovered balance so that the total amount lent and interest are recovered exactly with the last receipt. *Rate of return* describes both of these perspectives.

> **Rate of return (ROR) is the rate paid on the unpaid balance of borrowed money, or the rate earned on the unrecovered balance of an investment, so that the final payment or receipt brings the balance to exactly zero with interest considered.**

The rate of return is expressed as a percent per period, for example, $i = 10\%$ per year. It is stated as a positive percentage; the fact that interest paid on a loan is actually a negative rate of return from the borrower's perspective is not considered. The numerical value of i can range from -100% to infinity, that is, $-100\% < i < \infty$. In terms of an investment, a return of $i = -100\%$ means the entire amount is lost.

The definition above does not state that the rate of return is on the initial amount of the investment, rather it is on the *unrecovered balance,* which changes each time period. The example below illustrates this difference.

EXAMPLE 7.1

Wells Fargo Bank lent a newly graduated engineer $1000 at $i = 10\%$ per year for 4 years to buy home office equipment. From the bank's perspective (the lender), the investment in this young engineer is expected to produce an equivalent net cash flow of $315.47 for each of 4 years.

$$A = \$1000(A/P,10\%,4) = \$315.47$$

This represents a 10% per year rate of return on the bank's unrecovered balance. Compute the amount of the unrecovered investment for each of the 4 years using (*a*) the rate of return on the unrecovered balance (the correct basis) and (*b*) the return on the initial $1000 investment. (*c*) Explain why all the initial $1000 amount is not recovered by the final payment in part (*b*).

Solution
(*a*) Table 7–1 shows the unrecovered balance at the end of each year in column 6 using the 10% rate on the *unrecovered balance at the beginning of the year.* After 4 years the total $1000 is recovered, and the balance in column 6 is exactly zero.

(*b*) Table 7–2 shows the unrecovered balance if the 10% return is always figured on the *initial $1000.* Column 6 in year 4 shows a remaining unrecovered amount of $138.12, because only $861.88 is recovered in the 4 years (column 5).

TABLE 7–1   Unrecovered Balances Using a Rate of Return of 10% on the Unrecovered Balance

(1)	(2) Beginning Unrecovered	(3) = 0.10 × (2) Interest on Unrecovered	(4) Cash	(5) = (4) − (3) Recovered	(6) = (2) + (5) Ending Unrecovered
Year	Balance	Balance	Flow	Amount	Balance
0	—	—	$−1,000.00	—	$−1,000.00
1	$−1,000.00	$100.00	+315.47	$215.47	−784.53
2	−784.53	78.45	+315.47	237.02	−547.51
3	−547.51	54.75	+315.47	260.72	−286.79
4	−286.79	28.68	+315.47	286.79	0
		$261.88		$1,000.00	

TABLE 7–2   Unrecovered Balances Using a 10% Return on the Initial Amount

(1)	(2) Beginning Unrecovered	(3) = 0.10 × (2) Interest on	(4) Cash	(5) = (4) − (3) Recovered	(6) = (2) + (5) Ending Unrecovered
Year	Balance	Initial Amount	Flow	Amount	Balance
0	—	—	$−1,000.00	—	$−1,000.00
1	$−1,000.00	$100	+315.47	$215.47	−784.53
2	−784.53	100	+315.47	215.47	−569.06
3	−569.06	100	+315.47	215.47	−353.59
4	−353.59	100	+315.47	215.47	−138.12
		$400		$861.88	

(c)   A total of $400 in interest must be earned if the 10% return each year is based on the initial amount of $1000. However, only $261.88 in interest must be earned if a 10% return on the unrecovered balance is used. There is more of the annual cash flow available to reduce the remaining loan when the rate is applied to the unrecovered balance as in part (a) and Table 7–1. Figure 7–1 illustrates the correct interpretation of rate of return in Table 7–1. Each year the $315.47 receipt represents 10% interest on the unrecovered balance in column 2 plus the recovered amount in column 5.

Because rate of return is the interest rate on the unrecovered balance, the computations in *Table 7–1 for part (a) present a correct interpretation of a 10% rate of return.* Clearly, an interest rate applied only to the principal represents a higher rate than is stated. In practice, a so-called add-on interest rate is frequently based on principal only, as in part (b). This is sometimes referred to as the *installment financing* problem.

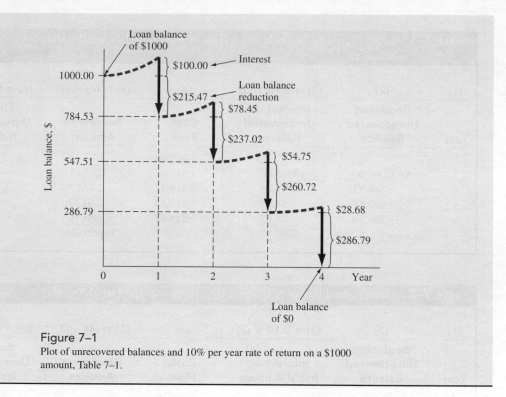

Figure 7–1
Plot of unrecovered balances and 10% per year rate of return on a $1000 amount, Table 7–1.

Installment financing can be discovered in many forms in everyday finances. One popular example is a "no-interest program" offered by retail stores on the sale of major appliances, audio and video equipment, furniture, and other consumer items. Many variations are possible, but in most cases, if the purchase is not paid for in full by the time the promotion is over, usually 6 months to 1 year later, *finance charges are assessed from the original date of purchase.* Further, the program's fine print may stipulate that the purchaser use a credit card issued by the retail company, which often has a higher interest rate than that of a regular credit card, for example, 24% per year compared to 18% per year. In all these types of programs, the one common theme is more interest paid over time by the consumer. Usually, the correct definition of *i* as interest on the unpaid balance does not apply directly; *i* has often been manipulated to the financial disadvantage of the purchaser.

7.2   RATE OF RETURN CALCULATION USING A PW OR AW EQUATION

To determine the rate of return of a cash flow series, set up the ROR equation using either PW or AW relations. The present worth of costs or disbursements PW_D is equated to the present worth of incomes or receipts PW_R. Equivalently, the two can be subtracted and set equal to zero. That is, solve for i using

$$PW_D = PW_R$$

$$0 = -PW_D + PW_R \qquad \text{[7.1]}$$

The annual worth approach utilizes the AW values in the same fashion to solve for i.

$$AW_D = AW_R$$

$$0 = -AW_D + AW_R \qquad \text{[7.2]}$$

The i value that makes these equations numerically correct is called i^*. It is the root of the ROR relation. To determine if the alternative's cash flow series is viable, compare i^* with the established MARR.

If $i^* \geq$ MARR, accept the alternative as economically viable.

If $i^* <$ MARR, the alternative is not economically viable.

In Chapter 2 the method for calculating the rate of return on an investment was illustrated when only one engineering economy factor was involved. Here the present worth equation is the basis for calculating the rate of return when several factors are involved. Remember that the basis for engineering economy calculations is *equivalence* in PW, FW, or AW terms for a stated $i \geq 0\%$. In rate of return calculations, the objective is to *find the interest rate i^* at which the cash flows are equivalent*. The calculations are the reverse of those made in previous chapters, where the interest rate was known. For example, if you deposit $1000 now and are promised payments of $500 three years from now and $1500 five years from now, the rate of return relation using PW factors is

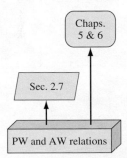

$$1000 = 500(P/F,i^*,3) + 1500(P/F,i^*,5) \qquad \text{[7.3]}$$

The value of i^* to make the equality correct is to be computed (see Figure 7–2). If the $1000 is moved to the right side of Equation [7.3], we have

$$0 = -1000 + 500(P/F,i^*,3) + 1500(P/F,i^*,5) \qquad \text{[7.4]}$$

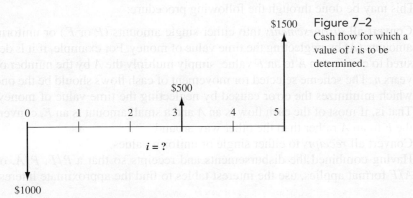

Figure 7–2
Cash flow for which a value of i is to be determined.

which is the general form of Equation [7.1]. The equation is solved for i to obtain $i^* = 16.9\%$ by hand using trial and error or by computer using spreadsheet functions. The rate of return will always be greater than zero if the total amount of receipts is greater than the total amount of disbursements, when the time value of money is considered. Using $i^* = 16.9\%$, a graph similar to Figure 7–1 can be constructed. It will show that the unrecovered balances each year, starting with -1000 in year 1, are exactly recovered by the $500 and $1500 receipts in years 3 and 5.

It should be evident that rate of return relations are merely a rearrangement of a present worth equation. That is, if the above interest rate is known to be 16.9%, and it is used to find the present worth of $500 three years from now and $1500 five years from now, the PW relation is

$$PW = 500(P/F,16.9\%,3) + 1500(P/F,16.9\%,5) = \$1000$$

This illustrates that rate of return and present worth equations are set up in exactly the same fashion. The only differences are what is given and what is sought.

There are two ways to determine i^* once the PW relation is established: solution via trial and error by hand and solution by spreadsheet function. The second is faster; the first helps in understanding how ROR computations work. We summarize both methods here and in Example 7.2.

i^* Using Trial and Error by Hand   The general procedure of using a PW-based equation is

1. Draw a cash flow diagram.
2. Set up the rate of return equation in the form of Equation [7.1].
3. Select values of i by trial and error until the equation is balanced.

When the trial-and-error method is applied to determine i^*, it is advantageous in step 3 to get fairly close to the correct answer on the first trial. If the cash flows are combined in such a manner that the income and disbursements can be represented by a *single factor* such as P/F or P/A, it is possible to look up the interest rate (in the tables) corresponding to the value of that factor for n years. The problem, then, is to combine the cash flows into the format of only one of the factors. This may be done through the following procedure:

1. Convert all *disbursements* into either single amounts (P or F) or uniform amounts (A) by neglecting the time value of money. For example, if it is desired to convert an A to an F value, simply multiply the A by the number of years n. The scheme selected for movement of cash flows should be the one which minimizes the error caused by neglecting the time value of money. That is, if most of the cash flow is an A and a small amount is an F, convert the F to an A rather than the other way around.
2. Convert all *receipts* to either single or uniform values.
3. Having combined the disbursements and receipts so that a P/F, P/A, or A/F format applies, use the interest tables to find the approximate interest

rate at which the P/F, P/A, or A/F value is satisfied. The rate obtained is a good estimate for the first trial.

It is important to recognize that this first-trial rate is only an *estimate* of the actual rate of return, because the time value of money is neglected. The procedure is illustrated in Example 7.2.

i* by Computer   The fastest way to determine an i^* value by computer, when there is a series of equal cash flows (A series), is to apply the RATE function. This is a powerful one-cell function, where it is acceptable to have a separate P value in year 0 and an F value in year n. The format is

$$\mathbf{RATE(}n,A,P,F\mathbf{)}$$

Q-Solv

The F value does not include the series A amount.

When cash flows vary from year to year (period to period), the best way to find i^* is to enter the net cash flows into contiguous cells (including any $0 amounts) and apply the IRR function in any cell. The format is

$$\mathbf{IRR(first\_cell:last\_cell,guess)}$$

where "guess" is the i value at which the computer starts searching for i^*.

The PW-based procedure for sensitivity analysis and a graphical estimation of the i^* value (or multiple i^* values, as discussed later) is as follows:

1. Draw the cash flow diagram.
2. Set up the ROR relation in the form of Equation [7.1].
3. Enter the cash flows onto the spreadsheet in contiguous cells.
4. Develop the IRR function to display i^*.
5. Use the NPV function to develop a chart of PW vs. i values. This graphically shows the i^* value at which PW = 0.

E-Solve

EXAMPLE 7.2

The HVAC engineer for a company constructing one of the world's tallest buildings (Shanghai Financial Center in the Peoples' Republic of China) has requested that $500,000 be spent now during construction on software and hardware to improve the efficiency of the environmental control systems. This is expected to save $10,000 per year for 10 years in energy costs and $700,000 at the end of 10 years in equipment refurbishment costs. Find the rate of return by hand and by computer.

Solution by Hand
Use the trial-and-error procedure based on a PW equation.

1. Figure 7–3 shows the cash flow diagram.
2. Use Equation [7.1] format for the ROR equation.

$$0 = -500,000 + 10,000(P/A,i^*,10) + 700,000(P/F,i^*,10) \qquad [7.5]$$

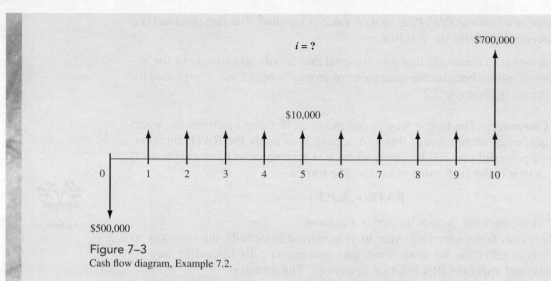

Figure 7–3
Cash flow diagram, Example 7.2.

3. Use the estimation procedure to determine i for the first trial. All income will be regarded as a single F in year 10 so that the P/F factor can be used. The P/F factor is selected because most of the cash flow ($700,000) already fits this factor and errors created by neglecting the time value of the remaining money will be minimized. Only for the first estimate of i, define $P = \$500,000$, $n = 10$, and $F = 10(10,000) + 700,000 = \$800,000$. Now we can state that

$$500,000 = 800,000(P/F,i,10)$$

$$(P/F,i,10) = 0.625$$

The roughly estimated i is between 4% and 5%. Use 5% as the first trial because this approximate rate for the P/F factor is lower than the true value when the time value of money is considered. At $i = 5\%$, the ROR equation is

$$0 = -500,000 + 10,000(P/A,5\%,10) + 700,000(P/F,5\%,10)$$

$$0 < \$6946$$

The result is positive, indicating that the return is more than 5%. Try $i = 6\%$.

$$0 = -500,000 + 10,000(P/A,6\%,10) + 700,000(P/F,6\%,10)$$

$$0 > \$-35,519$$

Since the interest rate of 6% is too high, linearly interpolate between 5% and 6%.

Sec. 2.4

Interpolation

$$i^* = 5.00 + \frac{6946 - 0}{6946 - (-35,519)}(1.0)$$

$$= 5.00 + 0.16 = 5.16\%$$

Solution by Computer

Enter the cash flows from Figure 7–3 into the RATE function. The entry RATE(10, 10000, −500000, 700000) displays $i^* = 5.16\%$. It is equally correct to use the IRR function. Figure 7–4, column B, shows the cash flows and IRR(B2:B12) function to obtain i^*.

Q-Solv

X Microsoft Excel   _ 🗗 ✕

File   Edit   View   Insert   Format   Tools   Data   Window   Help

D18   ▼   =

Example 7.2   _ ☐ ✕

	A	B	C	D	E	F	G	H	I	J
1	Year	Amount	Trial i	PW value						
2	0	-$500,000	4.00%	$54,004						
3	1	$10,000	4.20%	$44,204						
4	2	$10,000	4.40%	$34,603						
5	3	$10,000	4.60%	$25,198						
6	4	$10,000	4.80%	$15,984						
7	5	$10,000	5.00%	$6,957						
8	6	$10,000	5.20%	-$1,888						
9	7	$10,000	5.40%	-$10,555						
10	8	$10,000	5.60%	-$19,047						
11	9	$10,000	5.80%	-$27,368						
12	10	$710,000	6.00%	-$35,523						
13										
14	Rate of Return	5.16%								
15										
16					=NPV(C12,$B$3:$B$12)+$B$2					
17										
18				=IRR(B2:B12)						
19										
20										

Chart: PW value vs Interest rate i, values $60,000 $40,000 $20,000 $0 -$20,000 -$40,000 at 3.8% 4.2% 4.6% 5.0% 5.4% 5.8%

◄ ► ►│ **Sheet 1** ╱ Sheet2 ╱ Sheet3 ╱ Sheet4 ╱ Sheet5 ╱ Sheet6 ╱ Sheet7 ╱ Sheeti ◄

Ready   NUM

Figure 7–4

Spreadsheet solution for i^* and a plot of PW vs. i values, Example 7.2.

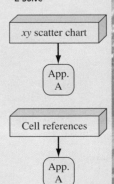
E-Solve

For a more thorough analysis, use the i^* by computer procedure above.

1, 2. The cash flow diagram and ROR relation are the same as in the by-hand solution.

3. Figure 7–4 shows the net cash flows in column B.

4. The IRR function in cell B14 displays $i^* = 5.16\%$.

5. In order to graphically observe i^*, column D displays PW for different i values (column C). The NPV function is used repeatedly to calculate PW for the Excel xy scatter chart of PW vs. i. The i^* is slightly less than 5.2%.

As indicated in the cell D12 tag, $ signs are inserted into the NPV functions. This provides _absolute cell referencing,_ which allows the NPV function to be correctly shifted from one cell to another (dragged with the mouse).

xy scatter chart → App. A

Cell references → App. A

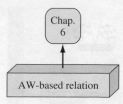

Chap. 6

AW-based relation

E-Solve

Just as i^* can be found using a PW equation, it may equivalently be determined using an AW relation. This method is preferred when uniform annual cash flows are involved. Solution by hand is the same as the procedure for a PW-based relation, except Equation [7.2] is used.

The procedure for solution by computer is exactly the same as outlined above using the IRR function. Internally, IRR calculates the NPV function at different i values until NPV $= 0$ is obtained. (There is no equivalent way to utilize the PMT function, since it requires a fixed value of i to calculate an A value.)

EXAMPLE 7.3

Use AW computations to find the rate of return for the cash flows in Example 7.2.

Solution
1. Figure 7–3 shows the cash flow diagram.
2. The AW relations for disbursements and receipts are formulated using Equation [7.2].

$$AW_D = -500,000(A/P,i,10)$$
$$AW_R = 10,000 + 700,000(A/F,i,10)$$
$$0 = -500,000(A/P,i^*,10) + 10,000 + 700,000(A/F,i^*,10)$$

3. Trial-and-error solution yields these results:

$$\text{At } i = 5\%, 0 < \$900$$
$$\text{At } i = 6\%, 0 > \$-4826$$

By interpolation, $i^* = 5.16\%$, as before.

In closing, to determine i^* by hand, choose the PW, AW, or any other equivalence equation. It is generally better to consistently use one of the methods in order to avoid errors.

7.3 CAUTIONS WHEN USING THE ROR METHOD

The rate of return method is commonly used in engineering and business settings to evaluate one project, as discussed in this chapter, and to select one alternative from two or more, as explained in the next chapter.

> **When applied correctly, the ROR technique will always result in a good decision, indeed, the same one as with a PW or AW (or FW) analysis.**

However, there are some assumptions and difficulties with ROR analysis that must be considered when calculating i^* and in interpreting its real-world meaning for a particular project. The summary provided below applies for solutions by hand and by computer.

- *Multiple i^* values.* Depending upon the sequence of net cash flow disbursements and receipts, there may be more than one real-number root to the ROR equation, resulting in more than one i^* value. This difficulty is discussed in the next section.

- *Reinvestment at i*.* Both the PW and AW methods assume that any net positive investment (i.e., net positive cash flows once the time value of money is
  considered) are reinvested at the MARR. But the ROR method assumes reinvestment at the *i** rate. When *i** is not close to the MARR (e.g., if *i** is substantially larger than MARR), this is an unrealistic assumption. In such cases,
  the *i** value is not a good basis for decision making. Though more involved
  computationally than PW or AW at the MARR, there is a procedure to use the
  ROR method and still obtain one unique *i** value. The concept of net positive
  investment and this method are discussed in Section 7.5.
- *Computational difficulty versus understanding.* Especially in obtaining a
  trial-and-error solution by hand for one or multiple *i** values, the computations rapidly become very involved. Spreadsheet solution is easier; however,
  there are no spreadsheet functions that offer the same level of understanding
  to the learner as that provided by hand solution of PW and AW relations.
- *Special procedure for multiple alternatives.* To correctly use the ROR method
  to choose from two or more mutually exclusive alternatives requires an analysis procedure significantly different from that used in PW and AW. Chapter 8
  explains this procedure.

In conclusion, from an engineering economic study perspective, the annual
worth or present worth method at a stated MARR should be used in lieu of the
ROR method. However, there is a strong appeal for the ROR method because
rate of return values are very commonly quoted. And it is easy to compare a proposed project's return with that of in-place projects.

> **When working with two or more alternatives, and when it is important to
> know the exact value of *i**, a good approach is to determine PW or AW at
> the MARR, then follow up with the specific *i** for the selected alternative.**

As an illustration, if a project is evaluated at MARR $= 15\%$ and has PW < 0,
there is no need to calculate *i**, because *i** $< 15\%$. However, if PW > 0, then
calculate the exact *i** and report it along with the conclusion that the project is
financially justified.

7.4 MULTIPLE RATE OF RETURN VALUES

In Section 7.2 a unique rate of return *i** was determined. In the cash flow series
presented thus far, the algebraic signs on the *net cash flows* changed only once,
usually from minus in year 0 to plus at some time during the series. This is called
a *conventional (or simple) cash flow series.* However, for many series the net
cash flows switch between positive and negative from one year to another, so
there is more than one sign change. Such a series is called *nonconventional (nonsimple).* As shown in the examples of Table 7–3, each series of positive or negative signs may be one or more in length. When there is more than one sign
change in the net cash flows, it is possible that there will be multiple *i** values in
the -100% to plus infinity range. There are two tests to perform in sequence on
the nonconventional series to determine if there is one unique or multiple *i** values that are real numbers. The first test is the *(Descartes') rule of signs* which

TABLE 7–3   Examples of Conventional and Nonconventional Net Cash Flow for a 6-year Project

Type of Series	Sign on Net Cash Flow							Number of Sign Changes
	0	1	2	3	4	5	6	
Conventional	−	+	+	+	+	+	+	1
Conventional	−	−	−	+	+	+	+	1
Conventional	+	+	+	+	+	−	−	1
Nonconventional	−	+	+	+	−	−	−	2
Nonconventional	+	+	−	−	−	+	+	2
Nonconventional	−	+	−	−	+	+	+	3

states that the total number of real-number roots is always less than or equal to the number of sign changes in the series. This rule is derived from the fact that the relation set up by Equation [7.1] or [7.2] to find i^* is an nth-order polynomial. (It is possible that imaginary values or infinity may also satisfy the equation.)

The second and more discriminating test determines if there is one, real-number, positive i^* value. This is the *cumulative cash flow sign test,* also known as *Norstrom's criterion.* It states that only one sign change in the series of cumulative cash flows which *starts negatively,* indicates that there is one positive root to the polynomial relation. To perform this test, determine the series

$$S_t = \text{cumulative cash flows through period } t$$

Observe the sign of S_0 and count the sign changes in the series $S_0, S_1, \ldots, S_n$. Only if $S_0 < 0$ and signs change one time in the series is there a single, real-number, positive i^*.

With the results of these two tests, the ROR relation is solved for either the unique i^* or the multiple i^* values, using trial and error by hand, or by computer using an IRR function that incorporates the "guess" option. The development of the PW vs. i graph is recommended, especially when using a spreadsheet. Example 7.4 illustrates the tests and solution for i^* by hand and by computer.

EXAMPLE 7.4

The engineering design and testing group for Honda Motor Corp. does contract-based work for automobile manufacturers throughout the world. During the last 3 years, the net cash flows for contract payments have varied widely, as shown below, primarily due to a large manufacturer's inability to pay its contract fee.

Year	0	1	2	3
Cash Flow ($1000)	+2000	−500	−8100	+6800

(a) Determine the maximum number of i^* values that may satisfy the ROR relation.
(b) Write the PW-based ROR relation and approximate the i^* value(s) by plotting PW vs. i by hand and by computer.
(c) Calculate the i^* values more exactly using the IRR function of the spreadsheet.

Solution by Hand

(a) Table 7–4 shows the annual cash flows and cumulative cash flows. Since there are two sign changes in the cash flow sequence, the rule of signs indicates a maximum of two real-number i^* values. The cumulative cash flow sequence starts with a positive number $S_0 = +2000$, indicating there is not just one positive root. The conclusion is that as many as two i^* values can be found.

TABLE 7–4   Cash Flow and Cumulative Cash Flow Sequences, Example 7.4

Year	Cash Flow ($1000)	Sequence Number	Cumulative Cash Flow ($1000)
0	+2000	S_0	+2000
1	−500	S_1	+1500
2	−8100	S_2	−6600
3	+6800	S_3	+200

(b) The PW relation is

$$PW = 2000 - 500(P/F,i,1) - 8100(P/F,i,2) + 6800(P/F,i,3)$$

Select values of i to find the two i^* values, and plot PW vs. i. The PW values are shown below and plotted in Figure 7–5 for i values of 0, 5, 10, 20, 30, 40, and 50%. The characteristic parabolic shape for a second-degree polynomial is obtained, with PW crossing the i axis at approximately $i_1^* - 8$ and $i_2^* - 41\%$.

$i\%$	0	5	10	20	30	40	50
PW ($1000)	+200	+51.44	−39.55	−106.13	−82.01	−11.83	+81.85

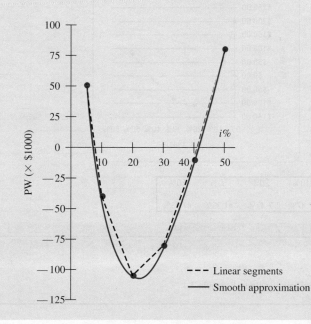

Figure 7–5
Present worth of cash flows at several interest rates, Example 7.4.

Solution by Computer

(*a*) See Figure 7–6. The NPV function is used in column D to determine the PW value at several *i* values (column C), as indicated by the cell tag. The accompanying Excel *xy* scatter chart presents the PW vs. *i* graph. The i^* values cross the PW = 0 line at approximately 8% and 40%.

(*b*) Row 19 in Figure 7–6 contains the ROR values (including a negative value) entered as guess into the IRR function to find the i^* root of the polynomial that is closest to the guess value. Row 21 includes the two resulting i^* values: $i_1^* = 7.47\%$ and $i_2^* = 41.35\%$.

   If "guess" is omitted from the IRR function, the entry IRR(B4:B7) will determine only the first value, 7.47%. As a check on the two i^* values, the NPV function can be set up to find PW at the two i^* values. Both NPV(7.47%,B5:B7)+B4 and NPV(41.35%,B5:B7)+B4 will display approximately $0.00.

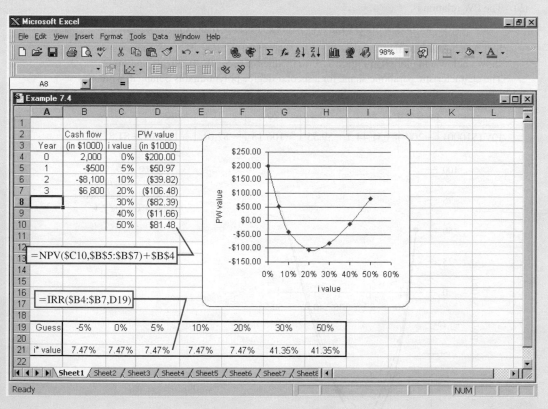

Figure 7–6
Spreadsheet showing PW vs. *i* graph and multiple i^* values, Example 7.4.

EXAMPLE 7.5

Two student engineers started a software development business during their junior year in college. One package in three dimensional modeling has now been licensed through IBM's Small Business Partners Program for the next 10 years. Table 7–5 gives the estimated net cash flows developed by IBM from the perspective of the small business. The negative values in years 1, 2, and 4 reflect heavy marketing costs. Determine the number of i^* values; estimate them graphically and by the IRR function of a spreadsheet.

E-Solve

TABLE 7–5   Net Cash Flow Series and Cumulative Cash Flow Series, Example 7.5

| | Cash Flow, $100 | | | Cash Flow, $100 | |
Year	Net	Cumulative	Year	Net	Cumulative
1	−2000	−2000	6	+500	−900
2	−2000	−4000	7	+400	−500
3	+2500	−1500	8	+300	−200
4	−500	−2000	9	+200	0
5	+600	−1400	10	+100	+100

Solution by Computer

The rule of signs indicates a nonconventional net cash flow series with up to three roots. The cumulative net cash flow series starts negatively and has only one sign change in year 10, thus indicating that one unique positive root can be found. (Zero values in the cumulative cash flow series are neglected when applying Norstrom's criterion.) A PW-based ROR relation is used to find i^*.

$$0 = -2000(P/F,i,1) - 2000(P/F,i,2) + \cdots + 100(P/F,i,10)$$

The PW of the right side is calculated for different values of i and plotted on the spreadsheet (Figure 7–7). The unique value $i^* = 0.77\%$ is obtained using the IRR function with the same "guess" values for i as in the PW vs. i graph.

Comment

Once the spreadsheet is set up as in Figure 7–7, the cash flows can be "tweaked" to perform sensitivity analysis on the i^* value(s). For example, if the cash flow in year 10 is changed only slightly from $\$+100$ to $\$-100$, the results displayed change across the spreadsheet to $i^* = -0.84\%$. Also, this simple change in cash flow substantially alters the cumulative cash flow sequence. Now $S_{10} = \$-100$, as can be confirmed in Table 7–5. There are now no sign changes in the cumulative cash flow sequence, so *no unique positive root* can be found. This is confirmed by the value $i^* = -0.84\%$. If other cash flows are altered, the two tests we have learned should be applied to determine whether multiple roots may now exist. This means that spreadsheet-based sensitivity analysis must be performed carefully when the ROR method is applied, because not all i^* values may be determined as cash flows are tweaked on the screen.

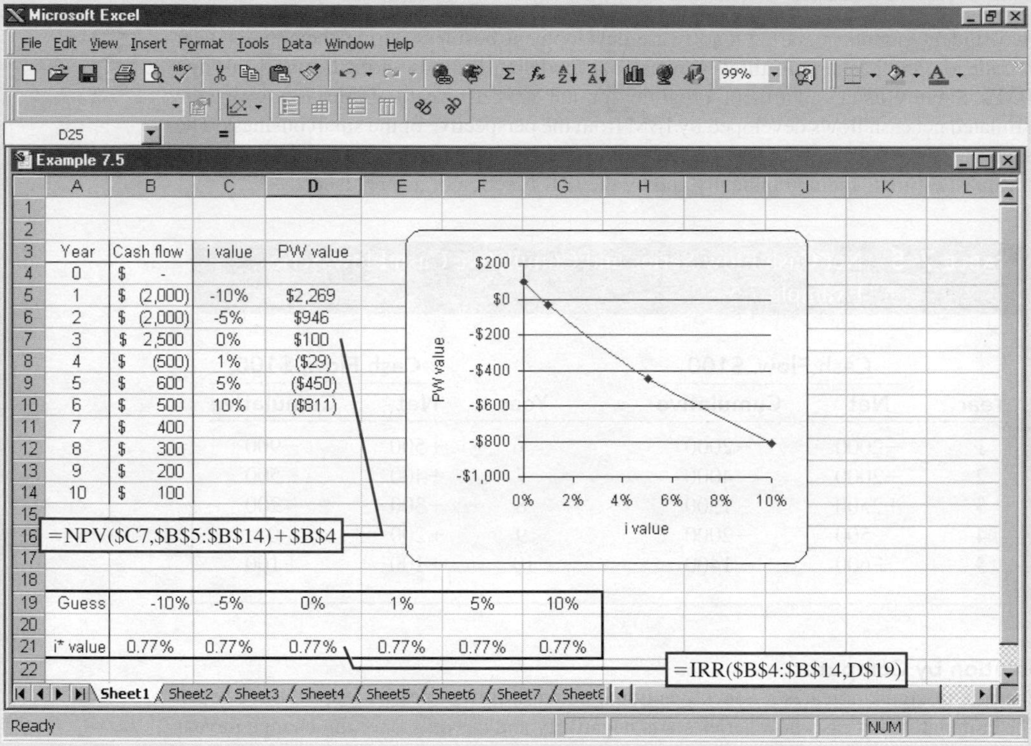

Figure 7–7
Spreadsheet solution to find *i**, Example 7.5.

In many cases some of the multiple *i** values will seem ridiculous because they are too large or too small (negative). For example, values of 10, 150, and 750% for a sequence with three sign changes are difficult to use in practical decision making. (Obviously, one advantage of the PW and AW methods for alternative analysis is that unrealistic rates do not enter into the analysis.) In determining which *i** value to select as *the* ROR value, it is common to neglect negative and large values or to simply never compute them. *Actually, the correct approach is to determine the unique composite rate of return,* as described in the next section.

If a standard spreadsheet system, such as Excel, is used, it will normally determine only one real-number root, unless different "guess" amounts are entered sequentially. This one *i** value determined from Excel is usually a realistically valued root, because the *i** which solves the PW relation is determined by the spreadsheet's built-in trial-and-error method. This method starts with a default value, commonly 10%, or with the user-supplied guess, as illustrated in the previous example.

7.5 COMPOSITE RATE OF RETURN: REMOVING MULTIPLE i^* VALUES

The rates of return we have computed thus far arc the rates that exactly balance plus and minus cash flows with the time value of money considered. Any method which accounts for the time value of money can be used in calculating this balancing rate, such as PW, AW, or FW. The interest rate obtained from these calculations is known as the *internal rate of return (IRR)*. Simply stated, the internal rate of return is the rate of return on the unrecovered balance of an investment, as defined earlier. The funds that remain unrecovered are still inside the investment, hence the name *internal rate of return*. The general terms rate of return and interest rate usually imply internal rate of return. The interest rates quoted or calculated in previous chapters are all internal rates.

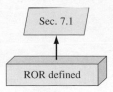

The concept of unrecovered balance becomes important when positive net cash flows are generated (thrown off) before the end of a project. A positive net cash flow, once generated, becomes *released as external funds to the project* and is not considered further in an internal rate of return calculation. These positive net cash flows may cause a nonconventional cash flow series and multiple i^* values to develop. However, there is a method to explicitly consider these funds, as discussed below. Additionally, the dilemma of multiple i^* roots is eliminated.

It is important to understand that the procedure detailed below is used to

Determine the rate of return for cash flow estimates when there are multiple i^* values indicated by both the cash flow rule of signs and the cumulative cash flow rule of signs, and net positive cash flows from the project will earn at a stated rate that is different from any of the multiple i^* values.

For example, assume a cash flow series has two i^* values that balance the ROR equation—10% and 60% per year—and any cash released by the project is invested by the company at a rate of return of 25% per year. The procedure below will find a single unique rate of return for the cash flow series. However, if it is known that released cash will earn exactly 10%, the unique rate is 10%. The same statement can be made using the 60% rate.

As before, if the exact rate of return for a project's cash flow estimates is not needed, it is much simpler, and equally correct, to use a PW or AW analysis at the MARR to determine if the project is financially viable. This is the normal mode of operation in an engineering economy study.

Consider the internal rate of return calculations for the following cash flows: $10,000 is invested at $t = 0$, $8000 is received in year 2, and $9000 is received in year 5. The PW equation to determine i^* is

$$0 = -10,000 + 8000(P/F,i,2) + 9000(P/F,i,5)$$
$$i^* = 16.815\%$$

If this rate is used for the unrecovered balances, the investment will be recovered exactly at the end of year 5. The procedure to verify this is identical to that used in Table 7–1, which describes how the ROR works to exactly remove the unrecovered balance with the final cash flow.

Unrecovered balance at end of year 2 immediately before $8000 receipt:

$$-10,000(F/P,16.815\%,2) = -10,000(1 + 0.16815)^2 = \$-13,646$$

Unrecovered balance at end of year 2 immediately after $8000 receipt:

$$-13,646 + 8000 = \$-5646$$

Unrecovered balance at end of year 5 immediately before $9000 receipt:

$$-5646(F/P,16.815\%,3) = \$-9000$$

Unrecovered balance at end of year 5 immediately after $9000 receipt:

$$\$-9000 + 9000 = \$0$$

In this calculation, no consideration is given to the $8000 available after year 2. What happens if funds released from a project *are* considered in calculating the overall rate of return of a project? After all, something must be done with the released funds. One possibility is to assume the money is reinvested at some stated rate. The ROR method assumes funds that are excess to a project earn at the i^* rate, but this may not be a realistic rate in everyday practice. Another approach is to simply assume that reinvestment occurs at the MARR. In addition to accounting for all the money released during the project period and reinvested at a realistic rate, the approach discussed below has the advantage of converting a nonconventional cash flow series (with multiple i^* values) to a conventional series with one root, which can be considered *the* rate of return for making a decision about the project.

The rate of earnings used for the released funds is called the *reinvestment rate* or *external rate of return* and is symbolized by c. This rate, established outside (external to) the cash flow estimates being evaluated, depends upon the market rate available for investments. If a company is making, say, 8% on its daily investments, then $c = 8\%$. It is common practice to set c equal to the MARR. The one interest rate that now satisfies the rate of return equation is called the *composite rate of return (CRR)* and is symbolized by i'. By definition

> The *composite rate of return i'* is the unique rate of return for a project that assumes that net positive cash flows, which represent money not immediately needed by the project, are reinvested at the reinvestment rate *c*.

The term *composite* is used here to describe this rate of return because it is derived using another interest rate, namely, the reinvestment rate c. If c happens to equal any one of the i^* values, then the composite rate i' will equal that i^* value. The CRR is also known by the term *return on invested capital (RIC)*. Once the unique i' is determined, it is compared to the MARR to decide on the project's financial viability, as outlined in Section 7.2.

The correct procedure to determine i' is called the *net-investment procedure*. The technique involves finding the future worth of the net investment amount 1 year in the future. Find the project's net-investment value F_t in year t from F_{t-1} by using the F/P factor for 1 year at the reinvestment rate c if the previous net investment F_{t-1} is positive (extra money generated by project), or at the CRR rate i' if F_{t-1} is negative (project used all available funds). To do this mathematically, for each year t set up the relation

$$F_t = F_{t-1}(1 + i) + C_t \qquad \text{[7.6]}$$

where $t = 1, 2, \ldots, n$

n = total years in project

C_t = net cash flow in year t

$$i = \begin{cases} c & \text{if } F_{t-1} > 0 \quad \text{(net positive investment)} \\ i' & \text{if } F_{t-1} < 0 \quad \text{(net negative investment)} \end{cases}$$

Set the net-investment relation for year n equal to zero ($F_n = 0$) and solve for i'. The i' value obtained is unique for a stated reinvestment rate c.

The development of F_1 through F_3 for the cash flow series below, which is graphed in Figure 7–8a, is illustrated for a reinvestment rate of c = MARR = 15%.

Year	Cash Flow, $
0	50
1	−200
2	50
3	100

The net investment for year $t = 0$ is

$$F_0 = \$50$$

which is positive, so it returns c = 15% during the first year. By Equation [7.6], F_1 is

$$F_1 = 50(1 + 0.15) - 200 = \$-142.50$$

This result is shown in Figure 7–8b. Since the project net investment is now negative, the value F_1 earns interest at the composite rate i' for year 2. Therefore, for year 2,

$$F_2 = F_1(1 + i') + C_2 = -142.50(1 + i') + 50$$

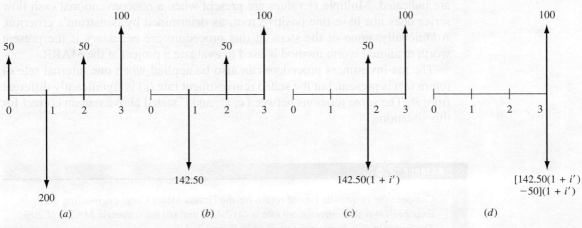

| (a) | (b) | (c) | (d) |

Figure 7–8

Cash flow series for which the composite rate of return i' is computed: (a) original form; equivalent form in (b) year 1, (c) year 2, and (d) year 3.

The i' value is to be determined (Figure 7–8c). Since F_2 will be negative for all $i' > 0$, use i' to set up F_3 as shown in Figure 7–8d.

$$F_3 = F_2(1 + i') + C_3 = [-142.50(1 + i') + 50](1 + i') + 100 \qquad [7.7]$$

Setting Equation [7.7] equal to zero and solving for i' will result in the unique composite rate of return i'. The resulting values are 3.13% and -168%, since Equation [7.7] is a quadratic relation (power 2 for i'). The value of $i' = 3.13\%$ is the correct i^* in the range -100% to ∞. The procedure to find i' may be summarized as follows:

1. Draw a cash flow diagram of the original net cash flow series.
2. Develop the series of net investments using Equation [7.6] and the c value. The result is the F_n expression in terms of i'.
3. Set $F_n = 0$ and find the i' value to balance the equation.

Several comments are in order. If the reinvestment rate c is equal to the internal rate of return i^* (or one of the i^* values when there are multiple ones), the i' that is calculated will be exactly the same as i^*; that is, $c = i^* = i'$. The closer the c value is to i^*, the smaller the difference between the composite and internal rates. As mentioned earlier, it is correct to assume that $c = $ MARR, if all throw-off funds from the project can realistically earn at the MARR rate.

A summary of the relations between c, i', and i^* follows, and the relations are demonstrated in Example 7.6.

Relation between Reinvestment Rate c and i^*	Relation between CRR i' and i^*
$c = i^*$	$i' = i^*$
$c < i^*$	$i' < i^*$
$c > i^*$	$i' > i^*$

Remember: This entire net-investment procedure is used when multiple i^* values are indicated. Multiple i^* values are present when a nonconventional cash flow series does not have one positive root, as determined by Norstrom's criterion. Additionally, none of the steps in this procedure are necessary if the present worth or annual worth method is used to evaluate a project at the MARR.

The net-investment procedure can also be applied when one internal rate of return (i^*) is present, but the stated reinvestment rate (c) is significantly different from i^*. The same relations between c, i^*, and i' stated above remain correct for this situation.

EXAMPLE **7.6**

Compute the composite rate of return for the Honda Motor Corp. engineering group in Example 7.4 if the reinvestment rate is (*a*) 7.47% and (*b*) the corporate MARR of 20%. The multiple i^* values are determined in Figure 7–6.

Solution

(a) Use the net-investment procedure to determine i' for $c = 7.47\%$.
   1. Figure 7–9 shows the original cash flow.
   2. The first net-investment expression is $F_0 = \$+2000$. Since $F_0 > 0$, use $c = 7.47\%$ to write F_1 by Equation [7.6].

$$F_1 = 2000(1.0747) - 500 = \$1649.40$$

Since $F_1 > 0$, use $c = 7.47\%$ to determine F_2.

$$F_2 = 1649.40(1.0747) - 8100 = \$-6327.39$$

Figure 7–10 shows the equivalent cash flow at this time. Since $F_2 < 0$, use i' to express F_3.

$$F_3 = -6327.39(1 + i') + 6800$$

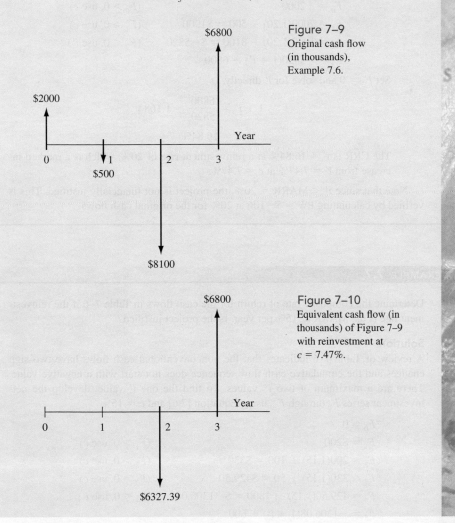

$\$6800$

Figure 7–9
Original cash flow (in thousands), Example 7.6.

$\$2000$

Year

0   1   2   3

$\$500$

$\$8100$

$\$6800$

Figure 7–10
Equivalent cash flow (in thousands) of Figure 7–9 with reinvestment at $c = 7.47\%$.

Year

0   1   2   3

$\$6327.39$

3.   Set $F_3 = 0$ and solve for i' directly.

$$-6327.39(1 + i') + 6800 = 0$$

$$1 + i' = \frac{6800}{6327.39} = 1.0747$$

$$i' = 7.47\%$$

The CRR is 7.47%, which is the same as c, the reinvestment rate, and the i_1^* value determined in Example 7.4, Figure 7–6. Note that 41.35%, which is the second i^* value, no longer balances the rate of return equation. The equivalent future worth result for the cash flow in Figure 7–10, if i' were 41.35%, is

$$6327.39(F/P,41.35\%,1) = \$8943.77 \neq \$6800$$

(b)   For MARR $= c = 20\%$, the net-investment series is

$F_0 = +2000$	$(F_0 > 0,\ \text{use } c)$
$F_1 = 2000(1.20) - 500 = \$1900$	$(F_1 > 0,\ \text{use } c)$
$F_2 = 1900(1.20) - 8100 = \$-5820$	$(F_2 < 0,\ \text{use } i')$
$F_3 = -5820(1 + i') + 6800$	

Set $F_3 = 0$ and solve for i' directly.

$$1 + i' = \frac{6800}{5820} = 1.1684$$

$$i' = 16.84\%$$

The CRR is $i' = 16.84\%$ at a reinvestment rate of 20%, which is a marked increase from $i' = 7.47\%$ at $c = 7.47\%$.

Note that since $i' < \text{MARR} = 20\%$, the project is not financially justified. This is verified by calculating PW $= \$-106$ at 20% for the original cash flows.

EXAMPLE 7.7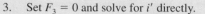

Determine the composite rate of return for the cash flows in Table 7–6 if the reinvestment rate is the MARR of 15% per year. Is the project justified?

Solution

A review of Table 7–6 indicates that the nonconventional cash flows have two sign changes and the cumulative cash flow sequence does not start with a negative value. There are a maximum of two i^* values. To find the one i' value, develop the net-investment series F_0 through F_{10} using Equation [7.6] and $c = 15\%$.

$F_0 = 0$	
$F_1 = \$200$	$(F_1 > 0,\ \text{use } c)$
$F_2 = 200(1.15) + 100 = \$330$	$(F_2 > 0,\ \text{use } c)$
$F_3 = 330(1.15) + 50 = \$429.50$	$(F_3 > 0,\ \text{use } c)$
$F_4 = 429.50(1.15) - 1800 = \$-1306.08$	$(F_4 < 0,\ \text{use } i')$
$F_5 = -1306.08(1 + i') + 600$	

| | **TABLE 7–6** | Cash Flow and Cumulative Cash Flow Sequences, Example 7.7 | | | | |

	Cash Flow, $				Cash Flow, $	
Year	Net	Cumulative		Year	Net	Cumulative
0	0	0		6	500	−350
1	200	+200		7	400	+50
2	100	+300		8	300	+350
3	50	+350		9	200	+550
4	−1800	−1450		10	100	+650
5	600	−850				

Since we do not know if F_5 is greater than zero or less than zero, all remaining expressions use i'.

$$F_6 = F_5(1 + i') + 500 = [-1306.08(1 + i') + 600](1 + i') + 500$$

$$F_7 = F_6(1 + i') + 400$$

$$F_8 = F_7(1 + i') + 300$$

$$F_9 = F_8(1 + i') + 200$$

$$F_{10} = F_9(1 + i') + 100$$

To find i', the expression $F_{10} = 0$ is solved by trial and error. Solution determines that $i' = 21.24\%$. Since $i' >$ MARR, the project is justified. In order to work more with this exercise and the net-investment procedure, do the case study in this chapter.

Comment
The two rates which balance the ROR equation are $i_1^* = 28.71\%$ and $i_2^* = 48.25\%$. If we rework this problem at either reinvestment rate, the i' value will be the same as this reinvestment rate; that is, if $c = 28.71\%$, then $i' = 28.71\%$.

There is a spreadsheet function called MIRR (modified IRR) which determines a unique interest rate when you input a reinvestment rate c for positive cash flows. However, the function does not implement the net-investment procedure for nonconventional cash flow series as discussed here, and the function requires that a finance rate for the funds used as the initial investment be supplied. So the formulas for MIRR and CRR computation are not the same. The MIRR will not produce exactly the same answer as Equation [7.6] unless all the rates happen to be the same and this value is one of the roots of the ROR relation.

7.6   RATE OF RETURN OF A BOND INVESTMENT

In Chapter 5 we learned the terminology of bonds and how to calculate the PW of a bond investment. The cash flow series for a bond investment is conventional and has one unique i^*, which is best determined by solving a PW-based rate of

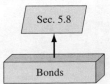

Sec. 5.8

Bonds

return equation in the form of Equation [7.1]. Examples 7.8 and 7.9 illustrate the procedure.

EXAMPLE 7.8

Allied Materials needs $3 million in debt capital for expanded composites manufacturing. It is offering small-denomination bonds at a discount price of $800 for a 4% $1000 bond that matures in 20 years with interest payable semiannually. What nominal and effective interest rates per year, compounded semiannually, will Allied Materials pay an investor?

Solution
The income that a purchaser will receive from the bond purchase is the bond interest $I = \$20$ every 6 months plus the face value in 20 years. The PW-based equation for calculating the rate of return

$$0 = -800 + 20(P/A,i^*,40) + 1000(P/F,i^*,40)$$

Solve by computer (IRR function) or by hand to obtain $i^* = 2.87\%$ semiannually. The nominal interest rate per year is computed by multiplying i^* by 2.

Nominal $i = 2.87\%(2) = 5.74\%$ per year, compounded semiannually

Using Equation [4.5], the effective annual rate is

$$i_a = (1.0287)^2 - 1 = 5.82\%$$

Sec. 4.2

Effective i

EXAMPLE 7.9

Gerry is an entry-level engineer at Boeing Aerospace in California. He took a financial risk and bought a bond from a different corporation that had defaulted on its interest payments. He paid $4240 for an 8% $10,000 bond with interest payable quarterly. The bond paid no interest for the first 3 years after Gerry bought it. If interest was paid for the next 7 years, and then Gerry was able to resell the bond for $11,000, what rate of return did he make on the investment? Assume the bond is scheduled to mature 18 years after he bought it. Perform hand and computer analysis.

Solution by Hand
The bond interest received in years 4 through 10 was

$$I = \frac{(10,000)(0.08)}{4} = \$200 \text{ per quarter}$$

The effective rate of return *per quarter* can be determined by solving the PW equation developed on a per quarter basis, since this basis makes PP = CP.

$$0 = -4240 + 200(P/A,i^* \text{ per quarter},28)(P/F,i^* \text{ per quarter},12)$$
$$+ 11,000(P/F,i^* \text{ per quarter},40)$$

The equation is correct for $i^* = 4.1\%$ per quarter, which is a nominal 16.4% per year, compounded quarterly.

Secs. 4.4 and 4.6

PP = CP

Solution by Computer

The solution is shown in Figure 7–11. The spreadsheet is set up to directly calculate an annual interest rate of 16.41% in cell E1. The quarterly bond interest receipts of $200 arc converted to equivalent annual receipts of $724.24 using the PV function in cell E6. A quarterly rate could be determined initially on the spreadsheet, but this approach would require four times as many entries of $200 each, compared to the six times $724.24 is entered here. (A circular reference may be indicated by Excel between cells E1, E6, and B6. However, clicking on OK will override and the solution $i^* = 16.41\%$ is displayed. A circular reference is avoided if all 40 quarters of $0 and $200 are entered in column B with necessary changes in the column E relations to find the quarterly rate.)

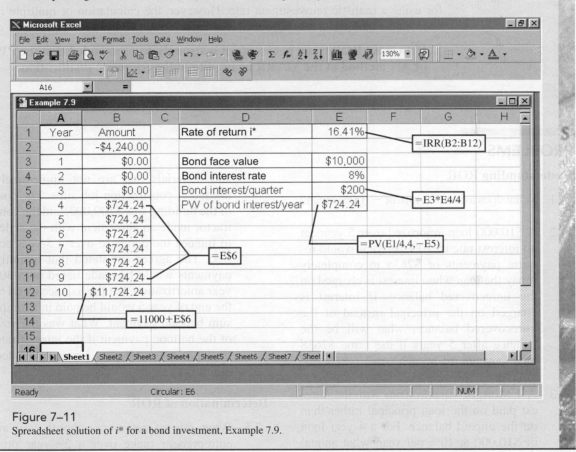

Figure 7–11
Spreadsheet solution of i^* for a bond investment, Example 7.9.

CHAPTER SUMMARY

Rate of return, or interest rate, is a term used and understood by almost everybody. Most people, however, can have considerable difficulty in calculating a rate of return i^* correctly for any cash flow series. For some types of series, more than one ROR possibility exists. The maximum number of i^* values is equal to the number of changes in the signs of the net cash flow series (Descartes' rule of signs). Also, a single positive rate can be found if the cumulative net cash flow series starts negatively and has only one sign change (Norstrom's criterion).

For all cash flow series where there is an indication of multiple roots, a decision must be made about whether to calculate the multiple i^* internal rates, or the one composite rate of return using an externally-determined reinvestment rate. This rate is commonly set at the MARR. While the internal rate is usually easier to calculate, the composite rate is the correct approach with two advantages: multiple rates of return are eliminated, and released project net cash flows are accounted for using a realistic reinvestment rate. However, the calculation of multiple i^* rates, or the composite rate of return, is often computationally involved.

If an exact ROR is not necessary, it is strongly recommended that the PW or AW method at the MARR be used to judge economic justification.

PROBLEMS

Understanding ROR

7.1 What does a rate of return of -100% mean?

7.2 A $10,000 loan amortized over 5 years at an interest rate of 10% per year would require payments of $2638 to completely repay the loan when interest is charged on the unrecovered balance. If interest is charged on the principal instead of the unrecovered balance, what will be the balance after 5 years if the same $2638 payments are made each year?

7.3 A-1 Mortgage makes loans with the interest paid on the loan principal rather than on the unpaid balance. For a 4-year loan of $10,000 at 10% per year, what annual payment would be required to repay the loan in 4 years if interest is charged on (a) the principal and (b) the unrecovered balance?

7.4 A small industrial contractor purchased a warehouse building for storing equipment and materials that are not immediately needed at construction job sites. The cost of the building was $100,000, and the contractor made an agreement with the seller to finance the purchase over a 5-year period. The agreement stated that monthly payments would be made based on a 30-year amortization, but the balance owed at the end of year 5 would be paid in a lump-sum balloon payment. What was the size of the balloon payment if the interest rate on the loan was 6% per year, compounded monthly?

Determination of ROR

7.5 What rate of return per month will an entrepreneur make over a 2½-year project period if he invested $150,000 to produce portable 12-volt air compressors? His estimated monthly costs are $27,000 with income of $33,000 per month.

7.6 The Camino Real Landfill was required to install a plastic liner to prevent leachate from migrating into the groundwater. The fill area was 50,000 square meters, and the installed liner cost was $8 per square meter. To recover the investment, the owner charged $10 for pickup loads, $25 for dump truck loads, and $70 for compactor truck loads. If the monthly distribution is 200 pickup loads, 50 dump truck loads, and 100 compactor truck loads, what rate of return will the landfill owner make on the investment if the fill area is adequate for 4 years?

7.7 Swagelok Enterprises is a manufacturer of miniature fittings and valves. Over a 5-year period, the costs associated with one product line were as follows: first cost of $30,000 and annual costs of $18,000. Annual revenue was $27,000, and the used equipment was salvaged for $4000. What rate of return did the company make on this product?

7.8 Barron Chemical uses a thermoplastic polymer to enhance the appearance of certain RV panels. The initial cost of one process was $130,000 with annual costs of $49,000 and revenues of $78,000 in year 1, increasing by $1000 per year. A salvage value of $23,000 was realized when the process was discontinued after 8 years. What rate of return did the company make on the process?

7.9 A graduate of New Mexico State University who built a successful business wanted to start an endowment in her name that would provide scholarships to IE students. She wanted the scholarships to amount to $10,000 per year, and she wanted the first one to be given on the day she made the donation (i.e., at time 0). If she planned to donate $100,000, what rate of return would the university have to make in order to be able to award the $10,000 per year scholarships forever?

7.10 PPG manufactures an epoxy amine that is used to protect the contents of polyethylene terephthalate (PET) containers from reacting with oxygen. The cash flow (in millions) associated with the process is shown below. Determine the rate of return.

Year	Cost, $	Revenue, $
0	−10	—
1	−4	2
2	−4	3
3	−4	9
4	−3	9
5	−3	9
6	−3	9

7.11 An entrepreneurial mechanical engineer started a tire shredding business to take advantage of a Texas state law that outlaws the disposal of whole tires in sanitary landfills. The cost of the shredder was $220,000. She spent $15,000 to get 460-volt power to the site and another $76,000 in site preparation. Through contracts with tire dealers, she was paid $2 per tire and handled an average of 12,000 tires per month for 3 years. The annual operating costs for labor, power, repairs, etc., amounted to $1.05 per tire. She also sold some of the tire chips to septic tank installers for use in drain fields. This endeavor netted $2000 per month. After 3 years, she sold the equipment for $100,000. What rate of return did she make (a) per month and (b) per year (nominal and effective)?

7.12 An Internet B to C company projected the cash flows (in millions) page 266.

What annual rate of return will be realized if the cash flows occur as projected?

Year	Expenses, $	Revenue, $
0	−40	—
1	−40	12
2	−43	15
3	−45	17
4	−46	51
5	−48	63
6–10	−50	80

7.13 The University of California at San Diego is considering a plan to build an 8-megawatt cogeneration plant to provide for part of its power needs. The cost of the plant is expected to be $41 million. The university consumes 55,000 megawatt-hours per year at a cost of $120 per megawatt-hour. (*a*) If the university will be able to produce power at one-half the cost that it now pays, what rate of return will it make on its investment if the power plant lasts 30 years? (*b*) If the university can sell an average of 12,000 megawatt-hours per year back to the utility at $90 per megawatt-hour, what rate of return will it make?

7.14 A new razor from Gillette called the M3Power emits pulses that cause the skin to prop up hair so that it can be cut off more easily. This might make the blades last longer because there would be less need to repeatedly shave over the same surface. The M3Power system (including batteries) sells for $14.99 at some stores. The blades cost $10.99 for a package of four. The more conventional M3Turbo blades cost $7.99 for a package of four. If the blades for the M3Power system last 2 months while the blades for the M3Turbo last only 1 month, what rate of return (*a*) per month and (*b*) per year (nominal and effective) will be made if a person purchases the M3Power system? Assume the person already has an M3Turbo razor but needs to purchase blades at time 0. Use a 1-year project period.

7.15 Techstreet.com is a small Web design business that provides services for two main types of websites: brochure sites and e-commerce sites. One package involves an up-front payment of $90,000 and monthly payments of 1.4¢ per "hit." A new CAD software company is considering the package. The company expects to have at least 6000 hits per month, and it hopes that 1.5% of the hits will result in a sale. If the average income from sales (after fees and expenses) is $150, what rate of return per month will the CAD software company realize if it uses the website for 2 years?

7.16 A plaintiff in a successful lawsuit was awarded a judgment of $4800 per month for 5 years. The plaintiff needs a fairly large sum of money now for an investment and has offered the defendant the opportunity to pay off the award in a lump-sum amount of $110,000. If the defendant accepts the offer and pays the $110,000 now, what rate of return will the defendant have made on the "investment"? Assume the next $4800 payment is due 1 month from now.

7.17 Army Research Laboratory scientists developed a diffusion-enhanced adhesion process that is expected to significantly improve the performance of multifunction hybrid composites. NASA engineers estimate that composites made using the new process will result in savings in many space exploration projects. The cash flows for one project are shown below. Determine the rate of return per year.

Year t	Cost ($1000)	Savings ($1000)
0	−210	—
1	−150	—
2–5	—	$100 + 60(t - 2)$

7.18 ASM International, an Australian steel company, claims that a savings of 40% of the cost of stainless steel threaded bar can be achieved by replacing machined threads with precision weld depositions. A U.S. manufacturer of rock bolts and grout-in fittings plans to purchase the equipment. A mechanical engineer with the company has prepared the following cash flow estimates. Determine the expected rate of return per quarter and per year (nominal).

Quarter	Cost, $	Savings, $
0	−450,000	—
1	−50,000	10,000
2	−40,000	20,000
3	−30,000	30,000
4	−20,000	40,000
5	−10,000	50,000
6−12	—	80,000

7.19 An indium-gallium-arsenide-nitrogen alloy developed at Sandia National Laboratory is said to have potential uses in electricity-generating solar cells. The new material is expected to have a longer life, and it is believed to have a 40% efficiency rate, which is nearly twice that of standard silicon solar cells. The useful life of a telecommunications satellite could be extended from 10 to 15 years by using the new solar cells. What rate of return could be realized if an extra investment now of $950,000 would result in extra revenues of $450,000 in year 11, $500,000 in year 12, and amounts increasing by $50,000 per year through year 15?

7.20 A permanent endowment at the University of Alabama is to award scholarships to engineering students. The awards are to be made beginning 5 years after the $10 million lump-sum donation is made. If the interest from the endowment is to fund 100 students each year in the amount of $10,000 each, what annual rate of return must the endowment fund earn?

7.21 A charitable foundation received a donation from a wealthy building contractor in the amount of $5 million. It specifies that $200,000 is to be awarded each year for 5 years starting *now* (i.e., 6 awards) to a university engaged in research pertaining to the development of layered composite materials. Thereafter, grants equal to the amount of interest earned each year are to be made. If the size of the grants from year 6 into the indefinite future is expected to be $1,000,000 per year, what annual rate of return is the foundation earning?

Multiple ROR Values

7.22 What is the difference between a conventional and a nonconventional cash flow series?

7.23 What cash flows are associated with Descartes' rule of signs and Norstrom's criterion?

7.24 According to Descartes' rule of signs, how many possible i^* values are there for net cash flows that have the following signs?
(a) − − − + + + − +
(b) − − − − − − + + + + +
(c) + + + + − − − − − − + − + − − −

7.25 The cash flow (in 1000s) associated with a new method of manufacturing box cutters is shown on page 268 for a 2-year period. (a) Use Descartes' rule to determine the maximum number of possible rate of return values. (b) Use Norstrom's criterion to determine if there is only one positive rate of return value.

Quarter	Expense, $	Revenue, $
0	−20	0
1	−20	5
2	−10	10
3	−10	25
4	−10	26
5	−10	20
6	−15	17
7	−12	15
8	−15	2

7.26 RKI Instruments manufactures a ventilation controller designed for monitoring and controlling carbon monoxide in parking garages, boiler rooms, tunnels, etc. The net cash flow associated with one phase of the operation is shown below. (a) How many possible rate of return values are there for this cash flow series? (b) Find all the rate of return values between 0% and 100%.

Year	Net Cash Flow, $
0	−30,000
1	20,000
2	15,000
3	−2,000

7.27 A manufacturer of heavy-tow carbon fibers (used for sporting goods, thermoplastic compounds, windmill blades, etc.) reported the net cash flows below. (a) Determine the number of possible rate of return values, and (b) find all rate of return values between −50% and 120%.

Year	Net Cash Flow, $
0	−17,000
1	20,000
2	−5,000
3	8,000

7.28 Arc-bot Technologies, manufacturers of six-axis, electric servo-driven robots, has experienced the cash flows below in a shipping department. (a) Determine the number of possible rate of return values.
(b) Find all i* values between 0% and 100%.

Year	Expense, $	Savings, $
0	−33,000	0
1	−15,000	18,000
2	−40,000	38,000
3	−20,000	55,000
4	−13,000	12,000

7.29 Five years ago, a company made a $5 million investment in a new high-temperature material. The product was not well accepted after the first year on the market. However, when it was reintroduced 4 years later, it did sell well during the year. Major research funding to broaden the applications has cost $15 million in year 5. Determine the rate of return for these cash flows (shown below in $1000s).

Year	Net Cash Flow, $
0	−5,000
1	4,000
2	0
3	0
4	20,000
5	−15,000

Composite Rate of Return

7.30 What is meant by the term *reinvestment rate*?

7.31 An engineer working for GE invested his bonus money each year in company stock. His bonus has been $5000 each year for the past 6 years (i.e., at the end of years 1 through 6). At the end of year 7, he *sold* $9000 worth of his stock to remodel his kitchen (he didn't purchase any stock that year). In years 8 through 10, he again invested his $5000 bonus. The engineer sold all his remaining stock for $50,000 immediately after the last investment at the end of year 10. (*a*) Determine the number of possible rate of return values in the net cash flow series. (*b*) Find the internal rate of return(s). (*c*) Determine the composite rate of return. Use a reinvestment rate of 20% per year.

7.32 A company that makes clutch disks for race cars had the cash flows shown below for one department. Calculate (*a*) the internal rate of return and (*b*) the composite rate of return, using a reinvestment rate of 15% per year.

Year	Cash Flow, $1000
0	−65
1	30
2	84
3	−10
4	−12

7.33 For the cash flow series below, calculate the composite rate of return, using a reinvestment rate of 14% per year.

Year	Cash Flow, $
0	3000
1	−2000
2	1000
3	−6000
4	3800

7.34 For the high-temperature material project in Problem 7.29, determine the composite rate of return if the reinvestment rate is 15% per year. The cash flows (repeated below) are in $1000 units.

Year	Cash Flow, $
0	−5,000
1	4,000
2	0
3	0
4	20,000
5	−15,000

Bonds

7.35 A municipal bond that was issued by the city of Phoenix 3 years ago has a face value of $25,000 and a bond interest rate of 6% per year payable semiannually. If the bond is due 25 years after it was issued, (*a*) what are the amount and frequency of the bond interest payments and (*b*) what value of *n* must be used in the P/A formula to find the present worth of the remaining bond interest payments? Assume the market interest rate is 8% per year, compounded semiannually.

7.36 A $10,000 mortgage bond with a bond interest rate of 8% per year, payable quarterly, was purchased for $9200. The bond was kept until it was due, a total of 7 years. What rate of return was made by the purchaser per 3 months and per year (nominal)?

7.37 A plan for remodeling the downtown area of the city of Steubenville, Ohio, required the city to issue $5 million worth of general obligation bonds for infrastructure replacement. The bond interest rate was set at 6% per year, payable quarterly, with the principal repayment date 30 years into the future. The brokerage fees for the transactions amounted to $100,000. If the city received $4.6 million (*before* paying the brokerage fees) from the bond issue, (*a*) what interest rate (per quarter) did the investors require to purchase the bonds and (*b*) what are the nominal and effective rates of return per year to the investors?

7.38 A collateral bond with a face value of $5000 was purchased by an investor for $4100. The bond was due in 11 years, and it had a bond interest rate of 4% per year, payable semiannually. If the investor kept the bond to maturity, what rate of return per semiannual period did she make?

7.39 An engineer planning for his child's college education purchased a zero coupon corporate bond (i.e., a bond that has no interest payments) for $9250. The bond has a face value of $50,000 and is due in 18 years. If the bond is held to maturity, what rate of return will the engineer make on the investment?

7.40 Four years ago, Texaco issued $5 million worth of debenture bonds having a bond interest rate of 10% per year, payable semiannually. Market interest rates dropped, and the company called the bonds (i.e., paid them off in advance) at a 10% premium on the face value (i.e., paid $5.5 million to retire the bonds). What semiannual rate of return did an investor make if he purchased one $5000 bond at face value 4 years ago and held it until it was called 4 years later?

7.41 Five years ago, GSI, an oil services company, issued $10 million worth of 12%, 30-year bonds with interest payable quarterly. The interest rate in the marketplace decreased enough that the company is considering calling the bonds. If the company buys the bonds back now for $11 million, (a) what rate of return per quarter will the company make on the $11 million expenditure and (b) what nominal rate of return per year will it make on the $11 million investment? *Hint:* By spending $11 million now, the company will not have to make the quarterly bond interest payments or pay the face value of the bonds when they come due 25 years from now.

FE REVIEW PROBLEMS

7.42 When the net cash flow for an alternative changes signs more than once, the cash flow is said to be
(a) Conventional
(b) Simple
(c) Extraordinary
(d) Nonconventional

7.43 According to Descartes' rule of signs, how many possible rate of return values are there for net cash flow that has the following signs?
$$+ + + + - - - - - - - + - + - - - + +$$
(a) 3
(b) 5
(c) 6
(d) Less than 3

7.44 A small manufacturing company borrowed $1 million and repaid the loan through monthly payments of $20,000 for 2 years plus a single lump-sum payment of $1 million at the end of 2 years. The interest rate on the loan was closest to
(a) 0.5% per month
(b) 2% per month
(c) 2% per year
(d) 8% per year

7.45 According to Norstrom's criterion, there is only one positive rate of return value in a cash flow series when
(a) The cumulative cash flow starts out positive and changes sign only once.
(b) The cumulative cash flow starts out negative and changes sign only once.
(c) The cumulative cash flow total is greater than zero.
(d) The cumulative cash flow total is less than zero.

7.46 An investment of $60,000 resulted in uniform income of $10,000 per year for 10 years. The rate of return on the investment was closest to
(a) 10.6% per year
(b) 14.2% per year
(c) 16.4% per year
(d) 18.6% per year

7.47 For the net cash flows shown below, the maximum number of possible rate of return solutions is
(a) 0
(b) 1
(c) 2
(d) 3

Year	Net Cash Flow, $
0	−60,000
1	20,000
2	22,000
3	15,000
4	35,000
5	13,000
6	−2,000

7.48 A bulk materials hauler purchased a used dump truck for $50,000. The operating cost was $5000 per month, with average revenues of $7500 per month. After 2 years, the truck was sold for $11,000. The rate of return was closest to
(a) 2.6% per month
(b) 2.6% per year
(c) 3.6% per month
(d) 3.6% per year

7.49 Assume you are told that by investing $100,000 now, you will receive $10,000 per year *starting in year 5* and continuing forever. If you accept the offer, the rate of return on the investment is
(a) Less than 10% per year
(b) 0% per year
(c) 10% per year
(d) Over 10% per year

7.50 Five years ago, an alumnus of a small university donated $50,000 to establish a permanent endowment for scholarships. The first scholarships were awarded 1 year after the money was donated. If the amount awarded each year (i.e., the interest) is $4500, the rate of return earned on the fund is closest to
(a) 7.5% per year
(b) 8.5% per year
(c) 9% per year
(d) 10% per year

7.51 When positive net cash flows are generated before the end of a project, and when these cash flows are reinvested at an interest rate that is greater than the internal rate of return,
(a) The resulting rate of return is equal to the internal rate of return.
(b) The resulting rate of return is less than the internal rate of return.
(c) The resulting rate of return is equal to the reinvestment rate of return.
(d) The resulting rate of return is greater than the internal rate of return.

7.52 A $10,000 mortgage bond that is due in 20 years pays interest of $250 every 6 months. The bond interest rate is closest to
(a) 2.5% per year, payable quarterly
(b) 5.0% per year, payable quarterly
(c) 5% per year, payable semiannually
(d) 10% per year, payable quarterly

7.53 A $10,000 bond that matures in 20 years with interest at 8% per year payable quarterly was issued 4 years ago. If the bond is purchased now for $10,000 and held to maturity, what will be the *effective rate of return per quarter* to the purchaser?
(a) 2%
(b) 2.02%
(c) 4%
(d) 8%

7.54 A person purchases a $5000, 5% per year bond, with interest payable semiannually, for the amount of $4000. The bond has a maturity date of 14 years from now. The equation for calculating how much the person must sell the bond for 6 years from now in order to make a rate of return of 12% per year, compounded semiannually, is
(a) $0 = -4000 + 125(P/A,6\%,12) + x(P/F,6\%,12)$
(b) $0 = -4000 + 100(P/A,6\%,12) + x(P/F,6\%,12)$
(c) $0 = -5000 + 125(P/A,6\%,12) + x(P/F,6\%,12)$

(d) $0 = -4000 + 125(P/A,12\%,6) + x(P/F,12\%,6)$

7.55 A $50,000 corporate bond due in 20 years with an interest rate of 10% per year, payable quarterly, is for sale for $50,000. If an investor purchases the bond and holds it to maturity, the rate of return will be closest to
(a) Nominal 10% per year, compounded quarterly
(b) 2.5% per quarter
(c) Both (a) and (b) are correct
(d) Effective 10% per year

EXTENDED EXERCISES

EXTENDED EXERCISE 1—THE COST OF A POOR CREDIT RATING

Two people each borrow $5000 at a 10% per year interest rate for 3 years. A portion of Charles's loan agreement states that interest ". . . is paid at the rate of 10% compounded each year on the declining balance." Charles is told his annual payment will be $2010.57, due at the end of each year of the loan.

Jeremy currently has a slightly degraded credit rating, which the bank loan officer discovered. Jeremy has a habit of paying his bills late. The bank approved the loan, but a part of his loan agreement states that interest ". . . is paid at a rate of 10% compounded each year on the original loan amount." Jeremy is told his annual payment will be $2166.67 due at the end of each year.

Questions

Answer the following by hand, by computer, or both.

1. Develop a table and a plot for Charles and for Jeremy of the unrecovered balances (total amount owed) just before each payment is due.
2. How much more total money and interest will Jeremy pay than Charles over the 3 years?

EXTENDED EXERCISE 2—WHEN IS IT BEST TO SELL A BUSINESS?

After Jeff finished medical school and Imelda completed a degree in engineering, the couple decided to put a substantial part of their savings into rental property.

With a hefty bank loan and a cash down payment of $120,000 of their own funds, they were able to purchase six duplexes from a person exiting the residential rental business. Net cash flow on rental income after all expenses and taxes for the first 4 years was good: $25,000 at the end of the first year, increasing by $5000 each year thereafter. A business friend of Jeff's introduced him to a potential buyer for all properties with an estimated $225,000 net cash-out after the 4 years of ownership. But they did not sell. They wanted to stay in the business for a while longer, given the increasing net cash flows they had experienced thus far.

During year 5, an economic downturn reduced net cash flow to $35,000. In response, an extra $20,000 was spent in improvements and advertising in each of years 6 and 7, but the net cash flow continued to decrease by $10,000 per year through year 7. Jeff had another offer to sell in year 7 for only $60,000. This was considered too much of a loss, so they did not take advantage of the opportunity.

In the last 3 years, they have expended $20,000, $20,000, and $30,000 each year in improvements and advertising costs, but the net cash flow from the business has been only $15,000, $10,000, and $10,000 each year.

Imelda and Jeff want out, but they have no offer to buy at any price, and they have most of their savings committed to the rental property.

Questions

Determine the rate of return for the following:

1. At the end of year 4, first, if the $225,000 purchase offer had been accepted; second, without selling.
2. After 7 years, first, if the $60,000 "sacrifice" offer had been accepted; and, second, without selling.
3. Now, after 10 years, with no prospect of sale.
4. If the houses are sold and given to a charity, assume a net cash infusion to Jeff and Imelda of $25,000 after taxes at the end of this year. What is the rate of return over the 10 years of ownership?

CASE STUDY

BOB LEARNS ABOUT MULTIPLE RATES OF RETURN[1]

Background

When Bob began a summer internship with VAC, an electricity distribution company in an Atlantic coast city of about 275,000, he was given a project on the first day by his boss, Kathy. Homeworth, one of the major corporate customers, just placed a request for a lower rate per kwh, once its minimum required usage

[1]Contributed by Dr. Tep Sastri (former Associate Professor, Industrial Engineering, Texas A&M University).

is exceeded each month. Kathy has an internal report from the Customer Relations Department that itemizes the net cash flows below for the Homeworth account during the last 10 years.

Year	Cash Flow ($1000)
1993	$200
1994	100
1995	50
1996	−1800
1997	600
1998	500
1999	400
2000	300
2001	200
2002	100

The report also states that the annual rate of return is between 25 and 50%, but no further information is provided. This information is not detailed enough for Kathy to evaluate the company's request.

Over the next few hours, Bob and Kathy had a series of discussions as Bob worked to answer Kathy's increasingly more specific questions. The following is an abbreviated version of these conversations. Luckily, both Bob and Kathy took an engineering economy course during their undergraduate work, and their professors covered the method to find a unique rate of return for any cash flow series.

Development of the Situation

1. Kathy asked Bob to do a preliminary study to find the correct rate of return. She wanted only one number, not a range, and not two or three possible values. She did, however, have a passing interest in initially knowing the values of multiple rates, if they do exist, in order to determine if the report from customer relations was correct or just a "shot in the dark."

   Kathy told Bob that the MARR for the company is 15% per year for these major clients. She also explained that the 1996 negative cash flow was caused by an on-site equipment upgrade

when Homeworth expanded its manufacturing capacity and increased power usage about 5-fold.

2. Once Bob had finished his initial analysis, Kathy told him that she had forgotten to tell him that the rate of return earned externally on the positive cash flows from these major clients is placed into a venture capital pool headquartered in Chicago. It has been making 35% per year for the last decade. She wanted to know if a unique return still existed and if the Homeworth account was financially viable at a MARR of 35%.

In response to this request, Bob developed the four-step procedure outlined below to closely estimate the composite rate of return i' for any reinvestment rate c and two multiple rates i_1^* and i_2^*. He plans to apply this procedure to answer this latest question and show the results to Kathy.

Step 1. Determine the i^* roots of the PW relation for the cash flow series.

Step 2. For a given reinvestment rate c and the two i^* values from step 1, determine which of the following conditions applies:

(a) If $c < i_1^*$, then $i' < i_1^*$.

(b) If $c > i_2^*$, then $i' > i_2^*$.

(c) If $i_1^* < c < i_2^*$, then i' can be less than c or greater than c, and $i_1^* < i' < i_2^*$.

Step 3. Guess a starting value for i' according to the result from step 2. Apply the net-investment method from periods 1 to n. Repeat this step until F_n is close to 0. If this F_n is a small positive value, guess another i' that will result in a small negative F_n value, and vice versa.

Step 4. Using the two F_n results from step 3, linearly interpolate i' such that the corresponding F_n is approximately zero. Of course, the final i' value can also be obtained directly in step 3, without interpolation.

3. Finally, Kathy asked Bob to reevaluate the cash flows for Homeworth at the MARR of 35%, but using a reinvestment rate of 45% to determine if the series is still justified.

Case Study Exercises

1, 2, and 3.   Answer the questions for Bob using spreadsheets.

4. If the i' approximating procedure Bob developed is not available, use the original cash flow data to apply the basic net-investment procedure, and answer Exercises 2 and 3, where c is 35 and 45%, respectively.

5. Kathy concluded from this exercise that any cash flow series is economically justified for any reinvestment rate that is larger than the MARR. Is this a correct conclusion? Explain why or why not.

CHAPTER 8

Rate of Return Analysis: Multiple Alternatives

This chapter presents the methods by which two or more alternatives can be evaluated using a rate of return (ROR) comparison based on the methods of the previous chapter. The ROR evaluation correctly performed will result in the same selection as the PW, AW, and FW analyses, but the computational procedure is considerably different for ROR evaluations.

The first case study involves multiple options for a business owned for many years by one person. The second case explores nonconventional cash flow series with multiple rates of return and the use of the PW method in this situation.

LEARNING OBJECTIVES

Purpose: Select the best mutually exclusive alternative on the basis of rate of return analysis of incremental cash flows.

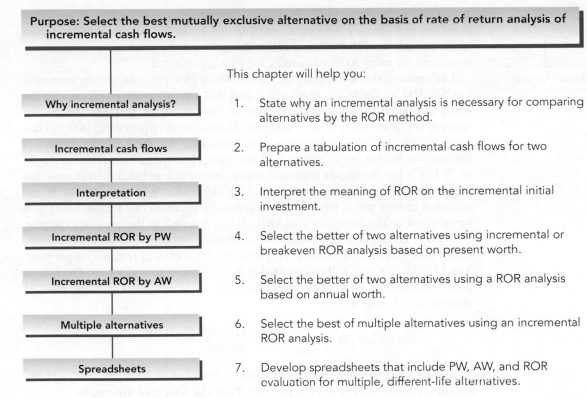

This chapter will help you:

Why incremental analysis?	1. State why an incremental analysis is necessary for comparing alternatives by the ROR method.
Incremental cash flows	2. Prepare a tabulation of incremental cash flows for two alternatives.
Interpretation	3. Interpret the meaning of ROR on the incremental initial investment.
Incremental ROR by PW	4. Select the better of two alternatives using incremental or breakeven ROR analysis based on present worth.
Incremental ROR by AW	5. Select the better of two alternatives using a ROR analysis based on annual worth.
Multiple alternatives	6. Select the best of multiple alternatives using an incremental ROR analysis.
Spreadsheets	7. Develop spreadsheets that include PW, AW, and ROR evaluation for multiple, different-life alternatives.

8.1 WHY INCREMENTAL ANALYSIS IS NECESSARY

When two or more mutually exclusive alternatives are evaluated, engineering economy can identify the one alternative that is the best economically. As we have learned, the PW, AW, and FW techniques can be used to do so. Now the procedure for using ROR to identify the best is presented.

Let's assume that a company uses a MARR of 16% per year, that the company has $90,000 available for investment, and that two alternatives (A and B) are being evaluated. Alternative A requires an investment of $50,000 and has an internal rate of return i_A^* of 35% per year. Alternative B requires $85,000 and has an i_B^* of 29% per year. Intuitively we may conclude that the better alternative is the one that has the larger return, A in this case. However, this is not necessarily so. While A has the higher projected return, it requires an initial investment that is much less than the total money available ($90,000). What happens to the investment capital that is left over? It is generally assumed that excess funds will be invested at the company's MARR, as we learned in the previous chapter. Using this assumption, it is possible to determine the consequences of the alternative investments. If alternative A is selected, $50,000 will return 35% per year. The $40,000 left over will be invested at the MARR of 16% per year. The rate of return on the total capital available, then, will be the weighted average. Thus, if alternative A is selected,

$$\text{Overall ROR}_A = \frac{50,000(0.35) + 40,000(0.16)}{90,000} = 26.6\%$$

If alternative B is selected, $85,000 will be invested at 29% per year, and the remaining $5000 will earn 16% per year. Now the weighted average is

$$\text{Overall ROR}_B = \frac{85,000(0.29) + 5000(0.16)}{90,000} = 28.3\%$$

These calculations show that even though the i^* for alternative A is higher, alternative B presents the better overall ROR for the $90,000. If either a PW or AW comparison is conducted using the MARR of 16% per year as i, alternative B will be chosen.

This simple example illustrates a major fact about the rate of return method for comparing alternatives:

> **Under some circumstances, project ROR values do not provide the same ranking of alternatives as do PW, AW, and FW analyses. This situation does not occur if we conduct an *incremental* cash flow ROR analysis (discussed in the next section).**

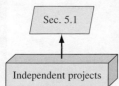

When independent projects are evaluated, no incremental analysis is necessary between projects. Each project is evaluated separately from others, and more than one can be selected. Therefore, the only comparison is with the do-nothing alternative for each project. The ROR can be used to accept or reject each independent project.

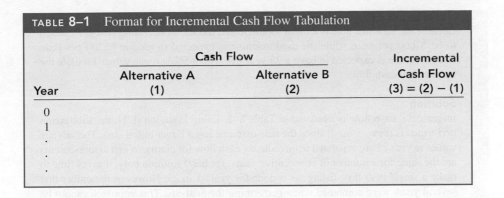

TABLE 8–1   Format for Incremental Cash Flow Tabulation

| | Cash Flow | | Incremental |
Year	Alternative A (1)	Alternative B (2)	Cash Flow (3) = (2) − (1)
0			
1			
.			
.			
.			

8.2   CALCULATION OF INCREMENTAL CASH FLOWS FOR ROR ANALYSIS

It is necessary to prepare an *incremental cash flow tabulation* between two alternatives in preparation for an incremental ROR analysis. A standardized format for the tabulation will simplify this process. The column headings are shown in Table 8–1. If the alternatives have *equal lives,* the year column will go from 0 to *n*. If the alternatives have *unequal lives,* the year column will go from 0 to the LCM (least common multiple) of the two lives. The use of the LCM is necessary because incremental ROR analysis requires equal-service comparison between alternatives. Therefore, all the assumptions and requirements developed earlier apply for any incremental ROR evaluation. When the LCM of lives is used, the salvage value and reinvestment in each alternative are shown at appropriate times. If a planning period is defined, the cash flow tabulation is for the specified period.

Only for the purpose of simplification, use the convention that between two alternatives, the one with the *larger initial investment* will be regarded as *alternative B.* Then, for each year in Table 8–1,

$$\text{Incremental cash flow} = \text{cash flow}_B - \text{cash flow}_A \qquad [8.1]$$

The initial investment and annual cash flows for each alternative (excluding the salvage value) occur in one of two patterns identified in Chapter 5:

Revenue alternative, where there are both negative and positive cash flows.

Service alternative, where all cash flow estimates are negative.

In either case, Equation [8.1] is used to determine the incremental cash flow series with the sign of each cash flow carefully determined. The next two examples illustrate incremental cash flow tabulation of service alternatives of equal and different lives. Later examples treat revenue alternatives.

Sec. 5.1

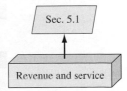

Revenue and service

EXAMPLE 8.1

A tool and die company in Pittsburgh is considering the purchase of a drill press with fuzzy-logic software to improve accuracy and reduce tool wear. The company has the opportunity to buy a slightly used machine for $15,000 or a new one for $21,000.

Because the new machine is a more sophisticated model, its operating cost is expected to be $7000 per year, while the used machine is expected to require $8200 per year. Each machine is expected to have a 25-year life with a 5% salvage value. Tabulate the incremental cash flow.

Solution

Incremental cash flow is tabulated in Table 8–2. Using Equation [8.1], the subtraction performed is (new − used) since the new machine has a larger initial cost. The salvage values in year 25 are separated from ordinary cash flow for clarity. When disbursements are the same for a number of consecutive years, for hand solution only, it saves time to make a single cash flow listing, as is done for years 1 to 25. However, remember that several years were combined when performing the analysis. This approach cannot be used for spreadsheets.

TABLE 8–2   Cash Flow Tabulation for Example 8.1

Year	Cash Flow Used Press	Cash Flow New Press	Incremental Cash Flow (New − Used)
0	$ −15,000	$ −21,000	$ −6,000
1–25	−8,200	−7,000	+1,200
25	+750	+1,050	+300
Total	$−219,250	$−194,950	$+24,300

Comment

When the cash flow columns are subtracted, the difference between the totals of the two cash flow series should equal the total of the incremental cash flow column. This only provides a check of the addition and subtraction in preparing the tabulation. It is not a basis for selecting an alternative.

EXAMPLE 8.2

Sandersen Meat Processors has asked its lead process engineer to evaluate two different types of conveyors for the bacon curing line. Type A has an initial cost of $70,000 and a life of 8 years. Type B has an initial cost of $95,000 and a life expectancy of 12 years. The annual operating cost for type A is expected to be $9000, while the AOC for type B is expected to be $7000. If the salvage values are $5000 and $10,000 for type A and type B, respectively, tabulate the incremental cash flow using their LCM.

Solution

The LCM of 8 and 12 is 24 years. In the incremental cash flow tabulation for 24 years (Table 8–3) note that the reinvestment and salvage values are shown in years 8 and 16 for type A and in year 12 for type B.

TABLE 8–3   Incremental Cash Flow Tabulation, Example 8.2

Year	Cash Flow Type A	Cash Flow Type B	Incremental Cash Flow (B − A)
0	$ −70,000	$ −95,000	$−25,000
1–7	−9,000	−7,000	+2,000
8	⎧ −70,000 ⎨ −9,000 ⎩ +5,000	−7,000	+67,000
9–11	−9,000	−7,000	+2,000
12	−9,000	⎧ −95,000 ⎨ −7,000 ⎩ +10,000	−83,000
13–15	−9,000	−7,000	+2,000
16	⎧ −70,000 ⎨ −9,000 ⎩ +5,000	−7,000	+67,000
17–23	−9,000	−7,000	+2,000
24	⎧ −9,000 ⎩ +5,000	⎧ −7,000 ⎩ +10,000	+7,000
	$−411,000	$−338,000	$+73,000

The use of a spreadsheet to obtain incremental cash flows requires one entry for each year through the LCM for each alternative. Therefore, some combining of cash flows may be necessary before the entry is made for each alternative. The incremental cash flow column results from an application of Equation [8.1]. As an illustration, the first 8 years of the 24 years in Table 8–3 should appear as follows when entered onto a spreadsheet. The incremental values in column D are obtained using a subtraction relation, for example, C4−B4.

E-Solve

Column A Year	Column B Type A	Column C Type B	Column D Incremental
0	$ −70,000	$ −95,000	$ −25,000
1	−9,000	−7,000	+2,000
2	−9,000	−7,000	+2,000
3	−9,000	−7,000	+2,000
4	−9,000	−7,000	+2,000
5	−9,000	−7,000	+2,000
6	−9,000	−7,000	+2,000
7	−9,000	−7,000	+2,000
8	−74,000	−7,000	+67,000
etc.			

8.3 INTERPRETATION OF RATE OF RETURN ON THE EXTRA INVESTMENT

The incremental cash flows in year 0 of Tables 8–2 and 8–3 reflect the *extra investment* or *cost* required if the alternative with the larger first cost is selected. This is important in an incremental ROR analysis in order to determine the ROR earned on the extra funds expended for the larger-investment alternative. If the incremental cash flows of the larger investment don't justify it, we must select the cheaper one. In Example 8.1 the new drill press requires an extra investment of $6000 (Table 8–2). If the new machine is purchased, there will be a "savings" of $1200 per year for 25 years, plus an extra $300 in year 25. The decision to buy the used or new machine can be made on the basis of the profitability of investing the extra $6000 in the new machine. If the equivalent worth of the savings is greater than the equivalent worth of the extra investment at the MARR, the extra investment should be made (i.e., the larger first-cost proposal should be accepted). On the other hand, if the extra investment is not justified by the savings, select the lower-investment proposal.

It is important to recognize that the rationale for making the selection decision is the same as if only *one alternative* were under consideration, that alternative being the one represented by the incremental cash flow series. When viewed in this manner, it is obvious that unless this investment yields a rate of return equal to or greater than the MARR, the extra investment should not be made. As further clarification of this extra-investment rationale, consider the following: The rate of return attainable through the incremental cash flow is an alternative to investing at the MARR. Section 8.1 states that any excess funds not invested in the alternative are assumed to be invested at the MARR. The conclusion is clear:

> **If the rate of return available through the incremental cash flow equals or exceeds the MARR, the alternative associated with the extra investment should be selected.**

Not only must the return on the extra investment meet or exceed the MARR, but also the return on the investment that is common to both alternatives must meet or exceed the MARR. Accordingly, before starting an incremental ROR analysis, it is advisable to determine the internal rate of return i^* for each alternative. (Of course, this is much easier with evaluation by computer than by hand.) This can be done only for revenue alternatives, because service alternatives have only cost (negative) cash flows and no i^* can be determined. The guideline is as follows:

> **For multiple revenue alternatives, calculate the internal rate of return i^* for each alternative, and eliminate all alternatives that have an $i^* <$ MARR. Compare the remaining alternatives incrementally.**

As an illustration, if the MARR = 15% and two alternatives have i^* values of 12 and 21%, the 12% alternative can be eliminated from further consideration. With only two alternatives, it is obvious that the second one is selected. If both alternatives have $i^* <$ MARR, no alternative is justified and the do-nothing alternative is the best economically. When three or more alternatives are evaluated, it is usually worthwhile, but not required, to calculate i^* for each alternative for preliminary screening. Alternatives that cannot meet the MARR may be eliminated from further evaluation using this option. This option is especially useful when performing the analysis by computer. The IRR function applied to each alternative's cash flow estimates can quickly indicate unacceptable alternatives, as demonstrated later in Section 8.6.

Q-Solv

When *independent projects* are evaluated, there is no comparison on the extra investment. The ROR value is used to accept all projects with $i^* \geq$ MARR, assuming there is no budget limitation. For example, assume MARR = 10%, and three independent projects are available with ROR values of:

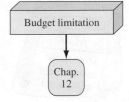

$$i_A^* = 12\% \qquad i_B^* = 9\% \qquad i_C^* = 23\%$$

Projects A and C are selected, but B is not because $i_B^* <$ MARR. Example 8.8 in Section 8.7 on spreadsheet applications illustrates selection from independent projects using ROR values.

8.4 RATE OF RETURN EVALUATION USING PW: INCREMENTAL AND BREAKEVEN

In this section we discuss the primary approach to making mutually exclusive alternative selections by the incremental ROR method. A PW-based relation like Equation [7.1] is developed for the incremental cash flows. Use hand or computer means to find Δi_{B-A}^*, the internal ROR for the series. Placing Δ (delta) before i_{B-A}^* distinguishes it from the ROR values i_A^* and i_B^*.

Since incremental ROR requires equal-service comparison, the LCM of lives must be used in the PW formulation. Because of the reinvestment requirement for PW analysis for different-life assets, the incremental cash flow series may contain several sign changes, indicating multiple Δi^* values. Though incorrect, this indication is usually neglected in actual practice. The correct approach is to establish the reinvestment rate c and follow the approach of Section 7.5. This means that the unique composite rate of return $(\Delta i')$ for the incremental cash flow series is determined. These three required elements—incremental cash flow series, LCM, and multiple roots—are the primary reasons that the ROR method is often applied incorrectly in engineering economy analyses of multiple alternatives. As stated earlier, it is always possible, and generally advisable, to use a PW or AW analysis *at an established MARR* in lieu of the ROR method when multiple rates are indicated.

The complete procedure by hand or computer for an incremental ROR analysis for two alternatives is as follows:

1. Order the alternatives by initial investment or cost, starting with the smaller one, called A. The one with the larger initial investment is in the column labeled B in Table 8–1.
2. Develop the cash flow and incremental cash flow series using the LCM of years, assuming reinvestment in alternatives.
3. Draw an incremental cash flow diagram, if needed.

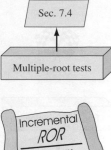

4. Count the number of sign changes in the incremental cash flow series to determine if multiple rates of return may be present. If necessary, use Norstrom's criterion on the cumulative incremental cash flow series to determine if a single positive root exists.

5. Set up the PW equation for the incremental cash flows in the form of Equation [7.1], and determine Δi^*_{B-A} using trial and error by hand or spreadsheet functions.

6. Select the economically better alternative as follows:

If Δi^*_{B-A} < MARR, select alternative A.

If Δi^*_{B-A} ≥ MARR, the extra investment is justified; select alternative B.

If the incremental i^* is exactly equal to or very near the MARR, noneconomic considerations will most likely be used to help in the selection of the "best" alternative.

In step 5, if trial and error is used to calculate the rate of return, time may be saved if the Δi^*_{B-A} value is bracketed, rather than approximated by a point value using linear interpolation, provided that a single ROR value is not needed. For example, if the MARR is 15% per year and you have established that Δi^*_{B-A} is in the 15 to 20% range, an exact value is not necessary to accept B since you already know that Δi^*_{B-A} ≥ MARR.

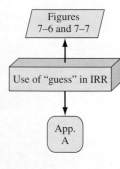

The IRR function on a spreadsheet will normally determine one Δi^* value. Multiple guess values can be input to find multiple roots in the range −100% to ∞ for a nonconventional series, as illustrated in Examples 7.4 and 7.5. If this is not the case, to be correct, the indication of multiple roots in step 4 requires that the net-investment procedure, Equation [7.6], be applied in step 5 to make $\Delta i' = \Delta i^*$. If one of these multiple roots is the same as the expected reinvestment rate c, this root can be used as the ROR value, and the net-investment procedure is not necessary. In this case only, $\Delta i' = \Delta i^*$, as concluded at the end of Section 7.5.

EXAMPLE 8.3

In 2000, Bell Atlantic and GTE merged to form a giant telecommunications corporation named Verizon Communications. As expected, some equipment incompatibilities had to be rectified, especially for long distance and international wireless and video services. One item had two suppliers—a U.S. firm (A) and an Asian firm (B). Approximately 3000 units of this equipment were needed. Estimates for vendors A and B are given for each unit.

	A	B
Initial cost, $	−8,000	−13,000
Annual costs, $	−3,500	−1,600
Salvage value, $	0	2,000
Life, years	10	5

TABLE 8–4 Incremental Cash Flow Tabulation, Example 8.3

Year	Cash Flow A (1)	Cash Flow B (2)	Incremental Cash Flow (3) = (2) − (1)
0	$ −8,000	$−13,000	$ −5,000
1–5	−3,500	−1,600	+1,900
5	—	$\begin{cases} +2,000 \\ -13,000 \end{cases}$	−11,000
6–10	−3,500	−1,600	+1,900
10	—	+2,000	+2,000
	$−43,000	$−38,000	$ +5,000

Determine which vendor should be selected if the MARR is 15% per year. Show hand and computer solutions.

Solution by Hand

These are service alternatives, since all cash flows are costs. Use the procedure described above to determine Δi_{B-A}^*.

1. Alternatives A and B are correctly ordered with the higher first-cost alternative in column (2).
2. The cash flows for the LCM of 10 years are tabulated in Table 8–4.
3. The incremental cash flow diagram is shown in Figure 8–1.
4. There are three sign changes in the incremental cash flow series, indicating as many as three roots. There are also three sign changes in the cumulative incremental series, which starts negatively at $S_0 = \$-5000$ and continues to $S_{10} = \$+5000$, indicating that more than one positive root may exist.
5. The rate of return equation based on the PW of incremental cash flows is

$$0 = -5000 + 1900(P/A,\Delta i,10) - 11,000(P/F,\Delta i,5) + 2000(P/F,\Delta i,10) \qquad [8.2]$$

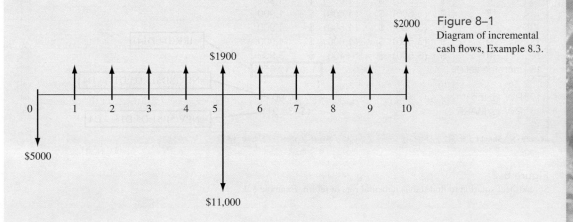

Figure 8–1
Diagram of incremental cash flows, Example 8.3.

Assume that the reinvestment rate is equal to the resulting Δi^*_{B-A} (or Δi^* for a shortened symbol). Solution of Equation [8.2] for the first root discovered results in Δi^* between 12 and 15%. By interpolation $\Delta i^* = 12.65\%$.

6. Since the rate of return of 12.65% on the extra investment is less than the 15% MARR, the lower-cost vendor A is selected. The extra investment of $5000 is not economically justified by the lower annual cost and higher salvage estimates.

Comment

In step 4, the presence of up to three i^* values is indicated. The preceding analysis finds one of the roots at 12.65%. When we state that the incremental ROR is 12.65%, we assume that any positive net-investments are reinvested at $c = 12.65\%$. If this is not a reasonable assumption, the net-investment procedure must be applied and an estimated reinvestment rate c must be used to find a different value $\Delta i'$ to compare with MARR = 15%.

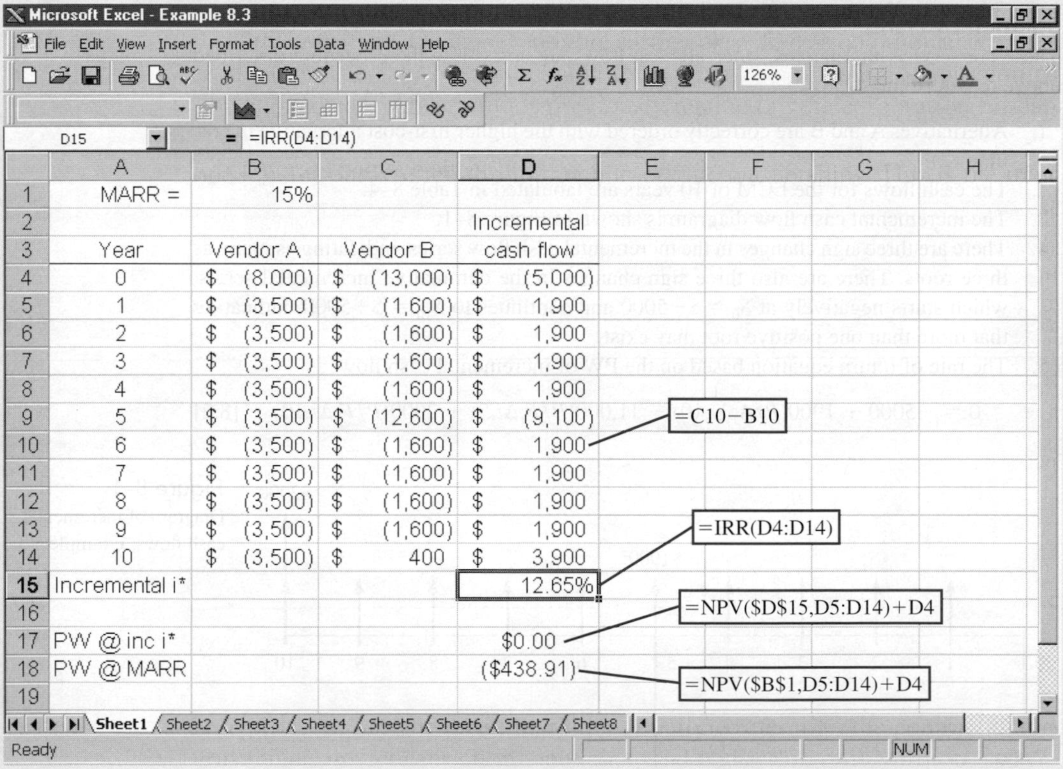

Figure 8–2
Spreadsheet solution to find the incremental rate of return, Example 8.3.

The other two roots are very large positive and negative numbers, as the IRR function of Excel reveals. So they are not useful to the analysis.

Solution by Computer

Steps 1 through 4 are the same as above.

5. Figure 8–2 includes the incremental net cash flows from Table 8–4 calculated in column D. Cell D15 displays the Δi^* value of 12.65% using the IRR function.
6. Since the rate of return on the extra investment is less than the 15% MARR, the lower-cost vendor A is selected.

Comment

Once the spreadsheet is set up, there are a wide variety of analyses that can be performed. For example, cell D17 uses the NPV function to verify that the present worth is zero at the calculated Δi^*. Cell D18 is the PW at MARR $= 15\%$, which is negative, thus indicating in yet another way that the extra investment does not return the MARR. Of course, any cash flow estimate and the MARR can be changed to determine what happens to Δi^*. A PW vs. Δi chart could easily be added, if two more columns are inserted, similar to those in Figures 7–6 and 7–7.

The rate of return determined for the incremental cash flow series can be interpreted as a *breakeven rate of return* value. If the incremental cash flow ROR (Δi^*) is greater than the MARR, the larger-investment alternative is selected. For example, if the PW vs. i graph for the incremental cash flows in Table 8–4 (and spreadsheet Figure 8–2) is plotted for various interest rates, the graph shown in Figure 8–3 is obtained. It shows the Δi^* breakeven at 12.65%. The conclusions are that

*   For MARR $< 12.65\%$, the extra investment for B is justified.
*   For MARR $> 12.65\%$, the opposite is true; the extra investment in B should not be made, and vendor A is selected.
*   If MARR is exactly 12.65%, the alternatives are equally attractive.

Figure 8–4, which is a breakeven graph of PW vs. i for the cash flows (not incremental) of each alternative in Example 8.3, provides the same results. Since all net cash flows are negative (service alternatives), the PW values are negative. Now, the same conclusions are reached using the following logic:

*   If MARR $< 12.65\%$, select B since its PW of cost cash flows is smaller (numerically larger).
*   If MARR $> 12.65\%$, select A since now its PW of costs is smaller.
*   If MARR is exactly 12.65%, either alternative is equally attractive.

The next example illustrates incremental ROR evaluation and breakeven rate of return graphs for revenue alternatives. More of breakeven analysis is covered in Chapter 13.

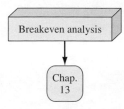

Figure 8–3
Plot of present worth of
incremental cash flows
for Example 8.3 at
various Δi values.

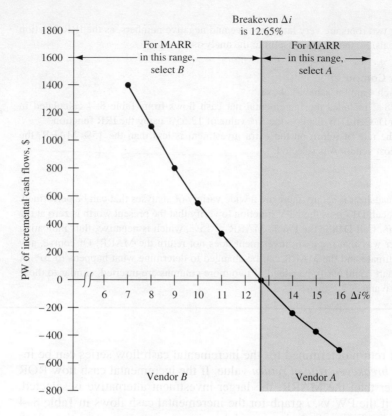

The rate of return on the incremental cash flow series may be interpreted as a breakeven rate of return value. If the incremental cash flow ROR (Δi^*) is greater than MARR, the larger investment alternative is selected. For example, if the PW vs. Δi graph for the incremental cash flow in Table 8–4 and spreadsheet (Figure 8–2) is plotted for various interest rates, the graph shown in Figure 8–3 is obtained. It shows the Δi^* breakeven at 12.65%. The conclusions are:

- For MARR < 12.65%, the extra investment for B is justified.
- For MARR > 12.65%, the opposite is true—the extra investment in B is not to be made, and vendor A is selected.
- If MARR is exactly 12.65%, the alternatives are equally attractive.

Figure 8–4, which is a breakeven graph of PW of the cash flows for each alternative (not of the incremental amounts), provides the same results. Since all net cash flows are negative (cost alternatives), the PW values are negative. Now the curves intersect at the breakeven ROR using the larger cash flows.

- If MARR < 12.65%, select A since its PW of cost cash flows is smaller (numerically larger).
- If MARR > 12.65%, select A since its PW of costs is smaller.
- If MARR is exactly 12.65%, either alternative is equally attractive.

The next example illustrates the incremental ROR analysis and breakeven rate of return graph for a cost alternatives. More than two alternatives is covered in Chapter 9.

Figure 8–4
Breakeven graph of
Example 8.3 cash flows
(not incremental).

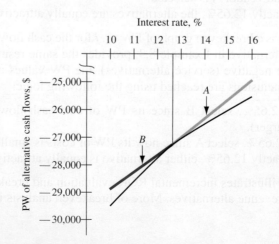

EXAMPLE 8.4

Bank of America uses a MARR of 30% on alternatives for its own business that are considered risky, that is, the response of the public to the service has not been well established by test marketing. Two alternative software systems and the marketing/delivery plans have been jointly developed by software engineers and the marketing department. They are for new online banking and loan services to passenger cruise ships and military vessels at sea internationally. For each system, start-up, annual net income, and salvage value (i.e., sell-out value to another financial corporation) estimates are summarized below.

E-Solve

(a) Perform the incremental ROR analysis by computer.
(b) Develop the PW vs. *i* graphs for each alternative and the increment. Which alternative, if either one, should be selected.?

	System A	System B
Initial investment, $1000	−12,000	−18,000
Estimated net income, $1000 per year	5,000	7,000
Salvage value, $1000	2,500	3,000
Estimated competitive life, years	8	8

Solution by Computer

(a) Refer to Figure 8–5a. The IRR function is used in cells B13 and E13 to display *i** for each alternative. We use the *i** values as a preliminary screening tool only to determine which alternatives exceed the MARR. If none do, the DN alternative is indicated automatically. In both cases *i** > 30%; they are both retained. The incremental cash flows are calculated (column G = column E − column B), and the IRR function results in $\Delta i* = 29.41\%$. This value is slightly less than the MARR; alternative A is selected as the better economic choice.

(b) Figure 8–5b contains plots of PW vs. *i* for all three cash flow series between the interest rates of 25 and 42%. The bottom curve (incremental analysis) indicates the breakeven ROR at 29.41%, which is where the two alternative PW curves cross. The conclusion is again the same; with MARR = 30%, select alternative A because its PW value ($2930 in cell D5 of Figure 8–5a) is slightly larger than that for B ($2841 in F5).

Comment

With this spreadsheet format, both a PW analysis and an incremental ROR analysis have been accomplished with graphic backup to demonstrate the conclusion of the engineering economy analysis.

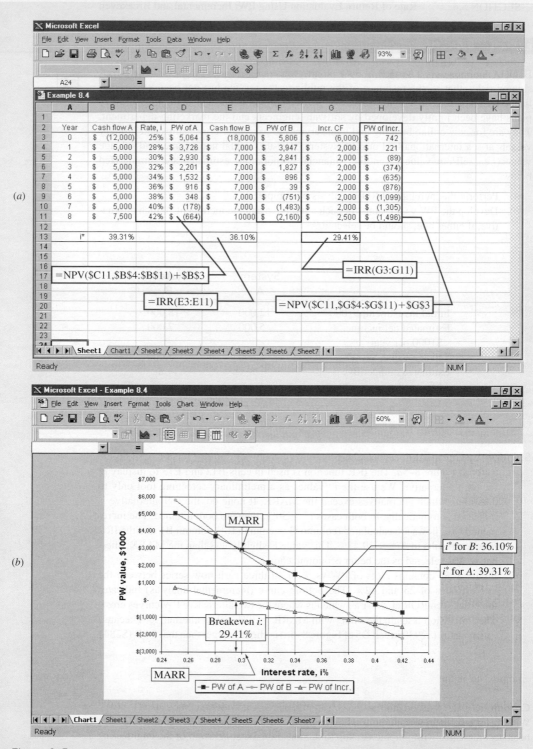

Figure 8–5
Spreadsheet solution to compare two alternatives: (*a*) incremental ROR analysis, (*b*) PW vs. *i* graphs, Example 8.4.

Figure 8–5*b* provides an excellent opportunity to see why the ROR method can result in selecting the wrong alternative when only $i*$ values are used to select between two alternatives. This is sometimes called the *ranking inconsistency problem* of the ROR method. *The inconsistency occurs when the MARR is set less than the breakeven rate between two revenue alternatives.* Since the MARR is established based on conditions of the economy and market, MARR is established external to any particular alternative evaluation. In Figure 8–5*b*, the breakeven rate is 29.41%, and the MARR is 30%. If the MARR were established lower than breakeven, say at 26%, the *incremental* ROR analysis results in correctly selecting B, because $\Delta i* = 29.41\%$, which exceeds 26%. But if only the $i*$ values were used, system A would be wrongly chosen, because its $i* = 39.31\%$. This error occurs because the rate of return method assumes reinvestment at the alternative's ROR value (39.31%), while PW and AW analyses use the MARR as the reinvestment rate. The conclusion is simple:

If the ROR method is used to evaluate two or more alternatives, use the *incremental cash flows* and $\Delta i*$ to make the decision between alternatives.

8.5   RATE OF RETURN EVALUATION USING AW

Comparing alternatives by the ROR method (correctly performed) always leads to the same selection as PW, AW, and FW analyses, whether the ROR is determined using a PW-based, an AW-based, or a FW-based relation. However, for the AW-based technique, there are two equivalent ways to perform the evaluation: using the *incremental cash flows* over the LCM of alternative lives, just as for the PW-based relation (previous section), or finding the AW for each alternative's *actual cash flows* and setting the difference of the two equal to zero to find the $\Delta i*$ value. Of course, there is no difference between the two approaches if the alternative lives are equal. Both methods are summarized here.

Since the ROR method requires comparison for equal service, *the incremental cash flows must be evaluated over the LCM of lives.* When solving by hand for $\Delta i*$, there may be no real computational advantage to using AW, as was found in Chapter 6. The same six-step procedure of the previous section (for PW-based calculation) is used, except in step 5 the AW-based relation is developed.

For comparison by computer with equal or unequal lives, the incremental cash flows must be calculated over the LCM of the two alternatives' lives. Then the IRR function is applied to find $\Delta i*$. This is the same technique developed in the previous section and used in the spreadsheet in Figure 8–2. *Use of the IRR function in this manner is the correct way to use Excel spreadsheet functions to compare alternatives using the ROR method.*

The second AW-based method takes advantage of the AW technique's assumption that the equivalent AW value is the same for each year of the first and all succeeding life cycles. Whether the lives are equal or unequal, set up the *AW relation for the cash flows of each alternative,* form the relation below, and solve for $i*$.

$$0 = AW_B - AW_A \qquad [8.3]$$

Equation [8.3] applies to solution by hand only, not solution by computer.

REMEMBER: COMPARE EQUAL SERVICE

E-Solve

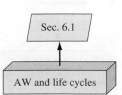

Sec. 6.1

AW and life cycles

For both methods, all equivalent values are on an AW basis, so the i^* that results from Equation [8.3] is the same as the Δi^* found using the first approach. Example 8.5 illustrates ROR analysis using AW-based relations for unequal lives.

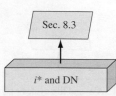

E-Solve

EXAMPLE 8.5

Compare the alternatives of vendors A and B for Verizon Communications in Example 8.3, using an AW-based incremental ROR method and the same MARR of 15% per year.

Solution

For reference, the PW-based ROR relation, Equation [8.2], for the incremental cash flow in Example 8.3 shows that vendor A should be selected with $\Delta i^* = 12.65\%$.

For the AW relation, there are two equivalent solution approaches. Write an AW-based relation on the *incremental* cash flow series over the *LCM of 10 years*, or write Equation [8.3] for the *two actual* cash flow series over *one life cycle* of each alternative.

For the incremental method, the AW equation is

$$0 = -5000(A/P,\Delta i,10) - 11{,}000(P/F,\Delta i,5)(A/P,\Delta i,10) + 2000(A/F,\Delta i,10) + 1900$$

It is easy to enter the incremental cash flows onto a spreadsheet, as in Figure 8–2, column D, and use the IRR(D4:D14) function to display $\Delta i^* = 12.65\%$.

For the second method, the ROR is found by Equation [8.3] using the respective lives of 10 years for A and 5 years for B.

$$AW_A = -8000(A/P,i,10) - 3500$$
$$AW_B = -13{,}000(A/P,i,5) + 2000(A/F,i,5) - 1600$$

Now develop $0 = AW_B - AW_A$.

$$0 = -13{,}000(A/P,i,5) + 2000(A/F,i,5) + 8000(A/P,i,10) + 1900$$

Solution again yields an interpolated value of $i^* = 12.65\%$.

Comment

It is very important to remember that when an incremental ROR analysis using an AW-based equation is made on the *incremental cash flows,* the least common multiple of lives must be used.

8.6   INCREMENTAL ROR ANALYSIS OF MULTIPLE, MUTUALLY EXCLUSIVE ALTERNATIVES

This section treats selection from multiple alternatives that are mutually exclusive, using the incremental ROR method. Acceptance of one alternative automatically precludes acceptance of any others. The analysis is based upon PW (or AW) relations for incremental cash flows between two alternatives at a time.

When the incremental ROR method is applied, the entire investment must return at least the MARR. When the i^* values on several alternatives exceed the MARR, incremental ROR evaluation is required. (For revenue alternatives, if

Sec. 8.3

i^* and DN

not even one $i^* \geq$ MARR, the do-nothing alternative is selected.) For all alternatives (revenue or service), the incremental investment must be separately justified. If the return on the extra investment equals or exceeds the MARR, then the extra investment should be made in order to maximize the total return on the money available, as discussed in Section 8.1.

Thus, for ROR analysis of multiple, mutually exclusive alternatives, the following criteria are used. Select the one alternative that

1.   Requires the *largest investment,* and
2.   Indicates that the *extra investment over another acceptable alternative is justified.*

An important rule to apply when evaluating multiple alternatives by the incremental ROR method is that *an alternative should never be compared with one for which the incremental investment is not justified.*

The incremental ROR evaluation procedure for multiple, equal-life alternatives is summarized below. Step 2 applies only to revenue alternatives, because the first alternative is compared to DN only when revenue cash flows are estimated. The terms *defender* and *challenger* are dynamic in that they refer, respectively, to the alternative that is currently selected (the defender) and the one that is challenging it for acceptance based on Δi^*. In every pairwise evaluation, there is one of each. The steps for solution by hand or by computer are as follows:

1.   Order the alternatives from *smallest to largest initial investment.* Record the annual cash flow estimates for each equal-life alternative.
2.   *Revenue alternatives only:* Calculate i^* for the first alternative. In effect, this makes DN the defender and the first alternative the challenger. If $i^* <$ MARR, eliminate the alternative and go to the next one. Repeat this until $i^* \geq$ MARR for the first time, and define that alternative as the defender. The next alternative is now the challenger. Go to step 3. (*Note:* This is where solution by computer spreadsheet can be a quick assist. Calculate the i^* for all alternatives first, using the IRR function, and select as the defender the first one for which $i^* \geq$ MARR. Label it the defender and go to step 3.)

Q-Solv

3.   Determine the incremental cash flow between the challenger and defender, using the relation

  Incremental cash flow = challenger cash flow − defender cash flow

  Set up the ROR relation.
4.   Calculate Δi^* for the incremental cash flow series using a PW-, AW-, or FW-based equation. (PW is most commonly used.)
5.   If $\Delta i^* \geq$ MARR, the challenger becomes the defender and the previous defender is eliminated. Conversely, if $\Delta i^* <$ MARR, the challenger is removed, and the defender remains against the next challenger.
6.   Repeat steps 3 to 5 until only one alternative remains. It is the selected one.

Note that only two alternatives are compared at any one time. It is vital that the correct alternatives be compared, or the wrong alternative may be selected.

EXAMPLE 8.6

Caterpillar Corporation wants to build a spare parts storage facility in the Phoenix, Arizona, vicinity. A plant engineer has identified four different location options. Initial cost of earthwork and prefab building, and annual net cash flow estimates are detailed in Table 8–5. The annual net cash flow series vary due to differences in maintenance, labor costs, transportation charges, etc. If the MARR is 10%, use incremental ROR analysis to select the one economically best location.

TABLE 8–5 Estimates for Four Alternative Building Locations, Example 8.6

	A	B	C	D
Initial cost, $	−200,000	−275,000	−190,000	−350,000
Annual cash flow, $	+22,000	+35,000	+19,500	+42,000
Life, years	30	30	30	30

Solution

All sites have a 30-year life, and they are revenue alternatives. The procedure outlined above is applied.

1. The alternatives are ordered by increasing initial cost in Table 8–6.
2. Compare location C with the do-nothing alternative. The ROR relation includes only the P/A factor.

$$0 = -190,000 + 19,500(P/A,i^*,30)$$

Table 8–6, column 1, presents the calculated $(P/A,\Delta i^*,30)$ factor value of 9.7436 and $\Delta i_c^* = 9.63\%$. Since $9.63\% < 10\%$, location C is eliminated. Now the comparison is A to DN, and column 2 shows that $\Delta i_A^* = 10.49\%$. This eliminates the do-nothing alternative; the defender is now A and the challenger is B.

TABLE 8–6 Computation of Incremental Rate of Return for Four Alternatives, Example 8.6

	C (1)	A (2)	B (3)	D (4)
Initial cost, $	−190,000	−200,000	−275,000	−350,000
Cash flow, $	+19,500	+22,000	+35,000	+42,000
Alternatives compared	C to DN	A to DN	B to A	D to B
Incremental cost, $	−190,000	−200,000	−75,000	−75,000
Incremental cash flow, $	+19,500	+22,000	+13,000	+7,000
Calculated $(P/A,\Delta i^*,30)$	9.7436	9.0909	5.7692	10.7143
$\Delta i^*,\%$	9.63	10.49	17.28	8.55
Increment justified?	No	Yes	Yes	No
Alternative selected	DN	A	B	B

3. The incremental cash flow series, column 3, and Δi^* for *B-to-A comparison* is determined from

$$0 = -275{,}000 - (-200{,}000) + (35{,}000 - 22{,}000)(P/A,\Delta i^*,30)$$
$$= -75{,}000 + 13{,}000(P/A,\Delta i^*,30)$$

4. From the interest tables, look up the P/A factor at the MARR, which is $(P/A,10\%,30) = 9.4269$. Now, any P/A value greater than 9.4269 indicates that the Δi^* will be less than 10% and is unacceptable. The P/A factor is 5.7692, so B is acceptable. For reference purposes, $\Delta i^* = 17.28\%$.

5. Alternative B is justified incrementally (new defender), thereby eliminating A.

6. Comparison D-to-B (steps 3 and 4) results in the PW relation $0 = -75{,}000 + 7000(P/A,\Delta i^*,30)$ and a P/A value of 10.7143 ($\Delta i^* = 8.55\%$). Location D is eliminated, and only alternative B remains; it is selected.

Comment

An alternative must *always* be incrementally compared with an acceptable alternative, and the do-nothing alternative can end up being the only acceptable one. Since C was not justified in this example, location A was not compared with C. Thus, *if* the B-to-A comparison had not indicated that B was incrementally justified, then the D-to-A comparison would be correct instead of D-to-B.

To demonstrate how important it is to apply the ROR method correctly, consider the following. If the i^* of each alternative is computed initially, the results by ordered alternatives are

Location	C	A	B	D
i^*, %	9.63	10.49	12.35	11.56

Now apply *only* the first criterion stated earlier; that is, make the largest investment that has a MARR of 10% or more. Location D is selected. But, as shown above, this is the wrong selection, because the extra investment of $75,000 over location B will not earn the MARR. In fact, it will earn only 8.55%.

For service alternatives (costs only), the incremental cash flow is the difference between costs for two alternatives. There is no do-nothing alternative and no step 2 in the solution procedure. Therefore, the lowest-investment alternative is the initial defender against the next-lowest investment (challenger). This procedure is illustrated in Example 8.7 using a spreadsheet solution for equal-life service alternatives.

EXAMPLE 8.7

As the film of an oil spill from an at-sea tanker moves ashore, great losses occur for aquatic life as well as shoreline feeders and dwellers, such as birds. Environmental engineers and lawyers from several international petroleum corporations and transport companies—Exxon-Mobil, BP, Shell, and some transporters for OPEC producers—have developed

TABLE 8–7   Costs for Four Alternative Machines, Example 8.7

	Machine 1	Machine 2	Machine 3	Machine 4
First cost, $	−5,000	−6,500	−10,000	−15,000
Annual operating cost, $	−3,500	−3,200	−3,000	−1,400
Salvage value, $	+500	+900	+700	+1,000
Life, years	8	8	8	8

a plan to strategically locate throughout the world newly developed equipment that is substantially more effective than manual procedures in cleaning crude oil residue from bird feathers. The Sierra Club, Greenpeace, and other international environmental interest groups are in favor of the initiative. Alternative machines from manufacturers in Asia, America, Europe, and Africa are available with the cost estimates in Table 8–7. Annual cost estimates are expected to be high to ensure readiness at any time. The company representatives have agreed to use the average of the corporate MARR values, which results in MARR = 13.5%. Use a computer and incremental ROR analysis to determine which manufacturer offers the best economic choice.

Solution by Computer

Follow the procedure for incremental ROR analysis outlined prior to Example 8.6. The spreadsheet in Figure 8–6 contains the complete solution.

1. The alternatives are already ordered by increasing first costs.
2. These are service alternatives, so there is no comparison to DN, since i^* values cannot be calculated.
3. Machine 2 is the first challenger to machine 1; the incremental cash flows for the 2-to-1 comparison are in column D.
4. The 2-to-1 comparison results in $\Delta i^* = 14.57\%$ in cell D17 by applying the IRR function.
5. This return exceeds MARR = 13.5%, so machine 2 is the new defender (cell D19).

The comparison continues for 3-to-2 in cell E17 where the return is very negative at $\Delta i^* = -18.77\%$; machine 2 is retained as the defender. Finally the 4-to-2 comparison has an incremental ROR of 13.60%, which is slightly larger than MARR = 13.5%. The conclusion is to purchase machine 4 because the extra investment is (marginally) justified.

Comment

As mentioned earlier, it is not possible to generate a PW vs. i graph for each service alternative because all cash flows are negative. However, it is possible to generate PW vs. i graphs for the incremental series in the same fashion as we have done previously. The curves will cross the PW = 0 line at the Δi^* values determined by the IRR functions.

The spreadsheet does not include logic to select the better alternative at each stage of the solution. This feature could be added at each comparison using the Excel IF-operator and the correct arithmetic operations for each incremental cash flow and Δi^* values. This is too time-consuming; it is faster for the analyst to make the decision and then develop the required functions for each comparison.

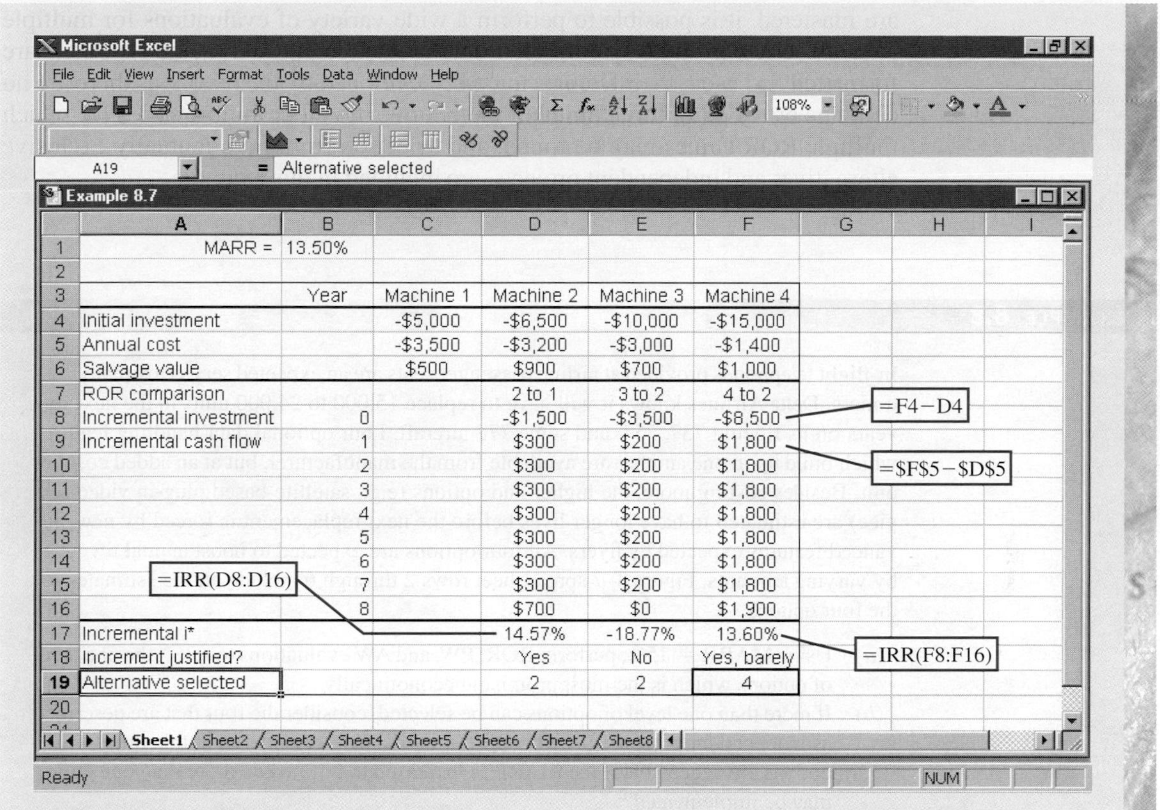

	A	B	C	D	E	F
1	MARR =	13.50%				
2						
3		Year	Machine 1	Machine 2	Machine 3	Machine 4
4	Initial investment		-$5,000	-$6,500	-$10,000	-$15,000
5	Annual cost		-$3,500	-$3,200	-$3,000	-$1,400
6	Salvage value		$500	$900	$700	$1,000
7	ROR comparison			2 to 1	3 to 2	4 to 2
8	Incremental investment	0		-$1,500	-$3,500	-$8,500
9	Incremental cash flow	1		$300	$200	$1,800
10		2		$300	$200	$1,800
11		3		$300	$200	$1,800
12		4		$300	$200	$1,800
13		5		$300	$200	$1,800
14		6		$300	$200	$1,800
15		7		$300	$200	$1,800
16		8		$700	$0	$1,900
17	Incremental i*			14.57%	-18.77%	13.60%
18	Increment justified?			Yes	No	Yes, barely
19	Alternative selected			2	2	4
20						

Figure 8–6
Spreadsheet solution to select from four service alternatives, Example 8.7.

Selection from multiple, mutually exclusive alternatives with *unequal lives* using Δi^* values requires that the incremental cash flows be evaluated over the LCM of the two alternatives being compared. This is another application of the principle of equal-service comparison. The spreadsheet application in the next section illustrates the computations.

It is always possible to rely on PW or AW analysis of the incremental cash flows at the MARR to make the selection. In other words, don't find Δi^* for each pairwise comparison; find PW or AW at the MARR instead. However, it is still necessary to make the comparison over the LCM number of years for an incremental analysis to be performed correctly.

8.7 SPREADSHEET APPLICATION—PW, AW, AND ROR ANALYSES ALL IN ONE

The following spreadsheet example combines many of the economic analysis techniques we have learned so far—(internal) ROR analysis, incremental ROR analysis, PW analysis, and AW analysis. Now that the IRR, NPV, and PV functions are mastered, it is possible to perform a wide variety of evaluations for multiple alternatives on a single spreadsheet. To better understand how the functions are formatted and used, their formats must be developed by the reader, as there are no cell tags provided in this example. A nonconventional cash flow series for which multiple ROR values may be found, and selection from both mutually exclusive alternatives and independent projects, are included in this example.

EXAMPLE 8.8

In-flight telephones provided at airline passenger seats are an expected service by many customers. Delta Airlines knows it will have to replace 15,000 to 24,000 units in the next few years on its Boeing 737, 757, and some 777 aircraft. Four optional data-handling features which build upon one another are available from the manufacturer, but at an added cost per unit. Besides costing more, the higher-end options (e.g., satellite-based plug-in video service) are estimated to have longer lives before the next replacement is forced by new, advanced features expected by flyers. All four options are expected to boost annual revenues by varying amounts. Figure 8–7 spreadsheet rows 2 through 6 include all the estimates for the four options.

(a)   Using MARR $= 15\%$, perform ROR, PW, and AW evaluations to select the one level of options which is the most promising economically.

(b)   If more than one level of options can be selected, consider the four that are described as independent projects. If no budget limitations are considered at this time, which options are acceptable if the MARR is increased to 20% when more than one option may be implemented?

Solution by Computer

E-Solve

(a)   The spreadsheet (Figure 8–7) is divided into six sections:

Section 1 (rows 1, 2): MARR value and the alternative names (A through D) in increasing order of initial cost.

Section 2 (rows 3 to 6): Per-unit net cash flow estimates for each alternative. These are revenue alternatives with unequal lives.

Section 3 (rows 7 to 20): Actual and incremental cash flows are displayed here.

Section 4 (rows 21, 22): Because these are all revenue alternatives, i^* values are determined by the IRR function. If an alternative passes the MARR test ($i^* > 15\%$), it is retained and a column is added to the right of its actual cash flows so the incremental cash flows can be determined. Columns F and H were inserted to make space for the incremental evaluations. Alternative A does not pass the i^* test.

Section 5 (rows 23 to 25): The IRR functions display the Δi^* values in columns F and H. Comparison of C to B takes place over the LCM of 12 years. Since $\Delta i^*_{C-B} = 19.42\% > 15\%$, eliminate B; alternative C is the new defender and D is the next challenger. The final comparison of D to C over 12 years results in $\Delta i^*_{D-C} = 11.23\% < 15\%$, so D is eliminated. Alternative C is the chosen one.

	A	B	C	D	E	F	G	H
1		MARR =	15%					
2	Alternative		A	B	C		D	
3	Initial cost		-$6,000	-$7,000	-$9,000		-$17,000	
4	Annual cash flow		$2,000	$3,000	$3,000		$3,500	
5	Salvage value		$0	$200	$300		$1,000	
6	Life	Year	3	4	6		12	
7	Incr. ROR comparison		Actual CF	Actual CF	Actual CF	C to B	Actual CF	D to C
8	Incremental investment	0	-6000	-7000	-9000	-$2,000	-17 000	-$8,000
9	Incremental cash flow	1	2000	3000	3000	$0	3500	$500
10	over the LCM	2	2000	3000	3000	$0	3500	$500
11		3	2000	3000	3000	$0	3500	$500
12		4		3200	3000	$6,800	3500	$500
13		5			3000	$0	3500	$500
14		6			3300	-$8,700	3500	$9,200
15		7				$0	3500	$500
16		8				$6,800	3500	$500
17		9				$0	3500	$500
18		10				$0	3500	$500
19		11				$0	3500	$500
20		12				$100	4500	$1,200
21	i*		0.00%	26.32%	24.68%		17.87%	
22	Retain or eliminate?		Eliminate	Retain	Retain		Retain	
23	Incremental i*					19.42%		11.23%
24	Increment justified?					Yes		No
25	Alternative selected					C		C
26	AW at MARR		($628)	$588	$656		$398	
27	PW at MARR		($3,403)	$3,188	$3,557		$2,159	
28	Alterenative selected		No	No	Yes		No	
29	Alternative		A	B	C		D	

Figure 8–7
Spreadsheet analysis using ROR, PW, and AW methods for unequal-life, revenue alternatives, Example 8.8.

Section 6 (rows 26 to 29): These include the AW and PW analyses. The AW value over the life of each alternative is calculated using the PMT function at the MARR with an embedded NPV function. Also, the PW value is determined from the AW value for 12 years using the PV function. For both measures, alternative C has the numerically largest value, as expected.

Conclusion: All methods result in the same, correct choice of alternative C.

(b) Since each option is independent of the others, and there is no budget limitation at this time, each i* value in row 21 of Figure 8–7 is compared to MARR = 20%. This is a comparison of each option with the do-nothing alternative. Of the four, options B and C have i* > 20%. They are acceptable; the other two are not.

Comment

In part (a), we should have applied the two multiple-root sign tests to the incremental cash flow series for the C-to-B comparison. The series itself has three sign changes, and the cumulative cash flow series starts negatively and also has three sign changes. Therefore, up to three real-number roots may exist. The IRR function is applied in cell F23 to obtain

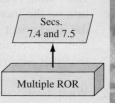

Secs. 7.4 and 7.5

Multiple ROR

$\Delta i^*_{C-B} = 19.42\%$ without using the net-investment procedure. This action assumes that the reinvestment assumption of 19.42% for positive net-investment cash flows is a reasonable one. If the MARR = 15%, or some other earning rate, were more appropriate, the net-investment procedure would have to be applied to determine the composite rate, which would be different from 19.42%. Depending upon the reinvestment rate chosen, alternative C may or may not be incrementally justified against B. Here, the assumption is made that the Δi^* value is reasonable, so C is justified.

CHAPTER SUMMARY

Just as present worth, annual worth, and future worth methods find the best alternative from among several, incremental rate of return calculations can be used for the same purpose. In using the ROR technique, it is necessary to consider the incremental cash flows when selecting between mutually exclusive alternatives. This was not necessary for the PW, AW, or FW methods. The incremental investment evaluation is conducted between only two alternatives at a time beginning with the lowest initial investment alternative. Once an alternative has been eliminated, it is not considered further.

If there is no budget limitation when independent projects are evaluated using the ROR method, the ROR value of each project is compared to the MARR. Any number, or none, of the projects can be accepted.

Rate of return values have a natural appeal to management, but the ROR analysis is often more difficult to set up and complete than the PW, AW, or FW analysis using an established MARR. Care must be taken to perform a ROR analysis correctly on the incremental cash flows; otherwise it may give incorrect results.

PROBLEMS

Understanding Incremental ROR

8.1 If alternative A has a rate of return of 10% and alternative B has a rate of return of 18%, what is known about the rate of return on the increment between A and B if the investment required in B is (a) larger than that required for A and (b) smaller than that required for A?

8.2 What is the overall rate of return on a $100,000 investment that returns 20% on the first $30,000 and 14% on the remaining $70,000?

8.3 Why is an incremental analysis necessary when you are conducting a rate of return analysis for service alternatives?

8.4 If all of the incremental cash flows are negative, what is known about the rate of return on the incremental investment?

8.5 Incremental cash flow is calculated as cash flow$_B$ − cash flow$_A$, where B represents the alternative with the larger initial investment. If the two cash flows were switched wherein B represents the one with the *smaller* initial investment, which alternative should be selected if the incremental rate of return is 20% per year and the company's MARR is 15% per year? Explain.

8.6 A food processing company is considering two types of moisture analyzers. The company expects an infrared model to yield a rate of return of 18% per year. A more expensive microwave model will yield a rate of return of 23% per year. If the company's MARR is 18% per year, can you determine which model(s) should be purchased solely on the basis of the rate of return information provided if (*a*) either one or both analyzers can be selected and (*b*) only one can be selected? Why or why not?

8.7 For each of the following scenarios, state whether an incremental investment analysis would be required to select an alternative and state why or why not. Assume that alternative Y requires a higher initial investment than alternative X and that the MARR is 20% per year.
   (*a*) X has a rate of return of 28% per year, and Y has a rate of return of 20% per year.
   (*b*) X has a rate of return of 18% per year, and Y has a rate of return of 23% per year.
   (*c*) X has a rate of return of 16% per year, and Y has a rate of return of 19% per year.
   (*d*) X has a rate of return of 30% per year, and Y has a rate of return of 26% per year.
   (*e*) X has a rate of return of 21% per year, and Y has a rate of return of 22% per year.

8.8 A small construction company has $100,000 set aside in a sinking fund to purchase new equipment. If $30,000 is invested at 30%, $20,000 at 25% and the remaining $50,000 at 20% per year, what is the overall rate of return on the entire $100,000?

8.9 A total of $50,000 was available for investing in a project to reduce insider theft in an appliance warehouse. Two alternatives identified as Y and Z were under consideration. The overall rate of return on the $50,000 was determined to be 40%, with the rate of return on the $20,000 increment between Y and Z at 15%. If Z is the higher first-cost alternative, (*a*) what is the size of the investment required in Y and (*b*) what is the rate of return on Y?

8.10 Prepare a tabulation of cash flow for the alternatives shown below.

	Machine A	Machine B
First cost, $	−15,000	−25,000
Annual operating cost, $/year	−1,600	−400
Salvage value, $	3,000	6,000
Life, years	3	6

8.11 A chemical company is considering two processes for making a cationic polymer. Process A will have a first cost of $100,000 and an annual operating cost of $60,000. Process B will have a first cost of $165,000. If both processes will be adequate for 4 years and the rate of return on the increment between the alternatives is 25%, what is the amount of the operating cost for process B?

Incremental ROR Comparison (Two Alternatives)

8.12 When the rate of return on the incremental cash flow between two alternatives is exactly equal to the MARR, which alternative should be selected—the one with the higher or lower initial investment? Why?

8.13 A consulting engineering firm is trying to decide whether it should purchase Ford Explorers or Toyota 4Runners for company principals. The models under consideration would cost $29,000 for the Ford and $32,000 for the Toyota. The annual operating cost of the Explorer is expected to be $200 per year less than that of the 4Runner. The trade-in values after 3 years are estimated to be 50% of first cost for the Explorer and 60% for the Toyota. (*a*) What is the rate of return relative to that of Ford, if the Toyota is selected? (*b*) If the firm's MARR is 18% per year, which make of vehicle should it buy?

8.14 A plastics company is considering two injection molding processes. Process X will have a first cost of $600,000, annual costs of $200,000, and a salvage value of $100,000 after 5 years. Process Y will have a first cost of $800,000, annual costs of $150,000, and a salvage value of $230,000 after 5 years. (*a*) What is the rate of return on the increment of investment between the two? (*b*) Which process should the company select on the basis of a rate of return analysis, if the MARR is 20% per year?

8.15 A company that manufactures amplified pressure transducers is trying to decide between the machines shown below. Compare them on the basis of rate of return, and determine which should be selected if the company's MARR is 15% per year.

	Variable Speed	Dual Speed
First cost, $	−250,000	−225,000
Annual operating cost, $/year	−231,000	−235,000
Overhaul in year 3, $	—	−26,000
Overhaul in year 4, $	−39,000	—
Salvage value, $	50,000	10,000
Life, years	6	6

8.16 The manager of a canned food processing plant is trying to decide between two labeling machines. Determine which should be selected on the basis of rate of return with a MARR of 20% per year.

	Machine A	Machine B
First cost, $	−15,000	−25,000
Annual operating cost, $/year	−1,600	−400
Salvage value, $	3,000	4,000
Life, years	2	4

8.17 A solid waste recycling plant is considering two types of storage bins. Determine which should be selected on the basis of rate of return. Assume the MARR is 20% per year.

	Alternative P	Alternative Q
First cost, $	−18,000	−35,000
Annual operating cost, $/year	−4,000	−3,600
Salvage value, $	1,000	2,700
Life, years	3	6

8.18 The incremental cash flow between alternatives J and K is estimated below. If the MARR is 20% per year, which alternative should be selected on the basis of rate of return? Assume K requires the extra $90,000 initial investment.

Year	Incremental Cash Flow, $(K − J)
0	−90,000
1–3	+10,000
4–9	+20,000
10	+5,000

8.19 A chemical company is considering two processes for isolating DNA material. The incremental cash flow between two alternatives J and S is as shown. The company uses a MARR of 50% per year. The rate of return on the incremental cash flow below is less than 50%, but the company CEO prefers the more expensive process. The CEO believes she can negotiate the initial cost of the more expensive process downward. By how much would she have to reduce the first cost of S, the higher-cost alternative, for it to have an incremental rate of return of exactly 50%?

Year	Incremental Cash Flow, $(S − J)
0	−900,000
1	400,000
2	600,000
3	850,000

8.20 Alternative R has a first cost of $100,000, annual M&O costs of $50,000, and a $20,000 salvage value after 5 years. Alternative S has a first cost of $175,000 and a $40,000 salvage value after 5 years, but its annual M&O costs are not known. Determine the M&O costs for alternative S that would yield an incremental rate of return of 20% per year.

8.21 The incremental cash flows for alternatives M and N are shown below. Determine which should be selected, using an AW-based rate of return analysis. The MARR is 12% per year, and alternative N requires the larger initial investment.

Year	Incremental Cash Flow, $(N − M)
0	−22,000
1–8	+4,000
9	+12,000

8.22 Determine which of the two machines below should be selected, using an AW-based rate of return analysis, if the MARR is 18% per year.

	Semiautomatic	Automatic
First cost, $	−40,000	−90,000
Annual cost, $/year	−100,000	−95,000
Salvage value, $	5,000	7,000
Life, years	2	4

8.23 The incremental cash flows for alternatives X and Y are shown below. Calculate the incremental rate of return per month, and determine which should be selected, using an AW-based rate of return analysis. The MARR is 24% per year, compounded monthly, and alternative Y requires the larger initial investment.

Month	Incremental Cash Flow, $(Y − X)
0	−62,000
1–23	+4,000
24	+10,000

8.24 The incremental cash flow between alternatives Z1 and Z2 is shown below (Z2 has the higher initial cost). Use an AW-based rate of return equation to determine the incremental rate of return and which alternative should be selected, if the MARR is 17% per year. Let k = year 1 through 10.

Year	Incremental Cash Flow, $(Z2 − Z1)
0	−40,000
1–10	9000 − 500k

8.25 Two roadway designs are under consideration for access to a permanent suspension bridge. Design 1A will cost $3 million to build and $100,000 per year to maintain. Design 1B will cost $3.5 million to build and $40,000 per year to maintain. Use an AW-based rate of return equation to determine which design is preferred. Assume n = 10 years and the MARR is 6% per year.

8.26 A manufacturing company is in need of 3000 square meters for expansion because of a new 3-year contract it just won. The company is considering the purchase of land for $50,000 and erecting a temporary metal structure on it at a cost of $90 per square meter. At the end of the 3-year period, the company expects to be able to sell the land for $55,000 and the building for $60,000. Alternatively, the company can lease space for $3 per square meter *per month*, payable at the beginning of each year. Use an AW-based rate of return equation to determine which alternative is preferred. The MARR is 28% per year.

8.27 Four mutually exclusive *service* alternatives are under consideration for automating a manufacturing operation. The alternatives were ranked in order of increasing initial investment and then compared by incremental investment rate of return analysis. The rate of return on each increment of investment was less than the MARR. Which alternative should be selected?

Multiple-Alternative Comparison

8.28 A metal plating company is considering four different methods for recovering by-product heavy metals from a manufacturing site's liquid waste. The investment costs and incomes associated with each method have been estimated. All methods have an 8-year life. The MARR is 11% per year. (a) If the methods are independent, because they can be implemented at different plants, which ones are acceptable? (b) If the methods are mutually exclusive, determine which one method should be selected, using a ROR evaluation.

Method	First Cost, $	Salvage Value, $	Annual Income, $/year
A	−30,000	+1,000	+4,000
B	−36,000	+2,000	+5,000
C	−41,000	+500	+8,000
D	−53,000	−2,000	+10,500

8.29 Mountain Pass Canning Company has determined that any one of five machines can be used in one phase of its canning operation. The costs of the machines are estimated below, and all machines have a 5-year life. If the minimum attractive rate of return is 20% per year, determine the one machine that should be selected on the basis of a rate of return analysis.

Machine	First Cost, $	Annual Operating Cost, $/year
1	−31,000	−18,000
2	−28,000	−19,500
3	−34,500	−17,000
4	−48,000	−12,000
5	−41,000	−15,500

8.30 An independent dirt contractor is trying to determine which size dump truck to buy. The contractor knows that as the bed size increases, the net income increases, but he is uncertain whether the incremental expenditure required for the larger trucks is justified. The cash flows associated with each size truck are estimated below. The contractor's MARR is 18% per year, and all trucks are expected to have a useful life of 5 years. (a) Determine which size truck should be purchased. (b) If two trucks of different size are to be purchased, what should be the size of the second truck?

Truck Bed Size, Cubic Meters	Initial Investment, $	Annual Operating Cost, $/year	Salvage Value, $	Annual Income, $/year
8	−30,000	−14,000	+2000	+26,500
10	−34,000	−15,500	+2500	+30,000
15	−38,000	−18,000	+3000	+33,500
20	−48,000	−21,000	+3500	+40,500
25	−57,000	−26,000	+4600	+49,000

8.31 An engineer at Anode Metals is considering the projects below, all of which can be considered to last indefinitely. If the company's MARR is 15% per year, determine which should be selected (a) if they are independent and (b) if they are mutually exclusive.

	First Cost, $	Annual Income, $/year	Alternative's Rate of Return, %
A	−20,000	+3,000	15
B	−10,000	+2,000	20
C	−15,000	+2,800	18.7
D	−70,000	+10,000	14.3
E	−50,000	+6,000	12

8.32 Only one of four different machines is to be purchased for a certain production process. An engineer performed the following analyses to select the best machine. All machines are assumed to have a 10-year life. Which machine, if any, should the company select if its MARR is (a) 12% per year and (b) 20% per year?

	Machine			
	1	2	3	4
Initial cost, $	−44,000	−60,000	−72,000	−98,000
Annual cost, $/year	−70,000	−64,000	−61,000	−58,000
Annual savings, $/year	+80,000	+80,000	+80,000	+82,000
ROR, %	18.6	23.4	23.1	20.8
Machines compared		2 to 1	3 to 2	4 to 3
Incremental investment, $		−16,000	−12,000	−26,000
Incremental cash flow, $/year		+6,000	+3,000	+5,000
ROR on increment, %		35.7	21.4	14.1

8.33 The four alternatives described below are being evaluated.
   (a) If the proposals are independent, which should be selected when the MARR is 16% per year?
   (b) If the proposals are mutually exclusive, which one should be selected when the MARR is 9% per year?
   (c) If the proposals are mutually exclusive, which one should be selected when the MARR is 12% per year?

Alternative	Initial Investment, $	Rate of Return, %	Incremental Rate of Return, %, When Compared with Alternative		
			A	B	C
A	−40,000	29			
B	−75,000	15	1		
C	−100,000	16	7	20	
D	−200,000	14	10	13	12

8.34 A rate of return analysis was initiated for the infinite-life alternatives below.
   (a) Fill in the blanks in the incremental rate of return column on the incremental cash flow portion of the table.
   (b) How much revenue is associated with each alternative?

   (c) What alternative should be selected if they are mutually exclusive and the MARR is 16%?
   (d) What alternative should be selected if they are mutually exclusive and the MARR is 11%?
   (e) Select the two best alternatives at a MARR of 19%.

Alternative	Alternative's Investment, $	Alternative's Rate of Return, %	Incremental Rate of Return, %, on Incremental Cash Flow When Compared with Alternative			
			E	F	G	H
E	−20,000	20	—			
F	−30,000	35		—		
G	−50,000	25			—	11.7
H	−80,000	20			11.7	—

8.35 A rate of return analysis was initiated for the infinite-life alternatives below.
   (a) Fill in the blanks in the alternative's rate of return column and incremental rate of return columns of the table.

   (b) What alternative should be selected if they are independent and the MARR is 21% per year?
   (c) What alternative should be selected if they are mutually exclusive and the MARR is 24% per year?

Alternative	Alternative's Investment, $	Alternative's Rate of Return, %	Incremental Rate of Return, %, on Incremental Cash Flow When Compared with Alternative			
			E	F	G	H
E	−10,000	25	—	20		
F	−25,000		20	—	4	
G	−30,000			4	—	
H	−60,000	30				—

FE REVIEW PROBLEMS

8.36 Alternative I requires an initial investment of $20,000 and will yield a rate of return of 15% per year. Alternative C, which requires a $30,000 investment, will yield 20% per year. Which of the following statements is true about the rate of return on the $10,000 increment?
   (a) It is greater than 20% per year.
   (b) It is exactly 20% per year.
   (c) It is between 15% and 20% per year.
   (d) It is less than 15% per year.

8.37 The rate of return for alternative X is 18% per year and for alternative Y is 17% per year, with Y requiring a larger initial investment. If a company has a minimum attractive rate of return of 16% per year,

(a) The company should select alternative X.

(b) The company should select alternative Y.

(c) The company should conduct an incremental analysis between X and Y to select the economically better alternative.

(d) The company should select the do-nothing alternative.

8.38 When one is conducting an ROR analysis of mutually exclusive service projects,

(a) All the projects must be compared against the do-nothing alternative.

(b) More than one project may be selected.

(c) An incremental investment analysis is necessary to identify the best one.

(d) The project with the highest incremental ROR should be selected.

8.39 When one is comparing two mutually exclusive alternatives by the ROR method, if the rate of return on the alternative with the higher first cost is less than that of the lower first-cost alternative,

(a) The rate of return on the increment between the two is greater than the rate of return for the lower first-cost alternative.

(b) The rate of return on the increment is less than the rate of return for the lower first-cost alternative.

(c) The higher first-cost alternative may be the better of the two alternatives.

(d) None of the above.

8.40 The incremental cash flow between two alternatives is shown below.

Year	Incremental Cash Flow, $
0	−20,000
1–10	+3,000
10	+400

The equation(s) that can be used to correctly solve for the incremental rate of return is (are)

(a) $0 = -20,000 + 3000(A/P,i,10) + 400(P/F,i,10)$

(b) $0 = -20,000 + 3000(A/P,i,10) + 400(A/F,i,10)$

(c) $0 = -20,000(A/P,i,10) + 3000 + 400(P/F,i,10)$

(d) $0 = -20,000 + 3000(P/A,i,10) + 400(P/F,i,10)$

Questions 8.41 through 8.43 are based on the following. The five alternatives are being evaluated by the rate of return method.

Alternative	Initial Investment, $	Alternative Rate of Return, %	Incremental Rate of Return, %, When Compared with Alternative				
			A	B	C	D	E
A	−25,000	9.6	—	28.9	19.7	36.7	25.5
B	−35,000	15.1		—	1.5	39.8	24.7
C	−40,000	13.4			—	49.4	28.0
D	−60,000	25.4				—	−0.6
E	−75,000	20.2					—

8.41 If the alternatives are independent and the MARR is 18% per year, the one(s) that should be selected is (are)
   (a) Only D
   (b) Only D and E
   (c) Only B, D, and E
   (d) Only E

8.42 If the alternatives are mutually exclusive and the MARR is 15% per year, the alternative to select is

   (a) B
   (b) D
   (c) E
   (d) None of them

8.43 If the alternatives are mutually exclusive and the MARR is 25% per year, the alternative to select is
   (a) A
   (b) D
   (c) E
   (d) None of them

EXTENDED EXERCISE

INCREMENTAL ROR ANALYSIS WHEN ESTIMATED ALTERNATIVE LIVES ARE UNCERTAIN

Make-to-Specs is a software system under development by ABC Corporation. It will be able to translate digital versions of three-dimensional computer models, containing a wide variety of part shapes with machined and highly finished (ultra-smooth) surfaces. The product of the system is the numerically controlled (NC) machine code for the part's manufacturing. Additionally, Make-to-Specs will build the code for super-fine finishing of surfaces with continuous control of the finishing machines. There are two alternative computers that can provide the server function for the software interfaces and shared database updates on the manufacturing floor while Make-to-Specs is operating in parallel mode. The server first cost and estimated contribution to annual net cash flow are summarized below.

	Server 1	Server 2	
First cost, $	$100,000	$200,000	
Net cash flow, $/year	$35,000	$50,000	year 1, plus $5000 per year for years 2, 3, and 4 (gradient).
		$70,000	maximum for years 5 on, even if the server is replaced.
Life, years	3 or 4	5 or 8	

The life estimates were developed by two different individuals: a design engineer and a manufacturing manager. They have asked that at this stage of the project, all analyses be performed using both life estimates for each system.

Questions

Use computer analysis to answer the following:

1. If the MARR = 12%, which server should be selected? Use the PW or AW method to make the selection.

2. Use incremental ROR analysis to decide between the servers at MARR = 12%.

3. Use any method of economic analysis to display on the spreadsheet the value of the incremental ROR between server 2 with a life estimate of 5 years and a life estimate of 8 years.

CASE STUDY 1

SO MANY OPTIONS. CAN A NEW ENGINEERING GRADUATE HELP HIS FATHER?[1]

Background

"I don't know whether to sell it, expand it, lease it, or what. But I don't think we can keep doing the same thing for many more years. What I really want to do is to keep it for 5 more years, then sell it for a bundle," Elmer Kettler said to his wife Janise, their son, John Kettler, and new daughter-in-law, Suzanne Gestory, as they were gathered around the dinner table. Elmer was sharing thoughts on Gulf Coast Wholesale Auto Parts, a company he has owned and operated for 25 years on the southern outskirts of Houston, Texas. The business has excellent contracts for parts supply with several national retailers operating in the area—NAPA, AutoZone, O'Reilly, and Advance. Additionally, Gulf Coast operates a rebuild shop serving these same retailers for major automobile components, such as carburetors, transmissions, and air conditioning compressors.

At his home after dinner, John decided to help his father with an important and difficult decision: What to do with his business? John graduated just last year with an engineering degree from a major state university in Texas, where he completed a course in engineering economy. Part of his job at Energcon Industries is to perform basic rate of return and present worth analyses on energy management proposals.

Options

Over the next few weeks, Mr. Kettler outlined five options, including his favorite of selling in 5 years. John summarized all the estimates over a 10-year horizon. The options and estimates were given to Elmer, and he agreed with them.

Option #1: *Remove rebuild.* Stop operating the rebuild shop and concentrate on selling wholesale parts. The removal of the rebuild operations and the switch to an "all parts house" is expected to cost $750,000 in the first year. Overall revenues will drop to $1 million the first year with an expected 4% increase per year thereafter. Expenses are projected at $0.8 million the first year, increasing 6% per year thereafter.

Option #2: *Contract rebuild operations.* To get the rebuild shop ready for an operations contractor to take over will cost $400,000 immediately. If expenses stay the same for 5 years, they will average $1.4 million per year, but they can be expected to rise to $2 million per year in year 6 and thereafter. Elmer thinks revenues under a contract arrangement can be $1.4 million the first year and rise 5% per year for the duration of a 10-year contract.

Option #3: *Maintain status quo and sell out after 5 years.* (Elmer's personal favorite.) There is no cost now, but the

[1]Based upon a study by Mr. Alan C. Stewart, Consultant, Communications and High Tech Solutions Engineering, Accenture LLP.

current trend of negative net profit will probably continue. Projections are $1.25 million per year for expenses and $1.15 million per year in revenue. Elmer had an appraisal last year, and the report indicated Gulf Coast Wholesale Auto Parts is worth a net $2 million. Elmer's wish is to sell out completely after 5 more years at this price, and to make a deal that the new owner pay $500,000 per year at the end of year 5 (sale time) and the same amount for the next 3 years.

Option #4: *Trade-out.* Elmer has a close friend in the antique auto parts business who is making a "killing," so he says, with e-commerce. Although the possibility is risky, it is enticing to Elmer to consider a whole new line of parts, but still in the basic business that he already understands. The trade-out would cost an estimated $1 million for Elmer immediately. The 10-year horizon of annual expenses and revenues is considerably higher than for his current business. Expenses are estimated at $3 million per year and revenues at $3.5 million each year.

Option #5: *Lease arrangement.* Gulf Coast could be leased to some turnkey company with Elmer remaining the owner and bearing part of the expenses for building, delivery trucks, insurance, etc. The first-cut estimates for this option are $1.5 million to get the business ready now, with annual expenses at $500,000 per year and revenues at $1 million per year for a 10-year contract.

Case Study Exercises

Help John with the analysis by doing the following:

1. Develop the actual cash flow series and incremental cash flow series (in $1000 units) for all five options in preparation for an incremental ROR analysis.

2. Discuss the possibility of multiple rate of return values for all the actual and incremental cash flow series. Find any multiple rates in the range of 0 to 100%.

3. If John's father insists that he make 25% per year or more on the selected option over the next 10 years, what should he do? Use all the methods of economic analysis you have learned so far (PW, AW, ROR) so John's father can understand the recommendation in one way or another.

4. Prepare plots of the PW vs. *i* for each of the five options. Estimate the breakeven rate of return between options.

5. What is the minimum amount that must be received in each of years 5 through 8 for option #3 (the one Elmer wants) to be best economically? Given this amount, what does the sale price have to be, assuming the same payment arrangement as presented in the description?

CASE STUDY 2

PW ANALYSIS WHEN MULTIPLE INTEREST RATES ARE PRESENT[2]

Background

Two engineering economy students, Jane and Bob, could not agree on what evaluation tool should be used to select one of the following investment plans. The cash flow series are identical except for their signs. They recall that a PW or AW equation should be set up to solve for a rate of return. It seems that the two investment plans should have identical ROR value(s). It may be that the two plans are equivalent and should both be acceptable.

[2]Contributed by Dr. Tep Sastri (former Associate Professor, Industrial Engineering, Texas A&M University).

Year	Plan A	Plan B
0	$+1900	$−1900
1	−500	+500
2	−8000	+8000
3	+6500	−6500
4	+400	−400

Up to this point in class, the professor has discussed the present worth and annual worth methods at a given MARR for evaluating alternatives. He explained the composite rate of return method during the last class. The two students remember that the professor said, "The calculation of the composite rate of return is often computationally involved. If an actual ROR is not necessary, it is strongly recommended that the PW or AW at MARR be used to decide on project acceptability."

Bob admitted that it is not very clear to him why the simplistic "PW at MARR" is strongly recommended. Bob is unsure how to determine if a rate of return is "not necessary." He said to Jane, "Since the composite ROR technique always yields a unique ROR value and every student has a calculator or a computer with a spreadsheet system on it, who cares about the computation problem? I would always perform the composite ROR method." Jane was more cautious and suggested that a good analysis starts with a simple, common-sense approach. She suggested that Bob inspect the cash flows and see if he could pick the better plan just through observation of the cash flows. Jane also proposed that they try every method they had learned so far. She said, "If we experiment with them, I think we may understand the real reason that the PW (or AW) at the MARR method is recommended over the composite rate of return method."

Case Study Exercises

Given their discussion, the following are some questions Jane and Bob need to answer. Help them develop the answers.

1. By simply inspecting the two cash flow patterns, determine which is the preferred plan. In other words, if someone is offering the two plans, which one do you think might obtain a higher rate of return?

2. Which plan is the better choice if the MARR is (a) 15% per year and (b) 50% per year? Two approaches should be taken here: First, evaluate the two options using PW analysis at the MARR, ignoring the multiple roots, whether they exist or not. Second, determine the internal rate of return of the two plans. Do the two cash flow series have the same ROR values?

3. Perform an incremental ROR analysis of the two plans. Are there still multiple roots to the incremental cash flow series that limit Bob's and Jane's ability to make a definitive choice? If so, what are they?

4. The students want to know if the composite ROR analysis will consistently yield a logical and unique decision as the MARR value changes. To answer this question, find out which plan should be accepted if any end-of-year released cash flows (excess project funds) earn at the following three reinvestment rates. The MARR rates change also.

   (a) Reinvestment rate is 15% per year; MARR is 15% per year.

   (b) Reinvestment is at 45% per year; MARR is 15% per year.

   (c) Reinvestment rate and MARR are both 50% per year.

   (d) Explain your findings about these three different rate combinations to Bob and Jane.

Benefit/Cost Analysis and Public Sector Economics

The evaluation methods of previous chapters are usually applied to alternatives in the private sector, that is, for-profit and not-for-profit corporations and businesses. Customers, clients, and employees utilize the installed alternatives. This chapter introduces *public sector alternatives* and their economic consideration. Here the owners and users (beneficiaries) are the citizens of the government unit—city, county, state, province, or nation. Government units provide the mechanisms to raise (investment) capital and operating funds for projects through taxes, user fees, bond issues, and loans. There are substantial differences in the characteristics of public and private sector alternatives and their economic evaluation, as outlined in the first section. Partnerships of the public and private sector have become increasingly common, especially for large infrastructure construction projects such as major highways, power generation plants, water resource developments, and the like.

The benefit/cost (B/C) ratio was developed, in part, to introduce objectivity into the economic analysis of public sector evaluation, thus reducing the effects of politics and special interests. However, there is always predictable disagreement among citizens (individuals and groups) about how the benefits of an alternative are defined and economically valued. The different formats of B/C analysis, and associated disbenefits of an alternative, are discussed here. The B/C analysis can use equivalency computations based on PW, AW, or FW values. Performed correctly, the benefit/cost method will always select the same alternative as PW, AW, and ROR analyses.

A public sector project to enhance freeway lighting is the subject of the case study.

LEARNING OBJECTIVES

Purpose: Understand public sector economics; evaluate a project and compare alternatives using the benefit/cost ratio method.

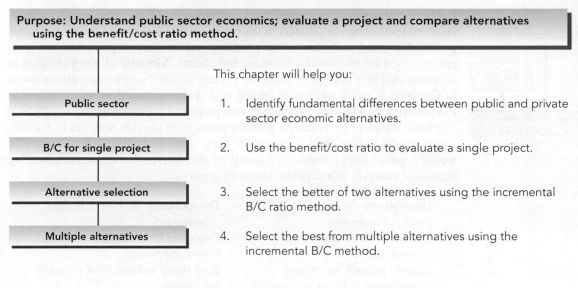

This chapter will help you:

Public sector

1. Identify fundamental differences between public and private sector economic alternatives.

B/C for single project

2. Use the benefit/cost ratio to evaluate a single project.

Alternative selection

3. Select the better of two alternatives using the incremental B/C ratio method.

Multiple alternatives

4. Select the best from multiple alternatives using the incremental B/C method.

9.1   PUBLIC SECTOR PROJECTS

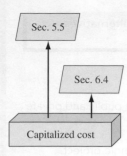

Sec. 5.5

Sec. 6.4

Capitalized cost

Public sector projects are owned, used, and financed by the citizenry of any government level, whereas projects in the private sector are owned by corporations, partnerships, and individuals. The products and services of private sector projects are used by individual customers and clients. Virtually all the examples in previous chapters have been from the private sector. Notable exceptions occur in Chapters 5 and 6 where capitalized cost was introduced as an extension to PW analysis for long-life alternatives and perpetual investments.

Public sector projects have a primary purpose to provide services to the citizenry for the public good at no profit. Areas such as health, safety, economic welfare, and utilities comprise a majority of alternatives that require engineering economic analysis. Some public sector examples are

Hospitals and clinics	Transportation: highways, bridges,
Parks and recreation	waterways
Utilities: water, electricity, gas,	Police and fire protection
sewer, sanitation	Courts and prisons
Schools: primary, secondary,	Food stamp and rent relief programs
community colleges, universities	Job training
Economic development	Public housing
Convention centers	Emergency relief
Sports arenas	Codes and standards

There are significant differences in the characteristics of private and public sector alternatives.

Characteristic	Public sector	Private sector
Size of investment	Larger	Some large; more medium to small

Often alternatives developed to serve public needs require large initial investments, possibly distributed over several years. Modern highways, public transportation systems, airports, and flood control systems are examples.

Life estimates	Longer (30–50+ years)	Shorter (2–25 years)

The long lives of public projects often prompt the use of the capitalized cost method, where infinity is used for n and annual costs are calculated as $A = P(i)$. As n gets larger, especially over 30 years, the differences in calculated A values become small. For example, at $i = 7\%$, there will be a very small difference in 30 and 50 years, because $(A/P,7\%,30) = 0.08059$ and $(A/P,7\%,50) = 0.07246$.

Annual cash flow estimates	No profit; costs, benefits, and disbenefits, are estimated	Revenues contribute to profits; costs are estimated

Public sector projects (also called publicly-owned) do not have profits; they do have costs that are paid by the appropriate government unit; and they benefit the citizenry. Public sector projects often have undesirable consequences, as stated by some portion of the public. It is these consequences that can cause public controversy about the projects. The economic analysis should consider these consequences in monetary terms to the degree estimable. (Often in private sector analysis, undesirable consequences are not considered, or they may be directly addressed as costs.) To perform an economic analysis of public alternatives, the costs (initial and annual), the benefits, and the disbenefits, if considered, must be estimated as accurately as possible in *monetary units*.

Costs—estimated expenditures *to the government entity* for construction, operation, and maintenance of the project, less any expected salvage value.

Benefits—advantages to be experienced *by the owners, the public*.

Disbenefits—expected undesirable or negative consequences *to the owners* if the alternative is implemented. Disbenefits may be indirect economic disadvantages of the alternative.

The following is important to realize:

It is difficult to estimate and agree upon the economic impact of benefits and disbenefits for a public sector alternative.

For example, assume a short bypass around a congested area in town is recommended. How much will it benefit a driver in *dollars per driving minute* to be able to bypass five traffic lights while averaging 35 miles per hour, as compared to currently driving through the lights averaging 20 miles per hour and stopping at an average of two lights for an average of 45 seconds each? The bases and standards for benefits estimation are always difficult to establish and verify. Relative to revenue cash flow estimates in the private sector, benefit estimates are much harder to make, and vary more widely around uncertain averages. And the disbenefits that accrue from an alternative are harder to estimate. In fact, the disbenefit itself may not be known at the time the evaluation is performed.

Funding	Taxes, fees, bonds, private funds	Stocks, bonds, loans, individual owners

The capital used to finance public sector projects is commonly acquired from taxes, bonds, and fees. Taxes are collected from those who are the owners—the citizens (e.g., federal gasoline taxes for highways are paid by all gasoline users). This is also the case for fees, such as toll road fees for drivers. Bonds are often issued: U.S. Treasury bonds, municipal bond issues, and special-purpose bonds, such as utility district bonds. Private lenders can provide up-front financing. Also, private donors may provide funding for museums, memorials, parks, and garden areas through gifts.

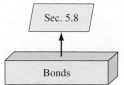

Interest rate	Lower	Higher, based on market cost of capital

Because many of the financing methods for public sector projects are classified as *low-interest,* the interest rate is virtually always lower than for private sector alternatives. Government agencies are exempt from taxes levied by higher-level units. For example, municipal projects do not have to pay state taxes. (Private corporations and individual citizens do pay taxes.) Many loans are very low-interest, and grants with no repayment requirement from federal programs may share project costs. This results in interest rates in the 4 to 8% range. It is common that a government agency will direct that all projects be evaluated at a specific rate. For example, the U.S. Office of Management and Budget (OMB) declared at one time that federal projects should be evaluated at 10% (with no inflation adjustment). As a matter of standardization, directives to use a specific interest rate are beneficial because different government agencies are able to obtain varying types of funding at different rates. This can result in projects of the same type being rejected in one city or county, but accepted in a neighboring district. Therefore, standardized rates tend to increase the consistency of economic decisions and to reduce gamesmanship.

The determination of the interest rate for public sector evaluation is as important as the determination of the MARR for a private sector analysis. The public sector interest rate is identified as *i*; however, it is referred to by other names to distinguish it from the private sector rate. The most common terms are *discount rate* and *social discount rate.*

Alternative selection criteria	Multiple criteria	Primarily based on rate of return

Multiple categories of users, economic as well as noneconomic interests, and special-interest political and citizen groups make the selection of one alternative over another much more difficult in public sector economics. Seldom is it possible to select an alternative on the sole basis of a criterion such as PW or ROR. It is important to describe and itemize the criteria and selection method prior to the analysis. This helps determine the perspective or viewpoint when the evaluation is performed. Viewpoint is discussed below.

Environment of the evaluation	Politically inclined	Primarily economic

There are often public meetings and debates associated with public sector projects to accommodate the various interests of citizens (owners). Elected officials commonly assist with the selection, especially when pressure is brought to bear by voters, developers, environmentalists, and others. The selection process is not as "clean" as in private sector evaluation.

The viewpoint of the public sector analysis must be determined before cost, benefit, and disbenefit estimates are made and before the evaluation is formulated and performed. There are several viewpoints for any situation, and the different perspectives may alter how a cash flow estimate is classified.

Some example perspectives are the citizen; the city tax base; number of students in the school district; creation and retention of jobs; economic development potential; a particular industry interest, such as agriculture, banking, or electronics manufacturing; and many others. In general, the viewpoint of the analysis should be as broadly defined as those who will bear the costs of the project and reap its benefits. Once established, the viewpoint assists in categorizing the costs, benefits, and disbenefits of each alternative, as illustrated in Example 9.1.

EXAMPLE 9.1

The citizen-based Capital Improvement Projects (CIP) Committee for the city of Dundee has recommended a $5 million bond issue for the purchase of greenbelt/floodplain land to preserve low-lying green areas and wildlife habitat on the east side of this rapidly expanding city of 62,000. The proposal is referred to as the Greenway Acquisition Initiative. Developers immediately opposed the proposal due to the reduction of available land for commercial development. The city engineer and economic development director have made the following preliminary estimates for some obvious areas, considering the Initiative's consequences in maintenance, parks, commercial development, and flooding over a projected 15-year planning horizon. The inaccuracy of these estimates is made very clear in the report to the Dundee City Council. The estimates are not yet classified as costs, benefits, or disbenefits. If the Greenway Acquisition Initiative is implemented, the estimates are as follows.

Economic Dimension	Estimate
1. Annual cost of $5 million in bonds over 15 years at 6% bond interest rate	$300,000 (years 1–14) $5,300,000 (year 15)
2. Annual maintenance, upkeep, and program management	$75,000 + 10% per year
3. Annual parks development budget	$500,000 (years 5–10)
4. Annual loss in commercial development	$2,000,000 (years 8–10)
5. State sales tax rebates not realized	$275,000 + 5% per year (years 8 on)
6. Annual municipal income from park use and regional sports events	$100,000 + 12% per year (years 6 on)
7. Savings in flood control projects	$300,000 (years 3–10) $1,400,000 (years 10–15)
8. Property damage (personal and city) not experienced due to flooding	$500,000 (years 10 and 15)

Identify three different viewpoints for the economic analysis of the proposal, and classify the estimates accordingly.

Solution
There are many perspectives to take; three are addressed here. The viewpoints and goals are identified and each estimate is classified as a cost, benefit, or disbenefit. (How the classification is made will vary depending upon who does the analysis. This solution offers only one logical answer.)

Viewpoint 1: Citizen of the city. Goal: Maximize the quality and wellness of citizens with family and neighborhood as prime concerns.

| Costs: 1, 2, 3 | Benefits: 6, 7, 8 | Disbenefits: 4, 5 |

Viewpoint 2: City budget. Goal: Ensure the budget is balanced and of sufficient size to fund rapidly growing city services.

Costs: 1, 2, 3, 5   Benefits: 6, 7, 8   Disbenefits: 4

Viewpoint 3: Economic development. Goal: Promote new commercial and industrial economic development for creation and retention of jobs.

Costs: 1, 2, 3, 4, 5   Benefits: 6, 7, 8   Disbenefits: none

Classification of estimates 4 (loss of commercial development) and 5 (loss of sales tax rebates) changes depending upon the view taken for the economic analysis. If the analyst favors the economic development goals of the city, commercial development losses are considered real costs, whereas they are undesirable consequences (disbenefits) from the citizen and budget viewpoints. Also, the loss of sales tax rebates from the state is interpreted as a real cost from the budget and economic development perspectives, but as a disbenefit from the citizen viewpoint.

Comment

Disbenefits may be included or disregarded in an analysis, as discussed in the next section. This decision can make a distinctive difference in the acceptance or rejection of a public sector alternative.

During the last several decades, larger public sector projects have been developed increasingly often through public-private partnerships. This is the trend in part because of the greater efficiency of the private sector and in part because of the sizable cost to design, construct, and operate such projects. Full funding by the government unit may not be possible using traditional means of government financing—fees, taxes, and bonds. Some examples of the projects are as follows:

Project	Some Purposes of the Project
Bridges and tunnels	Speed traffic flows; reduce congestion; improve safety
Ports and harbors	Increase cargo capacity; support industrial development
Airports	Increase capacity; improve passenger safety; support development
Water resources	Desalination for drinking water; meet irrigation and industrial needs; improve wastewater treatment

In these joint ventures, the public sector (government) is responsible for the cost and service to the citizenry, and the private sector partner (corporation) is responsible for varying aspects of the projects as detailed below. The government unit cannot make a profit, but the corporation(s) involved can realize a reasonable profit; in fact the profit margin is usually written into the contract that governs the design, construction, operation, and ownership of the project.

Traditionally, such construction projects have been designed for and financed by a government unit with a contractor doing the construction under either a lump-sum *(fixed-price)* contract or a cost reimbursement *(cost-plus)* contract that specifies the agreed upon margin of profit. In these cases, the contractor does not share the risk of the project's success with the government "owner." When a partnership of public and private interests is developed, the project is commonly contracted under an arrangement called *build-operate-transfer (BOT),* which may also be referred to as BOOT, where the first O is for *own.* The BOT-administered project may require that the contractor be responsible partially or completely for design and financing, and completely responsible for the construction (the build element), operation (operate), and maintenance activities for a specified number of years. After this time period, the owner becomes the government unit when the title of ownership is transferred (transfer) at no or very low cost. This arrangement may have several advantages, some of which are

- Better efficiency of resource allocation of private enterprise
- Ability to acquire funds (loans) based on financial record of the government and corporate partners
- Environmental, liability, and safety issues addressed by the private sector, where there usually is greater expertise
- Contracting corporation(s) able to realize a return on the investment during the operation phase

Many of the projects in international settings and in developing countries utilize the BOT form of partnership. There are, of course, disadvantages to this arrangement. One risk is that the amount of financing committed to the project may not cover the actual build cost because it is considerably higher than estimated. A second risk is that a reasonable profit may not be realized by the private corporation due to low usage of the facility during the operate phase. To plan against such problems, the original contract may provide for special loans guaranteed by the government unit and special subsidies. The subsidy may cover costs plus (contractually agreed-to) profit if usage is lower than a specified level. The level used may be the breakeven point with the agreed-to profit margin considered.

A variation of the BOT/BOOT method is BOO (build-own-operate), where the transfer of ownership never takes place. This form of public-private partnership may be used when the project has a relatively short life or the technology deployed is changing quickly.

9.2   BENEFIT/COST ANALYSIS OF A SINGLE PROJECT

The benefit/cost ratio is relied upon as a fundamental analysis method for public sector projects. The B/C analysis was developed to introduce more objectivity into public sector economics, and as one response to the U.S. Congress approving the Flood Control Act of 1936. There are several variations of the B/C ratio; however, the fundamental approach is the same. All cost and benefit estimates must be converted to a common equivalent monetary unit (PW, AW, or FW) at the discount rate (interest rate). The B/C ratio is then calculated using one of these relations:

$$\text{B/C} = \frac{\text{PW of benefits}}{\text{PW of costs}} = \frac{\text{AW of benefits}}{\text{AW of costs}} = \frac{\text{FW of benefits}}{\text{FW of costs}} \qquad [9.1]$$

Present worth and annual worth equivalencies are more used than future worth values. The sign convention for B/C analysis is positive signs, so *costs are preceded by a + sign.* Salvage values, when they are estimated, are subtracted from costs. Disbenefits are considered in different ways depending upon the model used. Most commonly, disbenefits are subtracted from benefits and placed in the numerator. The different formats are discussed below.

The decision guideline is simple:

If B/C ≥ 1.0, accept the project as economically acceptable for the estimates and discount rate applied.

If B/C < 1.0, the project is not economically acceptable.

If the B/C value is exactly or very near 1.0, noneconomic factors will help make the decision for the "best" alternative.

The *conventional B/C ratio*, probably the most widely used, is calculated as follows:

$$\text{B/C} = \frac{\text{benefits} - \text{disbenefits}}{\text{costs}} = \frac{B - D}{C} \qquad [9.2]$$

In Equation [9.2] disbenefits are subtracted from benefits, not added to costs. The B/C value could change considerably if disbenefits are regarded as costs. For example, if the numbers 10, 8, and 8 are used to represent the PW of benefits, disbenefits, and costs, respectively, the correct procedure results in B/C = $(10 - 8)/8 = 0.25$. The incorrect placement of disbenefits in the denominator results in B/C = $10/(8 + 8) = 0.625$, which is more than twice the correct B/C value of 0.25. Clearly, then, the method by which disbenefits are handled affects the magnitude of the B/C ratio. However, no matter whether disbenefits are (correctly) subtracted from the numerator or (incorrectly) added to costs in the denominator, a B/C ratio of less than 1.0 by the first method will always yield a B/C ratio less than 1.0 by the second method, and vice versa.

The *modified B/C ratio* includes maintenance and operation (M&O) costs in the numerator and treats them in a manner similar to disbenefits. The denominator includes only the initial investment. Once all amounts are expressed in PW, AW, or FW terms, the modified B/C ratio is calculated as

$$\text{Modified B/C} = \frac{\text{benefits} - \text{disbenefits} - \text{M\&O costs}}{\text{initial investment}} \qquad [9.3]$$

Salvage value is included in the denominator as a negative cost. The modified B/C ratio will obviously yield a different value than the conventional B/C method. However, as with disbenefits, *the modified procedure can change the magnitude of the ratio but not the decision to accept or reject the project.*

The *benefit and cost difference* measure of worth, which does not involve a ratio, is based on the difference between the PW, AW, or FW of benefits and costs, that is, $B - C$. If $(B - C) \geq 0$, the project is acceptable. This method has the advantage of eliminating the discrepancies noted above when disbenefits are regarded as costs, because B represents *net benefits*. Thus, for the numbers 10, 8, and 8 the same result is obtained regardless of how disbenefits are treated.

Subtracting disbenefits from benefits: $B - C = (10 - 8) - 8 = -6$

Adding disbenefits to costs: $B - C = 10 - (8 + 8) = -6$

Before calculating the B/C ratio by any formula, check whether the alternative with the larger AW or PW of costs also yields a larger AW or PW of benefits. It is possible for one alternative with larger costs to generate lower benefits than other alternatives, thus making it unnecessary to further consider the larger-cost alternative.

EXAMPLE 9.2

The Ford Foundation expects to award $15 million in grants to public high schools to develop new ways to teach the fundamentals of engineering that prepare students for university-level material. The grants will extend over a 10-year period and will create an estimated savings of $1.5 million per year in faculty salaries and student-related expenses. The Foundation uses a rate of return of 6% per year on all grant awards.

This grants program will share Foundation funding with ongoing activities, so an estimated $200,000 per year will be removed from other program funding. To make this program successful, a $500,000 per year operating cost will be incurred from the regular M&O budget. Use the B/C method to determine if the grants program is economically justified.

Solution

Use annual worth as the common monetary equivalent. All three B/C models are used to evaluate the program.

> **AW of investment cost.** $15,000,000(A/P,6%,10) = $2,038,050 per year
> **AW of benefit.** $1,500,000 per year
> **AW of disbenefit.** $200,000 per year
> **AW of M&O cost.** $500,000 per year

Use Equation [9.2] for conventional B/C analysis, where M&O is placed in the denominator as an annual cost.

$$\text{B/C} = \frac{1,500,000 - 200,000}{2,038,050 + 500,000} = \frac{1,300,000}{2,538,050} = 0.51$$

The project is not justified, since B/C < 1.0.

By Equation [9.3] the modified B/C ratio treats the M&O cost as a reduction to benefits.

$$\text{Modified B/C} = \frac{1,500,000 - 200,000 - 500,000}{2,038,050} = 0.39$$

The project is also not justified by the modified B/C method, as expected.

For the $(B - C)$ model, B is the net benefit, and the annual M&O cost is included with costs.

$$B - C = (1,500,000 - 200,000) - (2,038,050 + 500,000) = \$-1.24 \text{ million}$$

Since $(B - C) < 0$, the program is not justified.

EXAMPLE 9.3

Aaron is a new project engineer with the Arizona Department of Transportation (ADOT). After receiving a degree in engineering from Arizona State University, he decided to gain experience in the public sector before applying to master's degree programs. Based on annual worth relations, Aaron performed the conventional B/C analysis of the two separate proposals shown below.

Bypass proposal: new routing around part of Flagstaff to improve safety and decrease average travel time.

 Source of proposal: State ADOT office of major thoroughfare analysis.

 Initial investment in present worth: $P = \$40$ million.

 Annual maintenance: $1.5 million.

 Annual benefits to public: $B = \$6.5$ million.

 Expected life: 20 years.

 Funding: Shared 50–50 federal and state funding; federally required 8% discount rate applies.

Upgrade proposal: widening of roadway through parts of Flagstaff to alleviate traffic congestion and improve traffic safety.

 Source of proposal: Local Flagstaff district office of ADOT.

 Initial investment in present worth: $P = \$4$ million.

 Annual maintenance: $150,000.

 Annual benefits to public: $B = \$650,000$.

 Expected life: 12 years.

 Funding: 100% state funding required; usual 4% discount rate applies.

Aaron used a hand solution for the conventional B/C analysis in Equation [9.2] with AW values calculated at 8% per year for the bypass proposal and at 4% per year for the upgrade proposal.

 Bypass proposal: AW of investment = $\$40,000,000(A/P,8\%,20) = \$4,074,000$ per year

$$\text{B/C} = \frac{6,500,000}{4,074,000 + 1,500,000} = 1.17$$

 Upgrade proposal: AW of investment = $\$4,000,000(A/P,4\%,12) = \$426,200$ per year

$$\text{B/C} = \frac{650,000}{426,200 + 150,000} = 1.13$$

Both proposals are economically justified since B/C > 1.0.

(a) Perform the same analysis by computer, using a minimum number of computations.

(b) The discount rate for the upgrade proposal is not certain, because ADOT is thinking of asking for federal funds for it. Is the upgrade economically justified if the 8% discount rate also applies to it?

Solution by Computer

(a) See Figure 9–1a. The B/C values of 1.17 and 1.13 are in B4 and D4 ($1 million units). The function PMT($i\%,n,-P$) plus the annual maintenance cost calculates the AW of costs in the denominator. See the cell tags.

(b) Cell F4 uses an i value of 8% in the PMT function. There is a real difference in the justification decision. At the 8% rate, the upgrade proposal is no longer justified.

Q-Solv

Comment

Figure 9–1b presents a complete B/C spreadsheet solution. There are no differences in the conclusions from those in the Q-solv spreadsheet, but the proposal estimates and B/C results are shown in detail on this spreadsheet. Also, additional sensitivity analysis is easily performed on this expanded version, because of the use of cell reference functions.

E-Solve

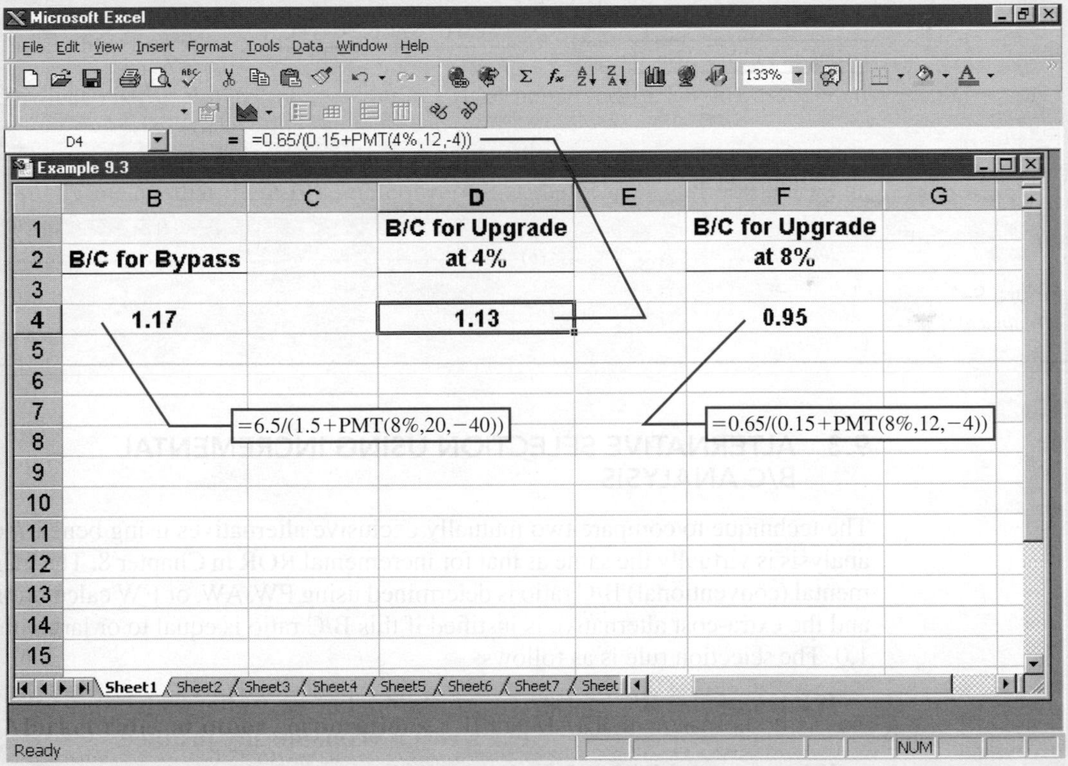

X Microsoft Excel

File Edit View Insert Format Tools Data Window Help

D4   = =0.65/(0.15+PMT(4%,12,-4))

Example 9.3

	B	C	D	E	F	G
1			**B/C for Upgrade**		**B/C for Upgrade**	
2	**B/C for Bypass**		**at 4%**		**at 8%**	
3						
4	1.17		1.13		0.95	
5						
6						
7		=6.5/(1.5+PMT(8%,20,−40))			=0.65/(0.15+PMT(8%,12,−4))	
8						
9						
10						
11						
12						
13						
14						
15						

Sheet1 / Sheet2 / Sheet3 / Sheet4 / Sheet5 / Sheet6 / Sheet7 / Sheet

Ready   NUM

(a)

Figure 9–1

Spreadsheet for B/C ratio of two proposals: (a) Q-solv solution and (b) expanded solution, Example 9.3.

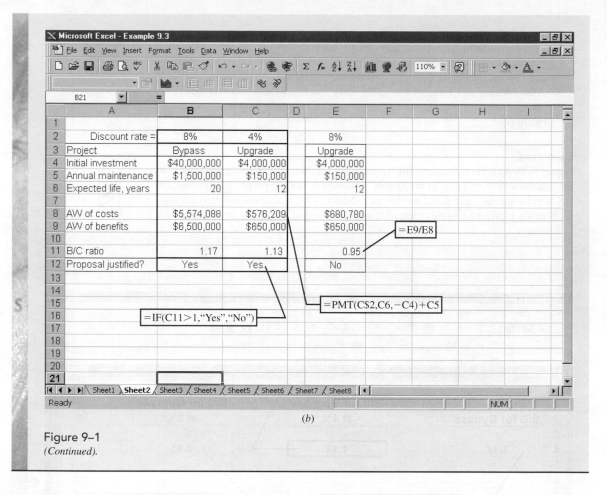

Figure 9–1
(Continued).

9.3 ALTERNATIVE SELECTION USING INCREMENTAL B/C ANALYSIS

The technique to compare two mutually exclusive alternatives using benefit/cost analysis is virtually the same as that for incremental ROR in Chapter 8. The incremental (conventional) B/C ratio is determined using PW, AW, or FW calculations, and the extra-cost alternative is justified if this B/C ratio is equal to or larger than 1.0. The selection rule is as follows:

> **If incremental B/C ≥ 1.0, choose the higher-cost alternative, because its extra cost is economically justified.**
>
> **If incremental B/C < 1.0, choose the lower-cost alternative.**
>
> **To perform a correct incremental B/C analysis, it is required that each alternative be compared only with another alternative for which the incremental cost is already justified. This same rule was used previously in incremental ROR analysis.**

There are several special considerations for B/C analysis that make it slightly different from that for ROR analysis. As mentioned earlier, all costs have a positive sign in the B/C ratio. Also, the *ordering of alternatives is done on the basis of total costs* in the denominator of the ratio. Thus, if two alternatives, A and B, have equal initial investments and lives, but B has a larger equivalent annual cost, then B must be incrementally justified against A. (This is illustrated in the next example.) If this convention is not correctly followed, it is possible to get a negative cost value in the denominator, which can incorrectly make B/C < 1 and reject a higher-cost alternative that is actually justified.

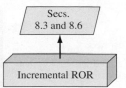

Follow these steps to correctly perform a conventional B/C ratio analysis of two alternatives. Equivalent values can be expressed in PW, AW, or FW terms.

1. Determine the total equivalent costs for both alternatives.
2. Order the alternatives by total equivalent cost; smaller first, then larger. Calculate the incremental cost (ΔC) for the larger-cost alternative. This is the denominator in B/C.
3. Calculate the total equivalent benefits and any disbenefits estimated for both alternatives. Calculate the incremental benefits (ΔB) for the larger-cost alternative. (This is $\Delta(B - D)$ if disbenefits are considered.)
4. Calculate the incremental B/C ratio using Equation [9.2], $(B - D)/C$.
5. Use the selection guideline to select the higher-cost alternative if B/C $\geq$ 1.0.

When the B/C ratio is determined for the lower-cost alternative, it is a comparison with the do-nothing (DN) alternative. If B/C < 1.0, then DN should be selected and compared to the second alternative. If neither alternative has an acceptable B/C value, the DN alternative must be selected. In public sector analysis, the DN alternative is usually the current condition.

EXAMPLE **9.4**

The city of Garden Ridge, Florida, has received designs for a new patient room wing to the municipal hospital from two architectural consultants. One of the two designs must be accepted in order to announce it for construction bids. The costs and benefits are the same in most categories, but the city financial manager decided that the three estimates below should be considered to determine which design to recommend at the city council meeting next week and to present to the citizenry in preparation for an upcoming bond referendum next month.

	Design A	Design B
Construction cost, $	10,000,000	15,000,000
Building maintenance cost, $/year	35,000	55,000
Patient usage cost, $/year	450,000	200,000

The patient usage cost is an estimate of the amount paid by patients over the insurance coverage generally allowed for a hospital room. The discount rate is 5%, and the life of the building is estimated at 30 years.

(a) Use conventional B/C ratio analysis to select design A or B.

(b) Once the two designs were publicized, the privately owned hospital in the directly adjacent city of Forest Glen lodged a complaint that design A will reduce its own municipal hospital's income by an estimated $500,000 per year because some of the day-surgery features of design A duplicate its services. Subsequently, the Garden Ridge merchants' association argued that design B could reduce its annual revenue by an estimated $400,000, because it will eliminate an entire parking lot used by their patrons for short-term parking. The city financial manager stated that these concerns would be entered into the evaluation as disbenefits of the respective designs. Redo the B/C analysis to determine if the economic decision is still the same as when disbenefits were not considered.

Solution

(a) Since most of the cash flows are already annualized, the incremental B/C ratio will use AW values. No disbenefit estimates are considered. Follow the steps of the procedure above:

1. The AW of costs is the sum of construction and maintenance costs.

$$AW_A = 10,000,000(A/P,5\%,30) + 35,000 = \$685,500$$

$$AW_B = 15,000,000(A/P,5\%,30) + 55,000 = \$1,030,750$$

2. Design B has the larger AW of costs, so it is the alternative to be incrementally justified. The incremental cost value is

$$\Delta C = AW_B - AW_A = \$345,250 \text{ per year}$$

3. The AW of benefits is derived from the patient usage costs, since these are consequences to the public. The benefits for the B/C analysis are not the costs themselves, but the *difference* if design B is selected. The lower usage cost each year is a positive benefit for design B.

$$\Delta B = \text{usage}_A - \text{usage}_B = \$450,000 - \$200,000 = \$250,000 \text{ per year}$$

4. The incremental B/C ratio is calculated by Equation [9.2].

$$B/C = \frac{\$250,000}{\$345,250} = 0.72$$

5. The B/C ratio is less than 1.0, indicating that the extra costs associated with design B are not justified. Therefore, design A is selected for the construction bid.

(b) The revenue loss estimates are considered disbenefits. Since the disbenefits of design B are $100,000 less than those of A, this positive difference is added to the $250,000 benefits of B to give it a total benefit of $350,000. Now

$$B/C = \frac{\$350,000}{\$345,250} = 1.01$$

Design B is slightly favored. In this case the inclusion of disbenefits has reversed the previous economic decision. This has probably made the situation more difficult politically. New disbenefits will surely be claimed in the near future by other special-interest groups.

Like other methods, B/C analysis requires *equal-service comparison* of alternatives. Usually, the expected useful life of a public project is long (25 or 30 or more years), so alternatives generally have equal lives. However, when alternatives do have unequal lives, the use of PW to determine the equivalent costs and benefits requires that the LCM of lives be used. This is an excellent opportunity to use the AW equivalency of costs and benefits, if the implied assumption that the project could be repeated is reasonable. Therefore, use AW-based analysis for B/C ratios when different-life alternatives are compared.

9.4   INCREMENTAL B/C ANALYSIS OF MULTIPLE, MUTUALLY EXCLUSIVE ALTERNATIVES

The procedure necessary to select one from three or more mutually exclusive alternatives using incremental B/C analysis is essentially the same as that of the last section. The procedure also parallels that for incremental ROR analysis in Section 8.6. The selection guideline is as follows:

Choose the largest-cost alternative that is justified with an incremental B/C ≥ 1.0 when this selected alternative has been compared with another justified alternative.

There are two types of benefit estimates—estimation of *direct benefits,* and implied benefits based on *usage cost* estimates. Example 9.4 is a good illustration of the second type of implied benefit estimation. *When direct benefits are estimated,* the B/C ratio for each alternative may be calculated first as an initial screening mechanism to eliminate unacceptable alternatives. At least one alternative must have B/C ≥ 1.0 to perform the incremental B/C analysis. If all alternatives are unacceptable, the DN alternative is indicated as the choice. (This is the same approach as that of step 2 for "revenue alternatives only" in the ROR procedure of Section 8.6. However, the term "revenue alternative" is not applicable to public sector projects.)

As in the previous section comparing two alternatives, selection from multiple alternatives by incremental B/C ratio utilizes total equivalent costs to initially order alternatives from smallest to largest. Pairwise comparison is then undertaken. Also, remember that all costs are considered positive in B/C calculations. The terms *defender* and *challenger alternative* are used in this procedure, as in a ROR-based analysis. The procedure for incremental B/C analysis of multiple alternatives is as follows:

1. Determine the total equivalent cost for all alternatives. (Use AW, PW, or FW equivalencies for equal lives; use AW for unequal lives.)
2. Order the alternatives by total equivalent cost, smallest first.
3. Determine the total equivalent benefits (and any disbenefits estimated) for each alternative.
4. *Direct benefits estimation only:* Calculate the B/C for the first ordered alternative. (In effect, this makes DN the defender and the first alternative the challenger.) If B/C < 1.0, eliminate the challenger, and go to the next

challenger. Repeat this until B/C ≥ 1.0. The defender is eliminated, and the next alternative is now the challenger. (For analysis by computer, determine the B/C for all alternatives initially and retain only acceptable ones.)

5. Calculate incremental costs (ΔC) and benefits (ΔB) using the relations

$$\Delta C = \text{challenger cost} - \text{defender cost} \qquad [9.4]$$

$$\Delta B = \text{challenger benefits} - \text{defender benefits} \qquad [9.5]$$

If relative *usage costs* are estimated for each alternative, rather than direct benefits, ΔB may be found using the relation

$$\Delta B = \text{defender usage costs} - \text{challenger usage costs} \qquad [9.6]$$

6. Calculate the incremental B/C for the first challenger compared to the defender.

$$B/C = \Delta B/\Delta C \qquad [9.7]$$

If incremental B/C ≥ 1.0 in Equation [9.7], the challenger becomes the defender and the previous defender is eliminated. Conversely, if incremental B/C < 1.0, remove the challenger and the defender remains against the next challenger.

7. Repeat steps 5 and 6 until only one alternative remains. It is the selected one.

In all the steps above, incremental disbenefits may be considered by replacing ΔB with $\Delta(B - D)$, as in the conventional B/C ratio, Equation [9.2].

EXAMPLE 9.5

The Economic Development Corporation (EDC) for the city of Bahia, California, and Moderna County is operated as a not-for-profit corporation. It is seeking a developer that will place a major water park in the city or county area. Financial incentives will be awarded. In response to a request for proposal (RFP) to the major water park developers in the country, four proposals have been received. Larger and more intricate water rides and increased size of the park will attract more customers, thus different levels of initial incentives are requested in the proposals. One of these proposals will be accepted by the EDC and recommended to the Bahia City Council and Moderna County Board of Trustees for approval.

Approved and in-place economic incentive guidelines allow entertainment industry prospects to receive up to $500,000 cash as a first-year incentive award and 10% of this amount each year for 8 years in property tax reduction. All the proposals meet the requirements for these two incentives. Each proposal includes a provision that residents of the city or county will benefit from reduced entrance (usage) fees when using the park. This fee reduction will be in effect as long as the property tax reduction incentive continues. The EDC has estimated the annual total entrance fees with the reduction included for local residents. Also, EDC estimated the extra sales tax revenue expected for the four park designs. These estimates and the costs for the initial incentive and annual 10% tax reduction are summarized in the top section of Table 9–1.

TABLE **9–1**   Estimates of Costs and Benefits, and the Incremental B/C Analysis for Four Water Park Proposals, Example 9.5

	Proposal 1	Proposal 2	Proposal 3	Proposal 4
Initial incentive, $	250,000	350,000	500,000	800,000
Tax incentive cost, $/year	25,000	35,000	50,000	80,000
Resident entrance fees, $/year	500,000	450,000	425,000	250,000
Extra sales taxes, $/year	310,000	320,000	320,000	340,000
Study period, years	8	8	8	8
AW of total costs, $	66,867	93,614	133,735	213,976
Alternatives compared		2-to-1	3-to-2	4-to-2
Incremental costs ΔC, $/year		26,747	40,120	120,360
Entrance fee reduction, $/year		50,000	25,000	200,000
Extra sales tax, $/year		10,000	0	20,000
Incremental benefits ΔB, $/year		60,000	25,000	220,000
Incremental B/C ratio		2.24	0.62	1.83
Increment justified?		Yes	No	Yes
Alternative selected		2	2	4

Utilize hand and computer analysis to perform an incremental B/C study to determine which park proposal is the best economically. The discount rate used by the EDC is 7% per year. Can the current incentive guidelines be used to accept the winning proposal?

Solution by Hand

The viewpoint taken for the economic analysis is that of a resident of the city or county. The first-year cash incentives and annual tax reduction incentives are real costs to the residents. Benefits are derived from two components: the decreased entrance fee estimates and the increased sales tax receipts. These will benefit each citizen indirectly through the increase in money available to those who use the park and through the city and county budgets where sales tax receipts are deposited. Since these benefits must be calculated indirectly from these two components, the initial proposal B/C values cannot be calculated to initially eliminate any proposals. A B/C analysis incrementally comparing two alternatives at a time must be conducted.

Table 9–1 includes the results of applying the procedure above. Equivalent AW values are used for benefit and cost amounts per year. Since the benefits must be derived indirectly from the entrance fee estimates and sales tax receipts, step 4 is not used.

1. For each alternative, the capital recovery amount over 8 years is determined and added to the annual property tax incentive cost. For proposal #1,

$$\text{AW of total costs} = \text{initial incentive}(A/P,7\%,8) + \text{tax cost}$$

$$= \$250,000(A/P,7\%,8) + 25,000 = \$66,867$$

2. The alternatives are ordered by the AW of total costs in Table 9–1.
3. The annual benefit of an alternative is the incremental benefit of the entrance fees and sales tax amounts. These are calculated in step 5.
4. This step is not used.
5. Table 9–1 shows incremental costs calculated by Equation [9.4]. For the 2-to-1 comparison,

$$\Delta C = \$93,614 - 66,867 = \$26,747$$

Incremental benefits for an alternative are the sum of the resident entrance fees compared to those of the next-lower-cost alternative, plus the increase in sales tax receipts over those of the next-lower-cost alternative. Thus, the benefits are determined incrementally for each pair of alternatives. For example, when proposal #2 is compared to proposal #1, the resident entrance fees decrease by \$50,000 per year and the sales tax receipts increase by \$10,000. Then the total benefit is the sum of these, that is, $\Delta B = \$60,000$ per year.
6. For the 2-to-1 comparison, Equation [9.7] results in

$$B/C = \$60,000/\$26,747 = 2.24$$

Alternative #2 is clearly incrementally justified. Alternative #1 is eliminated, and alternative #3 is the new challenger to defender #2.
7. This process is repeated for the 3-to-2 comparison, which has an incremental B/C of 0.62 because the incremental benefits are substantially less than the increase in costs. Therefore, proposal #3 is eliminated, and the 4-to-2 comparison results in

$$B/C = \$220,000/\$120,360 = 1.83$$

Since B/C > 1.0, proposal #4 is retained. Since proposal #4 is the one remaining alternative, it is selected.

The recommendation for proposal #4 requires an initial incentive of \$800,000, which exceeds the \$500,000 limit of the approved incentive limits. The EDC will have to request the City Council and County Trustees to grant an exception to the guidelines. If the exception is not approved, proposal #2 is accepted.

Solution by Computer

E-Solve

Figure 9–2 presents a spreadsheet using the same calculations as those in Table 9–1. Row 8 cells include the function PMT(7%,8,−initial incentive) to calculate the capital recovery for each alternative, plus the annual tax cost. These AW of total cost values are used to order the alternatives for incremental comparison.

The cell tags for rows 10 through 13 detail the formulas for incremental costs and benefits used in the incremental B/C computation (row 14). Note the difference in row 11 and 12 formulas, which find the incremental benefits for entrance fees and sales tax, respectively. The order of the subtraction between columns in row 11 (e.g., =B5 − C5, for the 2-to-1 comparison) must be correct to obtain the incremental entrance fees benefit. The IF operators in row 15 accept or reject the challenger, based upon the size of B/C. After the 3-to-2 comparison with B/C = 0.62 in cell D14, alternative #3 is eliminated. The final selection is alternative #4, as in the solution by hand.

	A	B	C	D	E	F	G
1	Discount rate =	7%					
2	Alternative	**#1**	**#2**	**#3**	**#4**		
3	Initial incentive, $	$ 250,000	$ 350,000	$ 500,000	$ 800,000		
4	Tax incentive cost, $/yr	$ 25,000	$ 35,000	$ 50,000	$ 80,000		
5	Resident entrance fees, $/yr	$ 500,000	$ 450,000	$ 425,000	$ 250,000		
6	Extra sales tax, $/yr	$ 310,000	$ 320,000	$ 320,000	$ 340,000		
7	Life	8	8	8	8		
8	AW of total costs	$ 66,867	$ 93,614	$ 133,734	$ 213,974		
9	Alternatives compared		2-to-1	3-to-2	4-to-2		
10	Incremental costs (delta C)		$ 26,747	$ 40,120	$ 120,360		
11	Entrance fees reduction, $/yr		50,000	25,000	200,000		
12	Extra sales taxes, $/yr		10,000	-	20,000		
13	Incremental benefits (delta B)		$ 60,000	$ 25,000	$ 220,000		
14	Incremental B/C ratio		2.24	0.62	1.83		
15	Increment justified?		Yes	No	Yes		
16	Alternative selected		#2	#2	#4		
17							
18	=PMT($B$1,C$7,−C3)+C4	=C$8−B$8		=IF(E14>1,"Yes","No")			
19		=B$5−C$5					
20		=C$6−B$6					
21		=C$11+C$12					

=E$13/E$10

Figure 9–2
Spreadsheet solution for an incremental B/C analysis of four mutually exclusive alternatives, Example 9.5.

When the lives of alternatives are so long that they can be considered infinite, the capitalized cost is used to calculate the equivalent PW or AW values for costs and benefits. Equation [5.3], $A = P(i)$, is used to determine the equivalent AW values in the incremental B/C analysis.

If two or more *independent projects* are evaluated using B/C analysis and there is no budget limitation, no incremental comparison is necessary. The only comparison is between each project separately with the do-nothing alternative. The project B/C values are calculated, and those with B/C $\geq$ 1.0 are accepted. This is the same procedure as that used to select from independent projects using the ROR method (Chapter 8). When a budget limitation is imposed, the capital budgeting procedure discussed in Chapter 12 must be applied.

EXAMPLE 9.6

The Army Corps of Engineers wants to construct a dam on a flood-prone river. The estimated construction cost and average annual dollar benefits are listed below. (*a*) If a 6% per year rate applies and dam life is infinite for analysis purposes, select the one best location using the B/C method. If no site is acceptable, other sites will be determined later. (*b*) If more than one dam site can be selected, which sites are acceptable, using the B/C method?

Site	Construction Cost, $ millions	Annual Benefits, $
A	6	350,000
B	8	420,000
C	3	125,000
D	10	400,000
E	5	350,000
F	11	700,000

Solution

(*a*) The capitalized cost $A = Pi$ is used to obtain AW values for annual capital recovery of the construction cost, as shown in the first row of Table 9–2. Since benefits are estimated directly, the site B/C ratio can be used for initial screening. Only sites E and F have B/C > 1.0, so they are evaluated incrementally. The E-to-DN comparison is performed because it is not required that one site must be selected. The analysis between the mutually exclusive alternatives in the lower portion of Table 9–2 is based on Equation [9.7].

$$\text{Incremental B/C} = \frac{\Delta \text{ annual benefits}}{\Delta \text{ annual costs}}$$

Since only site E is incrementally justified, it is selected.

(*b*) The dam site proposals are now independent projects. The site B/C ratio is used to select from none to all six sites. In Table 9–2, B/C > 1.0 for sites E and F only; they are acceptable, the rest are not.

TABLE 9–2   Use of Incremental B/C Ratio Analysis for Example 9.6 (Values in $1000)

	C	E	A	B	D	F
Capital recovery cost, $	180	300	360	480	600	660
Annual benefits, $	125	350	350	420	400	700
Site B/C	0.69	1.17	0.97	0.88	0.67	1.06
Decision	No	Retain	No	No	No	Retain
Comparison		E-to-DN				F-to-E
Δ Annual cost, $		300				360
Δ Annual benefits, $		350				350
Δ (B/C) ratio		1.17				0.97
Increment justified?		Yes				No
Site selected		E				E

Comment

In part (a), suppose that site G is added with a construction cost of $10 million and an annual benefit of $700,000. The site B/C is acceptable at B/C = 700/600 − 1.17. Now, incrementally compare G-to-E, the incremental B/C = 350/300 = 1.17, in favor of G. In this case, site F must be compared with G. Since the annual benefits are the same ($700,000), the B/C ratio is zero and the added investment is not justified. Therefore, site G is chosen.

CHAPTER SUMMARY

The benefit/cost method is used primarily to evaluate projects and to select from alternatives in the public sector. When one is comparing mutually exclusive alternatives, the incremental B/C ratio must be greater than or equal to 1.0 for the incremental equivalent total cost to be economically justified. The PW, AW, or FW of the initial costs and estimated benefits can be used to perform an incremental B/C analysis. If alternative lives are unequal, the AW values should be used, provided the assumption of project repetition is not unreasonable. For independent projects, no incremental B/C analysis is necessary. All projects with B/C ≥ 1.0 are selected provided there is no budget limitation.

Public sector economics are substantially different from those of the private sector. For public sector projects, the initial costs are usually large, the expected life is long (25, 35, or more years), and the sources for capital are usually a combination of taxes levied on the citizenry, user fees, bond issues, and private lenders. *It is very difficult to make accurate estimates of benefits for a public sector project.* The interest rates, called the discount rates in the public sector, are lower than those for corporate capital financing. Although the discount rate is as important to establish as the MARR, it can be difficult to establish, because various government agencies qualify for different rates. Standardized discount rates are established for some federal agencies.

PROBLEMS

Public Sector Economics

9.1 State the difference between public and private sector alternatives with respect to the following characteristics.
   (a) Size of investment
   (b) Life of project
   (c) Funding
   (d) MARR

9.2 Indicate whether the following characteristics are primarily associated with public sector or private sector projects.
   (a) Profits
   (b) Taxes
   (c) Disbenefits
   (d) Infinite life
   (e) User fees
   (f) Corporate bonds

9.3   Identify each cash flow as a benefit, disbenefit, or cost.
   (a) $500,000 annual income from tourism created by a freshwater reservoir
   (b) $700,000 per year maintenance by container ship port authority
   (c) Expenditure of $45 million for tunnel construction on an interstate highway
   (d) Elimination of $1.3 million in salaries for county residents based on reduced international trade
   (e) Reduction of $375,000 per year in car accident repairs because of improved lighting
   (f) $700,000 per year loss of revenue by farmers because of highway right-of-way purchases

9.4   During its 20 years in business, Deware Construction Company has always developed its contracts under a fixed-fee or cost-plus arrangement. Now it has been offered an opportunity to participate in a project to provide cross-country highway transportation in an international setting, specifically, a country in Africa. If accepted, Deware will work as subcontractor to a larger European corporation, and the BOT form of contracting will be used with the African country government. Describe for the president of Deware at least four of the significant differences that may be expected when the BOT format is utilized in lieu of its more traditional forms of contract making.

9.5   If a corporation accepts the BOT form of contracting, (a) identify two risks taken by a corporation and (b) state how these risks can be reduced by the government partner.

Project B/C Value

9.6   The estimated annual cash flows for a proposed city government project are costs of

$450,000 per year, benefits of $600,000 per year, and disbenefits of $100,000 per year. Determine the (a) B/C ratio and (b) value of B − C.

9.7   Use spreadsheet software such as Excel, PW analysis, and a discount rate of 5% per year to determine that the B/C value for the following estimates is 0.375, making the project not acceptable using the benefit/cost method. (a) Enter the values and equations on the spreadsheet so they may be changed for the purpose of sensitivity analysis.

First cost = $8 million
Annual cost = $800,000 per year
Benefit = $550,000 per year
Disbenefit = $100,000 per year

   (b) Do the following sensitivity analysis by changing only two cells on your spreadsheet. Change the discount rate to 3% per year, and adjust the annual cost estimate until B/C = 1.023. This makes the project just acceptable using benefit/cost analysis.

9.8   A proposed regulation regarding the removal of arsenic from drinking water is expected to have an annual cost of $200 per household per year. If it is assumed that there are 90 million households in the country and that the regulation could save 12 lives per year, what would the value of a human life have to be for the B/C ratio to be equal to 1.0?

9.9   The U.S. Environmental Protection Agency has established that 2.5% of the median household income is a reasonable amount to pay for safe drinking water. The median household income is $30,000 per year. For a regulation that would affect the health of people in 1% of the households, what would the health benefits have to equal in dollars per household (for that 1% of the households) for the B/C ratio to be equal to 1.0?

9.10 Use a spreadsheet to set up and solve Problem 9.9, and then apply the following changes. Observe the increases and decreases in the required economic value of the health benefits for each of these changes.

(a) Median income is $18,000 (poorer country), and percentage of household income is reduced to 2%.

(b) Median income is $30,000 and 2.5% is spent on safe water, but only 0.5% of the households are affected.

(c) What percentage of the households must be affected if the required health benefit and annual income both equal $18,000? Assume the 2.5% of income estimate is maintained.

9.11 The fire chief of a medium-sized city has estimated that the initial cost of a new fire station will be $4 million. Annual upkeep costs are estimated at $300,000. Benefits to citizens of $550,000 per year and disbenefits of $90,000 per year have also been identified. Use a discount rate of 4% per year to determine if the station is economically justified by (a) the conventional B/C ratio and (b) the B − C difference.

9.12 As part of the rehabilitation of the downtown area of a southern U.S. city, the Parks and Recreation Department is planning to develop the space below several overpasses into basketball, handball, miniature golf, and tennis courts. The initial cost is expected to be $150,000 for improvements which are expected to have a 20-year life. Annual maintenance costs are projected to be $12,000. The department expects 24,000 people per year to use the facilities an average of 2 hours each. The value of the recreation has been conservatively set at $0.50 per hour. At a discount

rate of 3% per year, what is the B/C ratio for the project?

9.13 The B/C ratio for a new flood control project along the banks of the Mississippi River is required to be 1.3. If the benefit is estimated at $600,000 per year and the maintenance cost is expected to total $300,000 per year, what is the allowed maximum initial cost of the project? The discount rate is 7% per year, and a project life of 50 years is expected. Solve in two ways: (a) by hand and (b) using a spreadsheet set up for sensitivity analysis.

9.14 Use the spreadsheet developed in Problem 9.13(b) to determine the B/C ratio if the initial cost is actually $3.23 million and the discount rate is now 5% per year.

9.15 The modified B/C ratio for a city-owned hospital heliport project is 1.7. If the initial cost is $1 million and the annual benefits are $150,000, what is the amount of the annual M&O costs used in the calculation, if a discount rate of 6% per year applies? The estimated life is 30 years.

9.16 Calculate the B/C ratio for the following cash flow estimates at a discount rate of 6% per year.

Item	Cash Flow
PW of benefits, $	3,800,000
AW of disbenefits, $/year	45,000
First cost, $	2,200,000
M&O costs, $/year	300,000
Life of project, years	15

9.17 Hemisphere Corp. is considering a BOT contract to construct and operate a large dam with a hydroelectric power generation facility in a developing nation in the southern hemisphere. The initial cost of the dam is expected to be $30 million, and it is expected to cost $100,000 per year to operate and maintain. Benefits from flood control, agricultural development, tourism, etc., are expected to be $2.8 million per year. At an interest rate of 8% per year, should the dam be constructed on the basis of its conventional B/C ratio? The dam is assumed to be a permanent asset for the country. (*a*) Solve by hand. (*b*) Using a spreadsheet, find the B/C ratio with only a single cell computation.

9.18 The U.S. Army Corps of Engineers is considering the feasibility of constructing a small flood control dam in an existing arroyo. The initial cost of the project will be $2.2 million, with inspection and upkeep costs of $10,000 per year. In addition, minor reconstruction will be required every 15 years at a cost of $65,000. If flood damage will be reduced from the present cost of $90,000 per year to $10,000 annually, use the benefit/cost method to determine if the dam should be constructed. Assume that the dam will be permanent and the interest rate is 12% per year.

9.19 A highway construction company is under contract to build a new roadway through a scenic area and two rural towns in Colorado. The road is expected to cost $18 million, with annual upkeep estimated at $150,000 per year. Additional income from tourists of $900,000 per year is estimated. If the road is expected to have a useful commercial life of 20 years, use one spreadsheet to determine if the highway should be constructed at an interest rate of 6% per year by applying (*a*) the B − C method, (*b*) the B/C method, and (*c*) the modified B/C method. (Additionally, if the instructor requests it: Set up the spreadsheet for sensitivity analysis and use the Excel IF operator to make the build–don't build decision in each part of the problem.)

9.20 The U.S. Bureau of Reclamation is considering a project to extend irrigation canals into a desert area. The initial cost of the project is expected to be $1.5 million, with annual maintenance costs of $25,000 per year. (*a*) If agricultural revenue is expected to be $175,000 per year, do a B/C analysis to determine whether the project should be undertaken, using a 20-year study period and a discount rate of 6% per year. (*b*) Rework the problem, using the modified B/C ratio.

9.21 (*a*) Set up the spreadsheet and (*b*) use hand calculations to calculate the B/C ratio for Problem 9.20 if the canal must be dredged every 3 years at a cost of $60,000 and there is a $15,000 per year disbenefit associated with the project.

Alternative Comparison

9.22 Apply incremental B/C analysis at an interest rate of 8% per year to determine which alternative should be selected. Use a 20-year study period, and assume the damage costs might occur in year 6 of the study period.

	Alternative A	Alternative B
Initial cost, $	600,000	800,000
Annual M&O costs, $/year	50,000	70,000
Potential damage costs, $	950,000	250,000

9.23 Two routes are under consideration for a new interstate highway segment. The long route would be 25 kilometers and would have an initial cost of $21 million. The short transmountain route would span 10 kilometers and would have an initial cost of $45 million. Maintenance costs are estimated at $40,000 per year for the long route and $15,000 per year for the short route. Additionally, a major overhaul and resurfacing will be required every 10 years at a cost of 10% of the first cost of each route. Regardless of which route is selected, the volume of traffic is expected to be 400,000 vehicles per year. If the vehicle operating expense is assumed to be $0.35 per kilometer and the value of reduced travel time for the short route is estimated at $900,000 per year, determine which route should be selected, using a conventional B/C analysis. Assume an infinite life for each road, an interest rate of 6% per year, and that one of the roads will be built.

9.24 A city engineer and economic development director of Buffalo are evaluating two sites for construction of a multipurpose sports arena. At the downtown site, the city already owns enough land for the arena. However, the land for construction of a parking garage will cost $1 million. The west side site is 30 kilometers from downtown, but the land will be donated by a developer who knows that an arena at this site will increase the value of the remainder of his land holdings by many times. The downtown site will have extra construction costs of about $10 million because of infrastructure relocations, the parking garage, and drainage improvements. However, because of its centralized location, there will be greater attendance at most of the events held there. This will result in more revenue to vendors and local merchants in the amount of $350,000 per year. Additionally, the average attendee will not have to travel as far, resulting in annual benefits of $400,000 per year. All other costs and revenues are expected to be the same at either site. If the city uses a discount rate of 8% per year, where should the arena be constructed? One of the two sites must be selected.

9.25 A country with rapid economic expansion has contracted for an economic evaluation of possibly building a new container port to augment the current port. The west coast site has deeper water so the dredging cost is lower than that for the east coast site. Also, the redredging of the west site will be required only every 6 years while the east site must be reworked each 4 years. Redredging, which is expected to increase in cost by 10% each time, will not take place in the last year of a port's commercial life. Disbenefit estimates vary from west (fishing revenue loss) to east (fishing and resort revenue losses). Fees to shippers per 20-foot STD equivalent are expected to be higher at the west site due to greater difficulty in handling ships because of the ocean currents present in the area and a higher cost of labor in this area of the country. All estimates are summarized below in $1 million, except annual revenue and life. Use spreadsheet analysis and a discount rate of 4% per year to determine if either port should be constructed. It is not necessary that the country build either port since one is already operating successfully.

	West Coast Site	East Coast Site
Initial cost, $		
Year 0	21	8
Year 1	0	8
Dredging cost, $, year 0	5	12
Annual M&O, $/year	1.5	0.8
Recurring dredging cost, $	2 each 6 years with increase of 10% each time	1.2 each 4 years with increase of 10% each time
Annual disbenefits, $/year	4	7
Annual fees: number of 20-foot. STD at $/container	5 million/year at $2.50 each	8 million/year at $2 each
Commercial life, years	20	12

9.26 A privately owned utility is considering two cash rebate programs to achieve water conservation. Program 1, which is expected to cost an average of $60 per household, would involve a rebate of 75% of the purchase and installation costs of an ultralow-flush toilet. This program is projected to achieve a 5% reduction in overall household water use over a 5-year evaluation period. This will benefit the citizenry to the extent of $1.25 per household per month. Program 2 would involve grass replacement with desert landscaping. This is expected to cost $500 per household, but it will result in reduced water cost at an estimated $8 per household per month (on average). At a discount rate of 0.5% per month, which program, if either, should the utility undertake? Use the B/C method.

9.27 Solar and conventional alternatives are available for providing energy at a remote space research site. The costs associated with each alternative are shown below. Use the B/C method to determine which should be selected at a discount rate of 0.75% per month over a 6-year study period.

	Conventional	Solar
Initial cost, $	2,000,000	4,500,000
M&O cost, $/month	50,000	10,000
Salvage value, $	0	150,000

9.28 The California Forest Service is considering two locations for a new state park. Location E would require an investment of $3 million and $50,000 per year in maintenance. Location W would cost $7 million to construct, but the Forest Service would receive an additional $25,000 per year in park use fees. The operating cost of location W will be $65,000 per year. The revenue to park concessionaires will be $500,000 per year at location E and $700,000 per year at location W. The disbenefits associated with each location are $30,000 per year for location E and $40,000 per year for location W. Use (a) the B/C method and (b) the modified B/C method to determine which location, if either, should be selected, using an interest rate of 12% per year. Assume that the park will be maintained indefinitely.

9.29 Three engineers made the estimates shown below for two optional methods by which new construction technology would be implemented at a site for public housing. Either one of the two options or the current method may be selected. Set up a spreadsheet for B/C sensitivity analysis, and determine if option 1, option 2, or the do-nothing option is selected by each of the three engineers. Use a life of 5 years and a discount rate of 10% per year for all analyses.

	Engineer Bob		Engineer Judy		Engineer Chen	
	Option 1	Option 2	Option 1	Option 2	Option 1	Option 2
Initial cost, $	50,000	90,000	75,000	90,000	60,000	70,000
Cost, $/year	3,000	4,000	3,800	3,000	6,000	3,000
Benefits, $/year	20,000	29,000	30,000	35,000	30,000	35,000
Disbenefits, $/year	500	1,500	1,000	0	5,000	1,000

Multiple Alternatives

9.30 One of four new techniques, or the current method, can be used to control mildly irritating chemical fume leakage into the surounding air from a mixing machine. The estimated costs and benefits (in the form of reduced employee health costs) are given below for each method. Assuming that all methods have a 10-year life with zero salvage value, determine which one should be selected, using a MARR of 15% per year and the B/C method.

	Technique			
	1	2	3	4
Installed cost, $	15,000	19,000	25,000	33,000
AOC, $/year	10,000	12,000	9,000	11,000
Benefits, $/year	15,000	20,000	19,000	22,000

9.31 Use a spreadsheet to perform a B/C analysis for the techniques in Problem 9.30, assuming they are independent projects. The benefits are cumulative if more than one technique is used in addition to the current method.

9.32 The Water Service Authority of Dubay is considering four sizes of pipe for a new water line. The costs per kilometer ($/km) for each size are given in the table. Assuming that all pipes will last 15 years and the MARR is 8% per year, which size pipe should be purchased based on a B/C analysis? Installation cost is considered a part of the initial cost.

	Pipe Size, Millimeters			
	130	150	200	230
Initial equipment cost, $/km	9,180	10,510	13,180	15,850
Installation cost, $/km	600	800	1,400	1,500
Usage cost, $/km per year	6,000	5,800	5,200	4,900

9.33 The federal government is considering three sites in the National Wildlife Preserve for mineral extraction. The cash flows (in millions) associated with each site are given below. Use the B/C method to determine which site, if any, is best, if the extraction period is limited to 5 years and the interest rate is 10% per year.

	Site A	Site B	Site C
Initial cost, $	50	90	200
Annual cost, $/year	3	4	6
Annual benefits, $/year	20	29	61
Annual disbenefits, $/year	0.5	1.5	2.1

9.34 Over the last several months, seven different toll bridge designs have been proposed and estimates made to connect a resort island to the mainland of an Asian country.

Location	Construction Cost, $ Millions	Annual Excess Fees Over Expenses, $100,000
A	14	4.0
B	8	6.1
C	22	10.8
D	9	8.0
E	12	7.5
F	6	3.9
G	18	9.3

A public-private partnership has been formed, and the national bank will be providing funding at a rate of 4% per year. Each bridge is expected to have a very long useful life. Use B/C analysis to answer the following. Solution by spreadsheet or by hand is acceptable.

(a) If one bridge design must be selected, determine which one is the best economically.

(b) An international bank has offered to fund as many as two additional bridges, since it is estimated that the trafffic and trade between the island and mainland will increase significantly. Determine which are the three best designs economically, if there is no budget restraint for the purpose of this analysis.

9.35 Three alternatives identified as X, Y, and Z were evaluated by the B/C method. The analyst, Joyce, calculated project B/C values of 0.92, 1.34, and 1.29. The alternatives are listed in order of increasing total equivalent costs. She isn't sure whether an incremental analysis is needed.

(a) What do you think? If no incremental analysis is needed, why not; if so, which alternatives must be compared incrementally?

(b) For what type of projects is incremental analysis never necessary? If X, Y, and Z are all this type of project, which alternatives are selected for the B/C values calculated?

9.36 The four mutually exclusive alternatives below are being compared using the B/C method. What alternative, if any, should be selected?

Alternative	Initial Investment, $ Millions	B/C Ratio	Incremental B/C When Compared with Alternative			
			J	K	L	M
J	20	1.10	—			
K	25	0.96	0.40	—		
L	33	1.22	1.42	2.14	—	
M	45	0.89	0.72	0.80	0.08	—

9.37 The city of Ocean View, California, is considering various proposals regarding the disposal of used tires. All the proposals involve shredding, but the charges for the service and handling of the tire shreds differ in each plan. An incremental B/C analysis was initiated, but the engineer conducting the study left recently. (a) Fill in the blanks in the incremental B/C portion of the table. (b) What alternative should be selected?

Alternative	Initial Investment, $ Millions	B/C Ratio	Incremental B/C When Compared with Alternative			
			P	Q	R	S
P	10	1.1	—	2.83		
Q	40	2.4	2.83	—		
R	50	1.4			—	
S	80	1.5				—

FE REVIEW PROBLEMS

9.38 When a B/C analysis is conducted,
  (a) The benefits and costs must be expressed in terms of their present worths.
  (b) The benefits and costs must be expressed in terms of their annual worths.
  (c) The benefits and costs must be expressed in terms of their future worths.
  (d) The benefits and costs can be expressed in terms of PW, AW, or FW.

9.39 In a conventional B/C ratio,
  (a) Disbenefits and M&O costs are subtracted from benefits.
  (b) Disbenefits are subtracted from benefits, and M&O costs are added to costs.
  (c) Disbenefits and M&O costs are added to costs.
  (d) Disbenefits are added to costs, and M&O costs are subtracted from benefits.

9.40 In a modified B/C ratio analysis,
  (a) Disbenefits and M&O costs are subtracted from benefits.
  (b) Disbenefits are subtracted from benefits, and M&O costs are added to costs.
  (c) Disbenefits and M&O costs are added to costs.
  (d) Disbenefits are added to costs, and M&O costs are subtracted from benefits.

9.41 An alternative has the following cash flows: benefits = $60,000 per year, disbenefits = $17,000 per year, and costs = $35,000 per year. The B/C ratio

is closest to
  (a) 0.92
  (b) 0.96
  (c) 1.23
  (d) 2.00

9.42 In evaluating three mutually exclusive alternatives by the B/C method, the alternatives were ranked in terms of increasing total equivalent cost (A, B, and C, respectively), and the following results were obtained for the project B/C ratios: 1.1, 0.9, and 1.3. On the basis of these results, you should
  (a) Select A.
  (b) Select C.
  (c) Select A and C.
  (d) Compare A and C incrementally.

9.43 Four independent projects are evaluated, using B/C ratios. The ratios are as follows:

Project	A	B	C	D
B/C ratio	0.71	1.29	1.07	2.03

On the basis of these results, you should
  (a) Reject B and D.
  (b) Select D only.
  (c) Reject A only.
  (d) Compare B, C and D incrementally.

9.44 If two mutually exclusive alternatives have B/C ratios of 1.5 and 1.4 for the lower first-cost and higher first-cost alternatives, respectively,
  (a) The B/C ratio on the increment between them is less than 1.4.
  (b) The B/C ratio on the increment between them is between 1.4 and 1.5.
  (c) The B/C ratio on the increment between them is greater than 1.4.
  (d) The lower-cost alternative is the better one.

EXTENDED EXERCISE

COSTS TO PROVIDE LADDER TRUCK SERVICE FOR FIRE PROTECTION

For many years, the city of Medford has paid a neighboring city (Brewster) for the use of its ladder truck when needed. The charges for the last few years have been $1000 per event when the ladder truck is only dispatched to a site in Medford, and $3000 each time the truck is activated. There has been no annual fee charged. With the approval of the Brewster city manager, the newly hired fire chief has presented a substantially higher cost to the Medford fire chief for the use of the ladder truck:

Annual flat fee	$30,000 with 5 years' fees paid up front (now)
Dispatch fee	$3000 per event
Activation fee	$8000 per event

The Medford chief has developed an alternative to purchase a ladder truck, with the following cost estimates for the truck and the fire station addition to house it:

Truck:

Initial cost	$850,000
Life	15 years
Cost per dispatch	$2000 per event
Cost per activation	$7000 per event

Building:

Initial cost	$500,000
Life	50 years

The chief has also taken data from a study completed last year and updated it. The study estimated the insurance premium and property loss reductions that the citizenry experienced by having a ladder truck available. The past savings and current estimates, if Medford had its own truck for more rapid response, are as follows:

	Past Average	Estimate If Truck Is Owned
Insurance premium reduction, $/year	100,000	200,000
Property loss reduction, $/year	300,000	400,000

Additionally, the Medford chief obtained the average number of events for the last 3 years and estimated the future use of the ladder truck. He believes there has been a reluctance to call for the truck from Brewster in the past.

	Past Average	Estimate If Truck Is Owned
Number of dispatches per year	10	15
Number of activations per year	3	5

Either the new cost structure must be accepted, or a truck must be purchased. The option to have no ladder truck service is not acceptable. Medford has a good rating for its bonds; a discount rate of 6% per year is used for all proposals.

Questions

Use a spreadsheet to do the following.

1. Perform an incremental B/C evaluation to determine if Medford should purchase a ladder truck.
2. Several of the new city council members are "up in arms" over the new annual fee and cost structure. However, they do not want to build more fire station capacity or own a ladder truck that will be used an average of only 20 times per year. They believe that Brewster can be convinced to reduce or remove the annual $30,000 fee. How much must the annual fee be reduced for the alternative to purchase the ladder truck to be rejected?
3. Another council member is willing to pay the annual fee, but wants to know how much the building cost can change from $500,000 to make the alternatives equally attractive. Find this first cost for the building.
4. Finally, a compromise proposal offered by the Medford mayor might be acceptable to Brewster. Reduce the annual fee by 50%, and reduce the per event charges to the same amount that the Medford fire chief estimates it will cost if the truck is owned. Then Medford will possibly adjust (if it seems reasonable) the sum of the insurance premium reduction and property loss reduction estimates to just make the arrangement with Brewster more attractive than owning the truck. Find this sum (for the estimates of premium reduction and property loss reduction). Does this new sum seem reasonable relative to the previous estimates?

CASE STUDY

FREEWAY LIGHTING

Introduction

A number of studies have shown that a disproportionate number of freeway traffic accidents occur at night. There are a number of possible explanations for this, one of which might be poor visibility. In an effort to determine whether freeway lighting was economically beneficial for reducing nighttime accidents, data were collected regarding accident frequency rates on lighted and unlighted sections of certain freeways. This case study is an analysis of part of those data.

Background

The Federal Highway Administration (FHWA) places value on accidents depending on the severity of the crash. There are a number of crash categories, the most severe of which is fatal. The cost of a fatal accident is placed at $2.8 million. The most common type of accident is not fatal or injurious and involves only property damage. The cost of this type of accident is placed at $4500. The ideal way to determine whether lights reduce traffic accidents is through before-and-after studies on a given section of free-way. However, this type of information is not readily available, so other methods must be used. One such method compares night to day accident rates for lighted and unlighted freeways. If lights are benefi-cial, the ratio of night to day accidents will be lower on the lighted section than on the unlighted one. If there is a difference, the reduced accident rate can be translated into benefits which can be compared to the cost of lighting to determine its economic feasibility. This technique is used in the following analysis.

Economic Analysis

The results of one particular study conducted over a 5-year period are presented in the table below. For il-lustrative purposes, only the property damage cate-gory will be considered.

The ratios of night to day accidents involving property damage for the unlighted and lighted freeway sections are $199/379 = 0.525$ and $839/2069 = 0.406$, respectively. These results indicate that the lighting was beneficial. To quantify the benefit, the accident-rate ratio from the unlighted section will be applied to the lighted section. This will yield the number of acci-dents that were prevented. Thus, there would have been $(2069)(0.525) = 1086$ accidents instead of the 839 if there had not been lights on the freeway. This is

a difference of 247 accidents. At a cost of $4500 per accident, this results in a net benefit of

$$B = (247)(\$4500) = \$1,111,500$$

To determine the cost of the lighting, it will be as-sumed that the light poles are center poles 67 me-ters apart with 2 bulbs each. The bulb size is 400 watts, and the installation cost is $3500 per pole. Since these data were collected over 87.8 kilometers (54.5 miles) of lighted freeway, the installed cost of the lighting is

$$\text{Installation cost} = \$3500\left(\frac{87.8}{0.067}\right)$$
$$= 3500(1310.4)$$
$$= \$4,586,400$$

The annual power cost based on 1310 poles is

Annual power cost

$$= 1310 \text{ poles}(2 \text{ bulbs/pole})(0.4 \text{ kilowatts/bulb})$$
$$\times (12 \text{ hours/day})(365 \text{ days/year})$$
$$\times (\$0.08/\text{kilowatt-hour})$$
$$= \$367,219 \text{ per year}$$

These data were collected over a 5-year period. There-fore, the annualized cost C at $i = 6\%$ per year is

$$\text{Total annual cost} = \$4,586,400(A/P,6\%,5)$$
$$+ 367,219$$
$$= \$1,456,030$$

Freeway Accident Rates, Lighted and Unlighted

Accident Type	Unlighted		Lighted	
	Day	Night	Day	Night
Fatal	3	5	4	7
Incapaciting	10	6	28	22
Evident	58	20	207	118
Possible	90	35	384	161
Property damage	379	199	2069	839
Totals	540	265	2697	1147

Source: Michael Griffin, "Comparison of the Safety of Lighting Options on Urban Free-ways," *Public Roads*, 58 (Autumn 1994), pp. 8–15.

The B/C ratio is

$$B/C = \frac{\$1,111,500}{\$1,456,030} = 0.76$$

Since B/C < 1, the lighting is not justified on the basis of property damage alone. To make a final determination about the economic viability of the lighting, the benefits associated with the other accident categories would obviously also have to be considered.

Case Study Exercises

1. What would the B/C ratio be if the light poles were twice as far apart as assumed above?

2. What is the ratio of night to day accidents for fatalities?

3. What would the B/C ratio be if the installation cost were only $2500 per pole?

4. How many accidents would be prevented on the unlighted portion of freeway if it were lighted? Consider the property damage category only.

5. Using only the category of property damage, what would the lighted night-to-day accident ratio have to be for the lighting to be economically justified?

Making Choices: The Method, MARR, and Multiple Attributes

This chapter broadens the capabilities of an engineering economy study. Some of the fundamental elements specified previously are unspecified here. As a result, many of the textbook aspects apparent in previous chapters are removed, thus coming closer to treating the more complex, real-world situations in which professional practice and decision making occur.

In all the previous chapters, the method for evaluating a project or comparing alternatives has been stated, or was obvious from the context of the problem. Also, when any method was used, the MARR was stated. Finally, only one dimension or attribute—the economic one—has been the judgment basis for the economic viability of one project, or the selection basis from two or more alternatives. In this chapter, the determination of all three of these parameters—evaluation method, MARR, and attributes—is discussed. Guidelines and techniques to determine each are developed and illustrated.

The case study examines the best balance between debt and equity capital, using a MARR-based analysis.

LEARNING OBJECTIVES

Purpose: Choose an appropriate method and MARR to compare alternatives economically, and using multiple attributes.

This chapter will help you:

Choose a method	1. Choose an appropriate method to compare mutually exclusive alternatives.
Cost of capital and MARR	2. Describe the cost of capital and its relation to the MARR, while considering reasons for MARR variation.
WACC	3. Understand the debt-to-equity mix and calculate the weighted average cost of capital (WACC).
Cost of debt capital	4. Estimate the cost of debt capital.
Cost of equity capital	5. Estimate the cost of equity capital, and explain how it compares to WACC and MARR.
High D-E mixes	6. Explain the relation of corporate risk to high debt-to-equity mixes.
Multiple attributes	7. Identify and develop weights for multiple attributes used in alternative selection.
Weighted attribute method	8. Use the weighted attribute method for multiple-attribute decision making.

10.1 COMPARING MUTUALLY EXCLUSIVE ALTERNATIVES BY DIFFERENT EVALUATION METHODS

In the previous five chapters, several equivalent evaluation techniques have been discussed. Any method—PW, AW, FW, ROR, or B/C—can be used to select one alternative from two or more and obtain the same, correct answer. Only one method is needed to perform the engineering economy analysis, because any method, correctly performed, will select the same alternative. Yet different information about an alternative is available with each different method. The selection of a method and its correct application can be confusing.

Table 10–1 gives a recommended evaluation method for different situations, if it is not specified by the instructor in a course or by corporate practice in professional work. The primary criteria for selecting a method are speed and ease of performing the analysis. Interpretation of the entries in each column follows.

Evaluation period: Most private sector alternatives (revenue and service) are compared over their equal or unequal estimated lives, or over a specific period of time. Public sector projects are commonly evaluated using the B/C ratio and usually have long lives that may be considered as infinite for economic computation purposes.

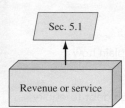

Type of alternatives: Private sector alternatives have cash flow estimates that are revenue-based (includes income and cost estimates) or service-based (cost estimates only). For service alternatives, the revenue cash flow series

TABLE 10–1	Recommended Method to Compare Mutually Exclusive Alternatives, Provided the Method Is Not Pre-selected		
Evaluation Period	**Type of Alternatives**	**Recommended Method**	**Series to Evaluate**
Equal lives of alternatives	Revenue or service	AW or PW	Cash flows
	Public sector	B/C, based on AW or PW	Incremental cash flows
Unequal lives of alternatives	Revenue or service	AW	Cash flows
	Public sector	B/C, based on AW	Incremental cash flows
Study period	Revenue or service	AW or PW	Updated cash flows
	Public sector	B/C, based on AW or PW	Updated incremental cash flows
Long to infinite	Revenue or service	AW or PW	Cash flows
	Public sector	B/C, based on AW	Incremental cash flows

is assumed to be equal for all alternatives. Public sector projects normally are service-based with the difference between costs and timing used to select one alternative over another.

Recommended method: Whether an analysis is performed by hand or by computer, the method(s) recommended in Table 10–1 will correctly select one alternative from two or more as rapidly as possible. Any other method can be applied subsequently to obtain additional information and, if needed, verification of the selection. For example, if lives are unequal and the rate of return is needed, it is best to first apply the AW method at the MARR and then determine the selected alternative's i^* using the same AW relation with i as the unknown.

Series to evaluate: The estimated cash flow series for one alternative and the incremental series between two alternatives are the only two options for present worth or annual worth evaluation. For spreadsheet analyses, this means that the NPV or PV functions (for present worth) or the PMT function (for annual worth) is applied. The word "updated" is added as a remainder that a study period analysis requires that cash flow estimates (especially salvage/market values) be reexamined and updated before the analysis is performed.

Once the evaluation method is selected, a specific procedure must be followed. These procedures were the primary topics of the last five chapters. Table 10–2 summarizes the important elements of the procedure for each method—PW, AW, ROR, and B/C. FW is included as an extension of PW. The meaning of the entries in Table 10–2 follows.

Equivalence relation: The basic equation written to perform any analysis is either a PW or an AW relation. The capitalized cost (CC) relation is a PW relation for infinite life, and the FW relation is likely determined from the PW equivalent value. Additionally, as we learned in Chapter 6, AW is simply PW times the A/P factor over the LCM of their lives.

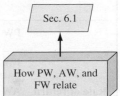

Sec. 6.1

How PW, AW, and FW relate

Lives of alternatives and *Time period for analysis:* The length of time for an evaluation (the n value) will always be one of the following: equal lives of the alternatives, LCM of unequal lives, specified study period, or infinity because the lives are so long.

> **PW analysis always requires the LCM of all alternatives.**
>
> **Incremental ROR and B/C methods require the LCM of the two alternatives being compared.**
>
> **The AW method allows analysis over the respective alternative lives.**

The one exception is for incremental ROR method for unequal-life alternatives using an AW relation for *incremental cash flows.* The LCM of the two alternatives compared must be used. This is equivalent to using an AW relation for the *actual cash flows* over the respective lives. Both approaches find the incremental rate of return Δi^*.

Series to evaluate: Either the estimated cash flow series or the incremental series is used to determine the PW value, the AW value, the i^* value, or the B/C ratio.

TABLE 10–2 Characteristics of an Economic Analysis of Mutually Exclusive Alternatives Once the Evaluation Method Is Determined

Evaluation Method	Equivalence Relation	Lives of Alternatives	Time Period for Analysis	Series to Evaluate	Rate of Return; Interest Rate	Decision Guideline: Select*
Present worth	PW	Equal	Lives	Cash flows	MARR	Numerically largest PW
	PW	Unequal	LCM	Cash flows	MARR	Numerically largest PW
	PW	Study period	Study period	Updated cash flows	MARR	Numerically largest PW
	CC	Long to infinite	Infinity	Cash flows	MARR	Numerically largest CC
Future worth	FW	Same as present worth for equal lives, unequal lives, and study period				Numerically largest FW
Annual worth	AW	Equal or unequal	Lives	Cash flows	MARR	Numerically largest AW
	AW	Study period	Study period	Updated cash flows	MARR	Numerically largest AW
	AW	Long to infinite	Infinity	Cash flows	MARR	Numerically largest AW
Rate of return	PW or AW	Equal	Lives	Incremental cash flows	Find Δi^*	Last $\Delta i^* \geq$ MARR
	PW or AW	Unequal	LCM of pair	Incremental cash flows	Find Δi^*	Last $\Delta i^* \geq$ MARR
	AW	Unequal	Lives	Cash flows	Find Δi^*	Last $\Delta i^* \geq$ MARR
	PW or AW	Study period	Study period	Updated incremental cash flows	Find Δi^*	Last $\Delta i^* \geq$ MARR
Benefit/cost	PW	Equal or unequal	LCM of pairs	Incremental cash flows	Discount rate	Last $\Delta B/C \geq$ 1.0
	AW	Equal or unequal	Lives	Incremental cash flows	Discount rate	Last $\Delta B/C \geq$ 1.0
	AW or PW	Long to infinite	Infinity	Incremental cash flows	Discount rate	Last $\Delta B/C \geq$ 1.0

*Lowest equivalent cost or largest equivalent income.

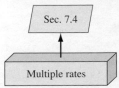

Rate of return (interest rate): The MARR value must be stated to complete the PW, FW, or AW method. This is also correct for the discount rate for public sector alternatives analyzed by the B/C ratio. The ROR method requires that the incremental rate be found in order to select one alternative.

It is here that the dilemma of multiple rates appears, if the sign tests indicate that a unique, real number root does not necessarily exist for a non-conventional series.

Decision guideline: The selection of one alternative is accomplished using the general guideline in the rightmost column. Always select the alternative with the *numerically largest PW, FW, or AW value*. This is correct for both revenue and service alternatives. The incremental cash flow methods—ROR and B/C—require that the largest initial cost and incrementally justified alternative be selected, provided it is justified against an alternative that is itself justified. This means that the incremental *i** exceeds MARR, or the incremental B/C exceeds 1.0.

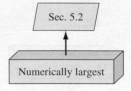

Table 10–2 is also printed on the inner-leaf sheet of the back cover of the text with references to the section(s) where the evaluation method is discussed.

EXAMPLE 10.1

Read through the problem statement of the following examples, neglecting the evaluation method used in the example. Determine which evaluation method is probably the fastest and easiest to apply to select between the two or more alternatives: (*a*) 8.6, (*b*) 6.5, (*c*) 5.6, (*d*) 5.12.

Solution
Refering to the contents of Table 10–1, the following methods should be applied first.

(*a*) Example 8.6 involves four revenue alternatives with equal lives. Use the AW or PW value at the MARR of 10%. (The incremental ROR method was applied in the example.)

(*b*) Example 6.5 requires selection between two public sector alternatives with unequal lives, one of which is very long. The B/C ratio of AW values is the best choice. (This is how the problem was solved.)

(*c*) Since Example 5.6 involves two service alternatives with one having a long life, either AW or PW can be used. Since one life is long, capitalized cost, as an extension of the PW method, is best in this case. (This is the method applied in the example.)

(*d*) Significantly unequal lives are present for two service alternatives in Example 5.12. The AW method is the clear choice here. (The PW method over the LCM was used.)

10.2 MARR RELATIVE TO THE COST OF CAPITAL

The MARR value used in alternative evaluation is one of the most important parameters of a study. In Chapter 1 the MARR was described relative to the weighted costs of debt and equity capital. This section and the next four sections explain how to establish a MARR under varying conditions.

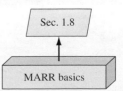

To form the basis for a realistic MARR, the cost of each type of capital financing is initially computed separately, and then the proportion from debt and equity sources is weighted to estimate the average interest rate paid for investment capital. This percentage is called the *cost of capital*. The MARR is then set relative to it. Additionally, the financial health of the corporation, the expected return on invested capital, and many other factors are considered when the MARR is established. If no specific MARR is established, the de facto MARR is set by the project's net cash flow estimates and the availability of capital funds. That is, in reality, the MARR is the *opportunity cost*, which is the i^* of the first project rejected due to the unavailability of funds.

Before we discuss cost of capital, the two primary sources of capital are reviewed.

Debt capital represents borrowing from outside the company, with the principal repaid at a stated interest rate following a specified time schedule. Debt financing includes borrowing via *bonds, loans,* and *mortgages.* The lender does not share in the profits made using the debt funds, but there is risk in that the borrower could default on part of or all the borrowed funds. The amount of outstanding debt financing is indicated in the liabilities section of the corporate balance sheet.

Equity capital is corporate money comprised of the funds of owners and retained earnings. Owners' funds are further classified as common and preferred stock proceeds or owners' capital for a private (non-stock-issuing) company. Retained earnings are funds previously retained in the corporation for capital investment. The amount of equity is indicated in the net worth section of the corporate balance sheet.

To illustrate the relation between cost of capital and MARR, assume a computer system project will be completely financed by a $5,000,000 bond issue (100% debt financing) and assume the dividend rate on the bonds is 8%. Therefore, the cost of debt capital is 8% as shown in Figure 10–1. This 8% is the minimum for MARR. Management may increase this MARR in increments that reflect its desire for added return and its perception of risk. For example, management may add an amount for all capital commitments in this area. Suppose this amount is 2%. This increases the expected return to 10% (Figure 10–1). Also, if

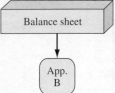

Figure 10–1
A fundamental relation between cost of capital and MARR used in practice.

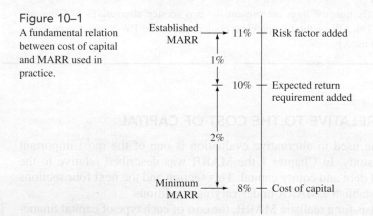

the risk associated with the investment is considered substantial enough to warrant an additional 1% return requirement, the final MARR is 11%.

The recommended approach does not follow the logic presented above. Rather the cost of capital (8% here) should be the established MARR. Then the i^* value is determined from the estimated net cash flows. Using this approach, suppose the computer system is estimated to return 11%. Now, additional return requirements and risk factors are considered to determine if 3% above the MARR of 8% is sufficient to justify the capital investment. After these considerations, if the project is rejected, the effective MARR is now 11%. This is the opportunity cost discussed previously—the rejected project i^* has established the effective MARR for computer system alternatives at 11%, not 8%.

The setting of the MARR for an economy study is not an exact process. The debt and equity capital mix changes over time and between projects. Also, the MARR is not a fixed value established corporatewide. It is altered for different opportunities and types of projects. For example, a corporation may use a MARR of 10% for evaluating the purchase of assets (equipment, cars) and a MARR of 20% for expansion investments, such as purchasing smaller companies.

The effective MARR varies from one project to another and through time because of factors such as the following:

Project risk.   Where there is greater risk (perceived or actual) associated with proposed projects, the tendency is to set a higher MARR. This is encouraged by the higher cost of debt capital for projects considered risky. This usually means that there is some concern that the project will not realize its projected revenue requirements.

Investment opportunity.   If management is determined to expand in a certain area, the MARR may be lowered to encourage investment with the hope of recovering lost revenue in other areas. This common reaction to investment opportunity can create havoc when the guidelines for setting on MARR are too strictly applied. Flexibility becomes very important.

Tax structure.   If corporate taxes are rising (due to increased profits, capital gains, local taxes, etc.), pressure to increase the MARR is present. Use of after-tax analysis may assist in eliminating this reason for a fluctuating MARR, since accompanying business expenses will tend to decrease taxes and after-tax costs.

Limited capital.   As debt and equity capital become limited, the MARR is increased. If the demand for limited capital exceeds supply, the MARR may tend to be set even higher. The opportunity cost has a large role in determining the MARR actually used.

Market rates at other corporations.   If the MARR increases at other corporations, especially competitors, a company may alter its MARR upward in response. These variations are often based on changes in interest rates for loans, which directly impact the cost of capital.

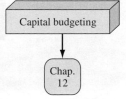

If the details of after-tax analysis are not of interest, but the effects of income taxes are important, the MARR may be increased by incorporating an effective

tax rate using the formula

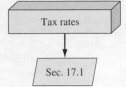

Tax rates

Sec. 17.1

$$\text{Before-tax MARR} = \frac{\text{after-tax MARR}}{1 - \text{tax rate}}$$

The total or effective tax rate, including federal, state, and local taxes, for most corporations is in the range of 30% to 50%. If an after-tax rate of return of 10% is required and the effective tax rate is 35%, the MARR for the before-tax economic analysis is $10\%/(1 - 0.35) = 15.4\%$.

EXAMPLE 10.2

Twin brother and sister, Carl and Christy graduated several years ago from college. Carl, an architect, has worked in home design with Bulte Homes since graduation. Christy, a civil engineer, works with Butler Industries in structural components and analysis. They both reside in Richmond, Virginia. They have started a creative e-commerce network through which Virginia-based builders can buy their "spec home" plans and construction materials much more cheaply. Carl and Christy want to expand into a regional e-business corporation. They have gone to the Bank of America (BA) in Richmond for a business development loan. Identify some factors that might cause the loan rate to vary when BA provides the quote. Also, indicate any impact on the established MARR when Carl and Christy make economic decisions for their business.

Solution

In all cases the direction of the loan rate and the MARR will be the same. Using the five factors mentioned above, some loan rate considerations are as follows:

Project risk: The loan rate may increase if there has been a noticeable downturn in housing starts, thus reducing the need for the e-commerce connection.

Investment opportunity: The rate could increase if other companies offering similar services have already applied for a loan at other BA branches regionally or nationwide.

Taxes: If the state recently removed house construction materials from the list of items subject to sales tax, the rate might be lowered slightly.

Capital limitation: Assume the computer equipment and software rights held by Carl and Christy were bought with their own funds and there are no outstanding loans. If additional equity capital is not available for this expansion, the rate for the loan (debt capital) should be lowered.

Market loan rates: The local BA branch probably obtains its development loan money from a large national pool. If market loan rates to this BA branch have increased, the rate for this loan will likely increase, because money is becoming "tighter."

10.3   DEBT-EQUITY MIX AND WEIGHTED AVERAGE COST OF CAPITAL

The *debt-to-equity (D-E) mix* identifies the percentages of debt and equity financing for a corporation. A company with a 40–60 D-E mix has 40% of its capital originating from debt capital sources (bonds, loans, and mortgages) and 60% derived from equity sources (stocks and retained earnings).

Most projects are funded with a combination of debt and equity capital made available specifically for the project or taken from a corporate *pool of capital.* The *weighted average cost of capital (WACC)* of the pool is estimated by the relative fractions from debt and equity sources. If known exactly, these fractions are used to estimate WACC; otherwise the historical fractions for each source are used in the relation

$$\textbf{WACC} = \textbf{(equity fraction)(cost of equity capital)}$$
$$\textbf{+ (debt fraction)(cost of debt capital)} \qquad \textbf{[10.1]}$$

The two *cost* terms are expressed as percentage interest rates.

Since virtually all corporations have a mixture of capital sources, the WACC is a value between the debt and equity costs of capital. If the fraction of each type of equity financing—common stock, preferred stock, and retained earnings—is known, Equation [10.1] is expanded.

$$WACC = \text{(common stock fraction)(cost of common stock capital)}$$
$$+ \text{(preferred stock fraction)(cost of preferred stock capital)}$$
$$+ \text{(retained earnings fraction)(cost of retained earnings capital)}$$
$$+ \text{(debt fraction)(cost of debt capital)} \qquad [10.2]$$

Figure 10–2 indicates the usual shape of cost of capital curves. If 100% of the capital is derived from equity or 100% is from debt sources, the WACC equals the cost of capital of that source of funds. There is virtually always a mixture of capital sources involved for any capitalization program. As an illustration only, Figure 10–2 indicates a minimum WACC at about 45% debt capital. Most firms operate over a range of D-E mixes. For example, a range of 30% to 50% debt

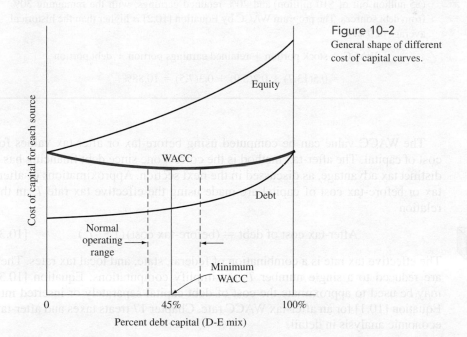

Figure 10–2
General shape of different cost of capital curves.

financing for some companies may be very acceptable to lenders, with no increases in risk or MARR. However, another company may be considered "risky" with only 20% debt capital. It takes knowledge about management ability, current projects, and the economic health of the specific industry to determine a reasonable operating range of the D-E mix for a particular company.

EXAMPLE 10.3

A new program in genetics engineering at Gentex will require $10 million in capital. The chief financial officer (CFO) has estimated the following amounts of financing at the indicated interest rates:

Common stock sales	$5 million at 13.7%
Use of retained earnings	$2 million at 8.9%
Debt financing through bonds	$3 million at 7.5%

Historically, Gentex has financed projects using a D-E mix of 40% from debt sources costing 7.5%, and 60% from equity sources costing 10.0%. Compare the historical WACC value with that for this current genetics program.

Solution

Equation [10.1] is used to estimate the historical WACC.

$$WACC = 0.6(10) + 0.4(7.5) = 9.0\%$$

For the current program, the equity financing is comprised of 50% common stock ($5 million out of $10 million) and 20% retained earnings, with the remaining 30% from debt sources. The program WACC by Equation [10.2] is higher than the historical average of 9%.

$$WACC = \text{stock portion} + \text{retained earnings portion} + \text{debt portion}$$

$$= 0.5(13.7) + 0.2(8.9) + 0.3(7.5) = 10.88\%$$

The WACC value can be computed using before-tax or after-tax values for cost of capital. The after-tax method is the correct one since debt financing has a distinct tax advantage, as discussed in the next section. Approximations of after-tax or before-tax cost of capital are made using the effective tax rate T_e in the relation

$$\text{After-tax cost of debt} = (\text{before-tax cost})(1 - T_e) \qquad [10.3]$$

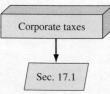

Corporate taxes

Sec. 17.1

The effective tax rate is a combination of federal, state, and local tax rates. They are reduced to a single number T_e to simplify computations. Equation [10.3] may be used to approximate the cost of debt capital separately or inserted into Equation [10.1] for an after-tax WACC rate. Chapter 17 treats taxes and after-tax economic analysis in detail.

10.4 DETERMINATION OF THE COST OF DEBT CAPITAL

Debt financing includes borrowing, primarily via bonds and loans. In the United States, bond interest and loan interest payments are tax-deductible as a corporate expense. This reduces the taxable income base upon which taxes are calculated, with the end result of less taxes paid. The cost of debt capital is, therefore, reduced because there is an annual *tax savings* of the expense cash flow times the effective tax rate T_e. This tax savings is subtracted from the debt-capital expense cash flow in order to calculate the cost of debt capital. In formula form,

$$\text{Tax savings} = \text{(expenses)(effective tax rate)} = \text{expenses } (T_e) \quad \text{[10.4]}$$

$$\text{Net cash flow} = \text{expenses} - \text{tax savings} = \text{expenses } (1 - T_e) \quad \text{[10.5]}$$

To find the cost of debt capital, develop a PW or AW-based relation of the net cash flow (NCF) series with i^* as the unknown. Find i^* by hand through trial and error, or by the RATE or IRR functions on a computer spreadsheet. This is the cost of debt capital percentage used in the WACC computation in Equation [10.1].

EXAMPLE **10.4**

AT&T will generate $5 million in debt capital by issuing five thousand $1000 8% per year 10-year bonds. If the effective tax rate of the company is 50% and the bonds are discounted 2% for quick sale, compute the cost of debt capital (*a*) before taxes and (*b*) after taxes from the company perspective. Obtain the answers by hand and by computer.

Solution by Hand

(*a*)   The annual bond dividend is $1000(0.08) = $80, and the 2% discounted sales price is $980 now. Using the company perspective, find the i^* in the PW relation

$$0 = 980 - 80(P/A,i^*,10) - 1000(P/F,i^*,10)$$

$$i^* = 8.3\%$$

The before-tax cost of debt capital is $i^* = 8.3\%$, which is slightly higher than the 8% bond interest rate, because of the 2% sales discount.

(*b*)   With the allowance to reduce taxes by deducting the bond interest, Equation [10.4] shows a tax savings of $80(0.5) = $40 per year. The bond dividend amount for the PW relation is now $80 - $40 = $40. Solving for i^* after taxes reduces the cost of debt capital by nearly one-half, to 4.25%.

Q-Solv

Solution by Computer

Figure 10–3 is a spreadsheet image for both before-tax (column B) and after-tax (column C) analysis using the IRR function. The after-tax net cash flow is calculated using Equation [10.5] with $T_e = 0.5$. See the cell tag.

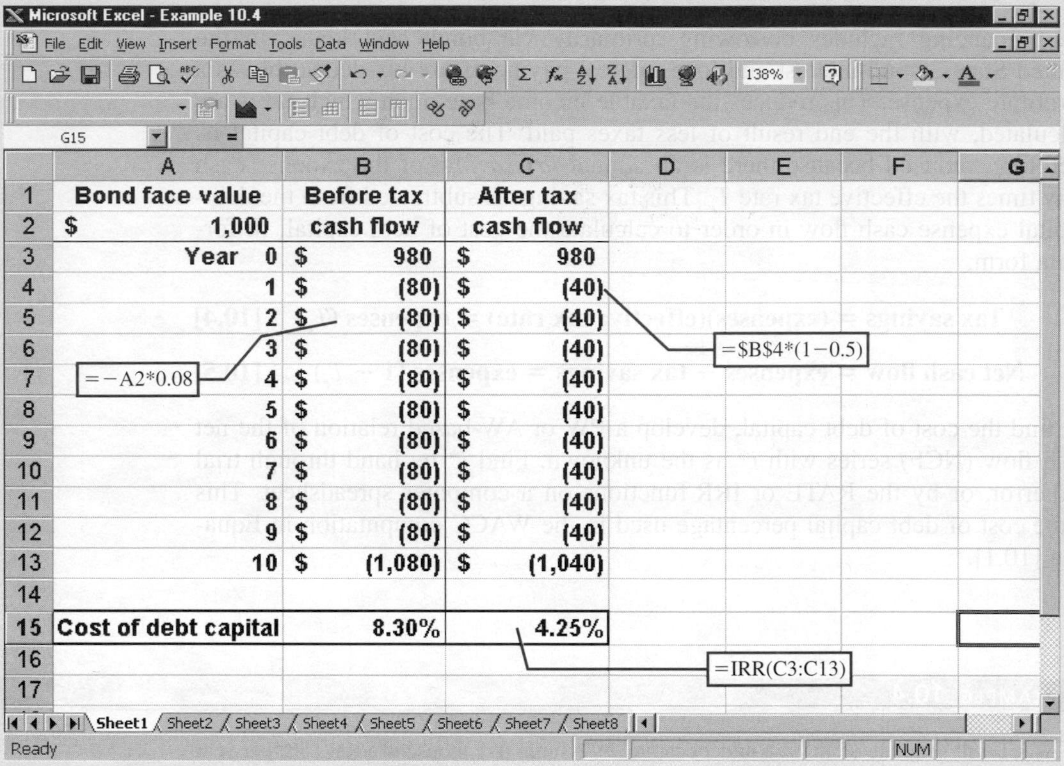

	A		B	C
1	Bond face value		Before tax	After tax
2	$	1,000	cash flow	cash flow
3	Year	0	$ 980	$ 980
4		1	$ (80)	$ (40)
5		2	$ (80)	$ (40)
6		3	$ (80)	$ (40)
7		4	$ (80)	$ (40)
8		5	$ (80)	$ (40)
9		6	$ (80)	$ (40)
10		7	$ (80)	$ (40)
11		8	$ (80)	$ (40)
12		9	$ (80)	$ (40)
13		10	$ (1,080)	$ (1,040)
14				
15	Cost of debt capital		8.30%	4.25%

Figure 10–3

Use of the IRR function to determine before-tax and after-tax cost of debt capital, Example 10.4.

EXAMPLE 10.5

The Hershey Company will purchase a $20,000 10-year-life asset. Company managers have decided to put $10,000 down now and borrow $10,000 at an interest rate of 6%. The simplified loan repayment plan is $600 in interest each year, with the entire $10,000 principal paid in year 10. What is the after-tax cost of debt capital if the effective tax rate is 42%?

Solution

The after-tax net cash flow for interest on the $10,000 loan is an annual amount of 600(1 − 0.42) = $348 by Equation [10.5]. The loan repayment is $10,000 in year 10. PW is used to estimate a cost of debt capital of 3.48%.

$$0 = 10{,}000 - 348(P/A,i^*,10) - 10{,}000(P/F,i^*,10)$$

10.5 DETERMINATION OF THE COST OF EQUITY CAPITAL AND THE MARR

Equity capital is usually obtained from the following sources:

Sale of preferred stock

Sale of common stock

Use of retained earnings

The cost of each type of financing is estimated separately and entered into the WACC computation. A summary of one commonly accepted way to estimate each source's cost of capital is presented here. There are additional methods for estimating the cost of equity capital via common stock. *There are no tax savings for equity capital, because dividends paid to stockholders are not tax-deductible.*

Issuance of *preferred stock* carries with it a commitment to pay a stated dividend annually. The cost of capital is the stated dividend percentage, for example, 10%, or the dividend amount divided by the price of the stock. A $20 dividend paid on a $200 share is a 10% cost of equity capital. Preferred stock may be sold at a discount to speed the sale, in which case the actual proceeds from the stock should be used as the denominator. For example, if a 10% dividend preferred stock with a value of $200 is sold at a 5% discount for $190 per share, there is a cost of equity capital of ($20/$190) × 100% = 10.53%.

Estimating the cost of equity capital for *common stock* is more involved. The dividends paid are not a true indication of what the stock issue will actually cost in the future. Usually a valuation of the common stock is used to estimate the cost. If R_e is the cost of equity capital (in decimal form),

$$R_e = \frac{\text{first-year dividend}}{\text{price of stock}} + \text{expected dividend growth rate}$$

$$R_e = \frac{DV_1}{P} + g \qquad\qquad [10.6]$$

The growth rate g is an estimate of the annual increase in returns that the shareholders receive. Stated another way, it is the compound growth rate on dividends that the company believes is required to attract stockholders. For example, assume a multinational corporation plans to raise capital through its U.S. subsidiary for a new plant in South America by selling $2,500,000 worth of common stock valued at $20 each. If a 5% or $1 dividend is planned for the first year and an appreciation of 4% per year is anticipated for future dividends, the cost of capital for this common stock issue from Equation [10.6] is 9%.

$$R_e = \frac{1}{20} + 0.04 = 0.09$$

The *retained earnings* cost of equity capital is usually set equal to the common stock cost, since it is the shareholders who will realize any returns from projects in which retained earnings are invested.

Once the cost of capital for all planned equity sources is estimated, the WACC is calculated using Equation [10.2].

A second method used to estimate the cost of common stock capital is the *capital asset pricing model (CAPM)*. Because of the fluctuations in stock prices and the higher return demanded by some corporations' stocks compared to others, this valuation technique is commonly applied. The cost of equity capital from common stock R_e, using CAPM, is

$$R_e = \text{risk-free return} + \text{premium above risk-free return}$$
$$= R_f + \beta(R_m - R_f) \qquad [10.7]$$

where β = volatility of a company's stock relative to other stocks in the market (β = 1.0 is the norm)

R_m = return on stocks in a defined market portfolio measured by a prescribed index

Sec. 5.8

Safe investments

The term R_f is usually the quoted U.S. Treasury bill rate, since it is considered a "safe investment." The term $(R_m - R_f)$ is the premium paid above the safe or risk-free rate. The coefficient β (beta) indicates how the stock is expected to vary compared to a selected portfolio of stocks in the same general market area, often the Standard and Poor's 500 stock index. If $\beta < 1.0$, the stock is less volatile, so the resulting premium can be smaller; when $\beta > 1.0$, larger price movements are expected, so the premium is increased.

Security is a word which identifies a stock, bond, or any other instrument used to develop capital. To better understand how CAPM works, consider Figure 10–4. This is a plot of a market security line, which is a linear fit by

Figure 10–4
Expected return on common stock issue using CAPM

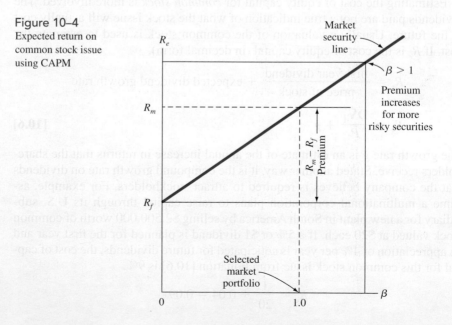

regression analysis to indicate the expected return for different β values. When $\beta = 0$, the risk-free return R_f is acceptable (no premium). As β increases, the premium return requirement grows. Beta values are published periodically for most stock-issuing corporations. Once complete, this estimated cost of common stock equity capital can be included in the WACC computation in Equation [10.2].

EXAMPLE 10.6

The lead software engineer at SafeSoft, a food industry service corporation, has convinced the president to develop new software technology for the meat and food safety industry. It is envisioned that processes for prepared meats can be completed more safely and faster using this automated control software. A common stock issue is a possibility to raise capital if the cost of equity capital is below 15%. SafeSoft, which has a historical beta value of 1.7, uses CAPM to determine the premium of its stock compared to other software corporations. The security market line indicates that a 5% premium above the risk-free rate is desirable. If U.S. Treasury bills are paying 4%, estimate the cost of common stock capital.

Solution
The premium of 5% represents the term $R_m - R_f$ in Equation [10.7].

$$R_e = 4.0 + 1.7(5.0) = 12.5\%$$

Since this cost is lower than 15%, SafeSoft should issue common stock to finance this new venture.

In theory, a correctly performed engineering economy study uses a MARR equal to the cost of the capital committed to the specific alternatives in the study. Of course, such detail is not known. For a combination of debt and equity capital, the calculated WACC sets the minimum for the MARR. The most rational approach is to set MARR between the cost of equity capital and the corporation's WACC. The risks associated with an alternative should be treated separately from the MARR determination, as stated earlier. This supports the guideline that the MARR should not be arbitrarily increased to account for the various types of risk associated with the cash flow estimates. Unfortunately, the MARR is often set above the WACC because management does want to account for risk by increasing the MARR.

EXAMPLE 10.7

The Engineering Products Division of 4M Corporation has two mutually exclusive alternatives A and B with ROR values of $i_A^* = 9.2\%$ and $i_B^* = 5.9\%$. The financing scenario is yet unsettled, but it will be one of the following: plan 1—use all equity funds, which are currently earning 8% for the corporation; plan 2—use funds from the corporate capital pool which is 25% debt capital costing 14.5% and the remainder from the same equity funds mentioned above. The cost of debt capital is currently high because the company has narrowly missed its projected revenue on common stock for the last two quarters, and banks have increased the borrowing rate for 4M. Make the economic decision on alternative A versus B under each financing scenario.

Solution
The capital is available for one of the two mutually exclusive alternatives. For plan 1, 100% equity, the financing is specifically known, so the cost of equity capital is the MARR, that is, 8%. Only alternative A is acceptable; alternative B is not since the estimated return of 5.9% does not exceed this MARR.

Under financing plan 2, with a D-E mix of 25–75,

$$\text{WACC} = 0.25(14.5) + 0.75(8.0) = 9.625\%$$

Now, neither alternative is acceptable since both ROR values are less than MARR = WACC = 9.625%. The selected alternative should be to do nothing, unless one alternative absolutely must be selected, in which case noneconomic attributes must be considered.

10.6   EFFECT OF DEBT-EQUITY MIX ON INVESTMENT RISK

The D-E mix was introduced in Section 10.3. As the proportion of debt capital increases, the calculated cost of capital decreases due to the tax advantages of debt capital. *However, the leverage offered by larger debt capital percentages increases the riskiness of projects undertaken by the company.* When large debts are already present, additional financing using debt (or equity) sources gets more difficult to justify, and the corporation can be placed in a situation where it owns a smaller and smaller portion of itself. This is sometimes referred to as a *highly leveraged* corporation. Inability to obtain operating and investment capital means increased difficulty for the company and its projects. Thus, a reasonable balance between debt and equity financing is important for the financial health of a corporation. Example 10.8 illustrates the disadvantages of unbalanced D-E mixes.

EXAMPLE 10.8

Three manufacturing companies have the following debt and equity capital amounts and D-E mixes. Assume all equity capital is in the form of common stock.

	Amount of Capital		
Company	Debt ($ in millions)	Equity ($ in millions)	D-E Mix (%−%)
A	10	40	20−80
B	20	20	50−50
C	40	10	80−20

Assume the annual revenue is $15 million for each one and that, after interest on debt is considered, the net incomes are $14.4, $13.4, and $10.0 million, respectively. Compute the return on common stock for each company, and comment on the return relative to the D-E mixes.

Solution
Divide the net income by the stock (equity) amount to compute the common stock return. In million dollars,

A: $\text{Return} = \dfrac{14.4}{40} = 0.36$ (36%)

B: $\text{Return} = \dfrac{13.4}{20} = 0.67$ (67%)

C: $\text{Return} = \dfrac{10.0}{10} = 1.00$ (100%)

As expected, the return is by far the largest for highly leveraged C, where only 20% of the company is in the hands of the ownership. The return is excellent, but the risk associated with this firm is high compared to A, where the D-E mix is only 20% debt.

The use of large percentages of debt financing greatly *increases the risk* taken by lenders and stock owners. Long-term confidence in the corporation diminishes, no matter how large the short-term return on stock.

The leverage of large D-E mixes does increase the return on *equity capital,* as shown in previous examples; but it can also work against the owners and investors. A small percentage decrease in asset value will more negatively affect a highly debt-leveraged investment compared to one with small leveraging. Example 10.9 illustrates this fact.

EXAMPLE 10.9

Two engineers place $10,000 each in different investments. Marylynn invests $10,000 in airline stock, and Carla leverages the $10,000 by purchasing a $100,000 residence to be used as rental property. Compute the resulting value of the $10,000 equity capital if there is a 5% decrease in the value of both the stock and the residence. Do the same for a 5% increase. Neglect any dividend, income, or tax considerations.

Solution
The airline stock value decreases by $10,000(0.05) = 500, and the house value decreases by $100,000(0.05) = 5000. The effect is that a smaller amount of the $10,000 is returned, if the investment must be sold immediately.

Marylynn's loss: $\dfrac{500}{10,000} = 0.05$ (5%)

Carla's loss: $\dfrac{5000}{10,000} = 0.50$ (50%)

The 10-to-1 leveraging by Carla gives her a 50% decrease in the equity position, while Marylynn has only the 5% loss since there is no leveraging.

> The opposite is correct for a 5% increase; Carla would benefit by a 50% gain on her $10,000, while Marylynn has only a 5% gain. The larger leverage is more risky. It offers a much *higher return for an increase* in the value of the investment and a much *larger loss for a decrease* in the value of the investment.

The same principles as discussed above for corporations are applicable to individuals. The person who is highly leveraged has large debts in terms of credit card balances, personal loans, and house mortgages. As an example, assume two engineers each have a take-home amount of $40,000 after all income tax, social security, and insurance premiums are deducted from their annual salaries. Further, assume that the cost of the debt (money borrowed via credit cards and loans) averages 15% per year and that the current debt principal is being repaid in equal amounts over 20 years. If Jamal has a total debt of $25,000 and Barry owes $100,000, the remaining amount of the annual take-home pay may be calculated as follows:

Person	Total Debt, $	Cost of Debt at 15%, $	Repayment of Debt over 20-Year Period, $	Amount Remaining from $40,000, $
Jamal	25,000	3,750	1,250	35,000
Barry	100,000	15,000	5,000	20,000

Jamal has 87.5% of his base available while Barry has only 50% available.

10.7 MULTIPLE ATTRIBUTE ANALYSIS: IDENTIFICATION AND IMPORTANCE OF EACH ATTRIBUTE

In Chapter 1 the role and scope of engineering economy in decision making were outlined. The decision-making process explained in that chapter included the seven steps listed on the right side of Figure 10–5. Step 4 is to identify the one or multiple attributes upon which the selection criteria are based. In all prior evaluations in this text, only one attribute—the economic one—has been identified

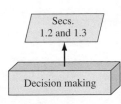

Figure 10–5
Expansion of the decision-making process to include multiple attributes.

Consider multiple attributes

4-1. Identify the attributes for decision making.
4-2. Determine the relative importance (weights) of attributes.
4-3. For each alternative, determine each attribute's value rating.

5. Evaluate each alternative using a multiple-attribute technique. Use sensitivity analysis for key attributes.

Emphasis on one attribute

1. Understand the problem; define the objective.
2. Collect relevant information.
3. Define alternatives; make estimates.
4. Identify the selection criteria (one or more attributes).

5. Evaluate each alternative; use sensitivity analysis.

6. Select the best alternative.
7. Implement the solution and monitor results.

and used to select the best alternative. The criterion has been the maximization of the equivalent value of PW, AW, ROR, or the B/C ratio. As we are all aware, most evaluations do and should take into account multiple attributes in decision making. These are the factors labeled as noneconomic at the bottom of Figure 1–1, which describes the primary elements in performing an engineering economy study. However, these noneconomic dimensions tend to be intangible and often difficult, if not impossible, to quantify directly with economic and other scales. Nonetheless, among the many attributes that can be identified, there are key ones that must be considered in earnest before the alternative selection process is complete. This and the next section describe some of the techniques that accommodate multiple attributes in an engineering study.

Multiple attributes enter into the decision-making process in many studies. Public sector projects are excellent examples of multiple-attribute problem solving. For example, the proposal to construct a dam to form a lake in a low-lying area or to widen the catch basin of a river usually has several purposes, such as flood control; drinking water; industrial use; commercial development; recreation; nature conservation for fish, plants, and birds; and possibly other less obvious purposes. High levels of complexity are introduced into the selection process by the multiple attributes thought to be important in selecting an alternative for the dam's location, design, environmental impact, etc.

The left side of Figure 10–5 expands steps 4 and 5 to consider multiple attributes. The discussion below concentrates on the expanded step 4, and the next section focuses on the evaluation measure and alternative selection of step 5.

4-1 Attribute Identification Attributes to be considered in the evaluation methodology can be identified and defined by several methods, some much better than others depending upon the situation surrounding the study itself. To seek input from individuals other than the analyst is important; it helps focus the study on key attributes. The following is an incomplete listing of ways in which key attributes are identified.

- Comparison with similar studies that include multiple attributes.
- Input from experts with relevant past experience.
- Surveys of constituencies (customers, employees, managers) impacted by the alternatives.
- Small group discussions using approaches such as focus groups, brainstorming, or nominal group technique.
- Delphi method, which is a progressive procedure to develop reasoned consensus from different perspectives and opinions.

As an illustration, assume that Continental Airlines has decided to purchase five new Boeing 777s for overseas flights, primarily between the North American west coast and Asian cities, principally Hong Kong, Tokyo, and Singapore. There are approximately 8000 options for each plane that must be decided upon by engineering, purchasing, maintenance, and marketing personnel at Continental before the order to Boeing is placed. Options range in scope from the material and color of the plane's interior to the type of latching devices used on the engine cowlings, and in function from maximum engine thrust to pilot instrument design.

An economic study based on the equivalent AW of the estimated passenger income per trip has determined that 150 of these options are clearly advantageous. But other noneconomic attributes are to be considered before some of the more expensive options are specified. A Delphi study was performed using input from 25 individuals. Concurrently, option choices for another, unidentified airline's recent order were shared with Continental personnel. From these two studies it was determined that there are 10 key attributes for options selection. Four of the most important attributes are

- *Repair time:* mean time to repair or replace (MTTR) if the option is or affects a flight-critical component.
- *Safety:* mean time to failure (MTTF) of flight-critical components.
- *Economic:* estimated extra revenue for the option. (Basically, this is the attribute evaluated by the economy study already performed.)
- *Crewmember needs:* some measure of the necessity and/or benefits of the option as judged by representative crewmembers—pilots and attendants.

The economic attribute of extra revenue may be considered an indirect measure of customer satisfaction, one that is more quantitative than customer opinion/satisfaction survey results. Of course, there are many other attributes that can be, and are, used. However, the point is that the economic study may directly address only one or a few of the key attributes vital to alternative decision making.

An attribute routinely identified by individuals and groups is *risk.* Actually, risk is not a stand-alone attribute, because it is a part of every attribute in one form or another. Considerations of variation, probabilistic estimates, etc., in the decision-making process are treated later in this text. Formalized sensitivity analysis, expected values, simulation, and decision trees are some of the techniques useful in handling the risk inherent in an attribute.

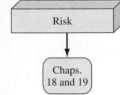

4-2 Importance (Weights) for the Attributes Determination of the *extent of importance* for each attribute i results in a weight W_i that is incorporated into the final evaluation measure. The weight, a number between 0 and 1, is based upon the experienced opinion of one individual or a group of persons familiar with the attributes, and possibly the alternatives. If a group is utilized to determine the weights, there must be consensus among the members for each weight. Otherwise, some averaging technique must be applied to arrive at one weight value for each attribute.

Table 10–3 is a tabular layout of attributes and alternatives used to perform a multiple attribute evaluation. Weights W_i for each attribute are entered on the left side. The remainder of the table is discussed as we proceed through steps 4 and 5 of the expanded decision-making process.

Attribute weights are usually normalized such that their sum over all the alternatives is 1.0. This normalizing implies that each attribute's importance score is divided by the sum S over all attributes. Expressed in formula form, these two properties of weights for attribute i ($i = 1, 2, \ldots, m$) are

$$\text{Normalized weights:} \quad \sum_{i=1}^{m} W_i = 1.0 \qquad [10.8]$$

		Alternatives				
Attributes	**Weights**	**1**	**2**	**3**	$\cdots$	**n**
1	W_1					
2	W_2					
3	W_3		Value ratings V_{ij}			
$\vdots$	$\vdots$					
m	W_m					

TABLE 10–3   Tabular Layout of Attributes and Alternatives Used for Multiple Attribute Evaluation

Weight calculation: $W_i = \dfrac{\text{importance score}_i}{\sum\limits_{i=1}^{m} \text{importance score}_i} = \dfrac{\text{importance score}_i}{S}$ [10.9]

Of the many procedures developed to assign weights to an attribute, an analyst is likely to rely upon one which is relatively simple, such as equal weighting, rank order, or weighted rank order. Each is briefly presented below.

Equal Weighting   All attributes are considered to be of approximately the same importance, or there is no rationale to distinguish the more important from the less important attribute. This is the default approach. Each weight in Table 10–3 will be $1/m$, according to Equation [10.9]. Alternatively, the normalizing can be omitted, in which case each weight is 1 and their sum is m. In this case, the final evaluation measure for an alternative will be the sum over all attributes.

Rank Order   The m attributes are placed (ranked) in order of increasing importance with a score of 1 assigned to the least important and m assigned to the most important. By Equation [10.9], the weights follow the pattern $1/S, 2/S, \ldots, m/S$. With this method, the difference in weights between attributes of increasing importance is constant.

Weighted Rank Order   The m attributes are again placed in the order of increasing importance. However, now differentiation between attributes is possible. The most important attribute is assigned a score, usually 100, and all other attributes are scored relative to it between 100 and 0. Now, define the score for each attribute as s_i, and Equation [10.9] takes the form

$$W_i = \frac{s_i}{\sum\limits_{i=1}^{m} s_i}$$ [10.10]

This is a very practical method to determine weights because one or more attributes can be heavily weighted if they are significantly more important than the remaining ones, and Equation [10.10] automatically normalizes the weights. For example, suppose the four key attributes in the previous aircraft purchase example are ordered: safety, repair time, crewmember needs, and economic. If repair time is only half as important as safety, and the last two attributes are each half as important as repair time, the scores and weights are as follows.

Attribute	Score	Weights
Safety	100	100/200 = 0.50
Repair time	50	50/200 = 0.25
Crewmember needs	25	25/200 = 0.125
Economic	25	25/200 = 0.125
Sum of scores and weights	200	1.000

There are other attribute weighting techniques, especially for group processes, such as utility functions, pairwise comparison, and others. These become increasingly sophisticated, but they are able to provide an advantage that these simple methods do not afford the analyst: *consistency of ranks and scores* between attributes and between individuals. If this consistency is important in that several decision makers with diverse opinions about attribute importance are involved in a study, a more sophisticated technique may be warranted. There is substantial literature on this topic.

4-3 Value Rating of Each Alternative, by Attribute   This is the final step prior to calculating the evaluation measure. Each alternative is awarded a value rating V_{ij} for each attribute i. These are the entries within the cells in Table 10–3. The ratings are appraisals by decision makers of how well an alternative will perform as each attribute is considered.

The scale for the value rating can vary depending upon what is easiest to understand for those who do the valuation. A scale of 0 to 100 can be used for attribute importance scoring. However, the most popular is a scale of 4 or 5 gradations about the perceived ability of an alternative to accomplish the intent of the attribute. This is called a *Likert scale,* which can have descriptions for the gradations (e.g., very poor, poor, good, very good), or numbers assigned between 0 and 10, or -1 to $+1$, or -2 to $+2$. The last two scales can give a negative impact to the evaluation measure for poor alternatives. An example numerical scale of 0 to 10 is as follows:

If You Value the Alternative As	Give It a Rating between the Numbers
Very poor	0–2
Poor	3–5
Good	6–8
Very good	7–10

It is preferable to have a Likert scale with four choices (an even number) so that the central tendency of "fair" is not overrated.

If we now build upon the aircraft purchase illustration to include value ratings, the cells are filled with ratings awarded by a decision maker. Table 10–4 includes example ratings V_{ij} and the weights W_i determined above. Initially, there will be one such table for each decision maker. Prior to calculating the final

TABLE 10–4   Completed Layout for Four Attributes and Three Alternatives for Multiple-Attribute Evaluation

Attributes	Weights	Alternatives		
		1	2	3
Safety	0.50	6	4	8
Repair	0.25	9	3	1
Crew needs	0.125	5	6	6
Economic	0.125	5	9	7

evaluation measure R_j, the ratings can be combined in some fashion; or a different R_j can be calculated using each decision maker's ratings. Determination of this evaluation measure is discussed below.

10.8   EVALUATION MEASURE FOR MULTIPLE ATTRIBUTES

The need for an evaluation measure that accommodates multiple attributes is indicated in step 5 of Figure 10–5. This measure should be a single-dimension number that effectively combines the different dimensions addressed by the attribute importance scores W_i and the alternative value ratings V_{ij}. The result is a formula to calculate an aggregated measure that can be used to select from two or more alternatives. The result is often referred to as a *rank-and-rate method*. This reduction process removes much of the complexity of trying to balance the different attributes; however, it also eliminates much of the robust information captured by the process of ranking attributes for their importance and rating each alternative's performance against each attribute.

There are additive, multiplicative, and exponential measures, but by far the most commonly applied is the additive model. The most used additive model is the *weighted attribute method*. The evaluation measure, symbolized by R_j for each alternative j, is defined as

$$R_j = \sum_{j=1}^{n} W_i V_{ij} \qquad \text{[10.11]}$$

The W_i numbers are the attribute importance weights, and V_{ij} is the value rating by attribute i for each alternative j. If the attributes are of equal weight (also called unweighted), all $W_i = 1/m$, as determined by Equation [10.9]. This means that W_i can be moved outside of the summation in the formula for R_j. (If an equal weight of $W_i = 1.0$ is used for all attributes, in lieu of $1/m$, then the R_j value is simply the sum of all ratings for the alternative.)

The selection guideline is as follows:

Choose the alternative with the largest R_j value. This measure assumes that increasing weights W_i mean more important attributes and increasing ratings V_{ij} mean better performance of an alternative.

Sensitivity analysis for any score, weight, or value rating is used to determine sensitivity of the decision to it.

EXAMPLE **10.10**

An interactive regional dispatching and scheduling system for trains has been in place for several years at MB+O Railroad. Management and dispatchers alike agree it is time for an updated software system, and possibly new hardware. Discussions have led to three alternatives:

1. Purchase new hardware and develop new customized software in-house.
2. Lease new hardware and use an outside contractor for software services.
3. Develop new software using an outside contractor, and upgrade specific hardware components for the new software.

Six attributes for alternative comparison have been defined using a Delphi process involving decision makers from dispatching, field operations, and train engineering.

1. Initial investment requirement.
2. Annual cost of hardware and software maintenance.
3. Response time to "collision conditions."
4. User interface for dispatching.
5. On-train software interface.
6. Software system interface with other company dispatching systems.

Attribute scores developed by the decision makers are shown in Table 10–5, using a weighted rank-order procedure with scores between 0 and 100. Attributes 2 and 3 are considered equally the most important attributes; a score of 100 is assigned to them. Once each alternative had been detailed enough to judge the capabilities from the system specifications, a three-person group placed value ratings on the three alternatives, again using a rating scale of 0 to 100 (Table 10–5). As an example, for alternative 3, the economics are excellent (scores of 100 for both attributes 1 and 2), but the on-train software interface is considered very poor, thus the low rating of 10. Use these scores and ratings to determine which alternative is the best to pursue.

TABLE **10–5**   Attribute Scores and Alternative Ratings for Multiple-Attribute Evaluation, Example 10.10

| | | Value Ratings (0 to 100), V_{ij} | | |
Attribute i	Importance Score	Alternative 1	Alternative 2	Alternative 3
1	50	75	50	100
2	100	60	75	100
3	100	50	100	20
4	80	100	90	40
5	50	85	100	10
6	70	100	100	75
Total	450			

TABLE 10–6 Results for the Weighted-Attribute Method, Example 10.10

Attribute i	Normalized Weight W_i	$R_j = W_i V_{ij}$ Alternative 1	Alternative 2	Alternative 3
1	0.11	8.3	5.5	11.0
2	0.22	13.2	16.5	22.0
3	0.22	11.0	22.0	4.4
4	0.18	18.0	16.2	7.2
5	0.11	9.4	11.0	1.1
6	0.16	16.0	16.0	12.0
Totals	1.00	75.9	87.2	57.7

Solution

Table 10–6 includes the normalized weights for each attribute determined by Equation [10.9]; the total is 1.0, as required. The evaluation measure R_j for the weighted attribute method is obtained by applying Equation [10.11] in each alternative column. For alternative 1,

$$R_1 = 0.11(75) + 0.22(60) + \cdots + 0.16(100) = 75.9$$

When the totals are reviewed for the largest measure, alternative 2 is the best choice at $R_2 = 87.2$. Further detailing of this alternative should be recommended to management.

Comment

Any economic measure can be incorporated into a multiple-attribute evaluation using this method. All measures of worth—PW, AW, ROR, B/C—can be included; however, their impact on the final selection will vary relative to the importance placed on the noneconomic attributes.

CHAPTER SUMMARY

The best method to economically evaluate and compare mutually exclusive alternatives is usually either the AW or PW method at the stated MARR. The choice depends, in part, upon the equal lives or unequal lives of the alternatives, and the pattern of the estimated cash flows, as summarized in Table 10–1. Public sector projects are best compared using the B/C ratio, but the economic equivalency is still AW or PW-based. Once the evaluation method is selected, Table 10–2 (also printed at the rear of the text with section references) can be used to determine the elements and decision guideline that must be implemented to correctly perform the study. If the estimated ROR for the selected alternative is needed, it is advisable to determine $i*$ by using the IRR function on a spreadsheet after the AW or PW method has indicated the best alternative.

The interest rate at which the MARR is established depends principally upon the cost of capital and the mix between debt and equity financing. The MARR should be set equal to the weighted average cost of capital (WACC). Risk, profit, and other factors can be considered after the AW, PW, or ROR analysis is completed and prior to final alternative selection.

If multiple attributes, which include more than the economic dimension of a study, are to be considered in making the alternative decision, first the attributes must be identified and their relative importance assessed. Then each alternative can be value-rated for each attribute. The evaluation measure is determined using a model such as the weighted attribute method, where the measure is calculated by Equation [10.11]. The largest value indicates the best alternative.

PROBLEMS

Choosing the Evaluation Method

10.1 When two or more alternatives are compared using the PW, AW, or B/C method, there are three circumstances for which the length of time of the evaluation period is the same for all alternatives. List these three circumstances.

10.2 For what evaluation methods is it mandatory that an incremental cash flow series analysis be performed to ensure that the correct alternative is selected?

10.3 Explain what is meant by the decision guideline to choose the "numerically largest value" when selecting the one best mutually exclusive alternative from two or more alternatives.

10.4 For the following situation, (a) determine which evaluation method is probably the easiest and fastest to apply by hand and by computer in order to select from the five alternatives, and (b) answer the two questions, using your chosen evaluation method.

An independent dirt contractor must determine which size dump truck to buy. The cash flows estimated with each size truck bed are tabulated. The MARR is 18% per year, and all alternatives are expected to have a useful life of 8 years. (1) What size truck bed should be purchased? (2) If two trucks are to be purchased, what should be the size of the second truck?

Truck Bed Size, Cubic Meters	Initial Investment, $	AOC, $/year	Salvage Value, $	Annual Revenue, $/year
8	−10,000	−4,000	+2,000	+6,500
10	−14,000	−5,500	+2,500	+10,000
15	−18,000	−7,000	+3,000	+14,000
20	−24,000	−11,000	+3,500	+20,500
25	−33,000	−16,000	+6,000	+26,500

10.5 Read Problem 9.26. (a) Determine which evaluation method is probably the easiest and fastest to apply by hand and by computer in order to select from the two alternatives. (b) If the evaluation method you chose is different from that used in Chapter 9, solve the problem using your chosen evaluation method.

10.6 For what type of alternatives should the capitalized cost method be used for comparison? Give several examples of these types of projects.

Working with MARR

10.7 After 15 years of employment in the airline industry, John started his own consulting company to use physical and computer simulation in the analysis of commercial airport accidents on runways. He estimates his average cost of new capital at 8% per year for physical simulation projects, that is, where he physically reconstructs the accident using scale versions of planes, buildings, vehicles, etc. He has established 12% per year as his MARR.

    (a) What (net) rate of return on capital investments for physical simulation does he expect?

    (b) John was recently offered an international project that he considers risky in that the information available is sketchy and the airport personnel do not appear to be willing to cooperate on the investigation. John considers this risk to be economically worth at least an added 5% return on his money. What is the recommended MARR in this situation, based upon what you have learned in this chapter? How should John consider the required return and perceived risk factors when evaluating this project opportunity?

10.8 State whether each of the following involves debt financing or equity financing.

    (a) A bond issue for $3,500,000 by a city-owned utility

    (b) An initial public offering (IPO) of $35,000,000 in common stock for a dot-com company

    (c) $25,000 taken from your retirement account to pay cash for a new car

    (d) A homeowner's equity loan for $25,000

10.9 Explain how the opportunity cost sets the effective MARR when, because of limited capital, only one alternative can be selected from two or more.

10.10 The board of directors of the Brazilia Group has less capital than needed to fund a $6.2 million project in the Middle East. Had it been funded, an estimated i^* of 18% per year would result. Corporate MARR = 15% has been applied for before-tax analysis. With the available $2.0 million in equity capital, a project with an estimated i^* of 16.6% is approved. The group president just asked you to estimate the after-tax opportunity cost that has been forgone. Assume you collect the following information in preparation to answer him:

Effective federal tax rate = 20% per year
Effective state tax rate = 6% per year
Equation for overall effective tax rate =
    state rate + (1 − state rate)(federal rate)

(*Hint:* First develop a drawing similar to Figure 1–6 for the Brazilia Group situation.)

10.11 The initial investment and incremental ROR values for four mutually exclusive alternatives are indicated below. Select the best alternative if a maximum of (a) $300,000, (b) $400,000, and (c) $700,000 in capital funds is available and the MARR is the cost of capital, which is estimated at 9% per year. (d) What is the de facto MARR for these alternatives if no specific MARR is stated, the available capital is $400,000, and the opportunity cost interpretation is applied?

Alternative	Initial Investment, $	Incremental Rate of Return, %	Alternative Rate of Return, %
1	−100,000	8.8 for 1 to DN	8.8
2	−250,000	12.5 for 2 to DN	12.5
3	−400,000	11.3 for 3 to 2	14.0
4	−550,000	8.1 for 4 to 3	10.0

10.12 State the recommended approach in establishing the MARR when other factors, such as alternative risk, taxes, and market fluctuations are considered in addition to the cost of capital.

10.13 A partnership of four engineers operates a duplex rental business. Five years ago they purchased a block of duplexes, using a MARR of 14% per year. The estimated return at that time for the duplexes was 15% per year, but the investment was considered very risky due to the poor rental business economy in the city and state overall. Nonetheless, the purchase was made with 100% equity financing at a cost of 10% per year. Fortunately, the return has averaged 18% per year over the 5 years. Another purchase opportunity for more duplexes has now presented itself, but a loan at 8% per year would have to be taken to make the investment. (a) If the economy for rental property has not changed significantly, is there likely to be a tendency to now make the MARR higher than, lower than, or the same as that used previously? Why? (b) What is the recommended way to consider the rental business economy risk now that debt capital would be involved?

D-E Mix and WACC

10.14 A new cross-country, transmountain range water pipeline needs to be built at an estimated first cost of $200,000,000. The consortium of cooperating companies has not fully decided the financial arrangements of this adventurous project. The WACC for similar projects has averaged 10% per year.

(a) Two financing alternatives have been identified. The first requires an investment of 60% equity funds at 12% and a loan for the balance at an interest rate of 9% per year. The second alternative requires only 20% equity funds and the balance obtained by a massive international loan estimated to carry an interest cost of 12.5% per year, which is, in part, based on the geographic location of the pipeline. Which financing plan will result in the smaller average cost of capital?

(b) If the consortium CFOs have decided that the WACC must not exceed the 5-year historical average of 10% per year, what is the maximum loan interest acceptable for each financing alternative?

10.15 A couple is planning ahead for their child's college education. They can fund part or all of the expected $100,000 tuition cost from their own funds (through an Education IRA) or borrow all or part of it. The expected return for their own funds is 8% per year, but the loan is expected to have higher interest rates as the amount of the loan increases. Use a spreadsheet-generated plot of the WACC curve and the estimated loan interest rates below to determine the best D-E mix for the couple.

Loan Amount, $	Estimated Interest Rate, % per year
10,000	7.0
25,000	7.5
50,000	9.0
60,000	10.0
75,000	12.0
100,000	18.0

10.16 Tiffany Baking Co. wants to arrange for $50 million in capital for manufacturing a new consumer product. The current financing plan is 60% equity capital and 40% debt financing. Compute the WACC for the following financing scenario.

Equity capital: 60%, or $35 million, via common stock sales for 40% of this amount that will pay dividends at a rate of 5% per year, and the remaining 60% from retained earnings, which currently earn 9% per year.

Debt capital: 40%, or $15 million, obtained through two sources— bank loans for $10 million borrowed at 8% per year, and the remainder in convertible bonds at an estimated 10% per year bond interest rate.

10.17 The possible D-E mixes and costs of debt and equity capital for a new project are summarized below. Use the data (a) to plot the curves for debt, equity, and weighted average costs of capital and (b) to determine what mix of debt and equity capital will result in the lowest WACC.

Plan	Debt Capital Percentage	Rate, %	Equity Capital Percentage	Rate, %
1	100	14.5		
2	70	13.0	30	7.8
3	65	12.0	35	7.8
4	50	11.5	50	7.9
5	35	9.9	65	9.8
6	20	12.4	80	12.5
7			100	12.5

10.18 For Problem 10.17, use a spreadsheet to (a) determine the best D-E mix and (b) determine the best D-E mix if the cost of debt capital increases by 10% per year.

10.19 A public corporation in which you own common stock reported a WACC of 10.7% for the year in its annual report to stockholders. The common stock that you own has averaged a total return of 6% per year over the last 3 years. The annual report also mentions that projects within the corporation are 80% funded by its own capital. Estimate the company's cost of debt capital. Does this seem like a reasonable rate for borrowed funds?

10.20 To understand the advantage of debt capital from a tax perspective in the United States, determine the before-tax and after-tax weighted average costs of capital if a project is funded 40%–60% with debt capital borrowed at 9% per year. A recent study indicates that corporate equity funds earn 12% per year and that the effective tax rate is 35% for the year.

Cost of Debt Capital

10.21 Bristol Myers Squibb, an international pharmaceutical company, is initiating a new project for which it requires $2.5 million in debt capital. The current plan is to sell 20-year bonds that pay 4.2% per year, payable quarterly, at a 3% discount on the face value. BMS has an effective tax rate of 35% per year. Determine (a) the total face value of the bonds required to obtain $2.5 million and (b) the effective annual after-tax cost of debt capital.

10.22 The Sullivans' plan to purchase a refurbished condo in their parents' hometown for investment purposes. The negotiated $200,000 purchase price will be financed with 20% of their savings which consistently make 6.5% per year after all relevant income taxes are paid. Eighty percent will be borrowed at 9% per year for 15 years with the principal repaid in equal annual installments. If their effective tax rate is 22% per year, based only on these data, answer the following. (Note: The 9% rate on the loan is a before-tax rate.)
(a)   What is the Sullivans' annual loan payment for each of the 15 years?

(b) What is the net present worth difference between the $200,000 now and the PW of the cost of the 80–20 D-E mix series of cash flows necessary to finance the purchase? What does this number mean?

(c) What is the Sullivans' after-tax WACC for this purchase?

10.23 An engineer is working on a design project for a plastics manufacturing company that has an after-tax cost of equity capital of 6% per year for retained earnings that may be used to 100% equity finance the project. An alternative financing strategy is to issue $4 million worth of 10-year bonds that will pay 8% per year interest on a quarterly basis. If the effective tax rate is 40%, which funding source has the lower cost of capital?

10.24 Tri-States Gas Processors expects to borrow $800,000 for field engineering improvements. Two methods of debt financing are possible—borrow it all from a bank or issue debenture bonds. The company will pay an effective 8% compounded per year for 8 years to the bank. The principal on the loan will be reduced uniformly over the 8 years, with the remainder of each annual payment going toward interest. The bond issue will be 800 10-year bonds of $1000 each that require a 6% per year interest payment.

(a) Which method of financing is cheaper after an effective tax rate of 40% is considered?

(b) What is the cheaper method using a before-tax analysis?

10.25 Charity Hospital, established in 1895 as a nonprofit corporation, pays no taxes on income and receives no tax advantage for interest paid. The board of directors has approved expanded cancer treatment equipment that will require $10 million

in debt capital to supplement $8 million in equity capital currently available. The $10 million can be borrowed at 7.5% per year through the Charity Hospital Corporation. Alternatively, 30-year trust bonds could be issued through the hospital's for-profit outpatient corporation, Charity Outreach, Inc. The interest on the bonds is expected to be 9.75% per year, which is tax-deductible. The bonds will be sold at a 2.5% discount for rapid sale. The effective tax rate of Charity Outreach is 32%. Which form of debt financing is less expensive after taxes?

Cost of Equity Capital

10.26 Common stocks issued by Henry Harmon Builders paid stockholders $0.93 per share on an average price of $18.80 last year. The company expects to grow the dividend rate at a maximum of 1.5% per year. The stock volatility of 1.19 is somewhat higher than that of other public firms in the construction industry, and other stocks in this market are paying an average of 4.95% per year dividend. U.S. Treasury bills are returning 4.5%. Determine the company's cost of equity capital last year, using (a) the dividend method and (b) the CAPM.

10.27 Government regulations from the U.S. Department of Agriculture (USDA) require that a Fortune 500 corporation implement the HACCP (Hazards Analysis and Critical Control Points) food safety program in its beef processing plants in 21 states. To finance the equipment and personnel training portions of this new program, Wholesome Chickens expects to use a D-E mix of 60%–40% to finance a $10 million effort for improved equipment, engineering, and quality control. After-tax cost of debt capital for loans is known to be 9.5% per year. However, obtaining sufficient equity capital will require the sale of common stock, as well as the commitment of corporate retained earnings. Use the following information to determine the WACC for the implementation of HACCP.

Common stock: 100,000 shares

Anticipated price = $32 per share
Initial dividend = $1.10 per share
Dividend growth
per share = 2% annually

Retained earnings: same cost of capital as for common stock

10.28 Last year a Japanese engineering materials corporation, Yamachi Inc., purchased some U.S. Treasury bonds that return an average of 4% per year. Now, Euro bonds are being purchased with a realized average return of 3.9% per year. The volatility factor of Yamachi stock last year was 1.10 and has increased this year to 1.18. Other publicly traded stocks in this same business are paying an average of 5.1% dividends per year. Determine the cost of equity capital for each year, and explain why the increase or decrease seems to have occurred.

10.29 An engineering graduate plans to purchase a new car. He has not decided how to pay the purchase price of $28,000 for the SUV he has selected. He has the total available in a savings account, so paying cash is an option; however, this would deplete virtually all his savings. These funds return an average of 6% per year, compounded every 6 months. Perform a before-tax analysis to determine which of the three financing plans below has the lowest WACC.

Plan 1: D-E is 50%–50%. Use $14,000 from the savings account and borrow $14,000 at a rate of 7% per year, compounded monthly. The difference between the payments

and the savings would be deposited at 6% per year, compounded semi-annually.

Plan 2: 100% equity. Take $28,000 from savings now.

Plan 3: 100% debt. Borrow $28,000 now from the credit union at an effective rate of 0.75% per month, and repay the loan at $581.28 per month for 60 months.

10.30 OILogistics.com has a total of 1.53 million shares of common stock outstanding at a market price of $28 per share. The before-tax cost of equity capital of common stock is 15% per year. Stocks fund 50% of the company's capital projects. The remaining capital is generated by equipment trust bonds and short-term loans. Thirty percent of the debt capital is from $5,000,000 worth of $10,000 6% per year 15-year bonds. The remaining 70% of debt capital is from loans repaid at an effective 10.5% before taxes. If the effective income tax rate is 35%, determine the weighted average cost of capital (*a*) before taxes and (*b*) after taxes.

10.31 Three projects have been identified. Capital will be developed 70% from debt sources at an average rate of 7.0% per year and 30% from equity sources at 10.34% per year. Set the MARR equal to WACC and make the economic decision, if the projects are (*a*) independent and (*b*) mutually exclusive.

Project	Initial Investment, $	Annual Net Cash Flow, $/year	Salvage Value, $	Life, Years
1	−25,000	6,000	4,000	4
2	−30,000	9,000	−1,000	4
3	−50,000	15,000	20,000	4

10.32 Shadowland, a manufacturer of airfreightable pet crates, has identified two projects that, though having a relatively high risk, are expected to move the company into new revenue markets. Utilize a spreadsheet solution (*a*) to select any combination of the projects if MARR = after-tax WACC and (*b*) to determine if the same projects should be selected if the risk factors are enough to require an additional 2% per year for the investment to be made.

Project	Initial Investment, $	Estimated After-Tax Cash Flow per Year, $/year	Life, Years
Wildlife (W)	−250,000	48,000	10
Reptiles (R)	−125,000	30,000	5

Financing will be developed using a D-E mix of 60–40 with equity funds costing 7.5% per year. Debt financing will be developed from $10,000 5% per year, paid quarterly, 10-year bonds. The effective tax rate is 30% per year.

10.33 The federal government imposes requirements upon industry in many areas, such as employee safety, pollution control, environmental protection, and noise control. One view of these regulations is that their compliance tends to decrease the return on investment and/or increase the cost of capital to the corporation. In many cases the economics of these regulated compliances cannot be evaluated as regular engineering economy alternatives. Use your knowledge of engineering economic analysis to explain how an engineer might economically evaluate alternatives that define the ways in which the company will comply with imposed regulations.

Different D-E Mixes

10.34 Why is it financially unhealthy for an individual to maintain a large percentage of debt financing over a long period of time, that is, to be highly debt-leveraged?

10.35 Fairmont Industries primarily relies on 100% equity financing to fund projects. A good opportunity is available that will require $250,000 in capital. The Fairmont owner can supply the money from personal investments that currently earn an average of 8.5% per year. The annual net cash flow from the project is estimated at $30,000 for the next 15 years. Alternatively, 60% of the required amount can be borrowed for 15 years at 9% per year. If the MARR is the WACC, determine which plan, if either, is better. This is a before-tax analysis.

10.36 Mrs. McKay has different methods by which a $600,000 project can be funded, using debt and equity capital. A net cash flow of $90,000 per year is estimated for 7 years.

Type of Financing	Financing Plan, %			Cost per Year, %
	1	2	3	
Debt	20	50	60	10
Equity	80	50	40	7.5

Determine the rate of return for each plan, and identify the ones that are economically acceptable if (*a*) MARR equals the cost of equity capital, (*b*) MARR equals the WACC, or (*c*) MARR is halfway between the cost of equity capital and the WACC.

10.37 Mosaic Software has an opportunity to invest $10,000,000 in a new engineering remote-control system for offshore drilling platforms. Financing for Mosaic will be split between common stock sales ($5,000,000) and a loan with an 8% per year interest rate. Mosaic's share of the annual net cash flow is estimated to be $2.0 million for each of the next 6 years. Mosaic is about to initiate CAPM as its common stock evaluation model. Recent analysis shows that it has a volatility rating of 1.05 and is paying a premium of 5% common stock dividend. The U.S. Treasury bills are currently paying 4% per year. Is the venture financially attractive if the MARR equals (a) the cost of equity capital and (b) the WACC?

10.38 Draw the general shape of the three cost of capital curves (debt, equity, and WACC), using the form of Figure 10–2. Draw them under the condition that a high D-E mix has been present for some time for the corporation. Explain via your graph and words the movement of the minimum WACC point under historically high leveraged D-E mixes. *Hint:* High D-E mixes cause the debt cost to increase substantially. This makes it harder to obtain equity funds, so the cost of equity capital also increases.

10.39 In a leveraged buyout of one company by another, the purchasing company usually obtains borrowed money and inserts as little of its own equity funds as possible into the purchase. Explain some circumstances under which such a buyout may put the purchasing company at economic risk.

Multiple-Attribute Evaluation

10.40 A committee of four people submitted the following statements about the attributes to be used in a weighted attribute method. Use the statements to determine the normalized weights if scores are assigned between 0 and 10.

Attribute	Comment
1. Flexibility	The most important factor
2. Safety	50% as important as uptime
3. Uptime	One-half as important as flexibility
4. Speed	As important as uptime
5. Rate of return	Twice as important as safety

10.41 Different types and capacities of crawler hoes are being considered for use in a major excavation on a pipe-laying project. Several supervisors on similar projects of the past have identified some of the attributes and their views of the importance of an attribute. The information has been shared with you. Determine the weighted rank order, using a 0 to 100 scale and the normalized weights.

Attribute	Comment
1. Truck versus hoe loading height	Vitally important factor
2. Type of topsoil	Usually only 10% of the problem
3. Type of soil below topsoil	One-half as important as matching trenching and laying speeds
4. Hoe cycle time	About 75% as important as soil type below topsoil
5. Match hoe trenching speed to pipe-laying speed	As important as attribute number one

10.42 You graduated 2 years ago, and you plan to purchase a new car. For three different models you have evaluated the initial cost and estimated annual costs for fuel and maintenance. You also evaluated the styling of each car in your role as a young engineering professional. List some additional factors (tangible and intangible) that might be used in your version of the weighted attribute method.

10.43 (*Note to instructor:* This and the next two problems may be assigned as a progressive exercise.) John, who works at Swatch, has decided to use the weighted attribute method to compare three systems for manufacturing a watchband. The vice president and her assistant have evaluated each of three attributes in terms of importance to them, and John has placed an evaluation from 0 to 100 on each alternative for the three attributes. John's *ratings* are as follows:

Attribute	Alternatives		
	1	2	3
Economic return > MARR	50	70	100
High throughput	100	60	30
Low scrap rate	100	40	50

Use the *weights* below to evaluate the alternatives. Are the results the same for the two persons' weights? Why?

Importance Score	VP	Assistant VP
Economic return > MARR	20	100
High throughput	80	80
Low scrap rate	100	20

10.44 In Problem 10.43 the vice president and assistant vice president are not consistent in their weights of the three attributes. Assume you are a consultant asked to assist John.

(*a*) What are some conclusions you can draw about the weighted attribute method as an alternative selection method, given the alternative ratings and results in Problem 10.43?

(*b*) Use the new alternative ratings below that you have developed yourself to select an alternative. Using the same scores as the vice president and her assistant given in Problem 10.43, comment on any differences in the alternative selected.

(*c*) What do your new alternative ratings tell you about the selections based on the importance scores of the vice president and assistant vice president?

Attribute	Alternatives		
	1	2	3
Economic return > MARR	30	40	100
High throughput	70	100	70
Low scrap rate	100	80	90

10.45 The watchband division discussed in Problems 10.43 and 10.44 has just been fined $1 million for environmental pollution due to the poor quality of its discharge water. Also, John has become the vice president, and there is no longer an assistant vice president. John always agreed with the importance scores of the former assistant vice president and the alternative ratings he developed earlier (those present initially in Problem 10.43). If he adds his own importance score of 80 to the new factor of environmental cleanliness and awards alternatives 1, 2, and 3 ratings of 80, 50, and 20, respectively, for this new factor, redo the evaluation to select the best alternative.

10.46 For Example 10.10, use an equal weighting of 1 for each attribute to choose the alternative. Did the weighting of attributes change the selected alternative?

10.47 The Athlete's Shop has evaluated two proposals for weight lifting and exercise equipment. A present worth analysis at $i = 15\%$ of estimated incomes and costs resulted in $PW_A = \$420,500$ and $PW_B = \$392,800$. In addition to this economic measure, three attributes were independently assigned a relative importance score from 0 to 100 by the shop manager and the lead trainer.

Attribute	Importance Score	
	Manager	Trainer
Economics	100	80
Durability	35	10
Flexibility	20	100
Maintainability	20	10

Separately, you have used the four attributes to rate the two equipment proposals on a scale of 0.0 to 1.0. The economic attribute was rated using the PW values.

Attribute	Proposal A	Proposal B
Economics	1.00	0.90
Durability	0.35	1.00
Flexibility	1.00	0.90
Maintainability	0.25	1.00

Select the better proposal, using each of the following methods.

(a) Present worth

(b) Weighted evaluations of the manager

(c) Weighted evaluations of the lead trainer

EXTENDED EXERCISE

EMPHASIZING THE RIGHT THINGS

A fundamental service provided to the citizens of a city is police protection. Increasing crime rates that include injury to persons have been documented in the close-in suburbs in Belleville, a densely populated historic area north of the capital. In phase I of the effort, the police chief has made and preliminarily examined four proposals of ways in which police surveillance and protection may be provided in the target residential areas. In brief, they are placing additional officers in cars, on bicycles, on foot, or on horseback. Each alternative has been evaluated separately to estimate annual costs. Placing six new officers on bicycles is clearly the least expensive option at an estimated $700,000 per year. The next best is on foot with 10 new officers at $925,000 per year. The other alternatives will cost slightly more than the "on foot" option.

Before entering phase II, which is a 3-month pilot study to test one or two of these approaches in the neighborhoods, a committee of five members (comprised of police staff and citizen-residents) has been asked to help determine and prioritize attributes that are important in this decision to them, as representatives of the residents and police officers. The five attributes agreed upon after 2 months of discussion are listed below, followed by each committee member's ordering of the attributes from the most important (a score of 1) to the least important (a score of 5).

Attribute	Committee Member					
	1	2	3	4	5	Sum
A. Ability to become 'close' to the citizenry	4	5	3	4	5	21
B. Annual cost	3	4	1	2	4	14
C. Response time upon call or dispatch	2	2	5	1	1	11
D. Number of blocks in coverage area	1	1	2	3	2	9
E. Safety of officers	5	3	4	5	3	20
Totals	15	15	15	15	15	75

Questions

1. Develop weights that can be used in the weighted attribute method for each attribute. The committee members have agreed that the simple average of their five ordered-attribute scores can be considered the indicator of how important each attribute is to them as a group.

2. One committee member recommended, and obtained committee approval for, reducing the attributes considered in the final selection to only those that were listed as number 1 by one or more committee members. Select these attributes and recalculate the weights as requested in question 1.

3. A crimes prevention analyst in the Police Department applied the weighted attribute method to the ordered attributes in question 1. The R_j values obtained using Equation [10.11] are listed below. Which two options should the police chief select for the pilot study?

Alternative	Car	Bicycles	Foot	Horse
R_j	62.5	50.5	47.2	35.4

CASE STUDY

WHICH WAY TO GO—DEBT OR EQUITY FINANCING?

The Opportunity

Pizza Hut Corporation has decided to enter the catering business in three states within its Southeastern U.S. Division, using the name Pizza Hut At-Your-Place. To deliver the meals and serving personnel, it is about to purchase 200 vans with custom interiors for a total of $1.5 million. Each van is expected to be used for 10 years and have a $1000 salvage value.

A feasibility study completed last year indicated that the At-Your-Place business venture could realize an estimated annual net cash flow of $300,000 before taxes in the three states. After-tax considerations would have to take into account the effective tax rate of 35% paid by Pizza Hut.

An engineer with Pizza Hut's Distribution Division has worked with the corporate finance office to determine how to best develop the $1.5 million capital

needed for the purchase of vans. There are two viable financing plans.

The Financing Options

Plan A is debt financing for 50% of the capital ($750,000) with the 8% per year compound interest loan repaid over 10 years with uniform year-end payments. (A simplifying assumption that $75,000 of the principal is repaid with each annual payment can be made.)

Plan B is 100% equity capital raised from the sale of $15 per share common stock. The financial manager informed the engineer that stock is paying $0.50 per share in dividends and that this dividend rate has been increasing at an average of 5% each year. This dividend pattern is expected to continue, based on the current financial environment.

Case Study Exercises

1. What values of MARR should the engineer use to determine the better financing plan?

2. The engineer must make a recommendation on the financing plan by the end of the day. He does not know how to consider all the tax angles for the debt financing in plan A. However, he does have a handbook that gives these relations for equity and debt capital about taxes and cash flows:

   Equity capital: no income tax advantages

   After-tax net cash flow
   = (before-tax net cash flow)(1 − tax rate)

   Debt capital: income tax advantage comes from interest paid on loans

   After-tax net cash flow
   = before-tax net cash flow − loan principal
   − loan interest − taxes

   Taxes
   = (taxable income) (tax rate)

   Taxable income
   = net cash flow − loan interest

   He decides to forget any other tax consequences and use this information to prepare a recommendation. Is A or B the better plan?

3. The division manager would like to know how much the WACC varies for different D-E mixes, especially about 15% to 20% on either side of the 50% debt financing option in plan A. Plot the WACC curve and compare its shape with that of Figure 10–2.

LEVEL THREE

MAKING DECISIONS ON REAL-WORLD PROJECTS

LEVEL ONE This Is How It All Starts	LEVEL TWO Tools for Evaluating Alternatives	LEVEL THREE Making Decisions on Real-World Projects	LEVEL FOUR Rounding Out the Study
Chapter 1 Foundations of Engineering Economy	Chapter 5 Present Worth Analysis	**Chapter 11 Replacement and Retention Decisions**	Chapter 14 Effects of Inflation
Chapter 2 Factors: How Time and Interest Affect Money	Chapter 6 Annual Worth Analysis	**Chapter 12 Selection from Independent Projects Under Budget Limitation**	Chapter 15 Cost Estimation and Indirect Cost Allocation
Chapter 3 Combining Factors	Chapter 7 Rate of Return Analysis: Single Alternative	**Chapter 13 Breakeven Analysis**	Chapter 16 Depreciation Methods
Chapter 4 Nominal and Effective Interest Rates	Chapter 8 Rate of Return Analysis: Multiple Alternatives		Chapter 17 After-Tax Economic Analysis
	Chapter 9 Benefit/Cost Analysis and Public Sector Economics		Chapter 18 Formalized Sensitivity Analysis and Expected Value Decisions
	Chapter 10 Making Choices: The Method, MARR, and Multiple Attributes		Chapter 19 More on Variation and Decision Making Under Risk

The chapters in this level extend the use of economic evaluation tools into real-world situations. A large percentage of economic evaluations involve other than selection from new assets or projects. Probably the most commonly performed evaluation is that of replacing or retaining an in-place asset. *Replacement analysis* applies the evaluation tools to make the correct economic choice.

Often the evaluation involves choosing from *independent projects* under the restriction of limited capital investment. This requires a special technique that is based on the previous chapters.

Future estimates are certainly not exact. Therefore, an alternative should not be selected on the basis of fixed estimates only. *Breakeven analysis* assists in the evaluation process of a range of estimates for *P*, *A*, *F*, *i*, or *n*, and operating variables such as production level, workforce size, design cost, raw material cost, and sales price. Spreadsheets speed up this important, but often detailed, analysis tool.

> ***Important note:* If asset depreciation and taxes are to be considered by an *after-tax analysis*, Chapters 16 and 17 should be covered before or in conjunction with these chapters. See the Preface for options.**

11

Replacement and Retention Decisions

CHAPTER

One of the most commonly performed engineering economy studies is that of replacement or retention of an asset or system that is currently installed. This differs from previous studies where all the alternatives are new. The fundamental question answered by a replacement study about a currently installed asset or system is, *Should it be replaced now or later?* When an asset is currently in use and its function is needed in the future, it will be replaced at some time. So, in reality, a replacement study answers the question of *when*, not *if*, to replace.

A replacement study is usually designed to first make the economic decision to retain or replace *now*. If the decision is to replace, the study is complete. If the decision is to retain, the cost estimates and decision will be revisited each year to ensure that the decision to retain is still economically correct. This chapter explains how to perform the initial year and follow-on year replacement studies.

A replacement study is an application of the AW method of comparing unequal-life alternatives, first introduced in Chapter 6. In a replacement study with no specified study period, the AW values are determined by a technique of cost evaluation called the *economic service life (ESL)* analysis. If a study period is specified, the replacement study procedure is different from that used when no study period is set. All these procedures are covered in this chapter.

The case study is a real-world replacement analysis involving in-place equipment and possible replacement by upgraded equipment.

If asset depreciation and taxes are to be considered in an *after-tax replacement analysis*, Chapters 16 and 17 should be covered before or in conjunction with this chapter. After-tax replacement analysis is included in Section 17.7.

LEARNING OBJECTIVES

Purpose: Perform a replacement study between an in-place asset or system and a new one that could replace it.

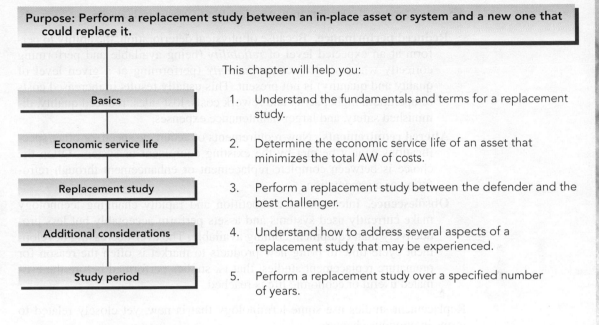

This chapter will help you:

Basics

1. Understand the fundamentals and terms for a replacement study.

Economic service life

2. Determine the economic service life of an asset that minimizes the total AW of costs.

Replacement study

3. Perform a replacement study between the defender and the best challenger.

Additional considerations

4. Understand how to address several aspects of a replacement study that may be experienced.

Study period

5. Perform a replacement study over a specified number of years.

11.1   BASICS OF THE REPLACEMENT STUDY

The need for a replacement study can develop from several sources:

Reduced performance. Because of physical deterioration, the ability to perform at an expected level of *reliability* (being available and performing correctly when needed) or *productivity* (performing at a given level of quality and quantity) is not present. This usually results in increased costs of operation, higher scrap and rework costs, lost sales, reduced quality, diminished safety, and larger maintenance expenses.

Altered requirements. New requirements of accuracy, speed, or other specifications cannot be met by the existing equipment or system. Often the choice is between complete replacement or enhancement through retrofitting or augmentation.

Obsolescence. International competition and rapidly changing technology make currently used systems and assets perform acceptably but less productively than equipment coming available. The ever-decreasing development cycle time to bring new products to market is often the reason for premature replacement studies, that is, studies performed before the estimated useful or economic life is reached.

Replacement studies use some terminology that is new, yet closely related to terms in previous chapters.

Defender and *challenger* are the names for two mutually exclusive alternatives. The defender is the currently installed asset, and the challenger is the potential replacement. A replacement study compares these two alternatives. The challenger is the "best" challenger because it has been selected as the one best challenger to possibly replace the defender. (This is the same terminology used earlier for incremental ROR and B/C analysis of two new alternatives.)

AW values are used as the primary economic measure of comparison between the defender and challenger. The term EUAC (equivalent uniform annual cost) may be used in lieu of AW, because often only costs are included in the evaluation; revenues generated by the defender or challenger are assumed to be equal. Since the equivalence calculations for EUAC are exactly the same as for AW, we use the term AW. Therefore, all values will be negative when only costs are involved. Salvage value, of course, is an exception; it is a cash inflow and carries a plus sign.

Economic service life (ESL) for an alternative is the *number of years* at which the lowest AW of cost occurs. The equivalency calculations to determine ESL establish the life n for the best challenger, and it also establishes the lowest cost life for the defender in a replacement study. (The next section of this chapter explains how to find the ESL by hand and by computer for any new or currently installed asset.)

Defender first cost is the initial investment amount P used for the defender. The *current market value* (MV) is the correct estimate to use for P for the defender in a replacement study. The fair market value may be obtained from professional appraisers, resellers, or liquidators who know the value of used assets. The estimated salvage value at the end of 1 year becomes the market value at the beginning of the next year, provided the estimates remain correct as the years pass. It is incorrect to use the following as MV for the defender first cost: trade-in value that *does not represent a fair market value,* or the depreciated book value taken from accounting records. If the defender must be upgraded or augmented to make it equivalent to the challenger (in speed, capacity, etc.), this cost is added to the MV to obtain the estimate of defender first cost. In the case of asset augmentation for the defender alternative, this separate asset and its estimates are included along with the installed asset estimates to form the complete defender alternative. This alternative is then compared with the challenger via a replacement study.

Challenger first cost is the amount of capital that must be recovered (amortized) when replacing a defender with a challenger. This amount is almost always equal to P, the first cost of the challenger. On occasion, an unrealistically high trade-in value may be offered for the defender compared to its fair market value. In this event, the *net* cash flow required for the challenger is reduced, and this fact should be considered in the analysis. The correct amount to recover and use in the economic analysis for the challenger is its first cost minus the difference between the trade-in value (TIV) and market value (MV) of the defender. In equation form, this is $P - (TIV - MV)$. This amount represents the actual cost to the company because it includes both the opportunity cost (i.e., market value of the defender) and the out-of-pocket cost (i.e., first cost − trade-in) to acquire the challenger. Of course, when the trade-in and market values are the same, the challenger P value is used in all computations.

The challenger first cost is the estimated initial investment necessary to acquire and install it. Sometimes, an analyst or manager will attempt to *increase* this first cost by an amount equal to the *unrecovered capital* remaining in the defender as shown on the accounting records for the asset. This is observed most often when the defender is working well and in the early stages of its life, but technological obsolescence, or some other reason, has forced consideration of a replacement. This unrecovered capital amount is referred to as a *sunk cost.* A sunk cost must not be added to the challenger's first cost, because it will make the challenger appear to be more costly than it is.

Sunk costs are capital losses and cannot be recovered in a replacement study. Sunk costs are correctly handled in the corporation's income statement and by tax law allowances.

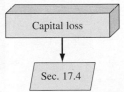

A replacement study is performed most objectively if the analyst takes the *viewpoint of a consultant* to the company or unit using the defender. In this way, the perspective taken is that neither alternative is currently owned, and the services provided by the defender could be purchased now with an "investment" that is equal to its first cost (market value). This is indeed correct because the market value will be a forgone opportunity of cash inflow if the question

"Replace now?" is answered with a no. Therefore, the consultant's viewpoint is a convenient way to allow the economic evaluation to be performed without bias for either alternative. This approach is also referred to as the *outsider's viewpoint*.

As mentioned in the introduction, a replacement study is an application of the annual worth method. As such, the fundamental assumptions for a replacement study parallel those of an AW analysis. If the *planning horizon is unlimited,* that is, a study period is not specified, the assumptions are as follows:

1. The services provided are needed for the indefinite future.
2. The challenger is the best challenger available now and in the future to replace the defender. When this challenger replaces the defender (now or later), it will be repeated for succeeding life cycles.
3. Cost estimates for every life cycle of the challenger will be the same.

As expected, none of these assumptions is precisely correct. We discussed this previously for the AW method (and the PW method). When the intent of one or more of the assumptions becomes incorrect, the estimates for the alternatives must be updated and a new replacement study conducted. The replacement procedure discussed in Section 11.3 explains how to do this. When the *planning horizon is limited to a specified study period, the assumptions above do not hold.* The procedure of Section 11.5 discusses how to perform the replacement study in this case.

EXAMPLE **11.1**

The Arkansas Division of ADM, a large agricultural products corporation, purchased a state-of-the-art ground-leveling system for rice field preparation 3 years ago for $120,000. When purchased, it had an expected service life of 10 years, an estimated salvage of $25,000 after 10 years, and AOC of $30,000. Current account book value is $80,000. The system is deteriorating rapidly; 3 more years of use and then salvaging it for $10,000 on the international used farm equipment network are now the expectations. The AOC is averaging $30,000.

A substantially improved, laser-guided model is offered today for $100,000 with a trade-in of $70,000 for the current system. The price goes up next week to $110,000 with a trade-in of $70,000. The ADM division engineer estimates the laser-guided system to have a useful life of 10 years, a salvage of $20,000, and an AOC of $20,000. A $70,000 market value appraisal of the current system was made today.

If no further analysis is made on the estimates, state the correct values to include if the replacement study is performed today.

Solution
Take the consultant's viewpoint and use the most current estimates.

Defender	Challenger
$P = \text{MV} = \$70,000$	$P = \$100,000$
$\text{AOC} = \$30,000$	$\text{AOC} = \$20,000$
$S = \$10,000$	$S = \$20,000$
$n = 3$ years	$n = 10$ years

> The defender's original cost, AOC, and salvage estimates, as well as its current book value, are all *irrelevant* to the replacement study. *Only the most current estimates should be used.* From the consultant's perspective, the services that the defender can provide could be obtained at a cost equal to the defender market value of $70,000. Therefore, this is the first cost of the defender for the study. The other values are as shown.

11.2 ECONOMIC SERVICE LIFE

Until now the estimated life *n* of an alternative or asset has been stated. In reality, the best life estimate to use in the economic analysis is not known initially. When a replacement study or an analysis between new alternatives is performed, the best value for *n* should be determined using current cost estimates. The best life estimate is called the *economic service life*.

> **The economic service life (ESL) is the number of years *n* at which the equivalent uniform annual worth (AW) of costs is the minimum, considering the most current cost estimates over all possible years that the asset may provide a needed service.**

The ESL is also referred to as the economic life or minimum cost life. Once determined, the ESL should be the estimated life for the asset used in an engineering economy study, if only economics are considered. When *n* years have passed, the ESL indicates that the asset should be replaced to minimize overall costs. To perform a replacement study correctly, it is important that the ESL of the challenger and ESL of the defender be determined, since their *n* values are usually not preestablished.

The ESL is determined by calculating the total AW of costs if the asset is in service 1 year, 2 years, 3 years, and so on, up to the last year the asset is considered useful. Total AW of costs is the sum of capital recovery (CR), which is the AW of the initial investment and any salvage value, and the AW of the estimated annual operating cost (AOC), that is,

$$\text{Total AW} = -\text{capital recovery} - \text{AW of annual operating costs}$$

$$= -\text{CR} - \text{AW of AOC} \qquad [11.1]$$

The ESL is the n value for the smallest total AW of costs. (Remember: These AW values are *cost* estimates, so the AW values are negative numbers. Therefore, $-200 is a lower cost than $-500.) Figure 11–1 shows the characteristic shape of a total AW of cost curve. The CR component of total AW decreases, while the AOC component increases, thus forming the concave shape. The two AW components are calculated as follows.

> **Decreasing cost of capital recovery.** The capital recovery is the AW of investment; it decreases with each year of ownership. Capital recovery is calculated by Equation [6.3], which is repeated here. The salvage value *S*,

Figure 11–1
Annual worth curves
of cost elements that
determine the economic
service life.

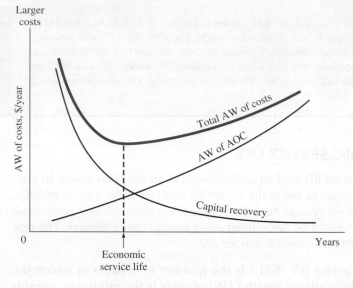

which usually decreases with time, is the estimated market value (MV) in
that year.

$$\text{Capital recovery} = -P(A/P,i,n) + S(A/F,i,n) \qquad [11.2]$$

Increasing cost of AW of AOC. Since the AOC estimates usually increase
over the years, the AW of AOC increases. To calculate the AW of the AOC
series for 1, 2, 3, . . . years, determine the present worth of each AOC value
with the P/F factor, then redistribute this P value over the years of ownership, using the A/P factor.

The complete equation for total AW of costs over k years is

$$\text{Total AW}_k = -P(A/P,i,k) + S_k(A/F,i,k) - \left[\sum_{j=1}^{j=k} \text{AOC}_j(P/F,i,j)\right](A/P,i,k)$$
$$[11.3]$$

where P = initial investment or current market value
          S_k = salvage value or market value after k years
          AOC_j = annual operating cost for year j (j = 1 to k)

The current MV is used for P when the asset is the defender, and the estimated
future MV values are substituted for the S values in years 1, 2, 3, ...

To determine ESL by computer, the PMT function (with embedded NPV
functions as needed) is used repeatedly for each year to calculate capital recovery and the AW of AOC. Their sum is the total AW for k years of ownership. The
PMT function formats for the capital recovery and AOC components for each
year k are as follows:

Capital recovery for the challenger: PMT(i%,years,P,−MV_in_year_k)

Capital recovery for the defender: PMT(i%,years,current_MV,−MV_in_year_k)

AW of AOC: −PMT(i%,years,NPV(i%,year_1_AOC:year_k_AOC)+0)

E-Solve

When the spreadsheet is developed, it is recommended that the PMT functions in year 1 be developed using cell-reference format, then drag down the function through each column. A final column summing the two PMT results displays total AW. Augmenting the table with an Excel *xy* scatter plot graphically displays the cost curves in the general form of Figure 11–1, and the ESL is easily identified. Example 11.2 illustrates ESL determination by hand and by computer.

EXAMPLE 11.2

A 3-year-old manufacturing process asset is being considered for early replacement. Its current market value is $13,000. Estimated future market values and annual operating costs for the next 5 years are given in Table 11–1, columns 2 and 3. What is the economic service life of this defender if the interest rate is 10% per year? Solve by hand and by computer.

TABLE 11–1   Computation of Economic Service Life

Year j (1)	MV_j (2)	AOC_j (3)	Capital Recovery (4)	AW of AOC (5)	Total AW_k (6) = (4) + (5)
1	$9000	$−2500	$−5300	$−2500	$−7800
2	8000	−2700	−3681	−2595	−6276
3	6000	−3000	−3415	−2717	−6132
4	2000	−3500	−3670	−2886	−6556
5	0	−4500	−3429	−3150	−6579

Solution by Hand

Equation [11.3] is used to calculate total AW_k for $k = 1, 2, \ldots, 5$. Table 11–1, column 4, shows the capital recovery for the $13,000 current market value ($j = 0$) plus 10% return. Column 5 gives the equivalent AW of AOC for k years. As an illustration, the computation of total AW for $k = 3$ from Equation [11.3] is

$$\text{Total AW}_3 = -P(A/P,i,3) + MV_3(A/F,i,3) - [\text{PW of } AOC_1, AOC_2, \text{ and } AOC_3](A/P,i,3)$$

$$= -13,000(A/P,10\%,3) + 6000(A/F,10\%,3) - [2500(P/F,10\%,1)$$

$$+ 2700(P/F,10\%,2) + 3000(P/F,10\%,3)](A/P,10\%,3)$$

$$= -3415 - 2717 = \$-6132$$

A similar computation is performed for each year 1 through 5. The lowest equivalent cost (numerically largest AW value) occurs at $k = 3$. Therefore, the defender ESL is $n = 3$ years, and the AW value is $−6132. In the replacement study, this AW will be compared with the best challenger AW determined by a similar ESL analysis.

Solution by Computer

See Figure 11–2 for the spreadsheet and chart for this example. (This format is a template for any ESL analysis; simply change the estimates and add rows for more years.) Contents of columns D and E are briefly described below. The PMT functions apply the formats for the defender as described above. Cell tags show detailed cell-reference format for year 5. The $ symbols are included for absolute cell referencing, needed when the entry is dragged down through the column.

E-Solve

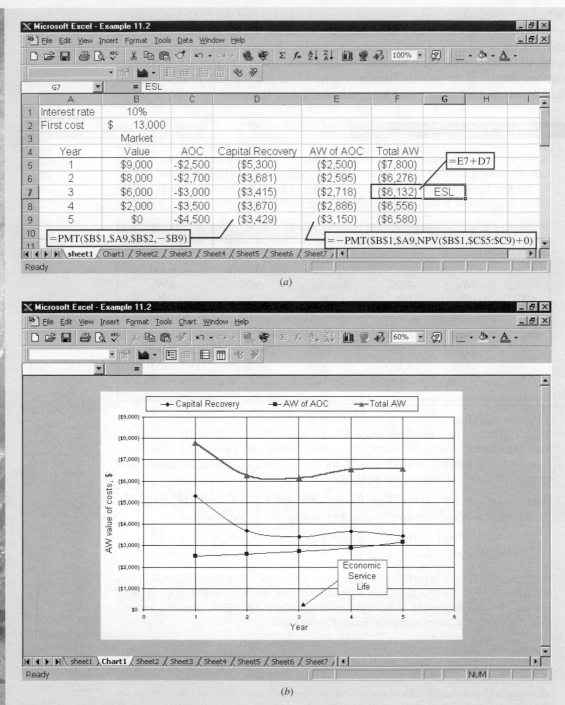

Figure 11–2

(*a*) Spreadsheet determination of the economic service life (ESL), and (*b*) plot of total AW and cost components, Example 11.2.

Column D: Capital recovery is the AW of the \$13,000 investment (cell B2) in year 0 for each year 1 through 5 with the estimated MV in that year. For example, in actual numbers, the cell-reference PMT function in year 5 shown on the spreadsheet reads PMT(10%,5,13000,−0), resulting in \$−3429. This series is plotted in Figure 11–2b as the middle curve, labeled capital recovery in the legend.

Column E: The NPV function embedded in the PMT function obtains the present worth in year 0 of all AOC estimates through year k. Then PMT calculates the AW of AOC over the k years. For example, in year 5, the PMT in numbers is −PMT(10%,5,NPV(10%,C5:C9)+0). The 0 is the AOC in year 0; it is optional. The graph plots the AW of AOC curve, which constantly increases in cost because the AOC estimates increase each year.

Comment

The *capital recovery curve* in Figure 11–2b (middle curve) is not a true concave shape because the estimated market value changes each year. If the same MV were estimated for each year, the curve would appear like Figure 11–1. When several total AW values are approximately equal, the curve will be flat over several periods. This indicates that the ESL is relatively insensitive to costs.

It is reasonable to ask about the difference between the ESL analysis above and the AW analyses performed in previous chapters. Previously we had an *estimated life of n years* with associated other estimates: first cost in year 0, possibly a salvage value in year n, and an AOC that remained constant or varied each year. For all previous analyses, the calculation of AW using these estimates determined the AW over n years. This is also the economic service life when n is fixed. In all previous cases, there were no year-by-year MV estimates applicable over the years. Therefore, we can conclude the following:

> **When the expected life n is known for the challenger or defender, determine its AW over n years, using the first cost or current market value, estimated salvage value after n years, and AOC estimates. This AW value is the correct one to use in the replacement study.**

It is not difficult to estimate a series of market/salvage values for a new or current asset. For example, an asset with a first cost of P can lose market value at 20% per year, so the market value series for years 0, 1, 2, . . . is P, $0.8P$, $0.64P$, . . . , respectively. (An overview of cost estimation approaches and techniques is presented in Chapter 15.) If it is reasonable to predict the MV series on a year-by-year basis, it can be combined with the AOC estimates to produce what is called the *marginal costs* for the asset.

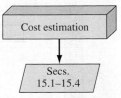

> **Marginal costs (MC) are year-by-year estimates of the costs to own and operate an asset for that year.**

There are three components to each annual marginal cost estimate:

- Cost of ownership (loss in market value is the best estimate of this cost).
- Forgone interest on the market value at the beginning of the year.
- AOC for each year.

Once the marginal costs are estimated for each year, their equivalent AW value can be calculated. The sum of the AW values of the first two of these components is identical to the capital recovery amount. Now, it should be clear that the total AW of all three marginal cost components over k years is the same value as the total annual worth for k years calculated in Equation [11.3]. That is, the following relation is

$$\text{AW of marginal costs} = \text{total AW of costs} \qquad [11.4]$$

Therefore, there is no need to perform a separate, detailed marginal cost analysis when yearly market values are estimated. The ESL analysis presented in Example 11.2 is sufficient in that it results in the same numerical values. This is demonstrated in Example 11.3 using the data of Example 11.2.

EXAMPLE 11.3

An engineer has determined that a 3-year-old manufacturing process asset has a market value of $13,000 now, and the estimated salvage/market values and AOC values shown in Table 11–1 (repeated in Figure 11–3, columns B and E). Determine the AW of the marginal

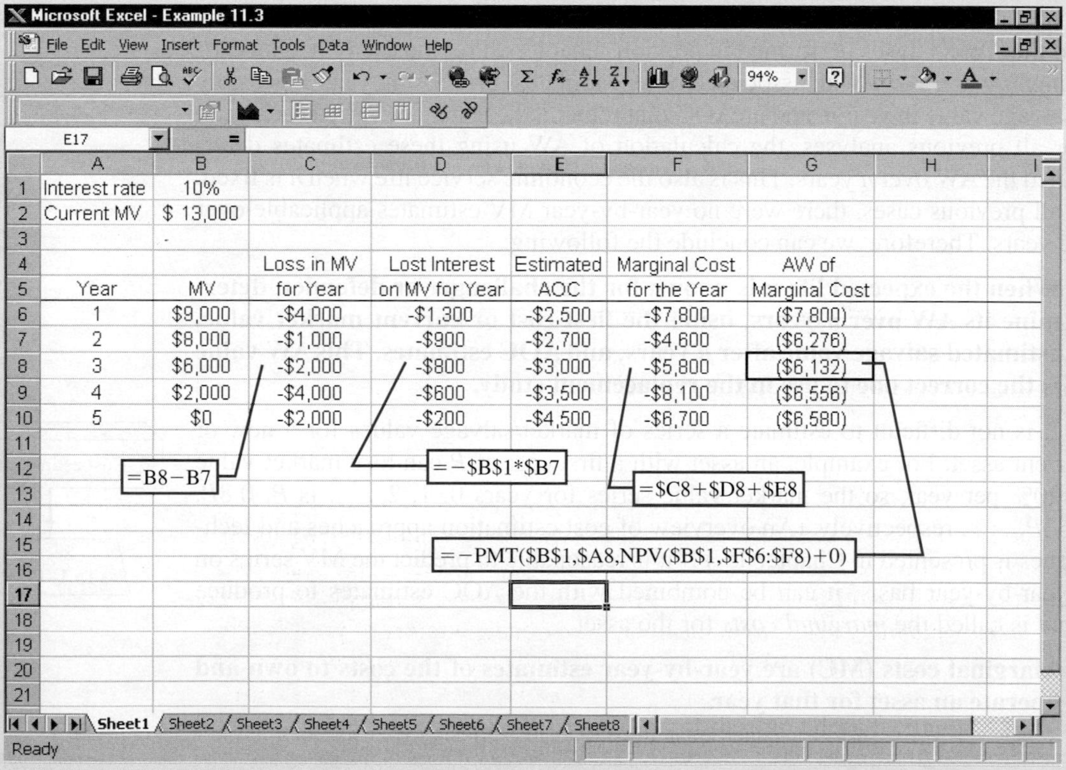

	X Microsoft Excel - Example 11.3								

File Edit View Insert Format Tools Data Window Help

E17   =

	A	B	C	D	E	F	G	H	I
1	Interest rate	10%							
2	Current MV	$ 13,000							
3		-							
4			Loss in MV	Lost Interest	Estimated	Marginal Cost	AW of		
5	Year	MV	for Year	on MV for Year	AOC	for the Year	Marginal Cost		
6	1	$9,000	-$4,000	-$1,300	-$2,500	-$7,800	($7,800)		
7	2	$8,000	-$1,000	-$900	-$2,700	-$4,600	($6,276)		
8	3	$6,000	-$2,000	-$800	-$3,000	-$5,800	($6,132)		
9	4	$2,000	-$4,000	-$600	-$3,500	-$8,100	($6,556)		
10	5	$0	-$2,000	-$200	-$4,500	-$6,700	($6,580)		
11									
12		=B8−B7		=−$B$1*$B7					
13					=$C8+$D8+$E8				
14									
15				=−PMT($B$1,$A8,NPV($B$1,$F$6:$F8)+0)					
16									
17									
18									
19									
20									
21									

Sheet1 / Sheet2 / Sheet3 / Sheet4 / Sheet5 / Sheet6 / Sheet7 / Sheet8

Ready

Figure 11–3
Calculation of AW of marginal cost series, Example 11.3.

cost values by computer, and compare it with the total AW values in Figure 11–2. Use the marginal cost series to determine the correct values for *n* and AW if the asset is the defender in a replacement study.

Solution by Computer

See Figure 11–3. The first marginal cost component is the loss in MV by year (column C). The 10% interest on the MV (column D) is the second component, the forgone interest on the MV. Their sum is the year-by-year capital recovery amount. Based on the description above, the marginal cost for each year is the sum of columns C, D, and E, as shown in the spreadsheet cell tags. The series AW of marginal cost values in column G is identical to those determined for total AW of costs using the ESL analysis in Figure 11–2*a*. The correct values for a replacement study are $n = 3$ years and AW = \$−6132; the same as for the ESL analysis in the previous example.

E-Solve

Now it is possible to draw two specific conclusions about the *n* and AW values to be used in a replacement study. These conclusions are based on the extent to which detailed annual estimates are made for the market value.

1. **Year-by-year market value estimates are made.** Use them to perform an ESL analysis, and determine the *n* value with the lowest total AW of costs. These are the best *n* and AW values for the replacement study.
2. **Yearly market value estimates are not made.** Here the only estimate available is market value (salvage value) in year *n*. Use it to calculate the AW over *n* years. These are the *n* and AW values to use; however, they may not be the "best" values in that they may not represent the best equivalent total AW of cost value.

Upon completion of the ESL analysis, the replacement study procedure in the next section is applied using the following values:

Challenger alternative (C): AW_C for n_C years
Defender alternative (D): AW_D for n_D years

11.3 PERFORMING A REPLACEMENT STUDY

Replacement studies are performed in one of two ways: without a study period specified or with one defined. Figure 11–4 gives an overview of the approach taken for each situation. The procedure discussed in this section applies when no study period (planning horizon) is specified. If a specific number of years is identified for the replacement study, for example, over the next 5 years, with no continuation considered after this time period in the economic analysis, the procedure in Section 11.5 is applied.

A replacement study determines when a challenger replaces the in-place defender. The complete study is finished if the challenger (C) is selected to replace the defender (D) now. However, if the defender is retained now, the study may extend over a number of years equal to the life of the defender n_D, after which a

Figure 11–4
Overview of replacement
study approaches.

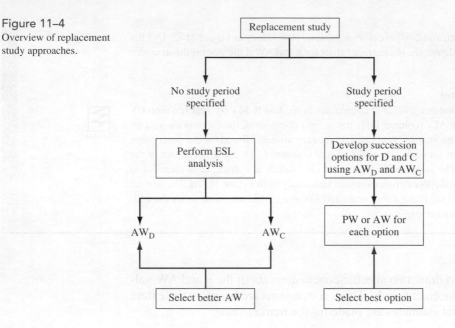

challenger replaces the defender. Use the annual worth and life values for C and D determined in the ESL analysis to apply the following replacement study procedure. This assumes the services provided by the defender could be obtained at the AW_D amount.

New replacement study:
1. On the basis of the better AW_C or AW_D value, select the challenger alternative (C) or defender alternative (D). When the challenger is selected, replace the defender now, and expect to keep the challenger for n_C years. This replacement study is complete. If the defender is selected, plan to retain it for up to n_D more years. (This is the left AW_D branch of Figure 11–4.) Next year, perform the following steps.

One-year-later analysis:
2. Are all estimates still current for both alternatives, especially first cost, market value, and AOC? If no, proceed to step 3. If yes and this is year n_D, replace the defender. If this is not year n_D, retain the defender for another year and repeat this same step. This step may be repeated several times.
3. Whenever the estimates have changed, update them and determine new AW_C and AW_D values. Initiate a new replacement study (step 1).

If the defender is selected initially (step 1), estimates may need updating after 1 year of retention (step 2). Possibly there is a new best challenger to compare with D. Either significant changes in defender estimates or availability of a new challenger indicates that a new replacement study is to be performed. In actuality, a replacement study can be performed each year to determine the advisability of replacing or retaining any defender, provided a competitive challenger is available.

Example 11.4 below illustrates the application of ESL analysis for a challenger and defender, followed by the use of the replacement study procedure. The planning horizon is unspecified in this example.

EXAMPLE 11.4

Two years ago, Toshiba Electronics made a $15 million investment in new assembly line machinery. It purchased approximately 200 units at $70,000 each and placed them in plants in 10 different countries. The equipment sorts, tests, and performs insertion-order kitting on electronic components in preparation for special-purpose printed circuit boards. This year, new international industry standards will require a $16,000 retro-fit on each unit, in addition to the expected operating cost. Due to the new standards, coupled with rapidly changing technology, a new system is challenging the retention of these 2-year-old machines. The chief engineer at Toshiba USA realizes that the economics must be considered, so he has asked that a replacement study be performed this year and each year in the future, if need be. At $i = 10\%$ and with the estimates below, use both hand and computer calculations to

(a) Determine the AW values and economic service lives necessary to perform the replacement study.

(b) Perform the replacement study now.

        Challenger: First cost: $50,000
                         Future market values: decreasing by 20% per year
                         Estimated retention period: no more than 5 years
                         AOC estimates: $5000 in year 1 with increases of $2000 per year
                             thereafter

        Defender: Current international market value: $15,000
                         Future market values: decreasing by 20% per year
                         Estimated retention period: no more than 3 more years
                         AOC estimates: $4000 next year, increasing by $4000 per year
                             thereafter, plus the $16,000 retro-fit next year

(c) After 1 year, it is time to perform the follow-up analysis. The challenger is making large inroads to the market for electronic components assembly equipment, especially with the new international standards features built in. The expected market value for the defender is still $12,000 this year, but it is expected to drop to virtually nothing in the future—$2000 next year on the worldwide market and zero after that. Also, this prematurely outdated equipment is more costly to keep serviced, so the estimated AOC next year has been increased from $8000 to $12,000 and to $16,000 two years out. Perform the follow-up replacement study analysis.

Solution by Hand

(a) The results of the ESL analysis, shown in Table 11–2, include all the MV and AOC estimates for the challenger in part (a) of the table. Note that $P = $50,000$ is also the MV in year 0. The total AW of costs is for each year, should the challenger be placed into service for that number of years. As an example, the year $k = 4$ amount of $-19,123$ is determined using Equation [11.3]. The A/G factor is applied in lieu of the P/F and A/P factors to find the AW of the arithmetic gradient

TABLE 11–2 Economic Service Life (ESL) Analysis of (a) Challenger and (b) Defender Costs, Example 11.4

		(a) challenger		
Challenger Year k	Market Value	AOC	Total AW If Owned k Years	
0	$50,000	—	—	
1	40,000	$ −5,000	$−20,000	
2	32,000	−7,000	−19,524	
3	25,600	−9,000	−19,245	
4	20,480	−11,000	−19,123	ESL
5	16,384	−13,000	−19,126	

		(b) defender		
Defender Year k	Market Value	AOC	Total AW If Retained k Years	
0	$15,000	—	—	
1	12,000	$−20,000	$−24,500	
2	9,600	−8,000	−18,357	
3	7,680	−12,000	−17,307	ESL

series in the AOC.

$$\text{Total AW}_4 = -50{,}000(A/P,10\%,4) + 20{,}480(A/F,10\%,4)$$
$$- [5000 + 2000(A/G,10\%,4)]$$
$$= \$-19{,}123$$

The defender costs are analyzed in the same way in Table 11–2b up to the maximum retention period of 3 years.

The lowest AW cost (numerically largest) values for the replacement study are

Challenger: $AW_C = \$-19{,}123$ for $n_C = 4$ years

Defender: $AW_D = \$-17{,}307$ for $n_D = 3$ years

If plotted, the challenger total AW of cost curve (Table 11–2a) would be relatively flat after 2 years; there is virtually no difference in the total AW for years 4 and 5. For the defender, note that the AOC values change substantially over the 3 years, and they do not constantly increase or decrease.

Parts b and c are solved below by computer.

Solution by Computer

E-Solve

(a)   Figure 11–5 includes the complete spreadsheet and total AW of cost graph for the challenger and defender. (The tables were generated by initially copying the spreadsheet developed in Figure 11–2a as a template. All the PMT functions in columns D and E, and summing function in column F, are identical. The first cost, market value, and AOC amounts are changed for this example.) Some critical functions are detailed in the cell tags. The xy charts show the total AW of cost curves. The AW and ESL values are the same as in the hand solution.

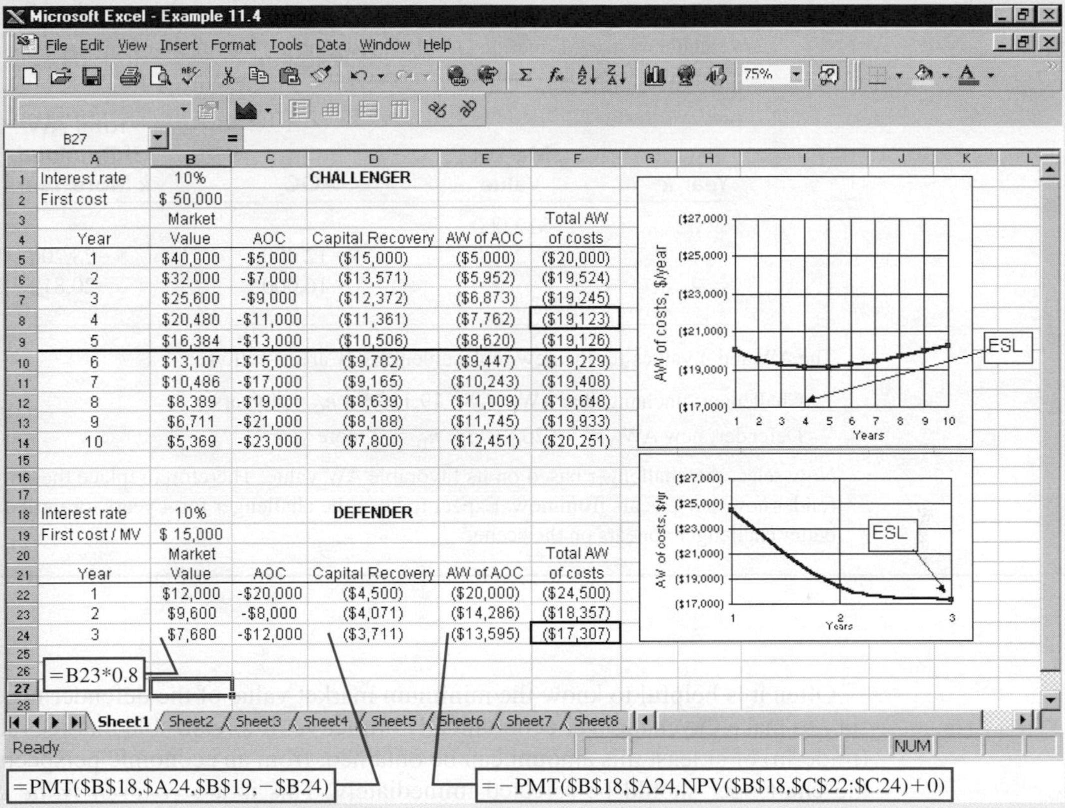

Figure 11–5
Economic service life (ESL) for the challenger and defender using a spreadsheet, Example 11.4.
(The tabular format and functions are the same as in Figure 11–2a.)

Spreadsheet (Microsoft Excel - Example 11.4), cell B27:

Year	Market Value	AOC	Capital Recovery	AW of AOC	Total AW of costs
Interest rate	10%		CHALLENGER		
First cost	$ 50,000				
1	$40,000	-$5,000	($15,000)	($5,000)	($20,000)
2	$32,000	-$7,000	($13,571)	($5,952)	($19,524)
3	$25,600	-$9,000	($12,372)	($6,873)	($19,245)
4	$20,480	-$11,000	($11,361)	($7,762)	($19,123)
5	$16,384	-$13,000	($10,506)	($8,620)	($19,126)
6	$13,107	-$15,000	($9,782)	($9,447)	($19,229)
7	$10,486	-$17,000	($9,165)	($10,243)	($19,408)
8	$8,389	-$19,000	($8,639)	($11,009)	($19,648)
9	$6,711	-$21,000	($8,188)	($11,745)	($19,933)
10	$5,369	-$23,000	($7,800)	($12,451)	($20,251)
Interest rate	10%		DEFENDER		
First cost / MV	$ 15,000				
1	$12,000	-$20,000	($4,500)	($20,000)	($24,500)
2	$9,600	-$8,000	($4,071)	($14,286)	($18,357)
3	$7,680	-$12,000	($3,711)	($13,595)	($17,307)

Cell formula box: `=B23*0.8`

Bottom formula callouts:
`=PMT($B$18,$A24,$B$19,-$B24)`
`=-PMT($B$18,$A24,NPV($B$18,$C$22:$C24)+0)`

Since it is very easy to add years to an ESL analysis, years 5 through 10 are appended to the challenger analysis in spreadsheet rows 10 to 14. Note that the total AW curve has a relatively flat bottom, and it returns to the early-life AW cost level (about $-20,000) after some number of years, 10 here. This is a classically shaped AW curve developed from constantly decreasing market values and constantly increasing AOC values. (Use of this tabular format and these functions is also recommended for an analysis where all the components of total AW need to be displayed.)

(b) To perform the replacement study now, apply only the first step of the procedure. Select the defender because it has the better AW of costs ($-17,307), and expect to retain it for 3 more years. Prepare to perform the one-year-later analysis 1 year from now.

(c) One year later, the situation has changed significantly for the equipment Toshiba retained last year. Apply the steps for the one-year-later analysis:

2. After 1 year of defender retention, the challenger estimates are still reasonable, but the defender market value and AOC estimates are substantially different. Go to step 3 to perform a new ESL analysis for the defender.

3. The defender estimates in Table 11–2*b* are updated below, and new AW values are calculated using Equation [11.3]. There is now a maximum of 2 more years of retention, 1 year less than the 3 years determined last year.

Year k	Market Value	AOC	Total AW If Retained k More Years
0	$12,000	—	—
1	2,000	$-12,000	$-23,200
2	0	-16,000	-20,819

The AW and *n* values for the new replacement study are as follows:

Challenger: unchanged at $AW_C = \$-19,123$ for $n_C = 4$ years

Defender: new $AW_D = \$-20,819$ for $n_D = 2$ more years

Now select the challenger based on its favorable AW value. Therefore, replace the defender now, not 2 years from now. Expect to keep the challenger for 4 years, or until a better challenger appears on the scene.

Often it is helpful to know the minimum market value of the defender necessary to make the challenger economically attractive. If a realizable market value (trade-in) of at least this amount can be obtained, from an economic perspective the challenger should be selected immediately. This is a *breakeven value* between AW_C and AW_D; it is referred to as the *replacement value (RV)*. Set up the relation $AW_C = AW_D$ with the market value for the defender substituted as RV, which is the unknown. The AW_C is known, so RV can be determined. The selection guideline is as follows:

If the actual market trade-in exceeds the breakeven replacement value, the challenger is the better alternative, and should replace the defender now.

For Example 11.4*b*, $AW_C = \$-19,123$, and the defender was selected. Therefore, RV should be larger than the estimated defender market value of $15,000. Equation [11.3] is set up for 3 years of defender retention and equated to $-19,123.

$$-RV(A/P,10\%,3) + 0.8^3\, RV(A/F,10\%,3) - [20,000(P/F,10\%,1)$$
$$+ 8,000(P/F,10\%,2) + 12,000(P/F,10\%,3)](A/P,10\%,3) = \$-19,123 \quad [11.5]$$
$$RV = \$22,341$$

Any market trade-in value above this amount is an economic indication to replace now with the challenger.

If the spreadsheet in Figure 11–5 has been developed for the ESL analysis, Excel's SOLVER (which is on the Tools toolbar) can find RV rapidly. It is important to understand what SOLVER does from an engineering economy perspective, so Equation [11.5] should be set up and understood. Cell F24 in

Q-Solv

Figure 11–5 is the "target cell" to equal $\$-19{,}123$ (the best AW_C in F8). This is how Excel sets up a spreadsheet equivalent of Equation [11.4]. SOLVER returns the RV value of $\$22{,}341$ in cell B19 with a new estimated market value of $\$11{,}438$ in year 3. Reflecting on the solution to Example 11.4(*b*), the current market value is $\$15{,}000$, which is less than RV $= \$22{,}341$. The defender is selected over the challenger. Use Appendix A or the Excel online help function to learn how to use SOLVER in an efficient way. SOLVER is used more extensively in Chapter 13 for breakeven analysis.

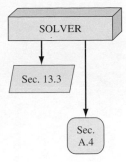

11.4 ADDITIONAL CONSIDERATIONS IN A REPLACEMENT STUDY

There are several additional aspects of a replacement study that may be introduced. Three of these are identified and discussed in turn.

- Future-year replacement decisions at the time of the initial replacement study.
- Opportunity-cost versus cash-flow approaches to alternative comparison.
- Anticipation of improved future challengers.

In most cases when management initiates a replacement study, the question is best framed as, "Replace now, 1 year from now, 2 years from now, etc.?" The procedure above does answer this question provided the estimates for C and D do not change as each year passes. In other words, *at the time it is performed, step 1 of the procedure does answer the replacement question for multiple years.* It is only when estimates change over time that the decision to retain the defender may be prematurely reversed in favor of the then-best challenger, that is, prior to n_D years.

The first costs (*P* values) for the challenger and defender have been correctly taken as the initial investment for the challenger C and current market value for the defender D. This is called the *opportunity-cost approach* because it recognizes that a cash inflow of funds equal to the market value is forgone if the defender is selected. This approach, also called the conventional approach, is correct for every replacement study. A second approach, called the *cash-flow approach,* recognizes that when C is selected, the market value cash inflow for the defender is received and, in effect, immediately reduces the capital needed to invest in the challenger. Use of the cash-flow approach is strongly discouraged for at least two reasons: possible violation of the equal-service assumption and incorrect capital recovery value for C. As we are aware, all economic evaluations must compare alternatives with equal service. Therefore, the cash-flow approach can work only when challenger and defender lives are exactly equal. This is commonly not the case; in fact, the ESL analysis and replacement study procedure are designed to compare two mutually exclusive, *unequal-life* alternatives via the annual worth method. If this equal-service comparison reason is not enough to avoid the cash flow approach, consider what happens to the challenger's capital recovery amount when its first cost is decreased by the market value of the defender. The capital recovery (CR) terms in Equation [11.3] will decrease, resulting in a falsely low value of CR for the challenger, were it selected. From the vantage point of the economic study itself, the decision for C or D will not change; but when C is selected and implemented, this CR value is not reliable. The conclusion is simple: *Use the initial investment of C and the MV of D as the first costs in the ESL analysis and in the replacement study.*

A basic premise of a replacement study is that some challenger will replace the defender at a future time, provided the service continues to be needed and a worthy challenger is available. The expectation of ever-improving challengers can offer strong encouragement to retain the defender until some situational elements—technology, costs, market fluctuations, contract negotiations, etc.—stabilize. This was the case in the previous example of the electronics assembly equipment. A large expenditure on equipment when the standards changed soon after purchase forced an early replacement consideration and a large loss of invested capital. The replacement study is no substitute for forecasting challenger availability. *It is important to understand trends, new advances, and competitive pressures that can complement the economic outcome of a good replacement study.* It is often better to compare a challenger with an augmented defender in the replacement study. Adding needed features to a currently installed defender may prolong its useful life and productivity until challenger choices are more appealing.

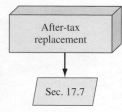

It is possible that a significant tax impact may occur when a defender is traded early in its expected life. If taxes should be considered, proceed now, or after the next section, to Chapter 17 and the after-tax replacement analysis in Section 17.7.

11.5   REPLACEMENT STUDY OVER A SPECIFIED STUDY PERIOD

When the time period for the replacement study is limited to a specified study period or planning horizon, for example, 6 years, the determinations of AW values for the challenger and for the remaining life of the defender are usually not based on the economic service life. What happens to the alternatives after the study period is not considered in the replacement analysis. Therefore, the services are not needed beyond the study period. In fact, a study period of fixed duration does not comply with the three assumptions stated in Section 11.1—service needed for indefinite future, best challenger available now, and estimates will be identical for future life cycles.

When performing a replacement study over a fixed study period, it is crucial that the estimates used to determine the AW values be accurate and used in the study. This is especially important for the defender. Failure to do the following violates the assumption of equal-service comparison.

When the defender's remaining life is shorter than the study period, the cost of providing the defender's services from the end of its expected remaining life to the end of the study period must be estimated as accurately as possible and included in the replacement study.

The right branch of Figure 11–4 presents an overview of the replacement study procedure for a stated study period.

1. *Succession options and AW values.* Develop all the viable ways to use the defender and challenger during the study period. There may be only one

option or many options; the longer the study period, the more complex this analysis becomes. The AW_D and AW_C values are used to build the equivalent cash flow series for each option.

2. *Selection of the best option.* The PW or AW for each option is calculated over the study period. Select the option with the lowest cost, or highest income if revenues are estimated. (As before, the best option will have the numerically largest PW or AW value.)

The following three examples use this procedure and illustrate the importance of making cost estimates for the defender alternative when its remaining life is less than the study period.

EXAMPLE 11.5

Claudia works with Lockheed-Martin (LMCO) in the aircraft maintenance division. She is preparing for what she and her boss, the division chief, hope to be a new 10-year defense contract with the U.S. Air Force on C-5A cargo aircraft. A key piece of equipment for maintenance operations is an avionics circuit diagnostics system. The current system was purchased 7 years ago on an earlier contract. It has no capital recovery costs remaining, and the following are reliable estimates: current market value = $70,000, remaining life of 3 more years, no salvage value, and AOC = $30,000 per year. The only options for this system are to replace it now or retain it for the full 3 additional years.

Claudia has found that there is only one good challenger system. Its cost estimates are: first cost of $750,000, life of 10 years, $S = 0$, and AOC = $50,000 per year.

Realizing the importance of accurate defender alternative cost estimates, Claudia asked the division chief what system would be a logical follow-on to the current one 3 years hence, if LMCO wins the contract. The chief predicted LMCO would purchase the very system she had identified as the challenger, because it is the best on the market. The company would keep it for the entire 10 additional years for use on an extension of this contract or some other application that could recover the remaining 3 years of invested capital. Claudia interpreted the response to mean that the last 3 years would also be capital recovery years, but on some project other than this one. Claudia's estimate of the first cost of this same system 3 years from now is $900,000. Additionally, the $50,000 per year AOC is the best estimate at this time.

The division chief mentioned any study had to be conducted using the interest rate of 10%, as mandated by the U.S. Office of Management and Budget (OMB). Perform a replacement study for the fixed contract period of 10 years.

Solution
The study period is fixed at 10 years, so the intent of the replacement study assumptions is not present. This means the defender follow-on estimates are very important to the analysis. Further, any analyses to determine the ESL values are unnecessary and incorrect since alternative lives are already set and no projected annual market values are available. The first step of the replacement study procedure is to define the options. Since the defender will be replaced now or in 3 years, there are only two options:

1. Challenger for all 10 years.
2. Defender for 3 years, followed by challenger for 7 years.

The AW values for C and D are calculated. For option 1, the challenger is used for all 10 years. Equation [11.3] is applied using the following estimates:

Challenger: $P = \$750,000$ $AOC = \$50,000$
    $n = 10$ years $S = 0$

$$AW_C = -750,000(A/P,10\%,10) - 50,000 = \$-172,063$$

The second option has more complex cost estimates. The AW for the in-place system is calculated over the first 3 years. *Added to this is the capital recovery for the defender follow-on for the next 7 years. However in this case, the CR amount is determined over its full 10-year life.* (It is not unusual for the recovery of invested capital to be moved between projects, especially for contract work.) Refer to the AW components as AW_{DC} (subscript DC for defender-current) and AW_{DF} (subscript DF for defender follow-on). The final cash flow diagrams are shown in Figure 11–6.

Defender current: market value $= \$70,000$ $AOC = \$30,000$
    $n = 3$ years $S = 0$

$$AW_{DC} = [-70,000 - 30,000(P/A,10\%,3)](A/P,10\%,10) = \$-23,534$$

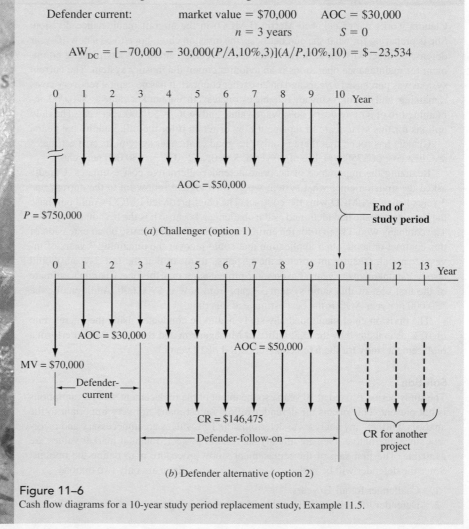

(a) Challenger (option 1)

(b) Defender alternative (option 2)

Figure 11–6
Cash flow diagrams for a 10-year study period replacement study, Example 11.5.

Defender follow-on: $P = \$900,000$, $n = 10$ years for capital recovery calculation only, AOC $= \$50,000$ for years 4 through 10, $S = 0$.

The CR and AW for all 10 years are

$$\text{CR}_{DF} = -900,000(A/P,10\%,10) = \$-146,475 \qquad [11.6]$$

$$\text{AW}_{DF} = (-146,475 - 50,000)(F/A,10\%,7)(A/F,10\%,10) = \$-116,966$$

Total AW_D for the defender is the sum of the two annual worth values above. This is the AW for option 2.

$$\text{AW}_D = \text{AW}_{DC} + \text{AW}_{DF} = -23,534 - 116,966 = \$-140,500$$

Option 2 has a lower cost ($\$-140,500$ versus $\$-172,063$). Retain the defender now and expect to purchase the follow-on system 3 years hence.

Comment

The capital recovery cost for the defender follow-on will be borne by some yet-to-be-identified project for years 11 through 13. If this assumption were not made, its capital recovery cost would be calculated over 7 years, not 10, in Equation [11.6], increasing CR to $\$-184,869$. This raises the annual worth to $\text{AW}_D = \$-163,357$. The defender alternative (option 1) is still selected.

EXAMPLE 11.6

Three years ago Chicago's O'hare Airport purchased a new fire truck. Because of flight increases, new fire-fighting capacity is needed once again. An additional truck of the same capacity can be purchased now, or a double-capacity truck can replace the current fire truck. Estimates are presented below. Compare the options at 12% per year using (a) a 12-year study period and (b) a 9-year study period.

	Presently Owned	New Purchase	Double Capacity
First cost P, \$	$-151,000$ (3 years ago)	$-175,000$	$-190,000$
AOC, \$	$-1,500$	$-1,500$	$-2,500$
Market value, \$	70,000	—	—
Salvage value, \$	10% of P	12% of P	10% of P
Life, years	12	12	12

Solution

Identify option 1 as retention of the presently owned truck and augmentation with a new same-capacity vehicle. Define option 2 as replacement with the double-capacity truck.

	Option 1		Option 2
	Presently Owned	Augmentation	Double Capacity
P, \$	$-70,000$	$-175,000$	$-190,000$
AOC, \$	$-1,500$	$-1,500$	$-2,500$
S, \$	15,100	21,000	19,000
n, years	9	12	12

(a) For a full-life 12-year study period of option 1,

$$AW_1 = (\text{AW of presently owned}) + (\text{AW of augmentation})$$
$$= [-70{,}000(A/P,12\%,9) + 15{,}100(A/F,12\%,9) - 1500]$$
$$\quad + [-175{,}000(A/P,12\%,12) + 21{,}000(A/F,12\%,12) - 1500]$$
$$= -13{,}616 - 28{,}882$$
$$= \$-42{,}498$$

This computation assumes the equivalent services provided by the current fire truck can be purchased at $\$-13{,}616$ per year for years 10 through 12.

$$AW_2 = -190{,}000(A/P,12\%,12) + 19{,}000(A/F,12\%,12) - 2500$$
$$= \$-32{,}386$$

Replace now with the double-capacity truck (option 2) at an advantage of $10,112 per year.

(b) The analysis for an abbreviated 9-year study period is identical, except that $n = 9$ in each factor; that is, 3 fewer years are allowed for the augmentation and double-capacity trucks to recover the capital investment plus a 12% per year return. The salvage values remain the same since they are quoted as a percentage of P for all years.

$$AW_1 = \$-46{,}539 \qquad AW_2 = \$-36{,}873$$

Option 2 is again selected; however, now the economic advantage is smaller. If the study period were abbreviated more severely, at some point the decision should reverse. If this example were solved by computer, the n values in the PMT functions could be decreased to determine if and when the decision reverses from option 2 to 1.

If there are several options for the number of years that the defender may be retained before replacement with the challenger, the first step of the replacement study—succession options and AW values—must include all the viable options. For example, if the study period is 5 years, and the defender will remain in service 1 year, or 2 years, or 3 years, cost estimates must be made to determine AW values for each defender retention period. In this case, there are four options; call them W, X, Y, and Z.

Option	Defender Retained	Challenger Serves
W	3 years	2 years
X	2	3
Y	1	4
Z	0	5

The respective AW values for defender retention and challenger use define the cash flows for each option. Example 11.7 illustrates the procedure.

EXAMPLE 11.7

Amoco Canada has oil field equipment placed into service 5 years ago for which a replacement study has been requested. Due to its special purpose, it has been decided that the current equipment will have to serve for either 2, 3, or 4 more years before replacement. The equipment has a current market value of $100,000, which is expected to decrease by $25,000 per year. The AOC is constant now, and is expected to remain so, at $25,000 per year. The replacement challenger is a fixed-price contract to provide the same services at $60,000 per year for a minimum of 2 years and a maximum of 5 years. Use MARR of 12% per year to perform a replacement study over a 6-year period to determine when to sell the current equipment and purchase the contract services.

Solution

Since the defender will be retained for 2, 3, or 4 years, there are three viable options (X, Y, and Z).

Option	Defender Retained	Challenger Serves
X	2 years	4 years
Y	3	3
Z	4	2

The defender annual worth values are identified with subscripts D2, D3, and D4 for the number of years retained.

$$AW_{D2} = -100,000(A/P,12\%,2) + 50,000(A/F,12\%,2) - 25,000 = \$-60,585$$
$$AW_{D3} = -100,000(A/P,12\%,3) + 25,000(A/F,12\%,3) - 25,000 = \$-59,226$$
$$AW_{D4} = -100,000(A/P,12\%,4) - 25,000 = \$-57,923$$

For all options, the challenger has an annual worth of

$$AW_C = \$-60,000$$

Table 11–3 presents the cash flows and PW values for each option over the 6-year study period. A sample PW computation for option Y is

$$PW_Y = -59,226(P/A,12\%,3) - 60,000(F/A,12\%,3)(P/F,12\%,6) = \$-244,817$$

Option Z has the lowest cost PW value ($-240,369). Keep the defender all 4 years, then replace it. Obviously, the same answer will result if the annual worth, or future worth, of each option is calculated at the MARR.

TABLE 11–3 Equivalent Cash Flows and PW Values for a 6-Year Study Period Replacement Analysis, Example 11.7

Option	Time in Service, Years Defender	Time in Service, Years Challenger	AW Cash Flows for Each Option, $/Year 1	2	3	4	5	6	Option PW, $
X	2	4	−60,585	−60,585	−60,000	−60,000	−60,000	−60,000	−247,666
Y	3	3	−59,226	−59,226	−59,226	−60,000	−60,000	−60,000	−244,817
Z	4	2	−57,923	−57,923	−57,923	−57,923	−60,000	−60,000	−240,369

> **Comment**
> If the study period is long enough, it is possible that the ESL of the challenger should be determined and its AW value used in developing the options and cash flow series. An option may include more than one life cycle of the challenger for its ESL period. Partial life cycles of the challenger can be included. Regardless, any years beyond the study period must be disregarded for the replacement study, or treated explicitly, in order to ensure that equal-service comparison is maintained, especially if PW is used to select the best option.

CHAPTER SUMMARY

It is important in a replacement study to compare the best challenger with the defender. *Best (economic) challenger is described as the one with the lowest annual worth (AW) of costs for some period of years.* If the expected remaining life of the defender and the estimated life of the challenger are specified, the AW values over these years are determined and the replacement study proceeds. However, if reasonable estimates of the expected market value (MV) and AOC for each year of ownership can be made, these year-by-year (marginal) costs help determine the best challenger.

The economic service life (ESL) analysis is designed to determine the best challenger's years of service and the resulting lowest total AW of costs. The resulting n_C and AW_C values are used in the replacement study procedure. The same analysis can be performed for the ESL of the defender.

Replacement studies in which no study period (planning horizon) is specified utilize the annual worth method of comparing two unequal-life alternatives. The better AW value determines how long the defender is retained before replacement.

When a study period is specified for the replacement study, it is vital that market value and cost estimates for the defender be as accurate as possible. When the defender's remaining life is shorter than the study period, it is critical that the cost for continuing service be estimated carefully. All the viable options for using the defender and challenger are enumerated, and their AW equivalent cash flows are determined. For each option, the PW or AW value is used to select the best option. This option determines how long the defender is retained before replacement.

PROBLEMS

Foundations of Replacement

11.1 Identify the basic assumptions made specifically about the challenger alternative when a replacement study is performed.

11.2 In a replacement analysis, what numerical value should be used as the first cost for the defender? How is this value best obtained?

11.3 Why is it important to take a consultant's viewpoint in a replacement analysis?

11.4 Chris is tired of driving the old used car she bought 2 years ago for $18,000. She estimates it is worth about $8000 now. A car salesman gave her this deal: "Look, I'll give you $10,000 in trade for this year's model. This is $2000 more than you expect, and it is $3000 more than the current official *Kelly Blue Book* value for your car. Our sales price for your new car is only $28,000, which is $6000 less than the manufacturer's sticker price of $34,000. Considering the extra $3000 on the trade-in and the $6000 reduction from sticker, you are paying $9000 less for the new car. So, I am giving you a great deal, and you get $2000 more for your old clunker than you estimated it was worth. So, let's trade now. Okay?" If Chris were to perform a replacement study at this moment, what is the correct first cost for (*a*) the defender and (*b*) the challenger?

11.5 New microelectronics testing equipment was purchased 2 years ago by Mytesmall Industries at a cost of $600,000. At that time, it was expected to be used for 5 years and then traded or sold for its salvage value of $75,000. Expanded business in newly developed international markets is forcing the decision to trade now for a new unit at a cost of $800,000. The current equipment could be retained, if necessary, for another 2 years, at which time it would have a $5000 estimated market value. The current unit is appraised at $350,000 on the international market, and if it is used for another 2 years, it will have M&O costs (exclusive of operator costs) of $125,000 per year. Determine the values of P, n, S, and AOC for this defender if a replacement analysis were performed today.

11.6 Buffett Enterprises installed a new fire monitoring and control system for its manufacturing process lines in California exactly 2 years ago for $450,000 with an expected life of 5 years. The market value was described then by the relation $400,000 - 50,000k^{1.4}$, where k was the years from time of purchase. Previous experience with fire monitoring equipment indicated that its annual operating costs follow the relation $10,000 + 100k^3$. If the relations are correct over time, determine the values of P, S, and AOC for this defender if a replacement analysis is performed (*a*) now with a study period of 3 years specified and (*b*) 2 years from now with no study period identified.

11.7 A machine purchased 1 year ago for $85,000 costs more to operate than anticipated. When purchased, the machine was expected to be used for 10 years with annual maintenance costs of $22,000 and a $10,000 salvage value. However, last year, it cost the company $35,000 to maintain it, and these costs are expected to escalate to $36,500 this year and increase by $1500 each year thereafter. The salvage value is now estimated to be $85,000 - $10,000k$, where k is the number of years since the machine was purchased. It is now estimated that this machine will be useful for a maximum of 5 more years. Determine the values of P, AOC, n, and S for a replacement study performed now.

Economic Service Life

11.8 Halcrow, Inc., expects to replace a downtime tracking system currently installed on CNC machines. The challenger system has a first cost of $70,000, an estimated annual operating cost of $20,000, a maximum useful life of 5 years, and a $10,000 salvage value anytime it is replaced. At an interest rate of 10% per year, determine its economic service life and corresponding AW value. Work this problem using a hand calculator.

11.9 Use a spreadsheet to work Problem 11.8 and plot the total AW curve and its components, (a) using the estimates originally made and (b) using new, more precise estimates, namely, an expected maximum life of 10 years, an AOC that will increase by 15% per year from the initial estimate of $20,000, and a salvage value that is expected to decrease by $1000 per year from the $10,000 estimated for the first year.

11.10 An asset with a first cost of $250,000 is expected to have a maximum useful life of 10 years and a market value that decreases $25,000 each year. The annual operating cost is expected to be constant at $25,000 per year for 5 years and to increase at a substantial 25% per year thereafter. The interest rate is a low 4% per year, because the company, Public Services Corp., is majority-owned by a municipality and regarded as a semiprivate corporation that enjoys public project interest rates on its loans. (a) Verify that the ESL is 5 years. Is the ESL sensitive to the changing market value and AOC estimates? (b) The engineer doing a replacement analysis determines that this asset should have an ESL of 10 years when it is pitted against any challenger. If the estimated AOC series has proved to be correct, determine the minimum market value that will make ESL equal 10 years. Solve by hand or spreadsheet as instructed.

11.11 A new gear grinding machine for composite materials has a first cost of $P = \$100,000$ and can be used for a maximum of 6 years. Its salvage value is estimated by the relation $S = P(0.85)^n$,

where n is the number of years after purchase. The operating cost will be $75,000 the first year and will increase by $10,000 per year thereafter. Use $i = 18\%$ per year.

(a) Determine the economic service life and corresponding AW of this challenger.

(b) It was hoped that the machine is economically justified for retention for all 6 years, but that is not the case since the ESL in part (a) is considerably less than 6 years. Determine the reduction in first cost that would have to be negotiated to make the equivalent annual cost for a full 6 years of ownership numerically equal to the AW estimate determined for the calculated ESL. Assume all other estimates remain the same, and neglect the fact that this lower P value will still not make a newly calculated ESL equal 6 years.

11.12 (a) Set up a general (cell reference format) spreadsheet that will indicate the ESL and associated AW value for any challenger asset that has a maximum useful life of 10 years. The relation for AW should be a single-cell formula to calculate AW for each year of ownership, using all the necessary estimates.

(b) Use your spreadsheet to find the ESL and AW values for the estimates tabulated. Assume $i = 10\%$ per year.

Year	Estimated Market Value, $	Estimated AOC, $
0	80,000	0
1	60,000	60,000
2	50,000	65,000
3	40,000	70,000
4	30,000	75,000
5	20,000	80,000

11.13 A piece of equipment has a first cost of $150,000, a maximum useful life of 7 years, and a salvage value described by $S = 120,000 - 20,000k$, where k is the number of years since it was purchased. The salvage value does not go below zero. The AOC series is estimated using $AOC = 60,000 + 10,000k$. The interest rate is 15% per year. Determine the economic service life (a) by hand solution, using regular AW computations, and (b) by computer, using annual marginal cost estimates.

11.14 Determine the economic service life and corresponding AW for a machine that has the following cash flow estimates. Use an interest rate of 14% per year and hand solution.

Year	Salvage Value, $	Operating Cost, $
0	100,000	—
1	75,000	−28,000
2	60,000	−31,000
3	50,000	−34,000
4	40,000	−34,000
5	25,000	−34,000
6	15,000	−45,000
7	0	−49,000

11.15 Use the annual marginal costs to find the economic service life for Problem 11.14 on a spreadsheet. Assume the salvage values are the best estimates of future market value. Develop an Excel chart of annual marginal costs (MC) and AW of MC over the 7 years.

Replacement Study

11.16 During a 3-year period Shanna, a project manager with Sherholme Medical Devices, performed replacement studies on microwave-based cancer detection equipment used in the diagnostic labs. She tabulated the ESL and AW values each year.

(a) What decision should be made each year?

(b) From the data, describe what changes took place in the defender and challenger over the 3 years.

	Maximum Life, Years	ESL Years	AW, $/Year
First Year, 200X			
Defender	3	3	−10,000
Challenger 1	10	5	−15,000
Second Year, 200X+1			
Defender	2	1	−14,000
Challenger 1	10	5	−15,000
Third Year, 200X+2			
Defender	1	1	−14,000
Challenger 2	5	3	−9,000

11.17 A consulting aerospace engineer at Aerospatial estimated AW values for a presently owned, highly accurate steel rivet inserter based on company records of similar equipment.

If Retained This Number of Years	AW Value, $/Year
1	−62,000
2	−50,000
3	−47,000
4	−53,000
5	−70,000

A challenger has ESL = 2 years and AW_C = $−49,000 per year. If the consultant must recommend a replace/retain decision today, should the company purchase the challenger? The MARR is 15% per year.

11.18 If a replacement study is performed and the defender is selected for retention for n_D years, explain what should be done 1 year later if a new challenger is identified.

11.19 BioHealth, a biodevice systems leasing company, is considering a new equipment purchase to replace a currently owned asset that was purchased 2 years ago for $250,000. It is appraised at a current market value of only $50,000. An upgrade is possible for $200,000 now that would be adequate for another 3 years of lease rights, after which the entire system could be sold on the international circuit for an estimated $40,000. The challenger can be purchased at a cost of $300,000, has an expected life of 10 years, and has a $50,000 salvage value. Determine whether the company should upgrade or replace at a MARR of 12% per year. Assume the AOC estimates are the same for both alternatives.

11.20   For the estimates in Problem 11.19, use a spreadsheet-based analysis to determine the maximum first cost for the augmentation of the current system that will make the defender and challenger break even. Is this a maximum or minimum for the upgrade, if the current system is to be retained?

11.21   A lumber company that cuts fine woods for cabinetry is evaluating whether it should retain the current bleaching system or replace it with a new one. The relevant costs for each system are known or estimated. Use an interest rate of 10% per year to (a) perform the replacement analysis and (b) determine the minimum resale price needed to make the challenger replacement choice now. Is this a reasonable amount to expect for the current system?

	Current System	New System
First cost 7 years ago, $	−450,000	
First cost, $		−700,000
Remaining life, years	5	10
Current market value, $	50,000	
AOC, $ per year	−160,000	−150,000
Future salvage, $	0	50,000

11.22   Five years ago, the Nuyork Port Authority purchased several containerized transport vehicles for $350,000 each. Last year a replacement study was performed with the decision to retain the vehicles for 2 more years. However, this year the situation has changed in that each transport vehicle is estimated to have a value of only $8000 now. If they are kept in service, upgrading at a cost of $50,000 will make them useful for up to 2 more years. Operating cost is expected to be $10,000 the first year and $15,000 the second year, with no salvage value at all. Alternatively, the company can purchase a new vehicle with an ESL of 7 years, no salvage value, and an equivalent annual cost of $−55,540 per year. The MARR is 10% per year. If the budget to upgrade the current vehicles is available this year, use these estimates to determine (a) when the company should replace the upgraded vehicles and (b) the minimum future salvage value of a new vehicle necessary to indicate that purchasing now is economically advantageous to upgrading.

11.23   Annabelle went to work this month for Caterpillar, a heavy equipment manufacturing company. When asked to verify the results of a replacement study that concluded in favor of the challenger, a new piece of heavy-duty metal forming equipment for the bulldozer processing plant, at first she concurred because the numerical results were in favor of this challenger.

	Challenger	Defender
Life, years	4	6 more
AW, $ per year	−80,000	−130,000

Curious about past decisions of this same kind, she learned that similar replacement analyses had been performed three previous times every 2 years for the same category of equipment. The decision was consistently to replace with the then-current challenger. During her study, Annabelle concluded that the ESL values were not determined prior to comparing AW values in the analyses made 6, 4, and 2 years ago. She reconstructed as best as possible the analyses for estimated life,

ESL, and associated AW values as tabulated. All cost amounts are rounded and in $1000-per-year units. Determine the two sets of replacement study conclusions (that is, life-based and ESL-based), and decide if Annabelle is correct in her initial conclusion that were the ESL and AW values calculated, the pattern of replacement decisions would have been significantly different.

Study Performed This Many Years Ago	Defender				Challenger			
	Life, Years	AW, $/Year	ESL, Years	AW, $/Year	Life, Years	AW, $/Year	ESL, Years	AW, $/Year
6	5	−140	2	−100	8	−130	7	−80
4	6	−130	5	−80	5	−120	3	−90
2	3	−140	3	−80	8	−130	8	−120
Now	6	−130	1	−100	4	−80	3 or 4	−80

11.24 Herald Richter and Associates 5 years ago purchased for $45,000 a microwave signal graphical plotter for corrosion detection in concrete structures. It is expected to have the market values and annual operating costs shown for the rest of its useful life of up to 3 years. It could be traded now at an appraised market value of $8000.

Year	Market Value at End of Year, $	AOC, $
1	6000	−50,000
2	4000	−53,000
3	1000	−60,000

A replacement plotter with new Internet-based, digital technology costing $125,000 has an estimated $10,000 salvage value after its 5-year life and an AOC of $31,000 per year. At an interest rate of 15% per year, determine how many more years Richter should retain

the present plotter. Solve (a) by hand and (b) using a spreadsheet.

11.25 What is meant by the opportunity-cost approach in a replacement study?

11.26 Why is it suggested that the cash flow approach not be used when one is performing a replacement study?

11.27 Two years ago, Geo-Sphere Spatial, Inc. (GSSI) purchased a new GPS tracker system for $1,500,000. The estimated salvage value was $50,000 after 9 years. Currently the expected remaining life is 7 years with an AOC of $75,000 per year. A French corporation, La Aramis, has developed a challenger that costs $400,000 and has an estimated 12-year life, $35,000 salvage value, and AOC of $50,000 per year. If the MARR = 12% per year, use a spreadsheet or hand solution (as instructed) to (a) find the minimum trade-in value necessary now to make the challenger economically advantageous, and (b) determine the number of years to retain the defender to just break even if the trade-in offer is $150,000. Assume the $50,000 salvage value can be realized for all retention periods up to 7 years.

11.28 Three years ago, Mercy Hospital significantly improved its hyperbaric oxygen (HBO) therapy equipment for advanced treatment of problem wounds, chronic bone infections, and radiation injury. The equipment cost $275,000 then and can be used for up to 3 years more. If the HBO system is replaced now, the hospital can realize $20,000. If retained, the market values and operating costs tabulated are estimated. A new system, made of a composite material, is cheaper to purchase initially at $150,000 and cheaper to operate during its initial years. It has a maximum life of 6 years, but market values and AOC change significantly after 3 years of use due to the projected deterioration of the composite material used in construction. Additionally, a recurring cost of $40,000 per year to inspect and rework the composite material is anticipated after 4 years of use. Market values, operating cost, and material rework estimates are tabulated. On the basis of these estimates and $i = 15\%$ per year, what are the ESL and AW values for the defender and challenger, and in what year should the current HBO system be replaced? Work this problem by hand. (See Problems 11.29 and 11.31 for more questions using these estimates.)

| | Current HBO System | | Proposed HBO System | | |
| | Market | | Market | | Material |
Year	Value, $	AOC, $	Value, $	AOC, $	Rework, $
1	10,000	−50,000	65,000	−10,000	
2	6,000	−60,000	45,000	−14,000	
3	2,000	−70,000	25,000	−18,000	
4			5,000	−22,000	
5			0	−26,000	−40,000
6			0	−30,000	−40,000

11.29 Refer to the estimates of Problem 11.28.
(a) Work the problem, using a spreadsheet.
(b) Use Excel's SOLVER to determine the maximum allowed rework cost of the challenger's composite material in years 5 and 6 such that the challenger's AW value for 6 years will exactly equal the defender's AW value at its ESL. Explain the impact of this lower rework cost on the conclusion of the replacement study.

Replacement Study over a Study Period

11.30 Consider two replacement studies to be performed using the same defenders and challengers and the same estimated costs. For the first study, no study period is specified; for the second, a study period of 5 years is specified.
(a) State the difference in the fundamental assumptions of the two replacement studies.

(b) Describe the differences in the procedures followed in performing the replacement studies for the conditions.

11.31 Reread the situation and estimates explained in Problem 11.28. (a) Perform the replacement study for a fixed study period of 5 years. (b) If, in lieu of the challenger purchase, a full-service contract for hyperbaric oxygen therapy were offered to Mercy Hospital for a total of $85,000 per year if contracted for 4 or 5 years or $100,000 for a 3-year or less contract, which option or combination is economically the best between the defender and the contract?

11.32 An in-place machine has an equivalent annual worth of $-200,000 for each year of its maximum remaining useful life of 2 years. A suitable replacement is determined to have equivalent annual worth values of $-300,000, $-225,000, and $-275,000 per year, if kept for 1, 2, or 3 years, respectively. When should the company replace the machine, if it uses a fixed 3-year planning horizon? Use an interest rate of 18% per year.

11.33 Use a spreadsheet to perform a replacement analysis for the following situation. An engineer estimates that the equivalent annual worth of an existing machine over its remaining useful life of 3 years is $-90,000 per year. It can be replaced now or after 3 years with a machine that will have an AW of $-90,000 per year if kept for 5 years or less and $-110,000 per year if kept for 6 to 8 years.
(a) Perform the analysis to determine the AW values for study periods of length 5 through 8 years at an interest rate of 10% per year. Select the study period with the lowest AW

value. How many years are the defender and challenger used?
(b) Can the PW values be used to select the best study period length and decide to retain or replace the defender? Why or why not?

11.34 Nabisco Bakers currently employs staff to operate the equipment used to sterilize much of the mixing, baking, and packaging facilities in a large cookie and cracker manufacturing plant in Iowa. The plant manager, who is dedicated to cutting costs but not sacrificing quality and cleanliness, has the projected data were the current system retained for up to its maximum expected life of 5 years. A contract company has proposed a turnkey sanitation system for $5.0 million per year if Nabisco signs on for 4 to 10 years and $5.5 million per year for a smaller number of years.
(a) At an MARR = 8% per year, perform a replacement study for the plant manager with a fixed planning horizon of 5 years, when it is anticipated that the plant will be shut down due to age of the facility and projected technological obsolescence. As you perform the study, take into account the fact that regardless of the number of years that the current sanitation system is retained, a one-time close-down cost will be incurred for personnel and equipment during the last year of operation.
(b) What is the percentage change in AW amount each year of the 5-year study period? If the decision to retain the current sanitation system is made, what is the economic disadvantage in AW amount compared to that of the best economic retention period?

| Current Sanitation System Estimates | | |
Years Retained	AW, $/Year	Close-down Expense Last Year of Retention, $
0		−3,000,000
1	−2,300,000	−2,500,000
2	−2,300,000	−2,000,000
3	−3,000,000	−1,000,000
4	−3,000,000	−1,000,000
5	−3,500,000	−500,000

11.35   A machine that was purchased 3 years ago for $140,000 is now too slow to satisfy increased demand. The machine can be upgraded now for $70,000 or sold to a smaller company for $40,000. The current machine will have an annual operating cost of $85,000 per year. If upgraded,

the presently owned machine will be kept in service for only 3 more years, then replaced with a machine that will be used in the manufacture of several other product lines. This replacement machine, which will serve the company now and for at least 8 years, will cost $220,000. Its salvage value will be $50,000 for years 1 through 5; $20,000 after 6 years; and $10,000 thereafter. It will have an estimated operating cost of $65,000 per year. The company asks you to perform an economic analysis at 20% per year, using a 5-year time horizon. Should the company replace the presently owned machine now, or do it 3 years from now? What are the AW values?

FE REVIEW PROBLEMS

11.36   Equipment purchased 2 years ago for $70,000 was expected to have a useful life of 5 years with a $5000 salvage value. Its performance was less than expected, and it was upgraded for $30,000 one year ago. Increased demand now requires that the equipment be upgraded again for an additional $25,000 or replaced with new equipment that will cost $85,000. If replaced, the existing equipment will be sold for $6000. In conducting a replacement study, the first cost that should be used for the presently owned machine is:
(a) $31,000
(b) $25,000
(c) $6,000
(d) $22,000

11.37   In a make/buy replacement study over a 4-year study period, a subcomponent is currently purchased under contract. The challenger system necessary to make the

component inhouse has an expected useful life of up to 6 years and an economic service life of 4 years. The current contract can be extended for up to 2 more years. The number of options available for the make/buy decision is:
(a)   None
(b)   One
(c)   Two
(d)   Three

11.38   The economic service life of an asset is:
(a)   The longest time that the asset will still perform the function that it was originally purchased for.
(b)   The length of time that will yield the lowest present worth of costs.
(c)   The length of time that will yield the lowest annual worth of costs.
(d)   The time required for its market value to reach the originally estimated salvage value.

11.39   In a replacement study conducted last year, it was determined that the defender should be kept for 3 more years. Now, however, it is clear that some of the estimates made last year for this year and next year have changed substantially. The proper course of action is to:
   (*a*)   Replace the existing asset now.
   (*b*)   Replace the existing asset 2 years from now, as was determined last year.
   (*c*)   Conduct a new replacement study using the new estimates.
   (*d*)   Conduct a new replacement study using last year's estimates.

11.40   The AW values calculated for a retain-or-replace decision with no stated study period are as shown.

Year	AW to Replace, $/Year	AW to Retain, $/Year
1	−25,500	−27,000
2	−25,500	−26,500
3	−26,900	−25,000
4	−27,000	−25,900

The defender should be replaced:
   (*a*)   After 4 more years.
   (*b*)   After 3 more years.
   (*c*)   After 1 more year.
   (*d*)   Now

EXTENDED EXERCISE

ECONOMIC SERVICE LIFE UNDER VARYING CONDITIONS

New pumper system equipment is under consideration by a Gulf Coast chemical processing plant. One crucial pump moves highly corrosive liquids from specially lined tanks on intercoastal barges into storage and preliminary refining facilities dockside. Because of the variable quality of the raw chemical and the high pressures imposed on the pump chassis and impellers, a close log is maintained on the number of hours per year that the pump operates. Safety records and pump component deterioration are considered critical control points for this system. As currently planned, rebuild and M&O cost estimates are increased accordingly when cumulative operating time reaches the 6000-hour mark. Estimates made for this pump are as follows:

First cost: $−800,000
Rebuild cost: $−150,000 whenever 6000 cumulative hours are logged. Each rework will cost 20% more than the previous one. A maximum of 3 rebuilds is allowed.
M&O costs: $25,000 for each year 1 through 4
                 $40,000 per year starting the year after the first rebuild, plus 15% per year thereafter
MARR: 10% per year

Based on previous logbook data, the current estimates for number of operating hours per year are as follows:

Year	Hours per Year
1	500
2	1500
3 on	2000

Questions

1. Determine the economic service life of the pump.
2. The plant superintendent told the new engineer on the job that only one re-build should be planned for, because these types of pumps usually have their minimum cost life before the second rebuild. Determine a market value for this pump that will force the ESL to be 6 years.
3. The plant superintendent also told the safety engineer that they should not plan for a rebuild after 6000 hours, because the pump will be replaced after a total of 10,000 hours of operation. The safety engineer wants to know what the base AOC in year 1 can be to make the ESL 6 years. The engineer assumes now that the 15% growth rate applies from year 1 forward. How does this base AOC value compare with the rebuild cost after 6000 hours?

CASE STUDY

REPLACEMENT ANALYSIS FOR QUARRY EQUIPMENT

Equipment used to move raw material from the quarry to the rock crushers was purchased 3 years ago by Tres Cementos, SA. When purchased, the equipment had $P = \$85,000$, $n = 10$ years, $S = \$5000$, with an annual capacity of 180,000 metric tons. Additional equipment with a capacity of 240,000 metric tons per year is now needed. Such equipment can be purchased for $P = \$70,000$, $n = 10$ years, $S = \$8000$.

However, a consultant has pointed out that the company can construct conveyor equipment to move the material from the quarry. This will cost an esti-mated $115,000 with a life of 15 years and no sig-nificant salvage value. It will carry 400,000 metric tons per year. The company needs some way to move material to the conveyor in the quarry. The presently owned equipment can be used, but it will have

excess capacity. If new smaller-capacity equipment is purchased, there is a $15,000 market value for the currently used equipment. The smaller-capacity equipment will require a capital outlay of $40,000 with an estimated life of $n = 12$ years and $S = \$3500$. The capacity is 400,000 metric tons per year over this short distance. Monthly operating, mainte-nance, and insurance costs will average $0.01 per ton-kilometer for the movers. Corresponding costs for the conveyor are expected to be $0.0075 per metric ton.

The company wants to make 12% per year on this investment. Records show that the equipment must move raw material an average of 2.4 kilometers from the quarry to the crusher pad. The conveyor will be placed to reduce this distance to 0.75 kilometer.

Case Study Exercises

1. You have been asked to determine if the old equipment should be augmented with new equipment or if the conveyor equipment should be considered as a replacement. If replacement is more economical, which method of moving the material in the quarry should be used?

2. Because of new safety regulations, the control of dust in the quarry and at the crusher site has become a real problem and implies that new capital must be invested to improve the environment for employees, or else large fines may be imposed. The Tres Cementos president has obtained an initial quote from a subcontractor which would take over the entire raw material movement operation being evaluated here for a base annual amount of $21,000 and a variable cost of 1 cent per metric ton moved. The 10 employees in the quarry operation would be employed elsewhere in the company with no financial impact upon the estimates for this evaluation. Should this offer be seriously considered if the best estimate is that 380,000 metric tons per year would be moved by the subcontractor? Identify any additional assumptions necessary to adequately address this new question posed by the president.

CHAPTER 12

Selection from Independent Projects Under Budget Limitation

In most of the previous economic comparisons, the alternatives have been mutually exclusive; only one could be selected. If the projects are not mutually exclusive, they are categorized as independent of each other, as discussed at the beginning of Chapter 5. Now we learn techniques to select from several independent projects. It is possible to select any number of projects from none (do nothing) to all viable projects.

There is virtually always some upper limit on the amount of capital available for investment in new projects. This limit is considered as each independent project is economically evaluated. The technique applied is called the *capital budgeting method*, also referred to as capital rationing. It determines the economically best rationing of initial investment capital among independent projects. The capital budgeting method is an application of the present worth method.

The case study takes a look at the project selection dilemmas of an engineering professional society striving to serve its membership with a limited budget in a technologically changing world.

LEARNING OBJECTIVES

Purpose: Select from several independent projects when there is a capital investment limit.

This chapter will help you:

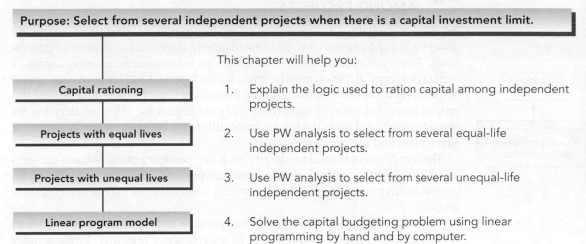

Capital rationing	1. Explain the logic used to ration capital among independent projects.
Projects with equal lives	2. Use PW analysis to select from several equal-life independent projects.
Projects with unequal lives	3. Use PW analysis to select from several unequal-life independent projects.
Linear program model	4. Solve the capital budgeting problem using linear programming by hand and by computer.

12.1   AN OVERVIEW OF CAPITAL RATIONING AMONG PROJECTS

Investment capital is a scarce resource for all corporations; thus there is virtually always a limited amount to be distributed among competing investment opportunities. When a corporation has several options for placing investment capital, a "reject or accept" decision must be made for each project. Effectively, each option is independent of other options, so the evaluation is performed on a project-by-project basis. Selection of one project does not impact the selection decision for any other project. This is the fundamental difference between mutually exclusive alternatives and independent projects.

The term *project* is used to identify each independent option. We use the term *bundle* to identify a collection of independent projects. The term mutually exclusive alternative continues to identify a project when only one may be selected from several.

There are two exceptions to purely independent projects: A *contingent project* is one that has a condition placed upon its acceptance or rejection. Two examples of contingent projects A and B are as follows: A cannot be accepted unless B is accepted; and A can be accepted in lieu of B, but both are not needed. A *dependent project* is one that must be accepted or rejected based on the decision about another project(s). For example, B must be accepted if both A and C are accepted. In practice, these complicating conditions can be bypassed by forming packages of related projects that are economically evaluated themselves as independent projects with the remaining, unconditioned projects.

The *capital budgeting* problem has the following characteristics:

1. Several independent projects are identified, and net cash flow estimates are available.
2. Each project is either selected entirely or not selected; that is, partial investment in a project is not possible.
3. A stated budgetary constraint restricts the total amount available for investment. Budget constraints may be present for the first year only or for several years. This investment limit is identified by the symbol b.
4. The objective is to maximize the return on the investments using a measure of worth, usually the PW value.

By nature, independent projects are usually quite different from one another. For example, in the public sector, a city government may develop several projects to choose from: drainage, city park, street widening, and an upgraded public bus system. In the private sector, sample projects may be: a new warehousing facility, expanded product base, improved quality program, upgraded information system, or acquisition of another firm. The typical capital budgeting problem is illustrated in Figure 12–1. For each independent project there is an initial investment, project life, and estimated net cash flows that can include a salvage value.

Present worth analysis is the recommended method to select projects. The selection guideline is as follows:

Accept projects with the best PW values determined at the MARR over the project life, provided the investment capital limit is not exceeded.

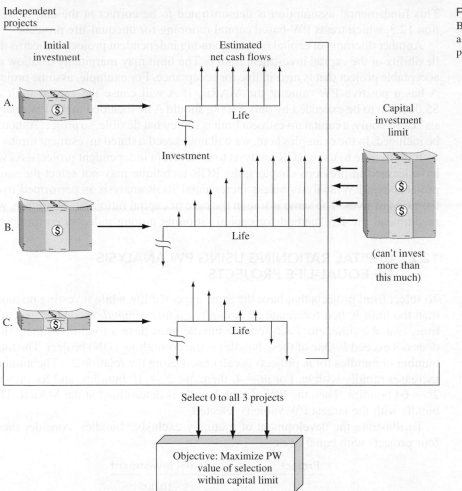

Independent projects

Figure 12–1
Basic characteristics of
a capital budgeting
problem.

This guideline is not different from that used for selection in previous chapters for independent projects. As before, each project is compared with the do-nothing project; that is, incremental analysis between projects is not necessary. The primary difference now is that the amount of money available to invest is limited. Therefore, a specific solution procedure that incorporates this constraint is needed.

Previously, PW analysis required the assumption of equal service between alternatives. This assumption is not valid for capital rationing, because there is no life cycle of a project beyond its estimated life. Yet, the selection guideline is based on the *PW over the respective life of each independent project*. This means there is an implied reinvestment assumption, as follows:

All positive net cash flows of a project are reinvested at the MARR from the time they are realized until the end of the longest-lived project.

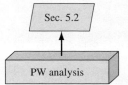

Sec. 5.2

PW analysis

This fundamental assumption is demonstrated to be correct at the end of Section 12.3, which treats PW-based capital rationing for unequal-life projects.

Another dilemma of capital rationing among independent projects concerns the flexibility of the capital investment limit b. The limit may marginally disallow an acceptable project that is next in line for acceptance. For example, assume project A has a positive PW value at the MARR. If A will cause the capital limit of $5,000,000 to be exceeded by only $1000, should A be included in the PW analysis? Commonly, a capital investment limit is somewhat flexible, so project A should be included. In the examples here, we will not exceed a stated investment limit.

It is possible to use a ROR analysis to select from independent projects. As we have learned in previous chapters, the ROR technique may not select the same projects as a PW analysis unless incremental ROR analysis is performed over the LCM of lives. The same is true in the case of capital rationing. Therefore, we recommend the PW method for capital rationing among independent projects.

12.2   CAPITAL RATIONING USING PW ANALYSIS OF EQUAL-LIFE PROJECTS

To select from projects that have the same expected life while investing no more than the limit b, first formulate all *mutually exclusive bundles*—one project at a time, two at a time, etc. Each feasible bundle must have a total investment that does not exceed b. One of these bundles is the do-nothing (DN) project. The total number of bundles for m projects is calculated using the relation 2^m. The number increases rapidly with m. For $m = 4$, there are $2^4 = 16$ bundles, and for $m = 6$, $2^6 = 64$ bundles. Then the PW of each bundle is determined at the MARR. The bundle with the largest PW value is selected.

To illustrate the development of mutually exclusive bundles, consider these four projects with equal lives.

Project	Initial Investment
A	$-10,000
B	-5,000
C	-8,000
D	-15,000

If the investment limit is $b = \$25,000$, of the 16 bundles, there are 12 feasible ones to evaluate. The bundles ABD, ACD, BCD, and ABCD have investment totals that exceed $25,000. The viable bundles are shown below.

Projects	Total Initial Investment	Projects	Total Initial Investment
A	$-10,000	AD	$-25,000
B	-5,000	BC	-13,000
C	-8,000	BD	-20,000
D	-15,000	CD	-23,000
AB	-15,000	ABC	-23,000
AC	-18,000	Do nothing	0

The procedure to solve a capital budgeting problem using PW analysis is as follows:

1. Develop all mutually exclusive bundles with a total initial investment that does not exceed the capital limit b.
2. Sum the net cash flows NCF_{jt} for all projects in each bundle j and each year t from 1 to the expected project life n_j. Refer to the initial investment of bundle j at time $t = 0$ as NCF_{j0}.
3. Compute the present worth value PW_j for each bundle at the MARR.

$$PW_j = \text{PW of bundle net cash flows} - \text{initial investment}$$

$$PW_j = \sum_{t=1}^{t=n_j} NCF_{jt}(P/F,i,t) - NCF_{j0} \qquad [12.1]$$

4. Select the bundle with the (numerically) largest PW_j value.

Selecting the maximum PW_j means that this bundle produces a return larger than any other bundle. Any bundle with $PW_j < 0$ is discarded, because it does not produce a return of MARR.

EXAMPLE 12.1

The projects review committee of Microsoft has $20 million to allocate next year to new software product development. Any or all of five projects in Table 12–1 may be accepted. All amounts are in $1000 units. Each project has an expected life of 9 years. Select the project if a 15% return is expected.

TABLE 12–1   Five Equal-Life Independent Projects ($1000 Units)

Project	Initial Investment	Annual Net Cash Flow	Project Life, Years
A	$-10,000	$2,870	9
B	-15,000	2,930	9
C	-8,000	2,680	9
D	-6,000	2,540	9
E	-21,000	9,500	9

Solution
Use the procedure above with $b = \$20,000$ to select one bundle that maximizes present worth. Remember the units are in $1000.

1. There are $2^5 = 32$ possible bundles. The eight bundles which require no more than $20,000 initial investments are described in columns 2 and 3 of Table 12–2. The $21,000 investment for E eliminates it from all bundles.

TABLE **12–2**   Summary of Present Worth Analysis of Equal-Life
Independent Projects ($1000 Units)

Bundle j (1)	Projects Included (2)	Initial Investment NCF$_{j0}$ (3)	Annual Net Cash Flow NCF$_j$ (4)	Present Worth PW$_j$ (5)
1	A	$-10,000	$2,870	$ +3,694
2	B	-15,000	2,930	-1,019
3	C	-8,000	2,680	+4,788
4	D	-6,000	2,540	+6,120
5	AC	-18,000	5,550	+8,482
6	AD	-16,000	5,410	+9,814
7	CD	-14,000	5,220	+10,908
8	Do nothing	0	0	0

2. The bundle net cash flows, column 4, are the sum of individual project net cash flows.

3. Use Equation [12.1] to compute the present worth for each bundle. Since the annual NCF and life estimates are the same for a bundle, PW$_j$ reduces to

$$PW_j = NCF_j(P/A,15\%,9) - NCF_{j0}$$

4. Column 5 of Table 12–2 summarizes the PW$_j$ values at $i = 15\%$. Bundle 2 does not return 15%, since PW$_2 < 0$. The largest is PW$_7 = \$10,908$; therefore, invest $14 million in C and D. This leaves $6 million uncommitted.

Comment

This analysis assumes that the $6 million not used in this initial investment will return the MARR by placing it in some other, unspecified investment opportunity. The return on bundle 7 exceeds 15% per year. The actual rate of return, using the relation $0 = -14,000 + 5220(P/A,i^*,9)$ is $i^* = 34.8\%$, which significantly exceeds MARR $= 15\%$.

12.3   CAPITAL RATIONING USING PW ANALYSIS OF UNEQUAL-LIFE PROJECTS

Usually independent projects do not have the same expected life. As stated in Section 12.1, the PW method for solution of the capital budgeting problem assumes that each project will last for the period of the longest-lived project n_L. *Additionally, reinvestment of any positive net cash flows is assumed to be at the MARR from the time they are realized until the end of the longest-lived project,* *that is, from year n_j through year n_L.* Therefore, the use of the least common multiple of lives is not necessary, and it is correct to use Equation [12.1] to select bundles of unequal-life projects by PW analysis using the procedure of the previous section.

EXAMPLE **12.2**

For a MARR = 15% per year and b = $20,000, select from the following independent projects. Solve by hand and by computer.

Project	Initial Investment	Annual Net Cash Flow	Project Life, Years
A	$ −8,000	$3,870	6
B	−15,000	2,930	9
C	−8,000	2,680	5
D	−8,000	2,540	4

Solution by Hand
The unequal-life values make the net cash flows vary over a bundle's life, but the selection procedure is the same as above. Of the $2^4 = 16$ bundles, 8 are economically feasible. Their PW values by Equation [12.1] are summarized in Table 12–3. As an illustration, for bundle 7:

$$PW_7 = -16,000 + 5220(P/A,15\%,4) + 2680(P/F,15\%,5) = \$235$$

Select bundle 5 (projects A and C) for a $16,000 investment.

TABLE **12–3**   Present Worth Analysis for Unequal-Life Independent Projects, Example 12.2

Bundle j (1)	Project (2)	Initial Investment, NCF_{j0} (3)	Net Cash Flows Year t (4)	Net Cash Flows NCF_{jt} (5)	Present Worth PW_j (6)
1	A	$ −8,000	1–6	$3,870	$+6,646
2	B	−15,000	1–9	2,930	−1,019
3	C	−8,000	1–5	2,680	+984
4	D	−8,000	1–4	2,540	−748
5	AC	−16,000	1–5	6,550	+7,630
			6	3,870	
6	AD	−16,000	1–4	6,410	+5,898
			5–6	3,870	
7	CD	−16,000	1–4	5,220	+235
			5	2,680	
8	Do nothing	0		0	0

Solution by Computer

Figure 12–2 presents a spreadsheet with the same information as in Table 12–3. It is necessary to initially develop the mutually exclusive bundles and total net cash flows each year. Bundle 5 (projects A and C) has the largest PW value (row 16 cells). The NPV function is used to determine PW for each bundle j over its respective life, using the format NPV(MARR,NCF_year_1: NCF_year_n_j)+investment.

E-Solve

Microsoft Excel - Example 12.2

	A	B	C	D	E	F	G	H	I
1	MARR =	15%							
2									
3	Bundle	1	2	3	4	5	6	7	8
4	Projects	A	B	C	D	AC	AD	CD	Do nothing
5	Year			Net cash flows, NCF(j,t)					
6	0	-$8,000	-$15,000	-$8,000	-$8,000	-$16,000	-$16,000	-$16,000	0
7	1	$3,870	$2,930	$2,680	$2,540	$6,550	$6,410	$5,220	0
8	2	$3,870	$2,930	$2,680	$2,540	$6,550	$6,410	$5,220	0
9	3	$3,870	$2,930	$2,680	$2,540	$6,550	$6,410	$5,220	0
10	4	$3,870	$2,930	$2,680	$2,540	$6,550	$6,410	$5,220	0
11	5	$3,870	$2,930	$2,680		$6,550	$3,870	$2,680	0
12	6	$3,870	$2,930			$3,870	$3,870		0
13	7		$2,930						0
14	8		$2,930						0
15	9		$2,930						0
16	PW Value	$6,646	-$1,019	$984	-$748	$7,630	$5,898	$235	$0

=D7+E7

=NPV($B$1,C7:C15)+C6

=NPV($B$1,H7:H15)+H6

Sheet1 / Sheet2 / Sheet3 / Sheet4 / Sheet5 / Sheet6 / Sheet7 / Sheet8

Figure 12–2
Spreadsheet analysis to select from independent projects of unequal life using the PW method of capital rationing, Example 12.2.

It is important to understand why solution of the capital budgeting problem by PW evaluation using Equation [12.1] is correct. The following logic verifies the assumption of reinvestment at the MARR for all net positive cash flows when project lives are unequal. Refer to Figure 12–3, which uses the general layout of a two-project bundle. Assume each project has the same net cash flow each year.

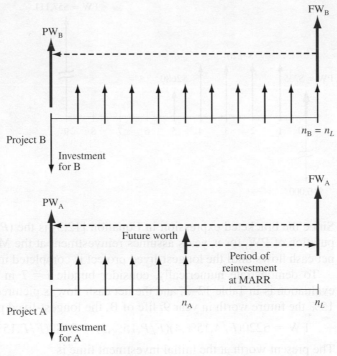

Figure 12–3
Representative cash flows used to compute PW for a bundle of two independent unequal-life projects by Equation [12.1].

The P/A factor is used for PW computation. Define n_L as the life of the longer lived project. At the end of the shorter-lived project, the bundle has a total future worth of $\mathrm{NCF}_j(F/A,\mathrm{MARR},n_j)$ as determined for each project. Now, assume reinvestment at the MARR from year n_{j+1} through year n_L (a total of $n_L - n_j$ years). The assumption of the return at the MARR is important; this PW approach does not necessarily select the correct projects if the return is not at the MARR. The results are the two future worth arrows in year n_L in Figure 12–3. Finally, compute the bundle PW value in the initial year. This is the bundle $\mathrm{PW} = \mathrm{PW_A} + \mathrm{PW_B}$. In general form, the bundle j present worth is

$$\mathrm{PW}_j = \mathrm{NCF}_j(F/A,\mathrm{MARR},n_j)(F/P,\mathrm{MARR},n_L-n_j)(P/F,\mathrm{MARR},n_L) \quad [12.2]$$

Substitute the symbol i for the MARR, and use the factor formulas to simplify.

$$\mathrm{PW}_j = \mathrm{NCF}_j \frac{(1+i)^{n_j}-1}{i}(1+i)^{n_L-n_j}\frac{1}{(1+i)^{n_L}}$$

$$= \mathrm{NCF}_j\left[\frac{(1+i)^{n_j}-1}{i(1+i)^{n_j}}\right] \quad [12.3]$$

$$= \mathrm{NCF}_j(P/A,i,n_j)$$

Figure 12–4
Initial investment and
cash flows for bundle 7,
projects C and D,
Example 17.2.

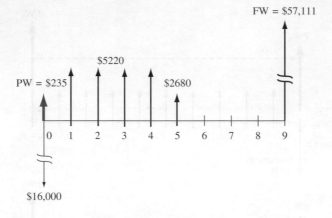

Since the bracketed expression in Equation [12.3] is the $(P/A,i,n_j)$ factor, computation of PW_j for n_j years assumes reinvestment at the MARR of all positive net cash flows until the longest-lived project is completed in year n_L.

To demonstrate numerically, consider bundle $j = 7$ in Example 12.2. The evaluation is in Table 12–3, and the net cash flow is pictured in Figure 12–4. At 15% the future worth in year 9, life of B, the longest-lived project of the four, is

$$FW = 5220(F/A,15\%,4)(F/P,15\%,5) + 2680(F/P,15\%,4) = \$57,111$$

The present worth at the initial investment time is

$$PW = -16,000 + 57,111(P/F,15\%,9) = \$235$$

The PW value is the same as PW_7 in Table 12–3 and Figure 12–2. This demonstrates the reinvestment assumption for positive net cash flows. If this assumption is not realistic, the PW analysis must be conducted using the *LCM of all project lives.*

Project selection can also be accomplished using the *incremental rate of return* procedure. Once all viable, mutually exclusive bundles are developed, they are ordered by increasing initial investment. Determine the incremental rate of return on the first bundle relative to the do-nothing bundle, and the return for each incremental investment and incremental net cash flow sequence on all other bundles. If any bundle has an incremental return less than the MARR, it is removed. The last justified increment indicates the best bundle. This approach results in the same answer as the PW procedure. There are a number of incorrect ways to apply the rate of return method, but the procedure of incremental analysis on mutually exclusive bundles ensures a correct result, as in previous applications of incremental rate of return.

12.4 CAPITAL BUDGETING PROBLEM FORMULATION USING LINEAR PROGRAMMING

The capital budgeting problem can be stated in the form of the linear programming model. The problem is formulated using the integer linear programming (ILP) model, which means simply that all relations are linear and that the variable x can take on only integer values. In this case, the variables can only take on the values 0 or 1, which makes it a special case called the 0-or-1 ILP model. The formulation in words follows.

Maximize: Sum of PW of net cash flows of independent projects.

Constraints:

- Capital investment constraint is that the sum of initial investments must not exceed a specified limit.
- Each project is completely selected or not selected.

For the math formulation, define b as the capital investment limit, and let x_k ($k = 1$ to m projects) be the variables to be determined. If $x_k = 1$, project k is completely selected; if $x_k = 0$, project k is not selected. Note that the subscript k represents each *independent project,* not a mutually exclusive bundle.

If the sum of PW of the net cash flows is Z, the math programming formulation is

Maximize:
$$\sum_{k=1}^{k=m} PW_k x_k = Z$$

$$\text{[12.4]}$$

Constraints:
$$\sum_{k=1}^{k=m} NCF_{k0} x_k \le b$$

$$x_k = 0 \text{ or } 1 \qquad \text{for } k = 1, 2, \ldots, m$$

The PW_k of each project is calculated using Equation [12.1] at MARR $= i$.

$$PW_k = \text{PW of project net cash flows for } n_k \text{ years}$$

$$= \sum_{t=1}^{t=n_k} NCF_{kt}(P/F,i,t) - NCF_{k0}$$

$$\text{[12.5]}$$

Computer solution is accomplished by a linear programming software package which treats the ILP model. Also, Excel and its optimizing tool SOLVER can be used to develop the formulation and select the projects, as illustrated in Example 12.3.

EXAMPLE 12.3

Review Example 12.2. (*a*) Formulate the capital budgeting problem using the math programming model presented in Equation [12.4], and insert the solution into the model to verify that it does indeed maximize present worth. (*b*) Set up and solve the problem using Excel.

Solution

(a) Define the subscript $k = 1$ through 4 for the four projects, which are relabeled as 1, 2, 3, and 4. The capital investment limit is $b = \$20,000$ in Equation [12.4].

Maximize:
$$\sum_{k=1}^{k=4} PW_k x_k = Z$$

Constraints:
$$\sum_{k=1}^{k=4} NCF_{k0} x_k \leq 20,000$$

$$x_k = 0 \text{ or } 1 \qquad \text{for } k = 1 \text{ through } 4$$

Calculate the PW_k for the estimated net cash flows using $i = 15\%$ and Equation [12.5].

Project k	Net Cash Flow NCF_{kt}	Life n_k	Factor $(P/A,15\%,n_k)$	Initial Investment NCF_{k0}	Project PW_k
1	$3,870	6	3.7845	$−8,000	$+6,646
2	2,930	9	4.7716	−15,000	−1,019
3	2,680	5	3.3522	−8,000	+984
4	2,540	4	2.8550	−8,000	−748

Now, substitute the PW_k values into the model, and put the initial investments in the budget constraint. Plus signs are used for all values in the capital investment constraint. We have the complete 0-or-1 ILP formulation.

Maximize: $6646x_1 - 1019x_2 + 984x_3 - 748x_4 = Z$

Constraints: $8000x_1 + 15,000x_2 + 8000x_3 + 8000x_4 < 20,000$

$$x_1, x_2, x_3, \text{ and } x_4 = 0 \text{ or } 1$$

Solution to select projects 1 and 3 is written:

$$x_1 = 1 \qquad x_2 = 0 \qquad x_3 = 1 \qquad x_4 = 0$$

for a PW value of $7630.

E-Solve

(b) Figure 12–5 presents a spreadsheet template developed to select from six or fewer independent projects with 12 years or less of net cash flow estimates per project. The spreadsheet template can be expanded in either direction if needed. Figure 12–6 shows the SOLVER parameters set to solve this example for four projects and an investment limit of $20,000. The descriptions below and the cell tags identify the contents of the rows and cells in Figure 12–5, and their linkage to SOLVER parameters.

Rows 4 and 5: Projects are identified by numbers in order to distinguish them from spreadsheet column letters. Cell I5 is the expression for Z, the sum of the PW values for the projects. This is the target cell for SOLVER to maximize (see Figure 12–6).

Rows 6 to 18: These are initial investments and net cash flow estimates for each project. Zero values that occur after the life of a project need not be entered; however, any $0 estimates that occur during a project's life must be entered.

Row 19: The entry in each cell is 1 for a selected project and 0 if not selected. These are the changing cells for SOLVER. Since each entry must be 0 or 1, a binary

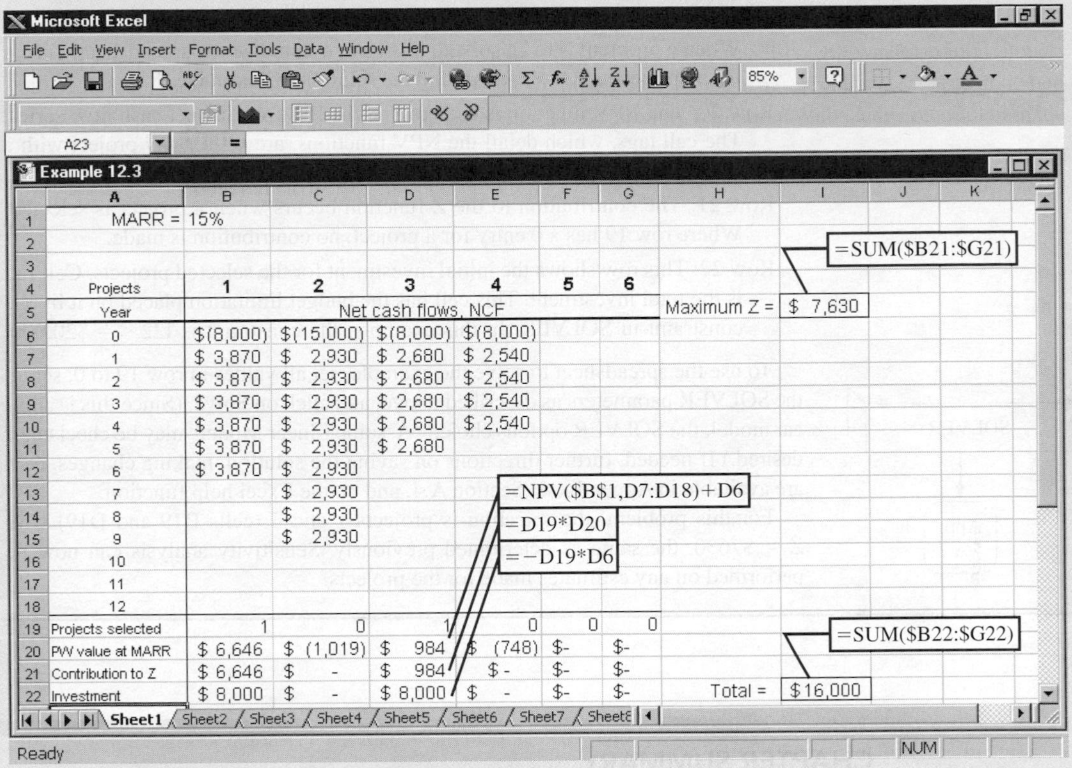

Figure 12–5
Excel spreadsheet configured to solve a capital budgeting problem, Example 12.3.

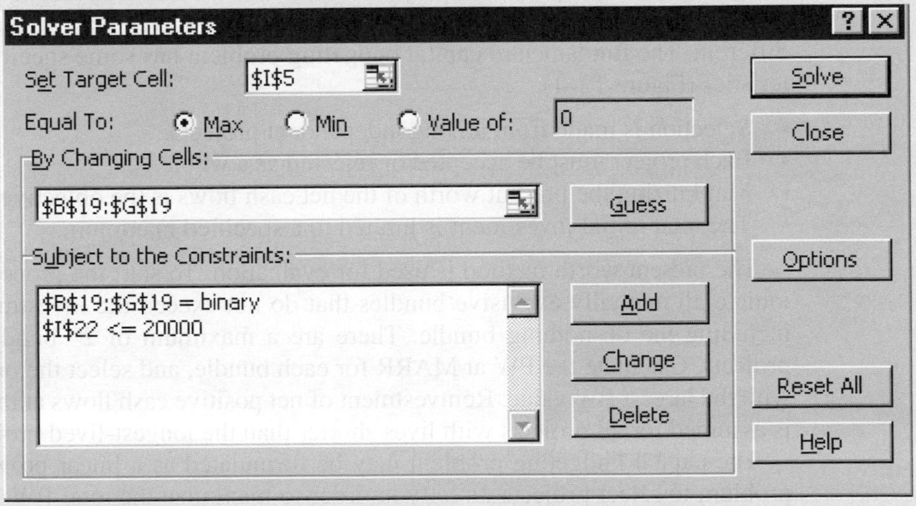

Figure 12–6
SOLVER parameters set to solve the capital budgeting problem in Example 12.3.

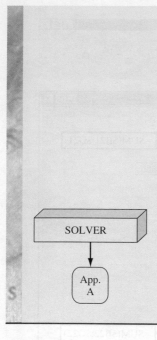

constraint is placed on all row 19 cells in SOLVER, as shown in Figure 12–6. When a problem is to be solved, it is best to initialize the spreadsheet with 0s for all projects. SOLVER will find the solution to maximize Z.

Row 20: The NPV function is used to find the PW for each net cash flow series. The cell tags, which detail the NPV functions, are set for any project with a life up to 12 years at the MARR entered in cell B1.

Row 21: The contribution to the Z function occurs when a project is selected. Where row 19 has a 0 entry for a project, no contribution is made.

Row 22: This row shows the initial investment for the selected projects. Cell I22 is the total investment. This cell has the budget limitation placed on it by the constraint in SOLVER. In this example, the constraint is I22 <= $20,000.

To use the spreadsheet to solve the example, set all values in row 19 to 0, set up the SOLVER parameters as described above, and click on Solve. (Since this is a linear model, the SOLVER options choice "Assume Linear Model" may be checked, if desired.) If needed, further directions on saving the solution, making changes, etc., are available in Appendix A, Section A.4, and on the Excel help function.

For this problem, the selection is projects 1 and 3 (cells B19 and D19) with Z = $7630, the same as determined previously. Sensitivity analysis can now be performed on any estimates made for the projects.

CHAPTER SUMMARY

Investment capital is always a scarce resource, so it must be rationed among competing projects using specific economic and noneconomic criteria. Capital budgeting involves proposed projects, each with an initial investment and net cash flows estimated over the life of the project. The lives may be the same or different. The fundamental capital budgeting problem has some specific characteristics (Figure 12–1).

- Selection is made from among independent projects.
- Each project must be accepted or rejected as a whole.
- Maximizing the present worth of the net cash flows is the objective.
- The total initial investment is limited to a specified maximum.

The present worth method is used for evaluation. To start the procedure, formulate all mutually exclusive bundles that do not exceed the investment limit, including the do-nothing bundle. There are a maximum of 2^m bundles for m projects. Calculate the PW at MARR for each bundle, and select the one bundle with the largest PW value. Reinvestment of net positive cash flows at the MARR is assumed for all projects with lives shorter than the longest-lived project.

The capital budgeting problem may be formulated as a linear programming problem to select projects directly in order to maximize the total PW. Mutually exclusive bundles are not developed using this solution approach. Excel and SOLVER can be used to solve this problem by computer.

PROBLEMS

Understanding the Capital Rationing Problem

12.1 Write a short paragraph that explains the problem of rationing investment capital among several projects that are independent of one another.

12.2 State the reinvestment assumption about project cash flows that is made when one is solving the capital budgeting problem.

12.3 Four independent projects (1, 2, 3, and 4) are to be evaluated for investment by Perfect Manufacturing. Develop all the acceptable mutually exclusive bundles based on the following selection restrictions developed by the department of engineering production:

Project 2 can be selected only if project 3 is selected.

Projects 1 and 4 should not both be selected; they are essentially duplicates.

12.4 Develop all acceptable mutually exclusive bundles for the four independent projects described below if the investment limit is $400 and the following project selection restriction applies: Project 1 can be selected only if both projects 3 and 4 are selected.

Project	Initial Investment, $
1	−250
2	−150
3	−75
4	−235

Selection from Independent Projects

12.5 (a) Determine which of the following independent projects should be selected for investment if $325,000 is available and the MARR is 10% per

year. Use the PW method to evaluate mutually exclusive bundles to make the selection.

Project	Initial Investment, $	Net Cash Flow, $/Year	Life, Years
A	−100,000	50,000	8
B	−125,000	24,000	8
C	−120,000	75,000	8
D	−220,000	39,000	8
E	−200,000	82,000	8

(b) If the five projects are mutually exclusive alternatives, perform the present worth analysis and select the best alternative.

12.6 Work Problem 12.5(a), using a spreadsheet.

12.7 The engineering department at General Tire has a total of $900,000 for no more than two projects in capital improvement for the year. Use a spreadsheet-based PW analysis and a minimum 12% per year return to answer the following.

(a) Which projects are acceptable from the three described below?

(b) What is the minimum required annual net cash flow necessary to select the bundle that expends as much as possible without violating either the budget limit or the two-project maximum restriction?

Project	Initial Investment, $	Estimated NCF, $/year	Life, Years	Salvage Value, $
A	−400,000	120,000	4	40,000
B	−200,000	90,000	4	30,000
C	−700,000	200,000	4	20,000

12.8 Jesse wants to choose exactly two independent projects from four opportunities.

Each project has an initial investment of $300,000 and a life of 5 years. The annual NCF estimates for the first three projects are available, but a detailed estimate for the fourth is not yet prepared and time has run out for the selection. Using MARR = 9% per year, determine the minimum NCF for the fourth project (Z) that will guarantee that it is part of the selected twosome.

Project	Annual NCF, $/year
W	90,000
X	50,000
Y	130,000
Z	at least 50,000

12.9 The engineer at Clean Water Engineering has established a capital investment limit of $800,000 for next year for projects that target improved recovery of highly brackish groundwater. Select any or all of the following projects, using a MARR of 10% per year. Present your solution by hand calculations, not Excel.

Project	Initial Investment, $	Annual NCF, $/Year	Life, Years	Salvage Value, $
A	−250,000	50,000	4	45,000
B	−300,000	90,000	4	−10,000
C	−550,000	150,000	4	100,000

12.10 Develop an Excel spreadsheet for the three projects in Problem 12.9. Assume that the engineer wants project C to be the only one selected. Considering the viable project options and b = $800,000, determine (a) the largest initial investment for C and (b) the largest MARR allowed to guarantee that C is selected.

12.11 Eight projects are available for selection at HumVee Motors. The listed PW values are determined at the corporate MARR of 10% per year and rounded to the nearest $1000. Project lives vary from 5 to 15 years.

Project	Initial Investment, $	PW value at 10%, $
1	−1,500,000	−50,000
2	−300,000	+35,000
3	−95,000	−9,000
4	−400,000	+75,000
5	−195,000	+125,000
6	−175,000	−27,000
7	−100,000	+62,000
8	−400,000	+110,000

Project selection guidelines:
1. No more than $400,000 in investment capital is available.
2. No negative PW project may be selected.
3. At least one project, but no more than three, must be selected.
4. The following selection restrictions apply to specific projects:
   - Project 4 can be selected only if project 1 is selected.
   - Projects 1 and 2 are duplicative; don't select both.
   - Projects 8 and 4 are also duplicative.
   - Project 7 requires that project 2 also be selected.

(a) Identify the viable project bundles and select the best economically justified projects. What is the investment assumption for any remaining capital funds?

(b) If as much of the $400,000 as possible *must* be invested, use the same restrictions and determine the project(s) to select. Is this a viable second choice for investing the $400,000? Why?

12.12 Use the analysis below of five independent projects to select the best, if the capital limitation is (a) $30,000, (b) $60,000, and (c) unlimited.

Project	Initial Investment, $	Life, Years	PW at 12% per Year, $
S	−15,000	6	8,540
A	−25,000	8	12,325
M	−10,000	6	3,000
E	−25,000	4	10
H	−40,000	12	15,350

			Estimated NCF, $/Year	
Project	Investment, $ Millions	Life, Years	Year 1	Gradient after Year 1
1	−0.9	6	250,000	−5000
2	−2.1	10	485,000	+5000
3	−1.0	5	200,000	+10%

12.13 The independent project estimates below have been developed by the engineering and finance managers. The corporate MARR is 15% per year, and the capital investment limit is $4 million.

(a) Use the PW method and hand solution to select the economically best projects.

(b) Use the PW method and computer solution to select the economically best projects.

Project	Project Cost, $ Millions	Life, Years	NCF, $/Year
1	−1.5	8	360,000
2	−3.0	10	600,000
3	−1.8	5	520,000
4	−2.0	4	820,000

12.14 The following capital rationing problem is defined. Three projects are to be evaluated at a MARR of 12.5% per year. No more than $3.0 million can be invested.

(a) Use a spreadsheet to select from the independent projects.

(b) Use SOLVER to determine the minimum year 1 NCF for project 3 alone to have the same PW as the best bundle in part (a) if project 3 life can be increased to 10 years for the same $1 million investment. All other estimates remain the same. With this increased NCF and life, what are the best projects for investment?

12.15 Use the PW method to evaluate four independent projects. Select as many as three of the four projects. The MARR is 12% per year, and an available capital investment limit is $16,000.

	Project			
	1	2	3	4
Investment, $	−5000	−8000	−9000	−10,000
Life, years	5	5	3	4
Year	NCF Estimates, $			
1	1000	500	5000	0
2	1700	500	5000	0
3	2400	500	2000	0
4	3000	500		17,000
5	3800	10,500		

12.16 Work Problem 12.15, using a spreadsheet.

12.17 Using the NCF estimates in Problem 12.15 for projects 3 and 4, demonstrate the reinvestment assumption made when the capital budgeting problem is solved for the four projects by using the PW method. (*Hint:* Refer to Equation [12.2].)

Linear Programming and Capital Budgeting

12.18 Formulate the linear programming model, develop a spreadsheet, and solve the capital rationing problem in Example 12.1 (a) as presented and (b) using an investment limit of $13 million.

12.19 For Problem 12.5, use Excel and SOLVER to (a) answer the question in part (*a*) and (*b*) select the projects if MARR = 12% per year and the investment limit is increased to $500,000.

12.20 Use SOLVER to work Problem 12.10.

12.21 Use SOLVER to find the minimum NCF required for project Z as detailed by Jesse in Problem 12.8.

12.22 Use linear programming and a spreadsheet-based solution technique to select from the independent unequal-life projects in Problem 12.13.

12.23 Solve the capital budgeting problem in Problem 12.14(*a*), using the linear programming model and Excel.

12.24 Solve the capital budgeting problem in Problem 12.15, using the linear programming model and Excel.

12.25 Using the data in Problem 12.15 and Excel solutions of the capital rationing problem for capital budget limits ranging from $b = \$5000$ to $b = \$25,000$, develop an Excel chart that plots b versus the value of Z.

CASE STUDY

LIFELONG ENGINEERING EDUCATION IN A WEB ENVIRONMENT

The Report

IME is a not-for-profit engineering professional society, headquartered in New York City with offices in several international sites. A task force was established last year with the charge to recommend ways to improve services to members in the area of lifelong learning. Overall sales to individuals, libraries, and businesses of technical journals, magazines, books, monographs, CDs, and videos have decreased by 35% over the last 3 years. IME, like virtually all for-profit corporations, is being negatively impacted by e-commerce. The just-published report of the task force contains the following conclusion and recommendations:

It is essential that IME take rapid, proactive steps to initiate web-based learning materials itself and/or in conjunction with other organizations. Topically, these materials should concentrate on areas such as:

Professional engineer certification and licensing.
Leading-edge technical topics.
Retooling topics for mature engineers.
Basic tools for individuals doing engineering analysis with inadequate training or education.

Projects should be started immediately and evaluated over the next 3 years to determine future directions of electronic learning materials for IME.

The Project Proposals

In the Action Items section of the report, four projects are identified, along with cost and net revenue estimates made on a 6-month basis. The project summaries that follow all require development and marketing of online learning materials.

Project A, niche markets. IME identifies several new technical areas and offers learning materials to members and nonmembers. An initial investment of $500,000 and a follow-up investment of another $500,000 after 18 months are necessary.

Project B, partnering. IME joins with several other professional societies to offer materials on a relatively wide spectrum. This business strategy could bring a larger investment to bear on lifelong learning materials. An initial investment by IME of $2 million is necessary. This project will require that a smaller project for network improvement be undertaken. This is project C, which follows.

Project C, web search engine. With an investment of only $200,000 six months from now, IME can offer members a web search engine for access to current publications of IME. An outside contractor can quickly install this capability on current equipment. This entry point to web-based learning is a stop-gap measure, which may increase services and revenue over the short-term only. This project is necessary if project B is pursued, but project C can be pursued separately from any other project.

Project D, service improvement. This project is a complete substitute for project B. This is a longer-term effort to improve the electronic publication and continuing education offerings of IME. Investments of $300,000 now with commitments of $400,000 in 6 months and another $300,000 after 6 more months will be necessary. Project D is slower-moving, but it will develop a firm base for most future web-based learning services of the IME.

The estimated net cash flows (in $1000) by 6-month period for IME are summarized as shown.

Period	Project			
	A	B	C	D
1	$ 0	$ 500	$ 0	$100
2	100	500	50	200
3	200	600	100	300
4	400	700	150	300
5	400	800	0	300
6	0	1000	0	300

The Finance Committee has responded that no more that $3.5 million can be committed to these projects. It also stated that the total amount per project should be committed up-front, regardless of when the initial and follow-on investment cash flows actually occur. The Finance Committee and a Board Committee will review progress each 3 months to determine if the selected projects should be continued, expanded, or discontinued. The IME capital, primarily in equity investments, has returned an average of 10% per 6-months over the last 5 years. There is no debt liability carried by IME at this time.

Case Study Exercises

1. Formulate all the investment opportunities for IME and the cash flow profiles, given the information in the task force report.

2. What projects should the Finance Committee recommend on a purely economic basis?

3. The Executive Director of IME has great interest in pursuing project D because of its perceived positive, longer-lasting effects upon the membership size of the Institute and future services offered to new and current members. Using a spreadsheet that details the project net cash flow estimates, determine some of the changes that the director may make to ensure that project D is accepted. No restrictions should be placed on this analysis; for example, investments and cash flows can be changed, and the restrictions between projects described in the task force report can be removed.

Breakeven Analysis

Breakeven analysis is performed to determine the value of a variable or parameter of a project or alternative that makes two elements equal, for example, the sales volume that will equate revenues and costs. A breakeven study is performed for two alternatives to determine when either alternative is equally acceptable, for example, the replacement value of the defender in a replacement study that makes the challenger an equally good choice (Section 11.3). Breakeven analysis is commonly applied in *make-or-buy decisions* when corporations and businesses must decide upon the source for manufactured components, services of all kinds, etc.

We have utilized the breakeven approach previously in payback analysis (Section 5.6) and for breakeven ROR analysis of two alternatives (Section 8.4). The Excel optimizing tool SOLVER, introduced and used most recently in Chapter 12 to select from independent projects, is a prime tool used to perform a computer-based breakeven analysis between two alternatives. This chapter expands our scope and understanding of performing a breakeven study.

Breakeven studies use estimates that are considered to be certain; that is, if the estimated values are expected to vary enough to possibly change the outcome, another breakeven study is necessary using different estimates. This leads to the observation that breakeven analysis is a part of the larger efforts of *sensitivity analysis.* If the variable of interest in a breakeven study is allowed to vary, the approaches of sensitivity analysis (Chapter 18) should be used. Additionally, if probability and risk assessment are considered, the tools of simulation (Chapter 19) can be used to supplement the static nature of a breakeven study.

This chapter's case study focuses on cost and efficiency measures in a public sector (municipal) setting.

LEARNING OBJECTIVES

Purpose: For one or more alternatives, determine the level of activity necessary or the value of a parameter to break even.

This chapter will help you:

| Breakeven point | 1. | Determine the breakeven value for a single project. |

| Two alternative breakeven | 2. | Calculate the breakeven value between two alternatives and use it to select one alternative. |

| Spreadsheets | 3. | Develop a spreadsheet that uses the Excel tool SOLVER to perform breakeven analysis. |

13.1 BREAKEVEN ANALYSIS FOR A SINGLE PROJECT

When one of the engineering economy symbols—P, F, A, i, or n—is not known or not estimated, a breakeven quantity can be determined by setting an equivalence relation for PW or AW equal to zero. This form of breakeven analysis has been used many times so far. For example, we have solved for the rate of return i^*, found the payback period n_p, and determined the P, F, A, or salvage value S at which a series of cash flow estimates return a specific MARR. Methods used to determine the quantity are

Direct solution by hand if only one factor is present (say, P/A) or only single amounts are estimated (for example, P and F).

Trial and error by hand when multiple factors are present.

Computer spreadsheet when cash flow and other estimates are entered into spreadsheet cells and used in resident functions, such as PV, FV, RATE, IRR, NPV, PMT, and NPER.

We now concentrate on the determination of the *breakeven quantity for a decision variable*. For example, the variable may be a design element to minimize cost, or the production level needed to realize revenues that exceed costs by 10%. This quantity, called the *breakeven point* Q_{BE}, is determined using relations for revenue and cost at different values of the variable Q. The size of Q may be expressed in units per year, percentage of capacity, hours per month, and many other dimensions.

Figure 13–1a presents different shapes of a revenue relation identified as R. A linear revenue relation is commonly assumed, but a nonlinear relation is often more realistic. It can model an increasing per unit revenue with larger volumes (curve 1 in Figure 13–1a), or a decreasing per unit price that usually prevails at higher quantities (curve 2).

Costs, which may be linear or nonlinear, usually include two components— fixed and variable—as indicated in Figure 13–1b.

Fixed costs (FC). Includes costs such as buildings, insurance, fixed overhead, some minimum level of labor, equipment capital recovery, and information systems.

Variable costs (VC). Includes costs such as direct labor, materials, indirect costs, contractors, marketing, advertisement, and warranty.

The fixed cost component is essentially constant for all values of the variable, so it does not vary for a large range of operating parameters, such as production level or workforce size. Even if no units are produced, fixed costs are incurred at some threshold level. Of course, this situation cannot last long before the plant must shut down to reduce fixed costs. Fixed costs are reduced through improved equipment, information systems and workforce utilization, less costly fringe benefit packages, subcontracting specific functions, and so on.

Variable costs change with production level, workforce size, and other parameters. It is usually possible to decrease variable costs through better product design, manufacturing efficiency, improved quality and safety, and higher sales volume.

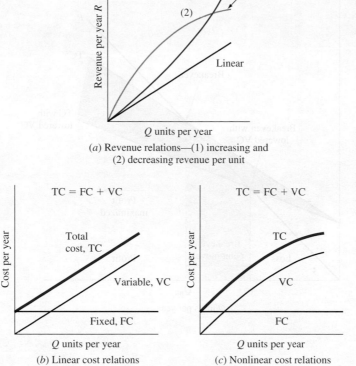

(a) Revenue relations—(1) increasing and (2) decreasing revenue per unit

(b) Linear cost relations

(c) Nonlinear cost relations

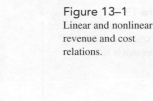

Figure 13–1
Linear and nonlinear revenue and cost relations.

When FC and VC are added, they form the total cost relation TC. Figure 13–1*b* illustrates the TC relation for linear fixed and variable costs. Figure 13–1*c* shows a general TC curve for a nonlinear VC in which unit variable costs decrease as the quantity level rises.

At a specific but unknown value Q of the decision variable, the revenue and total cost relations will intersect to identify the breakeven point Q_{BE} (Figure 13–2). If $Q > Q_{BE}$, there is a predictable profit; but if $Q < Q_{BE}$, there is a loss. For linear models of R and VC, the greater the quantity, the larger the profit. Profit is calculated as

$$\textbf{Profit = revenue} - \textbf{total cost}$$

$$= R - \text{TC} \qquad\qquad \textbf{[13.1]}$$

A relation for the breakeven point may be derived when revenue and total cost are linear functions of quantity Q by setting the relations for R and TC equal to each other, indicating a profit of zero.

$$R = \text{TC}$$

$$rQ = \text{FC} + \text{VC} = \text{FC} + vQ$$

where r = revenue per unit
         v = variable cost per unit

Figure 13–2
Effect on the breakeven point when the variable cost per unit is reduced.

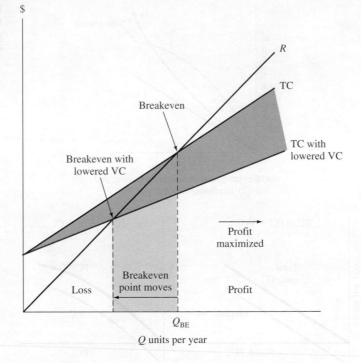

Figure 13–3
Breakeven points and maximum profit point for a nonlinear analysis.

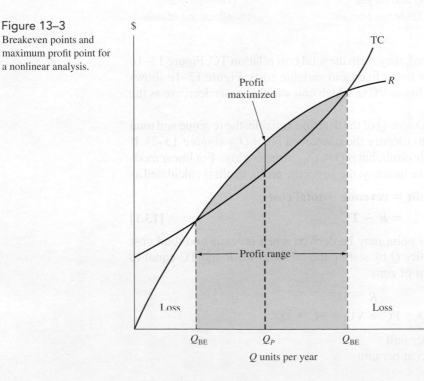

Solve for the breakeven quantity Q_{BE} to obtain

$$Q_{BE} = \frac{FC}{r - v} \qquad [13.2]$$

The breakeven graph is an important management tool because it is easy to understand and may be used in decision making and analysis in a variety of ways. For example, if the variable cost per unit is reduced, the TC line has a smaller slope (Figure 13–2), and the breakeven point will decrease. This is an advantage because the smaller the value of Q_{BE}, the greater the profit for a given amount of revenue.

If nonlinear R or TC models are used, there may be more than one breakeven point. Figure 13–3 presents this situation for two breakeven points. The maximum profit occurs at Q_P between the two breakeven points where the distance between the R and TC relations is greatest.

Of course, no static R and TC relations—linear or nonlinear—are able to estimate exactly the revenue and cost amounts over an extended period of time. But the breakeven point is an excellent target for planning purposes.

EXAMPLE 13.1

Lufkin Trailer Corporation assembles up to 30 trailers per month for 18-wheel trucks in its east coast facility. Production has dropped to 25 units per month over the last 5 months due to a worldwide economic slow down in transportation services. The following information is available.

Fixed costs	$FC = \$750{,}000$ per month
Variable cost per unit	$v = \$35{,}000$
Revenue per unit	$r = \$75{,}000$

(a) How does the reduced production level of 25 units per month compare with the current breakeven point?

(b) What is the current profit level per month for the facility?

(c) What is the difference between the revenue and variable cost per trailer that is necessary to break even at a monthly production level of 15 units, if fixed costs remain constant?

Solution

(a) Use Equation [13.2] to determine the breakeven number of units. All dollar amounts are in $1000 units.

$$Q_{BE} = \frac{FC}{r - v}$$

$$= \frac{750}{75 - 35} = 18.75 \text{ units per month}$$

Figure 13–4 is a plot of R and TC lines. The breakeven value is 18.75, or 19 in integer trailer units. The reduced production level of 25 units is above the breakeven value.

(b) To estimate profit in $1000 at $Q = 25$ units per month, use Equation [13.1].

$$\begin{aligned}
\text{Profit} &= R - TC = rQ - (FC + vQ) \\
&= (r - v)Q - FC \\
&= (75 - 35)25 - 750 \qquad [13.3]\\
&= \$250
\end{aligned}$$

There is a profit of $250,000 per month currently.

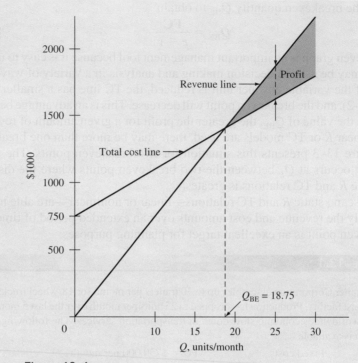

Figure 13–4
Breakeven graph, Example 13.1.

(c)   To determine the required difference $r - v$, use Equation [13.3] with profit = 0, $Q = 15$, and FC = $750,000. In $1000 units,

$$0 = (r - v)(15) - 750$$

$$r - v = \frac{750}{15} = \$50 \text{ per unit}$$

The spread between r and v must be $50,000. If v stays at $35,000, the revenue per trailer must increase from $75,000 to $85,000 just to break even at a production level of $Q = 15$ per month.

In some circumstances, breakeven analysis performed on a per unit basis is more meaningful. The value of Q_{BE} is still calculated using Equation [13.2], but the TC relation is divided by Q to obtain an expression for cost per unit, also termed *average cost per unit* C_u.

$$C_u = \frac{TC}{Q} = \frac{FC + vQ}{Q} = \frac{FC}{Q} + v \qquad [13.4]$$

At the breakeven quantity $Q = Q_{BE}$, the revenue per unit is exactly equal to the cost per unit. If graphed, the FC per unit term in Equation [13.4] takes on the shape of a hyperbola.

In Chapter 5, payback period analysis was discussed. Payback is the number of years n_p necessary to recover an initial investment. Payback analysis at a zero

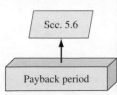
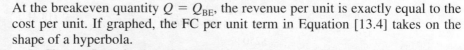

interest rate is performed only when there is no requirement to earn a rate of return greater than zero in addition to recovering the initial investment. (As discussed earlier, the technique is best used as a supplement to PW analysis at the MARR.) If payback analysis is coupled with breakeven, the quantity of the decision variable for different payback periods can be determined, as illustrated in the next example.

EXAMPLE **13.2**

The president of a local company, Online Ontime, Inc., expects a product to have a profitable life of between 1 and 5 years. He wants to know the breakeven number of units that must be sold annually to realize payback within each of the time periods of 1 year, 2 years, and so on up to 5 years. Find the answers using hand and computer solutions. The cost and revenue estimates are as follows:

Fixed costs: initial investment of $80,000 with $1000 annual operating cost.

Variable cost: $8 per unit.

Revenue: Twice the variable cost for the first 5 years and 50% of the variable cost thereafter.

Solution by Hand

Define X as the units sold per year to break even and n_p as the payback period, where $n_p = 1, 2, 3, 4,$ and 5 years. There are two unknowns and one relation, so it is necessary to establish values of one variable and solve for the other. The following approach is used: Establish annual cost and revenue relations with no time value of money considered, then use n_p values to find the breakeven value of X.

$$\text{Fixed costs} \qquad \frac{80{,}000}{n_p} + 1000$$

$$\text{Variable cost} \qquad 8X$$

$$\text{Revenue} \qquad \begin{cases} 16X & \text{years 1 through 5} \\ 4X & \text{year 6 and thereafter} \end{cases}$$

Set revenue equal to total cost, and solve for X.

$$\text{Revenue} = \text{total cost}$$

$$16X = \frac{80{,}000}{n_p} + 1000 + 8X \qquad\qquad [13.5]$$

$$X = \frac{10{,}000}{n_p} + 125$$

Insert the values 1 through 5 for n_p and solve for X (Figure 13–5). For example, payback in 2 years requires sales of 5125 units per year in order to break even. There is no consideration of interest in this solution; that is, $i = 0\%$.

Solution by Computer

Solve Equation [13.5] for X, retaining the symbols r and v for use in the spreadsheet.

$$X = \frac{80{,}000/n_p + 1000}{r - v} \qquad\qquad [13.6]$$

E-Solve

The spreadsheet in Figure 13–6 includes the breakeven Equation [13.6] in cells C9 through C13 as detailed in the cell tag. Column C and the *xy* scatter chart display the results. For example, payback in 1 year requires sales of $X = 10{,}125$, while a 5-year payback requires only 2125 units per year.

Breakeven

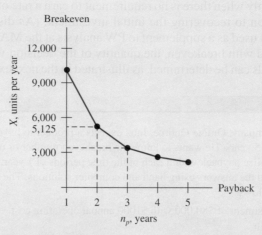

Figure 13–5
Breakeven sales volumes for different payback periods,
Example 13.2.

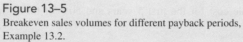

Figure 13–6
Spreadsheet solution of breakeven values for different payback years, Example 13.2.

13.2   BREAKEVEN ANALYSIS BETWEEN TWO ALTERNATIVES

Breakeven analysis involves the determination of a common variable or economic parameter between two alternatives. The parameter can be the interest rate i, first cost P, annual operating cost (AOC), or any parameter. We have already performed breakeven analysis between alternatives on several parameters. For example, the incremental ROR value (Δi^*) is the breakeven rate between alternatives. If the MARR is lower than Δi^*, the extra investment of the larger-investment alternative is justified. In Section 11.3, the replacement value (RV) of a defender was determined. If the market value is larger than RV, the decision should favor the challenger.

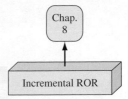

Often breakeven analysis involves revenue or cost variables common to both alternatives, such as price per unit, operating cost, cost of materials, and labor cost. Figure 13–7 illustrates this concept for two alternatives with linear cost relations. The fixed cost of alternative 2 is greater than that of alternative 1. However, alternative 2 has a smaller variable cost, as indicated by its lower slope. The intersection of the total cost lines locates the breakeven point. Thus, if the number of units of the common variable is greater than the breakeven amount, alternative 2 is selected, since the total cost will be lower. Conversely, an anticipated level of operation below the breakeven point favors alternative 1.

Instead of plotting the total costs of each alternative and estimating the breakeven point graphically, it may be easier to calculate the breakeven point numerically using engineering economy expressions for the PW or AW at the MARR. The AW is preferred when the variable units are expressed on a yearly basis, and AW calculations are simpler for alternatives with unequal lives. The following steps may be used to determine the breakeven point of the common variable and to select an alternative:

1. Define the common variable and its dimensional units.
2. Use AW or PW analysis to express the total cost of each alternative as a function of the common variable.
3. Equate the two relations and solve for the breakeven value of the variable.

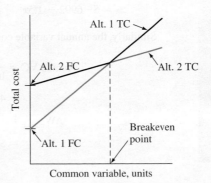

Figure 13–7
Breakeven between two alternatives with linear cost relations.

4. If the anticipated level is below the breakeven value, select the alternative with the higher variable cost (larger slope). If the level is above the breakeven point, select the alternative with the lower variable cost. Refer to Figure 13–7.

EXAMPLE 13.3

A small aerospace company is evaluating two alternatives: the purchase of an automatic feed machine and a manual feed machine for a finishing process. The auto-feed machine has an initial cost of $23,000, an estimated salvage value of $4000, and a predicted life of 10 years. One person will operate the machine at a rate of $12 per hour. The expected output is 8 tons per hour. Annual maintenance and operating cost is expected to be $3500.

The alternative manual feed machine has a first cost of $8000, no expected salvage value, a 5-year life, and an output of 6 tons per hour. However, three workers will be required at $8 per hour each. The machine will have an annual maintenance and operation cost of $1500. All projects are expected to generate a return of 10% per year. How many tons per year must be finished in order to justify the higher purchase cost of the auto-feed machine?

Solution

Use the steps above to calculate the breakeven point between the two alternatives.

1. Let x represent the number of tons per year.
2. For the auto-feed machine the annual variable cost is

$$\text{Annual VC} = \frac{\$12}{\text{hour}} \cdot \frac{1 \text{ hour}}{8 \text{ tons}} \cdot \frac{x \text{ tons}}{\text{year}}$$

$$= 1.5x$$

The VC is developed in dollars per year. The AW expression for the auto-feed machine is

$$\text{AW}_{\text{auto}} = -23{,}000(A/P{,}10\%{,}10) + 4000(A/F{,}10\%{,}10) - 3500 - 1.5x$$
$$= \$-6992 - 1.5x$$

Similarly, the annual variable cost and AW for the manual feed machine are

$$\text{Annual VC} = \frac{\$8}{\text{hour}} (3 \text{ operators}) \frac{1 \text{ hour}}{6 \text{ tons}} \frac{x \text{ tons}}{\text{year}}$$

$$= 4x$$

$$\text{AW}_{\text{manual}} = -8000(A/P{,}10\%{,}5) - 1500 - 4x$$

$$= \$\ 3610 - 4x$$

3. Equate the two cost relations and solve for x.

$$\text{AW}_{\text{auto}} = \text{AW}_{\text{manual}}$$
$$-6992 - 1.5x = -3610 - 4x$$
$$x = 1353 \text{ tons per year}$$

4. If the output is expected to exceed 1353 tons per year, purchase the auto-feed machine, since its VC slope of 1.5 is smaller than the manual feed VC slope of 4.

The breakeven analysis approach is commonly used for make-or-buy decisions. The alternative to buy usually has no fixed cost and a larger variable cost than the option to make. Where the two cost relations cross is the make-buy decision quantity. Amounts above this indicate that the item should be made, not purchased outside.

EXAMPLE **13.4**

Guardian is a national manufacturing company of home health care appliances. It is faced with a make-or-buy decision. A newly engineered lift can be installed in a car trunk to raise and lower a wheelchair. The steel arm of the lift can be purchased for $0.60 per unit or made inhouse. If manufactured on site, two machines will be required. Machine A is estimated to cost $18,000, have a life of 6 years, and have a $2000 salvage value; machine B will cost $12,000, have a life of 4 years, and have a $-500 salvage value (carry-away cost). Machine A will require an overhaul after 3 years costing $3000. The annual operating cost for machine A is expected to be $6000 per year and for machine B $5000 per year. A total of four operators will be required for the two machines at a rate of $12.50 per hour per operator. In a normal 8-hour period, the operators and two machines can produce parts sufficient to manufacture 1000 units. Use a MARR of 15% per year to determine the following.

(a) Number of units to manufacture each year to justify the inhouse (make) option.
(b) The maximum capital expense justifiable to purchase machine A, assuming all other estimates for machines A and B are as stated. The company expects to produce 125,000 units per year.

Solution

(a) Use steps 1 to 3 stated previously to determine the breakeven point.

1. Define x as the number of lifts produced per year.
2. There are variable costs for the operators and fixed costs for the two machines for the make option.

$$\text{Annual VC} = (\text{cost per unit})(\text{units per year})$$
$$= \frac{4 \text{ operators}}{1000 \text{ units}} \frac{\$12.50}{\text{hour}} (8 \text{ hours})x$$
$$= 0.4x$$

The annual fixed costs for machines A and B are the AW amounts.

$$AW_A = -18,000(A/P,15\%,6) + 2000(A/F,15\%,6)$$
$$- 6000 - 3000(P/F,15\%,3)(A/P,15\%,6)$$

$$AW_B = -12,000(A/P,15\%,4) - 500(A/F,15\%,4) - 5000$$

Total cost is the sum of AW_A, AW_B, and VC.

3. Equating the annual costs of the buy option ($0.60x$) and the make option yields

$$-0.60x = AW_A + AW_B - VC$$

$$= -18,000(A/P,15\%,6) + 2000(A/F,15\%,6) - 6000$$
$$- 3000(P/F,15\%,3)(A/P,15\%,6) - 12,000(A/P,15\%,4)$$
$$- 500(A/F,15\%,4) - 5000 - 0.4x \qquad [13.7]$$

$$-0.2x = -20,352.43$$

$$x = 101,762 \text{ units per year}$$

A minimum of 101,762 lifts must be produced each year to justify the make option, which has the lower variable cost of $0.40x$.

(b) Substitute 125,000 for x and P_A for the to-be-determined first cost of machine A (currently $18,000) in Equation [13.7]. Solution yields $P_A = \$35,588$. This is approximately twice the estimated first cost of $18,000, because the production of 125,000 per year is larger than the breakeven amount of 101,762.

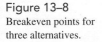

Figure 13–8
Breakeven points for
three alternatives.

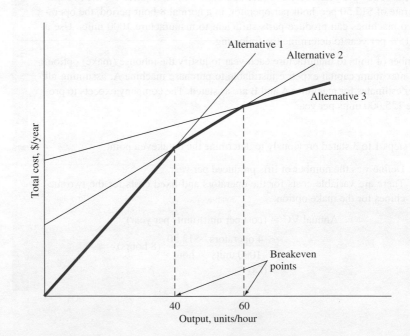

Even though the preceding examples treat two alternatives, the same type of analysis can be performed for three or more alternatives. To do so, compare the alternatives in pairs to find their respective breakeven points. The results are the ranges through which each alternative is more economical. For example, in Figure 13–8, if the output is less than 40 units per hour, alternative 1 should be selected. Between 40 and 60, alternative 2 is more economical; and above 60, alternative 3 is favored.

If the variable cost relations are nonlinear, analysis is more complicated. If the costs increase or decrease uniformly, mathematical expressions that allow direct determination of the breakeven point can be developed.

13.3   SPREADSHEET APPLICATION—USING EXCEL'S SOLVER FOR BREAKEVEN ANALYSIS

This section focuses on the utilization of Excel's optimizing tool SOLVER for basic breakeven analysis. We have used this tool previously with excellent success in speeding up the solution—Section 12.4 on the linear programming model formulation to select from independent projects, and Section 11.3 to find the replacement value for a defender asset in a replacement study. The latter is an application of SOLVER for breakeven analysis. Some discussion about SOLVER is included here in case it has not been used in previous chapters. Further discussion of its use can be found in Appendix A, Section A.4, and the Excel help system.

SOLVER is on the "Tools" toolbar of Excel. (If SOLVER is not on the computer, click "add-ins" on the Tools menu to receive installation directions.) Fundamentally, it is designed to perform breakeven and "what if?" analysis. This is a tool that demonstrates the real advantages of using a computer for engineering economic analysis. The SOLVER template is shown in Figure 13–9. Two primary cell designations of SOLVER are the target cell and changing cells.

Target cell—This is the cell that sets the objective. For example, to find the rate of return, we can set PW = 0. On the spreadsheet, the NPV function is set to 0. The target is the cell with the NPV function in it. The target cell must contain a formula or function. The entry may be maximized, minimized, or set to a specific value, such as 0. If specified, the value must be a number, not a cell reference. In breakeven analysis, the target cell is commonly set equal to the value of another cell, for example, equating the PW

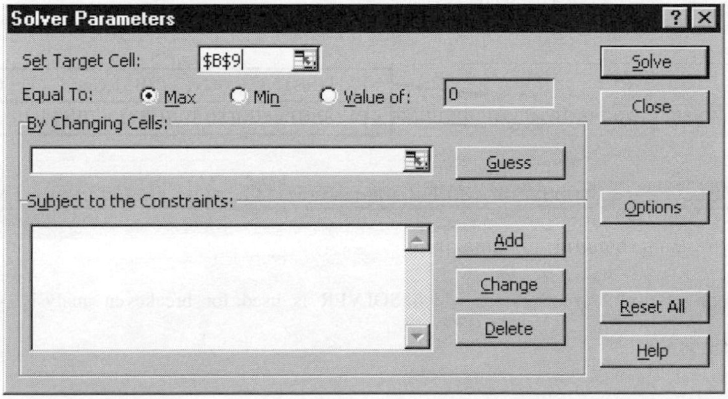

Figure 13–9
Excel's SOLVER template used to perform breakeven analysis and many other "what if?" analyses.

values of two alternatives. This is the spreadsheet equivalent of equating two relations for breakeven analysis.

Changing cells—This is the cell (or cells) where SOLVER will make the changes. These are the decision variables. One or more cells are identified that are directly or indirectly affected by the target cell. SOLVER changes the value of this cell until a solution is found that results in the target cell, as specified. Clicking "Guess" will display all the cells that can be changed once the target cell is identified.

Constraints—These are limits placed on the target and changing cells to set bounds on cell values while SOLVER is looking for a solution to match the target cell value.

EXAMPLE **13.5**

Cheryl is a project engineer in the ANCO Division of Federal-Mogul. She is searching for a replacement for an aging testing machine for after-market windshield wiper blades. One of her engineering technicians found two equivalent machines that have fundamentally the same estimates, except for the expected life.

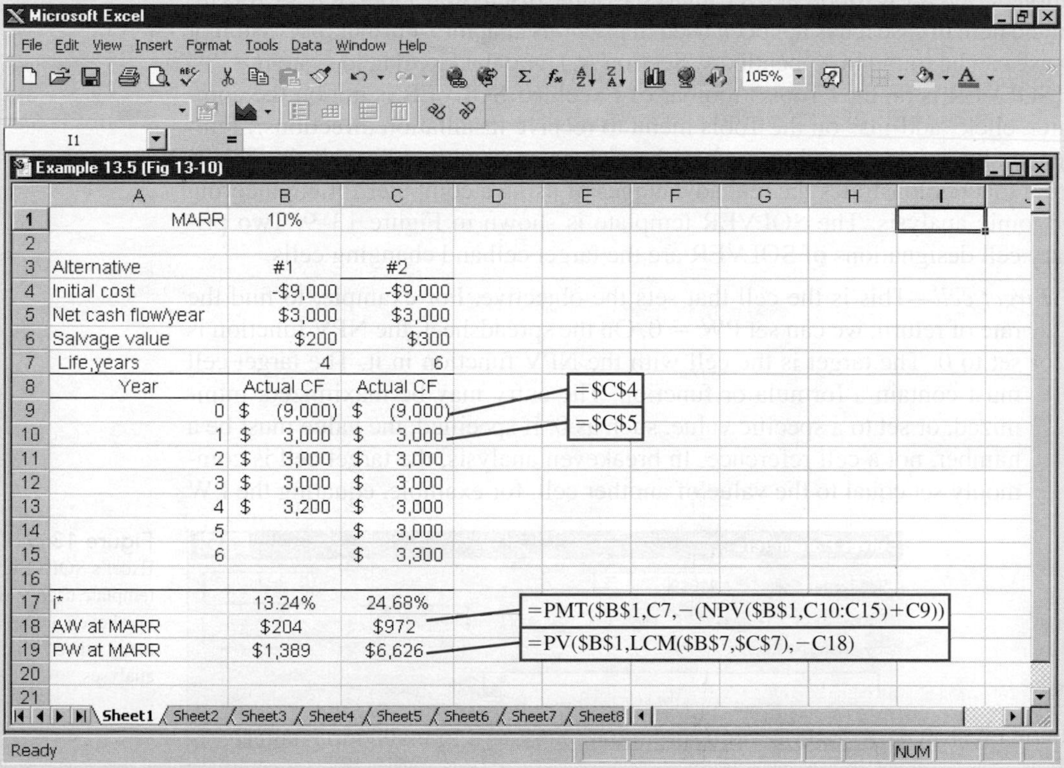

	A	B	C	D	E	F	G	H	I
1	MARR	10%							
2									
3	Alternative	#1	#2						
4	Initial cost	-$9,000	-$9,000						
5	Net cash flow/year	$3,000	$3,000						
6	Salvage value	$200	$300						
7	Life,years	4	6						
8	Year	Actual CF	Actual CF		=$C$4				
9	0	$ (9,000)	$ (9,000)		=$C$5				
10	1	$ 3,000	$ 3,000						
11	2	$ 3,000	$ 3,000						
12	3	$ 3,000	$ 3,000						
13	4	$ 3,200	$ 3,000						
14	5		$ 3,000						
15	6		$ 3,300						
16									
17	i*	13.24%	24.68%	=PMT($B$1,C7,−(NPV($B$1,C10:C15)+C9))					
18	AW at MARR	$204	$972						
19	PW at MARR	$1,389	$6,626	=PV($B$1,LCM($B$7,$C$7),−C18)					
20									
21									

Figure 13–10
Spreadsheet comparison of two alternatives for which SOLVER is used for breakeven analysis, - Example 13.5.

	Machine #1	Machine #2
First cost, $	$-9000	$-9000
Net cash flow, $/year	3000	3000
Salvage, $	200	300
Life, years	4	6

Cheryl knows that #2 is the better choice due to its longer life. However, she judges the features of alternative #1 as much better. She has decided to perform breakeven analysis to determine what alterations in estimated values are necessary to make the two machines economically indifferent. The MARR is 10% at the ANCO Division. Use the spreadsheet in Figure 13–10 and the SOLVER tool to explore breakeven values for (*a*) the first cost of machine #1 and (*b*) the net cash flow for machine #1.

Solution by Computer

The spreadsheet (Figure 13–10) is developed in cell reference format in preparation for any type of breakeven or sensitivity analysis. For example, the PV functions in row 19 use

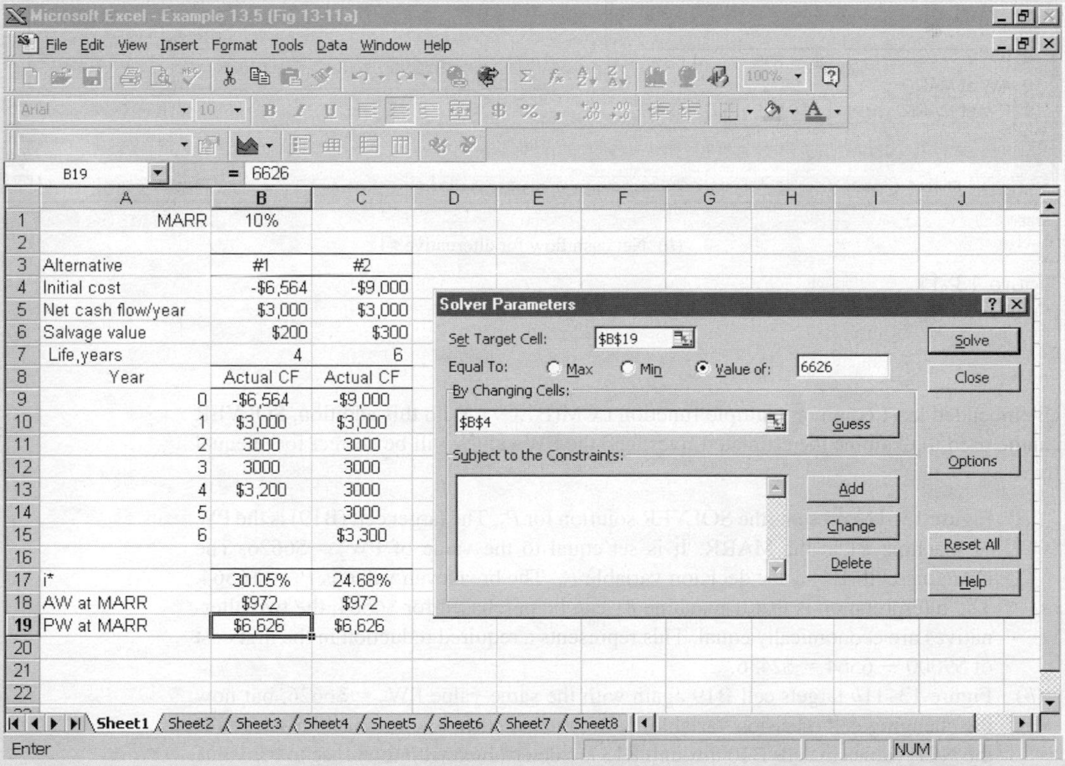

(*a*) Initial cost for alternative #1

Figure 13–11
Excel's SOLVER used to perform breakeven analysis on alternative #1: (*a*) first cost and (*b*) net cash flow, Example 13.5.

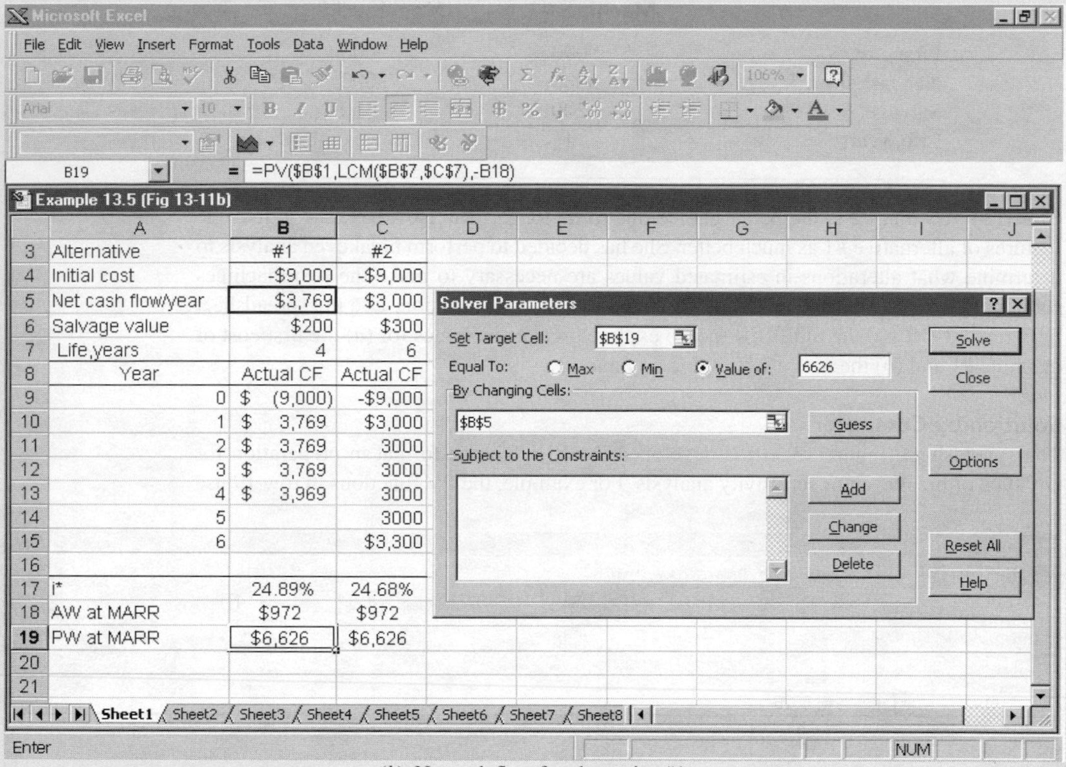

(b) Net cash flow for alternative #1

Figure 13–11
(Continued).

an embedded least common multiple function LCM(B7,C7). With this addition, SOLVER can be used to examine the estimated lives, and the PW values will be correct for unequal lives.

(a) Figure 13–11a presents the SOLVER solution for P_1. The target cell (B19) is the PW of machine #1 at the MARR. It is set equal to the value of $PW_2 = \$6626$. The changing cell (B4) is the decision variable P_1. The breakeven value is $P_1 = \$6564$. The interpretation is that if machine #1 can be purchased for $6564, the two alternatives are economically equal. This represents a required reduction in the first cost of $9000 − 6564 = \$2436$.

(b) Figure 13–11b targets cell B19 again with the same value $PW_2 = \$6626$, but now the changing cell (decision variable) is B5, the net cash flow for machine #1. Since the NCF values in cells B10 through B13 are determined with reference to the value in cell B5, they all change to the breakeven value of $3769. The salvage of $200 is added to NCF in cell B13. The interpretation is that if the estimated annual net cash flow can be raised from $3000 to $3769, the alternatives are economically equal with a first cost of $9000 for both.

CHAPTER SUMMARY

The breakeven point for a variable X for one project is expressed in terms such as units per year or hours per month. At the breakeven amount Q_{BE}, there is indifference to accept or reject the project. Use the following decision guideline:

Single Project   (Refer to Figure 13–2.)

Estimated quantity is *larger* than $Q_{BE} \rightarrow$ accept project

Estimated quantity is *smaller* than $Q_{BE} \rightarrow$ reject project

For two or more alternatives, determine the breakeven value of the common variable X. Use the following guideline to select an alternative:

Two Alternatives   (Refer to Figure 13–7.)

Estimated level of X is *below* breakeven $\rightarrow$ select alternative with the higher variable cost (larger slope)

Estimated level of X is *above* breakeven $\rightarrow$ select alternative with the lower variable cost (smaller slope)

   Breakeven analysis between two alternatives is accomplished by equating the PW or AW relations and solving for the parameter in question. The SOLVER tool is very effective in performing fast, accurate breakeven analyses on a computer.

PROBLEMS

Breakeven Analysis for a Project

13.1 The fixed costs at Harley Motors are $1 million annually. The main product has revenue of $8.50 per unit and $4.25 variable cost. Determine the following.
   (a) Breakeven quantity per year.
   (b) Annual profit if 200,000 units are sold and if 350,000 units are sold. Use both an equation and a plot of the revenue and total cost relations to answer.

13.2 If both linear and nonlinear revenue and total cost relations are considered, state at least one combination of mathematical relations for which there could easily be exactly two breakeven points.

13.3 A metallurgical engineer has estimated that the capital investment for recovering valuable metals (nickel, silver, gold, etc.) from a copper refinery's wastewater stream will be $15 million. The equipment will have a useful life of 10 years with no salvage value. The amount of metals currently discharged is 12,000 pounds per month. The monthly operating cost is represented by $(4,100,000)E^{1.8}$, where E is the efficiency of metal recovery in decimal form. Determine the minimum removal efficiency required for the company to break even, if the average selling price of the metals is $250 per pound. Use an interest rate of 1% per month.

13.4 For the estimates below, calculate the following.
   (a) Breakeven quantity per month.
   (b) Profit (loss) per unit at sales levels that are 10% above and 10% below breakeven.

(c) Plot average cost per unit for quantities ranging from 25% below to 30% above breakeven.

$$r = \$39.95 \text{ per unit}$$
$$v = \$24.75 \text{ per unit}$$
$$FC = \$4,000,000 \text{ per year}$$

13.5 Develop a plot of the average cost per unit versus production quantity for the houseware appliance assembly department of Ace-One, Inc., that has a fixed cost of $160,000 per year and a variable cost of $4.00 per unit; use it to answer the following questions.

(a) At what quantity is a $5 per unit average cost justified?

(b) If the fixed cost increases to $200,000, plot the new curve on the same graph and estimate the quantity that justifies an average cost of $6 per unit.

13.6 A call center in India used by U.S. and U.K. credit card holders has a capacity of 1,400,000 calls annually. The fixed cost of the center is $775,000 with an average variable cost of $2 and revenue of $3.50 per call.

(a) Find the percentage of the capacity that must be placed each year to break even.

(b) The center manager expects to dedicate the equivalent of 500,000 of the 1,400,000 capacity to a new product line. This is expected to increase the center's fixed cost to $900,000, of which 50% will be allocated to the new product line. Determine the average revenue per call necessary to make 500,000 calls the breakeven point for only the new product. How does this required revenue compare with the current center revenue of $3.50 per call?

13.7 For the last 2 years, the Homes-r-US company has experienced a fixed cost of $850,000 per year and an $(r - v)$ value of

$1.25 per unit. International competition has become severe enough that some financial changes must be made to keep market share at the current level. Perform a graphical analysis, using Excel, that estimates the effect on the breakeven point if the difference between revenue and variable cost per unit increases somewhere between 1% and 15% of its current value. If fixed costs and revenue per unit remain at their current values, what must change to make the breakeven point go down?

13.8 (This is an extension of Problem 13.7.) Expand the analysis performed in Problem 13.7, where a change in variable cost per unit is examined. The financial manager estimates that fixed costs will fall to $750,000 when the required production rate to break even is at or below 600,000 units. What happens to the breakeven points over the $r - v$ range of 1% to 15% increase as evaluated previously?

13.9 An automobile company is investigating the advisability of converting a plant that manufactures economy cars into one that will make retro sports cars. The initial cost for equipment conversion will be $200 million with a 20% salvage value anytime within a 5-year period. The cost of producing a car will be $21,000, but it is expected to have a selling price of $33,000 (to dealers). The production capacity for the first year will be 4000 units. At an interest rate of 12% per year, by what uniform amount will production have to increase each year in order for the company to recover its investment in 3 years?

13.10 Rod, an industrial engineer manager with Zema Corporation, has determined via the least squares method that the annual total cost per case of producing its top-selling drink can best be described by the quadratic relation $TC = 0.001Q^2 + 3Q + 2$, and that revenue is approximately linear with r = $25 per case. Rod asks you to do the following.

(a) Tabulate the profit function between the values of Q = 5000 and 25,000 cases. Estimate the maximum profit and the quantity at which it occurs.

(b) Find the answers to part (a) by using an Excel graph.

(c) Determine an equation for Q_{max}, the quantity at which the maximum profit should occur, and determine the amount of the profit at this point for the TC and r identified by the manager.

13.11 A civil engineer has been promoted to manager of engineered public systems. One of the products is an emergency intercept pump for potable water. If the tested water quality or volume varies by a preset percentage, the pump automatically switches to preselected options of treatments or water sources. The manufacturing process for the pump had the following fixed and variable costs over a 1-year period.

	Fixed Costs, $		Variable Costs, $/Unit
Administrative	30,000	Materials	2500
Salaries and		Labor	200
benefits: 20% of	350,000		
Equipment	100,000	Indirect labor	2000
Space, utilities, etc.	55,000	Subcontractors	800
Computers: 1/3 of	150,000		

(a) Determine the minimum revenue per unit to break even at the current production volume of 5000 units per year.

(b) If selling internationally and to large corporations is pursued, an increased production of 3000 additional units will be necessary. Determine the revenue per unit required if a profit goal of $500,000 is set for the entire product line. Assume the cost estimates above remain the same.

Breakeven Analysis Between Alternatives

13.12 A Yellow Pages directory company must decide whether it should compose the ads for its clients inhouse or pay a production company to compose them. To develop the ads inhouse, the company will have to purchase computers, printers, and other peripherals at a cost of $12,000. The equipment will have a useful life of 3 years, after which it will be sold for $2000. The employee who creates the ads will be paid $45,000 per year. In addition, each ad will have an average cost of $8 to prepare for delivery to the printer. A total of 4000 ads are anticipated for the next few years. Alternatively, the company can outsource ad development at a fee of $20 per ad regardless of the quantity. The current interest rate is 8% per year. What is the breakeven amount, and which alternative is economically better?

13.13 An engineering firm can lease a measurement system for $1000 per month or purchase one for $15,000. The leased system will have no monthly maintenance cost, but the purchased one will cost $80 per month. At an interest rate of 0.5% per month, how many months must the system be required to break even?

13.14 Two pumps can be used for pumping a corrosive liquid. A pump with a brass impeller costs $800 and is expected to last 3 years. A pump with a stainless steel impeller costs $1900 and will last 5 years. A rebuild costing $300 will be required after 2000 operating hours for the brass impeller pump while an overhaul costing $700 will be required for the stainless steel pump after 8000 hours. If the operating cost of each pump is $1 per hour, how many hours per year must the pump be required to justify the purchase of the more expensive pump? Use an interest rate of 10% per year.

13.15 Two bids have been received to repave a commercial parking lot. Proposal 1 includes new curbs, grading, and paving at an initial cost of $250,000. The life of the parking lot surface constructed in this manner is expected to be 4 years with annual costs for maintenance and repainting of pavement markings at $3000. Proposal 2 offers pavement of a significantly higher quality with an expected life of 8 years. The annual maintenance cost will be negligible for the pavement, but the markings will have to be repainted every 2 years at a cost of $3000. Markings are not repainted the last year of its expected life under proposal 2. If the company's current MARR is 12% per year, how much can it afford to spend on proposal 2 initially so the two break even?

13.16 Jeremy is evaluating the operational costs of the manufacturing processes for specific components of a wireless home security system. The same components are produced at plants in New York (NY) and Los Angeles (LA). The records for the last 3 years from NY report a fixed cost of $400,000 per year and a variable cost of $95 per unit in year 1, decreasing by $3 per unit per year. The LA reports indicate a fixed cost of $750,000 per year and a variable cost of $50 per unit, increasing by $4 per unit per year. If the trends continue, how many units must be produced in year 4 for the two processes break even? Use an interest rate of 10% per year.

13.17 Alfred Home Construction is considering the purchase of five dumpsters and a transport truck to store and transfer construction debris from building sites. The entire rig is estimated to have an initial cost of $125,000, a life of 8 years, a $5000 salvage value, an operating cost of $40 per day, and an annual maintenance cost of $2000. Alternatively, Alfred can obtain the same services from the city as needed at each construction site for an initial delivery cost of $125 per dumpster per site and a daily charge of $20 per day per dumpster. An estimated 45 construction sites will need debris storage throughout the average year. The minimum attractive rate of return is 12% per year. (a) How many days per year must the equipment be required to just break even? (b) If the expected usage is 75 days per year, which option—buy or lease—should be selected based on this economic analysis? Determine the expected annual cost of this decision.

13.18 Machine A has a fixed cost of $40,000 per year and a variable cost of $60 per unit. Machine B has an unknown fixed cost, but with this process 200 units can be produced each month at a total variable cost of $2000. If the total costs of the two machines break even at a production rate of 2000 units per year, what is the fixed cost of machine B?

13.19 A waste-holding lagoon situated near the main plant receives sludge daily. When the lagoon is full, it is necessary to remove the sludge to a site located 8.2 kilometers from the main plant. Currently, when the lagoon is full, the sludge is removed by pump into a tank truck and hauled away. This process requires the use of a portable pump that initially cost $800 and has an 8-year life. The company pays a contract individual to operate the pump and oversee environmental and safety factors at a rate of $100 per day, but the truck and driver must be rented for $200 per day.

The company has the option to install a pump and pipeline to the remote site. The pump would have an initial cost of $1600 and a life of 10 years and will cost $3 per day to operate. The company's MARR is 10% per year.

(a) If the pipeline will cost $12 per meter to construct and will have a 10-year life, how many days per year must the lagoon require pumping to justify construction of the pipeline?

(b) If the company expects to pump the lagoon once per week every week of the year, how much money can it afford to spend now on the 10-year-life pipeline to just break even?

Use of SOLVER for Breakeven Analysis

13.20 Develop a spreadsheet using cell reference format for the estimates of Problem 13.15 where two proposals for repaving a commercial parking lot are evaluated. Answer the following questions, using SOLVER.
(a) How much can the company afford to spend on proposal 2 initially so the two break even? (This is the same question posed in Problem 13.15.)
(b) Assume that $P_2 = \$-400{,}000$ is the actual cost of proposal 2 and that $P_1 = \$-250{,}000$ is correct. Use the results of the analysis above to determine if the annual costs of proposal 1 for maintenance and repainting can be reduced enough to make it a viable option.

13.21 A jeans manufacturing company is evaluating the purchase of a new automatic cutting machine with fuzzy-logic features. The machine will have a first cost of $40,000, a life of 10 years, and no salvage value. The maintenance cost of the machine is expected to be $2000 per year. The machine will require one operator at a total cost of $30 per hour. A total of 2500 yards of material can be cut each hour by the machine. Alternatively, if human labor is used, six workers, each earning $14 per hour, are required to cut 2500 yards per hour. The MARR is 8% per year. Determine the minimum yards per year to justify the purchase of the automatic machine. Use (a) hand and (b) computer solution.

13.22 This is an extension to Problem 13.21. Assume this is a North American company that elected to move some of its cutting operations to Asia when domestic commercial interest rates were high at 8% per year and Asian labor rates averaged $14 per hour. Now the machinery will cost $80,000 with the other estimates remaining the same, the interest rate is 6% per year, and Asian labor costs an average of $25 per hour. Further assume the company elected earlier to stay with human cutting operations since the garment line for which the analysis was done had an annual production rate of about 300,000 yards. Recalculate the new breakeven point, and determine if the earlier "human cutters" decision is still valid.

13.23 A 3-year-old house can be purchased for an excellent price of $100,000. Remodeling costs immediately after purchase are estimated to be $12,000. Taxes will be $3800 per year, utilities will cost $2500 per year, and the house must be repainted every 6 years at a cost of $1000. At present, resale houses are selling for $60 per square foot, but this price has been increasing, a trend that is expected to continue, by $1.50 per square foot per year. The 2500-square-foot house can be continually leased for $12,000 per year from this year until it is sold. A return of 8% per year is expected if the investment is made. (a) Determine how long the purchaser must own and lease the house to break even and the anticipated selling price at breakeven time. (b) Use Excel to approximate the breakeven years and estimated selling price.

13.24 Bovay Medical Labs is evaluating the alternatives of complete and partial in-house diabetes and associated blood sugar testing labs, instead of outsourcing samples to an independent laboratory for analysis. The alternatives and associated costs are as follows:

Complete lab inhouse: If the inhouse lab is completely equipped, the initial cost will be $50,000. A part-time technician will be employed at an annual equivalent salary of $26,000. Cost of chemicals and supplies is estimated at $10 per sample.

Partial lab inhouse: The lab can be partially equipped at an initial cost of $35,000. The part-time technician will have an annual equivalent salary of $10,000. The cost of in-house sample analysis will be only $3 per sample. However, since some tests cannot be conducted in-house, outside testing will be required at an average cost of $40 per sample for all samples.

Complete outsourcing: The cost averages $120 per sample analyzed.

Any laboratory equipment purchased will have a life of 6 years. If the MARR is 10% per year, determine the number of samples that must be tested each year to justify (*a*) the complete lab and (*b*) the partial lab. (*c*) Plot the total cost lines for all three options, and state the range of samples for which each option will have the lowest cost. (*d*) If Bovay expects to test 300 samples per year, which of the three options is the best economically?

13.25   The chief engineer of Domingo County is considering two methods for lining water-holding tanks. A bituminous coating can be applied at a cost of $8000. If the coating is touched up after 3 years at a cost of $1000, its life can be extended another 3 years. As an alternative, a plastic lining may be installed with a useful life of 15 years. If the discount rate is 4% per year, how much can be spent for the plastic lining so that the two methods just break even? Solve (*a*) by hand and (*b*) by computer.

13.26   A building contractor is evaluating two alternatives for improving the exterior appearance of a small commercial building that he is renovating. The building can be completely painted at a cost of $2800. The paint is expected to remain attractive for 4 years, at which time repainting is necessary. Every time the building is repainted, the cost will increase by 20% over the previous time. As an alternative, the building can be sandblasted now and every 10 years at a cost 40% greater than the previous time. The remaining life of the building is expected to be 38 years. The MARR is 10% per year. What is the maximum amount that can be spent now on the sandblasting alternative for the two alternatives to just break even? Use present worth analysis (*a*) by hand and (*b*) by computer to answer this question. (*c*) Use the spreadsheet to find the breakeven value if the cost increase for sandblasting can be reduced from 40% to 30%, or even 20%, each 10 years.

CASE STUDY

WATER TREATMENT PLANT PROCESS COSTS

Introduction

Aeration and sludge recirculation have been practiced for many years at municipal and industrial water treatment plants. Aeration is used primarily for the physical removal of gases or volatile compounds while sludge recirculation can be beneficial for turbidity removal and hardness reduction.

When the advantages of aeration and sludge recirculation in water treatment were first recognized, energy costs were so low that such considerations were seldom of concern in treatment plant design and operation. With the 10-fold increase in electricity cost that occurred in some localities, however, it became necessary to review the cost-effectiveness of all water treatment processes that consume significant amounts of energy. This study was conducted at a municipal water treatment plant for evaluating the cost-effectiveness of the pre-aeration and sludge recirculation practices.

Experimental Procedure

This study was conducted at a 106 m^3 per minute water treatment plant where, under normal operating circumstances, sludge from the secondary clarifiers is returned to the aerator and subsequently removed in the primary clarifiers. Figure 13–12 is a schematic of the process.

To evaluate the effect of sludge recirculation, the sludge pump was turned off, but aeration was continued. Next, the sludge pump was turned back on, and aeration was discontinued. Finally, both processes were discontinued. Results obtained during the test periods were averaged and compared to the values obtained when both processes were operational.

Results and Discussion

The results obtained from the four operating modes showed that the hardness decreased by 4.7% when both processes were in operation (i.e., sludge recirculation and aeration). When only sludge was recirculated, the reduction was 3.8%. There was no reduction due to aeration only, or when there was neither aeration nor recirculation. For turbidity, the reduction was 28% when both recirculation and aeration were used. The reduction was 18% when *neither* aeration nor recirculation was used. The reduction was also 18% when aeration alone was used, which means that aeration alone was of no benefit for turbidity reduction. With sludge recirculation alone, the turbidity reduction was only 6%, meaning that sludge recirculation alone actually resulted in an *increase* in turbidity—the difference between 18% and 6%.

Since aeration and sludge recirculation did cause readily identifiable effects on treated water quality (some good and others bad), the cost-effectiveness of each process for turbidity and hardness reduction was investigated. The calculations are based on the following data:

Aerator motor = 40 hp
Aerator motor efficiency = 90%
Sludge recirculation motor = 5 hp
Recirculation pump efficiency = 90%
Electricity cost = 9 ¢/kWh
Lime cost = 7.9 ¢/kg
Lime required = 0.62 mg/L per mg/L hardness
Coagulant cost = 16.5 ¢/kg
Days/month = 30.5

As a first step, the costs associated with aeration and sludge recirculation were calculated. In each case, costs are independent of flow rate.

Aeration cost:

40 hp × 0.75 kW/hp × 0.09 \$/kWh × 24 h/day
÷ 0.90 = \$72 per day or \$2196 per month

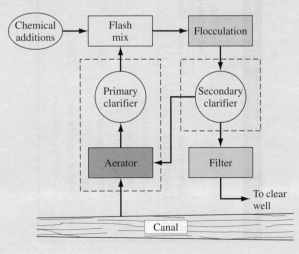

Figure 13–12
Schematic of water treatment plant.

TABLE 13–1 Cost Summary in Dollars per Month

Alt. I.D.	Alternative Description	Savings from Discontinuation of		Total Savings (3) = (1) + (2)	Extra Cost for Removal of		Total Extra Cost (6) = (4) + (5)	Net Savings (7) = (3) − (6)
		Aeration (1)	Recirculation (2)		Hardness (4)	Turbidity (5)		
1	Sludge recirculation and aeration			Normal operating condition				
2	Aeration only	—	275	275	1380	469	1849	− 1574
3	Sludge recirculation only	2196	—	2196	262	845	1107	+ 1089
4	Neither aeration nor sludge recirculation	2196	275	2471	1380	469	1849	+ 622

Sludge recirculation cost:

5 hp × 0.75 kW/hp × 0.09 $/kWh × 24 h/day
   ÷ 0.90 = $9 per day or $275 per month

The estimates appear in columns 1 and 2 of the cost summary in Table 13–1.

Costs associated with turbidity and hardness removal are a function of the chemical dosage required and the water flow rate. The calculations below are based on a design flow of 53 m³/minute.

As stated earlier, there was less turbidity reduction through the primary clarifier without aeration than there was with it (28% versus 6%). The extra turbidity reaching the flocculators could require further additions of the coagulating chemical. If it is assumed that, as a worst case, these chemical additions would be proportional to the extra turbidity, then 22 percent more coagulant would be required. Since the average dosage before discontinuation of aeration was 10 mg/L, the *incremental chemical cost* incurred because of the increased turbidity in the clarifier effluent would be

$(10 × 0.22)$ mg/L × 10^{-6}kg/mg × 53 m³/min
   × 1000 L/m³ × 0.165 $/kg × 60 min/h
   × 24 h/day = $27.70/day or $845/month

Similar calculations for the other operating conditions (i.e., aeration only, and neither aeration nor sludge recirculation) reveal that the additional cost for turbidity removal would be $469 per month in each case, as shown in column 5 of Table 13–1.

Changes in hardness affect chemical costs by virtue of the direct effect on the amount of lime required for water softening. With aeration and sludge recirculation, the average hardness reduction was 12.1 mg/L (that is, 258 mg/L × 4.7 percent). However, with sludge recirculation only, the reduction was 9.8 mg/L, resulting in a difference of 2.3 mg/L attributed to aeration. The *extra cost of lime* incurred because of the discontinuation of aeration, therefore, was

2.3 mg/L × 0.62 mg/L lime × 10^{-6} kg/mg
   × 53 m³/min × 1000 L/m³ × 0.079 $/kg
   × 60 min/h × 24 h/day = $8.60/day or
   $262/month

When sludge recirculation was discontinued, there was no hardness reduction through the clarifier, so that the extra lime cost would be $1380 per month.

The total savings and total costs associated with changes in plant operating conditions are tabulated in columns 3 and 6 of Table 13–1, respectively, with the net savings shown in column 7. Obviously, the optimum condition is represented by "sludge recirculation only." This condition would result in a net savings of $1089 per month, compared to a net savings of $622 per month when both processes are discontinued and a net *cost* of $1574 per month for aeration only. Since the calculations made here represent worst-case conditions, the actual savings that resulted from modifying the plant operating procedures were greater than those indicated.

In summary, the commonly applied water treatment practices of sludge recirculation and aeration can significantly affect the removal of some compounds in the primary clarifier. However, increasing energy and chemical costs warrant continued investigations on a case-by-case basis of the cost-effectiveness of such practices.

Case Study Exercises

1. What would be the monthly savings in electricity from discontinuation of aeration if the cost of electricity were 6 ¢/kWh?

2. Does a decrease in the efficiency of the aerator motor make the selected alternative of sludge recirculation only more attractive, less attractive, or the same as before?

3. If the cost of lime were to increase by 50%, would the cost difference between the best alternative and second-best alternative increase, decrease, or remain the same?

4. If the efficiency of the sludge recirculation pump were reduced from 90% to 70%, would the net savings difference between alternatives 3 and 4 increase, decrease, or stay the same?

5. If hardness removal were to be discontinued at the treatment plant, which alternative would be the most cost-effective?

6. If the cost of electricity decreased to 4 ¢/kWh, which alternative would be the most cost-effective?

7. At what electricity cost would the following alternatives just break even: (*a*) alternatives 1 and 2, (*b*) alternatives 1 and 3, (*c*) alternatives 1 and 4?

LEVEL FOUR

ROUNDING OUT THE STUDY

LEVEL ONE This Is How It All Starts	LEVEL TWO Tools for Evaluating Alternatives	LEVEL THREE Making Deci- sions on Real- World Projects	LEVEL FOUR Rounding Out the Study
Chapter 1 Foundations of Engineering Economy	Chapter 5 Present Worth Analysis	Chapter 11 Replacement and Retention Decisions	Chapter 14 Effects of Inflation
Chapter 2 Factors: How Time and Interest Affect Money	Chapter 6 Annual Worth Analysis	Chapter 12 Selection from Independent Projects Under Budget Limitation	Chapter 15 Cost Estimation and Indirect Cost Allocation
Chapter 3 Combining Factors	Chapter 7 Rate of Return Analysis: Single Alternative		Chapter 16 Depreciation Methods
Chapter 4 Nominal and Effective Interest Rates	Chapter 8 Rate of Return Analysis: Multiple Alternatives	Chapter 13 Breakeven Analysis	Chapter 17 After-Tax Economic Analysis
	Chapter 9 Benefit/Cost Analysis and Public Sector Economics		Chapter 18 Formalized Sensitivity Analysis and Expected Value Decisions
	Chapter 10 Making Choices: The Method, MARR, and Multiple Attributes		Chapter 19 More on Variation and Decision Making Under Risk

This level includes topics to enhance your ability to perform a thorough engineering economic study of one or more alternatives. The effects of inflation, depreciation, income taxes in all types of studies, and indirect costs are incorporated into the methods of previous chapters. Several techniques of cost estimation to better predict cash flows are treated in order to base alternative selection on more accurate estimates. The last two chapters include additional material on the use of engineering economics in decision making. An expanded version of sensitivity analysis is developed; it formalizes the approach to examine parameters that vary over a predictable range of values. Finally, the elements of risk and probability are explicitly considered using expected values, probabilistic analysis, and Monte Carlo–based computer simulation.

Several of these topics can be covered earlier in the text, depending on the objectives of the course. Use the chart in the Preface to determine appropriate points at which to introduce the material.

14

Effects of Inflation

This chapter concentrates upon understanding and calculating the effects of inflation in time value of money computations. Inflation is a reality that we deal with nearly everyday in professional and personal life.

The annual inflation rate is closely watched and historically analyzed by government units, businesses, and industrial corporations. An engineering economy study can have different outcomes in an environment in which inflation is a serious concern compared to one in which it is of minor consideration. In the last few years of the 20th century, and the beginning of the 21st century, inflation has not been a major concern in the U.S. or most industrialized nations. But the inflation rate is sensitive to real, as well as perceived, factors of the economy. Factors such as the cost of energy, interest rates, availability and cost of skilled people, scarcity of materials, political stability, and other, less tangible factors have short-term and long-term impacts on the inflation rate. In some industries, it is vital that the effects of inflation be integrated into an economic analysis. The basic techniques to do so are covered here.

LEARNING OBJECTIVES

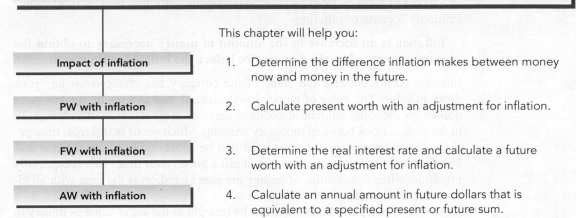

Purpose: Consider inflation in an engineering economy analysis.

This chapter will help you:

Impact of inflation

1. Determine the difference inflation makes between money now and money in the future.

PW with inflation

2. Calculate present worth with an adjustment for inflation.

FW with inflation

3. Determine the real interest rate and calculate a future worth with an adjustment for inflation.

AW with inflation

4. Calculate an annual amount in future dollars that is equivalent to a specified present or future sum.

14.1   UNDERSTANDING THE IMPACT OF INFLATION

We are all very well aware that $20 now does not purchase the same amount as $20 did in 1995 or 1996 and purchases significantly less than in 1980. Why? Primarily because of inflation.

Inflation is an increase in the amount of money necessary to obtain the same amount of product or service before the inflated price was present.

Inflation occurs because the value of the currency has changed—it has gone down in value. The value of money has decreased, and as a result, it takes more dollars for the same amount of goods or services. This is a sign of *inflation*. To make comparisons between monetary amounts which occur in different time periods, the different-valued dollars must first be converted to constant-value dollars in order to represent the same purchasing power over time. This is especially important when future sums of money are considered, as is the case with all alternative evaluations.

Money in one period of time t_1 can be brought to the same value as money in another period of time t_2 by using the equation

$$\text{Dollars in period } t_1 = \frac{\text{dollars in period } t_2}{\text{inflation rate between } t_1 \text{ and } t_2} \qquad [14.1]$$

Dollars in period t_1 are called *constant-value dollars* or *today's dollars*. Dollars in period t_2 are called *future dollars* or *then-current dollars*. If f represents the inflation rate per period (year) and n is the number of time periods (years) between t_1 and t_2, Equation [14.1] is

$$\textbf{Constant-value dollars} = \textbf{today's dollars} = \frac{\textbf{future dollars}}{\textbf{(1 + }f\textbf{)}^n} \qquad [14.2]$$

$$\textbf{Future dollars} = \textbf{today's dollars(1 + }f\textbf{)}^n \qquad [14.3]$$

It is correct to express future (inflated) dollars in terms of constant-value dollars, and vice versa, by applying the last two equations. This is how the Consumer Price Index (CPI) and cost estimation indices (of the next chapter) are determined. As an illustration, use the price of a McDonald's Big Mac in some parts of Texas.

$2.23   August 2004

If inflation averaged 4% during the last year, in *constant-value 2003 dollars*, this cost is last year's equivalent of

$2.23/(1.04) = $2.14   August 2003

A predicted price in 2005 is

$2.23(1.04) = $2.32   August 2005

If inflation averages 4% per year over the next 10 years, Equation [14.3] is used to predict a Big Mac price in 2014:

$2.23(1.04)^{10} = $3.30   August 2014

This is a 48% increase over the 2004 price at 4% inflation, which is considered low to average nationally and internationally. If inflation averages 6% per year, the Big Mac cost in 10 years will be $3.99, an increase of 79%. In some areas of the world, hyperinflation may average 50% per year. In such an unfortunate economy, the Big Mac in 10 years rises from the dollar equivalent of $2.23 to $128.59! This is why countries experiencing hyperinflation must devalue the currency by factors of 100 and 1000 when unacceptable inflation rates persist.

Placed into an industrial or business context, at a reasonably low inflation rate averaging 4% per year, equipment or services with a first cost of $209,000 will increase by 48% to $309,000 over a 10-year span. This is before any consideration of the rate of return requirement is placed upon the equipment's revenue-generating ability. *Make no mistake: Inflation is a formidable force in our economy.*

There are actually three different rates that are important: the real interest rate (i), the market interest rate (i_f), and the inflation rate (f). Only the first two are interest rates.

Real or inflation-free interest rate i. This is the rate at which interest is earned when the effects of changes in the value of currency (inflation) have been removed. Thus, the real interest rate presents an actual gain in purchasing power. (The equation used to calculate i, with the influence of inflation removed, is derived later in Section 14.3.) The real rate of return that generally applies for individuals is approximately 3.5% per year. This is the "safe investment" rate. The required real rate for corporations (and many individuals) is set above this safe rate when a MARR is established without an adjustment for inflation.

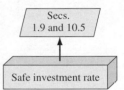

Secs. 1.9 and 10.5

Safe investment rate

Inflation-adjusted interest rate i_f. As its name implies, this is the interest rate that has been adjusted to take inflation into account. The *market interest rate,* which is the one we hear everyday, is an inflation-adjusted rate. This rate is a combination of the real interest rate i and the inflation rate f, and, therefore, it changes as the inflation rate changes. It is also known as the *inflated interest rate.*

Inflation rate f. As described above, this is a measure of the rate of change in the value of the currency.

A company's MARR adjusted for inflation is referred to as the inflation-adjusted MARR. The determination of this value is discussed in Section 14.3.

Deflation is the opposite of inflation in that when deflation is present, the purchasing power of the monetary unit is greater in the future than at present. That is, it will take fewer dollars in the future to buy the same amount of goods or services as it does today. Inflation occurs much more commonly than deflation, especially at the national economy level. In deflationary economic conditions, the market interest rate is always less than the real interest rate.

Temporary price deflation may occur in specific sectors of the economy due to the introduction of improved products, cheaper technology, or imported materials or products that force current prices down. In normal situations, prices equalize at a competitive level after a short time. However, deflation over a short time in a specific sector of an economy can be orchestrated through *dumping*. An example of dumping may be the importation of materials, such as steel, cement, or cars, into one country from international competitors at very low prices compared to current market prices in the targeted country. The prices will go down for the consumer, thus forcing domestic manufacturers to reduce their prices in order to compete for business. If domestic manufacturers are not in good financial condition, they may fail, and the imported items replace the domestic supply. Prices may then return to normal levels and, in fact, become inflated over time, if competition has been significantly reduced.

On the surface, having a moderate rate of deflation sounds good when inflation has been present in the economy over long periods. However, if deflation occurs at a more general level, say nationally, it is likely to be accompanied by the lack of money for new capital. Another result is that individuals and families have less money to spend due to fewer jobs, less credit, and fewer loans available; an overall "tighter" money situation prevails. As money gets tighter, less is available to be committed to industrial growth and capital investment. In the extreme case, this can evolve over time into a deflationary spiral that disrupts the entire economy. This has happened on occasion, notably in the United States during the Great Depression of the 1930s.

Engineering economy computations that consider deflation use the same relations as those for inflation. For basic equivalence between today's dollars and future dollars, Equations [14.2] and [14.3] are used, except the deflation rate is a $-f$ value. For example, if deflation is estimated to be 2% per year, an asset that costs $10,000 today would have a first cost 5 years from now determined by Equation [14.3].

$$10,000(1 - f)^n = 10,000(0.98)^5 = 10,000(0.9039) = \$9039$$

14.2   PRESENT WORTH CALCULATIONS ADJUSTED FOR INFLATION

When the dollar amounts in different time periods are expressed in *constant-value dollars,* the equivalent present and future amounts are determined using the real interest rate i. The calculations involved in this procedure are illustrated in Table 14–1 where the inflation rate is 4% per year. Column 2 shows the inflation-driven increase for each of the next 4 years for an item that has a cost of $5000 today. Column 3 shows the cost in future dollars, and column 4 verifies the cost in constant-value dollars via Equation [14.2]. When the future dollars of column 3 are converted to constant-value dollars (column 4), the cost is always $5000, the same as the cost at the start. This is predictably true when the costs are increasing by an amount *exactly equal* to the inflation rate. The actual (inflation-adjusted) cost of the item 4 years from now will be $5849, but in constant-value dollars the cost in 4 years will still amount to $5000. Column 5

TABLE 14–1 Inflation Calculations Using Constant-Value Dollars
$(f = 4\%, i = 10\%)$

Year n (1)	Cost Increase Due to 4% Inflation (2)	Cost in Future Dollars (3)	Future Cost in Constant-Value Dollars (4) = (3)/1.04^n	Present Worth at Real $i = 10\%$ (5) = (4)(P/F,10%,n)
0		$5000	$5000	$5000
1	$5000(0.04) = $200	5200	5200/(1.04)^1 = 5000	4545
2	5200(0.04) = 208	5408	5408/(1.04)^2 = 5000	4132
3	5408(0.04) = 216	5624	5624/(1.04)^3 = 5000	3757
4	5624(0.04) = 225	5849	5849/(1.04)^4 = 5000	3415

shows the present worth of future amounts of $5000 at a real interest rate of $i = 10\%$ per year.

Two conclusions can be made. At $f = 4\%$, $5000 today inflates to $5849 in 4 years. And $5000 four years from now has a PW of only $3415 constant-value dollars at a real interest rate of 10% per year.

Figure 14–1 shows the differences over a 4-year period of the constant-value amount of $5000, the future-dollar costs at 4% inflation, and the present worth at 10% real interest with inflation considered. The effect of compounded inflation and interest rates is large, as you can see by the shaded area.

An alternative, and less complicated, method of accounting for inflation in a present worth analysis involves adjusting the interest formulas themselves to account for inflation. Consider the P/F formula, where i is the real interest rate.

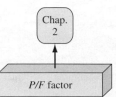

Chap. 2

P/F factor

$$P = F\frac{1}{(1 + i)^n}$$

The F is a future-dollar amount with inflation built in. And F can be converted into today's dollars by using Equation [14.2].

$$P = \frac{F}{(1 + f)^n}\frac{1}{(1 + i)^n}$$

$$= F\frac{1}{(1 + i + f + if)^n} \quad\quad [14.4]$$

If the term $i + f + if$ is defined as i_f, the equation becomes

$$P = F\frac{1}{(1 + i_f)^n} = F(P/F,i_f,n) \quad\quad [14.5]$$

Figure 14–1
Comparison of constant-value dollars, future dollars, and their present worth values.

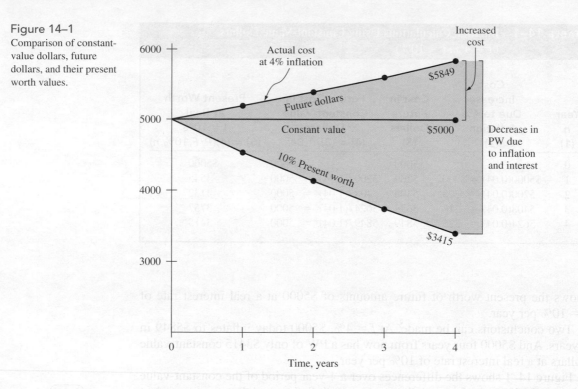

The symbol i_f is called the *inflation-adjusted interest rate* and is defined as

$$i_f = i + f + if \qquad \textbf{[14.6]}$$

where i = real interest rate
f = inflation rate

For a real interest rate of 10% per year and an inflation rate of 4% per year, Equation [14.6] yields an inflated interest rate of 14.4%.

$$i_f = 0.10 + 0.04 + 0.10(0.04) = 0.144$$

Table 14–2 illustrates the use of $i_f = 14.4\%$ in PW calculations for $5000 now, which inflates to $5849 in future dollars 4 years hence. As shown in column 4, the present worth for each year is the same as column 5 of Table 14–1.

The present worth of any series of cash flows—equal, arithmetic gradient, or geometric gradient—can be found similarly. That is, either i or i_f is introduced into the P/A, P/G, or P_g factors, depending upon whether the cash flow is expressed in constant-value (today's) dollars or future dollars. If the series is expressed in today's dollars, then its PW is simply the discounted value using the real interest rate i. If the cash flow is expressed in future dollars, the PW value is obtained using i_f. Alternatively, you can first convert all future dollars to today's dollars using Equation [14.2] and then find the PW at i.

TABLE 14–2   Present Worth Calculation Using an Inflated Interest Rate

Year n (1)	Cost in Future Dollars (2)	$(P/F,14.4\%,n)$ (3)	PW (4)
0	$5000	1	$5000
1	5200	0.8741	4545
2	5408	0.7641	4132
3	5624	0.6679	3757
4	5849	0.5838	3415

EXAMPLE 14.1

A former student of an engineering department wishes to donate to the department's scholarship fund. Three options are available:

Plan A. $60,000 now.
Plan B. $15,000 per year for 8 years beginning 1 year from now.
Plan C. $50,000 three years from now and another $80,000, five years from now.

From the department's perspective, it wants to select the plan that maximizes the buying power of the dollars received. The department head asked the engineering professor evaluating the plans to account for inflation in the calculations. If the donation earns a real 10% per year and the inflation rate is expected to average 3% per year, which plan should be accepted?

Solution
The quickest evaluation method is to calculate the present worth of each plan in today's dollars. For plans B and C, the easiest way to obtain the present worth is through the use of the inflated interest rate. By Equation [14.6],

$$i_f = 0.10 + 0.03 + 0.10(0.03) = 0.133$$

$$PW_A = \$60{,}000$$

$$PW_B = \$15{,}000(P/A,13.3\%,8) = \$15{,}000(4.7508) = \$71{,}262$$

$$PW_C = \$50{,}000(P/F,13.3\%,3) + 80{,}000(P/F,13.3\%,5)$$

$$= \$50{,}000(0.68756) + 80{,}000(0.53561) = \$77{,}227$$

Since PW_C is the largest in today's dollars, select plan C.

For spreadsheet analysis, the PV function is used to find PW_B and PW_C: PV(13.3%,8,−15000) in one cell, and PV(13.3%,3,,−50000) + PV(13.3%,5,,−80000) in another cell.

Q-Solv

Comment
The present worths of plans B and C can also be found by first converting the cash flows to today's dollars using $f = 3\%$ in Equation [14.2] and then using the real i of 10% in the P/F factors. This procedure is more time-consuming, but the answers are the same.

EXAMPLE 14.2

A 15-year \$50,000 bond that has a dividend rate of 10% per year, payable semiannually, is currently for sale. If the expected rate of return of the purchaser is 8% per year, compounded semiannually, and if the inflation rate is expected to be 2.5% each 6-month period, what is the bond worth now (*a*) without an adjustment for inflation and (*b*) when inflation is considered? Perform both hand and computer solutions.

Sec. 5.8

Bonds

Solution by Hand

(*a*) Without inflation adjustment: The semiannual dividend is $I = [(50{,}000)(0.10)]/2 =$ \$2500. At a nominal 4% per 6 months for 30 periods, the PW is

$$\text{PW} = 2500(P/A,4\%,30) + 50{,}000(P/F,4\%,30) = \$58{,}645$$

(*b*) With inflation: Use the inflated rate i_f.

$$i_f = 0.04 + 0.025 + (0.04)(0.025) = 0.066 \text{ per semiannual period}$$

$$\begin{aligned}\text{PW} &= 2500(P/A,6.6\%,30) + 50{,}000(P/F,6.6\%,30)\\ &= 2500(12.9244) + 50{,}000(0.1470)\\ &= \$39{,}660\end{aligned}$$

Q-Solv

Solution by Computer

(*a*) and (*b*) These both require simple, single-cell functions on a spreadsheet (Figure 14–2). Without inflation adjusted for, the PV function is developed in B2 at the nominal 4% rate for 30 periods; with inflation the rate is $i_f = 6.6\%$, as determined above. See the cell tags for the formats.

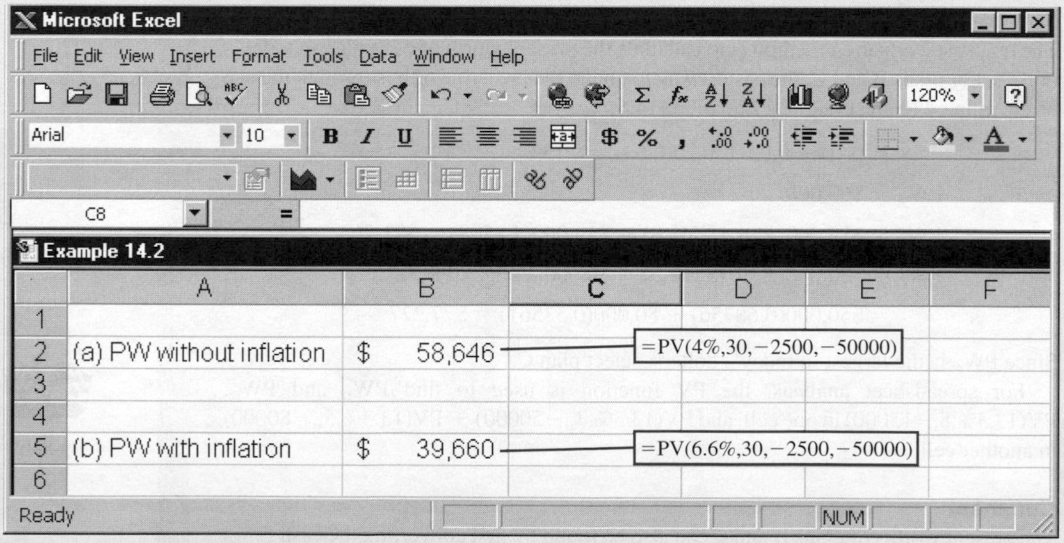
Microsoft Excel
File Edit View Insert Format Tools Data Window Help
120%
Arial
10
C8
Example 14.2
A B C D E F
1
2 (a) PW without inflation $ 58,646
=PV(4%,30,−2500,−50000)
3
4
5 (b) PW with inflation $ 39,660
=PV(6.6%,30,−2500,−50000)
6
Ready
NUM

Figure 14–2
Present worth of a bond without and with inflation adjustment, Example 14.2.

Comment

The $18,985 difference in PW values illustrates the tremendous negative impact made by only 2.5% inflation each 6 months (5.06% per year). Purchasing the $50,000 bond means receiving $75,000 in dividends over 15 years and the $50,000 principal in year 15. This is worth only $39,660 in constant-value (today's) dollars.

EXAMPLE 14.3

A self-employed chemical engineer is on contract with Dow Chemical, currently working in a relatively high-inflation country. She wishes to calculate a project's PW with estimated costs of $35,000 now and $7000 per year for 5 years beginning 1 year from now with increases of 12% per year thereafter for the next 8 years. Use a real interest rate of 15% per year to make the calculations (a) without an adjustment for inflation and (b) considering inflation at a rate of 11% per year.

Solution

(a) Figure 14–3 presents the cash flows. The PW without an adjustment for inflation is found using $i = 15\%$ and $g = 12\%$ in Equations [2.23] and [2.24] for the geometric series.

$$PW = -35,000 - 7000(P/A,15\%,4)$$

$$-\left\{ \frac{7000\left[1 - \left(\dfrac{1.12}{1.15}\right)^9\right]}{0.15 - 0.12} \right\}(P/F,15\%,4)$$

$$= -35,000 - 19,985 - 28,247$$

$$= \$-83,232$$

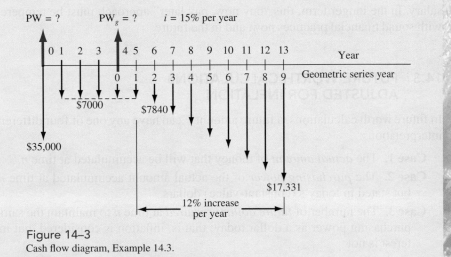

Figure 14–3
Cash flow diagram, Example 14.3.

In the P/A factor, $n = 4$ because the \$7000 cost in year 5 is the A_1 term in Equation [2.23].

(b) To adjust for inflation, calculate the inflated interest rate by Equation [14.6].

$$i_f = 0.15 + 0.11 + (0.15)(0.11) = 0.2765$$

$$PW = -35,000 - 7000(P/A,27.65\%,4)$$

$$- \left\{ \frac{7000 \left[1 - \left(\dfrac{1.12}{1.2765} \right)^9 \right]}{0.2765 - 0.12} \right\} (P/F,27.65\%,4)$$

$$= -35,000 - 7000(2.2545) - 30,945(0.3766)$$

$$= \$-62,436$$

This result demonstrates that in a high-inflation economy, when negotiating the amount of the payments to repay a loan, it is economically advantageous for the borrower to use future (inflated) dollars whenever possible to make the payments. The present value of future inflated dollars is significantly less when the inflation adjustment is included. And the higher the inflation rate, the larger the discounting because the P/F and P/A factors decrease in size.

The last two examples seem to add credence to the "buy now, pay later" philosophy of financial management. However, at some point, the debt-ridden company or individual will have to pay off the debts and the accrued interest with the inflated dollars. If cash is not readily available, the debts cannot be repaid. This can happen, for example, when a company unsuccessfully launches a new product, when there is a serious downturn in the economy, or when an individual loses a salary. In the longer term, this "buy now, pay later" approach must be tempered with sound financial practices now, and in the future.

14.3 FUTURE WORTH CALCULATIONS ADJUSTED FOR INFLATION

In future worth calculations, a future amount F can have any one of four different interpretations:

Case 1. The *actual amount* of money that will be accumulated at time n.

Case 2. The *purchasing power* of the actual amount accumulated at time n, but stated in today's (constant-value) dollars.

Case 3. The number of *future dollars required* at time n to maintain the same purchasing power as a dollar today; that is, inflation is considered, but interest is not.

Case 4. The number of dollars required at time n to *maintain purchasing power and earn a stated real interest rate.*

Depending upon which interpretation is intended, the F value is calculated differently, as described below. Each case is illustrated.

Case 1: Actual Amount Accumulated   It should be clear that F, the actual amount of money accumulated, is obtained using the inflation-adjusted (market) interest rate.

$$F = P(1 + i_f)^n = P(F/P,i_f,n) \qquad [14.7]$$

For example, when we are quoted a market rate of 10%, the inflation rate is included. Over a 7-year period, \$1000 will accumulate to

$$F = 1000(F/P,10\%,7) = \$1948$$

Case 2: Constant-Value with Purchasing Power   The purchasing power of future dollars is determined by first using the market rate i_f to calculate F and then deflating the future amount through division by $(1 + f)^n$.

$$F = \frac{P(1 + i_f)^n}{(1 + f)^n} = \frac{P(F/P,i_f,n)}{(1 + f)^n} \qquad [14.8]$$

This relation, in effect, recognizes the fact that inflated prices mean \$1 in the future purchases less than \$1 now. The percentage loss in purchasing power is a measure of how much less. As an illustration, consider the same \$1000 now, a 10% per year market rate, and an inflation rate of 4% per year. In 7 years, the purchasing power has risen, but only to \$1481.

$$F = \frac{1000(F/P,10\%,7)}{(1.04)^7} = \frac{\$1948}{1.3159} = \$1481$$

This is \$467 (or 24%) less than the \$1948 actually accumulated at 10% (case 1). Therefore, we conclude that 4% inflation over 7 years reduces the purchasing power of money by 24%.

Also for case 2, the future amount of money accumulated with today's buying power could equivalently be determined by calculating the real interest rate and using it in the F/P factor to compensate for the decreased purchasing power of the dollar. This *real interest rate* is the i in Equation [14.6].

$$i_f = i + f + if$$
$$= i(1 + f) + f$$
$$i = \frac{i_f - f}{1 + f} \qquad [14.9]$$

The real interest rate i represents the rate at which today's dollars expand with their *same purchasing power* into equivalent future dollars. An inflation rate larger than the market interest rate leads to a negative real interest rate. The use of this interest rate is appropriate for calculating the future worth of an investment (such as a savings account or money market fund) when the effect of inflation must be removed. For the example of \$1000 in today's dollars from Equation [14.9]

$$i = \frac{0.10 - 0.04}{1 + 0.04} = 0.0577, \text{ or } 5.77\%$$

$$F = 1000(F/P,5.77\%,7) = \$1481$$

The market interest rate of 10% per year has been reduced to a real rate that is less than 6% per year because of the erosive effects of inflation.

Case 3: Future Amount Required, No Interest   This case recognizes that prices increase when inflation is present. Simply put, future dollars are worth less, so more are needed. No interest rate is considered at all in this case. This is the situation present if someone asks, How much will a car cost in 5 years if its current cost is $20,000 and its price will increase by 6% per year? (The answer is $26,765.) No interest rate, only inflation, is involved. To find the future cost, substitute f for the interest rate in the F/P factor.

$$F = P(1 + f)^n = P(F/P,f,n) \qquad [14.10]$$

Reconsider the $1000 used previously. If it is escalating at exactly the inflation rate of 4% per year, the amount 7 years from now will be

$$F = 1000(F/P,4\%,7) = \$1316$$

Case 4: Inflation and Real Interest   This is the case applied when a MARR is established. Maintaining purchasing power and earning interest must account for both increasing prices (case 3) and the time value of money. If the growth of capital is to keep up, funds must grow at a rate equal to or above the real interest rate i plus a rate equal to the inflation rate f. Thus, to make a *real rate of return of 5.77%* when the inflation rate is 4%, i_f is the market (inflation-adjusted) rate that must be used. For the same $1000 amount,

$$i_f = 0.0577 + 0.04 + 0.0577(0.04) = 0.10$$
$$F = 1000(F/P,10\%,7) = \$1948$$

This calculation shows that $1948 seven years in the future will be equivalent to $1000 now with a real return of $i = 5.77\%$ per year and inflation of $f = 4\%$ per year.

Table 14–3 summarizes which rate is used in the equivalence formulas for the different interpretations of F. The calculations made in this section reveal that $1000 now at a market rate of 10% per year would accumulate to $1948 in 7 years; the $1948 would have the purchasing power of $1481 of today's dollars if $f = 4\%$ per year; an item with a cost of $1000 now would cost $1316 in 7 years at an inflation rate of 4% per year; and it would take $1948 of future dollars to be equivalent to the $1000 now at a real interest rate of 5.77% with inflation considered at 4%.

Most corporations evaluate alternatives at a MARR large enough to cover inflation plus some return greater than their cost of capital, and significantly higher than the safe investment return of approximately 3.5% mentioned earlier. Therefore, for case 4, the resulting MARR will normally be higher than the market rate i_f. Define the symbol MARR_f as the inflation-adjusted MARR, which is calculated in a fashion similar to i_f.

$$\text{MARR}_f = i + f + i(f) \qquad [14.11]$$

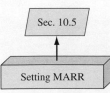

Sec. 10.5

Setting MARR

| | | **TABLE 14–3** Calculation Methods for Various Future Worth Interpretations | | |

Future Worth Desired	Method of Calculation	Example for $P = \$1000$, $n = 7$, $i_f = 10\%$, $f = 4\%$
Case 1: Actual dollars accumulated	Use stated market rate i_f in equivalence formulas	$F = 1000(F/P,10\%,7)$
Case 2: Purchasing power of accumulated dollars in terms of today's dollars	Use market rate i_f in equivalence and divide by $(1 + f)^n$	$F = \dfrac{1000(F/P,10\%,7)}{(1.04)^7}$
	or	or
	Use real i	$F = 1000(F/P,5.77\%,7)$
Case 3: Dollars required for same purchasing power	Use f in place of i in equivalence formulas	$F = 1000(F/P,4\%,7)$
Case 4: Future dollars to maintain purchasing power and to earn a return	Calculate i_f and use in equivalence formulas	$F = 1000(F/P,10\%,7)$

The real rate of return i used here is the required rate for the corporation relative to its cost of capital. Now the future worth F, or FW, is calculated as

$$F = P(1 + \text{MARR}_f)^n = P(F/P,\text{MARR}_f,n) \qquad [14.12]$$

For example, if a company has a WACC (weighted average cost of capital) of 10% per year and requires that a project return 3% per year above its WACC, the real return is $i = 13\%$. The inflation-adjusted MARR is calculated by including the inflation rate of, say, 4% per year. Then, the project PW, AW, or FW will be determined at the rate obtained from Equation [14.11].

$$\text{MARR}_f = 0.13 + 0.04 + 0.13(0.04) = 17.52\%$$

A similar computation can be made for an individual using i as the expected real rate that is above the safe investment rate. When an individual is satisfied with a required real return equal to a safe investment rate, approximately $i = 3.5\%$, or a corporation is satisfied with a real return equal to a safe investment rate, Equations [14.11] and [14.6] have the same result; that is, $\text{MARR}_f = i_f$, for the corporation or the individual.

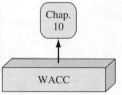

Chap. 10

WACC

EXAMPLE 14.4

Abbott Mining Systems wants to determine whether it should "buy" now or "buy" later for upgrading a piece of equipment used in deep mining operations in one of its international operations. If the company selects plan A, the equipment will be purchased now for $200,000. However, if the company selects plan I, the purchase will be deferred for 3 years when the cost is expected to rise rapidly to $340,000. Abbott is ambitious; it expects a real MARR of 12% per year. The inflation rate in the country has averaged 6.75% per year. From only an economic perspective, determine whether the company should purchase now or later (a) when inflation is not considered and (b) when inflation is considered.

Solution

(a) Inflation not considered: The real rate, or MARR, is $i = 12\%$ per year. The cost of plan I is $340,000 three years hence. Calculate the FW value for plan A three years from now and select the lower cost.

$$FW_A = -200,000(F/P,12\%,3) = \$-280,986$$

$$FW_I = \$-340,000$$

Select plan A (purchase now).

(b) Inflation considered: This is case 4; there is a real rate (12%), and inflation of 6.75% must be accounted for. First, compute the inflation-adjusted MARR by Equation [14.11].

$$i_f = 0.12 + 0.0675 + 0.12(0.0675) = 0.1956$$

Use i_f to compute the FW value for plan A in future dollars.

$$FW_A = -200,000(F/P,19.56\%,3) = \$-341,812$$

$$FW_I = \$-340,000$$

Purchase later (plan I) is now selected, because it requires fewer equivalent future dollars. The inflation rate of 6.75% per year has raised the equivalent future worth of costs by 21.6% to $341,812. This is the same as an increase of 6.75% per year, compounded over 3 years, or $(1.0675)^3 - 1 = 21.6\%$.

Most countries have inflation rates in the range of 2% to 8% per year, but *hyperinflation* is a problem in countries where political instability, overspending by the government, weak international trade balances, etc., are present. Hyperinflation rates may be very high—10% to 100% *per month*. In these cases, the government may take drastic actions: redefine the currency in terms of the currency of another country, control banks and corporations, and control the flow of capital into and out of the country in order to decrease inflation.

In a hyperinflated environment, people usually spend all their money immediately since the cost will be so much higher the next month, week, or day. To appreciate the disastrous effect of hyperinflation on a company's ability to keep up, we can rework Example 14.4b using an inflation rate of 10% per month, that is, a nominal 120% per year (not considering the compounding of inflation). The FW_A amount skyrockets and plan I is a clear choice. Of course, in such an environment the $340,000 purchase price for plan I three years hence would obviously not be guaranteed, so the entire economic analysis is unreliable. Good economic decisions in a hyperinflated economy are very difficult to make using traditional engineering economy methods, since the estimated future values are totally unreliable and the future availability of capital is uncertain.

14.4 CAPITAL RECOVERY CALCULATIONS ADJUSTED FOR INFLATION

It is particularly important in capital recovery calculations used for AW analysis to include inflation because current capital dollars must be recovered with future inflated dollars. Since future dollars have less buying power than today's dollars, it is obvious that more dollars will be required to recover the present investment. This suggests the use of the inflated interest rate in the A/P formula. For example, if $1000 is invested today at a real interest rate of 10% per year when the inflation rate is 8% per year, the equivalent amount that must be recovered each year for 5 years in future dollars is

$$A = 1000(A/P,18.8\%,5) = \$325.59$$

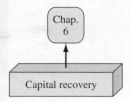

Chap. 6

Capital recovery

On the other hand, the decreased value of dollars through time means that investors can spend fewer present (higher-value) dollars to accumulate a specified amount of future (inflated) dollars. This suggests the use of a higher interest rate, that is, the i_f rate, to produce a lower A value in the A/F formula. The annual equivalent (with adjustment for inflation) of $F = \$1000$ five years from now in future dollars is

$$A = 1000(A/F,18.8\%,5) = \$137.59$$

This method is illustrated in the next example.

For comparison, the equivalent annual amount to accumulate $F = \$1000$ at a real $i = 10\%$, (without adjustment for inflation) is $1000(A/F,10\%,5) = \$163.80$. Thus, when F is fixed, uniformly distributed future costs should be spread over as long a time period as possible so that the leveraging effect of inflation will reduce the payment ($137.59 versus $163.80 here).

EXAMPLE 14.5

What annual deposit is required for 5 years to accumulate an amount of money with the same purchasing power as $680.58 today, if the market interest rate is 10% per year and inflation is 8% per year?

Solution
First, find the actual number of future (inflated) dollars required 5 years from now. This is case 3.

$$F = (\text{present buying power})(1 + f)^5 = 680.58(1.08)^5 = \$1000$$

The actual amount of the annual deposit is calculated using the market (inflated) interest rate of 10%. This is case 4 using A instead of P.

$$A = 1000(A/F,10\%,5) = \$163.80$$

Comment
The real interest rate is $i = 1.85\%$ as determined using Equation [14.9]. To put these calculations into perspective, if the inflation rate is zero when the real interest rate is 1.85%, the future amount of money with the same purchasing power as \$680.58 today is obviously \$680.58. Then the annual amount required to accumulate this future amount in 5 years is $A = 680.58(A/F,1.85\%,5) = \$131.17$. This is \$32.63 lower than the \$163.80 calculated above for $f = 8\%$. This difference is due to the fact that during inflationary periods, dollars deposited have more buying power than the dollars returned at the end of the period. To make up the buying power difference, more lower-value dollars are required. That is, to maintain equivalent purchasing power at $f = 8\%$ per year, an extra \$32.63 per year is required.

The logic discussed here explains why, in times of increasing inflation, lenders of money (credit card companies, mortgage companies, and banks) tend to further increase their market interest rates. People tend to pay off less of their incurred debt at each payment because they use any excess money to purchase additional items before the price is further inflated. Also, the lending institutions must have more dollars in the future to cover the expected higher costs of lending money. All this is due to the spiraling effect of increasing inflation. Breaking this cycle is difficult to do at the individual level and much more difficult to alter at a national level.

CHAPTER SUMMARY

Inflation, treated computationally as an interest rate, makes the cost of the same product or service increase over time due to the decreased value of money. There are several ways to consider inflation in engineering economy computations in terms of today's (constant-value) dollars and in terms of future dollars. Some important relations are:

Inflated interest rate: $i_f = i + f + if$

Real interest rate: $i = (i_f - f)/(1 + f)$

PW of a future amount with inflation considered: $P = F(P/F,i_f,n)$

Future worth of a present amount in constant-value dollars with the same purchasing power: $F = P(F/P,i,n)$

Future amount to cover a current amount with no interest: $F = P(F/P,f,n)$

Future amount to cover a current amount with interest: $F = P(F/P,i_f,n)$

Annual equivalent of a future dollar amount: $A = F(A/F,i_f,n)$

Annual equivalent of a present amount in future dollars: $A = P(A/P,i_f,n)$

Hyperinflation implies very high f values. Available funds are expended immediately because costs increase so rapidly that larger cash inflows cannot offset the fact that the currency is losing value. This can, and usually does, cause a national financial disaster when it continues over extended periods of time.

PROBLEMS

Adjusting for Inflation

14.1 Describe how to convert inflated dollars into constant-value dollars.

14.2 What is the inflation rate if something costs exactly twice as much as it did 10 years earlier?

14.3 In an effort to reduce pipe breakage, water hammer, and product agitation, a chemical company plans to install several chemically resistant pulsation dampeners. The cost of the dampeners today is $106,000, but the chemical company has to wait until a permit is approved for its bidirectional port-to-plant product pipeline. The permit approval process will take at least 2 years because of the time required for preparation of an environmental impact statement. Because of intense foreign competition, the manufacturer plans to increase the price only by the inflation rate each year. If the inflation rate is 3% per year, estimate the cost of the dampeners in 2 years in terms of (*a*) then-current dollars and (*b*) today's dollars.

14.4 Convert $10,000 present dollars into then-current dollars of year 10 if the inflation rate is 7% per year.

14.5 Convert $10,000 future dollars in year 10 into *constant-value dollars* (not equivalent dollars) of today if the inflation-adjusted (market) interest rate is 11% per year and the inflation rate is 7% per year.

14.6 Convert $10,000 future dollars in year 10 into *constant-value dollars* (not equivalent dollars) today if the inflation-adjusted (market) interest rate is 12% per year and the real interest rate is 3% per year.

14.7 Estimated costs for maintenance and operation of a certain machine are expected to be $13,000 per year (then-current dollars) in years 1 to 3. At an inflation rate of 6% per year, what is the constant-value amount (in terms of today's dollars) of *each year's* future dollar amount?

14.8 If the market interest rate is 12% per year and the inflation rate is 5% per year, determine the number of future dollars in year 5 that have the *same buying power* as $2000 now.

14.9 Ford Motor Company announced that the price of its F-150 pickup trucks is going to increase by only the inflation rate for the next 2 years. If the current price of a truck is $21,000 and the inflation rate is expected to average 2.8% per year, what is the expected *price* of a comparably equipped truck 2 years from now?

14.10 A headline in the *Chronicle of Higher Education* reads "College Costs Rise Faster than Inflation." The article states that tuition at public colleges and universities increased by 56% over the past 5 years. (*a*) What was the average annual percentage increase over that time? (*b*) If the inflation rate was 2.5% per year, how many percentage points over the inflation rate was the annual tuition increase?

14.11 A machine purchased by Holtzman Industries had a cost of $45,000 four years ago. If a similar machine costs $55,000 now and its price increased only by the inflation rate, what was the annual inflation rate over that 4-year period?

Real and Market Interest Rates

14.12 State the conditions under which the market interest is (*a*) higher than, (*b*) lower than, and (*c*) the same as the real interest rate.

14.13 Calculate the inflation-adjusted interest rate when the annualized inflation rate is 27% per year (Caracas, 2004) and the real interest rate is 4% per year.

14.14 What annual inflation rate is implied from a market interest rate of 15% per year when the real interest rate is 4% per year?

14.15 What market interest rate per quarter would be associated with a quarterly inflation rate of 5% and a real interest rate of 2% per quarter?

14.16 When the market interest rate is 48% per year, compounded monthly (due to hyperinflation), what is the monthly inflation rate if the real interest rate is 6% per year, compounded monthly?

14.17 What real rate of return will an investor make on a rate of return of 25% per year when the inflation rate is 10% per year?

14.18 What is the real interest rate per semiannual period when the market interest rate is 22% per year, compounded semiannually, and the inflation rate is 7% per 6 months?

14.19 A cash-value life insurance policy will pay a sum of $1,000,000 when the insured reaches the age of 65. If the insured will be 65 years old 27 years from today, what will be the value of the $1,000,000 in terms of dollars with today's buying power if the inflation rate is 3% per year over that time period?

Alternative Comparison with Adjustment for Inflation

14.20 A regional infrastructure building and maintenance contractor is trying to decide whether to buy a new compact horizontal directional drilling (HDD) machine now or wait to buy it 2 years from now (when a large pipeline contract will require the new equipment). The HDD machine will include an innovative pipe loader design and a maneuverable undercarriage system. The cost of the system is $68,000 if purchased now or $81,000 if purchased 2 years from now. At a real interest MARR of 10% per year and an inflation rate of 5% per year, determine if the company should buy now or later (a) without any adjustment for inflation and (b) with inflation considered.

14.21 As an innovative way to pay for various software packages, a new high-technology service company has offered to pay your company in one of three ways: (1) pay $400,000 now, (2) pay $1.1 million 5 years from now, or (3) pay an amount of money 5 years from now that has the same *buying power* as $750,000 now. If you want to earn a real interest rate of 10% per year and the inflation rate is 6% per year, which offer should you accept?

14.22 Consider alternatives A and B on the basis of their present worth values, using a real interest rate of 10% per year and an inflation rate of 3% per year, (a) without any adjustment for inflation and (b) with inflation considered.

	Machine A	Machine B
First cost, $	−31,000	−48,000
Annual operating cost, $/year	−28,000	−19,000
Salvage value, $	5,000	7,000
Life, years	5	5

14.23 Compare the alternatives below on the basis of their capitalized costs with adjustments made for inflation. Use $i_r = 12\%$ per year and $f = 3\%$ per year.

	Alternative X	Alternative Y
First cost, $	−18,500,000	−9,000,000
Annual operating cost, $/year	−25,000	−10,000
Salvage value, $	105,000	82,000
Life, years	∞	10

14.24 An engineer must recommend one of two machines for integration into an upgraded manufacturing line. She obtains estimates from two salespeople. Salesman A gives her the estimates in future (then-current) dollars, while saleswoman B provides the estimates in today's (constant-value) dollars. The company has a MARR of a real 15% per year, and it expects inflation to be 5% per year. Use PW analysis to determine which machine the engineer should recommend.

	Salesman A, Future $	Saleswoman B, Today's $
First cost, $	−60,000	−95,000
AOC, $/year	−55,000	−35,000
Life, years	10	10

Future Worth and Other Calculations with Inflation

14.25 An engineer purchased an inflation-linked corporate bond (i.e., bond interest changes with inflation) issued by Household Finance Bank that has a face value of $25,000. At the time the bond was purchased, the yield on the bond was 2.16% per year *plus* inflation, payable monthly. The bond interest rate is adjusted each month based on the change in the Consumer Price Index (CPI) from the same month of the previous year. In one particular month, the CPI was 3.02% higher than it was in the same month of the previous year.
 (a) What is the new yield on the bond?
 (b) If interest is paid monthly, how much interest did the engineer

receive that month (i.e., after the adjustment)?

14.26 An engineer deposits $10,000 into an account when the market interest rate is 10% per year and the inflation rate is 5% per year. The account is left undisturbed for 5 years.
 (a) How much money will be in the account?
 (b) What will be the buying (purchasing) power in terms of today's dollars?
 (c) What is the real rate of return that is made on the account?

14.27 A chemical company wants to set aside money now so that it will be able to purchase new data loggers 3 years from now. The price of the data loggers is expected to increase only by the inflation rate of 3.7% per year for each of the next 3 years. If the total cost of the data loggers now is $45,000, determine (a) their expected cost 3 years from now and (b) how much the company will have to set aside now, if it earns interest at a rate of 8% per year.

14.28 The cost of constructing a certain highway exit ramp was $625,000 seven years ago. An engineer designing another one that is almost exactly the same estimates the cost today will be $740,000. If the cost had increased only by the inflation rate over that time period, what was the inflation rate per year?

14.29 If you make an investment in commercial real estate that is guaranteed to net you $1.5 million 25 years from now, what will be the *buying power* of that money with respect to today's dollars if the market interest rate is 8% per year and the inflation rate stays at 3.8% per year over that time period?

14.30   Goodyear Tire and Rubber Corporation can purchase a piece of equipment now for $80,000 or buy it 3 years from now for $128,000. The MARR requirement for the plant is a real return of 15% per year. If an inflation rate of 4% per year must be accounted for, should the company buy the machine now or later?

14.31   In a period of 3% per year inflation, how much will a machine cost 3 years from now in terms of *constant-value dollars,* if the cost today is $40,000 and the cost of the machine is expected to increase only by the inflation rate?

14.32   In a period of 4% per year inflation, how much will a machine cost 3 years from now in terms of *constant-value dollars,* if the cost today is $40,000 and the manufacturer plans to raise the price so that the manufacturer will make a real rate of return of 5% per year over that time period?

14.33   Convert $100,000 of today's dollars into then-current dollars in year 10 when the *deflation rate* is 1.5% per year.

14.34   A company has been invited to invest $1 million in a partnership and receive a guaranteed total amount of $2.5 million after 4 years. By corporate policy, the MARR is always established at 4% above the real cost of capital. If the real interest rate paid on capital is currently 10% per year and the inflation rate during the 4-year period is expected to average 3% per year, is the investment economically justified?

14.35   The first Nobel Prize was awarded in 1901 in the amount of $150,000. In 1996, the award was raised from $489,000 to $653,000. (*a*) At what inflation rate would an award of $653,000 in 1996 be equivalent (in purchasing power) to the original award in 1901? (*b*) If the foundation expects inflation to average 3.5% per year from 1996 through 2010, how large will the award have to be in 2010 to make it worth the same as in 1996?

14.36   Factors that increase costs and prices—especially for materials and manufacturing costs sensitive to market, technology, and labor availability—can be considered separately using the real interest rate i, the inflation rate f, and additional increases that grow at a geometric rate g. The future amount is calculated based on a current estimate by using the relation

$$F = P(1 + i)^n(1 + f)^n(1 + g)^n$$
$$= P[(1 + i)(1 + f)(1 + g)]^n$$

The product of the first two terms enclosed in parentheses results in the inflated interest rate i_f. The geometric rate is the same one used in the geometric gradient series (Chapter 2). It commonly applies to maintenance and repair cost increases as machinery ages. This is over and above the inflation rate. If the current cost to manufacture an electronic subcomponent is $250,000 per year, what is the equivalent value in 5 years, if average annual rates are estimated to be $i = 5\%$, $f = 3\%$, and $g = 2\%$ per year?

Capital Recovery with Inflation

14.37   Aquatech Microsystems spent $183,000 for a communications protocol to achieve interoperability among its utility systems. If the company uses a real interest rate of 15% per year on such investments and a recovery period of 5 years, what is the annual worth of the expenditure in then-current dollars at an inflation rate of 6% per year?

14.38   A DSL company has made an equipment investment of $40 million with the expectation that it will be recovered in 10 years. The company has a MARR based on a real rate of return of 12% per year. If inflation is 7% per year, how much must the company make each year (a) in constant-value dollars and (b) in future dollars to meet its expectation?

14.39 What is the annual worth in then-current dollars in years 1 through 5 of a receipt of $750,000 now, if the *market interest rate* is 10% per year and the inflation rate is 5% per year?

14.40 A recently graduated mechanical engineer wants to build a reserve fund as a safety net to pay his expenses in the unlikely event that he is without work for a short time. His aim is to have $15,000 developed over the next 3 years, with the proviso that the amount have the same purchasing power as $15,000 today. If the expected market rate on investments is 8% per year and inflation is averaging 2% per year, find the annual amount necessary to meet his goal.

14.41 A European-based cattle genetics engineering research lab is planning for a major expenditure on research equipment. The lab needs $5 million of today's dollars so it can make the acquisition 4 years from now. The inflation rate is steady at 5% per year. (a) How many future dollars will be needed when the equipment is purchased, if purchasing power is maintained? (b) What is the

required amount of the annual deposit into a fund that earns the market rate of 10% per year to ensure that the amount calculated in part (a) is accumulated?

14.42 (a) Calculate the perpetual equivalent annual worth in future dollars (for years 1 through ∞) for income of $50,000 now and $5000 per year thereafter. Assume the market interest rate is 8% per year and inflation averages 4% per year. All amounts are quoted as future dollars.

(b) If the amounts are quoted in *constant-value dollars,* how do you find the annual worth in *future dollars*?

14.43 The two machines detailed are being considered for a chip manufacturing operation. Assume the company's MARR is a real return of 12% per year and that the inflation rate is 7% per year. Which machine should be selected on the basis of an annual worth analysis, if the estimates are in (a) constant-value dollars and (b) future dollars?

	Machine A	Machine B
First cost, $	−150,000	−1,025,000
Annual M&O cost, $/year	−70,000	−5,000
Salvage value, $	40,000	200,000
Life, years	5	∞

FE REVIEW PROBLEMS

14.44 For a real interest rate of 12% per year and an inflation rate of 7% per year, the market interest rate per year is closest to
(a) 4.7%
(b) 7%
(c) 12%
(d) 19.8%

14.45 When all future cash flows are expressed in then-current dollars, the rate that should be used to find the present worth is the
(a) Real MARR
(b) Inflation rate
(c) Inflated interest rate
(d) Real interest rate

14.46 To convert constant-value dollars into inflated dollars, it is necessary to
   (a) Divide by $(1 + i_f)^n$
   (b) Divide by $(1 + f)^n$
   (c) Divide by $(1 + i)^n$
   (d) Multiply by $(1 + f)^n$

14.47 To convert inflated dollars into constant-value dollars, it is necessary to
   (a) Divide by $(1 + i_f)^n$
   (b) Divide by $(1 + f)^n$
   (c) Divide by $(1 + i)^n$
   (d) Multiply by $(1 + f)^n$

14.48 When the market interest rate is less than the real interest rate, then
   (a) The inflated interest rate is higher than the real interest rate.
   (b) The real interest rate is zero.
   (c) A deflationary condition exists.
   (d) All the above.

14.49 When future dollars are expressed in terms of constant-value dollars, the rate that should be used in present worth calculations is the
   (a) Real interest rate
   (b) Market interest rate
   (c) Inflation rate
   (d) Market rate less the inflation rate

EXTENDED EXERCISE

FIXED-INCOME INVESTMENTS VERSUS THE FORCES OF INFLATION

The savings and investments that an individual maintains should have some balance between equity (e.g., corporate stocks that rely on market growth and dividend income) and fixed-income investments (e.g., bonds that pay dividends to the purchaser). When inflation is moderately high, conventional bonds offer a low return relative to stocks, because the potential for market growth is not present with bonds. Additionally, the force of inflation makes the dividends worth less in future years, because there is no inflation adjustment made in the amount of dividend paid as time passes. However, bonds do offer a steady income that may be important to an individual. And, they serve to preserve the principal invested in the bond, because the face value is returned at maturity.

Harold is an engineer who wants a predictable flow of money for travel and vacations. He has a salary high enough that he is in a relatively high tax bracket (28% or above). As a first step, he has decided to purchase a municipal bond, because of the predictable income and the fact that the dividend is completely free of federal and state income taxes. He plans to purchase a municipal tax-exempt bond that has a face value of $25,000, a coupon rate of 5.9% paid annually, and a maturity of 12 years.

Questions

Help Harold with some analysis by answering the following, using a spreadsheet.

1. What is the overall rate of return if the bond is kept until maturity? Does this return value have any of the inherent effects of inflation included in it?

2. Harold may decide to sell the bond immediately after the third annual dividend. What is his minimum selling price if he wants to make a 7% real return and wants to adjust for 4% per year inflation?

3. If Harold were in need of money immediately after the third dividend payment, what would be the minimum selling price in future dollars of the bond, if he will sell for an amount that is equivalent to the purchasing power of the original price?

4. As a follow-on to question #3, what happens to the selling price (in future dollars) 3 years after purchase, if Harold is willing to include the then-current purchasing power of each of the dividends in the computation to determine the selling price? Assume Harold spent the dividends immediately after receiving them.

5. Harold plans to retain the bond through its maturity of 12 years, but he requires a return of 7% per year adjusted for 4% per year inflation. He may be able to purchase the bond at a discount, that is, pay less than $25,000 now. What is the maximum he should pay for the bond?

CHAPTER 15

Cost Estimation and Indirect Cost Allocation

Up to this point, cost and revenue cash flow values have been stated or assumed as known. In reality, they are not; they must be estimated. This chapter explains what cost estimation involves, and applies cost estimation techniques. *Cost estimation* is important in all aspects of a project, but especially in the stages of project conception, preliminary design, detailed design, and economic analysis. When a project is developed in the private or the public sector, questions about costs and revenues will be posed by individuals representing many different functions: management, engineering, construction, production, quality, finance, safety, environmental, legal, and marketing, to name some. In engineering practice, the estimation of costs receives much more attention than revenue estimation; costs are the topic of this chapter.

Unlike direct costs for labor and materials, indirect costs are not easily traced to a specific department, machine, or processing line. Therefore, *allocation of indirect costs* for functions such as utilities, safety, management and administration, purchasing, and quality is made using some rational basis. Both the traditional method of allocation and the Activity-Based Costing (ABC) method are covered in this chapter. Comparison between these two approaches is made.

There are two case studies; the first concentrates on cost estimate sensitivity analysis, while the second examines the allocation of indirect costs in a manufacturing setting.

LEARNING OBJECTIVES

Purpose: Make cost estimates and include the dimension of indirect cost allocation in an engineering economy study.

This chapter will help you learn to:

Approaches	1. Describe different approaches to cost estimation.
Cost indexes	2. Use a cost index to estimate present cost based on historic data.
Cost-capacity equations	3. Estimate the cost of a component, system, or plant by using a cost-capacity equation.
Factor method	4. Estimate total plant cost using the factor method.
Indirect cost rates and allocation	5. Allocate indirect costs using traditional indirect cost rates.
ABC allocation	6. Allocate indirect costs using the Activity-Based Costing (ABC) method.

15.1   UNDERSTANDING HOW COST ESTIMATION IS ACCOMPLISHED

Cost estimation is a major activity performed in the initial stages of virtually every effort in industry, business, and government. In general, most cost estimates are developed for either a *project* or a *system;* however, combinations of these are very common. A project usually involves physical items, such as a building, bridge, manufacturing plant, and offshore drilling platform, to name just a few. A system is usually an operational design that involves processes, software, and other nonphysical items. Examples might be a purchase order system, a software package, and an Internet-based remote-control system. Additionally the cost estimates are usually made for the initial development of the project or system, with the life-cycle costs of maintenance and upgrade estimated as a percentage of first cost. Of course, many projects will have major elements that are not physical, so estimates of both types must be developed. For example, consider a computer network system. There would be no operational system if only the computer hardware plus wire and wireless connectors were estimated; it is equally important to estimate the software, personnel, and maintenance costs. Much of the discussion that follows concentrates on physical-based projects. However, the logic is widely applicable to cost estimation for project and system designs.

Thus far virtually all cash flow estimates in the examples, problems, and exercises of this text were stated or assumed to be known. In real-world practice, the cash flows for costs and revenues must be estimated prior to the evaluation of a project or comparison of alternatives. We concentrate on cost estimation because costs are the primary values estimated for the economic analysis. Revenue estimates utilized by engineers are usually developed in marketing, sales, and other departments.

Costs are comprised of *direct costs* and *indirect costs*. Normally direct costs are estimated with some detail, then the indirect costs are added using standard rates and factors. However, direct costs in many industries, including manufacturing and assembly settings, have become a small percentage of overall product cost, while indirect costs have become much larger. Accordingly, many industrial settings require some estimating for indirect costs as well. Indirect cost allocation is discussed in detail in later sections of this chapter. Primarily, direct costs are discussed here.

Because cost estimation is a complex activity, the following questions form a structure for our discussion.

- What cost components must be estimated?
- What approach to cost estimation will be applied?
- How accurate should the estimates be?
- What estimation techniques will be utilized?

Costs to Estimate   If a project revolves around a single piece of equipment, for example, an industrial robot, the *cost components* will be significantly simpler and fewer than the components for a complete system such as the manufacturing and testing line for a new product. Therefore, it is important to know up front how much the cost estimation task will involve. Examples of cost components are the first cost P and the annual operating cost (AOC), also called the M&O costs (maintenance and operating) of equipment. Each component will have several *cost elements,* some that are directly estimated, others that require examination of records of similar projects, and still others that must be modeled using an estimation technique. Listed below are sample elements of the first cost and AOC components.

First cost component P:

    Elements: Equipment cost

                  Delivery charges

                  Installation cost

                  Insurance coverage

                  Initial training of personnel for equipment use

Delivered-equipment cost is the sum of the first two elements; installed-equipment cost adds the third element. Capital recovery (CR) for the first cost is determined using the MARR and the A/P factor over the estimated life of the equipment.

AOC component, a part of the equivalent annual cost A:

    Elements: Direct labor cost for operating personnel

                  Direct materials

                  Maintenance (daily, periodic, repairs, etc.)

                  Rework and rebuild

Some of these elements, such as equipment cost, can be determined with high accuracy; others, such as maintenance costs, are harder to estimate. When costs for an entire system must be estimated, the number of cost components and elements is likely to be in the hundreds. It is then necessary to prioritize the estimation tasks.

For familiar projects (houses, office buildings, highways, and some chemical plants) there are standard cost estimation packages available in paper or software form. For example, state highway departments utilize software packages that prompt for the correct cost components (bridges, pavement, cut-and-fill profiles, etc.) and estimate costs with time-proven, built-in relations. Once these components are estimated, exceptions for the specific project are added. However, there are no "canned" software packages for a large percentage of industrial, business, and public sector cost estimation jobs.

Cost Estimation Approach   Traditionally in industry, business, and the public sector, a "bottom-up" approach to cost estimation is applied. For a simple rendition of this approach, see Figure 15–1 (left). The progression is as follows: cost components and their elements are identified, cost elements are estimated, and estimates are summed to obtain total direct cost. The price is then determined by adding indirect costs and the profit margin, which is usually a percentage of the total cost. This approach works well when competition is not the dominant factor in pricing the product or service.

The bottom-up approach treats the required price as an output variable and the cost estimates as input variables.

Figure 15–1
Simplified cost estimation
processes for bottom-up
and top-down
approaches.

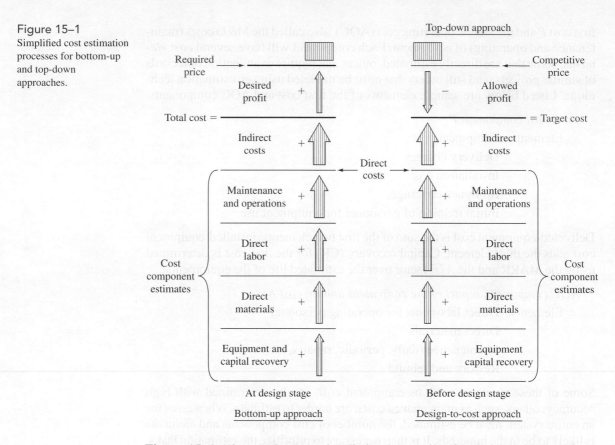

This is the figure showing the bottom-up and top-down (design-to-cost) approaches.

Figure 15–1 (above) shows a simplistic progression for the design-to-cost, or top-down, approach. The competitive price establishes the target cost.

The design-to-cost, or top-down, approach treats the competitive price as an input variable and the cost estimates as output variables.

This approach places greater emphasis on the accuracy of the price estimation activity. The target cost must be realistic, or it can be a disincentive to design and engineering staff.

The design-to-cost approach is best applied in the early stages of a new or enhanced product design. The detailed design and specific equipment options are not yet known, but the price estimates assist in establishing target costs for different components. This approach is useful in encouraging innovation, new design, manufacturing process improvement, and efficiency. These are some of the essentials of *value engineering* and *value-added* systems engineering.

Usually, the resulting approach is some combination of these two cost estimation philosophies. However, it is helpful to understand up front what approach is to be emphasized. Historically, the bottom-up approach is more predominant in Western engineering cultures, especially in the United States and Canada. The design-to-cost approach is considered routine in Eastern engineering cultures, especially in industrialized nations such as Japan and other Asian countries.

Accuracy of the Estimates   No cost estimates are expected to be exact; however, they are expected to be reasonable and accurate enough to support economic scrutiny. The accuracy required increases as the project progresses from preliminary design to detailed design and on to economic evaluation. Cost estimates made before and during the preliminary design stage are expected to be good "first-cut" estimates that serve as input to the project budget. Estimation techniques, such as the unit method, are applicable at this stage.

The *unit method* is a popular, very preliminary estimation technique. The total estimated cost is obtained by multiplying the number of units by a per unit cost factor. Sample cost factors are

Cost of automobile operation per mile, including fuel, insurance, wear and tear (e.g., 34.5 cents per mile)

Cost of house construction per livable square foot (e.g., $150 per square foot)

Cost of buried electrical cable per mile

Cost per parking space in a parking garage

Cost of constructing a standard-width suburban street per mile

Instances of the unit method are evident in everyday business activities. If house construction costs average $150 per square foot, a preliminary cost estimate for a 2000-square-foot home is $300,000. If car travel is expensed at $0.345 per mile, a 200-mile business trip should cost approximately $70 for the car.

When utilized at early and conceptual design stages, the estimates above are often referred to as *order-of-magnitude* estimates. At the detailed design stage, cost estimates are expected to be accurate enough to support economic evaluation for a go-no go decision. Every project setting has its own characteristics, but a range on estimates of $\pm5\%$ to $\pm15\%$ of actual costs is expected at the detailed design stage.

Cost Estimation Techniques   Methods such as expert opinion and comparison with comparable installations serve as excellent estimators. The use of *cost indexes* bases the present cost estimate on past cost experiences, with inflation considered. Models such as *cost-capacity equations* and the *factor method* are simple mathematical techniques applied at the preliminary design stage. These *cost-estimating relationships (CER)* are presented in the following sections. There are many additional methods discussed in the handbooks and publications of different industry and business sectors.

15.2   COST INDEXES

A *cost index* is a ratio of the cost of something today to its cost sometime in the past. As such, the index is a dimensionless number that shows the relative cost change over time. One such index that most people are familiar with is the Consumer Price Index (CPI), which shows the relationship between present and past costs for many of the things that "typical" consumers must buy. This index includes such items as rent, food, transportation, and certain services. Other indexes track the costs of equipment, and goods and services that are more pertinent to the engineering disciplines. Table 15–1 is a listing of some of the more common indexes.

TABLE 15–1 Types and Sources of Various Cost Indexes

Type of Index	Source
Overall prices	
Consumer (CPI)	Bureau of Labor Statistics
Producer (wholesale)	U.S. Department of Labor
Construction	
Chemical plant overall	*Chemical Engineering*
Equipment, machinery, and supports	
Construction labor	
Buildings	
Engineering and supervision	
Engineering News Record overall	*Engineering News Record* (*ENR*)
Construction	
Building	
Common labor	
Skilled labor	
Materials	
EPA treatment plant indexes	Environmental Protection Agency, *ENR*
Large-city advanced treatment (LCAT)	
Small-city conventional treatment (SCCT)	
Federal highway	
Contractor cost	
Equipment	
Marshall and Swift (M&S) overall	Marshall & Swift
M&S specific industries	
Labor	
Output per man-hour by industry	U.S. Department of Labor

The general equation for updating costs through the use of a cost index over a period from time $t = 0$ (base) to another time t is

$$C_t = C_0\left(\frac{I_t}{I_0}\right) \qquad \text{[15.1]}$$

where C_t = estimated cost at present time t
C_0 = cost at previous time t_0
I_t = index value at time t
I_0 = index value at time t_0

Generally, the indexes for equipment and materials are made up of a mix of components which are assigned certain weights, with the components sometimes

TABLE 15–2	Values for Selected Indexes		
Year	*CE* Plant Cost Index	*ENR* Construction Cost Index	M&S Equipment Cost Index
1985	325.3	4195	789.6
1986	318.4	4295	797.6
1987	323.8	4406	813.6
1988	342.5	4519	852.0
1989	355.4	4615	895.1
1990	357.6	4732	915.1
1991	361.3	4835	930.6
1992	358.2	4985	943.1
1993	359.2	5210	964.2
1994	368.1	5408	993.4
1995	381.1	5471	1027.5
1996	381.8	5620	1039.2
1997	386.5	5826	1056.8
1998	389.5	5920	1061.9
1999	390.6	6059	1068.3
2000	394.1	6221	1089.0
2001	394.3	6343	1093.9
2002	395.6	6538	1104.2
2003	401.7	6694	1123.6
2004	434.6	7064	1136.0
	(Mid-year)	(Mid-year)	(Estimated)

further subdivided into more basic items. For example, the equipment, machinery, and support component of the chemical plant cost index is subdivided into process machinery, pipes, valves and fittings, pumps and compressors, and so forth. These subcomponents, in turn, are built up from even more basic items such as pressure pipe, black pipe, and galvanized pipe. Table 15–2 shows the *Chemical Engineering* plant cost index, the *Engineering News Record (ENR)* construction cost index, and the Marshall and Swift (M&S) equipment cost index for several years. The base period of 1957 to 1959 is assigned a value of 100 for the *Chemical Engineering (CE)* plant cost index, 1913 = 100 for the *ENR* index, and 1926 = 100 for the M&S equipment cost index.

Current and past values of several of the indexes may be obtained from the Internet. For example, the *CE* plant cost index is available at www.che.com/pindex. The *ENR* construction cost index is found at www.construction.com. This latter site offers a comprehensive series of construction-related resources, including several *ENR* cost indexes and cost estimation systems. A website used by many engineering professionals in the form of a "technical chat room" for all types of topics, including estimation, is www.eng-tips.com.

EXAMPLE 15.1

In evaluating the feasibility of a major construction project, an engineer is interested in estimating the cost of skilled labor for the job. The engineer finds that a project of similar complexity and magnitude was completed 5 years ago at a skilled labor cost of $360,000. The *ENR* skilled labor index was 3496 then and is now 4038. What is the estimated skilled labor cost for the new project?

Solution

The base time t_0 is 5 years ago. Using Equation [15.1], the present cost estimate is

$$C_t = 360{,}000\left(\frac{4038}{3496}\right)$$

$$= \$415{,}812$$

In the manufacturing and service industries, tabulated cost indexes are not readily available. The cost index will vary, perhaps with the region of the country, the type of product or service, and many other factors. When estimating costs for a manufacturing system, it is often necessary to develop the cost index for high-priority cost variables such as subcontracted components, selected materials, and labor costs. The development of the cost index requires the actual cost at different times for a prescribed quantity and quality of the item. The *base period* is a selected time when the index is defined with a basis value of 100 (or 1). The index each year (period) is determined as the cost divided by the base-year cost and multiplied by 100. Future index values may be forecast using simple extrapolation or more refined mathematical techniques, such as time-series analysis. The development of cost indexes is illustrated in the next example.

EXAMPLE 15.2

A manufacturing engineer with Hughes Industries is in the process of estimating costs for a plant expansion. Two important items used in the manufacturing process are a subcontracted circuit board and a preprocessed platinum alloy. Spot checks on the contracted prices through the Purchasing Department at 6-month intervals (first and third quarters, or Q1 and Q3) show the following historical costs. Make the first quarter of 2001 the base period, and determine the cost indexes using a basis of 100.

Year	1999		2000		2001		2002
Quarter	Q1	Q3	Q1	Q3	Q1	Q3	Q1
Circuit board, $/unit	57.00	56.90	56.90	56.70	56.60	56.40	56.25
Platinum alloy, $/ounce	446	450	455	575	610	625	635

Solution

For each item, the index (I_t/I_0) is calculated with the first quarter 2001 cost used for the I_0 value. As indicated by the cost indexes shown, the index for the circuit board is stable, while the platinum alloy index is steadily rising.

Year	1999		2000		2001		2002
Quarter	Q1	Q3	Q1	Q3	Q1	Q3	Q1
Circuit board cost index	100.71	100.53	100.53	100.17	100.00	99.65	99.38
Platinum alloy cost index	73.11	73.77	74.59	94.26	100.00	102.46	104.10

Comment
Use of the cost index for forecasting costs should be performed with a good understanding of the variable itself. The cost index of the platinum alloy is rising, but platinum is much more susceptible to economic market trends and conditions than the
circuit board cost. Accordingly, cost indexes are often more reliable to estimate present
and short-term future costs.

Cost indexes are sensitive over time to technological change. The predefined
quantity and quality used to obtain cost values may be difficult to retain through
time, so "index creep" may occur. Updating of the index and its definition is necessary when identifiable changes occur.

15.3 COST-ESTIMATING RELATIONSHIPS: COST-CAPACITY EQUATIONS

Design variables (speed, weight, thrust, physical size, etc.) for plants, equipment,
and construction are determined in the early design stages. Cost-estimating relationships (CER) use these design variables to predict costs. Thus, a CER is
generically different from the cost index method, because the index is based on
the cost history of a defined quantity and quality of a variable.

One of the most widely used CER models is a *cost-capacity equation*. As the
name implies, an equation relates the cost of a component, system, or plant to its
capacity. This is also known as the *power law and sizing model*. Since many cost-
capacity equations plot as a straight line on log-log paper, a common form is

$$C_2 = C_1 \left(\frac{Q_2}{Q_1} \right)^x \qquad\qquad [15.2]$$

where C_1 = cost at capacity Q_1
C_2 = cost at capacity Q_2
x = correlating exponent

The value of the exponent for various components, systems, or entire plants can
be obtained or derived from a number of sources, including *Plant Design and
Economics for Chemical Engineers, Preliminary Plant Design in Chemical Engineering, Chemical Engineers' Handbook,* technical journals (especially *Chemical Engineering*), the U.S. Environmental Protection Agency, professional or
trade organizations, consulting firms, handbooks, and equipment companies.
Table 15–3 is a partial listing of typical values of the exponent for various units.
When an exponent value for a particular unit is not known, it is common practice

to use the average value of 0.6. In fact, in the chemical processing industry, Equation [15.2] is referred to as the six-tenth model. Commonly, $0 < x \leq 1$. For values $x < 1$, the economies of scale are taken advantage of; if $x = 1$, a linear relationship is present. When $x > 1$, there are diseconomies of scale in that a larger size is expected to be more costly than that of a purely linear relation.

It is especially powerful to combine the time adjustment of the cost index (I_t/I_0) from Equation [15.1] with a cost-capacity equation to estimate costs that change over time. If the index is embedded into the cost-capacity computation in Equation [15.2], the cost at time t and capacity level 2 may be written as the product of two independent terms.

$$C_{2,t} = (\text{cost at time 0 of level 2}) \times (\text{time adjustment cost index})$$

$$= \left[C_{1,0} \left(\frac{Q_2}{Q_1} \right)^x \right] \left(\frac{I_t}{I_0} \right)$$

This is commonly expressed without the time subscripts. Thus,

$$C_2 = C_1 \left(\frac{Q_2}{Q_1} \right)^x \left(\frac{I_t}{I_0} \right) \qquad [15.3]$$

The following example illustrates the use of this relation.

TABLE 15–3   Sample Exponent Values for Cost-Capacity Equations

Component/System/Plant	Size Range	Exponent
Activated sludge plant	1–100 MGD	0.84
Aerobic digester	0.2–40 MGD	0.14
Blower	1000–7000 ft/min	0.46
Centrifuge	40–60 in	0.71
Chlorine plant	3000–350,000 tons/year	0.44
Clarifier	0.1–100 MGD	0.98
Compressor, reciprocating (air service)	5–300 hp	0.90
Compressor	200–2100 hp	0.32
Cyclone separator	20–8000 ft³/min	0.64
Dryer	15–400 ft²	0.71
Filter, sand	0.5–200 MGD	0.82
Heat exchanger	500–3000 ft²	0.55
Hydrogen plant	500–20,000 scfd	0.56
Laboratory	0.05–50 MGD	1.02
Lagoon, aerated	0.05–20 MGD	1.13
Pump, centrifugal	10–200 hp	0.69
Reactor	50–4000 gal	0.74
Sludge drying beds	0.04–5 MGD	1.35
Stabilization pond	0.01–0.2 MGD	0.14
Tank, stainless	100–2000 gal	0.67

NOTE: MGD = million gallons per day; hp = horsepower; scfd = standard cubic feet per day.

EXAMPLE 15.3

The total design and construction cost for a digester to handle a flow rate of 0.5 million gallons per day (MGD) was $1.7 million in 2000. Estimate the cost today for a flow rate of 2.0 MGD. The exponent from Table 15–3 for the MGD range of 0.2 to 40 is 0.14. The cost index in 2000 of 131 has been updated to 225 for this year.

Solution
Equation [15.2] estimates the cost of the larger system in 2000, but it must be updated by the cost index to today's dollars. Equation [15.3] performs both operations at once. The estimated cost in current-value dollars is

$$C_2 = 1,700,000 \left(\frac{2.0}{0.5}\right)^{0.14} \left(\frac{225}{131}\right)$$

$$= 1,700,000(1.214)(1.718) = \$3,546,178$$

15.4 COST ESTIMATING RELATIONSHIPS: FACTOR METHOD

Another widely used model for preliminary cost estimates of process plants is called the *factor method*. While the methods discussed above can be used to estimate the costs of major items of equipment, processes, and the total plant costs, the factor method was developed specifically for total plant costs. The method is based on the premise that fairly reliable total plant costs can be obtained by multiplying the cost of the major equipment by certain factors. Since major equipment costs are readily available, rapid plant estimates are possible if the appropriate factors are known. These factors are commonly referred to as Lang factors after Hans J. Lang, who first proposed the method in 1947.

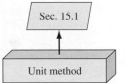

In its simplest form, the factor method is expressed in the same form as the unit method

$$C_T = hC_E \qquad [15.4]$$

where C_T = total plant cost
       h = overall cost factor or sum of individual cost factors
       C_E = total cost of major equipment

The h may be one overall cost factor or, more realistically, the sum of individual cost components such as construction, maintenance, direct labor, materials, and indirect cost elements. This follows the cost estimation approaches presented in Figure 15–1.

In his original work, Lang showed that direct cost factors and indirect cost factors can be combined into one overall factor for some types of plants as follows: solid process plants, 3.10; solid-fluid process plants, 3.63; and fluid process plants, 4.74. These factors reveal that the total installed-plant cost is many times the first cost of the major equipment.

EXAMPLE 15.4

An engineer with Phillips Petroleum has learned that an expansion of the solid-fluid process plant is expected to have a delivered equipment cost of $1.55 million. If the overall cost factor for this type of plant is 3.63, estimate the plant's total cost.

Solution
The total plant cost is estimated by Equation [15.4].

$$C_T = 3.63(1,550,000)$$

$$= \$5,626,500$$

Subsequent refinements of the factor method have led to the development of separate factors for direct and indirect cost components. Direct costs as discussed in Section 15.1, are specifically identifiable with a product, function, or process. Indirect costs are not directly attributable to a single function, but are shared by several because they are necessary to perform the overall objective. Examples of indirect costs are general administration, computer services, quality, safety, taxes, security, and a variety of support functions. The factors for both direct and indirect costs are sometimes developed for use with delivered-equipment costs and other times for installed-equipment costs. In this text, we assume that all factors apply to delivered-equipment costs, unless otherwise specified.

For indirect costs, some of the factors apply to equipment costs only, while others apply to the total direct cost. In the former case, the simplest procedure is to add the direct and indirect cost factors before multiplying by the delivered-equipment cost. The overall cost factor h can be written as

$$h = 1 + \sum_{i=1}^{n} f_i \qquad [15.5]$$

where f_i = factor for each cost component
i = 1 to n components, including indirect cost

If the indirect cost factor is applied to the total direct cost, only the direct cost factors are added to obtain h. Therefore, Equation [15.4] is rewritten.

$$C_T = \left[C_E \left(1 + \sum_{i=1}^{n} f_i \right) \right] (1 + f_I) \qquad [15.6]$$

where f_I = indirect cost factor
f_i = factors for direct cost components

Examples 15.5 and 15.6 illustrate these equations.

EXAMPLE 15.5

The delivered-equipment cost for a small chemical process plant is expected to be $2 million. If the direct cost factor is 1.61 and the indirect cost factor is 0.25, determine the total plant cost.

Solution

Since all factors apply to the delivered-equipment cost, they are added to obtain h, the total cost factor in Equation [15.5].

$$h = 1 + 1.61 + 0.25 = 2.86$$

From Equation [15.4], the total plant cost is

$$C_T = 2.86(2,000,000) = \$5,720,000$$

EXAMPLE 15.6

An activated sludge wastewater treatment plant is expected to have the following delivered-equipment first costs:

Equipment	Cost
Preliminary treatment	$30,000
Primary treatment	40,000
Activated sludge	18,000
Clarification	57,000
Chlorination	31,000
Digestion	70,000
Vacuum filtration	27,000
Total cost	$273,000

The cost factor for the installation of piping, concrete, steel, insulation, supports, etc., is 0.49. The construction factor is 0.53, and the indirect cost factor is 0.21. Determine the total plant cost if (*a*) all cost factors are applied to the cost of the delivered-equipment and (*b*) the indirect cost factor is applied to the total direct cost.

Solution

(*a*) Total equipment cost is $273,000. Since both the direct and indirect cost factors are applied to only the equipment cost, the overall cost factor from Equation [15.5] is

$$h = 1 + 0.49 + 0.53 + 0.21 = 2.23$$

The total plant cost is

$$C_T = 2.23(273,000) = \$608,790$$

(*b*) Now the total direct cost is calculated first, and Equation [15.6] is used to estimate the total plant cost.

$$h = 1 + \sum_{i=1}^{n} f_i = 1 + 0.49 + 0.53 = 2.02$$

$$C_T = [273,000(2.02)](1.21) = \$667,267$$

Comment

Note the decrease in estimated plant cost when the indirect cost is applied to the equipment cost only in part (*a*). This illustrates the importance of determining exactly what the factors apply to before they are used.

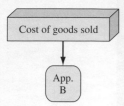

15.5 TRADITIONAL INDIRECT COST RATES AND ALLOCATION

Costs incurred in the production of an item or delivery of a service are tracked and assigned by a *cost accounting system*. For the manufacturing environment, it can be stated generally that the *statement of cost of goods sold* is one end product of this system. The cost accounting system accumulates material costs, labor costs, and indirect costs (also called overhead costs or factory expenses) by using *cost centers*. All costs incurred in one department or process line are collected under a cost center title, for example, Department 3X. Since direct materials and direct labor are usually directly assignable to a cost center, the system need only identify and track these costs. Of course, this in itself is no easy chore, and the cost of the tracking system may prohibit collection of all direct cost data to the level of detail desired.

One of the primary and more difficult tasks of cost accounting is the allocation of *indirect costs* when it is necessary to allocate them separately to departments, processes, and product lines. The costs associated with property taxes, service and maintenance departments, personnel, legal, quality, supervision, purchasing, utilities, software development, etc., must be allocated to the using cost center. Detailed collection of these data is cost-prohibitive and often impossible; thus, allocation schemes are utilized to distribute the expenses on a reasonable basis. A listing of possible bases is included in Table 15–4. Historically common bases have been direct labor cost, direct labor hours, machine-hours, number of employees, space, and direct materials.

Most allocation is accomplished utilizing a predetermined *indirect cost rate,* computed by using the general relation.

$$\text{Indirect cost rate} = \frac{\text{estimated indirect costs}}{\text{estimated basis level}}$$ [15.7]

The estimated indirect cost is the amount allocated to a cost center. For example, if a division has two producing departments, the total indirect cost allocated to a department is used as the numerator in Equation [15.7] to determine the department rate. Example 15.7 illustrates allocation when the cost center is a machine.

TABLE **15–4**   Indirect Cost Allocation Bases

Cost Category	Possible Allocation Basis
Taxes	Space occupied
Heat, light	Space, usage, number of outlets
Power	Space, direct labor hours, horsepower-hours, machine hours
Receiving, purchasing	Cost of materials, number of orders, number of items
Personnel, machine shop	Direct labor hours, direct labor cost
Building maintenance	Space occupied, direct labor cost
Software	Number of accesses
Quality control	Number of inspections

EXAMPLE 15.7

EnviroTech, Inc., is computing indirect cost rates for the production of glass products. The following information is obtained from last year's budget for the three machines used in production.

Cost Source	Allocation Basis	Estimated Activity Level
Machine 1	Direct labor cost	$100,000
Machine 2	Direct labor hours	2000 hours
Machine 3	Direct material cost	$250,000

Determine rates for each machine if the estimated annual indirect cost budget is $50,000 per machine.

Solution

Applying Equation [15.7] for each machine, the annual rates are

$$\text{Machine 1 rate} = \frac{\text{indirect budget}}{\text{direct labor cost}} = \frac{50,000}{100,000}$$

$$= \$0.50 \text{ per direct labor dollar}$$

$$\text{Machine 2 rate} = \frac{\text{indirect budget}}{\text{direct labor hours}} = \frac{50,000}{2000}$$

$$= \$25 \text{ per direct labor hour}$$

$$\text{Machine 3 rate} = \frac{\text{indirect budget}}{\text{material cost}} = \frac{50,000}{250,000}$$

$$= \$0.20 \text{ per direct material dollar}$$

Comment

Once the product has been manufactured, and actual direct labor costs and hours and material costs are computed, each dollar of direct labor cost spent on machine 1 implies that $0.50 in indirect cost will be added to the cost of the product. Similarly, indirect costs are added for machines 2 and 3.

When the same allocation basis is used to distribute indirect costs to several cost centers, an *overall* or *blanket rate* may be determined. For example, if direct materials are the basis for allocation to four separate processing lines, the blanket rate is

$$\text{Indirect cost rate} = \frac{\text{total indirect costs}}{\text{total direct materials cost}}$$

If totals of $500,000 in indirect costs and $3 million in materials are estimated for next year for the four lines, the blanket indirect rate to apply is 500,000/3,000,000 = $0.167 per materials cost dollar. Blanket rates are easier to calculate and apply, but they do not account for differences in the type of activities accomplished in each cost center.

In most cases, machinery or processes add value to the end product at different rates per unit or hour of use. For example, light machinery may contribute less per hour than heavy, more expensive machinery. This is especially true when advanced technology processing, for example, an automated manufacturing cell is used along with traditional methods, for example, nonautomated finishing equipment. The use of blanket or overall rates in these cases is not recommended as the indirect cost will be incorrectly allocated. The lower-value-contribution machinery will accumulate too much of the indirect cost. The approach to indirect cost allocation should be the application of different bases for different machines, activities, etc., as discussed earlier and illustrated in Example 15.7. The use of different, appropriate bases is often called the *productive hour rate method* since the cost rate is determined based on the value added, not a uniform or blanket rate. Realization that more than one basis should be normally used in allocating indirect costs has led to the use of activity-based costing methods, as discussed in the next section.

Once a period of time (month, quarter, or year) has passed, the indirect cost rates are applied to determine the indirect cost *charge,* which is then added to direct costs. This results in the total cost of production, which is called the cost of goods sold, or *factory cost.* These costs are all accumulated by *cost center.*

If the total indirect cost budget is correct, the indirect costs charged to all cost centers for the period of time should equal this budget amount. However, since some error in budgeting always exists, there will be overallocation or underallocation relative to actual charges, which is termed *allocation variance.* Experience in indirect cost estimation assists in reducing the variance at the end of the accounting period. Example 15.8 illustrates indirect cost allocation and variance computation.

EXAMPLE 15.8

Since we determined indirect cost rates for EnviroTech (Example 15.7), we can now compute the total factory cost for a month. Perform the computations using the data in Table 15–5. Also calculate the variance for indirect cost allocation for the month.

TABLE 15–5 Actual Monthly Data Used for Indirect Cost Allocation

Cost Source	Machine Number	Actual Cost	Actual Hours
Material	1	$3,800	
	3	19,550	
Labor	1	2,500	650
	2	3,200	750
	3	2,800	720

Solution

Start with the cost of goods sold (factory cost) relation given by Equation [B.1] in Appendix B, which is

Cost of goods sold = direct materials + direct labor + indirect costs

To determine indirect cost, the rates from Example 15.7 are applied:

Machine 1 indirect = (labor cost)(rate) = 2500(0.50) = $1250

Machine 2 indirect = (labor hours)(rate) = 750(25.00) = $18,750

Machine 3 indirect = (material cost)(rate) = 19,550(0.20) = $3910

Total charged indirect cost = $23,910

Factory cost is the sum of actual material and labor costs from Table 15–5 and the indirect cost charge for a total of $55,760.

Based on the annual indirect cost budget of $50,000 per machine, one month represents 1/12 of the total or

$$\text{Monthly budget} = \frac{3(50,000)}{12}$$

$$= \$12,500$$

The variance for total indirect cost is

Variance = 12,500 − 23,910 = $−11,410

This is a large budget underallocation, since much more was charged than allocated. The $12,500 budgeted for the three machines represents a 91.3% underallocation of indirect costs. This analysis for only one month of a year will most likely prompt a rapid review of the rates and the indirect cost budget for EnviroTech.

Once estimates of indirect costs are determined, it is possible to perform an economic analysis of the present operation versus a proposed operation. Such a study is described in Example 15.9.

EXAMPLE 15.9

For several years the Cuisinart Corporation has purchased the carafe assembly of its major coffee-maker line at an annual cost of $1.5 million. The suggestion to make the component inhouse has been made. For the three departments involved the annual indirect cost rates, estimated material, labor, and hours are found in Table 15–6. The allocated hours column is the time necessary to produce the carafes for a year.

Equipment must be purchased with the following estimates: first cost of $2 million, salvage value of $50,000, and life of 10 years. Perform an economic analysis for the make alternative, assuming that a market rate of 15% per year is the MARR.

TABLE 15–6 Production Cost Estimates for Example 15.9

	Indirect Costs				
Department	Basis, Hours	Rate Per Hour	Allocated Hours	Material Cost	Direct Labor Cost
A	Labor	$10	25,000	$200,000	$200,000
B	Machine	5	25,000	50,000	200,000
C	Labor	15	10,000	50,000	100,000
				$300,000	$500,000

Solution

For making the components inhouse, the AOC is comprised of direct labor, direct material, and indirect costs. Use the data of Table 15–6 to calculate the indirect cost allocation.

$$\text{Department A:} \quad 25,000(10) = \$250,000$$

$$\text{Department B:} \quad 25,000(5) \;= \; 125,000$$

$$\text{Department C:} \quad 10,000(15) = \underline{\;150,000\;}$$

$$\$525,000$$

$$\text{AOC} = 500,000 + 300,000 + 525,000 = \$1,325,000$$

The make alternative annual worth is the total of capital recovery and AOC.

$$\text{AW}_{make} = -P(A/P,i,n) + S(A/F,i,n) - \text{AOC}$$

$$= -2,000,000(A/P,15\%,10) + 50,000(A/F,15\%,10) - 1,325,000$$

$$= \$-1,721,037$$

Currently, the carafes are purchased with an AW of

$$\text{AW}_{buy} = \$-1,500,000$$

It is cheaper to purchase, because the AW of costs is less.

15.6   ACTIVITY-BASED COSTING (ABC) FOR INDIRECT COSTS

As automation, software, and manufacturing technologies have advanced, the number of direct labor hours necessary to manufacture a product has decreased substantially. Where once as much as 35% to 45% of the final product cost was represented in labor, now the labor component is commonly 5% to 15% of total manufacturing cost. However, the indirect cost may represent as much as 35% of the total manufacturing cost. The use of bases, such as direct labor hours, to allocate indirect cost is not accurate enough for automated and technologically advanced environments. This has led to the development of methods that supplement traditional cost allocations that rely upon one form or another of Equation [15.7]. Also, allocation bases different from traditional ones are commonly utilized.

It is important from an engineering economy viewpoint to realize when traditional indirect cost allocation systems should be augmented with better methods. A product that by traditional methods may have contributed a large portion to profit may actually be a loser when indirect costs are allocated more correctly. Companies that have a wide variety of products and produce some in small lots may find that traditional allocation methods have a tendency to underallocate the indirect cost to small-lot products. This may indicate that they are profitable, when in actuality they are losing money. Additionally, the productive hour rate method, which uses different allocation bases dependent on the value added per hour of operation, should be used, as discussed in Section 15.5.

An augmentation technique for indirect cost allocation is *Activity-Based Costing,* or *ABC* for short. By design, its goal is to develop a cost center, called a *cost pool,* for each event, or *activity,* which acts as a *cost driver.* In other words, cost drivers actually *drive* the consumption of a shared resource and are charged accordingly. Cost pools are usually departments or functions—purchasing, inspection, maintenance, and information technology. Activities are events such as purchase orders, reworks, repairs, software activations, machine setups, material movement, wait time, and engineering changes.

Some proponents of the ABC method recommend discarding the traditional cost accounting methods of a company and utilizing ABC exclusively. This is not a good approach, since ABC is not a complete cost system. The ABC method provides information that assists in *cost control,* while the traditional method emphasizes cost allocation and cost estimation. The two systems work well together with the traditional methods allocating costs that have identifiable direct bases, for example, direct labor. The ABC method can then be utilized to further allocate support service costs using activity bases such as those mentioned above.

The ABC methodology involves a two-step process:

1. *Define cost pools.* **Usually these are support functions.**
2. *Identify cost drivers.* **These help trace costs to the cost pools.**

As an illustration, a company which produces an industrial laser has three primary support departments identified as cost pools in step 1: A, B, and C. The annual support cost for the purchasing cost driver (step 2) is allocated to these departments based on the number of purchase orders each department issues to support its laser production functions. Example 15.10 illustrates the process of applying activity-based costing.

EXAMPLE 15.10

A multinational aerospace firm uses traditional methods to allocate manufacturing and management support costs for its European division. However, accounts such as business travel have historically been allocated on the basis of the number of employees at the plants in France, Italy, Germany, and Greece.

The president recently stated that some product lines are likely generating much more management travel than others. The ABC system is chosen to augment the traditional method to more precisely allocate travel costs to major product lines at each plant.

(a) First, assume that allocation of total observed travel expenses of $500,000 to the plants using a traditional basis of workforce size is sufficient. If total employment of 29,100 is distributed as follows, allocate the $500,000.

Paris, France plant	12,500 employees
Florence, Italy plant	8,600 employees
Hamburg, Germany plant	4,200 employees
Athens, Greece plant	3,800 employees

(b) Now, assume that corporate management wants to know more about travel expenses based on product line, not merely plant location and workforce size. The ABC method will be applied to allocate travel costs to major product lines. Annual plant support budgets indicate that the following percentages are expended for travel:

Paris	5% of $2 million
Florence	15% of $500,000
Hamburg	17.5% of $1 million
Athens	30% of $500,000

Further, the study indicates that in 1 year a total of 500 travel vouchers were processed by the management of the major five product lines produced at the four plants. The distribution is as follows:

Paris.   Product lines—1 and 2; number of vouchers—50 for line 1, 25 for 2.

Florence.   Product lines—1, 3, and 5; vouchers—80 for line 1, 30 for 3, 30 for 5.

Hamburg.   Product lines—1, 2 and 4; vouchers—100 for line 1, 25 for 2, 20 for 4.

Athens.   Product line—5; vouchers—140 for line 5.

Use the ABC method to determine how the product lines drive travel costs at the plants.

Solution

(a) Equation [15.7] takes the form of a blanket rate per employee.

$$\text{Indirect cost rate} = \frac{\text{travel budget}}{\text{total workforce}}$$

$$= \frac{\$500,000}{29,100} = \$17.1821 \text{ per employee}$$

Using this traditional basis of rate times workforce size results in a plant-by-plant allocation.

Paris:	$17.1821(12,500) = $214,777
Florence:	$147,766
Hamburg:	$72,165
Athens:	$65,292

(b) The ABC method is more involved since it requires the definition of the cost pool and its size (step 1) and the allocation to product lines using the cost driver (step 2). The by-plant amounts will be different from those in part (a) since completely different bases are being applied.

     Step 1. The cost pool is travel activity, and the size of the cost pool is determined from the percentages of each plant's support budget devoted to travel. Using the travel expense information in the problem statement, a total cost pool of $500,000 is to be allocated to the five product lines. This number is determined from the percent-of-budget data as follows:

$$0.05(2,000,000) + \cdots + 0.30(500,000) = \$500,000$$

     Step 2. The cost driver for the ABC method is the number of travel vouchers submitted by the management unit responsible for each product line at each plant. The allocation will be to the products directly, not to the plants. However, the travel allocation to the plants can be determined afterward since we know what product lines are produced at each plant. For the cost driver of travel vouchers, the format of Equation [15.7] can be used to determine an ABC allocation rate.

$$
\begin{aligned}
\text{ABC allocation per travel voucher} &= \frac{\text{total travel cost pool}}{\text{total number of vouchers}} \\
&= \frac{\$500,000}{500} \\
&= \$1000 \text{ per voucher}
\end{aligned}
$$

Table 15–7 summarizes the vouchers and allocation by product line and by city. Product 1 ($230,000) and product 5 ($170,000) drive the travel costs based on the ABC analysis. Comparison of the by-plant totals in Table 15–7 with the respective totals in part (a) indicates a substantial difference in the amounts allocated, especially to Paris, Hamburg, and Athens. This comparison verifies the president's suspicion that product lines, not plants, drive travel requirements.

TABLE 15–7 ABC Allocation of Travel Cost ($ in Thousands), Example 15.10

| | Product Line | | | | | |
	1	2	3	4	5	Total
Paris	50	25				75
Florence	80		30		30	140
Hamburg	100	25		20		145
Athens					140	140
Total	$230	$50	$30	$20	$170	$500

> **Comment**
> Let's assume that product 1 has been produced in small lots at the Hamburg plant for a number of years. This analysis, when compared to the traditional cost allocation method in part (*a*), reveals a very interesting fact. In the ABC analysis, Hamburg has a total of $145,000 travel dollars allocated, $100,000 from product 1. In the traditional analysis based on workforce size, Hamburg was allocated only $72,165—about 50% of the more precise ABC analysis amount. This indicates to management the need to examine the manufacturing lot size practices at Hamburg and possibly other plants, especially when a product is currently manufactured at more than one plant.

ABC analysis is usually more expensive and time-consuming than a traditional cost allocation system, but in many cases it can assist in understanding the economic impact of management decisions and in controlling certain types of indirect costs. Often the combination of traditional and ABC analyses reveals areas where further economic analysis is warranted.

CHAPTER SUMMARY

Cost estimates are not expected to be exact, but they should be accurate enough to support a thorough economic analysis using an engineering economy approach. There are bottom-up and top-down approaches; each treats price and cost estimates differently.

Costs can be updated via a cost index, which is a ratio of costs for the same item at two separate times. The Consumer Price Index (CPI) is an often-quoted example of cost indexing.

Cost estimating may also be accomplished with a variety of models called Cost-Estimating Relationships. Two of them are

Cost-capacity equation. Good for estimating costs from design variables for equipment, materials, and construction.

Factor method. Good for estimating total plant cost.

Traditional cost allocation uses an indirect cost rate determined for a machine, department, product line, etc. Bases such as direct labor cost, direct material cost, and direct labor hours are used. With increased automation and information technology, different techniques of indirect cost allocation have been developed. The Activity-Based Costing method is an excellent technique to augment the traditional allocation method.

The ABC method uses the rationale that cost drivers are activities—purchase orders, inspections, machine setups, reworks. These activities *drive* the costs accumulated in cost pools, which are commonly departments or functions, such as quality, purchasing, accounting, and maintenance. Improved understanding of how the company or plant actually accumulates indirect costs is a major byproduct of implementing the ABC method.

PROBLEMS

Cost Estimation Approaches

15.1 List three elements of the following costs for a new computer integrated manufacturing system.
 (a) First cost of equipment
 (b) AOC

15.2 Identify a primary difference between the bottom-up and design-to-cost approaches to cost estimation.

15.3 Identify each of the following costs associated with owning an automobile as direct or indirect. Assume a direct cost of ownership is one that keeps the car in your possession and running to provide you transportation at the time that you want it. If you are undecided, state the conditions under which the cost is direct and indirect.
 (a) Gasoline
 (b) Highway toll fee
 (c) Cost of repairs after a serious collision
 (d) Annual inspection fee
 (e) Federal gasoline tax
 (f) Monthly loan payment

15.4 Estimate the cost of purchasing a suburban lot, constructing a house, and furnishing the house, using the following unit cost estimates:

 Size of property: 100 × 150 feet

 House approximate size: 6 rooms, 50 × 46 feet with 75% livable space

 Price of property in the suburban area: $2.50 per square foot

 Average cost of construction: $125 per usable square foot

 Furniture and appointments: $3000 per room

15.5 Two people developed first-cut cost estimates to construct a new 130,000-square-foot building on a university campus. Person A applied a general-purpose per unit cost estimate of $120 per square foot for the estimate. Individual B was more specific; she used the area estimates and per unit cost factors shown below. What are the cost estimates developed by the two people? How do they compare?

Type of Usage	Percent of Area	Cost per Square Foot, $
Classroom	30	125
Laboratory	40	185
Offices	30	110
Furnishings—labs	25	150
Furnishings—all other	75	25

Cost Indexes

15.6 The cost of a standard cooling system in 1995 was $78,000. If the M&S equipment cost index applies, what will be the estimated cost for a similar system when the index is 1200?

15.7 Use the *ENR* construction cost index to determine the cost of constructing a section of highway similar to one built in 1995 at a cost of $2.3 million. Use the most current index value from the *ENR* website.

15.8 On the website containing the *ENR* construction cost index (enr.construction.com), two indexes are reported—the construction cost index and the building cost index. Locate the section that explains their use and discuss the difference between the two indexes and under what conditions each one is appropriate to make cost estimates.

15.9 If the cost of a certain piece of equipment was $20,000 when the M&S index was 915.1, what was the index value when the same equipment was estimated to cost $30,000?

15.10 (a) Estimate the value of the *ENR* construction cost index for the year 2002 by using the average (compound) percentage change in its value between 1990 and 2000 to predict the 2002 value. (b) How much difference is there between the estimated and historical 2002 values?

15.11 A particular type of labor cost index had a value of 720 in 1985 and 1315 in 2004. If the labor cost for constructing a building was $1.6 million in 2004, what would the labor cost have been in 1985?

15.12 Use the *ENR* construction cost index (Table 15–2) to update a cost of $325,000 in 1990 to a 2004 figure.

15.13 Chemical processing plant equipment was newly purchased in 1998 at a cost of $2.5 million. Similar equipment was purchased in 1994 at another site and again in 2002 at a third site. The plant engineer wants to know the compound rate of cost growth over the time span of the three purchases. Determine this annual rate. The *CE* plant cost index applies.

15.14 If a person makes 1990 the base year with an index value equal to 100 for the *CE* plant cost index (Table 15–2), what is the projected value of the index in (a) 2002 and (b) current calendar month? (*Hint:* Use the index website to find the most current index value.)

15.15 Determine the average (compound) percentage increase per year between 1990 and 2002 for the *CE* plant cost index.

15.16 Estimate the value of the M&S equipment cost index in 2005 if it was 1068.3 in 1999 and it increases by 2% per year.

15.17 A mass spectrometer can be purchased for $60,000 today. The owner of a mineral analysis laboratory expects the cost to increase exactly by the equipment inflation rate over the next 10 years. (a) The inflation rate is estimated to be 2% per year for the next 3 years and 5% per year thereafter. How much will the spectrometer cost 10 years from now if the lab's MARR is 10% per year? (b) If the applicable equipment cost index is at 1203 now, what will it be 10 years from now?

Cost-Estimating Relationships

15.18 What is the fundamental difference between estimating a cost using a CER and a cost index?

15.19 The cost of a high-quality 250-horsepower compressor was $13,000 when recently purchased. What would a 450-horsepower compressor be expected to cost?

15.20 Janus Co. purchased a 100-horsepower centrifugal pump and associated gear last year for $20,000. Two additional pump systems are needed at other plant sites, one rated at 200 horsepower and the other at 75 horsepower. (a) Estimate the cost of the two new pumps. (b) If the 200-horsepower pump is delayed for 3 years, estimate its future cost provided the cost index is expected to rise 20% from its current value of 185 over these years.

15.21 The cost for implementing a manufacturing process that has a capacity of 6000 units per day was $550,000. If the cost for a plant with a capacity of 100,000 units per day was $3 million, what is the value of the exponent in the cost-capacity equation?

15.22 The estimated cost for a multitube cyclone system with a capacity of 60,000 cubic feet per minute is $450,000. (a) If the $200,000 actual cost for a 35,000 cubic feet per minute system was inserted into the cost-capacity equation, what value of the exponent in the estimation equation was used? (b) What can be concluded about the economy of scale of the costs between the two systems?

15.23 The cost for construction of a desulfurization system for flue gas from utility boilers at a 600-megawatt plant was estimated to be $250 million. If a smaller plant has a cost of $55 million and the exponent in the cost-capacity equation is 0.67, what was the size of the smaller plant that served as the basis for the cost projection?

15.24 The net annual operating cost for a filtration plant treating water for a semiconductor fabrication line was estimated to be $1.5 million per year. The estimate was based on the $200,000 per year cost of a 1-MGD plant. If the exponent in the cost-capacity equation is 0.80, what was the size of the larger plant?

15.25 In the year 2002, new technology IP telephony equipment was installed at the IDS Building headquarters at a total cost of $1 million. In the same year, an estimate was made that a system with 3 times the capacity would be needed in 2 years, that economies of scale and technological development warranted a cost-capacity exponent of 0.2, and that a 10% increase in the cost index would be sufficient. In fact, a 3× capacity system was installed in 2004 at a cost of $2 million, and the cost index had gone up by 25% instead of the anticipated 10%. (*a*) What is the difference between the estimate made in 2002 and the actual 2004 cost? (*b*) What value of the cost-capacity equation exponent should have been used to correctly estimate the actual $2 million cost?

15.26 Estimate the cost in 2002 of processing equipment if the cost of a unit one-half its size was $50,000 in 1998. The exponent in the cost-capacity equation is 0.24. Use the tabulated *CE* plant cost index to update the cost.

15.27 Estimate the cost in 2002 of a 1000-horsepower steam turbine air compressor if a 200-horsepower unit cost $160,000 in 1995. The exponent in the cost-capacity

equation is 0.35. The equipment cost index increased by 35% between the two years.

15.28 In 1990, a 10,000-square-meter facility was constructed at a food processing plant in Chicago for in-process handling at a cost of $220,000. In 2002, an engineer was asked to estimate the cost of a similar structure, but for 5000 square meters in a new plant in London. What was the 2002 estimate if the six-tenths model was applied?

15.29 The equipment cost for phosphorus removal from wastewater at a 50-MGD plant will be $16 million. If the overall cost factor for this type of plant is 2.97, what is the total plant cost expected to be?

15.30 The delivered-equipment cost for a fabric filter particulate collection system is $1.6 million. The direct cost factor is 1.52, and the indirect cost factor is 0.31. Estimate the total plant cost if the indirect cost factor applies (*a*) to delivered equipment cost only and (*b*) to the total direct cost.

15.31 During major expansion in 1994, Douwalla' a Import Company developed a new processing line for which the delivered-equipment cost was $1.75 million. Now, 11 years later, the board of directors has decided to expand into new markets and expects to build the current version of the same line. Estimate the cost if the following factors are applicable: Construction cost factor is 0.20, installation cost factor is 0.50, indirect cost factor applied against equipment is 0.25, and the total plant cost index has risen from 2509 to 3713 over the years.

15.32 Josephina is an engineer on temporary assignment at a refining operation in Seaside. She has reviewed a cost estimate for $450,000, which covers some new processing equipment for the ethylene line. The equipment itself is estimated at $250,000 with a construction cost factor of 0.30 and an installation cost factor of 0.30. No indirect cost factor is listed, but she knows from other sites that indirect cost is a sizable amount that increases the total direct cost of the line's equipment. (*a*) If the indirect cost factor should be 0.40, determine whether the current estimate includes a factor comparable to this value. (*b*) Determine the cost estimate if the 0.40 indirect cost factor is used.

Indirect Cost Allocation

15.33   Direct labor hours are used as the allocation basis for indirect costs per quarter. A total amount of $450,000 is to be allocated at each plant in each quarter.

(*a*)   Determine the indirect cost rates for the Humboldt plant for each quarter if each type of machining is allocated 50% of the indirect cost.

(*b*)   Determine the blanket rate for Q1 for Humboldt. Calculate the amount of indirect cost charged to light machining only using this blanket quarterly rate, and the amount charged using the rate determined in part (*a*). If the rate that is sensitive to the type of machining is correct, by how much is light machining overcharged or undercharged when the blanket rate is used?

(*c*)   Determine the blanket indirect cost rate for the Concourse plant for each quarter.

Direct Labor Hours

Machining	Quarter Q1		Quarter Q2	
	Heavy	Light	Heavy	Light
Humboldt	2000	800	1500	1500
Concourse	1000	800	800	2000

15.34   A department has four processing lines, each one considered a separate cost center for indirect cost allocation purposes. Machine operating hours are used as the allocation basis for all lines. A total of $500,000 is allocated to the department for next year. Use the data collected this year to determine the indirect cost rate for each line.

Cost Center	Indirect Cost Allocated, $	Estimated Operating Hours
1	50,000	600
2	100,000	200
3	150,000	800
4	200,000	1200

15.35   Dirk, the department manager of Chassis Fabrication, has obtained from finance and accounting the records that indicate indirect cost allocation rates and the actual indirect charges for the prior 3 months and their estimates for this month and next month (September and October). The basis of allocation is not indicated. The finance and accounting manager says there is no record of the basis used. However, he tells Dirk to not be concerned about the total allocation because the rate is now constant at $1.25 and lower than previous rates.

		Indirect Cost, $	
Month	Rate	Allocated	Charged
June	1.50	20,000	22,000
July	1.33	34,000	38,000
August	1.37	35,000	35,000
September	1.25	36,000	
October	1.25	36,250	

During his evaluation, Dirk finds this additional information from departmental and accounting records.

Month	Direct Labor Hours	Cost, $	Material Cost, $	Dept. Space, Square Feet
June	13,330	53,000	54,000	20,000
July	6,400	25,560	46,000	20,000
August	6,400	25,560	57,000	29,000
September	6,400	27,200	63,000	29,000
October	8,000	33,200	65,000	29,000

(a) For each month, determine the allocation basis, and (b) comment on the statement of the finance and accounting manager that the rate is now constant and lower than previous rates.

15.36 A manufacturer serving the sea transport industry has five departments. Indirect cost allocations for 1 month are detailed below, along with space assigned, direct labor hours, and direct labor costs for each department that directly manufactures radar and sonar equipment.

		Actual Data for 1 Month		
Department	Indirect Cost Allocation, $	Space, Square Feet	Direct Labor Hours	Direct Labor Cost, $
Housing	20,000	10,000	480	31,680
Subassemblies	45,000	18,000	1,000	103,250
Final assembly	10,000	10,000	600	12,460
Testing	15,000	1,200		
Engineering	19,000	2,000		

Determine the manufacturing department allocation rates for redistributing the indirect cost allocation for testing and engineering ($34,000) to the other departments. Use the following bases to determine the rates: (a) space, (b) direct labor hours, and (c) direct labor costs.

15.37 For Problem 15.36, determine the actual indirect cost charges, using the rates determined. For actual charges, use bases of direct labor hours for the housing and subassemblies departments and direct labor cost for the final assembly department.

15.38 Use the individual cost center rates in Problem 15.34 to compute (a) the actual indirect cost charges and allocation variances for each line and (b) the total for all lines. The actual hours credited to each center are as follows: 1 has 700 hours; 2 has 350 hours; 3 has 650 hours; 4 has 1400 hours.

15.39 Indirect cost rates and bases for six producing departments at Haycrow Industries are listed below. (a) Use them to distribute indirect costs to the departments. (b) Determine the allocation variance relative to a total indirect cost allocation budget of $800,000.

Department	Allocation Basis*	Rate, $	Direct Labor Hours	Direct Labor Cost, $	Machine Hours
1	DLH	2.50	5,000	20,000	3,500
2	MH	0.95	5,000	35,000	25,000
3	DLC	1.25	10,500	44,100	5,000
4	DLC	5.75	12,000	84,000	40,000
5	DLC	3.45	10,200	54,700	10,200
6	DLH	0.75	19,000	69,000	60,500

*DLH = direct labor hours; MH = machine hours; DLC = direct labor cost.

15.40 A new plant manager has been assigned to Haycrow Industries. This individual has reviewed the information contained in the table in Problem 15.39 and determined that it is too complex to have more than one indirect cost allocation basis. The selected basis is direct labor cost. Therefore, for this year, the simple average of the DLC rates in departments 3, 4, and 5 will be used to calculate actual indirect charges. Determine the amount of charges for all six departments and the total variance relative to the indirect cost allocation budget of $800,000.

15.41 Tocomo Industries serves the sea transport industry. Indirect cost allocations for 1 month are detailed, along with space assigned, direct labor hours, and direct labor costs for the three departments that directly manufacture radar and sonar equipment. (For reference, this is the same information presented in Problem 15.36, but you do not need to have worked it to complete this problem.)

Actual Data for 1 Month

Department	Indirect Cost Allocation, $	Space, Square Feet	Direct Labor Hours	Cost, $
Housing	20,000	10,000	480	31,680
Subassemblies	45,000	18,000	1,000	103,250
Final assembly	10,000	10,000	600	12,460
Testing	15,000	1,200		
Engineering	19,000	2,000		

The company presently makes all the components required by the housing department. The company is considering buying rather than making these components. An outside contractor has offered to make the items for $87,500 per month.

(a) If the costs for housing for the particular month shown are considered good estimates for an engineering economy study, and if $41,000 worth of materials is charged to housing, do a comparison of the make-versus-buy alternatives. Assume that the housing department's share of the testing and engineering departments' costs is a total of $3500 per month.

(b) A third alternative for the company is to purchase new equipment for the housing department and continue to make the components. The machinery will cost $375,000 and will have a 5-year life, no salvage value, and a monthly operating cost of $5000. This purchase is expected to reduce monthly costs in testing and engineering by $2000 and $3000, respectively, and also reduce monthly direct labor hours to 200 and monthly direct labor costs to

$20,000 for the housing department. The redistribution of the indirect costs from testing and engineering to the three production departments is on the basis of direct labor hours. If other costs remain the same, compare three costs: the present cost of making the components, the estimated cost if the new equipment is purchased, and the outside contractor cost. Select the most economic alternative. A market MARR of 12% per year, compounded monthly, is used for capital investments.

15.42 A corporation operates three plants in one state. They all manufacture the same lines of precision and high-pressure fittings of a wide variety for the oil, gas, and chemical processing industries. The corporate offices market and ship the finished products. Additionally, the three sites share the same support services for purchasing, computing, design engineering, process engineering, human resources, safety, and many other functions, the costs for which are distributed annually to the three plants as an indirect cost allocation. This allocation reduces the total plant income as determined by the finance department. One of the primary measures of performance for each plant manager is the plant's net income contribution to corporate income. Therefore, the annual indirect cost allocation is a direct reduction to the plant's bottom line.

For the last 5 years a total of $10 million per year has been allocated to the three plants on the basis of direct labor hours (DLH), which have the following annual averages.

Plant	A	B	C
DLH per year	200,000	100,000	1,800,000

Employment and DLH have been relatively constant over the 5-year period. Therefore, the indirect cost allocation is determined by using a rate known by each plant manager.

$$\text{Indirect cost rate} = \frac{\text{total indirect costs}}{\text{total DLH}}$$

$$= \frac{\$10 \text{ million}}{2.1 \text{ million}}$$

$$= \$4.762 \text{ per DLH}$$

Historically, the oldest plant, A, has set the standard for plant capacity per year. It is 500 units per day for each 200,000 direct labor hours, which is exactly the capacity of plant A. Thus, at 250 days per work-year, the capacity for each plant that is used by the corporate office is as follows:

Plant	A	B	C
Capacity, units/year	125,000	62,500	1,125,000

The manager of plant B has been diligent about quality improvement, minimizing scrap and rework, worker incentives, etc. He believes the standard of DLH for indirect allocation is not representative of plant B when compared with the statistics for *number shipped* from plants A and C.

(a) Allocate the $10 million indirect cost on the basis of DLH.

(b) Plant A manager has proposed that a new blanket indirect cost rate using the basis of total output capacity of the corporation in units per year be used. Determine this rate and allocate the $10 million indirect cost on this basis.

(c) Records indicate that the number of quality-checked units shipped in the last 2 years has averaged 100,000 for plant A, 60,000 for B, and 900,000 for C. Plant B manager is convinced his plant is, and will continue to be, allocated more indirect cost than it should be, in part based on the fact that plant B consistently ships a higher percentage of its capacity than plants A and C. He will propose that an incentive be developed to ship a number of quality-checked units as close as possible to the capacity of each plant. The

allocation formula will use the rate based on plant capacity from part (b) but modified by a dimensionless ratio that measures the actual output relative to plant capacity.

Indirect cost allocation

$$= \frac{(\text{rate})(\text{plant capacity})}{\text{actual output/plant capacity}}$$

Determine the allocation by plant using this method, and compare it with the allocated amounts determined by the previous two methods.

ABC Method

15.43 Use Equation [15.7] and the bases listed in Table 15–4 to explain why a decrease in direct labor hours, coupled with an increase in indirect labor hours due to automation on a production line, may require the use of new bases to allocate indirect costs.

15.44 If the traditional method of indirect cost allocation assists in estimating the cost to produce a unit of product, what does the ABC method often assist with concerning the cost to produce a unit of product?

15.45 SNTTA Travel distributes food costs for its four hotel sites in Europe based on the size of the budget. For this year, in round numbers, the budgets and allocation of $1 million indirect costs for food are distributed at the rate of 10% of total hotel budget.

	Site			
	A	B	C	D
Budget, $	2 million	3 million	4 million	1 million
Allocation, $	200,000	300,000	400,000	100,000

(a) Use the ABC allocation method with a cost pool of $1 million in food costs. The activity is the number of guests during the year.

Site	A	B	C	D
Guests	3500	4000	8000	1000

(b) Again use the ABC method, but now the activity is the number of guest-nights. The average number of lodging-nights for guests at each site is as follows:

Site	A	B	C	D
Length of stay, nights	3.0	2.5	1.25	4.75

(c) Comment on the distribution of food costs, using the two methods. Identify any other activities (cost drivers) that might be considered for the ABC approach that may reflect a realistic allocation of indirect costs.

(d) If a new distribution scheme for the $1 million total indirect cost is instituted (that is, other than the 10% of total hotel budget), explain the difference it will make in the final actual amounts charged in parts (a) and (b). Will it make a difference in the allocation variances if, for example, the distribution is 30% to budgets of $3 million or more and only 20% to budgets of less than $3 million per year?

15.46 Indirect costs are allocated each calendar quarter to three processing lines, using direct labor hours as the basis. New automated equipment has decreased direct labor and the time to produce a unit significantly, so the division manager plans to use cycle time per unit produced as the basis. However, the manager wants, initially, to determine what the allocation would have been had the cycle time been the basis prior to the automation. Use the data below to determine the allocation rate and actual indirect charges for the three different situations, if the amount to be allocated in an average quarter is $400,000. Comment on changes in the amount of actual charged indirect cost per processing line.

Processing Line	10	11	12
Direct labor hours per quarter	20,000	12,700	18,600
Cycle time per unit now, seconds	3.9	17.0	24.8
Cycle time per unit previously, seconds	13.0	55.8	28.5

15.47 This problem consists of three parts that build upon each other. The object is to compare and comment upon the amount of indirect cost allocated to the electricity-generating facilities located in two states for the different situations described in each part.

(a) Historically, Mesa Power Authority has allocated the indirect costs associated with its employee safety program to its plants in California and Arizona based on the number of employees. Information to allocate a $200,200 budget for this year follows:

State	Workforce Size
California	900
Arizona	500

(b) The head of the department of accounting recommends that the traditional method be abandoned and that activity-based costing be used to allocate the $200,200, using expenditures on the safety program as the cost pool and the number of accidents as the activities which drive the costs. Accident statistics indicate the following for the year.

State	Number of Accidents
California	425
Arizona	135

(c) Further study indicates that 80% of the safety program indirect costs is expended for employees in generation areas, and the remaining 20%

goes to office area employees. Because of this apparent imbalance in expenditures, a split allocation of the $200,200 amount is proposed: 80% of total dollars allocated via ABC with the activities being the number of accidents occurring in generation areas and the cost pool being 80% of the total safety program expenditures; and 20% of the total dollars using the traditional indirect cost

allocation method with a basis of number of office area employees. The following data have been collected.

State	Number of Employees		Number of Accidents	
	Generation Area	Office Area	Generation Area	Office Area
California	300	600	405	20
Arizona	200	300	125	10

FE REVIEW PROBLEMS

15.48   The cost of constructing a certain building in 1999 was $400,000. The *ENR* construction cost index was 6059 at that time. If the *ENR* index now is 6950, the cost of constructing a similar building is closest to
(a)   Less than $450,000
(b) $508,300
(c) $458,800
(d)   More than $600,000

15.49   An assembly line robot with a first cost of $75,000 in 1995 had a cost of $89,750 in 2004. If the M&S equipment cost index was 1027 in 1995 and the robot cost increased exactly in proportion to the index, the value of the index in 2004 was closest to
(a)   Slightly less than 1250
(b)   1105.2
(c)   914.6
(d)   Slightly more than 1400

15.50   A 50-horsepower turbine pump was purchased for $2100. If the exponent in the cost capacity equation has a value of 0.76, a 200-horsepower turbine could be expected to cost about
(a)   Less than $5000
(b) $6020
(c) $5975
(d)   More than $6100

15.51   The cost of a certain machine was $15,000 in 2000 when the M&S equipment cost index was 1092. If the current index is 1164 and the exponent in the cost-capacity equation had a value of 0.65, the cost now of a similar machine twice as large would be estimated to be about
(a)   Less than $24,000
(b) $25,100
(c) $28,500
(d)   More than $30,000

CASE STUDY

TOTAL COST ESTIMATES FOR OPTIMIZING COAGULANT DOSAGE

Background

There are several processes involved in the treatment of drinking water, but three of the most important ones are associated with removal of suspended matter, which is known as turbidity. Turbidity removal is brought about by adding chemicals that cause the small suspended solids to clump together (coagulation), forming larger particles which can be removed through settling (sedimentation). The few particles that remain after sedimentation are filtered out in sand, carbon, or coal filters (filtration).

In general, as the dosage of chemicals is increased, more "clumping" occurs (up to a point), so there is increased removal of particles through the settling process. This means that fewer particles have to be removed through filtration, which obviously means that the filter will not have to be cleaned as often through backwashing. Thus, more chemicals mean less backwash water and vice versa. Since backwash water and chemicals both have costs, a primary question is, What amount of chemicals will result in the lowest overall cost when the chemical coagulation and filtration processes are considered together?

Formulation

To minimize the total cost associated with coagulation and filtration, it is necessary to obtain the relationship between the chemical dosage and water turbidity after coagulation and sedimentation, but before filtration. This allows the chemical costs for different operating strategies to be determined. This cost relationship, derived using polynomial regression analysis, is shown in Figure 15–2 and is described by the equation

$$T = 37.0893 - 7.7390F + 0.7263F^2 - 0.0233\,F^3$$
$$[15.8]$$

where T = settled water turbidity and F = coagulant dosage in milligrams per liter (mg/L).

Similarly, the backwash water data are described by the equation

$$B = -0.549 + 1.697T \qquad [15.9]$$

where B = backwash water rate, $m^3/1000\ m^3$ product water.

By substituting Equation [15.8] into Equation [15.9] and multiplying by the unit water cost of $0.0608/m^3$, the cost of washwater versus turbidity C_B is found.

$$C_B = -0.0024F^3 + 0.0749F^2 - 0.798F + 3.791$$
$$[15.10]$$

where C_B = cost of washwater, $/1000\ m^3$ product water.

The chemical cost for coagulation C_C is $0.183 per kilogram or

$$C_C = 0.183F \qquad [15.11]$$

The total cost C_T of backwash water and chemicals is obtained by adding the last two equations.

$$C_T = C_B + C_C \qquad [15.12]$$
$$= -0.0024F^3 + 0.0749F^2 - 0.615F + 3.791$$

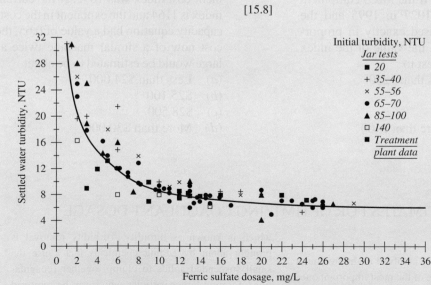

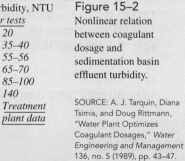

Initial turbidity, NTU
Jar tests
■ *20*
+ *35–40*
× *55–56*
● *65–70*
▲ *85–100*
□ *140*
■ *Treatment plant data*

Figure 15–2

Nonlinear relation between coagulant dosage and sedimentation basin effluent turbidity.

SOURCE: A. J. Tarquin, Diana Tsimis, and Doug Rittmann, "Water Plant Optimizes Coagulant Dosages," *Water Engineering and Management* 136, no. 5 (1989), pp. 43–47.

Equations [15.10] through [15.12] are plotted in Figure 15–3.

Results

As Figure 15–3 shows, the minimum cost occurs at a dosage of 5.6 mg/L. At this dosage the total cost for coagulation and filtration is approximately $C_T = \$2.27$ per 1000 m³ of product water. Prior to this analysis, the plant was using 12 mg/L. The costs at 5.6 and 12 mg/L are shown in Table 15–8. As an illustration, at the average flow rate of 189,250 m³/day, a 26% savings represents an annual dollar savings of over $53,000.

Case Study Exercises

1. What effect does an increase in chemical cost have on the optimum dosage?
2. What effect does an increase in backwash water cost have on the optimum dosage?
3. What is the chemical cost at a dosage of 10 mg/L?
4. What is the backwash water cost at a dosage of 14 mg/L?
5. If the chemical cost changes to $0.21 per kg, what will be the total coagulation and filtration cost at a dosage of 6 mg/L?
6. At what chemical cost will the minimum total cost occur at 8 mg/L?

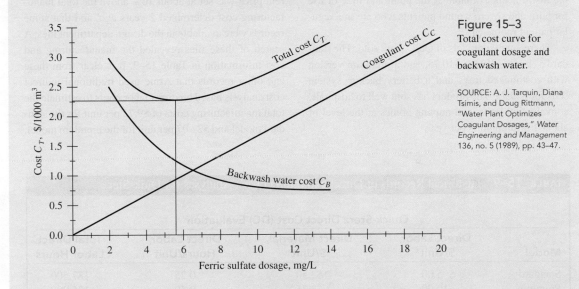

Figure 15–3
Total cost curve for coagulant dosage and backwash water.

SOURCE: A. J. Tarquin, Diana Tsimis, and Doug Rittmann, "Water Plant Optimizes Coagulant Dosages," *Water Engineering and Management* 136, no. 5 (1989), pp. 43–47.

TABLE 15–8 Operational Costs at 5.6 and 12 mg/L Coagulant Dosages

Coagulant Dosage, mg/L	Coagulant Cost, $/1000 m³	No. of Backwashes per Day*	Washwater Cost, $/1000 m³	Total Cost, $/1000 m³	Cost Savings
5.6	1.25	5.93	1.02	2.27	26%
12.0	2.20	4.12	0.85	3.05	

*Average amount of washwater per backwash is 305 m³, and average flow rate per day is 94,625 m³/day (25 MGD).

CASE STUDY

INDIRECT COST COMPARISON OF MEDICAL EQUIPMENT STERILIZATION UNIT

The Product

Three years ago Medical Dynamics, a medical equipment unit of Johnson and Sons, Inc., initiated the manufacture and sales of a portable sterilization unit (Quik-Sterz) that can be placed in the hospital room of a patient. This unit sterilizes and makes available at the bedside some of the reusable instruments that nurses and doctors usually obtain by walking to or having delivered from a centralized area. This new unit makes the instruments available at the point and time of use for burn and severe wound patients who are in a regular patient room.

There are two models of Quik-Sterz sold. The standard version sells for $10.75, and a premium version with customized trays and a battery backup system sells for $29.75. The product has sold well to hospitals, convalescent units, and nursing homes at the level of about 1 million units per year.

Cost Allocation Procedures

Medical Dynamics has historically used an indirect cost allocation system based upon direct hours to manufacture for all its other product lines. The same was applied when Quik-Sterz was priced. However, Arnie, the person who performed the indirect cost analysis and set the sales price, is no longer at the company, and the detailed analysis is no longer available. Through e-mail and telephone conversations, Arnie said the current price was set at about 10% above the total manufacturing cost determined 2 years ago, and that some records were available in the design department files. A search of these files revealed the manufacturing and cost information in Table 15–9. It is clear from these and other records that Arnie used traditional indirect cost analysis based on direct labor hours to estimate the total manufacturing costs of $9.73 per unit for the standard model and $27.07 per unit for the premium model.

TABLE 15–9 Historical Records of Direct and Indirect Cost Analyses for Quik-Sterz

	Quick-Sterz Direct Cost (DC) Evaluation			
Model	Direct Labor, $/Unit*	Direct Material, $/Unit	Direct Labor, Hours/Unit	Total Direct Labor Hours
Standard	$ 5.00	$2.50	0.25	187,500
Premium	10.00	3.75	0.50	125,000

	Quick-Sterz Indirect Cost (IDC) Evaluation			
Model	Direct Labor, Hours/Unit	Fraction IDC Allocated	Allocated IDC	Sales, Units/Year
Standard	0.25	$\frac{1}{3}$	$1.67 million	750,000
Premium	0.50	$\frac{2}{3}$	3.33 million	250,000

*Average direct labor rate is $20 per hour.

TABLE 15–10 Quik-Sterz Cost Pools, Cost Drivers, and Activity Levels for ABC-based Indirect Cost Allocation

Cost Pool Function	Cost Driver Activity	Activity, Volume/Year	Actual Cost, $/Year
Quality	Inspections	20,000 inspections	$ 800,000
Purchasing	Purchase orders	40,000 orders	1,200,000
Scheduling	Change orders	1,000 orders	800,000
Production setup	Setups	5,000 setups	1,000,000
Machine operations	Machine-hours	10,000 hours	1,200,000

Activity	Activity Level for the Year	
	Standard	Premium
Quality inspections	8,000	12,000
Purchase orders	30,000	10,000
Scheduling change orders	400	600
Production setups	1,500	3,500
Machine hours	7,000	3,000

Last year management decided to place the entire plant on the ABC system of indirect cost allocation. The costs and sales figures collected for Quik-Sterz the year before were still accurate. Five cost pools and their cost drivers were identified for the Medical Dynamics manufacturing operations (Table 15–10). Also, the number of activities for each model is summarized in this table.

The ABC method will be used henceforth, with the intention of determining the total cost and price based on its results. The first impression of production personnel is that the new system will show that indirect costs for Quik-Sterz are about the same as they have been for other products over the last several years when a standard and an upgrade (premium) model were sold. Predictably, they state, the standard model will receive about 1/3 of the indirect cost, and the premium will receive the remaining 2/3. Fundamentally, there are three reasons why production management does not like to produce premium versions: They cost too much in terms of both direct and indirect cost, they are less profitable for the company, and they require significantly more time and operations to manufacture.

Case Study Exercises

1. Use traditional indirect cost allocation to verify Arnie's cost and price estimates.

2. Use the ABC method to estimate the indirect cost allocation and total cost for each model.

3. If the prices and number of units sold are the same next year (750,000 standard and 250,000 premium), and all other costs remain constant, compare the profit from Quik-Sterz under the ABC method with the profit using the traditional indirect cost allocation method.

4. What prices should Medical Dynamics charge next year based on the ABC method and a 10% markup over cost? What is the total profit from Quik-Sterz predicted to be if sales hold steady?

5. Using the results above, comment on the first-impression observations and predictions of production management personnel about Quik-Sterz when the ABC method implementation was announced.

Depreciation Methods

The capital investments of a corporation in tangible assets—equipment, computers, vehicles, buildings, and machinery—are commonly recovered on the books of the corporation through *depreciation*. Although the depreciation amount is not an actual cash flow, the process of depreciating an asset, also referred to as *capital recovery*, accounts for the decrease in an asset's value because of age, wear, and obsolescence. Even though an asset may be in excellent working condition, the fact that it is worth less through time is taken into account in economic evaluation studies. An introduction to the classical depreciation methods is followed by a discussion of the *Modified Accelerated Cost Recovery System (MACRS)*, which is the standard in the United States for tax purposes. Other countries commonly use the classical methods for tax computations.

Why is depreciation important to engineering economy? Depreciation is a *tax-allowed deduction* included in tax calculations in virtually all industrialized countries. Depreciation lowers income taxes via the relation

$$\text{Taxes} = (\text{income} - \text{deductions})(\text{tax rate})$$

Income taxes are discussed further in Chapter 17.

This chapter concludes with an introduction to two methods of *depletion*, which are used to recover capital investments in deposits of natural resources such as minerals, ores, and timber.

***Important note:* To consider depreciation and after-tax analysis early in a course, cover this chapter and the next one (After-Tax Economic Analysis) after Chapter 6 (AW), Chapter 9 (B/C), or Chapter 11 (Replacement Analysis). Consult the Preface for more options on subject ordering.**

LEARNING OBJECTIVES

Purpose: Use classical and government-approved methods to reduce the value of the capital investment in an asset or natural resource.

This chapter will help you:

Depreciation terms	1.	Understand and use the basic terminology of depreciation.
Straight line	2.	Apply the straight line model of depreciation.
Declining balance	3.	Apply the declining balance and double declining balance models of depreciation.
MACRS	4.	Apply the Modified Accelerated Cost Recovery System (MACRS) of depreciation for U.S. corporations.
Recovery period	5.	Select the recovery period of an asset for MACRS depreciation.
Depletion	6.	Utilize the cost depletion and percentage depletion methods for natural resource investments.

Chapter Appendix

A1. Apply the sum-of-year digits method of depreciation.

A2. Determine when to switch from one depreciation model to another.

A3. Compute MACRS depreciation rates using depreciation model switching.

16.1 DEPRECIATION TERMINOLOGY

Primary terms used in depreciation are defined here. Most terms are applicable to corporations as well as individuals who own depreciable assets.

Depreciation is the reduction in value of an asset. The method used to depreciate an asset is a way to account for the decreasing value of the asset to the owner *and* to represent the diminishing value (amount) of the capital funds invested in it. The annual depreciation amount D_t does not represent an actual cash flow, nor does it necessarily reflect the actual usage pattern of the asset during ownership.

Book depreciation and **tax depreciation** are terms used to describe the purpose for reducing asset value. Depreciation may be performed for two reasons:

1. Use by a corporation or business for internal financial accounting. This is book depreciation.
2. Use in tax calculations per government regulations. This is tax depreciation.

The methods applied for these two purposes may or may not utilize the same formulas, as is discussed later. *Book depreciation* indicates the reduced investment in an asset based upon the usage pattern and expected useful life of the asset. There are classical, internationally accepted depreciation methods used to determine book depreciation: straight line, declining balance, and the infrequently used sum-of-year digits method. The amount of *tax depreciation* is important in an after-tax engineering economy study because of the following:

> **In the United States and many industrialized countries, the annual tax depreciation is tax deductible; that is, it is subtracted from income when calculating the amount of taxes due each year. However, the tax depreciation amount must be calculated using a government-approved method.**

Tax depreciation may be calculated and referred to differently in countries outside the United States. For example, in Canada the equivalent is CCA (capital cost allowance), which is calculated based on the undepreciated value of all corporate properties that form a particular class of assets, whereas in the United States, depreciation may be determined for each asset separately.

Where allowed, tax depreciation is usually based on an accelerated method, whereby the depreciation for the first years of use is larger than that for later years. In the United States, this method is called MACRS, as covered in later sections. In effect, accelerated methods defer some of the income tax burden to later in the asset's life; they do not reduce the total tax burden.

First cost or **unadjusted basis** is the delivered and installed cost of the asset including purchase price, delivery and installation fees, and other depreciable direct costs incurred to prepare the asset for use. The term unadjusted basis B, or simply basis, is used when the asset is new, with the term adjusted basis used after some depreciation has been charged.

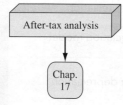

After-tax analysis

Chap. 17

Book value represents the remaining, undepreciated capital investment on the books after the total amount of depreciation charges to date have been subtracted from the basis. The book value BV_t is usually determined at the end of each year, which is consistent with the end-of-year convention.

Recovery period is the depreciable life n of the asset in years. Often there are different n values for book and tax depreciation. Both of these values may be different from the asset's estimated productive life.

Market value, a term also used in replacement analysis, is the estimated amount realizable if the asset were sold on the open market. Because of the structure of depreciation laws, the book value and market value may be substantially different. For example, a commercial building tends to increase in market value, but the book value will decrease as depreciation charges are taken. However, a computer workstation may have a market value much lower than its book value due to rapidly changing technology.

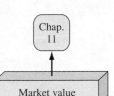

Salvage value is the estimated trade-in or market value at the end of the asset's useful life. The salvage value, S expressed as an estimated dollar amount or as a percentage of the first cost, may be positive, zero, or negative due to dismantling and carry-away costs.

Depreciation rate or **recovery rate** is the fraction of the first cost removed by depreciation each year. This rate, denoted by d_t, may be the same each year, which is called the straight-line rate, or different for each year of the recovery period.

Personal property, one of the two types of property for which depreciation is allowed, is the income-producing, tangible possessions of a corporation used to conduct business. Included is most manufacturing and service industry property—vehicles, manufacturing equipment, materials handling devices, computers and networking equipment, telephone equipment, office furniture, refining process equipment, construction assets, and much more.

Real property includes real estate and all improvements—office buildings, manufacturing structures, test facilities, warehouses, apartments, and other structures. *Land itself is considered real property, but it is not depreciable.*

Half-year convention assumes that assets are placed in service or disposed of in midyear, regardless of when these events actually occur during the year. This convention is utilized in this text and in most U.S.-approved tax depreciation methods. There are also midquarter and midmonth conventions.

As mentioned before, there are several models for depreciating assets. The straight line (SL) model is used, historically and internationally. Accelerated models, such as the declining balance (DB) model, decrease the book value to zero (or to the salvage value) more rapidly than the straight line method, as shown by the general book value curves in Figure 16–1.

For the classical methods—straight line, declining balance, and sum-of-year digits (SYD)—there are Excel functions available to determine annual depreciation. Each function is introduced and illustrated as the method is explained. Since SYD is applied less frequently, it is summarized in the appendix to this chapter.

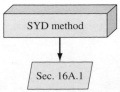

Figure 16–1
General shape of book
value curves for different
depreciation models.

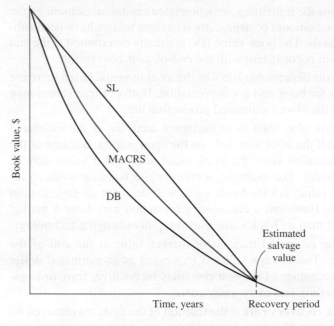

As expected, there are many rules and exceptions to the depreciation laws of a country. One that may be of interest to a U.S.-based *small business* performing an economic analysis is the *Section 179 Deduction.* This is an economic incentive that changes rapidly and is primarily intended for small businesses to invest capital in equipment directly used in the company. Up to a specified amount, the entire basis of an asset is treated as a business expense in the year of purchase. This tax treatment reduces federal income taxes, just as depreciation does, but it is allowed in lieu of depreciating the first cost over several years. The limit changes with time; it was $24,000 in 2002, but increased to $100,000 in 2003 and $102,000 in 2004. Investments above these limits are depreciated using MACRS. Tax law for 2004 stated that for capital investments above $410,000, the Section 179 Deduction is reduced dollar for dollar.

In the 1980s the U.S. government standardized accelerated methods for *federal tax depreciation* purposes. In 1981, all classical methods, including straight line, declining balance, and sum-of-year digits depreciation, were disallowed as tax deductible and replaced by the Accelerated Cost Recovery System (ACRS). In a second round of standardization, MACRS (Modified ACRS) was made the required tax depreciation method in 1986. To this date, the following is the law in the U.S.

Tax depreciation must be calculated using MACRS; book depreciation may be calculated using any classical method or MACRS.

MACRS has the DB and SL methods, in slightly different forms, embedded in it, but these two methods cannot be used directly if the annual depreciation is to be tax deductible. Many U.S. companies still apply the classical methods for keeping their own books, because these methods are more representative of how the usage patterns of the asset reflect the remaining capital invested in it. Additionally, most other countries still recognize the classical methods of straight line and declining balance for tax or book purposes. Because of the continuing importance of the SL and DB methods, they are explained in the next two sections prior to MACRS.

Tax law revisions occur often, and depreciation rules are changed from time to time in the United States and other countries. For example, in 2003 and 2004, a special depreciation allowance of either 30% or 50% of a new asset's first cost could be taken in its initial year in service. That was in addition to the Section 179 Deduction. This was an effort to promote capital investment. Although the tax rates and depreciation guidelines at the time you read this material may be slightly different, the general principles and equations are applicable to all U.S. corporations. For more depreciation and tax law information, consult the U.S. Department of the Treasury, Internal Revenue Service (IRS), website at www.irs.gov. Pertinent publications can be downloaded via Acrobat Reader. Publication 946, How to Depreciate Property, is especially applicable to this chapter. MACRS and most corporate tax depreciation laws are discussed in it.

16.2   STRAIGHT LINE (SL) DEPRECIATION

Straight line depreciation derives its name from the fact that the book value decreases linearly with time. The depreciation rate $d = 1/n$ is the same each year of the recovery period n.

Straight line is considered the standard against which any depreciation model is compared. For *book depreciation* purposes, it offers an excellent representation of book value for any asset that is used regularly over an estimated number of years. For *tax depreciation,* as mentioned earlier, it is not used directly in the United States, but it is commonly used in most countries for tax purposes. However, the U.S. MACRS method includes a version of SL depreciation with a larger n value than that allowed by regular MACRS (see Section 16.5).

The annual SL depreciation is determined by multiplying the first cost minus the salvage value by d. In equation form,

$$D_t = (B - S)d$$
$$= \frac{B - S}{n} \qquad \text{[16.1]}$$

where t = year $(t = 1, 2, \ldots, n)$
   D_t = annual depreciation charge
   B = first cost or unadjusted basis
   S = estimated salvage value
   n = recovery period
   d = depreciation rate = $1/n$

Since the asset is depreciated by the same amount each year, the book value after t years of service, denoted by BV_t, will be equal to the first cost B minus the annual depreciation times t.

$$\text{BV}_t = B - tD_t \qquad \text{[16.2]}$$

Earlier we defined d_t as a depreciation rate for a specific year t. However, the SL model has the same rate for all years, that is,

$$d = d_t = \frac{1}{n} \qquad \text{[16.3]}$$

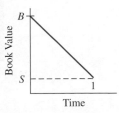

Q-Solv

The format for the Excel function to display the annual depreciation D_t in a single-cell operation is

$$\text{SLN}(B,S,n)$$

EXAMPLE 16.1

If an asset has a first cost of $50,000 with a $10,000 estimated salvage value after 5 years, (a) calculate the annual depreciation and (b) compute and plot the book value of the asset after each year, using straight line depreciation.

Solution

(a)   The depreciation each year for 5 years can be found by Equation [16.1].

$$D_t = \frac{B - S}{n} = \frac{50,000 - 10,000}{5} = \$8000$$

Enter the function SLN(50000,10000,5) in any cell to display the D_t of $8000.

(b)   The book value after each year t is computed using Equation [16.2]. The BV_t values are plotted in Figure 16–2. For years 1 and 5, for example,

$$BV_1 = 50,000 - 1(8000) = \$42,000$$

$$BV_5 = 50,000 - 5(8000) = \$10,000 = S$$

Q-Solv

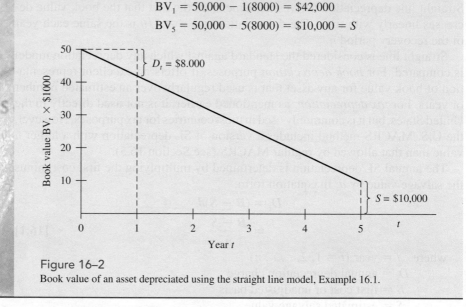

Figure 16–2
Book value of an asset depreciated using the straight line model, Example 16.1.

16.3   DECLINING BALANCE (DB) AND DOUBLE DECLINING BALANCE (DDB) DEPRECIATION

The declining balance method is commonly applied as the book depreciation method. Like the SL method, DB is embedded in the MACRS method, but the DB method itself cannot be used to determine the annual tax-deductible depreciation in the United States. This method is used routinely in most other countries for tax and book depreciation purposes.

Declining balance is also known as the fixed percentage or uniform percentage method. DB depreciation accelerates the write-off of asset value because

the annual depreciation is determined by multiplying the *book value at the beginning of a year* by a fixed (uniform) percentage d, expressed in decimal form. If $d = 0.1$, then 10% of the book value is removed each year. Therefore, the depreciation amount decreases each year.

The maximum annual depreciation rate for the DB method is twice the straight line rate, that is,

$$d_{max} = 2/n \qquad [16.4]$$

In this case the method is called *double declining balance (DDB)*. If $n = 10$ years, the DDB rate is $2/10 = 0.2$; so 20% of the book value is removed annually. Another commonly used percentage for the DB method is 150% of the SL rate, where $d = 1.5/n$.

The depreciation for year t is the fixed rate d times the book value at the end of the previous year.

$$D_t = (d)BV_{t-1} \qquad [16.5]$$

The actual depreciation rate for each year t, relative to the first cost B, is

$$d_t = d(1 - d)^{t-1} \qquad [16.6]$$

If BV_{t-1} is not known, the depreciation in year t can be calculated using B and d_t from Equation [16.6].

$$D_t = dB(1 - d)^{t-1} \qquad [16.7]$$

Book value in year t is determined in one of two ways: by using the rate d and first cost B, or by subtracting the current depreciation charge from the previous book value. The equations are

$$BV_t = B(1 - d)^t \qquad [16.8]$$
$$BV_t = BV_{t-1} - D_t \qquad [16.9]$$

It is important to understand that the book value for the DB method never goes to zero, because the book value is always decreased by a fixed percentage. The implied salvage value after n years is the BV_n amount, that is,

$$\text{Implied salvage value} = \text{implied } S = BV_n = B(1 - d)^n \qquad [16.10]$$

If a salvage value is estimated for the asset, this *estimated S value is not used in the DB or DDB method* to calculate annual depreciation. However, if the implied $S <$ estimated S, it is correct to stop charging further depreciation when the book value is at or below the estimated salvage value. In most cases, the estimated S is in the range of zero to the implied S value. (This guideline is important when the DB method can be used directly for tax depreciation purposes.)

If the fixed percentage d is not stated, it is possible to determine an implied fixed rate using the estimated S value, if $S > 0$. The range for d is $0 < d < 2/n$.

$$\text{Implied } d = 1 - \left(\frac{S}{B}\right)^{1/n} \qquad [16.11]$$

The Excel functions DDB and DB are used to display depreciation amounts for specific years (or any other unit of time). The function is repeated in consecutive

E-Solve

spreadsheet cells because the depreciation amount D_t changes with t. For the double declining balance method the format is

$$\text{DDB}(B,S,n,t,d)$$

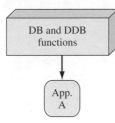

DB and DDB functions

App. A

The entry d is the fixed rate expressed as a number between 1 and 2. If omitted, this optional entry is assumed to be 2 for DDB. An entry of $d = 1.5$ makes the DDB function display 150% declining balance method amounts. The DDB function automatically checks to determine when the book value equals the estimated S value. No further depreciation is charged when this occurs. (In order to allow *full* depreciation charges to be made, ensure that the S entered is between zero and the implied S from Equation [16.10].) Note that $d = 1$ is the same as the straight line rate $1/n$, but D_t *will not be the SL amount* because declining balance depreciation is determined as a fixed percentage of the previous year's book value, which is completely different from the SL calculation in Equation [16.1].

The DB function must be used carefully. Its format is DB(B,S,n,t). The fixed rate d is not entered in the DB function; d is an embedded calculation using a spreadsheet equivalent of Equation [16.11]. Also, only three significant digits are maintained for d, so the book value may go below the estimated salvage value due to round-off errors. Therefore, *if the depreciation rate is known, always use the DDB function to ensure correct results.* The next two examples illustrate DB and DDB depreciation and these spreadsheet functions.

EXAMPLE 16.2

A fiber optics testing device is to be DDB depreciated. It has a first cost of $25,000 and an estimated salvage of $2500 after 12 years. (*a*) Calculate the depreciation and book value for years 1 and 4. Write the Excel functions to display depreciation for years 1 and 4. (*b*) Calculate the implied salvage value after 12 years.

Solution

(*a*) The DDB fixed depreciation rate is $d = 2/n = 2/12 = 0.1667$ per year. Use Equations [16.7] and [16.8].

Year 1: $\qquad D_1 = (0.1667)(25,000)(1 - 0.1667)^{1-1} = \$4167$

$\qquad\qquad\quad BV_1 = 25,000(1 - 0.1667)^1 = \$20,833$

Year 4: $\qquad D_4 = (0.1667)(25,000)(1 - 0.1667)^{4-1} = \$2411$

$\qquad\qquad\quad BV_4 = 25,000(1 - 0.1667)^4 = \$12,054$

The DDB functions for D_1 and D_4 are, respectively, DDB(25000,2500,12,1) and DDB(25000,2500,12,4).

Q-Solv

(*b*) From Equation [16.10], the implied salvage value after 12 years is

$$\text{Implied } S = 25,000(1 - 0.1667)^{12} = \$2803$$

Since the estimated $S = \$2500$ is less than $2803, the asset is not fully depreciated when its 12-year expected life is reached.

EXAMPLE 16.3

Freeport-McMoRan Mining Company has purchased a computer-controlled gold ore grading unit for $80,000. The unit has an anticipated life of 10 years and a salvage value of $10,000. Use the DB and DDB methods to compare the schedule of depreciation and book values for each year. Solve by hand and by computer.

Solution by Hand

An implied DB depreciation rate is determined by Equation [16.11].

$$d = 1 - \left(\frac{10,000}{80,000} \right)^{1/10} = 0.1877$$

Note that $0.1877 < 2/n = 0.2$, so this DB model does not exceed twice the straight line rate. Table 16–1 presents the D_t values using Equation [16.5] and the BV_t values from Equation [16.9] rounded to the nearest dollar. For example, in year $t = 2$, the DB results are

$$D_2 = d(BV_1) = 0.1877(64,984) = \$12,197$$

$$BV_2 = 64,984 - 12,197 = \$52,787$$

Because we round off to even dollars, $2312 is calculated for depreciation in year 10, but $2318 is deducted to make $BV_{10} = S = \$10,000$ exactly. Similar calculations for DDB with $d = 0.2$ result in the depreciation and book value series in Table 16–1.

TABLE 16–1 D_t and BV_t Values for DB and DDB Depreciation, Example 16.3

Year t	Declining Balance		Double Declining	
	D_t	BV_t	D_t	BV_t
0	—	$80,000	—	$80,000
1	$15,016	64,984	$16,000	64,000
2	12,197	52,787	12,800	51,200
3	9,908	42,879	10,240	40,960
4	8,048	34,831	8,192	32,768
5	6,538	28,293	6,554	26,214
6	5,311	22,982	5,243	20,972
7	4,314	18,668	4,194	16,777
8	3,504	15,164	3,355	13,422
9	2,846	12,318	2,684	10,737
10	2,318	10,000	737	10,000

Solution by Computer

The spreadsheet in Figure 16–3 displays the results for the DB and DDB methods. The xy scatter chart plots book values for each year. Since the fixed rates are close—0.1877 for DB and 0.2 for DDB—the annual depreciation and book value series are approximately equal for the two methods.

E-Solve

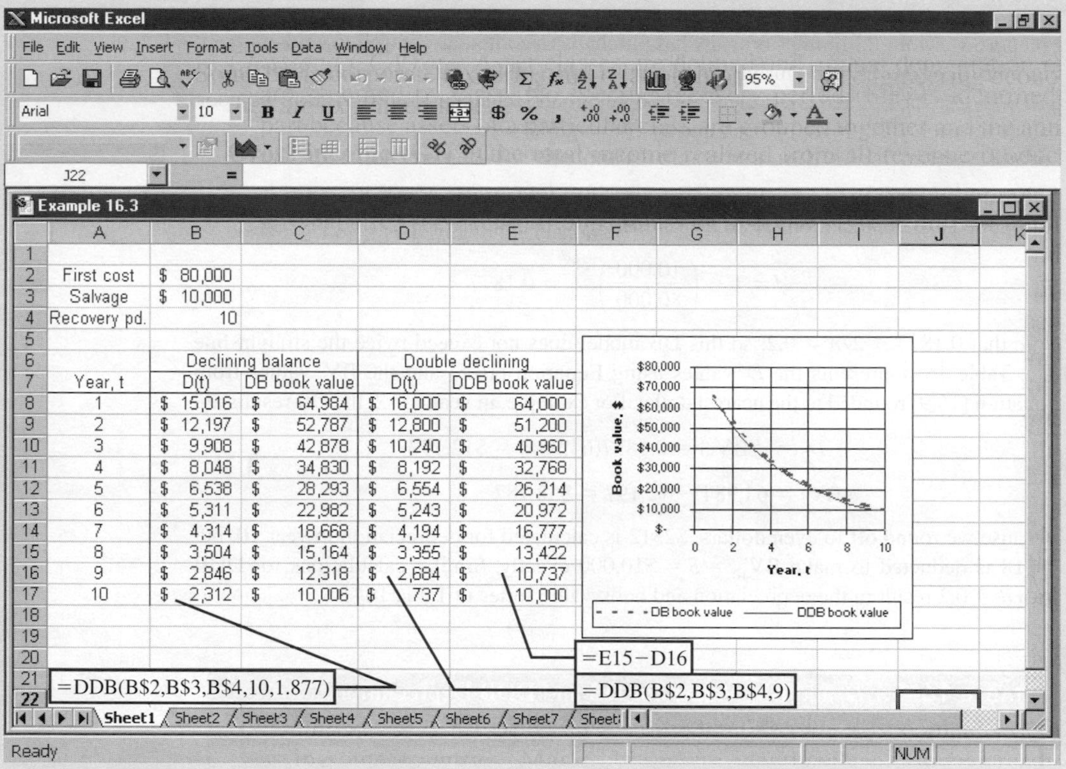

	A	B	C	D	E
1					
2	First cost	$ 80,000			
3	Salvage	$ 10,000			
4	Recovery pd.	10			
5					
6		Declining balance		Double declining	
7	Year, t	D(t)	DB book value	D(t)	DDB book value
8	1	$ 15,016	$ 64,984	$ 16,000	$ 64,000
9	2	$ 12,197	$ 52,787	$ 12,800	$ 51,200
10	3	$ 9,908	$ 42,878	$ 10,240	$ 40,960
11	4	$ 8,048	$ 34,830	$ 8,192	$ 32,768
12	5	$ 6,538	$ 28,293	$ 6,554	$ 26,214
13	6	$ 5,311	$ 22,982	$ 5,243	$ 20,972
14	7	$ 4,314	$ 18,668	$ 4,194	$ 16,777
15	8	$ 3,504	$ 15,164	$ 3,355	$ 13,422
16	9	$ 2,846	$ 12,318	$ 2,684	$ 10,737
17	10	$ 2,312	$ 10,006	$ 737	$ 10,000

Figure 16–3
Spreadsheet solution for annual depreciation and book values for DB and DDB depreciation, Example 16.3.

Comment

Note in the cell tags that the DDB function is used in both columns B and D to determine annual depreciation. The DDB function for the declining balance method has the *d* rate entered as 1.877. This is done for accuracy. As mentioned earlier, the DB function automatically calculates the implied rate by Equation [16.11] and maintains it to only three significant digits. Therefore, if the DB function were used in column B (Figure 16–3), the fixed rate applied would be 0.188. The resulting D_t and BV_t values for years 8, 9, and 10 would be as follows:

t	D_t	BV_t
8	$3,501	$15,120
9	2,842	12,277
10	2,308	9,969

Also noteworthy is the fact that the DB function uses the implied rate without a check to halt the book value at the estimated salvage value. Thus, BV_{10} will go slightly below S = $10,000, as shown above. However, the DDB function uses a relation different from that of the DB function to determine annual depreciation—one that correctly stops depreciating at the estimated salvage value, as shown in Figure 16–3, cell E17.

16.4 MODIFIED ACCELERATED COST RECOVERY SYSTEM (MACRS)

In the 1980s, the U.S. introduced MACRS as the required tax depreciation method for all depreciable assets. Through MACRS, the 1986 Tax Reform Act defined statutory depreciation rates that take advantage of the accelerated DB and DDB methods. Corporations are free to apply any of the classical methods for book depreciation.

Many aspects of MACRS deal with the specific depreciation accounting aspects of tax law. This chapter covers only the elements that materially affect after-tax economic analysis. Additional information on how the DDB, DB, and SL methods are embedded into MACRS and how to derive the MACRS depreciation rates is presented and illustrated in the chapter appendix, Sections 16A.2 and 16A.3.

MACRS determines annual depreciation amounts using the relation

$$D_t = d_t B \qquad \text{[16.12]}$$

where the depreciation rate d_t is provided in tabulated form. As for other methods, the book value in year t is determined by subtracting the depreciation amount from the previous year's book value,

$$\text{BV}_t = \text{BV}_{t-1} - D_t \qquad \text{[16.13]}$$

or by subtracting the total depreciation from the first cost.

$$\textbf{BV}_t = \textbf{first cost} - \textbf{sum of accumulated depreciation}$$

$$= B - \sum_{j=1}^{j=t} D_j \qquad \text{[16.14]}$$

The first cost B is always completely depreciated, since MACRS assumes that $S = 0$, even though there may be a positive salvage that is realizable.

The MACRS recovery periods are standardized to the values of 3, 5, 7, 10, 15, and 20 years for personal property. The real property recovery period for structures is commonly 39 years, but it is possible to justify a 27.5-year recovery for residential rental property. Section 16.5 explains how to determine an allowable MACRS recovery period. The MACRS personal property depreciation rates (d_t values) for $n = 3, 5, 7, 10, 15$, and 20 for use in Equation [16.12] are included in Table 16–2. (These values are used often throughout the remainder of this text, so you may wish to tab this page.)

MACRS depreciation rates incorporate the DDB method ($d = 2/n$) and switch to SL depreciation during the recovery period as an inherent component for *personal property* depreciation. The MACRS rates start with the DDB rate or the 150% DB rate and switch when the SL method offers faster write-off.

For *real property,* MACRS utilizes the SL method for $n = 39$ throughout the recovery period. The annual percentage depreciation rate is $d = 1/39 = 0.02564$.

	TABLE 16–2	Depreciation Rates, d_t, Applied to the First Cost B for the MACRS Method				

	Depreciation Rate (%) for Each MACRS Recovery Period in Years					
Year	$n = 3$	$n = 5$	$n = 7$	$n = 10$	$n = 15$	$n = 20$
1	33.33	20.00	14.29	10.00	5.00	3.75
2	44.45	32.00	24.49	18.00	9.50	7.22
3	14.81	19.20	17.49	14.40	8.55	6.68
4	7.41	11.52	12.49	11.52	7.70	6.18
5		11.52	8.93	9.22	6.93	5.71
6		5.76	8.92	7.37	6.23	5.29
7			8.93	6.55	5.90	4.89
8			4.46	6.55	5.90	4.52
9				6.56	5.91	4.46
10				6.55	5.90	4.46
11				3.28	5.91	4.46
12					5.90	4.46
13					5.91	4.46
14					5.90	4.46
15					5.91	4.46
16					2.95	4.46
17–20						4.46
21						2.23

However, MACRS forces partial-year recovery in years 1 and 40. The MACRS real property rates in percentage amounts are

Year 1	$100d_1 = 1.391\%$
Years 2–39	$100d_t = 2.564\%$
Year 40	$100d_{40} = 1.177\%$

Note that all MACRS depreciation rates are presented for 1 year longer than the stated recovery period. Also note that the extra-year rate is one-half of the previous year's rate. This is because a built-in *half-year convention* is imposed by MACRS. This convention assumes that all property is placed in service at the midpoint of the tax year of installation. Therefore, only 50% of the first-year DB depreciation applies for tax purposes. This removes some of the accelerated depreciation advantage and requires that one-half year of depreciation be taken in year $n + 1$.

There is no Excel function for MACRS; it is necessary to input the d_t rates and set up a function for $D_t = d_t B$.

E-Solve

EXAMPLE 16.4

Baseline, a nationwide franchise for environmental engineering services, has acquired new workstations and 3D modeling software for its 100 affiliate sites at a cost of $4000 per site. The estimated salvage for each system after 3 years is expected to be 5% of the first cost. The franchise manager in the home office in San Francisco wants to compare the depreciation for a 3-year MACRS model (tax depreciation) with that for a 3-year DDB model (book depreciation). He is especially curious about the depreciation over the next 2 years. Using hand and computer solution,

(a) Determine which model offers the larger total depreciation after 2 years.
(b) Determine the book value for each model after 2 years and at the end of the recovery period.

Solution by Hand

The basis is $B = \$400,000$, and the estimated $S = 0.05(400,000) = \$20,000$. The MACRS rates for $n = 3$ are taken from Table 16–2, and the depreciation rate for DDB is $d_{max} = 2/3 = 0.6667$. Table 16–3 presents the depreciation and book values. Year 3 depreciation for DDB would be $\$44,444(0.6667) = \$29,629$, except that this would make $BV_3 < \$20,000$. Only the remaining amount of $24,444 is removed.

(a) The 2-year accumulated depreciation values from Table 16–3 are

MACRS: $D_1 + D_2 = \$133,320 + 177,800 = \$311,120$

DDB: $D_1 + D_2 = \$266,667 + 88,889 = \$355,556$

The DDB depreciation is larger. (Remember that for tax purposes, Baseline does not have the choice in the United States of the DDB model as applied here.)

(b) After 2 years the book value for DDB at $44,444 is 50% of the MACRS book value of $88,880. At the end of recovery (4 years for MACRS due to the built-in half-year convention, and 3 years for DDB), the MACRS book value is $BV_4 = 0$ and for DDB, $BV_3 = \$20,000$. This occurs because MACRS always removes the entire first cost, regardless of the estimated salvage value. This is a tax depreciation advantage of the MACRS method (unless the asset is disposed of for more than the MACRS-depreciated book value, as discussed in Section 17.4).

TABLE 16–3   Comparing MACRS and DDB Depreciation, Example 16.4

Year	Rate	MACRS Tax Depreciation	MACRS Book Value	DDB Book Depreciation	DDB Book Value
0			$400,000		$400,000
1	0.3333	$133,320	266,680	$266,667	133,333
2	0.4445	177,800	88,880	88,889	44,444
3	0.1481	59,240	29,640	24,444	20,000
4	0.0741	29,640	0		

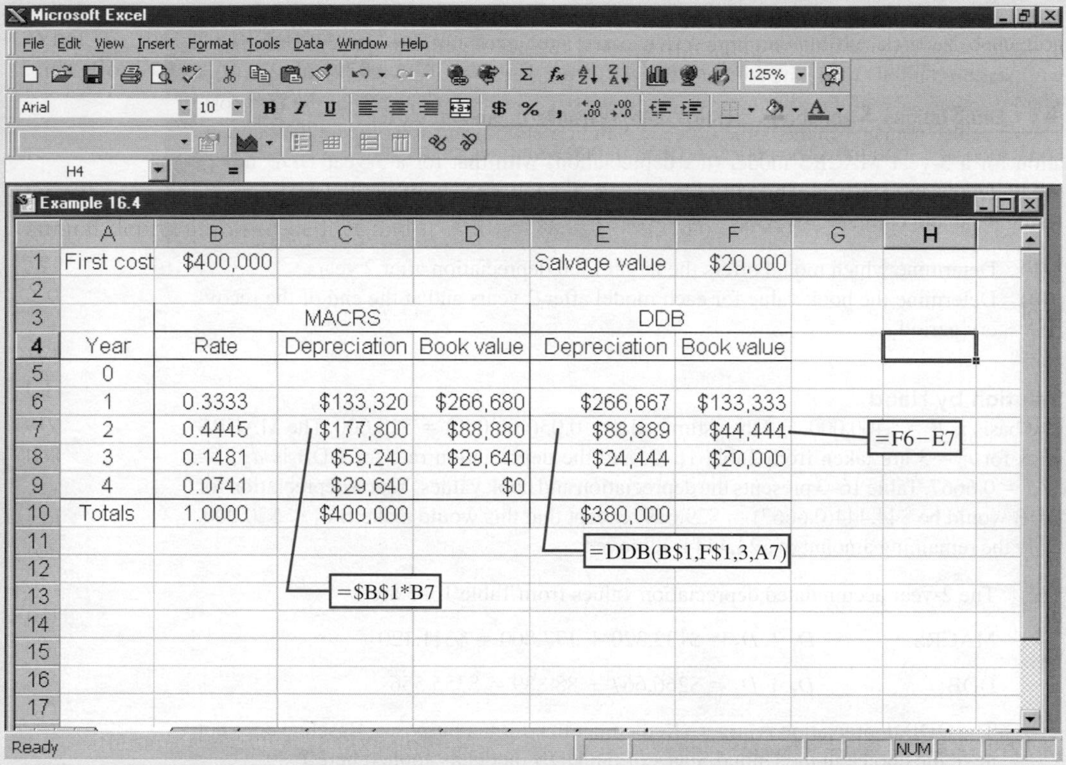

X Microsoft Excel							_	8	X

File Edit View Insert Format Tools Data Window Help

Arial ▼ 10 ▼ B I U ≡ ≡ ≡ ⊞ $ % , ⁺.⁰.⁰⁰ ⁺.⁰⁰.⁰ ⊞ ▼ ◇ ▼ A ▼

H4 ▼ =

Example 16.4

	A	B	C	D	E	F	G	H
1	First cost	$400,000			Salvage value	$20,000		
2								
3			MACRS		DDB			
4	Year	Rate	Depreciation	Book value	Depreciation	Book value		
5	0							
6	1	0.3333	$133,320	$266,680	$266,667	$133,333		
7	2	0.4445	$177,800	$88,880	$88,889	$44,444		=F6−E7
8	3	0.1481	$59,240	$29,640	$24,444	$20,000		
9	4	0.0741	$29,640	$0				
10	Totals	1.0000	$400,000		$380,000			
11					=DDB(B$1,F$1,3,A7)			
12								
13			=$B$1*B7					
14								
15								
16								
17								

Ready   NUM

Figure 16–4
Comparison of MACRS and DDB depreciation methods using a spreadsheet, Example 16.4.

Solution by Computer

Figure 16–4 presents the spreadsheet solution using the DDB function, in column E, the MACRS rates in column B, and Equation [16.12] in column C.

E-Solve

(a) The 2-year accumulated depreciation values are

MACRS, add cells C6 + C7: $133,320 + 177,800 = $311,120

DDB, add cells E6 + E7: $266,667 + 88,889 = $355,556

(b) Book values after 2 years are

MACRS, cell D7: $88,880

DDB, cell F7: $44,444

The book values at the end of the recovery periods are in cells D9 and F8.

Comment

It is advised to set up a *spreadsheet template* for use with depreciation problems in this and future chapters. The format and functions of Figure 16–4 are a good template for MACRS and DDB methods.

MACRS simplifies depreciation computations, but it removes the flexibility of model selection for a business or corporation. In general, an economic comparison that includes depreciation may be performed more rapidly and usually without altering the final decision by applying the classical straight-line model, in lieu of MACRS, to the cash flow estimates.

16.5   DETERMINING THE MACRS RECOVERY PERIOD

The expected useful life of property is estimated in years and used as the n value in alternative evaluation and in depreciation computations. For book depreciation the n value should be the expected useful life. However, when the depreciation will be claimed as tax deductible, the n value should be lower. The advantage of a recovery period shorter than the anticipated useful life is leveraged by the accelerated depreciation models which write off more of the basis B in the initial years. There are tables which assist in determining the life and recovery period for tax purposes.

The U.S. government requires that all depreciable property be classified into a *property class* which identifies its MACRS-allowed recovery period. Table 16–4, a summary of material from IRS Publication 946, gives examples of assets and the MACRS n values. Virtually any property considered in an economic analysis has a MACRS n value of 3, 5, 7, 10, 15, or 20 years.

Table 16–4 provides two MACRS n values for each property. The first is the *general depreciation system (GDS)* value, which we use in examples and problems. The depreciation rates in Table 16–2 correspond to the n values for the GDS column and provide the fastest write-off allowed. The rates utilize the DDB method or the 150% DB method with a switch to SL depreciation. Note that any asset not in a stated class is automatically assigned a 7-year recovery period under GDS.

The far right column of Table 16–4 lists the *alternative depreciation system* (*ADS*) recovery period range. This alternative method allows the use of *SL depreciation over a longer recovery period* than the GDS. The half-year convention applies, and any salvage value is neglected, as it is in regular MACRS. The use of ADS is generally a choice left to a company, but it is required for some special asset situations. Since it takes longer to depreciate the asset, and since the SL model is required (thus removing the advantage of accelerated depreciation), ADS is usually not considered an option for the economic analysis. This electable SL option is, however, sometimes chosen by businesses that are young and do not need the tax benefit of accelerated depreciation during the first years of operation and asset ownership. If ADS is selected, tables of d_t rates are available.

16.6   DEPLETION METHODS

Up to this point, we have discussed depreciation for assets that can be replaced. Depletion, though similar to depreciation, is applicable only to natural resources. When the resources are removed, they cannot be replaced or repurchased in the same manner as can a machine, computer, or structure. Depletion is applicable to natural deposits removed from mines, wells, quarries, geothermal deposits, forests, and the like. There are two methods of depletion—*cost depletion* and

TABLE **16–4**   Example MACRS Recovery Periods for Various Asset Descriptions

Asset Description (Personal and Real Property)	MACRS n Value, Years	
	GDS	ADS Range
Special manufacturing and handling devices, tractors, racehorses	3	3–5
Computers and peripherals, oil and gas drilling equipment, construction assets, autos, trucks, buses, cargo containers, some manufacturing equipment	5	6–9.5
Office furniture; some manufacturing equipment; railroad cars, engines, tracks; agricultural machinery; petroleum and natural gas equipment; *all property not in another class*	7	10–15
Equipment for water transportation, petroleum refining, agriculture product processing, durable-goods manufacturing, ship building	10	15–19
Land improvements, docks, roads, drainage, bridges, landscaping, pipelines, nuclear power production equipment, telephone distribution	15	20–24
Municipal sewers, farm buildings, telephone switching buildings, power production equipment (steam and hydraulic), water utilities	20	25–50
Residential rental property (house, mobile home)	27.5	40
Nonresidential real property attached to the land, but not the land itself	39	40

percentage depletion. Details for U.S. taxes are provided in IRS Publication 535, *Business Expenses.*

Cost depletion, sometimes referred to as factor depletion, is based on the level of activity or usage, not time, as in depreciation. It may be applied to most types of natural resources. The cost depletion factor for year t, denoted by p_t, is the ratio of the first cost of the resource to the estimated number of units recoverable.

$$p_t = \frac{\text{first cost}}{\text{resource capacity}} \qquad [16.15]$$

The annual depletion charge is p_t times the year's usage or volume. *The total cost depletion cannot exceed the first cost of the resource.* If the capacity of the property is reestimated some year in the future, a new cost depletion factor is determined based upon the undepleted amount and the new capacity estimate.

EXAMPLE 16.5

Temple-Inland Corporation has negotiated the rights to cut timber on privately held forest acreage for $700,000. An estimated 350 million board feet of lumber are harvestable.

(a) Determine the depletion amount for the first 2 years if 15 million and 22 million board feet are removed.

(b) After 2 years the total recoverable board feet was reestimated to be 450 million from the time the rights were purchased. Compute the new cost depletion factor for years 3 and later.

Solution

(a) Use Equation [16.15] for p_t in dollars per million board feet.

$$p_t = \frac{\$700{,}000}{350} = \$2000 \text{ per million board feet}$$

Multiply p_t by the annual harvest to obtain depletion of $30,000 in year 1 and $44,000 in year 2. Continue using p_t until a total of $700,000 is written off.

(b) After 2 years, a total of $74,000 has been depleted. A new p_t value must be calculated based on the remaining $700,000 − 74,000 = $626,000 investment. Additionally, with the new estimate of 450 million board feet, a total of 450 − 15 − 22 = 413 million board feet remain. For years $t = 3, 4, \ldots$, the cost depletion factor is

$$p_t = \frac{\$626{,}000}{413} = \$1516 \text{ per million board feet}$$

Percentage depletion, the second depletion method, is a special consideration given for natural resources. A constant, stated percentage of the resource's gross income may be depleted each year *provided it does not exceed 50% of the company's taxable income.* For oil and gas property, the limit is 100% of taxable income. The annual depletion amount is calculated as

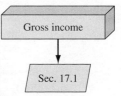

Gross income

Sec. 17.1

Percentage depletion amount = percentage
$\times$ **gross income from property**   **[16.16]**

Using percentage depletion, total depletion charges may exceed first cost with no limitation. The U.S. government does not generally allow percentage depletion to be applied to oil and gas wells (except small independent producers) or timber.

The depletion amount each year may be determined using either the cost method or the percentage method, as allowed by law. Usually, the percentage depletion amount is chosen because of the possibility of writing off more than the original cost. However, the law also requires that the cost depletion amount be chosen if the percentage depletion is smaller in any year. The guideline is:

Calculate both amounts—cost depletion ($Depl) and percentage depletion (%Depl)—and apply the following logic each year.

$$\text{Annual depletion} = \begin{cases} \%\text{Depl} & \text{if } \%\text{Depl} \geq \$\text{Depl} \\ \$\text{Depl} & \text{if } \%\text{Depl} < \$\text{Depl} \end{cases} \quad [16.17]$$

The annual percentage depletion for some common natural deposits is listed below for U.S. tax law. These percentages may change from time to time.

Deposit	Percentage
Sulfur, uranium, lead, nickel, zinc, and some other ores and minerals	22%
Gold, silver, copper, iron ore, and some oil shale	15
Oil and natural gas wells (varies)	15–22
Coal, lignite, sodium chloride	10
Gravel, sand, peat, some stones	5
Most other minerals, metallic ores	14

EXAMPLE **16.6**

A gold mine was purchased for $10 million. It has an anticipated gross income of $5.0 million per year for years 1 to 5 and $3.0 million per year after year 5. Assume that depletion charges do not exceed 50% of taxable income. Compute annual depletion amounts for the mine. How long will it take to recover the initial investment at $i = 0\%$?

Solution

A 15% depletion applies to gold. Depletion amounts are

Years 1 to 5:   0.15(5.0 million) = $750,000

Years thereafter: 0.15(3.0 million) = $450,000

A total of $3.75 million is written off in 5 years, and the remaining $6.25 million is written off at $450,000 per year. The total number of years is

$$5 + \frac{\$6.25 \text{ million}}{\$450,000} = 5 + 13.9 = 18.9$$

In 19 years, the initial investment could be fully depleted.

CHAPTER SUMMARY

Depreciation may be determined for internal company records (book depreciation) or for income tax purposes (tax depreciation). In the U.S., the MACRS method is the only one allowed for tax depreciation. In many other countries, straight line and declining balance methods are applied for both tax and book depreciation. Depreciation does not result in actual cash flow directly. It is a book method by which the capital investment in tangible property is recovered. The annual depreciation amount is tax deductible, which can result in actual cash flow changes.

Some important points about the straight line, declining balance, and MACRS models are presented below. Common relations for each method are summarized in Table 16–5.

Straight Line (SL)
- It writes off capital investment linearly over n years.
- The estimated salvage value is always considered.
- This is the classical, nonaccelerated depreciation model.

Declining Balance (DB)
- The model accelerates depreciation compared to straight line.
- The book value is reduced each year by a fixed percentage.
- The most used rate is twice the SL rate, which is called double declining balance (DDB).
- It has an implied salvage that may be lower than the estimated salvage.
- It is not an approved tax depreciation method in the United States. It is frequently used for book depreciation purposes.

Modified Accelerated Cost Recovery System (MACRS)
- It is the only approved tax depreciation system in the United States.
- It automatically switches from DDB or DB to SL depreciation.
- It always depreciates to zero; that is, it assumes $S = 0$.
- Recovery periods are specified by property classes.
- Depreciation rates are tabulated.
- The actual recovery period is 1 year longer due to the imposed half-year convention.
- MACRS straight line depreciation is an option, but recovery periods are longer than for regular MACRS.

Cost and *percentage depletion methods* recover investment in natural resources. The annual cost depletion factor is applied to the amount of resource removed. No more than the initial investment can be recovered with cost depletion. Percentage depletion, which can recover more than the initial investment, reduces the investment value by a constant percentage of gross income each year.

TABLE 16–5 Summary of Common Depreciation Method Relations

Model	MACRS	SL	DDB
Fixed depreciation rate d	Not defined	$\dfrac{1}{n}$	$\dfrac{2}{n}$
Annual rate d_t	Table 16–2	$\dfrac{1}{n}$	$d(1 - d)^{t-1}$
Annual depreciation D_t	$d_t B$	$\dfrac{B - S}{n}$	$d(\mathrm{BV}_{t-1})$
Book value BV_t	$\mathrm{BV}_{t-1} - D_t$	$B - tD_t$	$B(1 - d)^t$

PROBLEMS

Fundamentals of Depreciation

16.1 Write another term that may be used in lieu of each of the following that has the same interpretation in asset depreciation: *depreciation rate, fair market value, recovery period,* and *tangible property.*

16.2 State the difference between book depreciation and tax depreciation.

16.3 Explain why the recovery period used for tax depreciation purposes may be different from the estimated n value in an engineering economy study.

16.4 Explain why in the United States the explicit consideration of depreciation and income taxes in an engineering economy study may make a difference in the decision to accept or reject an alternative to acquire a depreciable asset.

16.5 Visit the U.S. Internal Revenue Service website at www.irs.gov, and answer the following questions about depreciation and MACRS by consulting online the current version of Publication 946, *How to Depreciate Property.*
  (*a*) What is the definition of depreciation according to the IRS?
  (*b*) What is the description of the term *salvage value?*
  (*c*) What are the two depreciation systems within MACRS, and what are the major differences between them?
  (*d*) What are the properties listed that cannot be depreciated under MACRS?

16.6 Visit the Canada Customs and Revenue Agency website at ccra-adrc.gc.ca/tax and answer the following questions about depreciation in Canada, using the glossary under the taxes—businesses section.
  (*a*) What is the definition of depreciation?
  (*b*) What is capital cost allowance (CCA) and what is the closest equivalent term in the U.S. system of depreciation?
  (*c*) What is the complete definition of real property?

16.7 Status Corporation purchased for $350,000 a new numerical controller during the last month of 2002. Extra installation costs were $40,000. The recovery period was 7 years with an estimated salvage value of 10% of the original purchase price. Status sold the system at the end of 2005 for $45,000.
  (*a*) What are the values needed to develop a depreciation schedule at purchase time?
  (*b*) State the numerical values for the following: remaining life at sale time, market value in 2005, book value at sale time if 65% of the unadjusted basis had been depreciated.

16.8 A $100,000 piece of testing equipment was installed and depreciated for 5 years. Each year the end-of-year book value decreased at a rate of 10% of the book value at the beginning of the year. The system was sold for $24,000 at the end of 5 years.
  (*a*) Compute the amount of the annual depreciation.
  (*b*) What is the actual depreciation rate for each year?
  (*c*) At the time of sale, what is the difference between the book value and the market value?
  (*d*) Plot the book value for each of the 5 years.

16.9 An asset with an unadjusted basis of $50,000 was depreciated over $n_{tax} = 10$ years for tax depreciation purposes and over $n_{book} = 5$ years for book depreciation purposes. The annual depreciation was $1/n$ using the relevant life value. Use Excel to plot on one graph the annual book value for both methods of depreciation.

Straight Line Depreciation

16.10 Home Health Care, Inc. (HHCI) purchased a new sonargram imaging unit for $300,000 and had it mounted on a truck body for an additional $100,000, including the truck chassis. The unit-truck system will be depreciated as one asset. The functional life is 8 years, and salvage is estimated at 10% of the purchase price of the imaging unit. (*a*) Use classical straight line depreciation and hand calculations to determine the salvage value, annual depreciation, and book value after 4 years. (*b*) Develop the cell reference worksheet in Excel to obtain the answers in part (*a*) for the original data. (*c*) Use your Excel worksheet to determine the answers, if the sonargram unit cost goes up to $350,000 and the expected life is decreased to 5 years.

16.11 Air handling equipment that costs $12,000 has a life of 8 years with a $2000 salvage value. (*a*) Calculate the straight line depreciation amount for each year. (*b*) Determine the book value after 3 years. (*c*) What is the rate of depreciation?

16.12 As asset has an unadjusted basis of $200,000, a salvage value of $10,000, and a recovery period of 7 years. Write a single-cell Excel function to display the book value after 5 years of straight line depreciation.

16.13 Simpson and Jones Pharmaceuticals purchased a prescription drug tablet-forming machine in 2004 for $750,000. They had planned to use the machine for 10 years, but due to rapid obsolescence it should be retired after 4 years. Develop the spreadsheet for depreciation and book value amounts necessary to answer the following.

 (*a*) What is the amount of capital investment remaining when the asset is retired due to obsolescence?

 (*b*) If the asset is sold at the end of 4 years for $75,000, what is the amount of capital investment lost based on straight line depreciation?

 (*c*) If the new technology machine has an estimated cost of $300,000, how many more years should the company retain and depreciate the currently owned machine to make its book value and the first cost of the new machine equal?

16.14 A special-purpose computer workstation has $B = \$50,000$ with a 4-year recovery period. Tabulate and plot the values for SL depreciation, accumulated depreciation, and book value for each year if (*a*) there is no salvage value and (*b*) $S = \$16,000$. (*c*) Use a spreadsheet to solve this problem.

16.15 A company owns the same asset in a U.S. plant and in a EU plant. It has $B = \$2,000,000$ with a salvage value of 20% of B. For tax depreciation purposes, the United States allows a straight line write-off over 5 years, while the EU allows write-off over 8 years. The general managers of the two plants want to know the difference in (*a*) the depreciation amount for year 5 and (*b*) the book value after 5 years. Use Excel and write cell equations in *only* two cells to answer these questions.

Declining Balance Depreciation

16.16 For the declining balance method of depreciation, explain the differences between the three rates: fixed percentage rate d, d_{max}, and the annual recovery rate d_t.

16.17 New equipment to read 96-bit product codes that are replacing old bar codes is to be purchased by General Food Stores. As a trial, 1000 of the items will be initially purchased. The DDB method will be used to depreciate the total amount of $50,000 over a 3-year recovery period. Calculate and plot the accumulated depreciation and book value curves (a) by hand and (b) by computer.

16.18 An asset has a first cost of $12,000, a recovery period of 8 years, and an estimated salvage value of $2000. (a) Use a spreadsheet to develop the depreciation schedule for both the SL and DDB methods. Plot the book value for SL and DDB depreciation on a single xy scatter chart. (b) Calculate the DDB annual depreciation rate for each of the years 1 through 8.

16.19 A warehouse costs $800,000 to construct for Ace Hardware. It has a 15-year life with an estimated resale value of 80% of the construction cost. However, the building will be depreciated to zero over a recovery period of 30 years. Calculate the annual depreciation charge for years 5, 10, and 25, using (a) straight line depreciation and (b) DDB depreciation. (c) What is the implied salvage value for DDB?

16.20 Allison and Carl are civil engineers who own a soil and water analysis business for which they have purchased computer equipment for $25,000. They do not expect the computers to have a positive salvage or trade-in value after the anticipated 5-year life. For book depreciation purposes, they want book value schedules for the following methods: SL, DB, and DDB. They want to use a fixed depreciation rate of 25% annually for the

DB model. Use a spreadsheet or hand computation to develop the schedules.

16.21 Equipment for immersion cooling of electronic components has an installed value of $182,000 with an estimated trade-in value of $50,000 after 18 years. (a) For years 2 and 18, determine the annual depreciation charge using DDB depreciation and DB depreciation by hand. (b) Use a spreadsheet to answer the questions above and to determine the year in which the estimated salvage value of $50,000 is reached for DDB depreciation.

16.22 Use the estimates $B = \$182,000$, $S = \$50,000$, and $n = 18$ years (from Problem 16.21) to write the Excel DDB function to determine the depreciation in year 18, using the implied depreciation rate.

16.23 For book depreciation purposes, declining balance depreciation at a rate of 1.5 times the straight line rate is used for automated process control equipment with $B = \$175,000$, $n = 12$, and $S = \$32,000$. (a) Compute the depreciation and book value for years 1 and 12. (b) Compare the estimated salvage value and the book value after 12 years. (c) Write the DDB function for Excel to calculate the depreciation each year.

MACRS Depreciation

16.24 (a) Develop a spreadsheet template to calculate the MACRS depreciation for any value of B and all allowed recovery periods.
(b) Test your template by determining the depreciation schedule for $B = \$10,000$ for $n = 3$ and $n = 10$ years.

16.25 Give at least two specific examples each of personal property and real property that must be depreciated by the MACRS method.

16.26 Zahra is a civil engineer working for Halcrow Engineering Consultants in the Middle East where the straight line method and (in some countries) the declining balance method of depreciation are utilized. She is about to conduct an after-tax analysis that involves a $500,000, 10-year-life piece of equipment where accelerated write-off of the capital investment is very important. The estimated salvage value is $100,000. To show her the effect of different methods, calculate the first-year depreciation for the following methods: classical SL, DDB, 150% DB, and MACRS.

16.27 Claude is an engineering economist with Reynolds. A new $30,000 personal property asset is to be depreciated using MACRS over 7 years. The salvage value is expected to be $2000. (*a*) Compare the book values for MACRS depreciation and classical SL depreciation over 7 years. (*b*) Plot the book values, using a spreadsheet.

16.28 An asset for a U.S.-based commercial farm corporation was purchased for $50,000 and has a 7-year useful life and an expected resale value of 20% of the first cost. An abbreviated recovery period of 5 years is allowed by MACRS. (*a*) Prepare the MACRS annual depreciation schedule and book values. (*b*) Compare these results with DDB depreciation and book value over $n = 7$ years that may be determined in another country. (*c*) Develop two Excel *xy* scatter charts for the comparisons above.

16.29 An automated assembly robot that cost $450,000 installed has a depreciable life of 5 years and no salvage value. An analyst in the financial management department used classical SL depreciation to determine end-of-year book values for

the robot when the original economic evaluation was performed. You are now performing a replacement analysis after 3 years of service and realize that the robot should have been depreciated using MACRS with $n = 5$ years. What is the amount of the difference in the book value caused by the classical SL method after the 3 years?

16.30 Develop the MACRS depreciation schedule for a commercial building purchased by Alpha Enterprises for $1,800,000.

16.31 Bowlers.com has installed $100,000 worth of depreciable software and equipment that represents the latest in Internet teaming and bowling, intended to allow the bowler to enjoy the sport on the Web or at the alley. No salvage value is estimated. The company can depreciate using MACRS for a 5-year recovery period or opt for the ADS alternative system over 10 years using the straight line model. The SL rates require the half-year convention; that is, only 50% of the regular annual rate applies for years 1 and 11. (*a*) Construct the book value curves for both models on one graph. Answer by hand or by computer. (*b*) After 3 years of use, what is the percentage of the $100,000 basis removed for each model?

16.32 A company has purchased special-purpose equipment for the manufacture of rubber products (asset class 30.11 in IRS Publication 946) and expects to use it predominately outside the United States. In this case, the ADS alternative to MACRS is required for tax depreciation purposes. The manager wants to understand the difference in the yearly recovery rates for classical SL, MACRS, and the ADS alternative to MACRS. Using a recovery period of 3 years, except for the

ADS alternative which requires a 4-year SL recovery with the half-year convention included, prepare a single graph showing the annual recovery rates (in percent) for the three methods.

16.33 Explain why shortened recovery periods coupled with higher depreciation rates in the initial years of an asset's life may be financially advantageous to a corporation.

Depletion

16.34 When WTA Corporation purchased rights to extract silver from a mine for a total price of $1.1 million 3 years ago, an estimated 350,000 ounces of silver was to be removed over the next 10 years. A total of 175,000 ounces has been removed and sold thus far. (a) What is the total cost depletion allowed over the 3 years? (b) New exploratory tests indicate that only an estimated 100,000 ounces remain in the veins of the mine. What is the cost depletion factor applicable for the next year? (c) If an additional 35,000 ounces is extracted this year and sold at an average of $5.50 per ounce, determine if the cost depletion amount or percentage depletion amount is allowed for U.S. income tax purposes. Assume that 50% of the company's taxable income is not exceeded.

16.35 A company owns copper mining operations in several states. One mine has the taxable income and sales results

following. Determine the annual percentage depletion for the mine. Use 2000 pounds per ton.

Year	Taxable Income, $	Sales, Tons	Sales Price, $/Pound
1	1,500,000	2000	0.80
2	2,000,000	4500	0.78
3	1,000,000	2300	0.65

16.36 A highway construction company has operated a quarry for the past 5 years. During this time the following tonnage has been extracted each year: 60,000; 50,000; 58,000; 60,000; and 65,000 tons. The mine is estimated to contain a total of 2.5 million tons of usable stones and gravel. The quarry land had an initial cost of $3.2 million. The company had a per-ton gross income of $30 for the first year, $25 for the second year, $35 for the next 2 years, and $40 for the last year.

(a) Compute the depletion charges each year, using the larger of the values for the two depletion methods.

(b) Compute the percent of the initial cost that has been written off in these 5 years, using the depletion charges in part (a).

(c) Use a spreadsheet to answer parts (a) and (b).

(d) If the quarry operation is reevaluated after the first 3 years of operation and estimated to contain another 1.5 million tons, rework parts (a) and (b).

FE REVIEW PROBLEMS

16.37 A machine with a 5-year life has a first cost of $20,000 and a $2000 salvage value. Its annual operating cost is $8000 per year. According to the classical straight line method, the depreciation

charge in year 2 is nearest to
(a) $3600
(b) $4000
(c) $11,600
(d) $12,000

16.38 A machine with a useful life of 10 years is to be depreciated by the MACRS method over 7 years. The machine has a first cost of $35,000 with a $5000 salvage value. Its annual operating cost is $7000 per year. The depreciation charge in year 3 is nearest to
(*a*) $3600
(*b*) $4320
(*c*) $5860
(*d*) $6120

16.39 An asset with a first cost of $50,000 is to be depreciated by the straight line method over a 5-year period. The asset will have annual operating costs of $20,000 and a salvage value of $10,000. According to the straight line method, the book value at the end of year 3 will be closest to
(*a*) $8000
(*b*) $26,000
(*c*) $24,000
(*d*) $20,000

16.40 An asset with a first cost of $50,000 is depreciated by the MACRS method over a 5-year period. If the asset will have a $20,000 salvage value, its book value at the end of year 2 will be closest to
(*a*) $10,000
(*b*) $16,000
(*c*) $24,000
(*d*) $30,000

16.41 An asset with a first cost of $50,000 is depreciated by the straight line method over its 5-year life. Its annual operating cost is $20,000, and its salvage value is expected to be $10,000. The book value at the end of year 5 will be nearest to
(*a*) $0
(*b*) $8000
(*c*) $10,000
(*d*) $14,000

16.42 An asset had a first cost of $50,000, an estimated salvage value of $10,000 and was depreciated by the MACRS method. If its book value at the end of year 3 was $21,850 and its market value was $25,850, the amount of depreciation charged against the asset up to that time was closest to
(*a*) $18,850
(*b*) $21,850
(*c*) $25,850
(*d*) $28,150

16.43 The depreciation recovery rate used for comparison with any method's rate is most commonly
(*a*) Straight line rate
(*b*) MACRS rate
(*c*) $2/n$
(*d*) Sum-of-year digits rate

CHAPTER 16   APPENDIX

16A.1 SUM-OF-YEAR-DIGITS (SYD) DEPRECIATION

The SYD method is a classical accelerated depreciation technique that removes much of the basis in the first one-third of the recovery period; however, write-off is not as rapid as for DDB or MACRS. This technique may be used in an engineering economy analysis in the depreciation of multiple-asset accounts (group and composite depreciation).

The mechanics of the method involve the sum of the year's digits from 1 through the recovery period n. The depreciation charge for any given year is obtained by multiplying the basis of the asset, less any salvage value, by the ratio of the number of years remaining in the recovery period to the sum of the year's digits SUM.

$$D_t = \frac{\text{depreciable years remaining}}{\text{sum of year digits}} (\text{basis} - \text{salvage value})$$

$$D_t = \frac{n-t+1}{\text{SUM}}(B - S) \qquad\qquad \text{[16A.1]}$$

where SUM is the sum of the digits 1 through n.

$$\text{SUM} = \sum_{j=1}^{j=n} j = \frac{n(n+1)}{2} \qquad\qquad \text{[16A.2]}$$

the book value for any year t is calculated as

$$\text{BV}_t = B - \frac{t(n-t/2+0.5)}{\text{SUM}}(B - S) \qquad\qquad \text{[16A.3]}$$

The rate of depreciation decreases each year and equals the multiplier in Equation [16A.1].

$$d_t = \frac{n-t+1}{\text{SUM}} \qquad\qquad \text{[16A.4]}$$

The SYD spreadsheet function displays the depreciation for the year t. The function format is

$$\text{SYD}(B,S,n,t)$$

EXAMPLE 16A.1

Calculate the SYD depreciation charges for year 2 for electro-optics equipment with $B = \$25,000$, $S = \$4000$, and an 8-year recovery period.

Solution
The sum of the year's digits is 36, and the depreciation amount for the second year by Equation [16A.1] is

$$D_2 = \frac{7}{36}(21,000) = \$4083$$

The SYD function is SYD(25000,4000,8,2).

Q-Solv

Figure 16A–1 is a plot of the book values for an $80,000 asset with $S = \$10,000$ and $n = 10$ years using four depreciation methods that we have learned. The MACRS, DDB, and SYD curves track closely except for year 1 and years 9 through 11. A spreadsheet and xy scatter chart can confirm the results of Figure 16A–1.

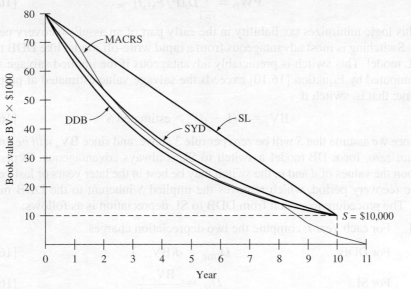

Figure 16A–1
Comparison of book values using SL, SYD, DDB, and MACRS depreciation.

16A.2 SWITCHING BETWEEN DEPRECIATION METHODS

Switching between depreciation models may assist in accelerated reduction of the book value. It also maximizes the present value of accumulated and total depreciation over the recovery period. Therefore, switching usually increases the tax advantage in years where the depreciation is larger. The approach below is an inherent part of MACRS.

Switching from a DB model to the SL method is the most common switch because it usually offers a real advantage, especially if the DB model is the DDB. General rules of switching are summarized here.

1. Switching is recommended when the depreciation for year t by the currently used model is less than that for a new model. The selected depreciation D_t is the larger amount.
2. Only one switch can take place during the recovery period.
3. Regardless of the (classical) depreciation models, the book value cannot go below the estimated salvage value. When switching from a DB model, the estimated salvage value, not the DB-implied salvage value, is used to compute the depreciation for the new method: we assume $S = 0$ in all cases. (This does not apply to MACRS, since it already includes switching.)
4. The undepreciated amount, that is, BV_t, is used as the new adjusted basis to select the larger D_t for the next switching decision.

In all situations, the criterion is to *maximize the present worth of the total depreciation* PW_D. The combination of depreciation models that results in the largest present worth is the best switching strategy.

$$PW_D = \sum_{t=1}^{t=n} D_t(P/F,i,t) \qquad [16A.5]$$

This logic minimizes tax liability in the early part of an asset's recovery period.

Switching is most advantageous from a rapid write-off model like DDB to the SL model. This switch is predictably advantageous if the implied salvage value computed by Equation [16.10] exceeds the salvage value estimated at purchase time; that is, switch if

$$BV_n = B(1 - d)^n > \text{estimated } S \qquad [16A.6]$$

Since we assume that S will be zero per rule 3 above, and since BV_n *will be greater than zero,* for a DB model a switch to SL is always advantageous. Depending upon the values of d and n, the switch may be best in the later years or last year of the recovery period, which removes the implied S inherent to the DDB model.

The procedure to switch from DDB to SL depreciation is as follows:

1. For each year t, compute the two depreciation charges.

    For DDB: $$D_{\text{DDB}} = d(BV_{t-1}) \qquad [16A.7]$$

    For SL: $$D_{\text{SL}} = \frac{BV_{t-1}}{n-t+1} \qquad [16A.8]$$

2. Select the larger depreciation value. The depreciation for each year is

    $$D_t = \max[D_{\text{DDB}}, D_{\text{SL}}] \qquad [16A.9]$$

3. If needed, compute the present worth of total depreciation, using Equation [16A.5].

It is acceptable, though not usually financially advantageous, to state that a switch will take place in a particular year, for example, a mandated switch from DDB to SL in year 7 of a 10-year recovery period. This approach is usually not taken, but the switching technique will work correctly for all depreciation models.

To use a spreadsheet for switching, first understand the depreciation model switching rules and practice the switching procedure from declining balance to straight line. Once these are understood, the mechanics of the switching can be speeded up by applying the spreadsheet function VDB (variable declining balance). This is a quite powerful function that determines the depreciation for 1 year or the total over several years for the DB-to-SL switch. The function format is

VDB(B,S,n,start_t,end_t,d,no_switch)

Appendix A explains all the fields in detail, but for simple applications, where the DDB and SL annual D_t values are needed, the following are correct entries:

    start_t is the year (t-1)

    end_t is year t

    d is optional; 2 for DDB is assumed, the same as in the DDB function

Income tax basics

Chap. 17

E-Solve

VDB function

App. A

no_switch is an optional logical value:

> FALSE or omitted—switch to SL occurs, if advantageous
>
> TRUE—DDB or DB model is applied with no switching to SL depreci-
> ation considered.

Entering TRUE for the no_switch option obviously causes the VDB function to display the same depreciation amounts as the DDB function. This is discussed in Example 16A.2(*d*).

EXAMPLE 16A.2

The Outback Steakhouse main office has purchased a $100,000 online document imaging system with an estimated useful life of 8 years and a tax depreciation recovery period of 5 years. Compare the present worth of total depreciation for (*a*) the SL method, (*b*) the DDB method, and (*c*) DDB-to-SL switching. (*d*) Perform the DDB-to-SL switch using a computer and plot the book values. Use a rate of $i = 15\%$ per year.

Solution by Hand
The MACRS method is not involved in this solution.

(*a*) Equation [16.1] determines the annual SL depreciation.

$$D_t = \frac{100,000 - 0}{5} = \$20,000$$

Since D_t is the same for all years, the P/A factor replaces P/F to compute PW_D.

$$\text{PW}_D = 20,000(P/A,15\%,5) = 20,000(3.3522) = \$67,044$$

(*b*) For DDB, $d = 2/5 = 0.40$. The results are shown in Table 16A–1. The value $\text{PW}_D = \$69,915$ exceeds the $67,044 for SL depreciation. As is predictable, the accelerated depreciation of DDB increases PW_D.

(*c*) Use the DDB-to-SL switching procedure.

1. The DDB values for D_t in Table 16A–1 are repeated in Table 16A–2 for compar-
    ison with the D_{SL} values from Equation [16A.8]. The D_{SL} values change each
    year because BV_{t-1} is different. Only in year 1 is $D_{\text{SL}} = \$20,000$, the same as
    computed in part (*a*). For illustration, compute D_{SL} values for years 2 and 4. For
    $t = 2, \text{BV}_1 = \$60,000$ by the DDB method and

    $$D_{\text{SL}} = \frac{60,000 - 0}{5-2+1} = \$15,000$$

    For $t = 4, \text{BV}_3 = \$21,600$ by the DDB method and

    $$D_{\text{SL}} = \frac{21,600 - 0}{5-4+1} = \$10,800$$

2. The column "Larger D_t" indicates a switch in year 4 with $D_4 = \$10,800$. The
    $D_{\text{SL}} = \$12,960$ in year 5 would apply only if the switch occurred in year 5.
    Total depreciation with switching is $100,000 compared to the DDB amount of
    $92,224.

3. With switching, $\text{PW}_D = \$73,943$, which is an increase over both the SL and
    DDB models.

TABLE 16A–1   DDB Model Depreciation and Present Worth Computations, Example 16A.2b

Year t	D_t	BV_t	(P/F,15%,t)	Present Worth of D_t
0		$100,000		
1	$40,000	60,000	0.8696	$34,784
2	24,000	36,000	0.7561	18,146
3	14,400	21,600	0.6575	9,468
4	8,640	12,960	0.5718	4,940
5	5,184	7,776	0.4972	2,577
Totals	$92,224			$69,915

TABLE 16A–2   Depreciation and Present Worth for DDB-to-SL Switching, Example 16A.2c

| Year t | DDB Model | | SL Model | Larger | P/F | Present Worth of |
	D_{DDB}	BV_t	D_{SL}	D_t	Factor	D_t
0	—	$100,000				
1	$40,000	60,000	$20,000	$ 40,000	0.8696	$34,784
2	24,000	36,000	15,000	24,000	0.7561	18,146
3	14,400	21,600	12,000	14,400	0.6575	9,468
4*	8,640	12,960	10,800	10,800	0.5718	6,175
5	5,184	7,776	12,960	10,800	0.4972	5,370
Totals	$92,224			$100,000		$73,943

*Indicates year of switch from DDB to SL depreciation.

Solution by Computer

E-Solve

(d) Figure 16A–2, column D, entries are the VDB functions to determine that the DDB-to-SL switch should take place in year 4. The entries "2,FALSE" at the end of the VDB function are optional (see the VDB function description). If TRUE were entered, the declining balance model is maintained throughout the recovery period, and the annual depreciation amounts are equal to those in column B. The plot in Figure 16A–2 indicates another difference in depreciation methods. The terminal book value in year 5 for the DDB model is $BV_5 = 7776, while the DDB-to-SL switch reduces the book value to zero.

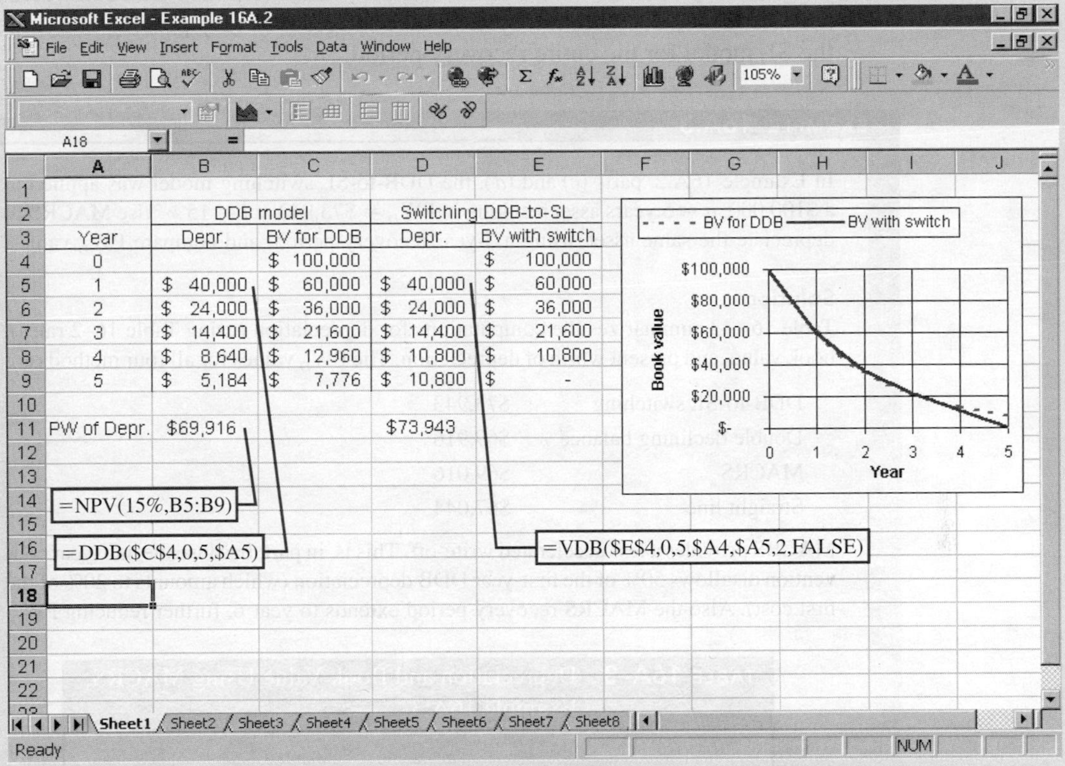

	A	B	C	D	E	F	G H I J
1							
2		DDB model		Switching DDB-to-SL			
3	Year	Depr.	BV for DDB	Depr.	BV with switch		
4	0		$ 100,000		$ 100,000		
5	1	$ 40,000	$ 60,000	$ 40,000	$ 60,000		
6	2	$ 24,000	$ 36,000	$ 24,000	$ 36,000		
7	3	$ 14,400	$ 21,600	$ 14,400	$ 21,600		
8	4	$ 8,640	$ 12,960	$ 10,800	$ 10,800		
9	5	$ 5,184	$ 7,776	$ 10,800	$ -		
10							
11	PW of Depr.	$69,916		$73,943			

=NPV(15%,B5:B9)

=DDB($C$4,0,5,$A5)

=VDB($E$4,0,5,$A4,$A5,2,FALSE)

Figure 16A–2
DDB-to-SL switch between depreciation models using Excel's VDB function, Example 16A.2*d*.

The NPV functions in row 11 determine the PW of depreciation. The results here are the same as in parts (*b*) and (*c*) above. The DDB-to-SL switch has the larger PW_D value.

In MACRS, recovery periods of 3, 5, 7, and 10 years apply DDB depreciation with half-year convention switching to SL. When the switch to SL takes place, which is usually in the last 1 to 3 years of the recovery period, any remaining basis is charged off in year $n + 1$ so that the book value reaches zero. Usually 50% of the applicable SL amount remains after the switch has occurred. For recovery periods of 15 and 20 years, 150% DB with the half-year convention and the switch to SL apply.

The present worth of depreciation PW_D will always indicate which model is the most advantageous. Only the MACRS rates for the GDS recovery periods

(Table 16–4) utilize the DDB-to-SL switch. The alternative MACRS rates for the alternative depreciation system have longer recovery periods and impose the SL model for the entire recovery period.

EXAMPLE 16A.3

In Example 16A.2, parts (c) and (d), the DDB-to-SL switching model was applied to a $100,000, $n = 5$ years asset resulting in $PW_D = \$73,943$ at $i = 15\%$. Use MACRS to depreciate the same asset, using a 5-year recovery period, and compare PW_D values.

Solution
Table 16A–3 summarizes the computations for depreciation (using Table 16–2 rates), book value, and present worth of depreciation. The PW_D values for all four methods are

DDB-to-SL switching	$73,943
Double declining balance	$69,916
MACRS	$69,016
Straight line	$67,044

MACRS provides a less accelerated write-off. This is, in part, because the half-year convention disallows 50% of the first-year DDB depreciation (which amounts to 20% of the first cost). Also the MACRS recovery period extends to year 6, further reducing PW_D.

TABLE 16A–3 Depreciation and Book Value Using MACRS, Example 16A.3

t	d_t	D_t	BV_t
0	—	—	$100,000
1	0.20	$ 20,000	80,000
2	0.32	32,000	48,000
3	0.192	19,200	28,800
4	0.1152	11,520	17,280
5	1.1152	11,520	5,760
6	0.0576	5,760	0
	1.000	$100,000	

$$PW_D = \sum_{t=1}^{t=6} D_t(P/F,15\%,t) = \$69,016$$

16A.3 DETERMINATION OF MACRS RATES

The depreciation rates for MACRS incorporate the DB-to-SL switching for all GDS recovery periods from 3 to 20 years. In the first year, some adjustments have been made to compute the MACRS rate. The adjustments vary and are not usually considered in detail in economic analyses. The half-year convention is always imposed, and any remaining book value in year n is removed in year $n + 1$. The value $S = 0$ is assumed for all MACRS schedules.

Since different DB depreciation rates apply for different n values, the following summary may be used to determine D_t and BV_t values. The symbols D_{DB} and D_{SL} are used to identify DB and SL depreciation, respectively.

For n = 3, 5, 7, and 10   Use DDB depreciation with the half-year convention, switching to SL depreciation in year t when $D_{\text{SL}} \geq D_{\text{DB}}$. Use the switching rules of Section 16A.2, and add one-half year when computing D_{SL} to account for the half-year convention. The yearly depreciation rates are

$$d_t = \begin{cases} \dfrac{1}{n} & t = 1 \\ \dfrac{2}{n} & t = 2, 3, \ldots \end{cases} \qquad [16A.10]$$

Annual depreciation values for each year t applied to the adjusted basis, allowing for the half-year convention, are

$$D_{\text{DB}} = d_t(\text{BV}_{t-1}) \qquad [16A.11]$$

$$D_{\text{SL}} = \begin{cases} \dfrac{1}{2}\left(\dfrac{1}{n}\right)B & t = 1 \\ \dfrac{\text{BV}_{t-1}}{n - t + 1.5} & t = 2, 3, \ldots, n \end{cases} \qquad [16A.12]$$

After the switch to SL depreciation takes place—usually in the last 1 to 3 years of the recovery period—any remaining book value in year n is removed in year $n + 1$.

For n = 15 and 20   Use 150% DB with the half-year convention and the switch to SL when $D_{\text{SL}} \geq D_{\text{DB}}$. Until SL depreciation is more advantageous, the annual DB depreciation is computed using the form of Equation [16A.7]

$$D_{\text{DB}} = d_t(\text{BV}_{t-1})$$

where

$$d_t = \begin{cases} \dfrac{0.75}{n} & t = 1 \\ \dfrac{1.50}{n} & t = 2, 3, \ldots \end{cases} \qquad [16A.13]$$

EXAMPLE **16A.4**

A wireless tracking system for shop floor control with a MACRS 5-year recovery period has been purchased for $10,000. (*a*) Use Equations [16A.10] through [16A.12] to obtain the annual depreciation and book value. (*b*) Determine the resulting annual depreciation rates and compare them with the MACRS rates in Table 16–2 for $n = 5$.

Solution

(*a*) With $n = 5$ and the half-year convention, use the DDB-to-SL switching procedure to obtain the results in Table 16A–4. The switch to SL depreciation, which occurs in year 4 when both depreciation values are equal, is indicated by

$$D_{DB} = 0.4(2880) = \$1152$$

$$D_{SL} = \frac{2880}{5 - 4 + 1.5} = \$1152$$

The SL depreciation of $1000 in year 1 results from applying the half-year convention included in the first relation of Equation [16A.12]. Also, the SL depreciation of $576 in year 6 is the result of the half-year convention.

(*b*) The actual rates are computed by dividing the "Larger D_t" column values by the first cost of $10,000. The rates below are the same as the Table 16–2 rates.

t	1	2	3	4	5	6
d_t	0.20	0.32	0.192	0.1152	0.1152	0.0576

TABLE 16A–4 Depreciation Amounts Used to Determine MACRS Rates for $n = 5$, Example 16A.4

Years	DDB		SL Depreciation	Larger	
t	d_t	D_{DB}	D_{SL}	D_t	BV_t
0	—	—	—	—	$10,000
1	0.2	$2,000	$1,000	$2,000	8,000
2	0.4	3,200	1,777	3,200	4,800
3	0.4	1,920	1,371	1,920	2,880
4	0.4	1,152	1,152	1,152	1,728
5	0.4	691	1,152	1,152	576
6	—	—	576	576	0
				$10,000	

It is clearly easier to use the rates in Table 16–2 than to determine each MACRS rate using the switching logic above. But the logic behind the MACRS rates is described here for those interested. The annual MACRS rates may be derived by using the applicable rate for the DB method. The subscripts DB and SL have been inserted along with the year t. For the first year $t = 1$,

$$d_{DB,1} = \frac{1}{n} \quad \text{or} \quad d_{SL,1} = \frac{1}{2}\frac{1}{n}$$

For summation purposes only, we introduce the subscript i ($i = 1, 2, \ldots, t$) on d. Then the depreciation rates for years $t = 2, 3, \ldots, n$ are

$$d_{DB,t} = d\left(1 - \sum_{i=1}^{i=t-1} d_i\right) \qquad [16A.14]$$

$$d_{SL,t} = \frac{\left(1 - \sum_{i=1}^{i=t-1} d_i\right)}{n - t + 1.5} \qquad [16A.15]$$

Also, for year $n + 1$, the MACRS rate is one-half the SL rate of the previous year n.

$$d_{SL,n+1} = \frac{1}{2d_{SL,n}} \qquad [16A.16]$$

The DB and SL rates are compared each year to determine which is larger and when the switch to SL depreciation should occur.

EXAMPLE 16A.5

Verify the MACRS rates in Table 16–2 for a 3-year recovery period. The rates in percent are 33.33, 44.45, 14.81, and 7.41.

Solution
The fixed rate for DDB with $n = 3$ is $d = 2/3 = 0.6667$. Using the half-year convention in year 1 and Equations [16A.14] through [16A.16], the results are as follows:

d_1. $d_{DB,1} = 0.5d = 0.5(0.6667) = 0.3333$

d_2.   Cumulative depreciation rate is 0.3333.

$$d_{DB,2} = 0.6667(1 - 0.3333) = 0.4445 \qquad \text{(larger value)}$$

$$d_{SL,2} = \frac{1 - 0.3333}{3 - 2 + 1.5} = 0.2267$$

d_3.   Cumulative depreciation rate is $0.3333 + 0.4445 = 0.7778$.

$$d_{DB,3} = 0.6667(1 - 0.7778) = 0.1481$$

$$d_{SL,3} = \frac{1 - 0.7778}{3 - 3 + 1.5} = 0.1481$$

Both values are the same; switch to straight line depreciation.

d_4.   This rate is 50% of the last SL rate.

$$d_4 = 0.5(d_{SL,3}) = 0.5(0.1481) = 0.0741$$

APPENDIX PROBLEMS

Sum-of-Year-Digits Depreciation

16A.1   A European manufacturing company has new equipment with a first cost of 12,000 euro, an estimated salvage value of 2000 euro, and a recovery period of 8 years. Use the SYD method to tabulate annual depreciation and book value.

16A.2   Earthmoving equipment with a first cost of $150,000 is expected to have a life of 10 years. The salvage value is expected to be 10% of the first cost. Calculate (*a*) by hand and (*b*) by computer the depreciation charge and book value for years 2 and 7 using the SYD method.

16A.3   If $B = \$12,000$, $n = 6$ years, and S is estimated at 15% of B, use the SYD method to determine (*a*) the book value after 3 years and (*b*) the rate of depreciation and the depreciation amount in year 4.

Switching Methods

16A.4   An asset has a first cost of $45,000, a recovery period of 5 years, and a $3000 salvage value. Use the switching procedure from DDB to SL depreciation, and calculate the present worth of depreciation at $i = 18\%$ per year.

16A.5   If $B = \$45,000$, $S = \$3000$, and $n =$ 5-year recovery period, use a spreadsheet and $i = 18\%$ per year to maximize the present worth of depreciation, using the following methods: DDB-to-SL switching (this was determined in Problem 16A.4) and MACRS. Given

that MACRS is the required depreciation system in the United States, comment on the results.

16A.6   Hempstead Industries has a new milling machine with $B = \$110,000$, $n = 10$ years, and $S = \$10,000$. Determine the depreciation schedule and present worth of depreciation at $i = 12\%$ per year, using the 175% DB method for the first 5 years and switching to the classical SL method for the last 5 years. Use a spreadsheet to solve this problem.

16A.7   Reliant Electric Company has erected a large portable building with a first cost of $155,000 and an anticipated salvage of $50,000 after 25 years. (*a*) Should the switch from DDB to SL depreciation be made? (*b*) For what values of the uniform depreciation rate in the DB method would it be advantageous to switch from DB to SL depreciation at some point in the life of the building?

MACRS Rates

16A.8   Verify the 5-year recovery period rates for MACRS given in Table 16–2. Start with the DDB model in year 1, and switch to SL depreciation when it offers a larger recovery rate.

16A.9   A video recording system was purchased 3 years ago at a cost of $30,000. A 5-year recovery period and MACRS depreciation have been used to write off the basis. The system is to be prematurely replaced with a trade-in value of $5000. Determine the MACRS depreciation, using the switching rules to find the difference between the book value and the trade-in value after the 3 years.

16A.10 Use the computations in Equations [16A.10] through [16A.12] to determine the MACRS annual depreciation for the following asset data: $B = \$50,000$ and a recovery period of 7 years.

16A.11 The 3-year MACRS recovery rates are 33.33%, 44.45%, 14.81%, and 7.41%, respectively. (a) What are the corresponding rates for the alternative MACRS straight line ADS model with the half-year convention imposed? (b) Compare the PW_D values for these two methods if $B = \$80,000$ and $i = 15\%$ per year.

17

After-Tax Economic Analysis

This chapter provides an overview of tax terminology, income tax rates, and tax equations pertinent to an after-tax economic analysis. The transfer from estimating cash flow before taxes (CFBT) to cash flow after taxes (CFAT) involves a consideration of significant tax effects that may alter the final decision, as well as estimate the magnitude of the tax effect on cash flow over the life of the alternative.

Mutually exclusive alternative comparisons using after-tax PW, AW, and ROR methods are explained with major tax implications considered. Replacement studies are discussed with tax effects that occur at the time that a defender is replaced. Also, the after-tax *economic value added* by an alternative is discussed in the context of annual worth analysis. All these methods use the procedures learned in earlier chapters, except now with tax effects considered.

An after-tax evaluation using any method requires more computations than those in previous chapters. The computer greatly reduces the analysis time due to the power of spreadsheet formatting and functions. Templates for tabulation of cash flow after taxes by hand and by computer are developed. Additional information on U.S. federal taxes—tax law, and annually updated tax rates—is available through Internal Revenue Service publications and, more readily, on the IRS website www.irs.gov. Publications 542, Corporations, and 544, Sales and Other Dispositions of Assets, are especially applicable to this chapter. Some differences in tax considerations outside the United States are summarized in the last section.

The case study provides the opportunity to perform a complete after-tax analysis of debt versus equity financing, with asset depreciation included. It is an application of the *generalized cash flow analysis* model.

LEARNING OBJECTIVES

Purpose: Perform an economic evaluation of one or more alternatives considering the effect of income taxes and other pertinent tax regulations.

This chapter will help you:

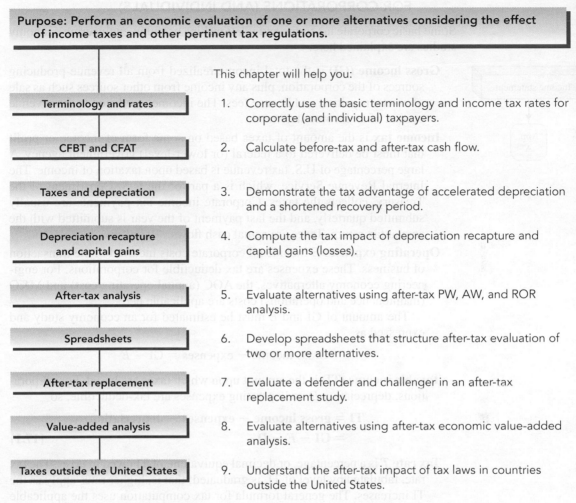

Terminology and rates	1. Correctly use the basic terminology and income tax rates for corporate (and individual) taxpayers.
CFBT and CFAT	2. Calculate before-tax and after-tax cash flow.
Taxes and depreciation	3. Demonstrate the tax advantage of accelerated depreciation and a shortened recovery period.
Depreciation recapture and capital gains	4. Compute the tax impact of depreciation recapture and capital gains (losses).
After-tax analysis	5. Evaluate alternatives using after-tax PW, AW, and ROR analysis.
Spreadsheets	6. Develop spreadsheets that structure after-tax evaluation of two or more alternatives.
After-tax replacement	7. Evaluate a defender and challenger in an after-tax replacement study.
Value-added analysis	8. Evaluate alternatives using after-tax economic value-added analysis.
Taxes outside the United States	9. Understand the after-tax impact of tax laws in countries outside the United States.

17.1 INCOME TAX TERMINOLOGY AND RELATIONS FOR CORPORATIONS (AND INDIVIDUALS)

Some basic corporate tax terms and relationships useful in engineering economy studies are explained here.

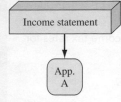

Gross income (GI) is the total income realized from all revenue-producing sources of the corporation, plus any income from other sources such as sale of assets, royalties, and license fees. The incomes are listed in the revenue section of an income statement.

Income tax is the amount of taxes based on some form of income or profit that must be delivered to a federal (or lower-level) government agency. A large percentage of U.S. tax revenue is based upon taxation of income. The Internal Revenue Service, which is a part of the U.S. Department of the Treasury, collects the taxes. Corporate income tax payments are usually submitted quarterly, and the last payment of the year is submitted with the annual tax return. Taxes are actual cash flows.

Operating expenses E include all corporate costs incurred in the transaction of business. These expenses are tax deductible for corporations. For engineering economy alternatives, the AOC (annual operating cost) and M&O (maintenance and operations) costs are applicable here.

The amount of GI and E must be estimated for an economy study and expressed as

$$\text{Gross income} - \text{expenses} = \text{GI} - E$$

Taxable income (TI) is the amount upon which taxes are based. For corporations, depreciation D and operating expenses are tax-deductible, so

$$\text{TI} = \text{gross income} - \text{expenses} - \text{depreciation}$$
$$= \text{GI} - E - D \qquad [17.1]$$

Tax rate T is a percentage, or decimal equivalent, of TI owed in taxes. The tax rate, tabulated by level of TI, is graduated; that is, higher rates apply as the TI increases. The general formula for tax computation uses the applicable T value.

$$\text{Taxes} = (\text{taxable income}) \times (\text{applicable tax rate})$$
$$= (\text{TI})(T) \qquad [17.2]$$

Net profit after taxes (NPAT) is the amount remaining each year when income taxes are subtracted from taxable income.

$$\text{NPAT} = \text{taxable income} - \text{taxes} = \text{TI} - (\text{TI})(T)$$
$$= (\text{TI})(1 - T) \qquad [17.3]$$

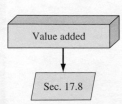

This is the amount of money returned to the corporation as a result of the capital invested during the year. It is a component of after-tax value-added analysis. NPAT is also referred to as net income (NI) and net operating profit after taxes (NOPAT).

TABLE 17–1 U.S. Corporate Federal Income Tax Rate Schedule (2003) (mil = million $)				
TI Limits (1)	TI Range (2)	Tax Rate T (3)	Maximum Tax for TI Range (4) = (2)T	Maximum Tax Incurred (5) = Sum of (4)
$1–$50,000	$ 50,000	0.15	$ 7,500	$ 7,500
$50,001–75,000	25,000	0.25	6,250	13,750
$75,001–100,000	25,000	0.34	8,500	22,250
$100,001–335,000	235,000	0.39	91,650	113,900
$335,001–10 mil	9.665 mil	0.34	3.2861 mil	3.4 mil
Over $10–15 mil	5 mil	0.35	1.75 mil	5.15 mil
Over $15–18.33 mil	3.33 mil	0.38	1.267 mil	6.417 mil
Over $18.33 mil	Unlimited	0.35	Unlimited	Unlimited

A variety of different bases may be used by federal, state, and local agencies for tax revenue. Some bases (and taxes) other than income are the total sales (sales tax); appraised value of property (property tax); value-added tax (VAT); net capital investment (asset tax); winnings from gambling (part of income tax); and retail value of items imported (import tax). Similar and different bases are utilized by different countries, provinces, and local tax districts. Governments that have no income tax must utilize bases other than income to develop revenue. No government entity survives for long without some form of tax revenue.

The annual federal tax rate T is based upon the principle of *graduated tax rates,* which means that corporations pay higher rates for larger taxable incomes. Table 17–1 presents T values for U.S. corporations. The rates and TI limits change based on legislation and IRS interpretation of tax law and economic conditions. Additionally each year the IRS reviews and/or alters the TI limits to account for inflation and other factors. This action is called *indexing.* The portion of each new dollar of TI is taxed at what is called the *marginal tax rate.* As an illustration, scan the tax rates in Table 17–1. A business with an annual TI of $50,000 has a marginal rate of 15%. However, a business with TI = $100,000 pays 15% for the first $50,000, 25% on the next $25,000, and 34% on the remainder.

Taxes = 0.15(50,000) + 0.25(75,000 − 50,000) + 0.34(100,000 − 75,000)

= $22,250

The graduated tax rate system gives businesses with small taxable incomes a slight advantage. The marginal rates are in the middle to upper thirty percentages for TI values above (roughly) $100,000, while smaller TIs have rates in the 15% to 25% range. Example 17.1 further illustrates the use of graduated tax rates.

Because the marginal tax rates change with TI, it is not possible to quote directly the percent of TI paid in income taxes. Alternatively, a single-value number, the *average tax rate,* is calculated as

$$\text{Average tax rate} = \frac{\text{total taxes paid}}{\text{taxable income}} = \frac{\text{taxes}}{\text{TI}} \qquad [17.4]$$

For a small business with TI = $100,000, the federal income tax burden averages $22,250/100,000 = 22.25%. If TI = $15 million, the average tax rate (Table 17–1) is $5.15 million/15 million = 34.33%.

As mentioned earlier, there are federal, state, and local taxes imposed. For the sake of simplicity, the tax rate used in an economy study is often a single-figure *effective tax rate T_e,* which accounts for all taxes. Effective tax rates are in the range of 35% to 50%. One reason to use the effective tax rate is that state taxes are deductible for federal tax computation. The effective tax rate and taxes are calculated as

$$T_e = \text{state rate} + (1 - \text{state rate})(\text{federal rate}) \qquad [17.5]$$

$$\textbf{Taxes} = \textbf{(TI)}(T_e) \qquad [17.6]$$

EXAMPLE 17.1

If the security division of OnStar has an annual gross income of $2,750,000 with expenses and depreciation totaling $1,950,000, (*a*) compute the company's exact federal income taxes. (*b*) Estimate total federal and state taxes if a state tax rate is 8% and a 34% federal average tax rate applies.

Solution

(*a*)   Compute the TI by Equation [17.1] and the income taxes using Table 17–1 rates.

$$\text{TI} = 2,750,000 - 1,950,000 = \$800,000$$

$$\text{Taxes} = 50,000(0.15) + 25,000(0.25) + 25,000(0.34)$$
$$+ 235,000(0.39) + (800,000 - 335,000)(0.34)$$
$$= 7500 + 6250 + 8500 + 91,650 + 158,100$$
$$= \$272,000$$

A faster approach uses the amount in column 5 of Table 17–1 that is closest to the total TI and adds the tax for the next TI range.

$$\text{Taxes} = 113,900 + (800,000 - 335,000)(0.34) = \$272,000$$

(*b*)   Equation [17.5] determines the effective tax rate.

$$T_e = 0.08 + (1 - 0.08)(0.34) = 0.3928$$

Estimate total taxes by Equation [17.6].

$$\text{Taxes} = (\text{TI})(T_e) = (800,000)(0.3928) = \$314,240$$

These two amounts are not comparable, because the tax in part (*a*) does not include state taxes.

TABLE **17–2** U.S. Individual Federal Income Tax Rate Schedule for Single and Married Filing Jointly (2003)		
	Taxable Income, $	
Tax Rate *T* (1)	Filing Single (2)	Filing Married and Jointly (3)
0.10	0–7,000	0–14,000
0.15	7,001–28,400	14,001–56,800
0.25	28,401–68,800	56,801–114,650
0.28	68,801–143,500	114,651–174,700
0.33	143,501–311,950	174,701–311,950
0.35	Over 311,950	Over 311,950

It is interesting to understand how corporate tax and individual tax computations differ. Gross income for an individual taxpayer is comparable, with business revenue replaced by salaries and wages. However, for an individual's taxable income, most of the expenses for living and working are not tax-deductible to the same degree as business expenses are for corporations. For individual taxpayers,

$$\text{Gross income} = \text{GI} = \text{salaries} + \text{wages} + \text{interest and dividends} + \text{other income}$$

$$\text{Taxable income} = \text{GI} - \text{personal exemption} - \text{standard or itemized deductions}$$

$$\text{Taxes} = (\text{taxable income})(\text{applicable tax rate}) = (\text{TI})(T)$$

For TI, corporate operating expenses are replaced by individual exemptions and specific deductions. Exemptions are yourself, spouse, children, and other dependents. Each exemption reduces TI by approximately $3000 per year, depending upon current exemption allowances.

Like the corporate tax structure, tax rates for individuals are graduated by TI level. As shown in Table 17–2, they range from 15% to 35% of TI.

EXAMPLE 17.2

Josh and Allison submit a married-filing-jointly return to the IRS. During the year their two jobs provided them with a combined income of $82,000. They had their second child during the year, and they plan to use the standard deduction of $9500 applicable for the year. Dividends and interest amounted to $3550, and an investment in a stock mutual fund had capital gains of $2500. Personal exemptions are $3100 currently. (*a*) Compute their exact federal tax liability. (*b*) Compute their average tax rate. (*c*) What percent of gross income is consumed by federal taxes?

Solution

(a) Josh and Allison have four personal exemptions and the standard deduction of $9500.

$$\text{Gross income} = \text{salaries} + \text{interest and dividends} + \text{capital gains}$$
$$= 82{,}000 + 3550 + 2500 = \$88{,}050$$

$$\text{Taxable income} = \text{gross income} - \text{exemptions} - \text{deductions}$$
$$= 88{,}050 - 4(3100) - 9500$$
$$= \$66{,}150$$

Table 17–2 indicates the 25% marginal rate. Using columns 1 and 3, federal taxes are

$$\text{Taxes} = 14{,}000(0.10) + (56{,}800 - 14{,}000)(0.15)$$
$$+ (66{,}150 - 56{,}800)(0.25)$$
$$= \$10{,}158$$

(b) From Equation [17.4],

$$\text{Average tax rate} = \frac{10{,}158}{66{,}150} = 15.4\%$$

This indicates that about 1 in 6 dollars of taxable income is paid to the U.S. government.

(c) Of the total $88,050, the percent paid in federal taxes is 10,158/88,050 = 11.5%.

Comment

There has been longstanding debate in the U.S. Congress about the advisability of changing from a graduated tax structure to a flat tax structure for individual taxpayers. (This structure is applied in some countries presently.) There are many ways to legislate taxes, and the amount to be chosen for the flat rate can be a real controversy.

For example, the flat tax structure may allow no standard or itemized deductions and have only the personal exemption allowance. In this example, were there a flat tax rate as low as 15% on gross income reduced only by the four personal exemptions, the computations would be

$$\text{Gross income} = \$88{,}050$$

$$\text{Flat rate taxable income} = 88{,}050 - 4(3100) = \$75{,}650$$

$$\text{Flat rate taxes} = 75{,}650(0.15) = \$11{,}348$$

In this case, a 15% flat tax rate would require that 11.7% more taxes be paid by this family—$11,348—compared to $10,158 using the graduated tax rates.

17.2   BEFORE-TAX AND AFTER-TAX CASH FLOW

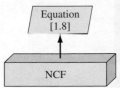

Equation [1.8]

NCF

Early in the text, the term *net cash flow (NCF)* was identified as the best estimate of actual cash flow each year. The NCF is calculated as cash inflows minus cash outflows. Since then, the annual NCF amounts have been used many times to perform alternative evaluations via the PW, AW, ROR, and B/C methods. Now that the impact on cash flow of depreciation and related taxes will be considered, it is time to expand our terminology. NCF is replaced by the term *cash flow before taxes* (*CFBT*), and we introduce the new term *cash flow after taxes* (*CFAT*).

CFBT and CFAT are *actual cash flows;* that is, they represent the estimated actual flow of money in and out of the corporation that will result from the alternative. The remainder of this section explains how to transition from before-tax to after-tax cash flows for solutions by hand and by computer, using income tax rates and other pertinent tax regulations described in the next few sections.

Once the CFAT estimates are developed, the economic evaluation is performed using the same methods and selection guidelines applied previously. However, the analysis is performed on the CFAT estimates.

The annual CFBT estimate must include the initial capital investment and salvage value for the years in which they occur. Incorporating the definitions of gross income and operating expenses, CFBT for any year is defined as

CFBT = gross income − expenses − initial investment + salvage value

$$= \text{GI} - E - P + S \qquad \qquad [17.7]$$

As in previous chapters, P is the initial investment (usually in year 0) and S is the estimated salvage value in year n. Once all taxes are estimated, the annual after-tax cash flow is simply

$$\text{CFAT} = \text{CFBT} - \text{taxes} \qquad \qquad [17.8]$$

where taxes are estimated using the relation $(\text{TI})(T)$ or $(\text{TI})(T_e)$, as discussed earlier.

We know from Equation [17.1] that depreciation D is subtracted to obtain TI. It is very important to understand the different roles of depreciation for income tax computations and in CFAT estimation.

Depreciation is a *non*cash flow. Depreciation is tax-deductible for determining the amount of income taxes only, but it does not represent a direct, after-tax cash flow to the corporation. Therefore, the after-tax engineering economy study must be based on actual cash flow estimates, that is, annual CFAT estimates that do not include depreciation as a negative cash flow.

Accordingly, if the CFAT expression is determined using the TI relation, depreciation must not be included outside of the TI component. Equations [17.7] and [17.8] are now combined as

$$\text{CFAT} = \text{GI} - E - P + S - (\text{GI} - E - D)(T_e) \qquad \qquad [17.9]$$

Suggested table column headings for CFBT and CFAT calculations by hand or by computer are shown in Table 17–3. The equations are shown in column numbers, with the effective tax rate T_e used for income taxes. Expenses E and initial investment P will be negative values.

The TI value in some years may be negative due to a depreciation amount that is larger than $(\text{GI} - E)$. It is possible to account for this in a detailed after-tax analysis using carry-forward and carry-back rules for operating losses. It is the exception that the engineering economy study will consider this level of

TABLE 17–3   Table Column Headings for Calculation of (*a*) CFBT and (*b*) CFAT

(a) CFBT table headings

Year	Gross Income GI	Operating Expenses E	Investment P and Salvage S	CFBT (4) = (1) + (2) + (3)
	(1)	(2)	(3)	

(b) CFAT table headings

Year	Gross Income GI	Operating Expenses E	Investment P and Salvage S	Depreciation D	Taxable Income TI (5) = (1) + (2) − (4)	Taxes $(TI)(T_e)$	CFAT (7) = (1) + (2) + (3) − (6)
	(1)	(2)	(3)	(4)		(6)	

detail. Rather, *the associated negative income tax is considered as a tax savings for the year.* The assumption is that the negative tax will offset taxes for the same year in other income-producing areas of the corporation.

EXAMPLE 17.3

TransAmerica Insurance expects to initiate a new outreach service next year. Small facilities will be constructed in about 35 high-risk cities across the continent. Company personnel from the area will offer training and consulting services to the citizens and officials of the cities and counties in fire avoidance, theft deterrence, and similar topics. Current and prospective customers will be invited to visit the facility. Each facility is expected to cost $550,000 initially with a resale (salvage) value of $150,000 after 6 years—the time period for which the TransAmerica Board of Directors approved this activity. MACRS depreciation allows a 5-year recovery period. A team of safety engineers, actuaries, and financial personnel estimate bottom-line results at annual net increases to the corporation of $200,000 in revenue and $90,000 in costs. Using an effective tax rate of 35%, tabulate the CFBT and CFAT estimates.

Solution by Hand and by Computer

The spreadsheet in Figure 17–1 presents before-tax and after-tax cash flows. The table format, identical for hand and computer calculations, uses the format of Table 17–3. Discussion and sample calculations follow.

   CFBT: The expenses and initial investment are shown as negative cash flows. The $150,000 salvage (resale) is a positive cash flow in year 6. CFBT is calculated by Equation [17.7]. In year 6, for example, the cell tag indicates that

$$CFBT_6 = 200,000 - 90,000 + 150,000 = \$260,000$$

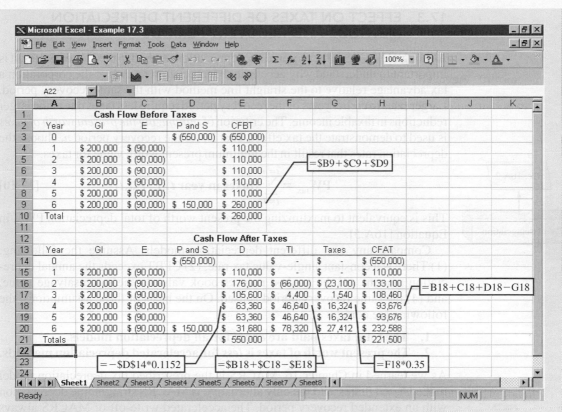

Figure 17–1

Computation of CFBT and CFAT using MACRS depreciation and a 35% effective tax rate, Example 17.3.

CFAT: Column E for MACRS depreciation (rates in Table 16–2 for $n = 5$) over the 6-year period writes off the entire \$550,000 investment. Taxable income, taxes, and CFAT, illustrated in the cell tags for year 4, are computed as

$$TI_4 = GI - E - D = 200,000 - 90,000 - 63,360 = \$46,640$$

$$Taxes = (TI)(0.35) = (46,640)(0.35) = \$16,324$$

$$CFAT_4 = GI - E - taxes = 200,000 - 90,000 - 16,324 = \$93,676$$

In year 2, MACRS depreciation is large enough to cause TI to be negative (\$−66,000). As mentioned above, the negative tax (\$−23,100) is considered a *tax savings* in year 2, thus increasing CFAT.

Comment

MACRS depreciates to a salvage value of $S = 0$. Later we will learn about a tax implication due to "recapturing of depreciation" when an asset is sold for an amount larger than zero, and MACRS was applied to fully depreciate the asset to zero.

17.3 EFFECT ON TAXES OF DIFFERENT DEPRECIATION METHODS AND RECOVERY PERIODS

Although MACRS is the required *tax* depreciation method in the United States, it is important to understand why accelerated depreciation rates give the corporation a tax advantage relative to the straight line method with the same recovery period. Larger rates in earlier years of the recovery period require less taxes due to the larger reductions in taxable income. The criterion of *minimizing the present worth of taxes* is used to demonstrate the tax effect. That is, for the recovery period *n*, choose the depreciation rates that result in the minimum present worth value for taxes.

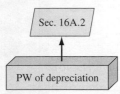

Sec. 16A.2

PW of depreciation

$$\text{PW}_{\text{tax}} = \sum_{t=1}^{t=n} (\text{taxes in year } t)(P/F,i,t) \qquad [17.10]$$

This is equivalent to maximizing the present worth of total depreciation PW_D, in Equation [16A.5].

Compare any two different depreciation models. Assume the following: (1) There is a constant single-value tax rate, (2) CFBT exceeds the annual depreciation amount, (3) the method reduces book value to the same salvage value, and (4) the same recovery period is used. On the basis of these assumptions, the following are correct:

1. **The total taxes paid are *equal* for all depreciation models.**
2. **The present worth of taxes is *less* for accelerated depreciation methods.**

As we learned in Chapter 16, MACRS is the prescribed tax depreciation model in the United States, and the only alternative is MACRS straight line depreciation with an extended recovery period. The accelerated write-off of MACRS always provides a smaller PW_{tax} compared to less accelerated models. If the DDB model were still allowed directly, rather than embedded in MACRS, DDB would not fare as well as MACRS. This is because DDB does not reduce the book value to zero. This is illustrated in Example 17.4.

EXAMPLE 17.4

An after-tax analysis for a new $50,000 machine proposed for a fiber optics manufacturing line is in process. The CFBT for the machine is estimated at $20,000. If a recovery period of 5 years applies, use the present-worth-of-taxes criterion, an effective tax rate of 35%, and a return of 8% per year to compare the following: classical straight line, classical DDB, and required MACRS depreciation. Use a 6-year period for the comparison to accommodate the half-year convention imposed by MACRS.

Solution

Table 17–4 presents a summary of annual depreciation, taxable income, and taxes for each model. For classical straight line depreciation with $n = 5$, $D_t = \$10,000$ for 5 years and $D_6 = 0$ (column 3). The CFBT of $20,000 is fully taxed at 35% in year 6.

The classical DDB percentage of $d = 2/n = 0.40$ is applied for 5 years. The implied salvage value is $\$50,000 - 46,112 = \$3888$, so not all $50,000 is tax deductible. The taxes using classical DDB would be $3888 (0.35) = \$1361$ larger than for the classical SL model.

TABLE 17–4 Comparison of Taxes and Present Worth of Taxes for Different Depreciation Models

		Classical Straight Line			Classical Double Declining Balance			MACRS		
(1) Year t	(2) CFBT	(3) D_t	(4) TI	(5) = 0.35(4) Taxes	(6) D_t	(7) TI	(8) = 0.35(7) Taxes	(9) D_t	(10) TI	(11) = 0.35(10) Taxes
1	+20,000	$10,000	$10,000	$ 3,500	$20,000	$ 0	$ 0	$10,000	$10,000	$ 3,500
2	+20,000	10,000	10,000	3,500	12,000	8,000	2,800	16,000	4,000	1,400
3	+20,000	10,000	10,000	3,500	7,200	12,800	4,480	9,600	10,400	3,640
4	+20,000	10,000	10,000	3,500	4,320	15,680	5,488	5,760	14,240	4,984
5	+20,000	10,000	10,000	3,500	2,592	17,408	6,093	5,760	14,240	4,984
6	+20,000	0	20,000	7,000	0	20,000	7,000	2,880	17,120	5,992
Totals		$50,000		$24,500	$46,112		$25,861*	$50,000		$24,500
PW_{tax}				$18,386			$18,549			$18,162

*Larger than other values since there is an implied salvage value of $3888 not recovered.

MACRS writes off the $50,000 in 6 years using the rates of Table 16–2. Total taxes are $24,500, the same as for classical SL depreciation.

The annual taxes (columns 5, 8, and 11) are accumulated year by year in Figure 17–2. Note the pattern of the curves, especially the lower total taxes relative to the SL model after year 1 for MACRS and in years 1 through 4 for DDB. These higher tax values for SL cause PW_{tax} for SL depreciation to be larger. The PW_{tax} values at the bottom of Table 17–4 are calculated using Equation [17.10]. The MACRS PW_{tax} value is the smallest at $18,162.

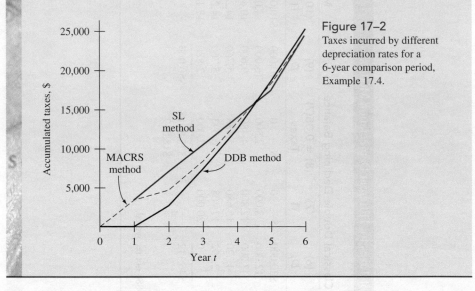

Figure 17–2
Taxes incurred by different depreciation rates for a 6-year comparison period, Example 17.4.

To compare taxes for different recovery periods, change only assumption four at the beginning of this section to read: The same depreciation method is applied. It can be shown that a shorter recovery period will offer a tax advantage over a longer period using the criterion to minimize PW_{tax}. Comparison will indicate that

1. **The total taxes paid are *equal* for all *n* values.**
2. **The present worth of taxes is *less* for smaller *n* values.**

This is why corporations want to use the shortest MACRS recovery period allowed for income tax purposes. Example 17.5 demonstrates these conclusions for classical straight line depreciation, but the conclusions are correct for MACRS, or any other tax depreciation method, were others available.

EXAMPLE 17.5

Grupo Grande Maquinaría, a diversified manufacturing corporation based in Mexico, maintains parallel records for depreciable assets in its European operations in Germany. This is common for multinational corporations. One set is for corporate use that reflects the estimated useful life of assets. The second set is for foreign government purposes, such as depreciation.

The company just purchased an asset for $90,000 with an estimated useful life of 9 years; however, a shorter recovery period of 5 years is allowed by German tax law. Demonstrate the tax advantage for the smaller n if $(GI - E) = \$30,000$ per year, an effective tax rate of 35% applies, invested money is returning 5% per year after taxes, and classical SL depreciation is allowed. Neglect the effect of any salvage value.

Solution

Determine the annual taxes by Equations [17.1] and [17.2], and the present worth of taxes using Equation [17.10] for both n values.

Useful life $n = 9$ years:

$$D = \frac{90{,}000}{9} = \$10{,}000$$

$$TI = 30{,}000 - 10{,}000 = \$20{,}000 \text{ per year}$$

$$\text{Taxes} = 20{,}000(0.35) = \$7000 \text{ per year}$$

$$PW_{tax} = 7000(P/A,5\%,9) = \$49{,}755$$

$$\text{Total taxes} = 7000(9) = \$63{,}000$$

Recovery period $n = 5$ years:

Use the same comparison period of 9 years, but depreciation occurs only during the first 5 years.

$$D_t = \begin{cases} \dfrac{90{,}000}{5} = \$18{,}000 & t = 1 \text{ to } 5 \\ 0 & t = 6 \text{ to } 9 \end{cases}$$

$$\text{Taxes} = \begin{cases} (30{,}000 - 18{,}000)(0.35) = \$4200 & t = 1 \text{ to } 5 \\ (30{,}000)(0.35) = \$10{,}500 & t = 6 \text{ to } 9 \end{cases}$$

$$PW_{tax} = 4200(P/A,5\%,5) + 10{,}500(P/A,5\%,4)(P/F,5\%,5)$$
$$= \$47{,}356$$

$$\text{Total taxes} = 4200(5) + 10{,}500(4) = \$63{,}000$$

A total of $63,000 in taxes is paid in both cases. However, the more rapid write-off for $n = 5$ results in a present worth tax savings of nearly $2400 ($49,755 − 47,356).

17.4 DEPRECIATION RECAPTURE AND CAPITAL GAINS (LOSSES): for corporations

All the tax implications discussed here are the result of disposing of a depreciable asset before, at, or after its recovery period. In an after-tax economic analysis of large investment assets, these tax effects should be considered. The key is the size of the selling price (or salvage or market value) relative to the book value at disposal time, and relative to the first cost. There are three relevant terms.

Capital gain (CG) is an amount incurred when the selling price exceeds its first cost. See Figure 17–3. At the time of asset disposal,

$$\textbf{Capital gain = selling price − first cost}$$

$$\textbf{CG = SP − P} \qquad \textbf{[17.11]}$$

Figure 17–3
Summary of calculations
and tax treatment for
depreciation recapture
and capital gains (losses).

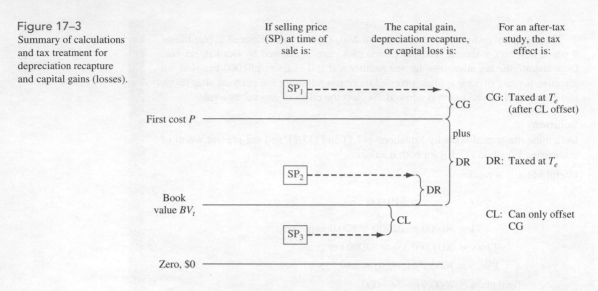

Since future capital gains are difficult to predict, they are usually not detailed in an after-tax economy study. An exception is for assets that historically increase in value, such as buildings and land. *If included, the gain is taxed as ordinary taxable income at the effective tax rate T_e.* (In actual tax law, there is a distinction between a short-term gain and long-term gain, where the asset is retained for at least 1 year.)

Depreciation recapture (DR) occurs when a depreciable asset is sold for more than the current book value BV_t. As shown in Figure 17–3,

Depreciation recapture = selling price − book value

$$DR = SP - BV_t \qquad [17.12]$$

Depreciation recapture is often present in the after-tax study. In the United States, an amount equal to the estimated salvage value can always be anticipated as DR when the asset is disposed of after the recovery period. This is correct simply because MACRS depreciates every asset to zero in $n + 1$ years. The amount is treated as ordinary taxable income in the year of asset disposal.

When the selling price exceeds first cost, a capital gain is also incurred and the TI due to the sale is the gain *plus* the depreciation recapture, as shown in Figure 17–3. The DR is now the total amount of depreciation taken thus far.

Capital loss (CL) occurs when a depreciable asset is disposed of for less than its current book value.

Capital loss = book value − selling price

$$CL = BV_t - SP \qquad [17.13]$$

An economic analysis does not commonly account for capital loss, simply because it is not estimable for a specific alternative. However, an after-tax replacement study should account for any capital loss if the defender must be traded at a "sacrifice" price. For the purposes of the economic study, this provides a tax savings in the year of replacement. Use the effective tax rate to estimate the tax savings. These savings are assumed to be offset elsewhere in the corporation by other income-producing assets that generate taxes.

Most depreciable corporate assets are retained in use for more than 1 year. When such an asset is sold, disposed of, or traded after the 1-year point, the U.S. tax consideration is referred to as a *Section 1231 transaction,* named after the IRS rules section of the same number. To determine the associated TI, all capital losses of the corporation are netted against all capital gains, because losses do not directly reduce taxes. On the other hand, *net capital gains* are taxed as ordinary TI. Additionally complicating the analysis may be the different tax treatment of long-term gains and losses (Section 1231 transactions) compared to short-term dispositions. Additional considerations are special, time-limited incentives offered by government agencies to boost capital, and possibly foreign investment, through allowances of increased depreciation and reduced taxes. These benefits come and go depending on the "health of the economy." Only if multiple-asset sales and/or exchanges are involved in an alternative's after-tax study may it be necessary to include this level of detail. These details are usually left to the accountants and finance personnel. (Reference to IRS Publications 334 and 544 may be of interest.) For most after-tax studies, it is *sufficient to apply the effective tax rate T_e to the alternative's TI in the year that the DR, CG, or CL occurs,* with a tax savings generated by the CL.

Finally, it is important to realize that this description and this tax treatment are for *corporations,* not individuals. Individual taxpayers use essentially the same calculations when they sell investments or depreciated assets, but the tax rates vary significantly from those for corporations, especially for capital gains. Also, tax laws and rates for individual taxpayers change more frequently. Refer to the IRS website and publications for details.

Equation [17.1] and the expression for TI in Equation [17.9] can now be expanded to include the additional cash flow estimates for asset disposal.

TI = gross income − expenses − depreciation + depreciation recapture + capital gain − capital loss

$$= \text{GI} - E - D + \text{DR} + \text{CG} - \text{CL} \qquad\qquad \textbf{[17.14]}$$

EXAMPLE **17.6**

Biotech, a medical imaging and modeling company, must purchase a bone cell analysis system for use by a team of bioengineers and mechanical engineers studying bone density in athletes. This particular part of a 3-year contract with the NBA will provide additional gross income of $100,000 per year. The effective tax rate is 35%. Estimates for two alternatives are summarized below.

	Analyzer 1	Analyzer 2
First cost, $	150,000	225,000
Operating expenses, $ per year	30,000	10,000
MACRS recovery, years	5	5

Answer the following questions, solving by hand and by computer.

(a) The Biotech president, who is very tax conscious, wishes to use a criterion of minimizing total taxes incurred over the 3 years of the contract. Which analyzer should be purchased?

(b) Assume that 3 years have now passed, and the company is about to sell the analyzer. Using the same total tax criterion, did either analyzer have an advantage? Assume the selling price is $130,000 for analyzer 1, or $225,000 for analyzer 2, the same as its first cost.

Solution by Hand

(a) Table 17–5 details the tax computations. First, the yearly MACRS depreciation is determined; rates are in Table 16–2. Equation [17.1], $TI = GI - E - D$, is used to calculate TI, after which the 35% tax rate is applied each year. Taxes for the 3-year period are summed, with no consideration of the time value of money.

$$\text{Analyzer 1 tax total: } \$36{,}120 \qquad \text{Analyzer 2 tax total: } \$38{,}430$$

The two analyzers are very close, but analyzer 1 wins with $2310 less in total taxes.

(b) When the analyzer is sold after 3 years of service, there is a depreciation recapture (DR) that is taxed at the 35% rate. This tax is in addition to the third-year tax in Table 17–5. For each analyzer, account for the DR by Equation [17.12]; then determine the TI, using Equation [17.14], $TI = GI - E - D + DR$. Again, find the total taxes for 3 years, and select the analyzer with the smaller total.

TABLE 17–5   Comparison of Total Taxes for Two Alternatives, Example 17.6a

Year	Gross Income GI	Operating Expenses E	First Cost P	MACRS Depreciation D	Book Value BV	Taxable Income TI	Taxes at 0.35TI
				Analyzer 1			
0			$150,000		$150,000		
1	$100,000	$30,000		$30,000	120,000	$40,000	$14,000
2	100,000	30,000		48,000	72,000	22,000	7,700
3	100,000	30,000		28,800	43,200	41,200	14,420
							$36,120
				Analyzer 2			
0			$225,000		$225,000		
1	$100,000	$10,000		$45,000	180,000	$45,000	$15,750
2	100,000	10,000		72,000	108,000	18,000	6,300
3	100,000	10,000		43,200	64,800	46,800	16,380
							$38,430

Analyzer 1:   DR = 130,000 − 43,200 = $86,800

Year 3 TI = 100,000 − 30,000 − 28,800 + 86,800 = $128,000

Year 3 taxes = 128,000(0.35) = $44,800

Total taxes = 14,000 + 7700 + 44,800 = $66,500

Analyzer 2:   DR = 225,000 − 64,800 − $160,200

Year 3 TI = 100,000 − 10,000 − 43,200 + 160,200 = $207,000

Year 3 taxes = 207,000(0.35) = $72,450

Total taxes = 15,750 + 6300 + 72,450 = $94,500

Now, analyzer 1 has a considerable advantage in total taxes ($94,500 versus $66,500).

Solution by Computer

(a) The Excel solution is concentrated in rows 5 through 9 (analyzer 1) and rows 15 through 19 (analyzer 2) of Figure 17–4. The cell tags reflect that the same equations discussed above are applied. The total taxes are in cells H9 and H19 for analyzers 1 and 2, respectively. As expected, analyzer 1 has a slight edge in total taxes.

E-Solve

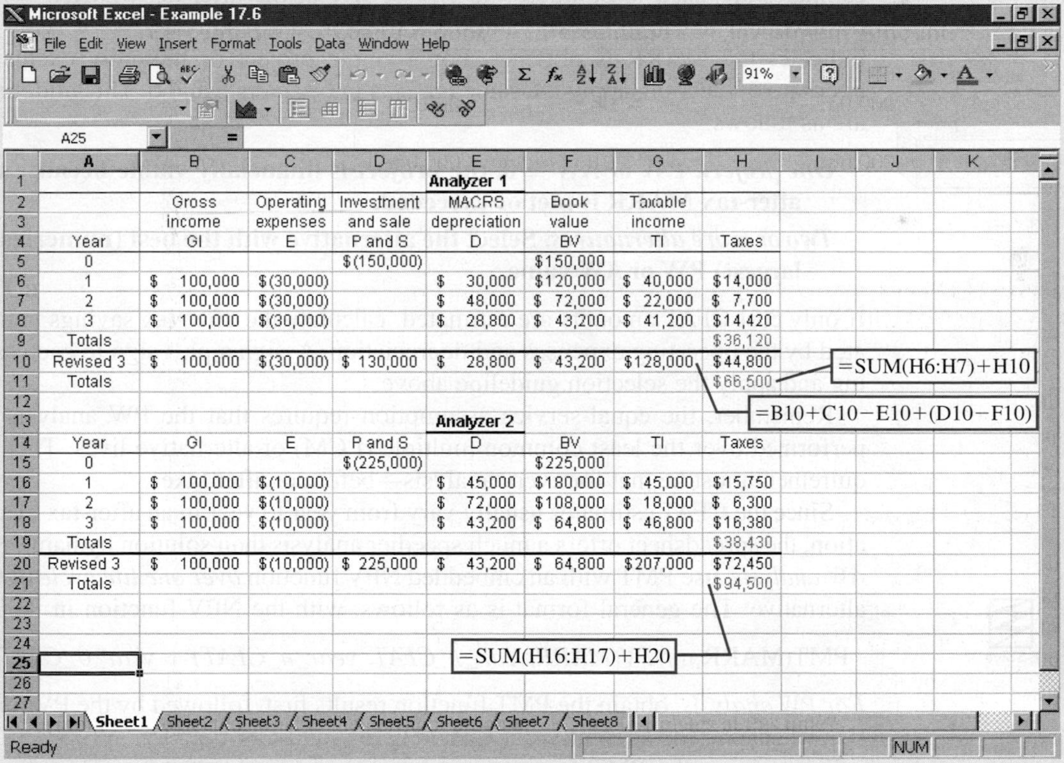

	A	B	C	D	E	F	G	H	I	J	K
1					**Analyzer 1**						
2		Gross	Operating	Investment	MACRS	Book	Taxable				
3		income	expenses	and sale	depreciation	value	income				
4	Year	GI	E	P and S	D	BV	TI	Taxes			
5	0			$(150,000)		$150,000					
6	1	$ 100,000	$(30,000)		$ 30,000	$120,000	$ 40,000	$14,000			
7	2	$ 100,000	$(30,000)		$ 48,000	$ 72,000	$ 22,000	$ 7,700			
8	3	$ 100,000	$(30,000)		$ 28,800	$ 43,200	$ 41,200	$14,420			
9	Totals							$ 36,120			
10	Revised 3	$ 100,000	$(30,000)	$ 130,000	$ 28,800	$ 43,200	$128,000	$44,800			
11	Totals							$66,500			
12											
13					**Analyzer 2**						
14	Year	GI	E	P and S	D	BV	TI	Taxes			
15	0			$(225,000)		$225,000					
16	1	$ 100,000	$(10,000)		$ 45,000	$180,000	$ 45,000	$15,750			
17	2	$ 100,000	$(10,000)		$ 72,000	$108,000	$ 18,000	$ 6,300			
18	3	$ 100,000	$(10,000)		$ 43,200	$ 64,800	$ 46,800	$16,380			
19	Totals							$38,430			
20	Revised 3	$ 100,000	$(10,000)	$ 225,000	$ 43,200	$ 64,800	$207,000	$72,450			
21	Totals							$94,500			
22											
23											
24											
25											
26											
27											

=SUM(H6:H7)+H10

=B10+C10−E10+(D10−F10)

=SUM(H16:H17)+H20

Figure 17–4
Analysis of impact of depreciation recapture on total taxes, Example 17.6.

(b)   When the purchased analyzer is sold in year 3 (at $130,000 in D10 for analyzer 1 and at $225,000 in D20 for analyzer 2), the depreciation recapture, TI, and taxes are recalculated. Rows 10 and 11, and rows 20 and 21, present the increased TI, and taxes. Now, analyzer 1 has a substantial $28,000 advantage in total taxes.

Comment
Note that no time value of money is considered in these analyses, as we have used in previous alternative evaluations. In the next Section we will rely upon PW, AW, and ROR analyses at an established MARR to make an after-tax decision based upon CFAT values.

17.5   AFTER-TAX PW, AW, AND ROR EVALUATION

Sec. 1.9

Sec. 10.2

Setting MARR

The required after-tax MARR is established using the market interest rate, the corporation's effective tax rate, and its average cost of capital. The CFAT estimates are used to compute the PW or AW at the after-tax MARR. When positive and negative CFAT values are present, the result of PW or AW < 0 indicates the MARR is not met. For a single project or mutually exclusive alternative selection, apply the same logic as in Chapters 5 and 6. The guidelines are as follows:

> *One project.* **PW or AW ≥ 0, the project is financially viable because the after-tax MARR is met or exceeded.**

> *Two or more alternatives.* **Select the alternative with the best (numerically largest) PW or AW value.**

If only cost CFAT amounts are estimated, calculate the after-tax savings generated by the operating expenses and depreciation. Assign a plus sign to each saving and apply the selection guideline above.

Remember, the equal-service assumption requires that the PW analysis be performed over the least common multiple (LCM) of alternative lives. This requirement must be met for every analysis—before or after taxes.

Since the CFAT estimates usually vary from year to year in an after-tax evaluation, the spreadsheet offers a much speedier analysis than solution by hand. *For AW analysis,* use PMT with an embedded NPV function *over one life cycle* of the alternative. The general format is as follows, with the NPV function in italics.

PMT(MARR,n,*NPV(MARR,year_1_CFAT: year_n_CFAT) + year_0_CFAT*)

For PW analysis, obtain the PMT function results first, followed by the PV function taken over the alternatives' LCM. (There is an LCM function in Excel.) The cell containing the PMT function result is entered as the A value. The general format is

PV(MARR, LCM_years, *PMT_result_cell*)

EXAMPLE 17.7

Paul is designing the interior walls of an industrial building. In some places, it is important to reduce noise transmission across the wall. Two construction options—stucco on metal lath (S) and bricks (B)—each have about the same transmission loss, approximately 33 decibels. This will reduce noise attenuation costs in adjacent office areas. Paul has estimated the first costs and after-tax savings each year for both designs. Use the CFAT values and an after-tax MARR of 7% per year to determine which is economically better. Solve both by hand and by computer.

	Plan S		Plan B
Year	CFAT	Year	CFAT
0	$-28,800	0	$-50,000
1–6	5,400	1	14,200
7–10	2,040	2	13,300
10	2,792	3	12,400
		4	11,500
		5	10,600

Solution by Hand

In this example, both AW and PW analyses are shown. Develop the AW relations using the CFAT values over each plan's life. Select the larger value.

$$AW_S = [-28,800 + 5400(P/A,7\%,6) + 2040(P/A,7\%,4)(P/F,7\%,6)$$
$$+ 2792(P/F,7\%,10)](A/P,7\%,10)$$
$$= \$422$$

$$AW_B = [-50,000 + 14,200(P/F,7\%,1) + \cdots + 10,600(P/F,7\%,5)](A/P,7\%,5)$$
$$= \$327$$

Both plans are financially viable; select plan S because AW_S is larger.

For the PW analysis, the LCM is 10 years. Use the AW values and the P/A factor for the LCM of 10 years to select stucco on metal lath, plan S.

$$PW_S = AW_S(P/A,7\%,10) = 422(7.0236) = \$2964$$
$$PW_B = AW_B(P/A,7\%,10) = 327(7.0236) = \$2297$$

Solution by Computer

The AW and PW values are displayed in rows 17 and 18 of Figure 17–5. The functions have been set up differently here than in previous examples because of the unequal lives. Follow along in the cell tags, which are shown in the order of development for each alternative. For plan S, first the NPV function in B18 is developed for PW, followed by the PMT function in B17 for the AW value. Note the minus sign in PMT to ensure that the function results in the correct sign on the PW value. This is not necessary for the NPV function, because it takes the cash flow sign from the cell entry itself.

E-Solve

The opposite order of function development is used for plan B. The PMT in C17 uses an embedded NPV function over the 5-year life. Again, note the minus sign. Finally, the PV function in C18 displays the PW value over 10 years.

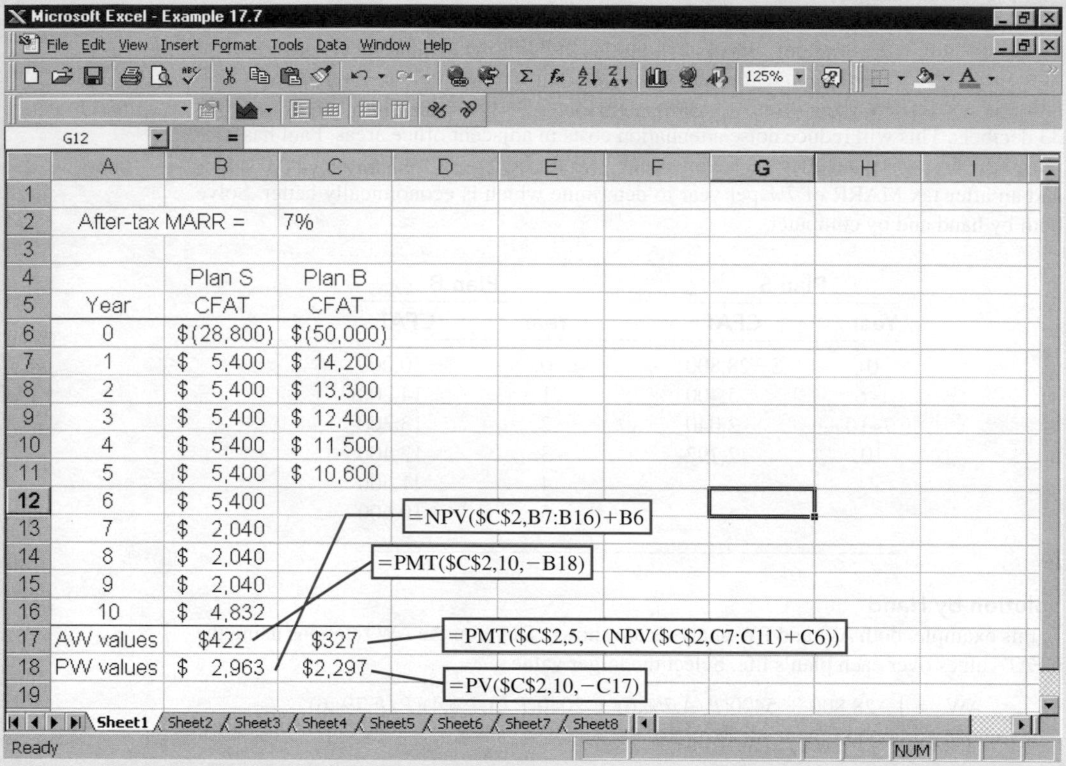

	A	B	C
1			
2	After-tax MARR =		7%
3			
4		Plan S	Plan B
5	Year	CFAT	CFAT
6	0	$(28,800)	$(50,000)
7	1	$ 5,400	$ 14,200
8	2	$ 5,400	$ 13,300
9	3	$ 5,400	$ 12,400
10	4	$ 5,400	$ 11,500
11	5	$ 5,400	$ 10,600
12	6	$ 5,400	
13	7	$ 2,040	
14	8	$ 2,040	
15	9	$ 2,040	
16	10	$ 4,832	
17	AW values	$422	$327
18	PW values	$ 2,963	$2,297
19			

Figure 17–5
After-tax PW and AW evaluation, Example 17.7.

Comment

It is important to remember the minus signs in PMT and PV functions when utilizing them to obtain the corresponding PW and AW values, respectively. If the minus is omitted, the AW and PW values are the opposite of the correct cash flow direction. Then it may appear that the plans are not financially viable in that they do not return at least the after-tax MARR. This is what would happen in this example. However, we know they are financially viable, based on the previous solution by hand. (Refer to Excel online help for more details on sign convention in PMT, PV, and NPV functions.)

To utilize the *ROR method* apply exactly the same procedures as in Chapter 7 (single project) and Chapter 8 (two or more alternatives) to the CFAT series. A PW or AW relation is developed to estimate the rate of return i^* for a project, or Δi^* for the incremental CFAT between two alternatives. Multiple roots may exist in the CFAT series, as they can for any cash flow series. For a single project, set the PW or AW equal to zero and solve for i^*.

Present worth: $$0 = \sum_{t=1}^{t=n} \text{CFAT}_t (P/F,i^*,t)$$ [17.15]

Annual worth: $$0 = \sum_{t=1}^{t=n} \text{CFAT}_t (P/F,i^*,t)(A/P,i^*,n)$$ [17.16]

Spreadsheet solution for i^* may be helpful for relatively complex CFAT series. It is performed using the IRR function with the general format

$$\text{IRR(year\_0\_CFAT:year\_n\_CFAT)}$$

If the after-tax ROR is important to the analysis, but the details of an after-tax study are not of interest, the before-tax ROR (or MARR) can be adjusted with the effective tax rate T_e using the *approximating* relation

$$\textbf{Before-tax ROR} = \frac{\textbf{after-tax ROR}}{1 - T_e}$$ [17.17]

For example, assume a company has an effective tax rate of 40% and normally uses an after-tax MARR of 12% per year for economic analyses that consider taxes explicitly. To *approximate* the effect of taxes without performing the details of an after-tax study, the before-tax MARR can be estimated as

$$\text{Before-tax MARR} = \frac{0.12}{1 - 0.40} = 20\% \text{ per year}$$

If the decision concerns the economic viability of a project, and the resulting PW or AW value is close to zero, the details of an after-tax analysis should be developed.

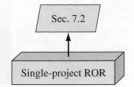

Sec. 7.2

Single-project ROR

E-Solve

EXAMPLE 17.8

A fiber optics manufacturing company operating in Hong Kong has spent $50,000 for a 5-year-life machine that has a projected $20,000 annual CFBT and annual depreciation of $10,000. The company has a T_e of 40%. (*a*) Determine the after-tax rate of return. (*b*) Approximate the before tax return.

Solution

(*a*) The CFAT in year 0 is $-50,000$. For years 1 through 5, combine Equations [17.8] and [17.9] to estimate the CFAT.

$$\text{CFAT} = \text{CFBT} - \text{taxes} = \text{CFBT} - (GI - E - D)(T_e)$$
$$= 20,000 - (20,000 - 10,000)(0.40)$$
$$= \$16,000$$

Since the CFAT for years 1 through 5 has the same value, use the P/A factor in Equation [17.15].

$$0 = -50,000 + 16,000(P/A,i^*,5)$$

$$(P/A,i^*,5) = 3.125$$

Solution gives $i^* = 18.03\%$ as the after-tax rate of return.

(b) Use Equation [17.17] for the before-tax return estimate.

$$\text{Before-tax ROR} = \frac{0.1803}{1 - 0.40} = 30.05\%$$

The actual before-tax i^* using CFBT = $20,000 for 5 years is 28.65% from the relation

$$0 = -50,000 + 20,000(P/A,i^*,5)$$

The tax effect will be slightly overestimated if a MARR of 30.05% is used in a before-tax analysis.

A rate of return evaluation performed by hand on two or more alternatives must utilize a PW or AW relation to determine the incremental return Δi^* of the incremental CFAT series between two alternatives. Solution by computer is accomplished using the incremental CFAT values and the IRR function. The equations and procedures applied are the same as in Chapter 8 for selection from mutually exclusive alternatives using the ROR method. You should review and understand the following sections before proceeding further with this section.

Section 8.4   ROR evaluation using PW: incremental and breakeven
Section 8.5   ROR evaluation using AW
Section 8.6   Incremental ROR analysis of multiple, mutually exclusive alternatives

From this review, several important facts should be recalled:

Selection guideline: The fundamental rule of incremental ROR evaluation at a stated MARR is as follows:

Select the one alternative that requires the largest initial investment, provided the extra investment is justified relative to another justified alternative.

Incremental ROR: Incremental analysis must be performed. Overall i^* values cannot be depended upon to select the correct alternative, unlike the PW or AW method at the MARR, which will always indicate the correct alternative.

Equal-service assumption: Incremental ROR analysis requires that the alternatives be evaluated over equal time periods. The LCM of the two alternative lives must be used to find the PW or AW of incremental cash flows. (The only exception, mentioned in Section 8.5, occurs when the AW analysis is performed on *actual cash flows, not the increments;* then one-life-cycle analysis is acceptable over the respective alternative lives.)

Revenue and service alternatives: Revenue alternatives (positive and negative cash flows) may be treated differently from service alternatives (cost-only cash flow estimates). For revenue alternatives, the overall i^* may be used to perform an initial screening. Alternatives that have $i^* < \text{MARR}$ can be removed from further evaluation. An i^* for cost-only (service) alternatives cannot be determined, so incremental analysis is required with all alternatives included.

These principles and the same procedures developed in Chapter 8 are applied to the CFAT series. The summary table in the rear of the book (which is also Table 10–2) details the requirements of all evaluation techniques. For the ROR method, in the column labeled "Series to Evaluate" change the words *cash flows* to *CFAT values*. Additionally, use the after-tax MARR as the decision guideline (far right column). Now, all entries for the ROR method are correct for an after-tax analysis.

Once the CFAT series are developed, the *breakeven ROR* can be obtained using a plot of PW vs. i^*. Solution of the PW relation for each alternative over the LCM at several interest rates can be accomplished by hand or by using the NPV spreadsheet function. For any after-tax MARR greater than the breakeven ROR, the extra investment is not justified.

Example 17.9 illustrates an after-tax ROR evaluation of two alternatives solved by hand. The next section includes additional examples solved by computer using incremental ROR analysis and the breakeven plot of PW vs. i^*.

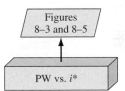

Figures 8–3 and 8–5

PW vs. i^*

EXAMPLE 17.9

Johnson Controls must decide between two alternatives in its northeast plant: system 1—a single robot assembly system for ICs will require a $100,000 investment now; and system 2—a combination of two robots requires a total of $130,000. Management intends to implement one of the plans. This manufacturer expects a 20% after-tax return on technology investments. Select one of the systems, if the following series of cost CFAT values have been estimated for the next 4 years.

			Year		
	0	1	2	3	4
System 1 CFAT, $	−100,000	−35,000	−30,000	−20,000	−15,000
System 2 CFAT, $	−130,000	−20,000	−20,000	−10,000	−5,000

Solution

System 2 is the alternative with the extra investment that must be justified. Since lives are equal, select PW analysis to estimate Δi^* for the incremental CFAT series shown here. All cash flows have been divided by $1000.

Year	0	1	2	3	4
Incremental CFAT, $1000	−30	+15	+10	+10	+10

A PW relation is set up to estimate the after-tax incremental return.

$$-30 + 15(P/F,\Delta i^*,1) + 10(P/A,\Delta i^*,3)(P/F,\Delta i^*,1) = 0$$

Solution indicates an incremental after-tax return of 20.10%, which just exceeds the 20% MARR. The extra investment in system 2 is marginally justified.

E-Solve

17.6 SPREADSHEET APPLICATIONS—AFTER-TAX INCREMENTAL ROR ANALYSIS

Two spreadsheet examples of after-tax ROR analysis are presented in this section. Both examples build upon earlier solutions in this chapter. The first example determines the incremental rate of return Δi^* and highlights the use of a PW vs. i plot. The second example illustrates incremental ROR analysis when the CFAT estimates must be calculated using MACRS depreciation.

The procedure outlined in Section 8.6 for performing an incremental ROR analysis of two or more alternatives is used in each example below. This procedure should be reviewed and understood before proceeding.

EXAMPLE 17.10

In Example 17.7, Paul estimated the CFAT for interior wall materials to reduce sound transmission; plan S is to construct with stucco on metal lath, and plan B is to construct using brick. Figure 17–5 presented both a PW analysis over 10 years and an AW analysis over the respective lives. Alternative S was selected. After reviewing this earlier solution, perform an ROR evaluation at the after-tax MARR of 7% per year.

Solution by Computer

The LCM is 10 years for the incremental ROR analysis, and plan B requires the extra investment that must be justified. Apply the procedure in Section 8.6 for incremental ROR analysis. Figure 17–6 shows the estimated CFAT for each alternative and the incremental CFAT series. Since these are revenue alternatives, the i^* is calculated first to ensure that they both make at least the MARR of 7%. Cells C14 and D14 indicate they do. The IRR function in cell E14 is applied to the incremental CFAT, indicating that $\Delta i^* = 6.35\%$. Since this is lower than the MARR, the extra investment in brick walls is not justified. Plan S is selected, the same as with the PW and AW methods.

In rows 17 through 21, Figure 17–6, the NPV function is used to find the PW of the alternatives' CFAT series at various i values. The graph indicates that the breakeven i^* occurs at 6.35%—the same result as with the IRR function above. Whenever the after-tax MARR is above 6.35%, as is the case here with MARR = 7%, the extra investment in plan B is not justified.

Microsoft Excel

File Edit View Insert Format Tools Data Window Help

A27

Example 17.10

	A	B	C	D	E
1			Plan S	Plan B	B - S
2		Year	CFAT	CFAT	Incr. CFAT
3		0	-$28,800	-$50,000	-$21,200
4		1	$5,400	$14,200	$8,800
5		2	$5,400	$13,300	$7,900
6		3	$5,400	$12,400	$7,000
7		4	$5,400	$11,500	$6,100
8		5	$5,400	-$39,400	-$44,800
9		6	$5,400	$14,200	$8,800
10		7	$2,040	$13,300	$11,260
11		8	$2,040	$12,400	$10,360
12		9	$2,040	$11,500	$9,460
13		10	$4,832	$10,600	$5,768
14	ROR		9.49%	8.05%	6.35%
15					
16		Rate	Plan S	Plan B	
17	PW @	5%	$5,721	$7,251	
18	PW @	6%	$4,296	$4,673	
19	PW @	7%	$2,963	$2,297	
20	PW @	8%	$1,715	$103	
21	PW @	9%	$544	($1,925)	

Chart: PW of CFAT, $ vs. After-tax rate of return, %
Plan S, Plan B
MARR
Breakeven: 6.35%

=IRR(E3:E13)

=NPV(B21,C4:C13)+C3   =NPV(B21,D4:D13)+D3

Sheet1 / Sheet2 / Sheet3 / Sheet4 / Sheet5 / Sheet6 / Sheet7 / Sheet8

Ready   NUM

Figure 17–6
After-tax evaluation of two alternatives with a plot of PW vs. i, Example 17.10.

Comment
Note that the incremental CFAT series has three sign changes. The cumulative series also has three sign changes (Norstrom's criterion). Accordingly, there may be multiple Δi^* values. The application of the IRR function using the "guess" option finds no other real number roots in the normal rate of return range.

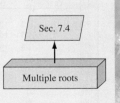

Sec. 7.4

Multiple roots

EXAMPLE 17.11

In Example 17.6 an after-tax analysis of two bone cell analyzers was initiated due to a new 3-year NBA contract. The criterion used to select analyzer 1 was the total taxes for the 3 years. The complete spreadsheet solution is in Figure 17–4.

Continue the spreadsheet analysis by performing an after-tax ROR evaluation, assuming the analyzers are sold after 3 years at the amounts estimated in Example 17.6: $130,000 for analyzer 1 and $225,000 for analyzer 2. The after-tax MARR is 10% per year.

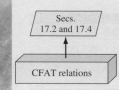

Solution by Computer

Figure 17–7 is an updated copy of the spreadsheet in Figure 17–4 to include the sale of the analyzer in year 3. The CFAT series (column I) is determined by the relation CFAT = CFBT − taxes, with the taxable income determined using Equation [17.14]. For example, in year 3 when analyzer 2 is sold for $S = \$225,000$, the CFAT calculation is

$$\text{CFAT}_3 = \text{CFBT} - (\text{TI})(T_e) = \text{GI} - E - P + S - (\text{GI}-E-D+\text{DR})(T_e)$$

The depreciation recapture DR is the amount above the year 3 book value received at sale time. By Equation [17.12], using the book value in cell F18,

$$\text{DR} = \text{selling price} - \text{BV}_3 = 225,000 - 64,800 = \$160,200$$

Now, the CFAT in year 3 for analyzer 2 can be determined.

$$\text{CFAT}_3 = 100,000 - 10,000 + 0 + 225,000$$
$$- (100,000-10,000-43,200 + 160,200)(0.35)$$
$$= 315,000 - 207,000(0.35) = \$242,550$$

The cell tags in row 18 of Figure 17–7 follow this same progression. The incremental CFAT is calculated in column J.

Figure 17–7
Incremental ROR analysis of CFAT with depreciation recapture in the last year, Example 17.11.

The procedure to compare alternatives outlined in Section 8.6 is applied. These are revenue alternatives, so the overall i^* values (cells I9 and I19) indicate that both CFAT series are acceptable. The value $\Delta i^* = 23.6\%$ (cell J19) also exceeds MARR = 10%, so *analyzer 2 is selected.* This decision applies the ROR method guideline: Select the alternative that requires the largest, *incrementally* justified investment.

Comment

In Section 8.4, Figure 8–5b demonstrated the fallacy of selecting an alternative based solely upon the ROR. The incremental ROR must be used. The same fact is demonstrated in this example. If the larger i^* alternative is chosen, analyzer 1 is incorrectly selected. When Δi^* exceeds the MARR, the larger investment is correctly chosen—analyzer 2 in this case. For verification, the PW at 10% is calculated for each analyzer (I10 and I20). Again, analyzer 2 is the winner, based on its larger PW of $93,905.

17.7   AFTER-TAX REPLACEMENT STUDY

When a currently installed asset (the defender) is challenged with possible replacement, the effect of taxes can have an impact upon the decision of the replacement study. The final decision may not be reversed by taxes, but the difference between before-tax AW values may be significantly different from the after-tax difference. There may be tax considerations in the year of the possible replacement due to *depreciation recapture* or *capital gain,* or there may be tax savings due to a sizable *capital loss,* if it is necessary to trade the defender at a sacrifice price. Additionally, the after-tax replacement study considers tax-deductible *depreciation* and *operating expenses* not accounted for in a before-tax analysis. The effective tax rate T_e is used to estimate the amount of annual taxes (or tax savings) from TI. The same procedure as the before-tax replacement study in Chapter 11 is applied here, but for CFAT estimates. The procedure should be thoroughly understood before proceeding. Special attention to Sections 11.3 and 11.5 is recommended.

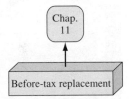

Example 17.12 presents a solution by hand of an after-tax replacement study using a simplifying assumption of classical SL (straight line) depreciation. Example 17.13 solves the same problem by computer, but includes the detail of MACRS depreciation. This provides an opportunity to observe the difference in the AW values between the two depreciation assumptions.

EXAMPLE 17.12

Midcontinent Power Authority purchased coal extraction equipment 3 years ago for $600,000. Management has discovered that it is technologically outdated now. New equipment has been identified. If the market value of $400,000 is offered as the trade-in for the current equipment, perform a replacement study using (a) a before-tax MARR of 10% per year and (b) a 7% per year after-tax MARR. Assume an effective tax rate of 34%. As a simplifying assumption, use classical straight line depreciation with $S = 0$ for both alternatives.

	Defender	Challenger
Market value, $	400,000	
First cost, $		1,000,000
Annual cost, $/year	−100,000	−15,000
Recovery period, years	8 (originally)	5

Solution

Assume that an ESL (economic service life) analysis has determined the best life values to be 5 more years for the defender and 5 years total for the challenger.

(a) For the *before-tax replacement study,* find the AW values. The defender AW uses the market value as the first cost, $P_D = \$-400,000$.

$$AW_D = -400,000(A/P,10\%,5) - 100,000 = \$-205,520$$

$$AW_C = -1,000,000(A/P,10\%,5) - 15,000 = \$-278,800$$

Applying step 1 of the replacement study procedure (Section 11.3), we select the better AW value. The defender is retained now with a plan to keep it for the 5 remaining years. The defender has a $73,280 lower equivalent annual cost compared to the challenger. This complete solution is included in Table 17–6 (left) for comparison with the after-tax study.

(b) For the *after-tax replacement study,* there are no tax effects other than income tax for the defender. The annual SL depreciation is $75,000, determined when the equipment was purchased 3 years ago.

$$D_t = 600,000/8 = \$75,000 \qquad t = 1 \text{ to } 8 \text{ years}$$

Table 17–6 shows the TI and taxes at 34%. The taxes are actually tax savings of $59,500 per year, as indicated by the minus sign. (Remember that for tax savings in an economic analysis it is assumed there is positive taxable income elsewhere in the corporation to offset the saving.) Since only costs are estimated, the annual CFAT is negative, but the $59,500 tax savings has reduced it. The CFAT and AW at 7% per year are

$$\text{CFAT} = \text{CFBT} - \text{taxes} = -100,000 - (-59,500) = \$-40,500$$

$$AW_D = -400,000(A/P,7\%,5) - 40,500 = \$-138,056$$

For the challenger, depreciation recapture on the defender occurs when it is replaced, because the trade-in amount of $400,000 is larger than the current book value. In year 0 for the challenger, Table 17–6 includes the following computations to arrive at a tax of $8500.

Defender book value, year 3:	$BV_3 = 600,000 - 3(75,000) = \$375,000$
Depreciation recapture:	$DR_3 = TI = 400,000 - 375,000 = \$25,000$
Taxes on the trade-in, year 0:	Taxes $= 0.34(25,000) = \$8500$

TABLE 17–6   Before-Tax and After-Tax Replacement Analyses, Example 17.12

Defender Age	Year	Before Taxes Expenses E	P and S	CFBT	After Taxes Depreciation D	Taxable Income TI	Taxes* at 0.34TI	CFAT
				DEFENDER				
3	0		$−400,000	$−400,000				$−400,000
4	1	$−100,000		−100,000	$75,000	$−175,000	$−59,500	−40,500
5	2	−100,000		−100,000	75,000	−175,000	−59,500	−40,500
6	3	−100,000		−100,000	75,000	−175,000	−59,500	−40,500
7	4	−100,000		−100,000	75,000	−175,000	−59,500	−40,500
8	5	−100,000	0	−100,000	75,000	−175,000	−59,500	−40,500
AW at 10%				$−205,520	AW at 7%			$−138,056
				CHALLENGER				
	0		$−1,000,000	$−1,000,000		$+25,000†	$ 8,500	$−1,008,500
	1	$−15,000		−15,000	$200,000	−215,000	−73,100	+58,100
	2	−15,000		−15,000	200,000	−215,000	−73,100	+58,100
	3	−15,000		−15,000	200,000	−215,000	−73,100	+58,100
	4	−15,000		−15,000	200,000	−215,000	−73,100	+58,100
	5	−15,000	0	−15,000	200,000	−215,000‡	−73,100	+58,100
AW at 10%				$−278,800	AW at 7%			$−187,863

* Minus sign indicates a tax savings for the year.
† Depreciation recapture on defender trade-in.
‡ Assumes challenger's salvage actually realized is $S = 0$; no tax.

The SL depreciation is $1,000,000/5 = \$200,000$ per year. This results in tax saving and CFAT as follows:

$$\text{Taxes} = (-15,000 - 200,000)(0.34) = \$-73,100$$

$$\text{CFAT} = \text{CFBT} - \text{taxes} = -15,000 - (-73,100) = \$+58,100$$

In year 5, it is assumed the challenger is sold for $0; there is no depreciation recapture. The AW for the challenger at the 7% after-tax MARR is

$$AW_C = -1,000,000(A/P,7\%,5) + 58,100 = \$-187,863$$

The defender is again selected; however, the equivalent annual advantage has decreased from $73,280 before taxes to $49,807 after taxes.

Conclusion: By either analysis, retain the defender now and plan to keep it for 5 more years. Additionally, plan to evaluate the estimates for both alternatives 1 year hence. If and when cash flow estimates change significantly, perform another replacement analysis.

Comment

If the market value (trade-in) had been less than the current defender book value of $375,000, a capital loss, rather than depreciation recapture, would occur in year 0. The resulting tax savings would decrease the CFAT (which is to reduce costs if CFAT is negative). For example, a trade-in amount of $350,000 would result in a TI of $350,000 − 375,000 = $−25,000 and a tax savings of $−8500 in year 0. The CFAT is then $−1,000,000 − (−8500) = $−991,500.

EXAMPLE **17.13**

Repeat the after-tax replacement study of the previous example (17.12*b*) using 7-year MACRS depreciation for the defender and 5-year MACRS depreciation for the challenger. Assume either asset is sold after the 5 years for exactly its book value. Determine if the answers are significantly different from those obtained when the simplifying assumption of classical SL depreciation was made.

Solution

Figure 17–8 shows the complete analysis. Refer to the cell tags for details of the computations. MACRS requires substantially more computation than SL depreciation, but this effort is easily reduced by the use of a spreadsheet. *Again the defender is selected for retention, but now by an advantage of $44,142 annually.* This compares to the $49,807 advantage using classical SL depreciation and the $73,280 before-tax advantage of the defender. Therefore, taxes and MACRS have both reduced the defender's economic advantage, but not enough to reverse the decision to retain it.

Several other differences in the results between SL and MACRS depreciation are worth noting. There is depreciation recapture in year 0 of the challenger due to trade in of the defender at $400,000, a value larger than the book value of the 3-year-old defender. This amount, $137,620 in cell G18, is treated as ordinary taxable income. The calculations for the DR and associated tax, by hand, are as follows:

$$BV_3 = \text{first cost} - \text{MACRS depreciation for 3 years}$$

$$= \text{total MACRS depreciation for years 4 through 8}$$

$$= \$262,380 \qquad \qquad \text{(cell F11)}$$

$$DR = TI_0 = \text{trade in} - BV_3$$

$$= 400,000 - 262,380 = \$137,620 \qquad \text{(cell G18)}$$

$$\text{Taxes} = 137,620(0.34) = \$46,790 \qquad \text{(cell H18)}$$

See the cell tags for the spreadsheet relations that duplicate this logic.

The assumption that the challenger is sold after 5 years at its book value implies a positive cash flow in year 5. The entry $57,600 in cell C23 of Figure 17–8 reflects this assumption, since the foregone MACRS depreciation in year 6 would be $1,000,000(0.0576) = $57,600. The spreadsheet relation B15-F24 in cell C23 determines

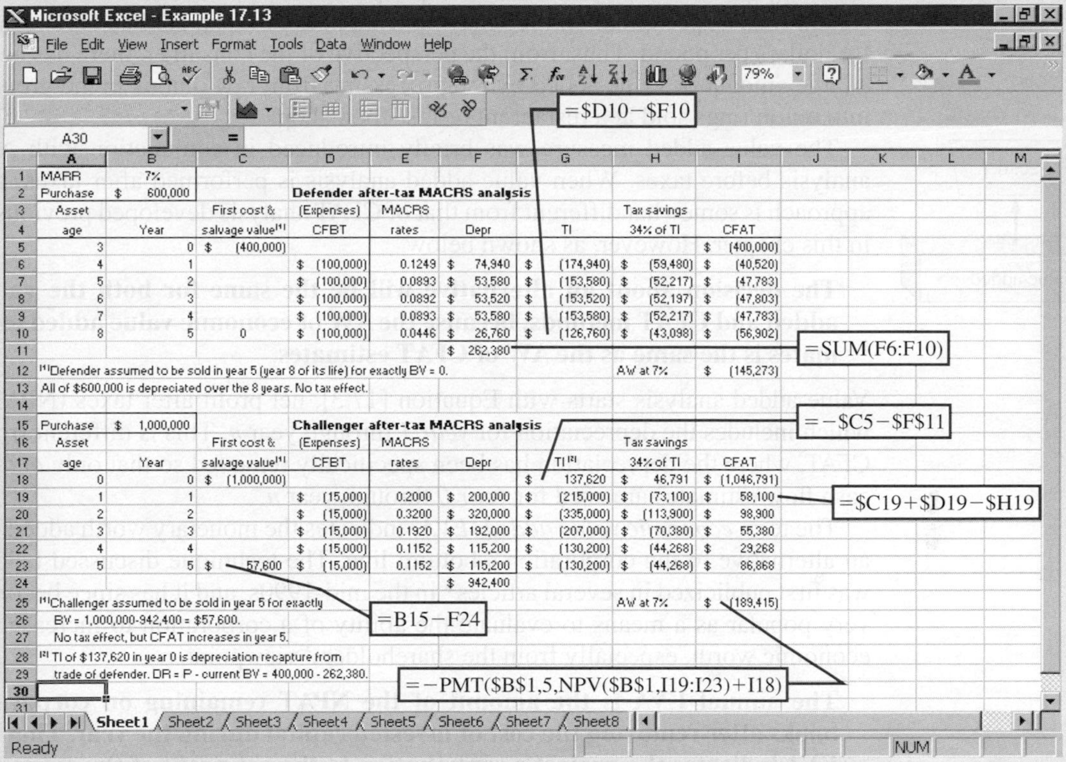

The spreadsheet shows:

```
X Microsoft Excel - Example 17.13                                          _ 8 X
 S] File  Edit  View  Insert  Format  Tools  Data  Window  Help            _ 8 X

 D  B  H   S  Q  ✓   X  B  B  ✓   K? -  □ -   Q  F   Σ  fx  A↓  Z↓  M  Q  A   79%  -  ?    -  ⬥ - A -

                      ▼  ⬛ -  ▤ ▦  ▤ ▦  ✂  ✐          =$D10-$F10

       A30        ▼        =
       A        B        C        D        E        F        G        H        I        J    K    L    M
```

	A	B	C	D	E	F	G	H	I
1	MARR	7%							
2	Purchase	$ 600,000		**Defender after-tax MACRS analysis**					
3	Asset		First cost &	(Expenses)	MACRS			Tax savings	
4	age	Year	salvage value[1]	CFBT	rates	Depr	TI	34% of TI	CFAT
5	3	0	$ (400,000)						$ (400,000)
6	4	1		$ (100,000)	0.1249	$ 74,940	$ (174,940)	$ (59,480)	$ (40,520)
7	5	2		$ (100,000)	0.0893	$ 53,580	$ (153,580)	$ (52,217)	$ (47,783)
8	6	3		$ (100,000)	0.0892	$ 53,520	$ (153,520)	$ (52,197)	$ (47,803)
9	7	4		$ (100,000)	0.0893	$ 53,580	$ (153,580)	$ (52,217)	$ (47,783)
10	8	5	0	$ (100,000)	0.0446	$ 26,760	$ (126,760)	$ (43,098)	$ (56,902)
11						$ 262,380			
12	[1]Defender assumed to be sold in year 5 (year 8 of its life) for exactly BV = 0.							AW at 7%	$ (145,273)
13	All of $600,000 is depreciated over the 8 years. No tax effect.								
14									
15	Purchase	$ 1,000,000		**Challenger after-tax MACRS analysis**					
16	Asset		First cost &	(Expenses)	MACRS			Tax savings	
17	age	Year	salvage value[1]	CFBT	rates	Depr	TI [2]	34% of TI	CFAT
18	0	0	$ (1,000,000)				$ 137,620	$ 46,791	$ (1,046,791)
19	1	1		$ (15,000)	0.2000	$ 200,000	$ (215,000)	$ (73,100)	$ 58,100
20	2	2		$ (15,000)	0.3200	$ 320,000	$ (335,000)	$ (113,900)	$ 98,900
21	3	3		$ (15,000)	0.1920	$ 192,000	$ (207,000)	$ (70,380)	$ 55,380
22	4	4		$ (15,000)	0.1152	$ 115,200	$ (130,200)	$ (44,268)	$ 29,268
23	5	5	$ 57,600	$ (15,000)	0.1152	$ 115,200	$ (130,200)	$ (44,268)	$ 86,868
24						$ 942,400			
25	[1]Challenger assumed to be sold in year 5 for exactly							AW at 7%	$ (189,415)
26	BV = 1,000,000-942,400 = $57,600.								
27	No tax effect, but CFAT increases in year 5.								
28	[2] TI of $137,620 in year 0 is depreciation recapture from								
29	trade of defender. DR = P - current BV = 400,000 - 262,380.								
30									

Cell tags shown: `=SUM(F6:F10)`, `=-$C5-$F$11`, `=$C19+$D19-$H19`, `=B15-F24`, `=-PMT($B$1,5,NPV($B$1,I19:I23)+I18)`

```
 ▶ ▶ H \Sheet1 / Sheet2 / Sheet3 / Sheet4 / Sheet5 / Sheet6 / Sheet7 / Sheet8  ◀
 Ready                                                                    NUM
```

Figure 17–8
After-tax replacement study using MACRS depreciation, Example 17.13.

this value using the accumulated depreciation in F24. [*Note:* If the amount $S = 0$ is definitely anticipated after 5 years, then a capital loss of $57,600 would be incurred. This would imply an additional tax saving of $57,600(0.34) = \$19,584$ in year 5. Conversely, if the salvage value exceeds the book value, a depreciation recapture and associated tax would be estimated.]

The cell tag indicates that the CFAT estimates are developed using columns C, D, and H. For example, in cell I19, the CFAT is determined as C19 + D19 − H19. Even though columns C and D do not usually both contain entries, this general form for CFAT calculation works correctly when the relation is dragged down through column I. Therefore, the CFAT computation in cell I23 correctly reflects the challenger sale in year 5 without altering the spreadsheet formula.

17.8 AFTER-TAX VALUE-ADDED ANALYSIS

Value added is a term used to indicate that a product or service has added worth from the perspective of the owner, an investor, or a consumer. It is possible to highly leverage the value-added activities of a process. For example, onions are grown and sold at the farm level for cents per pound. They may be purchased by the shopper in a store at 25 to 50 cents per pound. But when onions are cut and coated with a special batter, they may be fried in hot oil and sold as onion rings for dollars per pound. Thus, from the perspective of the consumer, there has been a large amount of value added by the processing from raw onions in the ground into onion rings sold at a restaurant or fast-food shop.

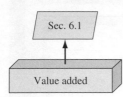

The value added measure was briefly introduced in conjunction with AW analysis before taxes. When value added analysis is performed after taxes, the approach is somewhat different from that of CFAT analysis developed previously in this chapter. However, as shown below,

> **The decision about an alternative will be the same for both the value added and CFAT methods, because the AW of economic value added estimates is the same as the AW of CFAT estimates.**

Value added analysis starts with Equation [17.3], net profit after taxes (NPAT), which includes the depreciation for *year 1* through year *n*. This is different from CFAT, where the depreciation has been specifically removed so that only *actual* cash flow estimates are used for *year 0* through year *n*.

The term *economic value added (EVA)* indicates the monetary worth added by an alternative to the corporation's bottom line. The technique discussed below was first publicized in several articles* in the mid-1990s, and it has since become very popular as a means to evaluate the ability of a corporation to increase its economic worth, especially from the shareholders' viewpoint.

> **The annual EVA is the amount of the NPAT remaining on corporate books after removing the cost of invested capital during the year. That is, EVA indicates the project's contribution to the net profit of the corporation after taxes.**

The *cost of invested capital* is the after-tax rate of return (usually the MARR value) multiplied by the book value of the asset during the year. This is the interest incurred by the current level of capital invested in the asset. (If different tax and book depreciation methods are used, the *book depreciation value is used* here, because it more closely represents the remaining capital invested in the asset from the corporation's perspective.) Computationally,

$$\text{EVA} = \text{NPAT} - \text{cost of invested capital}$$
$$= \text{NPAT} - (\text{after-tax interest rate})(\text{book value in year } t - 1)$$
$$= \text{TI}(1 - T_e) - (i)(\text{BV}_{t-1}) \qquad [17.18]$$

Since both TI and the book value consider depreciation, EVA is a measure of worth that mingles actual cash flow with noncash flows to determine the estimated financial worth contribution to the corporation. This financial worth is the amount used in public documents of the corporation (balance sheet, income statement, stock reports, etc.). Because corporations want to present the largest

---

*A. Blair, "EVA Fever," *Management Today,* Jan. 1997, pp. 42–45; W. Freedman, "How Do You Add Up?" *Chemical Week,* Oct. 9, 1996, pp. 31–34.

value possible to the stockholders and other owners, the EVA method may be more appealing from the financial perspective than the AW method.

The result of an EVA analysis is a series of annual EVA estimates. Two or more alternatives are compared by calculating the AW of EVA estimates and selecting the alternative with the larger AW value. If only one project is evaluated, AW $>$ 0 means the after-tax MARR is exceeded, thus making the project value adding.

Sullivan and Needy* have demonstrated that the AW of EVA and the AW of CFAT are identical in amount. Thus, either method can be used to make a decision. The annual EVA estimates indicate added worth to the corporation generated by the alternative, while the annual CFAT estimates describe how cash will flow. This comparison is made in Example 17.14.

EXAMPLE 17.14

Biotechnics Engineering has developed two mutually exclusive plans for investing in new capital equipment with the expectation of increased revenue from its medical diagnostic services to cancer patients. The estimates are summarized below. (*a*) Use classical straight line depreciation, an after-tax MARR of 12%, an effective tax rate of 40%, and solution by computer to perform two annual worth after-tax analyses: EVA and CFAT. (*b*) Explain the fundamental difference between the results of the two analyses.

	Plan A	Plan B
Initial investment	$500,000	$1,200,000
Gross income − expenses	$170,000 per year	$600,000 in year 1, decreasing by $50,000 per year thereafter
Estimated life	4 years	4 years
Salvage value	None	None

Solution by Computer

(*a*)   Refer to the spreadsheet and cell tags in Figure 17–9.

   EVA evaluation: All the necessary information for EVA estimation is determined in columns B through G. The net profit after taxes (NPAT) in column H is calculated by Equation [17.3], TI − taxes. The book values (column E) are used to determine the cost of invested capital in column I, using the second term in Equation [17.18], that is, $i(BV_{t-1})$, where i is the 12% after-tax MARR. This represents the amount of interest at 12% per year, after taxes, for the currently invested capital as reflected by the book value at the beginning of the year. The annual EVA estimate is the sum of columns H and I (Equation [17.18]) for years 1 through 4. *Notice there is no EVA estimate for year 0,* since NPAT and the cost of invested capital are estimated for years 1 through *n*. Finally, the larger AW of the EVA value is selected (J21), which indicates that plan B is better, and that plan A does not make the 12% return (J10).

E-Solve

---

* W. G. Sullivan and K. L. Needy, "Determination of Economic Value Added for a Proposed Investment in New Manufacturing." *The Engineering Economist,* vol. 45, no. 2 (2000), pp. 166–181.

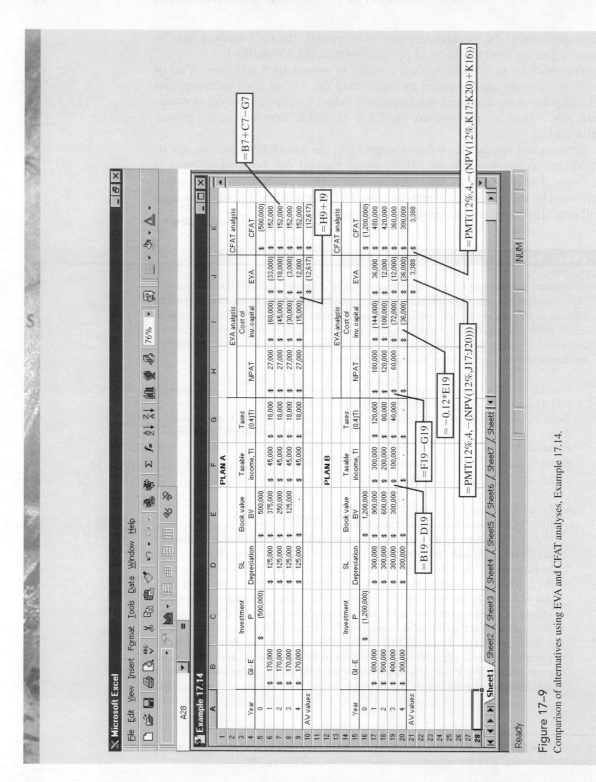

Figure 17–9
Comparison of alternatives using EVA and CFAT analyses, Example 17.14.

CFAT evaluation: As shown in the cell tag, CFAT estimates (column K) are calculated as GI − E − P − taxes, Equation [17.8]. The AW of CFAT (cell K21) again concludes that plan B is better and that plan A does not return the after-tax MARR of 12% (K10).

(b) What is the fundamental difference between the EVA and CFAT series in columns J and K? They are clearly equivalent from the time value of money perspective since the AW values are numerically the same. To answer the question, consider plan A, which has a constant CFAT estimate of $152,000 per year. To obtain the AW of EVA estimate of $−12,617 for years 1 through 4, the initial investment of $500,000 is distributed over the 4-year life using the A/P factor at 12%. That is, an equivalent amount of $500,000(A/P,12\%,4) = \$164,617$ is "charged" against the cash inflows in each of years 1 through 4. In effect, the yearly CFAT is reduced by this charge.

$$\text{CFAT} - (\text{initial investment})(A/P,12\%,4) = \$152,000 - 500,000(A/P,12\%,4)$$

$$152,000 - 164,617 = \$-12,617$$

$$= \text{AW of EVA}$$

This is the AW value for both series, demonstrating that the two methods are economically equivalent. However, the EVA method indicates an alternative's yearly estimated contribution to the *value of the corporation,* whereas the CFAT method estimates the actual cash flows to the corporation. This is why the EVA method is often more popular than the cash flow method with corporate executives.

Comment

The calculation $P(A/P,i,n) = \$500,000(A/P,12\%,4)$ is exactly the same as the capital recovery in Equation [6.3], assuming an estimated salvage value of zero. Thus, the cost of invested capital for EVA is the same as the capital recovery discussed in Chapter 6. This further demonstrates why the AW method is economically equivalent to the EVA evaluation.

17.9 AFTER-TAX ANALYSIS FOR INTERNATIONAL PROJECTS

Primary questions to be answered prior to performing a corporate-based after-tax analysis for international settings revolve around tax-deductible allowances—depreciation, business expenses, capital asset evaluation—and the effective tax rate needed for Equation [17.6], taxes $= TI(T_e)$. As discussed in Chapter 16, most governments of the world recognize and use the straight line (SL) and declining balance (DB) methods of depreciation with some variations to determine the annual tax-deductible allowance. Expense deductions vary widely from country to country. By way of example, some of these are summarized here.

Canada

Depreciation: This is deductible and is normally based on DB calculations, although SL may be used. An equivalent of the half-year convention is applied in the first year of ownership. The annual tax-deductible allowance is termed *capital cost allowance (CCA)*. As in the U.S. system, recovery rates are standardized, so the depreciation amount does not necessarily reflect the useful life of an asset.

Class and CCA rate: Asset classes are defined and annual depreciation rates are specified by class. No specific recovery period (life) is identified, in part because assets of a particular class are grouped together and the annual CCA is determined for the entire class, not individual assets. There are some 44 classes, and CCA rates vary from 4% per year (the equivalent of a 25-year-life asset) for buildings (class 1) to 100% (1-year life) for applications software, chinaware, dies, etc. (class 12). Most rates are in the range of 10% to 30% per year.

Expenses: Business expenses are deductible in calculating TI. Expenses related to capital investments are not deductible, since they are accommodated through the CCA.

Internet: Further details are available on the Revenue Canada website at www.ccra-adrc.gc.ca in the Forms and Publications section.

Mexico

Depreciation: This is a fully deductible allowance for calculating TI. The SL method is applied with an index for inflation considered each year. Monthly prorated depreciation is used for partial-year use. For some asset types, an immediate deduction of a percentage of the first cost is allowed. (This is a close equivalent to the capital expense deduction in the United States.)

Class and rates: Asset types are identified, though not as specifically defined as in some countries. Major classes are identified, and annual recovery rates vary from 5% for buildings (the equivalent of a 20-year life) to 100% for environmental machinery. Most rates range from 10% to 30% per year.

Profit tax: The income tax is levied on profits on income earned from carrying on business in Mexico. Most business expenses are deductible. Corporate income is taxed only once, at the federal level; no state-level taxes are imposed.

Tax on Net Assets (TNA): A tax of 1.8% of the average value of assets located in Mexico is paid annually in addition to income taxes, but income taxes paid are credited toward the TNA due.

Internet: The best information is via websites for corporations that assist international corporations located in Mexico. One example is PriceWaterhouseCoopers at www.pwcglobal.com/mx/eng.

Japan

Depreciation: This is fully tax-deductible and is based on classical SL or DB methods. A total of 95% of the unadjusted basis or first cost, as defined in Chapter 16, may be recovered through depreciation, but an asset's salvage value is assumed to be 10% of its acquisition cost. The capital investment is recovered for each asset or groups of similarly classified assets.

Class and life: A statutory useful life ranging from 4 to 24 years, with a 50-year life for reinforced concrete buildings, is specified.

Expenses: Business expenses are deductible in calculating TI.

TABLE 17–7	Summary of International Corporate Tax Rates*	

Tax Rate Levied on Taxable Income, %	For These Countries
≥ 40	United States, Japan, Saudi Arabia
36 to < 40	Canada, Germany, South Africa
32 to < 36	China, France, India, Mexico, New Zealand, Spain, Turkey
28 to < 32	Australia, Indonesia, United Kingdom, Republic of Korea
24 to < 28	Russia, Taiwan
20 to < 24	Singapore
< 20	Hong Kong, Iceland, Ireland, Hungary, Poland

*Sources: Extracted from KPMG's Corporate Tax Rates Survey, January 2004 (available at *www.kpmg.com/RUT2000_prod/documents/9/2004ctrs.pdf*), and from country websites on taxation for corporations.

Internet: Further details are available on the Japanese Ministry of Finance website at www.mof.go.jp.

The effective tax rate varies considerably among countries. Some countries levy taxes only at the federal level, while others impose taxes at several levels of government (federal, state or provincial, prefecture, county, and city). A summary of international corporate average tax rates is presented in Table 17–7 for a wide range of industrialized countries. These include income taxes at all reported levels of government within each country; however, other types of taxes may be imposed by a particular government. Although these average rates of taxation will vary from year to year, especially as tax reform is enacted, it can be surmised that most corporations face effective rates of about 20% to 40% of taxable income. If one were to examine the individual published rates over recent years, it is clear that countries, especially in Europe, have reduced their corporate tax rates considerably, as much as 15% to 20% in a single year in some cases. This reduction encourages corporate investment and business expansion within country borders.

CHAPTER SUMMARY

After-tax analysis does not usually change the decision to select one alternative over another; however, it does offer a much clearer estimate of the monetary impact of taxes. After-tax PW, AW, and ROR evaluations of one or more alternatives are performed on the CFAT series using exactly the same procedures as in previous chapters. The after-tax MARR is used in all PW and AW computations, and in deciding between two or more alternatives using incremental ROR analysis.

Income tax rates for U.S. corporations and individual taxpayers are graduated—higher taxable incomes pay higher income taxes. A single-value, effective tax rate T_e is usually applied in an after-tax economic analysis. Taxes are reduced because of tax-deductible items, such as depreciation and operating expenses. Because depreciation is a noncash flow, it is important to consider depreciation only in the TI computations, and not directly in the CFBT and CFAT calculations. Accordingly, key general cash flow after-tax relations for each year are

$$\text{CFBT} = \text{gross income} - \text{expenses} - \text{initial investment} + \text{salvage value}$$

$$\text{CFAT} = \text{CFBT} - \text{taxes} = \text{CFBT} - (\text{taxable income})(T_e)$$

$$\text{TI} = \text{gross income} - \text{expenses} - \text{depreciation} + \text{depreciation recapture}$$

If an alternative's estimated contribution to corporate financial worth is the economic measure, the economic value added (EVA) should be determined. Unlike CFAT, the EVA includes the effect of depreciation.

$$\text{EVA} = \text{net profit after taxes} - \text{cost of invested capital}$$

$$= \text{NPAT} - (\text{after-tax MARR})(\text{book value}) = \text{TI} - \text{taxes} - i(\text{BV})$$

The equivalent annual worths of CFAT and EVA estimates are the same numerically, due to the fact that they interpret the annual cost of the capital investment in different, but equivalent manners when the time value of money is taken into account.

In a replacement study, the tax impact of depreciation recapture or capital loss, either of which may occur when the defender is traded for the challenger, is accounted for in an after-tax analysis. The replacement study procedure of Chapter 11 is applied. The tax analysis may not reverse the decision to replace or retain the defender, but the effect of taxes will likely reduce (possibly by a significant amount) the economic advantage of one alternative over the other.

PROBLEMS

Basic Tax Computations

17.1 Write the equation to calculate TI and NPAT for a corporation, using only the following terms: *gross income, tax rate, business expenses,* and *depreciation.*

17.2 Describe the basic difference between an *income tax* and a *property tax* for an individual.

17.3 From the following list, select the tax-related term that is best described by each event below: *depreciation, operating expense, taxable income, income tax,* or *net profit after taxes.*

(a) A corporation reports that it had a negative $200,000 net profit on its annual income statement.

(b) An asset with a current book value of $80,000 was utilized on a new processing line to increase sales by $200,000 this year.

(c) A machine has an annual, straight line write-off equal to $21,000.

(d) The cost to maintain equipment during the past year was $3,680,200.

(e) A particular supermarket collected $23,550 in lottery ticket sales last year. Based on winnings paid to individuals holding these tickets, a rebate of $250 was sent to the store manager.

17.4 Two companies have the following values on their annual tax returns.

	Company 1	Company 2
Sales revenue, $	1,500,000	820,000
Interest revenue, $	31,000	25,000
Expenses, $	−754,000	−591,000
Depreciation, $	148,000	18,000

(a) Calculate the federal income tax for the year.

(b) Determine the percent of sales revenue each company will pay in federal income taxes.

(c) Estimate the taxes using an effective rate of 34% of the entire TI. Determine the percentage error made relative to the exact taxes in part (a).

17.5 Last year, one separate division of Compete.com, a dot-com sports industry service firm that provides real-time analysis of mechanical stress due to athlete injury, had $300,000 in taxable income. This year, TI is estimated to be $500,000. Calculate the federal income taxes and answer the following.

(a) What was the average federal tax rate paid last year?

(b) What is the marginal federal tax rate on the additional TI?

(c) What will be the average federal tax rate this year?

(d) What will be the NPAT on just the additional $200,000 in taxable income?

17.6 Yamachi and Nadler of Hawaii have a gross income of $6.5 million for the year. Depreciation and expenses total $4.1 million. If the combined state and local tax rate is 7.6%, use an effective federal rate of 34% to estimate the income taxes, using the effective tax rate equation.

17.7 Rotana Construction, Inc., has operated for the last 21 years in a northern U.S. state where the state income tax on corporate revenue is 6% per year. Rotana pays an average federal tax of 23% and reports taxable income of $7 million. Because of pressing labor cost increases, liability insurance premium increases, and other cost increases, the president wants to move to another state to reduce the total tax burden. The new state may have to be willing to offer tax allowances or an interest-free grant for the first couple of years in order to attract the company. You are an engineer with the company and are asked to do the following.

(a) Determine the effective tax rate for Rotana.

(b) Estimate the state tax rate that would be necessary to reduce the overall effective tax rate by 10% per year.

(c) Determine what the new state would have to do financially for Rotana to move there and to reduce its effective tax rate to 22% per year.

17.8 Workman Tools reported a TI of $80,000 last year. If the state income tax rate is 6%, determine the (a) average federal tax rate, (b) overall effective tax rate, (c) total taxes to be paid based on the effective tax rate, and (d) total taxes paid to the state and paid to the federal government.

17.9 Donald is a civil engineer with an annual salary of $98,000. He has dividends and interest of $7500 for the year. Total exemptions and deductions are $10,500.

(a) Calculate federal income taxes as a person filing single.

(b) Determine what percentage of his annual salary goes toward federal income taxes.

(c) Calculate by how much the total of exemptions and deductions has to increase for Donald's income taxes to go down by 10%.

CFBT and CFAT

17.10 What is the basic difference between cash flow after taxes (CFAT) and net profit after taxes (NPAT)?

17.11 Derive a general relation for calculating CFAT under the situation that there is no annual depreciation to deduct and it is a year in which no investment P or salvage S occurs.

17.12 Where is depreciation considered in the CFBT and CFAT expressions used to analyze an engineering economy alternative's cash flow estimates?

17.13 Four years ago ABB purchased an asset for $300,000 with an estimated salvage of $60,000. Depreciation was $60,000 per year. The following annual gross incomes and expenses were recorded. The asset was sold for $60,000 after 4 years.
- (a) Tabulate the cash flows by hand after an effective 32% tax rate is applied. Use the format of Table 17–3.
- (b) Continue the table above and calculate the net income (NI) estimates.
- (c) Set up the spreadsheet and determine the annual CFAT and NI values. Additionally, plot these values versus year of ownership.

Year of ownership	1	2	3	4
Gross income, $	80,000	150,000	120,000	100,000
Expenses, $	−20,000	−40,000	−30,000	−50,000

17.14 Four years ago Hartcourt-Banks purchased an asset for $200,000 with an estimated $S = \$40,000$. MACRS depreciation was charged over a 3-year recovery period. The following gross incomes and expenses were recorded, and an effective tax rate of 40% was applied. Tabulate CFAT under the assumption that the asset was (a) discarded for $0 after 4 years and (b) sold for $20,000 after 4 years. For this tabulation only, neglect any taxes that may be incurred on the sale of the asset.

Year of ownership	1	2	3	4
Gross income, $	80,000	150,000	120,000	100,000
Expenses, $	−20,000	−40,000	−30,000	−50,000

17.15 A petroleum engineer with Halstrom Exploration must estimate the minimum required cash flow before taxes if the CFAT is $2,000,000. The effective federal tax rate is 35%, and the state tax rate is 4.5%. A total of $1 million in tax-deductible depreciation will be charged this year. Estimate the required CFBT.

17.16 A division of Hanes has the following data at the end of a year.

$$\text{Total revenue} = \$48 \text{ million}$$
$$\text{Depreciation} = \$8.2 \text{ million}$$
$$\text{Operating expenses} = \$28 \text{ million}$$

For an effective federal tax rate of 35% and state tax rate of 6.5%, determine (a) CFAT, (b) percentage of total revenue expended on taxes, and (c) net income for the year.

17.17 Wal-Mart Distribution Centers has put into service forklifts and conveyors purchased for $250,000. Use a spreadsheet to tabulate the CFBT, CFAT, and NPAT for 6 years of ownership, using an effective tax rate of 40% and the estimated cash flow and depreciation amounts shown. Salvage is expected to be zero.

Year	Gross Income, $	Operating Expenses, $	MACRS Depreciation, $
1	90,000	−20,000	50,000
2	100,000	−20,000	80,000
3	60,000	−22,000	48,000
4	60,000	−24,000	28,800
5	60,000	−26,000	28,800
6	40,000	−28,000	14,400

17.18 A highway construction company purchased pipeline boring equipment for $80,000 and depreciated it using MACRS and a 5-year recovery period. It had no estimated salvage and produced annual $GI - E$ amounts of $50,000, which were taxed at an effective 38%. The company decided to prematurely sell the equipment after 2 full years of use.
(a) Find the price if the company wants to sell it for exactly the amount reflected by the current book value.
(b) Determine the CFAT values if the equipment was actually sold after 2 years for the amount determined in part (a) and no replacement was acquired.

Depreciation Method Effects on Taxes

17.19 An overland freight company has purchased new trailers for $150,000 and expects to realize a net $80,000 in gross income over operating expenses for each of the next 3 years. The trailers have a recovery period of 3 years. Assume an effective tax rate of 35% and an interest rate of 15% per year.
(a) Show the advantage of accelerated depreciation by calculating the present worth of taxes for the MACRS method versus the classical SL method. Since MACRS takes an additional year to fully depreciate the basis, assume no CFBT beyond year 3, but include any negative tax as a tax savings.
(b) Show that the total taxes are the same for both methods.

17.20 Work Problem 17.19 using one spreadsheet.

17.21 Perform an analysis using a spreadsheet between the following two depreciation options for a company operating in a South American country. Select the preferred method based on the better PW of tax value at $i = 8\%$ and $T_e = 40\%$.

	Option 1	Option 2
Initial investment, $	−100,000	−100,000
CFBT, $/year	40,000	40,000
Depreciation method	Straight line	Double declining balance
Recovery period, years	5	8

17.22 Imperial Chem, Inc., an international chemical products company headquartered in the United Kingdom, purchased two identical systems used in the manufacture of synthetic fibers. One system is located in Pennsylvania in the United States, the other in Genoa, Italy. Each is estimated to generate additional annual CFBT of $65,000 USD for the next 6 years. The company's division in Italy is incorporated there and may not use MACRS for depreciation. Assume the classical straight line method is the one approved for foreign corporations with incorporated units in Italy. The U.S. incorporated unit uses MACRS, the same as any U.S.-based corporation would. (a) What is the difference in the present worth of taxes and total taxes for the 6 years? Why are these two totals not equal? For this analysis, use the full 6 years as the evaluation period and an interest rate of 12% per year. Neglect any capital gains, losses, or depreciation recapture at sale time, and assume that any negative income tax is a tax savings.

	Asset Located in U.S.	Asset Located in Italy
First cost, $	−250,000	−250,000
Salvage value, $	25,000	25,000
Total annual CFBT, $/year	65,000	65,000
Tax rate, %	40	40
Depreciation method	MACRS	Classical SL (no half-year convention)
Recovery period, years	5	5

17.23 An asset with a first cost of $9000 is depreciated by MACRS over a 5-year recovery period. The CFBT is estimated at $10,000 the first 4 years and $5000 thereafter as long as the asset is retained. The effective tax rate is 40%, and money is worth 10% per year. In present worth dollars, how much of the cash flow generated by the asset over its recovery period is lost to taxes? Work using hand or spreadsheet calculations as directed by the instructor.

17.24 The effective tax savings in a year TS_t due to depreciation is calculated as

$$TS_t = \text{(depreciation)(effective tax rate)}$$
$$= (D_t)(T_e)$$

(a) Develop a relation for the present worth of tax savings PW_{TS}, and explain how it may be utilized rather than the PW_{tax} criterion in evaluating the effect of depreciation on taxes.

(b) Calculate PW_{TS} for an asset using MACRS with a first cost of $80,000, no salvage value, and a recovery period of 3 years. Use $i = 10\%$ per year and $T_e = 0.42$.

17.25 An engineering graduate has taken over his father's power tool manufacturing business, Hartley Tools. The company has purchased new equipment for $200,000 with an expected life of 10 years and no salvage value. For tax purposes, the depreciation choices are as follows:

- Alternative straight line depreciation with the half-year convention and a recovery period of 10 years
- MACRS depreciation with a recovery period of 5 years

The anticipated CFBT is $60,000 per year for only 10 years. The effective tax rate for Hartley is 42% including all taxes. Use a spreadsheet to determine the following:

(a) Rate of return before taxes over the total of 11 years.
(b) Rate of return after taxes over the total of 11 years.
(c) The depreciation method that minimizes the time value of taxes. Use an interest rate of 10% per year, and consider any negative tax as a tax savings in the year incurred.

Depreciation Recapture and Capital Gains (Losses)

17.26 Determine any depreciation recapture or capital gain or capital loss generated by each event described below. Use them to determine the amount of the income tax effect, if the effective tax rate is 30%.

(a) A strip of land zoned as 'Commercial A' purchased 8 years ago for $2.6 million was just sold at a 15% profit.

(b) Earthmoving equipment purchased for $155,000 was depreciated using MACRS over a 5-year recovery period. It was sold at the end of the fifth year of ownership for $10,000.

(c) A MACRS-depreciated asset with a 7-year recovery period has been sold after 8 years at an amount equal to 20% of its first cost, which was $150,000.

17.27 Determine any depreciation recapture or capital gain or capital loss generated by each event described below. Use them to determine the amount of the income tax effect, if the effective tax rate is 40%.

(a) A 21-year-old asset was removed from service and sold for $500. When purchased, the asset was entered on the books with a basis of $P = \$180,000$, $S = \$5000$, and $n = 18$ years. Classical straight line depreciation was used for the entire recovery period.

(b) A high-technology machine was sold internationally for $10,000 more than its purchase price just after it was in service 1 year. The asset had $P = \$100,000$, $S = \$1000$, and $n = 5$ years and was depreciated by the MACRS method for 1 year.

17.28 Set up a spreadsheet, using a modification of the column headings in Table 17-3(b) to (a) graphically compare the annual CFAT series and (b) numerically compare the total CFAT of two depreciation models—MACRS depreciation and SL depreciation with the half-year convention. Use the following asset situation.

First cost = \$10,000

Salvage estimate = \$500

Recovery period = 5 years

GI − E = \$5000 per year for 6 years

The asset was sold for \$500 after 6 years of use. The effective tax rate averaged 38% per year over the time of ownership.

17.29 Mercy Hospital, a for-profit corporation, purchased sterilization equipment at a cost of \$40,000. MACRS, with a recovery period of 5 years and an estimated salvage of \$5000, was used to write off the capital investment. The equipment increased gross income by \$20,000 and expenses by \$3000 per year. A full 2 years after purchasing it, the hospital sold the equipment to a newly organized clinic for \$21,000. The effective tax rate is 35%. Determine the (a) income taxes and (b) cash flow after taxes for the asset in the year of the sale.

17.30 A couple of years ago the company Health4All purchased land, a building, and two depreciable assets from another corporation. These have all recently been disposed of. Use the information shown

to determine the presence and amount of any capital gain, capital loss, or depreciation recapture.

Asset	Purchase Price, \$	Recovery Period, Years	Current Book Value, \$	Sales Price, \$
Land	−200,000	—		245,000
Building	−800,000	27.5	300,000	255,000
Cleaner	−50,500	3	15,500	18,500
Circulator	−10,000	3	5,000	10,500

17.31 In Problem 17.14(b), Hartcourt-Banks sold a 4-year-old asset for \$20,000. The asset was depreciated using the 3-year MACRS method. (a) Recalculate the CFAT in the year of sale, taking into account any additional tax effects caused by the \$20,000 sale price. (b) What is the change in the CFAT from the amount in Problem 17.14?

17.32 (Uses same asset data as in Problem 16A4.) An asset with a first cost of \$45,000 has a life of 5 years, a salvage value of \$3000, and an anticipated CFBT of \$15,000 per year. Determine the depreciation schedule for classical SL and for switching from DDB to SL to maximize depreciation. Use $i = 18\%$ and an effective tax rate of 50% to determine how much the present worth of taxes decreases when switching is allowed. Assume that the asset is sold for \$3000 in year 6 and that any negative tax or capital loss at sale time generates a tax savings.

17.33 Retrieve IRS Publication 544 from the Web and use the material in the chapter on reporting gains and losses to explain the calculations necessary to determine net capital gain and net capital loss on Section 1231 property transactions. Describe how they are to be reported on a corporate tax return.

After-Tax Economic Analysis

17.34 Compute the required before-tax return if an after-tax return of 9% per year is expected and the state and local tax rates total 6%. The effective federal tax rate is 35%.

17.35 A division of TexacoChevron has a TI of $8.95 million for a tax year. If the state tax rate averages 5% for all the states in which the corporation operates, find the equivalent after-tax ROR required of projects that are justified only if they can demonstrate a before-tax return of 22% per year.

17.36 John made an annual return of 8% after taxes on a stock investment. His sister told him this is equivalent to a 12% per year before-tax return. What percent of taxable income is she assuming will be taken by income taxes?

17.37 An engineer co-owns a real estate rental property business, which just purchased an apartment complex for $3,500,000, using all equity capital. For the next 8 years, an annual gross income before taxes of $480,000 is expected, offset by estimated annual expenses of $100,000. The owners hope to sell the property after 8 years for the currently appraised value of $4,050,000. The applicable tax rate for ordinary taxable income is 30%. The property will be straight line depreciated over a 20-year life with a salvage value of zero. Neglect the half-year convention in depreciation computations. (*a*) Tabulate the after-tax cash flows for the 8 years of ownership, and (*b*) determine the before-tax and after-tax rates of return. Use either hand or computer presentation of the CFAT tabulation template in Table 17–3, altered to accommodate this situation.

17.38 An asset has the following series of CFBT and CFAT estimates that have been entered into the indicated columns and rows of a spreadsheet. The company uses a rate of return of 14% per year before taxes and 9% per year after taxes. Write the spreadsheet functions for each series that will display the three results of PW, AW, and ROR. In solving this problem, use the spreadsheet functions NPV, PV, and IRR at a minimum.

Row	A	B	C
4	Year	CFBT, $	CFAT, $
5	0	−200,000	−200,000
6	1	75,000	62,000
7	2	75,000	60,000
8	3	75,000	52,000
9	4	75,000	53,000
10	5	90,000	65,000

Column header (spanning B and C): **Column**

17.39 NewsRecord, Inc., has owned two subsidiary companies for the last 4 years and expects to retain ownership for 4 more years of one company and sell the other company now. You have been asked to perform an economic analysis to determine which one to sell. When purchased, North Enterprises (NE) costs $20 million and The Southern Exchange (TSE) costs $40 million. Capital assets cost $10 million for NE and $20 million for TSE when purchased 4 years ago, and they will continue to be MACRS depreciated using $n = 7$ for the remaining 4 years of their lives. The NE company will require new investment funds of $500,000 immediately due to some bad decisions made previously. TSE does not require new investment funds.

Annual estimates of future gross income (revenue) and expenses are made. All values in the table are in $1000 units. The effective tax rate is 35% per year. The board of directors has set the corporate after-tax MARR at 25% per year.

Years from now	North Enterprises (NE)				The Southern Exchange (TSE)			
	1	2	3	4	1	2	3	4
Gross income, $	2000	2500	3000	3500	4000	3000	2000	1000
Expenses, $	−500	−800	−1100	−1400	−800	−1200	−1500	−2000

To make the retain/sell recommendation, consider only the next 4 years of after-tax cash flows to determine the breakeven ROR value. (*Note:* Since these are entire corporations being analyzed, a negative income tax should not be considered as a tax savings. In this case, the income tax amount is estimated to be zero for the year.)

17.40 A civil engineer must choose between two pieces of equipment used to supplement the pumping of concrete into foundation settings.

	Machine A	Machine B
First cost, $	−35,500	−19,000
Salvage, $	4,000	3,000
CFBT/year, $	8,000	6,500
Life, years	7	7

Both machines have an estimated 7-year useful life; however, MACRS depreciation is over a 5-year recovery period. The effective tax rate is 40%, and the after-tax MARR is 8% per year. Compare the two machines, using present worth after-tax analysis. Perform the analysis (a) by computer and (b) by hand.

17.41 Two alternatives are to be evaluated by Ned. His boss wants to know the rate of return value compared to the corporate after-tax MARR of 7% per year used to decide upon any new capital investment. Perform the analysis (a) before taxes and (b) after taxes with $T_e = 50\%$ and classical SL depreciation. (Develop the hand or spreadsheet solution per your instructor.)

	X	Y
First cost, $	−12,000	−25,000
AOC, $/year	−3,000	−1,500
Salvage, $	3,000	5,000
n, years	10	10

17.42 Two machines have the following estimates.

	Machine A	Machine B
First cost, $	−15,000	−22,000
Salvage, $	3,000	5,000
AOC, $/year	−3,000	−1,500
Life, years	10	10

Either machine is to be used for a total of 10 years, then sold for the estimated salvage value. The before-tax MARR is 14% per year, after-tax MARR is 7% per year, and T_e is 50%. Select a machine on the basis of (a) before-tax PW analysis, (b) after-tax PW analysis using classical SL depreciation over the 10-year life, and (c) after-tax PW analysis using MACRS depreciation with a 5-year recovery period.

17.43 A senior engineer at Tuskegee Industries developed per unit estimates for state-of-the-art truck tire balancing machines to be utilized for the next 3 years. Up to 1000 of these will be purchased for their 450 outlets. If after-tax MARR = 8%, MACRS depreciation (with no GI − E in year 4) and $T_e = 40\%$, (a) develop a PW versus i chart that shows the after-tax breakeven ROR, and (b) use SOLVER to determine the first cost of B to make the two break even if all other estimates are retained.

	Alternative A	Alternative B
First cost, $	−10,000	−13,000
Salvage, $	0	2,000
GI − E, $/year	4,500	5,000
Recovery period, years	3	3

17.44 A European candy manufacturing plant manager must select a new irradiation system to ensure the safety of specific products, while being economical. The two alternatives available have the following estimates.

	System A	System B
First cost, $	−150,000	−85,000
CFBT, $/year	60,000	20,000
Life, years	3	5

The company is in the 35% tax bracket and assumes classical straight line depreciation for alternative comparisons performed at an after-tax MARR of 6% per year. A salvage value of zero is used when depreciation is calculated. System B can be sold after 5 years for an estimated 10% of its first cost. System A has no anticipated salvage value. Determine which is more economical.

17.45 Use (a) hand calculations and (b) spreadsheet relations to find the after-tax rate of return for the following desalinization plant equipment over a 5-year time period. The equipment, designed for special jobs, will cost $2500, will have no salvage value, and will last no more than 5 years. Revenue minus expenses is estimated to be $1500 in year 1 and only $300 each additional year of use. The effective tax rate is 30%. (1) Use classical SL depreciation. (2) Use MACRS depreciation.

17.46 Automatic inspection equipment purchased for $78,000 by Stimson Engineering generated an average of $26,080 annually in before-tax cash flow during its 5-year estimated life. This represents a return of 20%. However, the corporate finance officer determined that the CFAT was $18,000 for the first year only and is decreasing by $1000 per year thereafter. If the president wants to realize an after-tax return of 12% per year, for how many more years must the equipment remain in service?

17.47 Solve Problem 17.46, using the NPV function in Excel.

17.48 Tabulate CFAT for Problem 17.42 estimates, using classical straight line depreciation over 10 years. (a) Estimate the breakeven rate of return using a PW versus i graph. (b) Select the better machine at each of the following after-tax MARR values: 5%, 9%, 10%, and 12% per year.

17.49 In Example 17.8, $P = \$50,000$, $S = 0$, $n = 5$, CFBT = \$20,000, and $T_e = 40\%$ for a fiber optics cable manufacturer. Straight line depreciation is used to compute an after-tax $i^* = 18.03\%$. If the owner requires an after-tax return of 20% per year, determine an estimate allowed for (a) the first cost and (b) the annual CFBT. When you determine one of these values, assume that the other parameter retains the value estimated in Example 17.8. Assume the effective tax rate remains at 40%. Solve this problem by hand.

17.50 Work Problem 17.49, using a spreadsheet and SOLVER for after-tax returns of (a) 20% and (b) 10% per year. Explain why the P and CFBT values have increased or decreased when a ROR value higher and lower than 18.03% is used.

After-Tax Replacement Study

17.51 Scotty Paper Company-Canada employee Stella Needleson was asked to determine if the current process of dying writing paper should be retained or a new, environmentally friendly process should be implemented. Estimates or actual values for the two processes are summarized below. She performed an after-tax replacement analysis at 10% per year and the corporation's effective tax rate of 32% to determine that economically, the new process should be chosen. Was she correct? Why or why not? (*Note:* Canadian tax law does not impose the half-year convention requirement.)

	Current Process	New Process
First cost 7 years ago, $	−450,000	
First cost, $		−700,000
Remaining life, years	5	10
Current market value, $	50,000	
AOC, $/year	−160,000	−150,000
Future salvage, $	0	50,000
Depreciation method	SL	SL

17.52 (a) The Los Angeles, California, city engineer is analyzing a for-profit public works project at the port authority, using an after-tax replacement analysis of the installed system (defender) and a challenger as detailed below. All values are in $1000 units. The effective state tax rate of 6% is applicable, but no federal taxes are imposed. The municipal after-tax return of 6% per year is required. Assume salvages in the future occur at the estimated amounts and use classical SL depreciation. Perform the analysis.

(b) Would the decision be different if a before-tax replacement analysis were performed at $i = 12\%$ per year?

	Defender	Challenger
First cost, $	−28,000	−15,000
AOC, $/year	−1,200	−1,500
Salvage estimate, $	2,000	3,000
Market value, $	15,000	
Life, years	10	8
Years in service	5	

17.53 In Problem 17.52(a), assume that 5 additional years have passed and the challenger has been in place all 5 years. A new city engineer decides to privatize the for-profit entities of the city government and sells the implemented challenger system for $10,000,000. Was the decision made 5 years earlier to select the challenger an economically correct one? Use the same values as before: 6% effective tax rate, 6% per year after-tax rate of return, and classical SL depreciation.

17.54 Apple Crisp Foods signed a contract some years ago for maintenance services on its fleet of trucks and cars. The contract is up for renewal now for a period of 1 year or 2 years only. The contract quote is $300,000 per year if taken for 1 year and $240,000 per year if taken for 2 years. The finance vice president wants to renew the contract for 2 years without further analysis, but the vice president for engineering believes it is more economical to perform the maintenance inhouse. Since much of the fleet is aging and must be replaced in the near future, a fixed 3-year study period has been agreed upon. The estimates for the inhouse (challenger) alternative are as follows:

First cost, $	−800,000
AOC, $/year	−120,000
Life, years	4
Estimated salvage, $	Loses 25% of *P* annually:
	End year 1 600,000
	End year 2 400,000
	End year 3 200,000
	End year 4 0
MACRS depreciation	3-year recovery period

The effective tax rate is 35%, and the after-tax MARR is 10% per year. Perform an after-tax AW analysis, and determine which vice president has the better economic strategy over the next 3 years.

17.55 Nuclear safety devices installed several years ago have been depreciated from a first cost of $200,000 to zero using MACRS. The devices can be sold on the used equipment market for an estimated $15,000. Or they can be retained in service for 5 more years with a $9000 upgrade now and an AOC of $6000 per year. The upgrade investment will be depreciated over 3 years with no salvage value. The challenger is a replacement with newer technology at a first cost of $40,000, $n = 5$ years, and $S = 0$. The new units will have operating expenses of $7000 per year.

(a) Use a 5-year study period, an effective tax rate of 40%, an after-tax MARR of 12% per year, and an assumption of classical straight line depreciation (no half-year convention) to perform an after-tax replacement study.

(b) If the challenger is known to be salable after 5 years for an amount between $2000 and $4000, will the challenger AW value become more or less costly? Why?

17.56 Develop a spreadsheet like the one in Table 17–8 for Example 17.13. Redo the after-tax replacement analysis, using the estimates that the market value of the defender is only $275,000 and that the challenger will be sold on the international market for $100,000. Salvage value is not considered in computing challenger depreciation.

17.57 Three years ago, Silver House Steel purchased a new quenching system for $550,000. The salvage value after 10 years at that time was estimated to be $50,000. Currently the expected remaining life is 7 years with an AOC of $27,000 per year. The new president has recommended early replacement of the system with one that costs $400,000 and has a 12-year life,

a $35,000 salvage value, and an estimated AOC of $50,000 per year. The MARR for the corporation is 12% per year. The president wishes to know the replacement value that will make the recommendation to replace now economically advantageous. Use a spreadsheet and the SOLVER tool to find the minimum trade-in value (a) before taxes and (b) after taxes, using an effective tax rate of 30%. For solution purposes, use classical SL depreciation for both systems. Comment on the difference in replacement value made by the consideration of taxes.

Economic Value Added

17.58 (a) What does the term *economic value added (EVA)* mean relative to the bottom line of a corporation?

(b) Why might an investor in a public corporation prefer to use the EVA estimates over the CFAT estimates for a project?

17.59 An asset has a first cost of $12,000, SL depreciation of $4000 for each year of its 3-year recovery period and no salvage value, and an estimated CFBT of $5000 per year. The effective tax rate is 50%, and the after-tax MARR of the Harriet Corporation is 10% per year. Use hand or spreadsheet solution as indicated by your instructor to demonstrate (a) that the present worth values of the EVA and CFAT series are identical and (b) that when the equivalent of the first cost is "charged" against annual CFAT, the resulting PW value is equal to the PW of EVA, as discussed in Example 17.14(b).

17.60 For Example 17.3 and an interest rate of 10% per year, do the following:

(a) Determine the EVA estimates for each year.

(b) Show that the annual worth of EVA estimates is numerically the same as the AW of the CFAT estimates, if the actual salvage value in year 6 is zero, not $150,000. (This makes the CFAT value in year 6 equal to $82,588.)

17.61 Sun Microsystems has developed partnerships with several large manufacturing corporations to use Java software in their consumer and industrial products. A new corporation will be formed to manage these applications. One major project involves using Java in commercial and industrial appliances that store and cook food. The gross income and expenses are expected to follow the relations shown for the estimated life of 6 years. For $t = 1$ to 6 years,

$$\text{Annual gross income} = 2,800,000 - 100,000t$$
$$\text{Annual expenses} = 950,000 + 50,000t$$

The effective tax rate is 35%, the interest rate is 12% per year, and the depreciation method chosen for the $3,000,000 in capital investment is the 5-year MACRS alternative that allows straight line write-off with the half-year convention in years 1 and 6. Using a spreadsheet, estimate (*a*) the annual economic contribution of the project to the new corporation and (*b*) the equivalent annual worth of these contributions.

17.62 Review the situations in Examples 17.6 and 17.11. Assume the NBA contract is now for 6 years, that the gross income and expenses continue for all 6 years, and that neither analyzer has any realized salvage value. Use an EVA analysis to choose between the two analyzers. The MARR is 10% per year, and $T_e = 35\%$.

CASE STUDY

AFTER-TAX EVALUATION OF DEBT AND EQUITY FINANCING

The Proposal

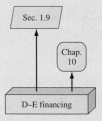

Young Brothers, Inc., a highway engineering company located in Seattle, Washington, wants to develop new business opportunities in Portland, Oregon. Brother Charles historically has looked after the financial end of the company. He is concerned about the method of financing the new office, work yard, and equipment at the planned Portland site. Debt financing (loan) from their Seattle bank and equity financing from company retained earnings are possible ways to finance the new office, but the best mix of debt and equity funds is unknown at this time. To help,

Charles has read a section of a handbook on *generalized cash flow analysis,* especially the part on after-tax analysis of the two financing methods—debt versus equity. What he learned and has summarized for himself follows.

Debt and Equity Financing
(Copied from the Handbook)

In a company uses loans and bonds to raise capital, it is called debt financing (DF). Loans require payment of periodic interest, and bonds require that periodic dividends be paid to the investor. The loan principal or bond face value is repaid after a stated number of years. These various loan and bond cash flows affect taxes and CFAT differently, as shown in the table.

Type of Debt	Cash Flow Involved	Tax Treatment	Effect on CFAT
Loan	Receive principal	No effect	Increases it
Loan	Pay interest	Deductible	Reduces it
Loan	Repay principal	Not deductible	Reduces it
Bond	Receive face value	No effect	Increases it
Bond	Pay dividend	Deductible	Reduces it
Bond	Repay face value	Not deductible	Reduces it

Only loan interest and bond dividends are tax-deductible. Use the symbol DF_I to identify the sum of these two. To develop a relation that explains the tax impact of debt financing, start with the fundamental net cash flow relation, that is, receipts minus disbursements. Identify receipts from debt financing as

$$DF_R = \text{loan principal receipt} + \text{bond sale receipt}$$

Define debt financing disbursements as

$$DF_D = \text{loan interest payment}$$
$$+ \text{bond dividend payment}$$
$$+ \text{loan principal repayment}$$
$$+ \text{bond face value repayment}$$

It is common that a loan or bond sale, not both, is involved in a single asset purchase. The two terms in the first line of the DF_D equation represent DF_I mentioned earlier.

If a company uses its own resources for capital investment, it is called *equity financing (EF)*. This includes (1) the use of a corporation's own funds, such as retained earnings; (2) the sale of corporation stock; and (3) the sale of corporate assets to raise capital. There are no direct tax advantages for equity financing. Expended retained earnings and stock dividends paid will reduce cash flow, but neither will reduce TI.

To explain the impact of equity financing, again start with the fundamental net cash flow relation—receipts minus disbursements. Equity financing disbursements, defined as EF_D, are the portion of the first cost of an asset covered by a corporation's own resources.

$$EF_D = \text{corporate owned funds}$$

Any equity financing receipts are

$$EF_R = \text{sale of corporate assets} + \text{stock sale receipts}$$

In EF_D, stock dividends are a part of disbursements, but they are small in comparison to other disbursements and their timing depends on the financial success of the corporation overall, so they can be neglected.

Combine the DF and EF terms to estimate annual CFAT. The initial investment is equal to the amount of corporate owned funds committed to the alternative's first cost; that is, $P = EF_D$.

$$\begin{aligned} CFAT = \ & -\text{equity financed investment} \\ & + \text{gross income} - \text{operating expenses} \\ & + \text{salvage value} - \text{taxes} \\ & + \text{debt financing receipts} \\ & - \text{disbursements} + \text{equity financing} \\ & \quad \text{receipts} \\ = \ & -EF_D + GI - E + S - TI(T_e) \\ & + (DF_R - DF_D) + EF_R \end{aligned}$$

Since DF_D includes DF_I, which is the tax-deductible portion of debt financing, taxes are

$$\begin{aligned} \text{Taxes} = \ & (TI)(T_e) \\ = \ & (\text{gross income} - \text{operating expenses} \\ & - \text{depreciation} - \text{loan interest} \\ & \quad \text{and bond dividends})(T_e) \\ = \ & (GI - E - D - DF_I)(T_e) \end{aligned}$$

These relations are easy to use when the investment involves only 100% equity financing or 100% debt financing, since only the relevant terms have nonzero values. Together they form the model for generalized cash flow analysis.

The Financial Picture

After consulting with their accountant, the brothers have agreed on the following estimates for the Portland branch, if it is pursued.

Initial investment = $1,500,000

Annual gross income = $700,000

Annual operating expenses = $100,000

Effective tax rate = 35%

All $1.5 million of the initial capital investment can be depreciated using MACRS with a 5-year recovery period.

Young Brothers is not a public corporation, so stock sales are not an option for equity financing; retained earnings must be used. And a loan is the only viable form of debt financing. The loan would be through the Seattle bank at 6% per year simple interest based on the initial loan principal. Repayment would be in five equal annual payments of interest and principal.

The percentage of debt financing is the real question for Young Brothers. If the Portland branch does not "make it," there is a loan commitment to repay that the Seattle operation will have to bear. However, if the financing is mostly from equity, the company will be "cash poor" for a while, thus limiting its ability to fund smaller projects as they arise. Therefore, a range of debt-equity (D-E) financing options to raise the $1,500,000 should be studied. These options have been identified:

Debt		Equity	
Percentage	Loan Amount	Percentage	Investment Amount
0%		100%	$1,500,000
50	$ 750,000	50	750,000
70	1,050,000	30	450,000
90	1,350,000	10	150,000

Case Study Exercises

1. For each funding option, perform a spreadsheet analysis that shows the total CFAT and its present worth over a 6-year period, the time it will take to realize the full advantage of MACRS depreciation. An after-tax return of 10% is expected. Which funding option is best for Young Brothers? (*Hint:* For the spreadsheet, sample column headings are: Year, GI − E, Loan interest, Loan principal, Equity investment, Depreciation rate, Depreciation, Book value, TI, Taxes, and CFAT.)

2. Observe the changes in the total 6-year CFAT as the D-E percentages change. If the time value of money is neglected, what is the constant amount by which this sum changes for every 10% increase in equity funding?

3. Charles's brother noticed that the CFAT total and PW values go in opposite directions as the equity percentage increases. He wants to know why this phenomenon occurs. How should Charles explain this to his brother?

4. The brothers have decided on the 50-50 split of debt and equity financing. Charles wants to know what additional bottom-line contributions to the economic worth of the company may be added by the new Portland office. What are the best estimates at this time?

CHAPTER 18

Formalized Sensitivity Analysis and Expected Value Decisions

This chapter includes several related topics about alternative evaluation. All these techniques build upon the methods and models used in previous chapters, especially those of the first eight chapters and the basics of breakeven analysis in Chapter 13. This chapter should be considered preparation for the topics of simulation and decision making under risk, presented in the next chapter.

The first two sections expand our capability to perform a *sensitivity analysis* of one or more parameters and of an entire alternative. Then the determination and use of the *expected value* of a cash flow series are treated. Finally, the technique of *decision trees* is covered. This approach helps an analyst make a series of economic decisions for alternatives that have different, but closely connected, stages.

The case study involves a thorough sensitivity analysis of a multiple-alternative, multiple-attribute (factor) project set in the public sector.

LEARNING OBJECTIVES

Purpose: Perform a formal sensitivity analysis of one or more parameters; and perform expected value and decision tree evaluations of alternatives.

This chapter will help you:

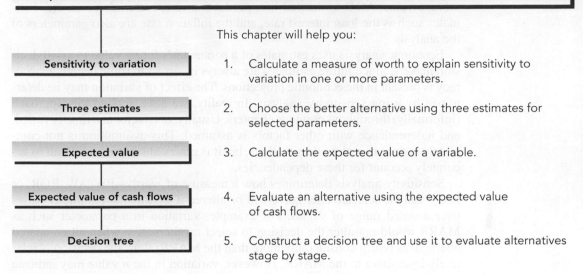

Sensitivity to variation

1. Calculate a measure of worth to explain sensitivity to variation in one or more parameters.

Three estimates

2. Choose the better alternative using three estimates for selected parameters.

Expected value

3. Calculate the expected value of a variable.

Expected value of cash flows

4. Evaluate an alternative using the expected value of cash flows.

Decision tree

5. Construct a decision tree and use it to evaluate alternatives stage by stage.

18.1   DETERMINING SENSITIVITY TO PARAMETER VARIATION

The term *parameter* is used in this chapter to represent any variable or factor for which an estimate or stated value is necessary. Example parameters are first cost, salvage value, AOC, estimated life, production rate, materials costs, etc. Estimates such as the loan interest rate, and the inflation rate are also parameters of the analysis.

Economic analysis uses estimates of a parameter's future value to assist decision makers. Since future estimates are always incorrect to some degree, inaccuracy is present in the economic projections. The effect of variation may be determined by using sensitivity analysis. In reality, we have applied this approach (informally) throughout previous chapters. Usually, one factor at a time is varied, and independence with other factors is assumed. This assumption is not completely correct in real world situations, but it is practical since it is difficult to accurately account for these dependencies.

Sensitivity analysis determines how a measure of worth—PW, AW, ROR, or B/C—and the selected alternative will be altered if a particular parameter varies over a stated range of values. For example, variation in a parameter such as MARR would not alter the decision to select an alternative when all compared alternatives return considerably more than the MARR; thus, the decision is relatively insensitive to the MARR. However, variation in the *n* value may indicate that selection from the same alternatives is very sensitive to the estimated life.

Usually the variations in life, annual costs, and revenues result from variations in selling price, operation at different levels of capacity, inflation, etc. For example, if an operating level of 90% of airline seating capacity for a domestic route is compared with 50% for a proposed international route, the operating cost and revenue per passenger mile will increase, but anticipated aircraft life will probably decrease only slightly. Usually several important parameters are studied to learn how the uncertainty of estimates affects the economic analysis.

Sensitivity analysis usually concentrates on the variation expected in estimates of *P*, AOC, *S*, *n*, unit costs, unit revenues, and similar parameters. These parameters are often the result of design questions and their answers, as discussed in Chapter 15. Parameters that are interest rate-based are not treated in the same manner.

Parameters such as MARR, and other interest rates (loan rates, inflation rate) are more stable from project to project. If performed, sensitivity analysis on them is for specific values or over a narrow range of values. Therefore, the sensitivity analysis is more constrained for interest rate parameters.

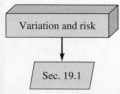

This point is important to remember if simulation is used for decision making under risk (Chapter 19).

Plotting the sensitivity of PW, AW, or ROR versus the parameter(s) studied is very helpful. Two alternatives can be compared with respect to a given parameter and the breakeven point. This is the value at which the two alternatives are economically equivalent. However, the breakeven chart commonly represents only one parameter per chart. Thus, several charts are constructed, and independence of each parameter is assumed. In previous uses of breakeven analysis, we computed the measure of worth at only two values of a parameter and connected the points with a straight line. However, if the results are sensitive to the parameter value, several intermediate points should be used to better evaluate the sensitivity, especially if the relationships are not linear.

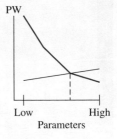

When several parameters are studied, sensitivity analysis can become quite complex. It may be performed one parameter at a time using a spreadsheet or computations by hand. The computer facilitates comparison of multiple parameters and multiple measures of worth, and the software can rapidly plot the results.

There is a general procedure to follow when conducting a thorough sensitivity analysis.

1. Determine which parameter(s) of interest might vary from the most likely estimated value.
2. Select the probable range and an increment of variation for each parameter.
3. Select the measure of worth.
4. Compute the results for each parameter, using the measure of worth as a basis.
5. To better interpret the sensitivity, graphically display the parameter versus the measure of worth.

This sensitivity analysis procedure should indicate the parameters that warrant closer study or require additional information. When there are two or more alternatives, it is better to use the PW or AW measure of worth in step 3. If ROR is used, it requires the extra efforts of incremental analysis between alternatives. Example 18.1 illustrates sensitivity analysis for one project.

EXAMPLE 18.1

Wild Rice, Inc., expects to purchase a new asset for automated rice handling. Most likely estimates are a first cost of $80,000, zero salvage value, and a cash flow before taxes (CFBT) per year t that follows the relation $27,000 - 2000t$. The MARR for the company varies from 10% to 25% per year for different types of investments. The economic life of similar machinery varies from 8 to 12 years. Evaluate the sensitivity of PW by varying (a) MARR, while assuming a constant n value of 10 years, and (b) n, while MARR is constant at 15% per year. Perform the analysis by hand and by computer.

Solution by Hand

(a) Follow the procedure above to understand the sensitivity of PW to MARR variation.
   1. MARR is the parameter of interest.
   2. Select 5% increments to evaluate sensitivity to MARR; the range is 10% to 25%.
   3. The measure of worth is PW.
   4. Set up the PW relation for 10 years. When MARR = 10%,

$$PW = -80,000 + 25,000(P/A,10\%,10) - 2000(P/G,10\%,10)$$
$$= \$27,830$$

The PW for all four values at 5% intervals is as follows:

MARR	PW
10%	$ 27,830
15	11,512
20	−962
25	−10,711

5. A plot of MARR versus PW is shown in Figure 18–1. The steep negative slope indicates that the decision to accept the proposal based on PW is quite sensitive to variations in the MARR. If the MARR is established at the upper end of the range, the investment is not attractive.

(b) 1. Asset life n is the parameter.
2. Select 2-year increments to evaluate PW sensitivity over the range 8 to 12 years.
3. The measure of worth is PW.

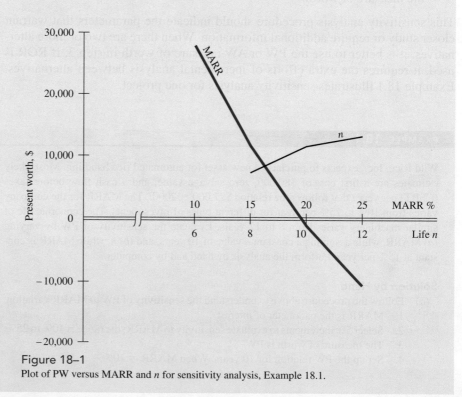

Figure 18–1
Plot of PW versus MARR and n for sensitivity analysis, Example 18.1.

4. Set up the same PW relation as in part (*a*) at *i* = 15%. The PW results are

n	PW
8	$ 7,221
10	11,511
12	13,145

5. Figure 18–1 presents the plot of PW versus *n*. Since the PW measure is positive for all values of *n*, the decision to invest is not materially affected by the estimated life. The PW curve levels out above *n* = 10. This insensitivity to changes in cash flow in the distant future is a predictable observation, because the *P/F* factor gets smaller as *n* increases.

Solution by Computer

Figure 18–2 presents two spreadsheets and accompanying plots of PW versus MARR (fixed *n*) and PW versus *n* (fixed MARR). The general relation for cash flow values is

$$\text{Cash flow}_t = \begin{cases} -80{,}000 & t = 0 \\ +27{,}000 - 2000t & t = 1, \ldots \end{cases}$$

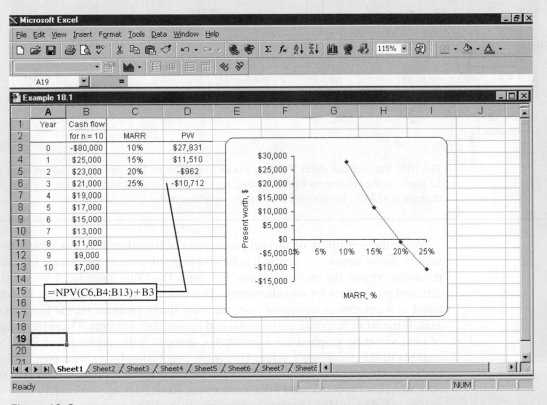

Figure 18–2
Sensitivity analysis of PW to variations in the MARR and estimated life, Example 18.1.

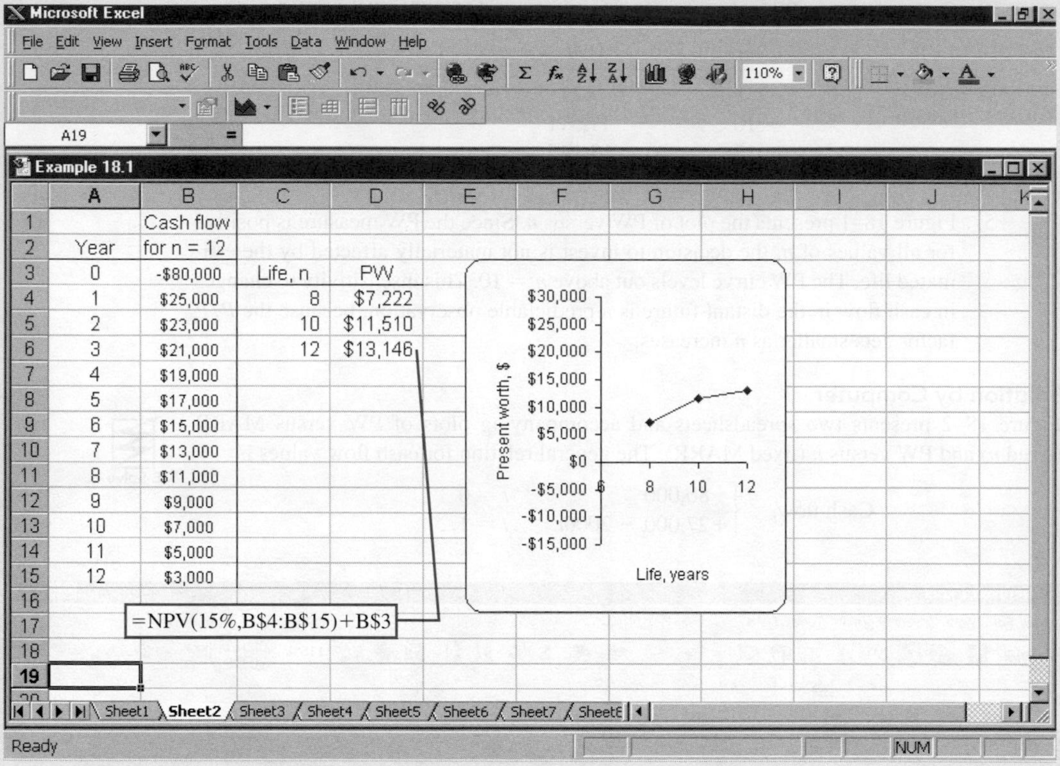

Figure 18–2
(Continued).

The NPV function calculates PW for *i* values from 10% to 25% and *n* values from 8 to 12 years. As the solution by hand indicated, so does the *xy* scatter chart; PW is sensitive to changes in MARR, but not very sensitive to variations in *n*.

When the sensitivity of *several parameters* is considered for *one alternative* using a *single measure of worth,* it is helpful to graph percentage change for each parameter versus the measure of worth. Figure 18–3 illustrates ROR versus six different parameters for one alternative. The variation in each parameter is indicated as a percentage deviation from the most likely estimate on the horizontal axis. If the ROR response curve is flat and approaches horizontal over the range of total variation graphed for a parameter, there is little sensitivity of ROR to changes in the parameter's value. This is the conclusion for indirect cost in Figure 18–3. On the other hand, ROR is very sensitive to sales price. A reduction of 30% from the expected sales price reduces the ROR from approximately 20% to −10%, whereas a 10% increase in price raises the ROR to about 30%.

If two *alternatives* are compared and the sensitivity to *one parameter* is sought, the graph may show quite nonlinear results. Observe the general shape of the sample sensitivity graphs in Figure 18–4. The plots are shown as linear

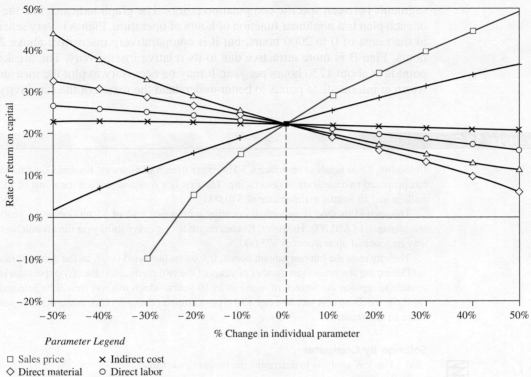

Parameter Legend

☐ Sales price   ✕ Indirect cost
◇ Direct material ○ Direct labor
+ Sales volume   △ Capital

Figure 18–3
Sensitivity analysis graph of percent variation from the most likely estimate.

SOURCE: L.T. Blank and A. J. Tarquin, Chap. 19, *Engineering Economy*, 4th ed., New York: McGraw-Hill, 1998.

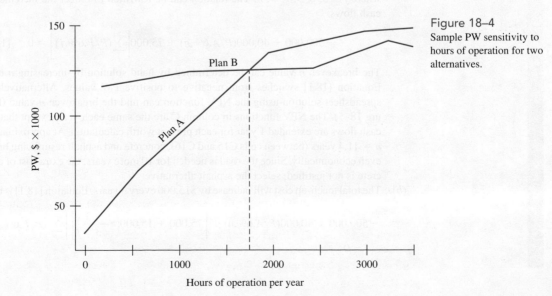

Figure 18–4
Sample PW sensitivity to hours of operation for two alternatives.

segments between specific computation points. The graph indicates that the PW of each plan is a nonlinear function of hours of operation. Plan A is very sensitive in the range of 0 to 2000 hours, but it is comparatively insensitive above 2000 hours. Plan B is more attractive due to its relative insensitivity. The breakeven point is at about 1750 hours per year. It may be necessary to plot the measure of worth at intermediate points to better understand the nature of the sensitivity.

EXAMPLE 18.2

Columbus, Ohio needs to resurface a 3-kilometer stretch of highway. Knobel Construction has proposed two methods of resurfacing. The first is a concrete surface for a cost of $1.5 million and an annual maintenance of $10,000.

The second method is an asphalt covering with a first cost of $1 million and a yearly maintenance of $50,000. However, Knobel requests that every third year the asphalt highway be touched up at a cost of $75,000.

The city uses the interest rate on bonds, 6% on its last bond issue, as the discount rate. (*a*) Determine the breakeven number of years of the two methods. If the city expects an interstate to replace this stretch of highway in 10 years, which method should be selected? (*b*) If the touch-up cost increases by $5000 per kilometer every 3 years, is the decision sensitive to this increase?

Solution by Computer

(*a*)   Use PW analysis to determine the breakeven n value.

$$\text{PW of concrete} = \text{PW of asphalt}$$

$$-1{,}500{,}000 - 10{,}000(P/A{,}6\%{,}n) = -1{,}000{,}000 - 50{,}000(P/A{,}6\%{,}n)$$

$$- 75{,}000\left[\sum_j (P/F{,}6\%{,}j)\right]$$

where $j = 3, 6, 9, \ldots, n$. The relation can be rewritten to reflect the incremental cash flows.

$$-500{,}000 + 40{,}000(P/A{,}6\%{,}n) + 75{,}000\left[\sum_j (P/F{,}6\%{,}j)\right] = 0 \qquad [18.1]$$

The breakeven n value can be determined by hand solution by increasing n until Equation [18.1] switches from negative to positive PW values. Alternatively, a spreadsheet solution using the NPV function can find the breakeven n value (Figure 18–5). The NPV functions in column C are the same each year, except that the cash flows are extended 1 year for each present worth calculation. At approximately $n = 11.4$ years (between cells C15 and C16), concrete and asphalt resurfacing break even economically. Since the road is needed for 10 more years, the extra cost of concrete is not justified; select the asphalt alternative.

(*b*)   The total touch-up cost will increase by $15,000 every 3 years. Equation [18.1] is now

$$-500{,}000 + 40{,}000(P/A{,}6\%{,}n) + \left[75{,}000 + 15{,}000\left(\frac{j-3}{3}\right)\right]\left[\sum_j (P/F{,}6\%{,}j)\right]$$

$$= 0$$

	A	B	C	D	E	F	G	H	I	J
1			Part (a)		Part (b)					
2		Incremental	PW for	Incremental	PW for					
3	Year, n	cash flow	n years	cash flow	n years					
4	0	$ (500,000)		$ (500,000)						
5	1	$ 40,000	$ (462,264)	$ 40,000	$ (462,264)					
6	2	$ 40,000	$ (426,664)	$ 40,000	$ (426,664)					
7	3	$ 115,000	$ (330,108)	$ 115,000	$ (330,108)					
8	4	$ 40,000	$ (298,424)	$ 40,000	$ (298,424)					
9	5	$ 40,000	$ (268,534)	$ 40,000	$ (268,534)					
10	6	$ 115,000	$ (187,464)	$ 130,000	$ (176,889)					
11	7	$ 40,000	$ (160,861)	$ 40,000	$ (150,287)					
12	8	$ 40,000	$ (135,765)	$ 40,000	$ (125,190)					
13	9	$ 115,000	$ (67,696)	$ 145,000	$ (39,365)		=NPV(6%,$D$5:$D14)+$D$4			
14	10	$ 40,000	$ (45,361)	$ 40,000	$ (17,029)		=NPV(6%,$D$5:$D15)+$D$4			
15	11	$ 40,000	$ (24,289)	$ 40,000	$ 4,042					
16	12	$ 115,000	$ 32,862	$ 160,000	$ 83,557					
17	13	$ 40,000	$ 51,616	$ 40,000	$ 102,311					
18	14	$ 40,000	$ 69,308	$ 40,000	$ 120,003					
19	15	$ 115,000	$ 117,293	$ 175,000	$ 193,024					
20	16	$ 40,000	$ 133,039	$ 40,000	$ 208,770					
21			=NPV(6%,$B$5:$B15)+$B$4							
22										
23			=NPV(6%,$B$5:$B16)+$B$4							
24										

Figure 18–5
Sensitivity of the breakeven point between two alternatives using PW analysis, Example 18.2.

Now the breakeven *n* value is between 10 and 11 years—10.8 years using linear interpolation (Figure 18–5, cells E14 and E15). The decision has become marginal for asphalt, since the interstate is planned for 10 years hence.

Noneconomic considerations may be used to determine if asphalt is still the better alternative. One conclusion is that the asphalt decision becomes more questionable as the asphalt alternative maintenance costs increase; that is, the PW value is sensitive to increasing touch-up costs.

18.2   FORMALIZED SENSITIVITY ANALYSIS USING THREE ESTIMATES

We can thoroughly examine the economic advantages and disadvantages among two or more alternatives by borrowing from the field of project scheduling the concept of making three estimates for each parameter: *a pessimistic, a most likely, and an optimistic estimate.* Depending upon the nature of a parameter, the pessimistic estimate may be the lowest value (alternative life is an example) or the largest value (such as asset first cost).

This formal approach allows us to study measure of worth and alternative selection sensitivity within a predicted range of variation for each parameter. Usually the most likely estimate is used for all other parameters when the measure of worth is calculated for one particular parameter or one alternative. This approach, essentially the same as the one-parameter-at-a-time analysis of Section 18.1, is illustrated by Example 18.3.

EXAMPLE 18.3

An engineer is evaluating three alternatives for which she has made three estimates for the salvage value, annual operating cost, and the life. The estimates are presented on an alternative-by-alternative basis in Table 18–1. For example, alternative B has pessimistic estimates of $S = \$500$, AOC = $\$-4000$, and $n = 2$ years. The first costs are known, so they have the same value. Perform a sensitivity analysis and determine the most economical alternative, using AW analysis at a MARR of 12% per year.

TABLE 18–1 Competing Alternatives with Three Estimates Made for Salvage Value, AOC, and Life Parameters

Strategy		First Cost, $	Salvage Value, $	AOC, $	Life n, Years
Alternative A					
Estimates	P	−20,000	0	−11,000	3
	ML	−20,000	0	−9,000	5
	O	−20,000	0	−5,000	8
Alternative B					
Estimates	P	−15,000	500	−4,000	2
	ML	−15,000	1,000	−3,500	4
	O	−15,000	2,000	−2,000	7
Alternative C					
Estimates	P	−30,000	3,000	−8,000	3
	ML	−30,000	3,000	−7,000	7
	O	−30,000	3,000	−3,500	9

P = pessimistic; ML = most likely; O = optimistic.

Solution

For each alternative in Table 18–1, calculate the AW value of costs. For example, the AW relation for alternative A, pessimistic estimates, is

$$AW = -20,000(A/P,12\%,3) - 11,000 = \$-19,327$$

Table 18–2 presents all AW values. Figure 18–6 is a plot of AW versus the three estimates of life for each alternative. Since the AW calculated using the ML estimates for

TABLE 18–2	Annual Worth Values, Example 18.3		
	Alternative AW Values		
Estimates	**A**	**B**	**C**
P	$-19,327	$-12,640	$-19,601
ML	-14,548	-8,229	-13,276
O	-9,026	-5,089	-8,927

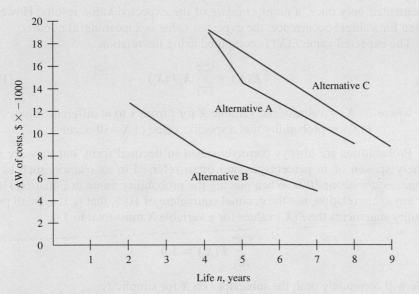

Figure 18–6
Plot of AW of costs for different-life estimates, Example 18.3.

alternative B ($-8229) is economically better than even the optimistic AW value for alternatives A and C, alternative B is clearly favored.

Comment

While the alternative that should be selected here is quite obvious, this is not normally the case. For example, in Table 18–2, if the pessimistic alternative B equivalent AW were much higher, say, $-21,000 per year (rather than $-12,640), and the optimistic AW values for alternatives A and C were less than that for B ($-5089), the choice of B is not apparent or correct. In this case, it would be necessary to select one set of estimates (P, ML, or O) upon which to base the decision. Alternatively, the different estimates can be used in an expected value analysis, which is introduced in the next section.

18.3 ECONOMIC VARIABILITY AND THE EXPECTED VALUE

Engineers and economic analysts usually deal with estimates about an uncertain future by placing appropriate reliance on past data, if any exist. This means that probability and samples are used. Actually the use of probabilistic analysis is not as common as might be expected. The reason for this is not that the computations are difficult to perform or understand, but that realistic probabilities associated with cash flow estimates are difficult to assign. Experience and judgment can often be used in conjunction with probabilities and expected values to evaluate the desirability of an alternative.

The *expected value* can be interpreted as a long-run average observable if the project is repeated many times. Since a particular alternative is evaluated or implemented only once, a *point estimate* of the expected value results. However, even for a single occurrence, the expected value is a meaningful number.

The expected value $E(X)$ is computed using the relation

$$E(X) = \sum_{i=1}^{i=m} X_i P(X_i) \qquad [18.2]$$

where X_i = value of the variable X for i from 1 to m different values
           $P(X_i)$ = probability that a specific value of X will occur

Probabilities are always correctly stated in decimal form, but they are routinely spoken of in percentages and often referred to as *chance,* such as *the chances are about 10%.* When placing the probability value in Equation [18.2] or any other relation, use the decimal equivalent of 10%, that is, 0.1. In all probability statements the $P(X_i)$ values for a variable X must total to 1.0.

$$\sum_{i=1}^{i=m} P(X_i) = 1.0$$

We will commonly omit the subscript i on X for simplicity.

If X represents the estimated cash flows, some will be positive and others negative. If a cash flow sequence includes revenues and costs, and the present worth at the MARR is calculated, the result is the expected value of the discounted cash flows $E(PW)$. If the expected value is negative, the overall outcome is expected to be a cash outflow. For example, if $E(PW) = \$-1500$, this indicates that the proposal is not expected to return the MARR.

EXAMPLE **18.4**

A downtown hotel is offering a new service for weekend travelers through its business and travel center. The manager estimates that for a typical weekend, there is a 50% chance of having a net cash flow of $5000 and a 35% chance of $10,000. He also estimates there is a small chance—5%—of no cash flow and a 10% chance of a loss of $500, which is the estimated extra personnel and utility costs to offer the service. Determine the expected net cash flow.

Solution

Let X be the net cash flow in dollars, and let $P(X)$ represent the associated probabilities. Using Equation [18.2],

$$E(X) = 5000(0.5) + 10,000(0.35) + 0(0.05) - 500(0.1) = \$5950$$

Although the "no cash flow" possibility does not increase or decrease $E(X)$, it is included because it makes the probability values sum to 1.0 and it makes the computation complete.

18.4   EXPECTED VALUE COMPUTATIONS FOR ALTERNATIVES

The expected value computation $E(X)$ is utilized in a variety of ways. Two ways are: (1) to prepare information for incorporation into a more complete engineering economy analysis and (2) to evaluate the expected viability of a fully formulated alternative. Example 18.5 illustrates the first situation, and Example 18.6 determines the expected PW when the cash flow series and probabilities are estimated.

EXAMPLE 18.5

An electric utility is experiencing a difficult time obtaining natural gas for electric generation. Fuels other than natural gas are purchased at an extra cost, which is transferred to the customer. Total monthly fuel expenses are now averaging \$7,750,000. An engineer with this city-owned utility has calculated the average revenue for the past 24 months using three fuel-mix situations—gas plentiful, less than 30% other fuels purchased, and 30% or more other fuels. Table 18–3 indicates the number of months that each fuel-mix situation occurred. Can the utility expect to meet future monthly expenses based on the 24 months of data, if a similar fuel-mix pattern continues?

TABLE 18–3   Revenue and Fuel-Mix Data, Example 18.5

Fuel-Mix Situation	Months in Past 24	Average Revenue, $ per Month
Gas plentiful	12	5,270,000
<30% other	6	7,850,000
≥30% other	6	12,130,000

Solution

Using the 24 months of data, estimate a probability for each fuel mix.

Fuel-Mix Situation	Probability of Occurrence
Gas plentiful	12/24 = 0.50
<30% other	6/24 = 0.25
≥30% other	6/24 = 0.25

Let the variable X represent average monthly revenue. Use Equation [18.2] to determine expected revenue per month.

$$E(\text{revenue}) = 5{,}270{,}000(0.50) + 7{,}850{,}000(0.25) + 12{,}130{,}000(0.25)$$
$$= \$7{,}630{,}000$$

With expenses averaging $7,750,000, the average monthly revenue shortfall is $120,000. To break even, other sources of revenue must be generated, or the additional costs must be transferred to the customer.

EXAMPLE 18.6

Lite-Weight Wheelchair Company has a substantial investment in tubular steel bending equipment. A new piece of equipment costs $5000 and has a life of 3 years. Estimated cash flows (Table 18–4) depend on economic conditions classified as receding, stable, or expanding. A probability is estimated that each of the economic conditions will prevail during the 3-year period. Apply expected value and PW analysis to determine if the equipment should be purchased. Use a MARR of 15% per year.

TABLE 18–4	Equipment Cash Flow and Probabilities, Example 18.6		
	Economic Condition		
Year	Receding (Prob. = 0.2)	Stable (Prob. = 0.6)	Expanding (Prob. = 0.2)
	Annual Cash Flow Estimates, $		
0	$−5000	$−5000	$−5000
1	+2500	+2000	+2000
2	+2000	+2000	+3000
3	+1000	+2000	+3500

Solution
First determine the PW of the cash flows in Table 18–4 for each economic condition, and then calculate $E(\text{PW})$, using Equation [18.2]. Define subscripts R for receding economy, S for stable, and E for expanding. The PW values for the three scenarios are

$$PW_R = -5000 + 2500(P/F,15\%,1) + 2000(P/F,15\%,2) + 1000(P/F,15\%,3)$$
$$= -5000 + 4344 = \$-656$$
$$PW_S = -5000 + 4566 = \$-434$$
$$PW_E = -5000 + 6309 = \$+1309$$

Only in an expanding economy will the cash flows return the 15% and justify the investment. The expected present worth is

$$E(\text{PW}) = \sum_{j=R,S,E} \text{PW}_j[P(j)]$$

$$= -656(0.2) - 434(0.6) + 1309(0.2)$$

$$= \$-130$$

At 15%, $E(\text{PW}) < 0$; the equipment is not justified using an expected value analysis.

Comment

It is also correct to calculate the E(cash flow) for each year and then determine PW of the E(cash flow) series, because the PW computation is a linear function of cash flows. Computing E(cash flow) first may be easier in that it reduces the number of PW computations. In this example, calculate $E(\text{CF}_t)$ for each year, then determine $E(\text{PW})$.

$E(\text{CF}_0) = \$-5000$

$E(\text{CF}_1) = 2500(0.2) + 2000(0.6) + 2000(0.2) = \$2100$

$E(\text{CF}_2) = \$2200$

$E(\text{CF}_3) = \$2100$

$E(\text{PW}) = -5000 + 2100(P/F,15\%,1) + 2200(P/F,15\%,2) + 2100(P/F,15\%,3)$

$\qquad = \$-130$

18.5 STAGED EVALUATION OF ALTERNATIVES USING A DECISION TREE

Alternative evaluation may require a series of decisions where the outcome from one stage is important to the next stage of decision making. When each alternative is clearly defined and probability estimates can be made to account for risk, it is helpful to perform the evaluation using a *decision tree*. A decision tree includes

- More than one stage of alternative selection.
- Selection of an alternative at one stage that leads to another stage.
- Expected results from a decision at each stage.
- Probability estimates for each outcome.
- Estimates of economic value (cost or revenue) for each outcome.
- Measure of worth as the selection criterion, such as $E(\text{PW})$.

The decision tree is constructed left to right and includes each possible decision and outcome. A square represents a *decision node* with the possible alternatives indicated on the *branches* from the decision node (Figure 18–7a). A circle represents a *probability node* with the possible outcomes and estimated probabilities on the branches (Figure 18–7b). Since outcomes always follow decisions, the treelike structure in Figure 18–7c results.

Usually each branch of a decision tree has some estimated economic value (often referred to as *payoff*) in cost, revenue, or benefit. These cash flows are

Figure 18–7
Decision and probability nodes used to construct a decision tree.

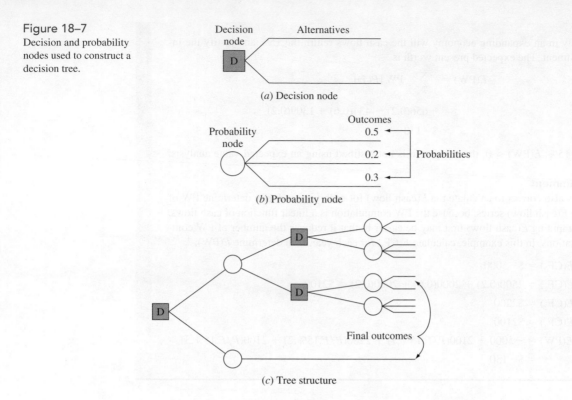

(a) Decision node

(b) Probability node

(c) Tree structure

expressed in terms of PW, AW, or FW values and are shown to the right of each final outcome branch. The cash flow and probability estimates on each outcome branch are used in calculating the expected economic value of each decision branch. This process, called *solving the tree* or *foldback,* is explained after Example 18.7, which illustrates the construction of a decision tree.

EXAMPLE **18.7**

Jerry Hill is president and CEO of an American-based food processing company, Hill Products and Services. He was recently approached by an international supermarket chain that wants to market in-country its own brand of frozen microwaveable dinners. The offer made to Jerry by the supermarket corporation requires that a series of two decisions be made, now and 2 years hence. The current decision involves two alternatives: (1) *Lease* a facility in the United Arab Emirates (UAE) from the supermarket chain, which has agreed to convert a current processing facility for immediate use by Jerry's company; or (2) *build and own* a processing and packaging facility in the UAE. Possible outcomes of this first decision stage are good market or poor market depending upon the public's response.

The decision choices 2 years hence are dependent upon the lease-or-own decision made now. If Hill *decides to lease,* good market response means that the future decision alternatives are to produce at twice, equal to, or one-half of the original volume. This

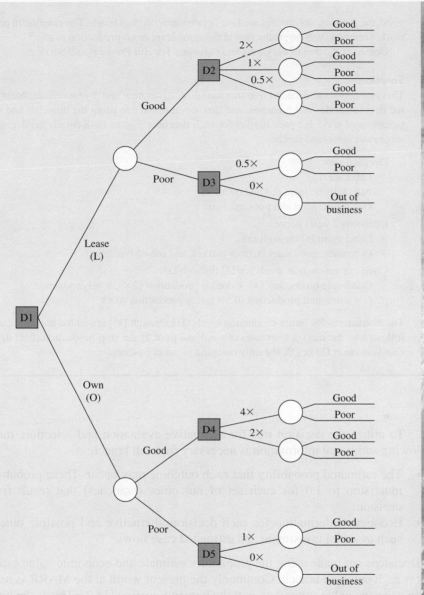

Figure 18–8
A two-stage decision tree identifying alternatives and possible outcomes.

will be a mutual decision between the supermarket chain and Jerry's company. A poor market response will indicate a one-half level of production, or complete removal from the UAE market. Outcomes for the future decisions are, again, good and poor market responses.

As agreed by the supermarket company, the current decision for Jerry *to own* the facility will allow him to set the production level 2 years hence. If market response is good, the decision alternatives are four or two times original levels. The reaction to poor market response will be production at the same level or no production at all.

Construct the tree of decisions and outcomes for Hill Products and Services.

Solution

This is a two-stage decision tree that has alternatives now and 2 years hence. Identify the decision nodes and branches, and then develop the tree using the branches and the outcomes of good and poor market for each decision. Figure 18–8 details the decision stages and outcome branches.

Decision now:
  Label it D1.
  Alternatives: lease (L) and own (O).
  Outcomes: good and poor markets.

Decisions 2 years hence:
  Label them D2 through D5
  Outcomes: good market, poor market, and out-of-business.

Choice of production levels for D2 through D5:
  Quadruple production (4×); double production (2×); level production (1×); one-half production (0.5×); stop production (0×)

The alternatives for future production levels (D2 through D5) are added to the tree and followed by the market responses of good and poor. If the stop-production (0×) decision is made at D3 or D5, the only outcome is out of business.

To utilize the decision tree for alternative evaluation and selection, the following additional information is necessary for each branch:

- The estimated probability that each outcome may occur. These probabilities must sum to 1.0 for each set of outcomes (branches) that result from a decision.
- Economic information for each decision alternative and possible outcome, such as initial investment and estimated cash flows.

Decisions are made using the probability estimate and economic value estimate for each outcome branch. Commonly the present worth at the MARR is used in an expected value computation of the type in Equation [18.2]. This is the general procedure to solve the tree using PW analysis:

1. Start at the top right of the tree. Determine the PW value for each outcome branch considering the time value of money.
2. Calculate the expected value for each decision alternative.

$$E(\text{decision}) = \sum (\text{outcome estimate})P(\text{outcome}) \qquad [18.3]$$

where the summation is taken over all possible outcomes for each decision alternative.

3. At each decision node, select the best E(decision) value—minimum cost or maximum value (if both costs and revenues are estimated).
4. Continue moving to the left of the tree to the root decision in order to select the best alternative.
5. Trace the best decision path back through the tree.

A decision is needed to either market or sell a new invention. If the product is marketed, the next decision is to take it international or national. Assume the details of the outcome branches result in the decision tree of Figure 18–9. The probabilities for each outcome and PW of CFBT (cash flow before taxes) are indicated. These payoffs are in millions of dollars. Determine the best decision at the decision node D1.

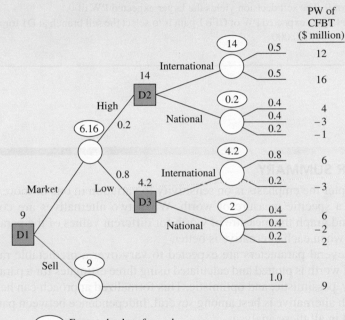

Expected values for each decision alternative

Figure 18–9
Solution of a decision tree with present worth of estimated CFBT values, Example 18.8.

Solution

Use the procedure above to determine that the D1 decision alternative to sell the invention should maximize E(PW of CFBT).

1. Present worth of CFBT is supplied.

2. Calculate the expected PW for alternatives from nodes D2 and D3, using Equation [18.3]. In Figure 18–9, to the right of decision node D2, the expected values of 14 and 0.2 in ovals are determined as

$$E(\text{international decision}) = 12(0.5) + 16(0.5) = 14$$
$$E(\text{national decision}) = 4(0.4) - 3(0.4) - 1(0.2) = 0.2$$

The expected PW values of 4.2 and 2 for D3 are calculated in a similar fashion.

3. Select the larger expected value at each decision node. These are 14 (international) at D2 and 4.2 (international) at D3.

4. Calculate the expected PW for the two D1 branches.

$$E(\text{market decision}) = 14(0.2) + 4.2(0.8) = 6.16$$
$$E(\text{sell decision}) = 9(1.0) = 9$$

The expected value for the sell decision is simple since the one outcome has a payoff of 9. The sell decision yields the larger expected PW of 9.

5. The largest expected PW of CFBT path is to select the sell branch at D1 for a guaranteed $9,000,000.

CHAPTER SUMMARY

In this chapter the emphasis is on sensitivity to variation in one or more parameters using a specific measure of worth. When two alternatives are compared, compute and graph the measure of worth for different values of the parameter to determine when each alternative is better.

When several parameters are expected to vary over a predictable range, the measure of worth is plotted and calculated using three estimates for a parameter—most likely, pessimistic, and optimistic. This formalized approach can help determine which alternative is best among several. Independence between parameters is assumed in all these analyses.

The combination of parameter and probability estimates results in the expected value relation

$$E(X) = \sum XP(X)$$

This expression is also used to calculate $E(\text{revenue})$, $E(\text{cost})$, $E(\text{cash flow})$, $E(\text{PW})$, and $E(i)$ for the entire cash flow sequence of an alternative.

Decision trees are used to make a series of alternative selections. This is a way to explicitly take risk into account. It is necessary to make several types of estimates for a decision tree: outcomes for each possible decision, cash flows, and probabilities. Expected value computations are coupled with those for the measure of worth to solve the tree and find the best alternatives stage by stage.

PROBLEMS

Sensitivity to Parameter Variation

18.1 The Central Drug Distribution Center wants to evaluate a new materials handling system for fragile products. The complete device will cost $62,000 and have an 8-year life and a salvage value of $1500. Annual maintenance, fuel, and overhead costs are estimated at $0.50 per metric ton moved. Labor cost will be $8 per hour for regular wages and $16 for overtime. A total of 20 tons can be moved in an 8-hour period. The center handles from 10 to 30 tons of fragile products per day. The center uses a MARR of 10%. Determine the sensitivity of present worth of costs to the annual volume moved. Assume the operator is paid regular wages for 200 days of work per year. Use a 10-metric-ton increment for the analysis.

18.2 An equipment alternative is being economically evaluated separately by three engineers at Raytheon. The first cost will be $77,000, and the life is estimated at 6 years with a salvage value of $10,000. The engineers disagree, however, on the estimated revenue the equipment will generate. Joe has made an estimate of $10,000 per year. Jane states that this is too low and estimates $14,000, while Carlos estimates $18,000 per year. If the before-tax MARR is 8% per year, use PW to determine if these different estimates will change the decision to purchase the equipment.

18.3 Perform the analysis in Problem 18.2 on a spreadsheet, and make it an after-tax consideration using 5-year MACRS depreciation and a 35% effective tax rate. Use estimated annual expenses of $2000. Determine the effective after-tax MARR from the before-tax MARR of 8%.

18.4 A manufacturing company needs 1000 square meters of storage space. Purchasing land for $80,000 and erecting a temporary metal building at $70 per square meter are one option. The president expects to sell the land for $100,000 and the building for $20,000 after 3 years. Another option is to lease space for $2.50 per square meter per month payable at the beginning of each year. The MARR is 20%. Perform a present worth analysis of the building and leasing alternatives to determine the sensitivity of the decision if construction costs go down 10% and the lease cost goes up to $2.75 per square meter per month.

18.5 A new demonstration system has been designed by Custom Baths & Showers. For the data shown, determine the sensitivity of the rate of return to the amount of the revenue gradient G for values from $1500 to $2500. If the MARR is 18% per year, would this variation in the revenue gradient affect the decision to build the demonstration system? Work this problem (a) by hand and (b) by computer.

$P = \$74,000 \qquad n = 10$ years $\qquad S = 0$

Expense: $30,000 first year, increasing $3000 per year thereafter

Revenue: $63,000 first year, decreasing by G per year thereafter

18.6 Consider the two air conditioning systems detailed below.

	System 1	System 2
First cost, $	−10,000	−17,000
Annual operating cost, $/year	−600	−150
Salvage value, $	−100	−300
New compressor and motor cost at midlife, $	−1,750	−3,000
Life, years	8	12

Use AW analysis to determine the sensitivity of the economic decision to MARR values of 4%, 6%, and 8%. Plot the sensitivity curve. Work this problem (a) by hand and (b) by computer.

18.7 Clint and Anne plan to either purchase a weekend home or buy a travel trailer and a four-wheel-drive vehicle to pull the trailer for vacations. They have found a 5-acre tract with a small house 25 miles from their home. It will cost them $130,000, and they estimate they can sell the place for $145,000 in 10 years when their children are grown. The insurance and upkeep costs are estimated at $1500 per year, but this weekend site is expected to save the family $150 every day they don't go on a traveling vacation. The couple estimates that even though the cabin is only 25 miles from home, they will travel 50 miles per day when at the cabin while working on it and visiting neighbors and local events. Their car averages 30 miles per gallon of gas.

The trailer and vehicle combination would cost $75,000 and could be sold for $20,000 in 10 years. Insurance and operating costs will average $1750 per year, but this alternative is expected to save $125 per vacation day. On a normal vacation, they travel 300 miles each day. Mileage per gallon for the vehicle and trailer is estimated at 60% that of the family car. Assume gas costs $1.20 per U.S. gallon. The money earmarked for this purchase currently earns 10% per year.
(a) Compute the breakeven number of vacation days per year for the two plans.
(b) Determine the sensitivity of AW for each plan if vacation time may vary as much as ±40% from the breakeven number.
(c) If Anne just started a new job and will have only 14 days for the next few years, which alternative is less costly?

18.8 (a) Calculate by hand and plot the sensitivity of rate of return versus the bond interest rate of a $50,000 fifteen-year bond that has bond interest paid quarterly and is discounted to $42,000. Consider bond rates of 5%, 7%, and 9%. (b) Use a spreadsheet to solve this problem.

18.9 Leona has been offered an investment opportunity that will require a cash outlay of $30,000 now for a cash inflow of $3500 for each year of investment. However, she must state now the number of years she plans to retain the investment. Additionally, if the investment is retained for 6 years, $25,000 will be returned to investors, but after 10 years the return is anticipated to be only $15,000, and after 12 years it is estimated to be $8000. If money is currently worth 8% per year, is the decision sensitive to the retention period?

18.10 An asset costs $8000 and has a maximum life of 15 years. Its AOC is expected to be $500 the first year and increase by an arithmetic gradient G between $60 and $140 per year thereafter. Determine the sensitivity of the economic service life to the cost gradient in increments of $40, and plot the results on the same graph. Use an interest rate of 5% per year.

18.11 For plans A and B, graph the sensitivity of PW values at 20% per year for the range −50% to +100% of the following single-point estimates for each of the parameters: (a) first cost, (b) AOC, and (c) annual revenue.

	Plan A	Plan B
First cost, $	−500,000	−375,000
AOC, $/year	−75,000	−80,000
Annual revenue, $/year	150,000	130,000
Salvage value, $	50,000	37,000
Expected life, years	5	5

18.12 Use a spreadsheet to determine and graph the sensitivity of the rate of return to a ±25% change in (a) purchase price and (b) selling price for the following investment. An engineer purchased an antique car for $25,000 with the plan to "make it original" and sell it at a profit. Improvements cost $5500 the first year, $1500 the second year, and $1300 the third year. He sold the car after 3 years for $35,000.

18.13 Use a spreadsheet to plot on one graph (similar to Figure 18–3) the sensitivity of AW over the range −30% to +50% of the parameters (a) first cost, (b) AOC, and (c) annual revenue. Use a MARR of 18% per year.

Process	Estimate
First cost, $	−80,000
Salvage value, $	10,000
Life, years	10
AOC, $/year	−15,000
Annual revenue, $/year	39,000

18.14 Graph the sensitivity of what a person should be willing to pay now for a 9% $10,000 bond due in 10 years if there is a ±30% change in (a) face value, (b) dividend rate, or (c) required nominal rate of return, which is expected to be 8% per year, compounded semiannually. The bond dividends are paid semiannually.

Three Estimates

18.15 An engineer must decide between two ways to pump concrete up to the top floors of a seven-story office building to be constructed. Plan 1 requires the purchase of equipment for $6000 which costs between $0.40 and $0.75 per metric ton to operate, with a most likely cost of $0.50 per metric ton. The asset is able to pump 100 metric tons per day. If

purchased, the asset will last for 5 years, have no salvage value, and be used from 50 to 100 days per year. Plan 2 is an equipment-leasing option and is expected to cost the company $2500 per year for equipment with a low cost estimate of $1800 and a high estimate of $3200 per year. In addition, an extra $5 per hour labor cost will be incurred for operating the leased equipment per 8-hour day. Plot the AW of each plan versus total annual operating cost or lease cost at $i = 12\%$. Which plan should the engineer recommend if the most likely estimate of use is (a) 50 days per year and (b) 100 days per year?

18.16 A meat packing plant must decide between two ways to cool cooked hams. Spraying cools to 30°C using approximately 80 liters of water for each ham. The immersion method uses 40 liters per ham, but an extra initial cost for equipment of $2000 and extra maintenance costs of $100 per year for the 10-year life are estimated. Ten million hams per year are cooked, and water costs $0.12 per 1000 liters. Another cost is $0.04 per 1000 liters for wastewater treatment, which is required for either method. The MARR is 15% per year.

If the spray method is selected, the amount of water used can vary from an optimistic value of 40 liters to a pessimistic value of 100 liters with 80 liters being the most likely amount. The immersion technique always takes 40 liters per ham. How will this varying use of water for the spray method affect the economic decision?

18.17 When the country's economy is expanding, AB Investment Company is optimistic and expects a MARR of 15% for new investments. However, in a receding economy the expected return is 8%. Normally a 10% return is required. An expanding economy causes the estimates of asset life to go down about 20%, and a receding economy makes the n values increase about 10%. Plot the sensitivity of present worth versus (a) the MARR and (b) the life values for the two plans detailed below, using the most likely estimates for the other factors. (c) Considering all the analyses, under which scenario, if any, should plan M or Q be rejected?

	Plan M	Plan Q
Initial investment, $	−100,000	−110,000
Cash flow, $/year	+15,000	+19,000
Life, years	20	20

Expected Value

18.18 Calculate the expected flow rate for each oil well using the estimated probabilities.

	Expected Flow, Barrels/Day			
	100	200	300	400
North well	0.15	0.75	0.10	—
East well	0.35	0.15	0.45	0.05

18.19 There are four estimates made for the anticipated cycle time to produce a subcomponent. The estimates, in seconds, are 10, 20, 30, and 70. (a) If equal weight is placed on each estimate what is the expected time to plan for? (b) If the largest time is disregarded, estimate the expected time. Does the large estimate seem to significantly increase the expected value?

18.20 The variable Y is defined as 3^n for $n = 1$, 2, 3, 4 with probabilities of 0.4, 0.3, 0.233, and 0.067, respectively. Determine the expected value of Y.

18.21 The AOC value for an alternative is expected to be one of two values. Your office partner told you that the low value is $2800 per year. If her computations show a probability of 0.75 for the high value and an expected AOC of $4575, what is the high AOC value she used in the computation of the average?

18.22 A total of 40 different proposals were evaluated by the IRAD (Industrial Research and Development) committee during the past year. Twenty were funded. Their rate of return estimates are summarized below with the i^* values rounded to the nearest integer. For the accepted proposals, calculate the expected rate of return $E(i)$.

Proposal Rate of Return, %	Number of Proposals
−8	1
−5	1
0	5
5	5
8	2
10	3
15	3
	20

18.23 Starbreak Foods has performed an economic analysis of proposed service in a new region of the country. The three-estimate approach to sensitivity analysis has been applied. The optimistic and pessimistic values each have an estimated 15% chance of occurring. Use the AW values shown to compute the expected AW.

	Optimistic	Most Likely	Pessimistic
AW value, $/year	+300,000	+50,000	−25,000

18.24 (a) Determine the expected present worth of the following cash flow series if each series may be realized with the probability shown at the head of each column. Let $i = 20\%$ per year.
   (b) Determine the expected AW value for the same cash flow series.

	Annual Cash Flow, $/year		
Year	Prob. = 0.5	Prob. = 0.2	Prob. = 0.3
0	−5000	−6000	−4000
1	1000	500	3000
2	1000	1500	1200
3	1000	2000	−800

18.25 A very successful health and recreation club wants to construct a mock mountain for climbing and exercise outside for its customers' use. Because of its location, there is a 30% chance of a 120-day season of good outdoor weather, a 50% chance of a 150-day season, and a 20% chance of a 165-day season. The mountain will be used by an estimated 350 persons each day of the 4-month (120-day) season, but by only 100 per day for each extra day the season lasts. The feature will cost $375,000 to construct and require a $25,000 rework each 4 years, and the annual maintenance and insurance costs will be $56,000. The climbing fee will be $5 per person. If a life of 10 years is anticipated and a 12% per year return is expected, determine if the addition is economically justified.

18.26 The owner of Ace Roofing may invest $200,000 in new equipment. A life of 6 years and a salvage value of 12% of first cost are anticipated. The annual extra revenue will depend upon the state of the housing and construction industry.

The extra revenue is expected to be only $20,000 per year if the current slump in the industry continues. Real estate economists estimate a 50% chance of the slump lasting 3 years and give it a 20% chance of continuing for 3 additional years. However, if the depressed market does improve, during either the first or second 3-year period, the revenue of the investment is expected to increase by a total of $35,000 per year. Can the company expect to make a return of 8% per year on its investment? Use present worth analysis.

18.27 Jeremy has $5000 to invest. If he puts the money in a certificate of deposit (CD), he is assured of receiving an effective 6.35% per year for 5 years. If he invests the money in stocks, he has a 50-50 chance of one of the following cash flow sequences for the next 5 years.

	Annual Cash Flow, $/year	
	Prob. = 0.5	Prob. = 0.5
Year	Stock 1	Stock 2
0	−5000	−5000
1–4	+250	+600
5	+6800	+4000

Finally, Jeremy can invest his $5000 in real estate for the 5 years with the following cash flow and probability estimates.

	Annual Cash Flow, $/year		
Year	Prob. = 0.3	Prob. = 0.5	Prob. = 0.2
0	−5000	−5000	−5000
1	−425	0	+500
2	−425	0	+600
3	−425	0	+700
4	−425	0	+800
5	+9500	+7200	+5200

Which of the three investment opportunities offers the best expected rate of return?

18.28   The California Company has $1 million in an investment pool which the board of directors plans to place in projects with different D-E mixes varying from 20–80 to 80–20. To assist with the decision, the plot shown below, prepared by the chief financial officer, of currently estimated annual equity rates of return (*i* on equity capital) versus various D-E mixes will be used. All investments will be for 10 years with no intermediate cash flows in or out of the projects. The motion passed by the board is to invest as follows:

D-E mix	20–80	50–50	80–20
Percent of pool	30%	50%	20%

(a)   What is the current estimate of the expected annual rate of return on the company's equity capital for the $1 million investments after 10 years?

(b)   What is the actual amount of equity capital invested now, and what is the expected total amount after 10 years for the board-approved investment plan?

(c)   If inflation is expected to average 4.5% per year over the next 10-year

period, determine both the real interest rates at which the equity investment funds will grow and the purchasing power in terms of today's (constant-value) dollars of the actual amount accumulated after 10 years.

18.29   A flagship hotel in Cedar Falls must construct a retaining wall next to its parking lot due to the widening of the city's main thoroughfare located in front of the hotel. The amount of rainfall experienced in a short period of time may cause damage in varying amounts, and the wall increases in cost in order to protect against larger and faster rainfalls. The probabilities of a specific amount of rainfall in a 30-minute period and wall cost estimates are as follows:

Rainfall, Inches/30 Minutes	Probability of Greater Rainfall	Estimated First Cost of Wall, $
2.0	0.3	−200,000
2.25	0.1	−225,000
2.5	0.05	−300,000
3.0	0.01	−400,000
3.25	0.005	−450,000

The wall will be financed through a 6% per year loan for the full amount that will be repaid over a 10-year period.

Records indicate an average damage of $50,000 has occurred with heavy rains, due to the relatively poor cohesive properties of the soil along the thoroughfare. A discount rate of 6% per year is applicable. Find the amount of rainfall to protect against by choosing the retaining wall with the smallest AW value over the 10-year period.

Decision Trees

18.30   For the decision tree branch shown, determine the expected values of the two

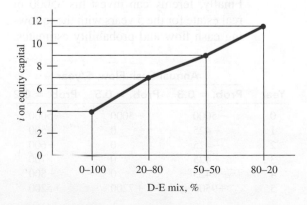

outcomes if decision D3 is already se-lected and the maximum outcome value is sought. (This decision branch is part of a larger tree.)

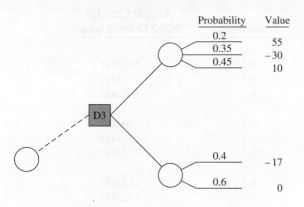

Probability	Value
0.2	
0.35	55
0.45	−30
	10
0.4	−17
0.6	0

18.31 A large decision tree has an outcome branch that is detailed for this problem. If decisions D1, D2, and D3 are all

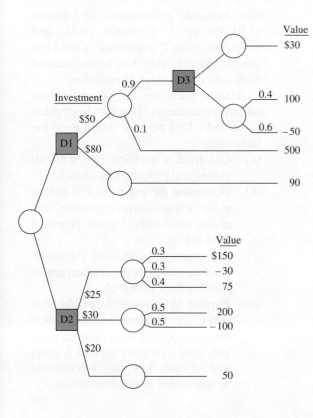

options in a 1-year time period, find the decision path which maximizes the out-come value. There are specific dollar in-vestments necessary for decision nodes D1, D2, and D3 as indicated on each branch.

18.32 Decision D4, which has three possible alternatives—*x, y,* or *z*—must be made in year 3 of a 6-year study period in order to maximize the expected value of present worth. Using a rate of return of 15% per year, the investment re-quired in year 3, and the estimated cash flows for years 4 through 6, determine which decision should be made in year 3. (This decision node is part of a larger tree.)

Investment required, year 3	Cash flow (× $1000) Year 4	Year 5	Year 6	Outcome probability
High −200,000	50	50	50	0.7
Low	40	30	20	0.3
High −75,000	30	40	50	0.45
Low	30	30	30	0.55
High −350,000	190	170	150	0.7
Low	−30	−30	−30	0.3

18.33 A total of 5000 mechanical subassem-blies are needed annually on a final as-sembly line. The subassemblies can be obtained in one of three ways: (1) *Make them* in one of three plants owned by the company; (2) *buy them off the shelf* from the one and only manufacturer; or (3) *contract to have them made* to speci-fications by a vendor.

The estimated annual cost for each alternative is dependent upon specific circumstances of the plant, producer, or contractor. The information shown details the circumstance, a probability of occurrence, and the estimated annual

cost. Construct and solve a decision tree to determine the least-cost alternative to provide the subassemblies.

Decision Alternative	Outcomes	Probability	Annual Cost for 5000 Units, $/year
1. Make	Plant:		
	A	0.3	−250,000
	B	0.5	−400,000
	C	0.2	−350,000
2. Buy off the shelf	Quantity:		
	<5000, pay premium	0.2	−550,000
	5000 available	0.7	−250,000
	>5000, forced to buy	0.1	−290,000
3. Contract	Delivery:		
	Timely delivery	0.5	−175,000
	Late delivery; then buy some off shelf	0.5	−450,000

18.34 The president of ChemTech is trying to decide whether to start a new product line or purchase a small company. It is not financially possible to do both. To make the product for a 3-year period will require an initial investment of $250,000. The expected annual cash flows with probabilities in parentheses are: $75,000 (0.5), $90,000 (0.4), and $150,000 (0.1).

To purchase the small company will cost $450,000 now. Market surveys indicate a 55% chance of increased sales for the company and a 45% chance of severe decreases with an annual cash flow of $25,000. If decreases are experienced in the first year, the company will be sold immediately (during year 1) at a price of $200,000. Increased sales could be $100,000 the first 2 years. If this occurs, a decision to expand after 2 years at an additional investment of $100,000 will be considered. This expansion could generate cash flows

with indicated probabilities as follows: $120,000 (0.3), $140,000 (0.3), and $175,000 (0.4). If expansion is not chosen, the current size will be maintained with anticipated sales to continue.

Assume there are no salvage values on any investments. Use the description given and a 15% per year return to do the following:

(a) Construct a decision tree with all values and probabilities shown.

(b) Determine the expected PW values at the "expansion/no expansion" decision node after 2 years provided sales are up.

(c) Determine what decision should be made now to offer the greatest return possible for ChemTech.

(d) Explain in words what would happen to the expected values at each decision node if the planning horizon were extended beyond 3 years and all cash flow values continued as forecasted in the description.

EXTENDED EXERCISE

LOOKING AT ALTERNATIVES FROM DIFFERENT ANGLES

Berkshire Controllers usually finances its engineering projects with a combination of debt and equity capital. The resulting MARR ranges from a low of 8% per year, if business is slow, to a high of 15% per year. Normally, a 10% per year return is expected. Also the life estimates for assets tend to go down about 20% from normal in a vigorous business environment, and up about 10% in a receding economy. The following estimates are the most likely values for two plans currently being evaluated. Use these data and a spreadsheet to answer the questions below.

| | Plan A | Plan B | |
		Asset 1	Asset 2
First cost, $	−10,000	−30,000	−5000
AOC, $ per year	−500	−100	−200
Salvage value, $	1000	5000	−200
Estimated life, years	40	40	20

Questions

1. Are the PW values for plans A and B sensitive to changes in the MARR?
2. Are the PW values sensitive to varying life estimates?
3. Plot the results above on separate charts for MARR and life estimates.
4. Is the breakeven point for the first cost of plan A sensitive to the changes in MARR as business goes from vigorous to receding?

CASE STUDY

SENSITIVITY ANALYSIS OF PUBLIC SECTOR PROJECTS—WATER SUPPLY PLANS

Introduction

One of the most basic services provided by municipal governments is the delivery of a safe, reliable water supply. As cities grow and extend their boundaries to outlying areas, they often inherit water systems that were not constructed according to city codes. The upgrading of these systems is sometimes more expensive than installing one correctly in the first place. To avoid these problems, city officials sometimes install water systems beyond the existing city limits in anticipation of future growth. This case study was extracted from such a countywide water and waste-water management plan and is limited to only some of the water supply alternatives.

Procedure

From about a dozen suggested plans, five methods were developed by an executive committee as alternative ways of providing water to the study area. These methods were then subjected to a preliminary evaluation to identify the most promising alternatives. Six attributes or factors used in the initial rating were: ability to serve the area, relative cost, engineering feasibility, institutional issues, environmental considerations, and lead time requirement. Each factor carried the same weighting and had values ranging from 1 to 5, with 5 being best. After the top three alternatives were identified, each was subjected to a detailed economic evaluation for selection of the best alternative. These detailed evaluations included an estimate of the capital investment of each alternative amortized over 20 years at 8% per year interest and the annual maintenance and operation (M&O) costs. The annual cost (an AW value) was then divided by the population served to arrive at a monthly cost per household.

Results of Preliminary Screening

Table 18–5 presents the results of the screening using the six factors rated on a scale of 1 to 5. Alternatives 1A, 3, and 4 were determined to be the three best and were chosen for further evaluation.

Detailed Cost Estimates for Selected Alternatives

All amounts are cost estimates.

Alternative 1A

Capital cost
Land with water rights: 1720 hectares
@ $5000 per hectare	$8,600,000
Primary treatment plant	2,560,000
Booster station at plant	221,425
Reservoir at booster station	50,325
Site cost	40,260
Transmission line from river	3,020,000
Transmission line right-of-way	23,350

Percolation beds	2,093,500
Percolation bed piping	60,400
Production wells	510,000
Well field gathering system	77,000
Distribution system	1,450,000
Additional distribution system	3,784,800
Reservoirs	250,000
Reservoir site, land, and development	17,000
Subtotal	22,758,060
Engineering and contingencies	5,641,940
Total capital investment	$28,400,000

Maintenance and operation costs (annual)
Pumping 9,812,610 kWh per year

@ $0.08 per kWh	$ 785,009
Fixed operating cost	180,520
Variable operating cost	46,730
Taxes for water rights	48,160
Total annual M&O cost	$1,060,419

$$\text{Total annual cost} = \text{equivalent capital investment} + \text{M\&O cost}$$
$$= 28,400,000(A/P,8\%,20) + 1,060,419$$
$$= 2,892,540 + 1,060,419$$
$$= \$3,952,959$$

Average monthly household cost to serve 95% of 4980 households is

$$\text{Household cost} = (3,952,959)\frac{1}{12}\frac{1}{4980}\frac{1}{0.95}$$
$$= \$69.63 \text{ per month}$$

Alternative 3

Total capital investment = $29,600,000

Total annual M&O cost = $867,119

$$\text{Total annual cost} = 29,600,000(A/P,8\%, 20) + 867,119$$
$$= 3,014,760 + 867,119$$
$$= \$3,881,879$$

Household cost = $68.38 per month

TABLE 18–5   Results of Rating Six Factors for Each Alternative, Case Study

Alternative	Description	Ability to Supply Area	Relative Cost	Engineering Feasibility	Institutional Issues	Environmental Considerations	Lead-Time Requirement	Total
					Factors			
1A	Receive city water and recharge wells	5	4	3	4	5	3	24
3	Joint city and county plant	5	4	4	3	4	3	23
4	County treatment plant	4	4	3	3	4	3	21
8	Desalt groundwater	1	2	1	1	3	4	12
12	Develop military water	5	5	4	1	3	1	19

Alternative 4

Total capital investment = $29,000,000

Total annual M&O cost = $1,063,449

Total annual cost = $29,000,000(A/P,8%, 20) + 1,063,449

= 2,953,650 + 1,063,449

= $4,017,099

Household cost = $70.76 per month

Conclusion

On the basis of the lowest monthly household cost, alternative 3 (joint city and county plant) is the most economically attractive.

Case Study Exercises

1. If the environmental considerations factor is to have a weighting of twice as much as any of the other five factors, what is its percentage weighting?

2. If the ability to supply area and relative cost factors were each weighted 20% and the other four factors 15% each, which alternatives would be ranked in the top three?

3. By how much would the capital investment of alternative 4 have to decrease in order to make it more attractive than alternative 3?

4. If alternative 1A served 100% of the households instead of 95%, by how much would the monthly household cost decrease?

5. (a) Perform a sensitivity analysis on the two parameters of M&O costs and number of households to determine if alternative 3 remains the best economic choice. Three estimates are made for each parameter in Table 18–6. M&O costs may vary up (pessimistic) or down (optimistic) from the most likely estimates presented in the case statement. The estimated number of households (4980) is determined to be the pessimistic

TABLE 18–6 Pessimistic, Most Likely, and Optimistic Estimates for Two Parameters

	Annual M&O Costs	Number of Households
Alternative 1A		
Pessimistic	+1%	4980
Most likely	$1,060,419	+2%
Optimistic	−1%	+5%
Alternative 3		
Pessimistic	+5%	4980
Most likely	$867,119	+2%
Optimistic	0%	+5%
Alternative 4		
Pessimistic	+2%	4980
Most likely	$1,063,449	+2%
Optimistic	−10%	+5%

estimate. Growth of 2% up to 5% (optimistic) will tend to lower the monthly cost per household.

(b) Consider the monthly cost per household for alternative 4, the optimistic estimate. The number of households is 5% above 4980, or 5230. What is the number of households that would have to be available in order for this option only to have exactly the same monthly household cost as that for alternative 3 at the optimistic estimate of 5230 households?

19

More on Variation and Decision Making Under Risk

C H A P T E R

This chapter further expands our ability to analyze variation in estimates, to consider probability, and to make decisions under risk. Fundamentals discussed include probability distributions, especially their graphs and properties of expected value and dispersion; random sampling; and the use of simulation to account for variation in engineering economy studies.

Through coverage of variation and probability, this chapter complements topics in the first sections of Chapter 1: the role of engineering economy in decision making and economic analysis in the problem-solving process. These techniques are more time-consuming than using estimates made with certainty, so they should be used primarily for critical parameters.

LEARNING OBJECTIVES

Purpose: Learn to incorporate decision making under risk into an engineering economy analysis using the basics of probability distributions, sampling, and simulation.

This chapter will help you:

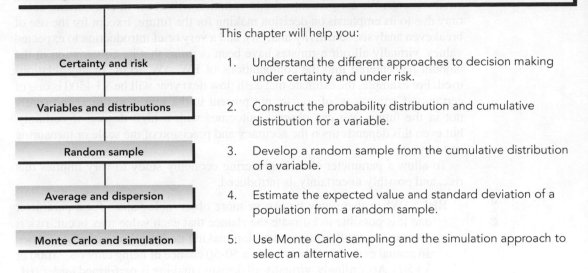

Certainty and risk

1. Understand the different approaches to decision making under certainty and under risk.

Variables and distributions

2. Construct the probability distribution and cumulative distribution for a variable.

Random sample

3. Develop a random sample from the cumulative distribution of a variable.

Average and dispersion

4. Estimate the expected value and standard deviation of a population from a random sample.

Monte Carlo and simulation

5. Use Monte Carlo sampling and the simulation approach to select an alternative.

19.1 INTERPRETATION OF CERTAINTY, RISK, AND UNCERTAINTY

All things in the world vary—one from another, over time, and with different environments. We are guaranteed that variation will occur in engineering economy due to its emphasis on decision making for the future. Except for the use of breakeven analysis, sensitivity analysis, and a very brief introduction to expected values, virtually all our estimates have been *certain;* that is, no variation in the amount has entered into the computations of PW, AW, ROR, or any relations used. For example, the estimate that cash flow next year will be $+4500 is one of certainty. Certainty is, of course, not present in the real world now and surely not in the future. We can observe outcomes with a high degree of certainty, but even this depends upon the accuracy and precision of the scale or measuring instrument.

To allow a parameter of an engineering economy study to vary implies that risk, and possibly uncertainty, is introduced.

Risk.   When there may be two or more observable values for a parameter *and* it is possible to estimate the chance that each value may occur, risk is present. As an illustration, decision making under risk is introduced when an annual cash flow estimate has a 50-50 chance of being either $−1000 or $+500. Accordingly, virtually all decision making is performed *under risk.*

Uncertainty.   Decision making under uncertainty means there are two or more values observable, but the chances of their occurring cannot be estimated or no one is willing to assign the chances. The observable values in uncertainty analysis are often referred to as *states of nature.* For example, consider the states of nature to be the rate of national inflation in a particular country during the next 2 to 4 years: remain low, increase 5% to 10% annually, or increase 20% to 50% annually. If there is absolutely no indication that the three values are equally likely, or that one is more likely than the others, this is a statement that indicates decision making under uncertainty.

Example 19.1 explains how a parameter can be described and graphed to prepare for decision making under risk.

EXAMPLE 19.1

Sue and Charles are both seniors in college and plan to be married next year. Based upon conversations with friends who have recently married, the couple has decided to make separate estimates of what each expects the ceremony to cost, with the chance that each estimate is actually observed expressed as a percentage. (*a*) Their separate estimates are tabulated at the top of Figure 19-1. Construct two graphs: one of Charles's estimated costs versus his chance estimates, and one for Sue. Comment on the shape of the plots relative to each other. (*b*) After some discussion, they decided the ceremony should cost somewhere between $7500 and $10,000. All values between the two limits are equally likely with a chance of 1 in 25. Plot these values versus chance.

Charles		Sue	
Estimated Cost, $	**Chance, %**	**Estimated Cost, $**	**Chance, %**
3,000	65	8,000	33.3
5,000	25	10,000	33.3
10,000	10	15,000	33.3

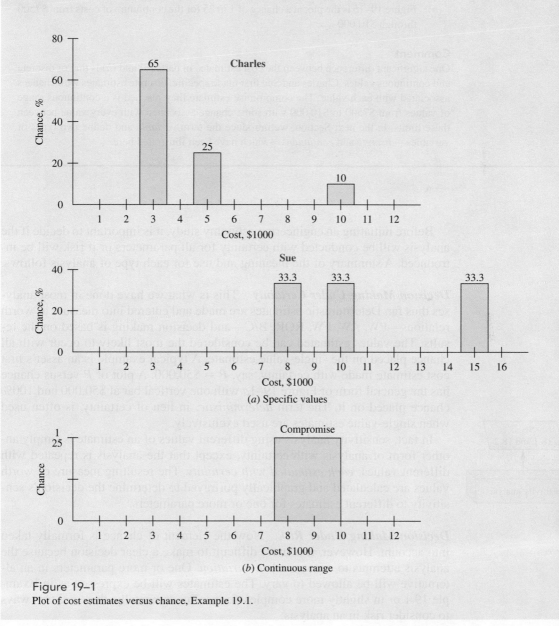

Figure 19–1
Plot of cost estimates versus chance, Example 19.1.

Solution

(a) Figure 19–1a presents the plot for Charles's and Sue's estimates, with the cost scales aligned. Sue expects the cost to be considerably higher than Charles. Additionally, Sue places equal (or uniform) chances on each value. Charles places a much higher chance on lower cost values; 65% of his chances are devoted to $3000, and only 10% to $10,000, which is Sue's middle cost estimate. The plots clearly show the different perceptions about their estimated wedding costs.

(b) Figure 19–1b is the plot at a chance of 1 in 25 for the continuum of costs from $7500 through $10,000.

Comment

One significant difference between the cost estimates in parts (a) and (b) is that of discrete and continuous values. Charles and Sue first made specific, discrete estimates with chances associated with each value. The compromise estimate they reached is a continuous range of values from $7500 to $10,000 with some chance associated with every value between these limits. In the next Section, we introduce the term *variable* and define two types of variables—*discrete* and *continuous*—which have been illustrated here.

Before initiating an engineering economy study, it is important to decide if the analysis will be conducted with certainty for all parameters or if risk will be introduced. A summary of the meaning and use for each type of analysis follows.

Decision Making Under Certainty   This is what we have done in most analyses thus far. Deterministic estimates are made and entered into measure of worth relations—PW, AW, FW, ROR, B/C—and decision making is based on the results. The values estimated can be considered the most likely to occur with all chance placed on the single-value estimate. A typical example is an asset's first cost estimate made with certainty, say, $P = \$50,000$. A plot of P versus chance has the general form of Figure 19–1a with one vertical bar at $50,000 and 100% chance placed on it. The term *deterministic,* in lieu of certainty, is often used when single-value estimates are used exclusively.

In fact, sensitivity analysis using different values of an estimate is simply another form of analysis with certainty, except that the analysis is repeated with different values, *each estimated with certainty.* The resulting measure of worth values are calculated and graphically portrayed to determine the decision's sensitivity to different estimates for one or more parameters.

Secs.
18.1 and 18.2

Sensitivity analysis

Decision Making Under Risk   Now the element of chance is formally taken into account. However, it is more difficult to make a clear decision because the analysis attempts to accommodate *variation.* One or more parameters in an alternative will be allowed to vary. The estimates will be expressed as in Example 19.1 or in slightly more complex forms. Fundamentally, there are two ways to consider risk in an analysis:

Expected value analysis. Use the chance and parameter estimates to calculate expected values, E(parameter) via formulas such as Equation [18.2]. Analysis results in E(cash flow), E(AOC), and the like; and the final result is the expected value for a measure of worth, such as E(PW), E(AW), E(ROR), E(B/C). To select the alternative, choose the most favorable expected value of the measure of worth. In an elementary form, this is what we learned about expected values in Chapter 18. The computations may become more elaborate, but the principle is fundamentally the same.

Simulation analysis. Use the chance and parameter estimates to generate repeated computations of the measure of worth relation by randomly sampling from a plot for each varying parameter similar to those in Figure 19–1. When a representative and random sample is complete, an alternative is selected utilizing a table or plot of the results. Usually, graphics are an important part of decision making via simulation analysis. Basically, this is the approach discussed in the rest of this chapter.

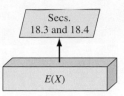

Decision Making Under Uncertainty   When chances are not known for the identified states of nature (or values) of the uncertain parameters, the use of expected value–based decision making under risk as outlined above is not an option. In fact, it is difficult to determine what criterion to use to even make the decision.

If it is possible to agree that each state is equally likely, then all states have the same chance, and the situation reduces to one of decision making under risk, because expected values can be determined.

Because of the relatively inconclusive approaches necessary to incorporate decision making under uncertainty into an engineering economy study, the techniques can be quite useful but are beyond the intended scope of this text.

In an engineering economy study, as well as all other forms of analysis and decision making, observed parameter values in the future will vary from the value estimated at the time of the study. However, when performing the analysis, not all parameters should be considered as probabilistic (or at risk). Those that are estimable with a relatively high degree of certainty should be fixed for the study. Accordingly, the methods of sampling, simulation, and statistical data analysis are selectively used on parameters deemed important to the decision-making process. As mentioned in Chapter 18, interest rate-based parameters (MARR, other interest rates, and inflation) are usually not treated as random variables in the discussions that follow. Parameters such as P, AOC, n, S, material and unit costs, revenues, etc., are the targets of decision making under risk and simulation. Anticipated and predictable variation in interest rates is more commonly addressed by the approaches of sensitivity analysis covered in the first two sections of Chapter 18.

The remainder of this chapter concentrates upon decision making under risk as applied in an engineering economy study. The next three sections provide foundation material necessary to design and correctly conduct a simulation analysis (Section 19.5).

19.2 ELEMENTS IMPORTANT TO DECISION MAKING UNDER RISK

Some basics of probability and statistics are essential to correctly perform decision making under risk via expected value or simulation analysis. These basics are explained here. (If you are already familiar with them, this section will provide a review.)

Random Variable (or Variable)   This is a characteristic or parameter that can take on any one of several values. Variables are classified as *discrete* or *continuous*. Discrete variables have several specific, isolated values, while continuous variables can assume any value between two stated limits, called the *range* of the variable.

The estimated life of an asset is a discrete variable. For example, n may be expected to have values of $n = 3, 5, 10$, or 15 years, and no others. The rate of return is an example of a continuous variable; i can vary from -100% to ∞, that is, $-100\% \le i < \infty$. The ranges of possible values for n (discrete) and i (continuous) are shown as the x axes in Figure 19–2a. (In probability texts, capital

Figure 19–2
(*a*) Discrete and continuous variable scales, and (*b*) scales for a variable versus its probability.

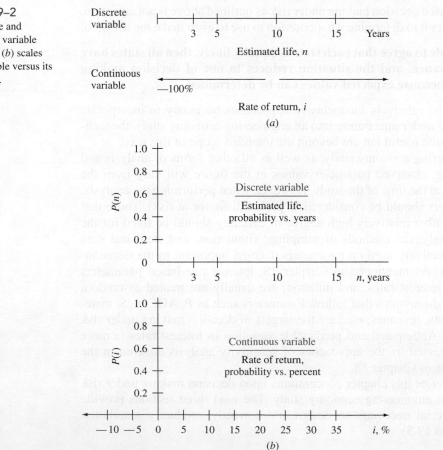

letters symbolize a variable, say X, and small letters, x, identify a specific value of the variable. Though correct, this level of rigor in terminology is not included in this chapter.)

Probability   This is a number between 0 and 1.0 which expresses the chance in decimal form that a random variable (discrete or continuous) will take on any value from those identified for it. Probability is simply the amount of chance, divided by 100. Probabilities are commonly identified by $P(X_i)$ or $P(X = X_i)$, which is read as the probability that the variable X takes on the value X_i. (Actually, for a continuous variable, the probability at a single value is zero, as shown in a later example.) The sum of all $P(X_i)$ for a variable must be 1.0, a requirement already discussed. The probability scale, like the percentage scale for chance in Figure 19–1, is indicated on the ordinate (y axis) of a graph. Figure 19–2b shows the 0 to 1.0 range of probability for the variables n and i.

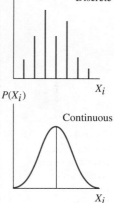

Probability Distribution   This describes how probability is distributed over the different values of a variable. Discrete variable distributions look significantly different from continuous variable distributions, as indicated by the inset at the right. The individual probability values are stated as

$$P(X_i) = \text{probability that } X \text{ equals } X_i \qquad [19.1]$$

The distribution may be developed in one of two ways: by listing each probability value for each possible variable value (see Example 19.2) or by a mathematical description or expression that states probability in terms of the possible variable values (Example 19.3).

Cumulative Distribution   Also called the cumulative probability distribution, this is the accumulation of probability over all values of a variable up to and including a specified value. Identified by $F(X_i)$, each cumulative value is calculated as

$$F(X_i) = \text{sum of all probabilities through the value } X_i$$
$$= P(X \le X_i) \qquad [19.2]$$

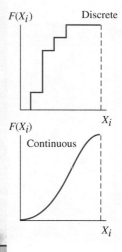

As with a probability distribution, cumulative distributions appear differently for discrete (stair-stepped) and continuous variables (smooth curve). The next two examples illustrate cumulative distributions that correspond to specific probability distributions. These fundamentals about $F(X_i)$ are applied in the next section to develop a random sample.

EXAMPLE **19.2**

Alvin is a medical doctor and biomedical engineering graduate who practices at Medical Center Hospital. He is planning to start prescribing an antibiotic that may reduce infection in patients with flesh wounds. Tests indicate the drug has been applied up to 6 times per day without harmful side effects. If no drug is used, there is always a positive probability that the infection will be reduced by a person's own immune system.

Published drug test results provide good probability estimates of positive reaction (i.e., reduction in the infection count) within 48 hours for different numbers of treatments per day. Use the probabilities listed below to construct a probability distribution and a cumulative distribution for the number of treatments per day.

Number of Treatments per Day	Probability of Infection Reduction
0	0.07
1	0.08
2	0.10
3	0.12
4	0.13
5	0.25
6	0.25

Solution

Define the random variable T as the number of treatments per day. Since T can take on only seven different values, it is a discrete variable. The probability of infection reduction is listed for each value in column 2 of Table 19–1. The cumulative probability $F(T_i)$ is determined using Equation [19.2] by adding all $P(T_i)$ values through T_i, as indicated in column 3.

Figure 19–3a and b shows plots of the probability distribution and cumulative distribution, respectively. The summing of probabilities to obtain $F(T_i)$ gives the cumulative distribution the stair-stepped appearance, and in all cases the final $F(T_i) = 1.0$, since the total of all $P(T_i)$ values must equal 1.0.

TABLE 19–1 Probability Distribution and Cumulative Distribution for Example 19.2

(1) Number per Day T_i	(2) Probability $P(T_i)$	(3) Cumulative Probability $F(T_i)$
0	0.07	0.07
1	0.08	0.15
2	0.10	0.25
3	0.12	0.37
4	0.13	0.50
5	0.25	0.75
6	0.25	1.00

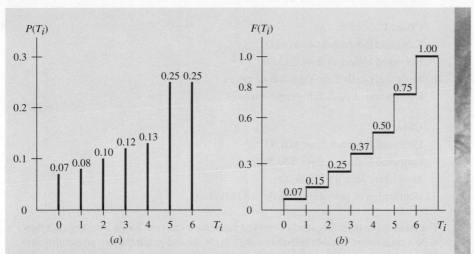

Figure 19–3
(a) Probability distribution $P(T_i)$, and (b) cumulative distribution $F(T_i)$ for Example 19.2.

Comment
Rather than use a tabular form as in Table 19-1 to state $P(T_i)$ and $F(T_i)$ values, it is possible to express them for each value of the variable.

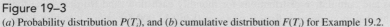

$$P(T_i) = \begin{cases} 0.07 & T_1 = 0 \\ 0.08 & T_2 = 1 \\ 0.10 & T_3 = 2 \\ 0.12 & T_4 = 3 \\ 0.13 & T_5 = 4 \\ 0.25 & T_6 = 5 \\ 0.25 & T_7 = 6 \end{cases} \qquad F(T_i) = \begin{cases} 0.07 & T_1 = 0 \\ 0.15 & T_2 = 1 \\ 0.25 & T_3 = 2 \\ 0.37 & T_4 = 3 \\ 0.50 & T_5 = 4 \\ 0.75 & T_6 = 5 \\ 1.00 & T_7 = 6 \end{cases}$$

In basic engineering economy situations, the probability distribution for a continuous variable is commonly expressed as a mathematical function, such as a *uniform distribution,* a *triangular distribution* (both discussed in Example 19.3 in terms of cash flow), or the more complex, but commonly used, *normal distribution.* For continuous variable distributions, the symbol $f(X)$ is routinely used instead of $P(X_i)$, and $F(X)$ is used instead of $F(X_i)$, simply because the point probability for a continuous variable is zero. Thus, $f(X)$ and $F(X)$ are continuous lines and curves.

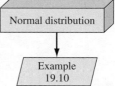

EXAMPLE 19.3

As president of a manufacturing systems consultancy, Sallie has observed the monthly cash flows that have occurred over the last 3 years into company accounts from two longstanding clients. Sallie has concluded the following about the distribution of these monthly cash flows:

Client 1

Estimated low cash flow: $10,000

Estimated high cash flow: $15,000

Most likely cash flow: same for all values

Distribution of probability: uniform

Client 2

Estimated low cash flow: $20,000

Estimated high cash flow: $30,000

Most likely cash flow: $28,000

Distribution of probability: mode at $28,000

The *mode* is the most frequently observed value for a variable. Sallie assumes cash flow to be a continuous variable referred to as *C*. (*a*) Write and graph the two probability distributions and cumulative distributions for monthly cash flow, and (*b*) determine the probability that monthly cash flow is no more than $12,000 for client 1 and no more than $25,000 for client 2.

Solution

All cash flow values are expressed in $1000 units.

Client 1: monthly cash flow distribution

(*a*)   The distribution of cash flows for client 1, identified by the variable C_1, follows the *uniform distribution*. Probability and cumulative probability take the following general forms.

$$f(C_1) = \frac{1}{\text{high} - \text{low}} \qquad \text{low value} \leq C_1 \leq \text{high value}$$

$$f(C_1) = \frac{1}{H - L} \qquad L \leq C_1 \leq H \qquad\qquad [19.3]$$

$$F(C_1) = \frac{\text{value} - \text{low}}{\text{high} - \text{low}} \qquad \text{low value} \leq C_1 \leq \text{high value}$$

$$F(C_1) = \frac{C_1 - L}{H - L} \qquad L \leq C_1 \leq H \qquad\qquad [19.4]$$

For client 1, monthly cash flow is uniformly distributed with $L = \$10$, $H = \$15$, and $\$10 \leq C_1 \leq \$15$. Figure 19–4 is a plot of $f(C_1)$ and $F(C_1)$ from Equations [19.3] and [19.4].

$$f(C_1) = \frac{1}{5} = 0.2 \qquad \$10 \leq C_1 \leq \$15$$

$$F(C_1) = \frac{C_1 - 10}{5} \qquad \$10 \leq C_1 \leq \$15$$

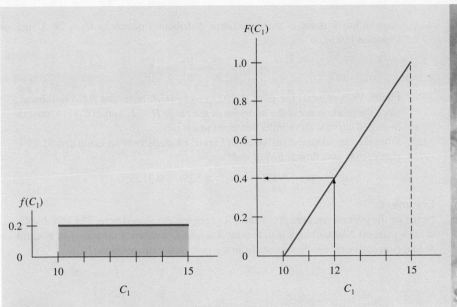

Figure 19–4
Uniform distribution for monthly cash flow, Example 19.3.

(b) The probability that client 1 has a monthly cash flow of less than $12 is easily
determined from the $F(C_1)$ plot as 0.4, or a 40% chance. If the $F(C_1)$ relation is
used directly, the computation is

$$F(\$12) = P(C_1 \leq \$12) - \frac{12 - 10}{5} = 0.4$$

Client 2: monthly cash flow distribution

(a) The distribution of cash flows for client 2, identified by the variable C_2, follows
the *triangular distribution*. This probability distribution has the shape of an
upward-pointing triangle with the peak at the mode M, and downward-sloping
lines joining the x axis on either side at the low (L) and high (H) values. The
mode of the triangular distribution has the maximum probability value.

$$f(\text{mode}) = f(M) = \frac{2}{H - L} \qquad [19.5]$$

The cumulative distribution is comprised of two curved line segments from 0 to
1 with a break point at the mode, where

$$F(\text{mode}) = F(M) = \frac{M - L}{H - L} \qquad [19.6]$$

For C_2, the low value is $L = \$20$, the high is $H = \$30$, and the most likely cash
flow is the mode $M = \$28$. The probability at M from Equation [19.5] is

$$f(28) = \frac{2}{30 - 20} = \frac{2}{10} = 0.2$$

and the break point in the cumulative distribution occurs at $C_2 = 28$. Using Equation [19.6],

$$F(28) = \frac{28 - 20}{30 - 20} = 0.8$$

Figure 19–5 presents the plots for $f(C_2)$ and $F(C_2)$. Note that $f(C_2)$ is skewed, since the mode is not at the midpoint of the range $H - L$, and $F(C_2)$ is a smooth S-shaped curve with an inflection point at the mode.

(b) From the cumulative distribution in Figure 19–5, there is an estimated 31.25% chance that cash flow is \$25 or less.

$$F(\$25) = P(C_2 \le \$25) = 0.3125$$

Comment

Note that the general relations $f(C_2)$ and $F(C_2)$ are not developed here. The variable C_2 is *not* a uniform distribution; it is triangular. Therefore, it requires the use of an integral to find cumulative probability values from the probability distribution $f(C_2)$.

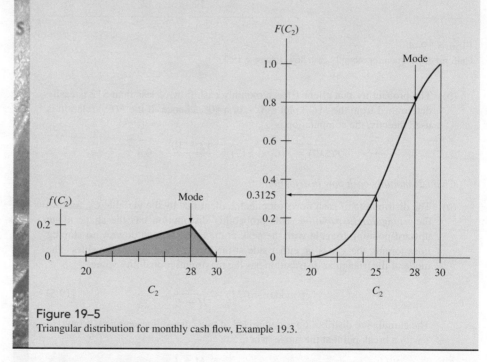

Figure 19–5
Triangular distribution for monthly cash flow, Example 19.3.

Additional Example 19.9

19.3 RANDOM SAMPLES

Estimating a parameter with a single value in previous chapters is the equivalent of taking a *random sample of size 1 from an entire population* of possible values. If all values in the population were known, the probability distribution and cumulative distribution would be known. Then a sample is not necessary. As an illustration, assume that estimates of first cost, annual operating cost, interest rate, and other parameters are used to compute one PW value in order to accept or reject an alternative. Each estimate is a sample of size 1 from an entire population of possible values for each parameter. Now, if a second estimate is made for each parameter and a second PW value is determined, a sample of size 2 has been taken.

Whenever we perform an engineering economy study and utilize decision making under certainty, we use one estimate for each parameter to calculate a measure of worth (i.e., a sample of size 1 for each parameter). The estimate is the most likely value, that is, one estimate of the expected value. We know that all parameters will vary somewhat; yet some are important enough, or will vary enough, that a probability distribution should be determined or assumed for it and the parameter treated as a random variable. This is using risk, and a sample from the parameter's probability distribution—$P(X)$ for discrete or $f(X)$ for continuous—helps formulate probability statements about the estimates. This approach complicates the analysis somewhat; however, it also provides a sense of confidence (or possibly a lack of confidence in some cases) about the decision made concerning the economic viability of the alternative based on the varying parameter. (We will further discuss this aspect later, after we learn how to correctly take a random sample from any probability distribution.)

A random sample of size n is the selection in a random fashion of n values from a population with an assumed or known probability distribution, such that the values of the variable have the same chance of occurring in the sample as they are expected to occur in the population.

Suppose Yvon is an engineer with 20 years of experience working for the Noncommercial Aircraft Safety Commission. For a two-crew aircraft, there are three parachutes on board. The safety standard states that 99% of the time, all three chutes must be "fully ready for emergency deployment." Yvon is relatively sure that nationwide the probability distribution of N, the number of chutes fully ready, may be described by the probability distribution

$$P(N = N_i) = \begin{cases} 0.005 & N = 0 \text{ chutes ready} \\ 0.015 & N = 1 \text{ chute ready} \\ 0.060 & N = 2 \text{ chutes ready} \\ 0.920 & N = 3 \text{ chutes ready} \end{cases}$$

This means that the safety standard is clearly not met nationwide. Yvon is in the process of sampling 200 (randomly selected) corporate and private aircraft across the nation to determine how many chutes are classified as fully ready. If the sample is truly random and Yvon's probability distribution is a correct representation of actual parachute readiness, the observed N values in the 200 aircraft will approximate the same proportions as the population probabilities, that is, 1 aircraft with 0 chutes ready, etc. Since this is a sample, it is likely that the results won't track the population exactly. However, if the results are relatively close, the study indicates that the sample results may be useful in predicting parachute safety across the nation.

TABLE 19–2 Random Digits Clustered into Two-Digit Numbers

51	82	88	18	19	81	03	88	91	46	39	19	28	94	70	76	33	15	64	20	14	52
73	48	28	59	78	38	54	54	93	32	70	60	78	64	92	40	72	71	77	56	39	27
10	42	18	31	23	80	80	26	74	71	03	90	55	61	61	28	41	49	00	79	96	78
45	44	79	29	81	58	66	70	24	82	91	94	42	10	61	60	79	30	01	26	31	42
68	65	26	71	44	37	93	94	93	72	84	39	77	01	97	74	17	19	46	61	49	67
75	52	14	99	67	74	06	50	97	46	27	88	10	10	70	66	22	56	18	32	06	24

To develop a random sample, use *random numbers* (*RN*) generated from a uniform probability distribution for the discrete numbers 0 through 9, that is,

$$P(X_i) = 0.1 \qquad \text{for } X_i = 0, 1, 2, \ldots, 9$$

In tabular form, the random digits so generated are commonly clustered in groups of two digits, three digits, or more. Table 19–2 is a sample of 264 random digits clustered into two-digit numbers. This format is very useful because the numbers 00 to 99 conveniently relate to the cumulative distribution values 0.01 to 1.00. This makes it easy to select a two-digit RN and enter $F(X)$ to determine a value of the variable with the same proportions as it occurs in the probability distribution. To apply this logic manually and develop a random sample of size n from a known discrete probability distribution $P(X)$ or a continuous variable distribution $f(X)$, the following procedure may be used.

1. Develop the cumulative distribution $F(X)$ from the probability distribution. Plot $F(X)$.
2. Assign the RN values from 00 to 99 to the $F(X)$ scale (the y axis) in the same proportion as the probabilities. For the parachute safety example, the probabilities from 0.0 to 0.15 are represented by the random numbers 00 to 14. Indicate the RNs on the graph.
3. To use a table of random numbers, determine the scheme or sequence of selecting RN values—down, up, across, diagonally. Any direction and pattern is acceptable, but the scheme should be used consistently for one entire sample.
4. Select the first number from the RN table, enter the $F(X)$ scale, and observe and record the corresponding variable value. Repeat this step until there are n values of the variable which constitute the random sample.
5. Use the n sample values for analysis and decision making under risk. These may include

- Plotting the sample probability distribution.
- Developing probability statements about the parameter.
- Comparing sample results with the assumed population distribution.
- Determining sample statistics (Section 19.4).
- Performing a simulation analysis (Section 19.5).

EXAMPLE **19.4**

Develop a random sample of size 10 for the variable N, number of months, as described by the probability distribution

$$P(N = N_i) = \begin{cases} 0.20 & N = 24 \\ 0.50 & N = 30 \\ 0.30 & N = 36 \end{cases} \qquad [19.7]$$

Solution

Apply the procedure above, using the $P(N = N_i)$ values in Equation [19.7]

1. The cumulative distribution, Figure 19–6, is for the discrete variable N, which can assume three different values.
2. Assign 20 numbers (00 through 19) to $N_1 = 24$ months, where $P(N = 24) = 0.2$; 50 numbers to $N_2 = 30$; and 30 numbers to $N_3 = 36$.
3. Initially select any position in Table 19–2, and go across the row to the right and onto the row below toward the left. (Any routine can be developed, and a different sequence for each random sample may be used.)
4. Select the initial number 45 (4th row, 1st column), and enter Figure 19–6 in the RN range of 20 to 69 to obtain $N = 30$ months.
5. Select and record the remaining nine values from Table 19–2 as shown below.

RN	45	44	79	29	81	58	66	70	24	82
N	30	30	36	30	36	30	30	36	30	36

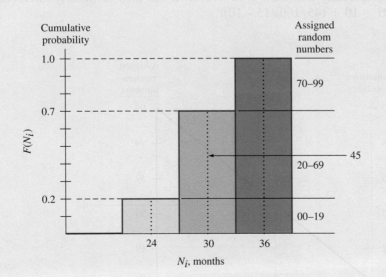

Figure 19–6
Cumulative distribution with random number values assigned in proportion to probabilities, Example 19.4.

Now, using the 10 values, develop the sample probabilities.

Months N	Times in Sample	Sample Probability	Equation [19.7] Probability
24	0	0.00	0.2
30	6	0.60	0.5
36	4	0.40	0.3

With only 10 values, we can expect the sample probability estimates to be different from the values in Equation [19.7]. Only the value $N = 24$ months is significantly different, since no RN of 19 or less occurred. A larger sample will definitely make the probabilities closer to the original data.

To take a *random sample of size n for a continuous variable,* the procedure above is applied, except the random number values are assigned to the cumulative distribution on a continuous scale of 00 to 99 corresponding to the $F(X)$ values. As an illustration, consider Figure 19–4, where C_1 is the *uniformly distributed* cash flow variable for client 1 in Example 19.3. Here $L = \$10$, $H = \$15$, and $f(C_1) = 0.2$ for all values between L and H (all values are divided by \$1000). The $F(C_1)$ is repeated as Figure 19–7 with the assigned random number values shown on the right scale. If the two-digit RN of 45 is chosen, the corresponding C_1 is graphically estimated to be \$12.25. It can also be linearly interpolated as $\$12.25 = 10 + (45/100)(15 - 10)$.

Figure 19–7
Random numbers assigned to the continuous variable of client 1 cash flows in Example 19.3.

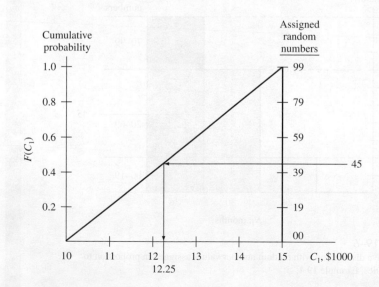

For greater accuracy when developing a random sample, especially for a continuous variable, it is possible to use 3-, 4-, or 5-digit RNs. These can be developed from Table 19–2 simply by combining digits in the columns and rows or obtained from tables with RNs printed in larger clusters of digits. In computer-based sampling, most simulation software packages have an RN generator built in which will generate values in the range of 0 to 1 from a continuous variable uniform distribution, usually identified by the symbol $U(0, 1)$. The RN values, usually between 0.00000 and 0.99999, are used to sample directly from the cumulative distribution employing essentially the same procedure explained here. The Excel functions RAND and RANDBETWEEN are described in Appendix A, Section A.3.

An initial question in random sampling usually concerns the *minimum size of n* required to ensure confidence in the results. Without detailing the mathematical logic, sampling theory, which is based upon the law of large numbers and the central-limit theorem (check a basic statistics book to learn about these), indicates that an n of 30 is sufficient. However, since reality does not follow theory exactly, and since engineering economy often deals with sketchy estimates, samples in the *range of 100 to 200* are the common practice. But samples as small as 10 to 25 provide a much better foundation for decision making under risk than the single-point estimate for a parameter that is known to vary widely.

E-Solve

19.4 EXPECTED VALUE AND STANDARD DEVIATION

Two very important measures or properties of a random variable are the expected value and standard deviation. If the entire population for a variable were known, these properties would be calculated directly. Since they are usually not known, random samples are commonly used to estimate them via the sample mean and the sample standard deviation, respectively. The following is a brief introduction to the interpretation and calculation of these properties using a random sample of size n from the population.

The usual symbols are Greek letters for the true population measures and English letters for the sample estimates.

	True Population Measure		Sample Estimate	
	Symbol	Name	Symbol	Name
Expected value	μ or $E(X)$	Mu or true mean	$\overline{X}$	Sample mean
Standard deviation	σ or $\sqrt{\text{Var}(X)}$ or $\sqrt{\sigma^2}$	Sigma or true standard deviation	s or $\sqrt{s^2}$	Sample standard deviation

The *expected value* is the long-run expected average if the variable is sampled many times.

The population expected value is not known exactly, since the population itself is not known completely, so μ is estimated either by $E(X)$ from a distribution or by $\overline{X}$, the sample mean. Equation [18.2], repeated here as Equation [19.8], is used to compute the $E(X)$ of a probability distribution, and Equation [19.9] is the sample mean, also called the sample average.

Population: μ

Probability distribution: $E(X) = \sum X_i P(X_i)$ [19.8]

Sample: $\overline{X} = \dfrac{\text{sum of sample values}}{\text{sample size}}$

$$= \frac{\sum X_i}{n} = \frac{\sum f_i X_i}{n}$$ [19.9]

The f_i in the second form of Equation [19.9] is the frequency of X_i, that is, the number of times each value occurs in the sample. The resulting $\overline{X}$ is not necessarily an observed value of the variable; it is the long-run average value and can take on any value within the range of the variable. (We omit the subscript i on X and f when there is no confusion introduced.)

EXAMPLE 19.5

Kayeu, an engineer with Pacific NW Utilities, is planning to test several hypotheses about residential electricity bills in North American and Asian countries. The variable of interest is X, the monthly residential bill in U.S. dollars (rounded to the nearest dollar). Two small samples have been collected from different countries of North America and Asia. Estimate the population expected value. Do the samples (from a nonstatistical viewpoint) appear to be drawn from one population of electricity bills or from two different populations?

North American, Sample 1, $	40	66	75	92	107	159	275
Asian, Sample 2, $	84	90	104	187	190		

Solution

Use Equation [19.9] for the sample mean.

Sample 1: $n = 7$ $\sum X_i = 814$ $\overline{X} = \$116.29$

Sample 2: $n = 5$ $\sum X_i = 655$ $\overline{X} = \$131.00$

Based solely on the small sample averages, the approximate $15 difference, which is less than 10% of the smaller average bill, does not seem sufficiently large to conclude that the two populations are different. There are several statistical tests available to determine if samples come from the same or different populations. (Check a basic statistics text to learn about them.)

Comment

There are three commonly used measures of central tendency for data. The sample average is the most popular, but the *mode* and the *median* are also good measures. The mode, which is the most frequently observed value, was utilized in Example 19.3 for a triangular distribution. There is no specific mode in Kayeu's two samples, since all values are different. The median is the middle value of the sample. It is not biased by extreme sample values, as is the mean. The two medians in the samples are $92 and $104. Based solely on the medians, the conclusion is still that the samples do not necessarily come from two different populations of electricity bills.

The *standard deviation* is the dispersion or spread of values about the expected value $E(X)$ or sample average $\overline{X}$.

The sample standard deviation s estimates the property σ, which is the population measure of dispersion about the expected value of the variable. A probability distribution for data with strong central tendency is more closely clustered about the center of the data, and has a smaller s than a wider, more dispersed distribution. In Figure 19–8, the samples with larger s values—s_1 and s_4—have a flatter, wider probability distribution.

Actually, the variance s^2 is often quoted as the measure of dispersion. The standard deviation is simply the square root of the variance, so either measure can be used. However, the s value is what we use routinely in making computations about risk and probability. Mathematically, the formulas and symbols for variance and standard deviation of a discrete variable and a random sample of size n are as follows:

Population: $\sigma^2 = \mathrm{Var}(X)$ and $\sigma = \sqrt{\sigma^2} = \sqrt{\mathrm{Var}(X)}$

Probability distribution: $\mathrm{Var}(X) = \sum [X_i - E(X)]^2 P(X_i)$ **[19.10]**

Sample: $s^2 = \dfrac{\textbf{sum of (sample value} - \textbf{sample average)}^2}{\textbf{sample size} - 1}$

$$= \frac{\Sigma(X_i - \overline{X})^2}{n - 1}$$ **[19.11]**

$$s = \sqrt{s^2}$$

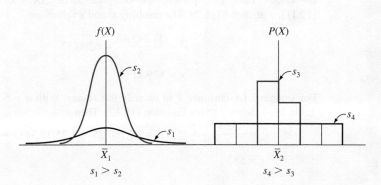

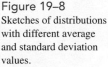

Figure 19–8
Sketches of distributions with different average and standard deviation values.

Equation [19.11] for sample variance is usually applied in a more computationally convenient form.

$$s^2 = \frac{\Sigma X_i^2}{n-1} - \frac{n}{n-1}\overline{X}^2 = \frac{\Sigma f_i X_i^2}{n-1} - \frac{n}{n-1}\overline{X}^2 \qquad [19.12]$$

The standard deviation uses the sample average as a basis about which to measure the spread or dispersion of data via the calculation $(X - \overline{X})$, which can have a minus or plus sign. To accurately measure the dispersion in both directions from the average, the quantity $(X - \overline{X})$ is squared. To return to the dimension of the variable itself, the square root of Equation [19.11] is extracted. The term $(X - \overline{X})^2$ is called the *mean-squared deviation,* and s has historically also been referred to as the *root-mean-square deviation.* The f_i in the second form of Equation [19.12] uses the frequency of each X_i value to calculate s^2.

One simple way to combine the average and standard deviation is to determine the percentage or fraction of the sample that is within ± 1, ± 2, or ± 3 standard deviations of the average, that is,

$$\overline{X} \pm ts \qquad \text{for } t = 1, 2, \text{ or } 3 \qquad [19.13]$$

In probability terms, this is stated as

$$P(\overline{X} - ts \le X \le \overline{X} + ts) \qquad [19.14]$$

Virtually all the sample values will always be within the $\pm 3s$ range of $\overline{X}$, but the percent within $\pm 1s$ will vary depending on how the data points are distributed about $\overline{X}$. The following example illustrates the calculation of s to estimate σ, and incorporates s with the sample average using $\overline{X} \pm ts$.

EXAMPLE 19.6

(*a*) Use the two samples of Example 19.5 to estimate population variance and standard deviation for electricity bills. (*b*) Determine the percentages of each sample that are inside the ranges of 1 and 2 standard deviations from the mean.

Solution

(*a*) For illustration purposes only, apply the two different relations to calculate s for the two samples. For sample 1 (North American) with $n = 7$, use X to identify the values. Table 19–3 presents the computation of $\Sigma(X - \overline{X})^2$ for Equation [19.11], with $X = \$116.29$. The resulting s^2 and s values are

$$s^2 = \frac{37,743.40}{6} = 6290.57$$

$$s = \$79.31$$

For sample 2 (Asian), use Y to identify the values. With $n = 5$ and $\overline{Y} = 131$, Table 19–4 shows ΣY^2 for Equation [19.12]. Then

$$s^2 = \frac{97,041}{4} - \frac{5}{4}(131)^2 = 42,260.25 - 1.25(17,161) = 2809$$

$$s = \$53$$

TABLE **19–3**	Computation of Standard Deviation Using Equation [19.11] with $\overline{X} =$ $116.29, Example 19.6	
X	**(X − X̄)**	**(X − X̄)²**
$ 40	−76.29	5,820.16
66	−50.29	2,529.08
75	−41.29	1,704.86
92	−24.29	590.00
107	−9.29	86.30
159	+42.71	1,824.14
275	+158.71	25,188.86
$814		$37,743.40

TABLE **19–4**	Computation of Standard Deviation Using Equation [19.12] with $\overline{Y} =$ $131, Example 19.6	
	Y	**Y²**
	$ 84	7,056
	90	8,100
	104	10,816
	187	34,969
	190	36,100
	$655	97,041

The dispersion is smaller for the Asian sample ($53) than for the North American sample ($79.31).

(b) Equation [19.13] determines the ranges of $\overline{X} \pm 1s$ and $\overline{X} \pm 2s$. Count the number of sample data points between the limits, and calculate the corresponding percentage. See Figure 19–9 for a plot of the data and the standard deviation ranges.

North American sample

$$\overline{X} \pm 1s = 116.29 \pm 79.31 \qquad \text{for a range of } \$36.98 \text{ to } \$195.60$$

Six out of seven values are within this range, so the percentage is 85.7%.

$$\overline{X} \pm 2s = 116.29 \pm 158.62 \qquad \text{for a range of } \$-42.33 \text{ to } \$274.91$$

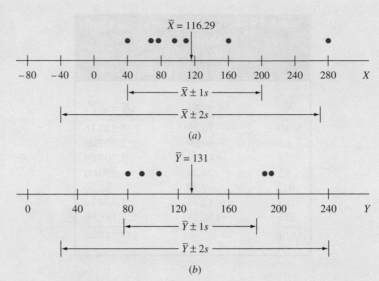

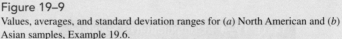

Figure 19–9
Values, averages, and standard deviation ranges for (*a*) North American and (*b*) Asian samples, Example 19.6.

There are still six of the seven values within the $\bar{X} \pm 2s$ range. The limit $-42.33 is meaningful only from the probabilistic perspective; from the practical viewpoint, use zero, that is, no amount billed.

Asian sample

$$\bar{Y} \pm 1s = 131 \pm 53 \qquad \text{for a range of \$78 to \$184}$$

There are three of five values, or 60%, within the range.

$$\bar{Y} \pm 2s = 131 \pm 106 \qquad \text{for a range of \$25 to \$237}$$

All five of the values are within the $\bar{Y} \pm 2s$ range.

Comment

A second common measure of dispersion is the *range,* which is simply the largest minus the smallest sample values. In the two samples here, the range estimates are $235 and $106.

Before we perform simulation analysis in engineering economy, it may be of use to summarize the expected value and standard deviation relations for a continuous variable, since Equations [19.8] through [19.12] address only discrete variables. The primary differences are that the sum symbol is replaced by the integral over the defined range of the variable, which we identify as *R*, and that

$P(X)$ is replaced by the differential element $f(X)dX$. For a stated continuous probability distribution $f(X)$, the formulas are

Expected value: $E(X) = \int_R Xf(X)\, dX$ [19.15]

Variance: $\text{Var}(X) = \int_R X^2 f(X)\, dX - [E(X)]^2$ [19.16]

For a numerical example, again use the uniform distribution in Example 19.3 (Figure 19–4) over the range R from \$10 to \$15. If we identify the variable as X, rather than C_1, the following are correct.

$$f(X) = \frac{1}{5} = 0.2 \qquad \$10 \le X \le \$15$$

$$E(X) = \int_R X(0.2)\, dX = 0.1X^2 \Big|_{10}^{15} = 0.1(225 - 100) = \$12.5$$

$$\text{Var}(X) = \int_R X^2(0.2)\, dX - (12.5)^2 = \frac{0.2}{3} X^3 \Big|_{10}^{15} - (12.5)^2$$

$$= 0.06667(3375 - 1000) - 156.25 = 2.08$$

$$\sigma = \sqrt{2.08} = \$1.44$$

Therefore, the uniform distribution between $L = \$10$ and $H = \$15$ has an expected value of \$12.5 (the midpoint of the range, as expected) and a standard deviation of \$1.44.

Additional Example 19.10

19.5   MONTE CARLO SAMPLING AND SIMULATION ANALYSIS

Up to this point, all alternative selections have been made using estimates with certainty, possibly followed by some testing of the decision via sensitivity analysis or expected values. In this section, we will use a simulation approach that incorporates the material of the previous sections to facilitate the engineering economy decision about one alternative or between two or more alternatives.

The random sampling technique discussed in Section 19.3 is called *Monte Carlo sampling*. The general procedure outlined below uses Monte Carlo sampling to obtain samples of size n for selected parameters of formulated alternatives. These parameters, expected to vary according to a stated probability distribution, warrant decision making under risk. All other parameters in an alternative are considered certain; that is, they are known, or they can be estimated with enough precision to consider them certain. An important assumption is made, usually without realizing it.

All parameters are independent; that is, one variable's distribution does not affect the value of any other variable of the alternative. This is referred to as the *property of independent random variables*.

The simulation approach to engineering economy analysis is summarized in the following basic steps.

Step 1. **Formulate alternative(s).**   Set up each alternative in the form to be considered using engineering economic analysis, and select the measure of worth upon which to base the decision. Determine the form of the relation(s) to calculate the measure of worth.

Step 2. **Parameters with variation.**   Select the parameters in each alternative to be treated as random variables. Estimate values for all other (certain) parameters for the analysis.

Step 3. **Determine probability distributions.**   Determine whether each variable is discrete or continuous, and describe a probability distribution for each variable in each alternative. Use standard distributions where possible to simplify the sampling process and to prepare for computer-based simulation.

Step 4. **Random sampling.**   Incorporate the random sampling procedure of Section 19.3 (the first four steps) into this procedure. This results in the cumulative distribution, assignment of RNs, selection of the RNs, and a sample of size n for each variable.

Step 5. **Measure of worth calculation.**   Compute n values of the selected measure of worth from the relation(s) determined in step 1. Use the estimates made with certainty and the n sample values for the varying parameters. (This is when the property of independent random variables is actually applied.)

Step 6. **Measure of worth description.**   Construct the probability distribution of the measure of worth, using between 10 and 20 cells of data, and calculate measures such as $\overline{X}$, s, $\overline{X} \pm ts$, and relevant probabilities.

Step 7. **Conclusions.**   Draw conclusions about each alternative, and decide which is to be selected. If the alternative(s) has (have) been previously evaluated under the assumption of certainty for all parameters, comparison of results may help with the final decision.

Example 19.7 illustrates this procedure using an abbreviated manual simulation analysis, and Example 19.8 utilizes spreadsheet simulation for the same estimates.

EXAMPLE **19.7**

Yvonne Ramos is the CEO of a chain of 50 fitness centers in the United States and Canada. An equipment salesperson has offered Yvonne two long term opportunities on new aerobic exercise systems, for which the usage is charged to customers on a per-use basis on top of the monthly fees paid by customers. As an enticement, the offer includes a guarantee of annual revenue for one of the systems for the first 5 years.

Since this is an entirely new and risky concept of revenue generation, Yvonne wants to do a careful analysis of each alternative. Details for the two systems follow:

System 1.   First cost is $P = \$12,000$ for a set period of $n = 7$ years with no salvage value. No guarantee for annual net revenue is offered.

System 2.   First cost is $P = \$8000$, there is no salvage value, and there is a guaranteed annual net revenue of $1000 for each of the first 5 years, but after this period, there is no guarantee. The equipment with updates may be useful up to 15 years, but the exact number is not known. Cancellation anytime after the initial 5 years is allowed, with no penalty.

For either system, new versions of the equipment will be installed with no added costs. If a MARR of 15% per year is required, use PW analysis to determine if neither, one, or both of the systems should be installed.

Solution by Hand

Estimates which Yvonne makes to correctly use the simulation analysis procedure are included in the following steps.

Step 1.   **Formulate alternatives.**   Using PW analysis, the relations for system 1 and system 2 are developed including the parameters known with certainty. The symbol NCF identifies the net cash flows (revenues), and NCF_G is the guaranteed NCF of $1000 for system 2.

$$PW_1 = -P_1 + NCF_1(P/A,15\%,n_1) \qquad\qquad [19.17]$$

$$PW_2 = -P_2 + NCF_G(P/A,15\%,5) \qquad\qquad [19.18]$$
$$+ NCF_2(P/A,15\%,n_2-5)(P/F,15\%,5)$$

Step 2.   **Parameters with variation.**   Yvonne summarizes the parameters estimated with certainty and makes distribution assumptions about three parameters treated as random variables.

> *System 1*
> **Certainty.** $P_1 = \$12,000$; $n_1 = 7$ years.
> **Variable.** NCF_1 is a continuous variable, uniformly distributed between $L = \$-4000$ and $H = \$6000$ per year, because this is considered a high-risk venture.
>
> *System 2*
> **Certainty.** $P_2 = \$8000$; $NCF_G = \$1000$ for first 5 years.
> **Variable.** NCF_2 is a discrete variable, uniformly distributed over the values $L = \$1000$ to $H = \$6000$ only in $1000 increments, that is, $1000, $2000, etc.
> **Variable.** n_2 is a continuous variable that is uniformly distributed between $L = 6$ and $H = 15$ years.

Now, rewrite Equations [19.17] and [19.18] to reflect the estimates made with certainty.

$$PW_1 = -12,000 + NCF_1(P/A,15\%,7)$$
$$= -12,000 + NCF_1(4.1604) \qquad\qquad [19.19]$$
$$PW_2 = -8000 + 1000(P/A,15\%,5)$$
$$+ NCF_2(P/A,15\%,n_2-5)(P/F,15\%,5)$$
$$= -4648 + NCF_2(P/A,15\%,n_2-5)(0.4972) \qquad\qquad [19.20]$$

Step 3.   **Determine probability distributions.**   Figure 19–10 (left side) shows the assumed probability distributions for NCF_1, NCF_2, and n_2.

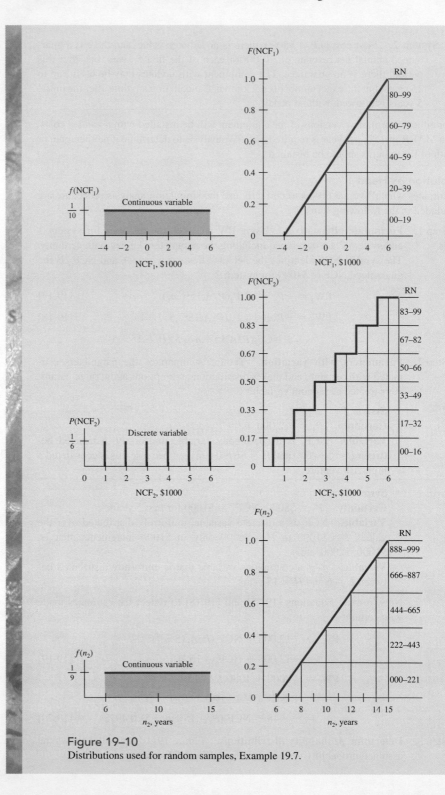

Figure 19–10
Distributions used for random samples, Example 19.7.

Step 4.   **Random sampling.**   Yvonne decides on a sample of size 30 and applies the first four of the random sample steps in Section 19.3. Figure 19–10 (right side) shows the cumulative distributions (step 1) and assigns RNs to each variable (step 2). The RNs for NCF_2 identify the x axis values so that all net cash flows will be in even $1000 amounts. For the continuous variable n_2, three-digit RN values are used in order to make the numbers come out evenly, and they are shown in cells only as "indexers" for easy reference when a RN is used to find a variable value. However, we round the number to the next higher value of n_2 because it is likely the contract may be canceled on an anniversary date. Also, now the tabulated compound interest factors for (n_2-5) years can be used directly (see Table 19–5).

      Once the first RN is selected randomly from Table 19–2, the sequence (step 3) used will be to proceed down the RN table column and then up the column to the left. Table 19–5 shows only the first five RN values selected for each sample and the corresponding variable values taken from the cumulative distributions in Figure 19–10 (step 4).

Step 5.   **Measure of worth calculation.**   With the five sample values in Table 19–5, calculate the PW values using Equations [19.19] and [19.20].

    1. $PW_1 = -12,000 + (-2200)(4.1604)$ $= \$-21,153$
    2. $PW_1 = -12,000 + 2000(4.1604)$ $= \$-3679$
    3. $PW_1 = -12,000 + (-1100)(4.1604)$ $= \$-16,576$
    4. $PW_1 = -12,000 + (-900)(4.1604)$ $= \$-15,744$
    5. $PW_1 = -12,000 + 3100(4.1604)$ $= \$+897$

    1. $PW_2 = -4648 + 1000(P/A,15\%,7)(0.4972)$ $= \$-2579$
    2. $PW_2 = -4648 + 1000(P/A,15\%,5)(0.4972)$ $= \$-2981$
    3. $PW_2 = -4648 + 5000(P/A,15\%,8)(0.4972)$ $= \$+6507$
    4. $PW_2 = -4648 + 3000(P/A,15\%,10)(0.4972) = \$+2838$
    5. $PW_2 = -4648 + 4000(P/A,15\%,3)(0.4972)$ $= \$-107$

TABLE 19–5   Random Numbers and Variable Values for NCF_1, NCF_2, and n_2, Example 19.7

NCF$_1$		NCF$_2$		n_2		
RN*	Value	RN†	Value	RN‡	Value	Rounded§
18	$-2200	10	$1000	586	11.3	12
59	+2000	10	1000	379	9.4	10
31	-1100	77	5000	740	12.7	13
29	-900	42	3000	967	14.4	15
71	+3100	55	4000	144	7.3	8

*Randomly start with row 1, column 4 in Table 19–2.
†Start with row 6, column 14.
‡Start with row 4, column 6.
§The n_2 value is rounded up.

Now, 25 more RNs are selected for each variable from Table 19–2 and the PW values are calculated.

Step 6.   **Measure of worth description.**   Figure 19–11a and b presents the PW_1 and PW_2 probability distributions for the 30 samples with 14 and 15 cells, respectively, as well as the range of individual PW values and the $\overline{X}$ and s values.

PW_1.   Sample values range from $\$-24{,}481$ to $\$+12{,}962$. The calculated measures of the 30 values are

$$\overline{X}_1 = \$-7729$$

$$s_1 = \$10{,}190$$

PW_2.   Sample values range from $\$-3031$ to $\$+10{,}324$. The sample measures are

$$\overline{X}_2 = \$2724$$

$$s_2 = \$4336$$

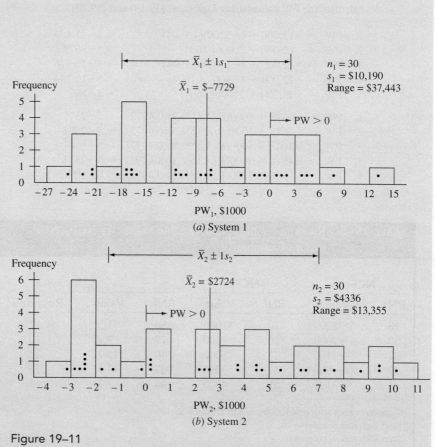

Figure 19–11
Probability distributions of simulated PW values for a sample of size 30, Example 19.7.

Step 7.   **Conclusions.** Additional sample values will surely make the central tendency of the PW distributions more evident and may reduce the s values, which are quite large. Of course, many conclusions are possible once the PW distributions are known, but the following seem clear.

> **System 1.**   Based on this small sample of 30 observations, *do not accept* this alternative. The likelihood of making the MARR = 15% is relatively small, since the sample indicates a probability of 0.27 (8 out of 30 values) that the PW will be positive, and $\overline{X}_1$ is a large negative. Though appearing large, the standard deviation may be used to determine that about 20 of the 30 sample PW values (two-thirds) are within the limits $\overline{X} \pm 1s$, which are $-17,919 and $2461. A larger sample may alter this analysis somewhat.

> **System 2.**   If Yvonne is willing to accept the longer-term commitment that may increase the NCF some years out, the sample of 30 observations indicates to *accept* this alternative. At a MARR of 15%, the simulation approximates the chance for a positive PW as 67% (20 of the 30 PW values in Figure 19–11b are positive). However, the probability of observing PW within the $\overline{X} \pm 1s$ limits ($-1612 and $7060) is 0.53 (16 of 30 sample values). This indicates that the PW sample distribution is more widely dispersed about its average than the system 1 PW sample.
>
> **Conclusion at this point.**   Reject system 1; accept system 2; and carefully watch net cash flow, especially after the initial 5-year period.

Comment

The estimates in Example 5.8 are very similar to those here, except all estimates were made with certainty (NCF$_1$ = $3000, NCF$_2$ = $3000, and n_2 = 14 years). The alternatives were evaluated by the payback period method at   MARR = 15%, and the first alternative was selected. However, the subsequent PW analysis in Example 5.8 selected alternative 2 based, in part, upon the anticipated larger cash flow in the later years.

Sec. 5.6

Payback analysis

EXAMPLE **19.8**

Help Yvonne Ramos set up an Excel spreadsheet simulation for the three random variables and PW analysis in Example 19.7. Does the PW distribution vary appreciably from that developed using manual simulation? Do the decisions to reject the system 1 proposal and accept the system 2 proposal still seem reasonable?

E-Solve

Solution by Computer

Figures 19–12 and 19–13 are spreadsheets that accomplish the simulation portion of the analysis described above in steps 3 (determine probability distribution) through 6 (measure of worth description). Most spreadsheet systems are limited in the variety of distributions they can accept for sampling, but common ones such as uniform and normal are available.

   Figure 19–12 shows the results of a small sample of 30 values (only a portion of the spreadsheet is printed here) from the three distributions using the RAND and IF functions. (See Section A.3 in Appendix A.)

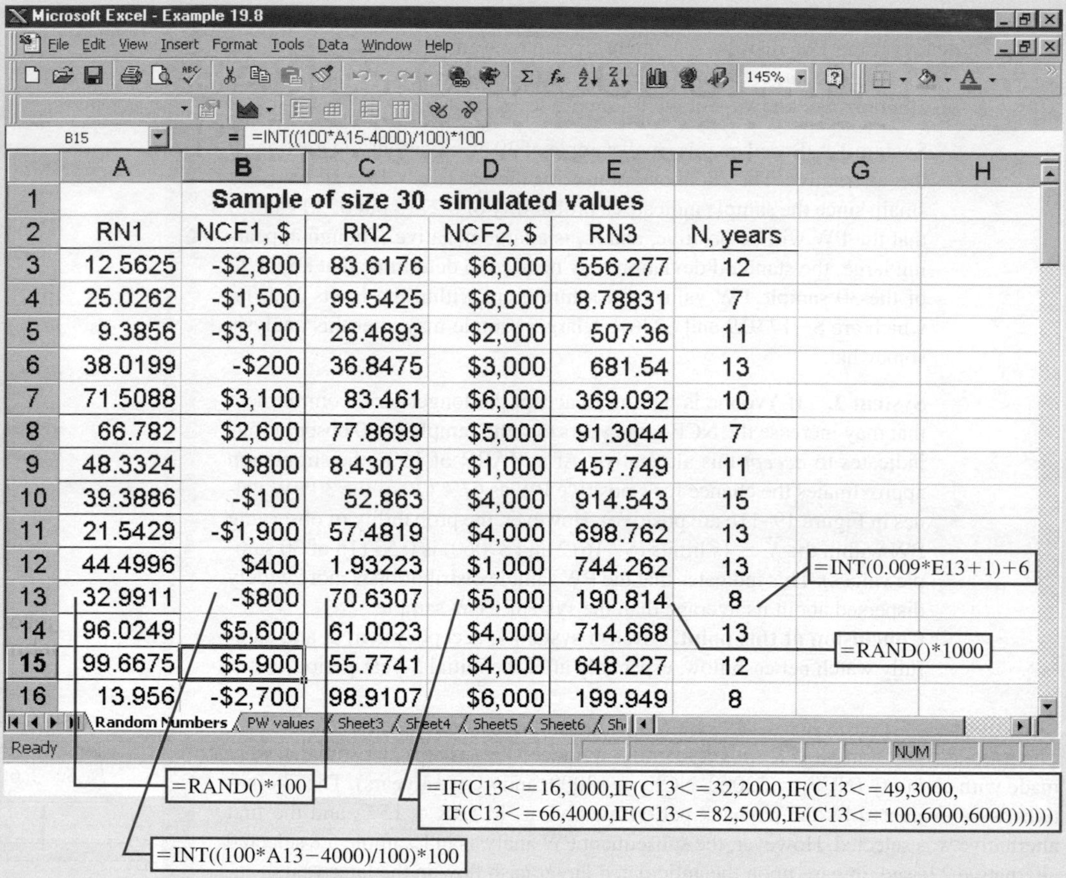

Microsoft Excel - Example 19.8

B15 = =INT((100*A15-4000)/100)*100

	A	B	C	D	E	F
1		Sample of size 30 simulated values				
2	RN1	NCF1, $	RN2	NCF2, $	RN3	N, years
3	12.5625	-$2,800	83.6176	$6,000	556.277	12
4	25.0262	-$1,500	99.5425	$6,000	8.78831	7
5	9.3856	-$3,100	26.4693	$2,000	507.36	11
6	38.0199	-$200	36.8475	$3,000	681.54	13
7	71.5088	$3,100	83.461	$6,000	369.092	10
8	66.782	$2,600	77.8699	$5,000	91.3044	7
9	48.3324	$800	8.43079	$1,000	457.749	11
10	39.3886	-$100	52.863	$4,000	914.543	15
11	21.5429	-$1,900	57.4819	$4,000	698.762	13
12	44.4996	$400	1.93223	$1,000	744.262	13
13	32.9911	-$800	70.6307	$5,000	190.814	8
14	96.0249	$5,600	61.0023	$4,000	714.668	13
15	99.6675	$5,900	55.7741	$4,000	648.227	12
16	13.956	-$2,700	98.9107	$6,000	199.949	8

Figure 19–12
Sample values generated using spreadsheet simulation, Example 19.8.

NCF_1: Continuous uniform from \$−4000 to \$6000. The relation in column B translates RN1 values (column A) into NCF1 amounts.

NCF_2: Discrete uniform in \$1000 increments from \$1000 to \$6000. Column D cells display NCF2 in the \$1000 increments using the logical IF function to translate from the RN2 values.

n_2: Continuous uniform from 6 to 15 years. The results in column F are integer values obtained using the INT function operating on the RN3 values.

Figure 19–13 presents the two alternatives' estimates in the top section. The PW1 and PW2 computations for the 30 repetitions of NCF1, NCF2, and n2 are the spreadsheet equivalent of Equations [19.19] and [19.20]. The tabular approach used here tallies the number of PW values below zero (\$0) and equal to or exceeding zero using the IF operator. For example, cell C17 contains a 1, indicating PW1 > 0 when NCF1 = \$3100 (in cell B7

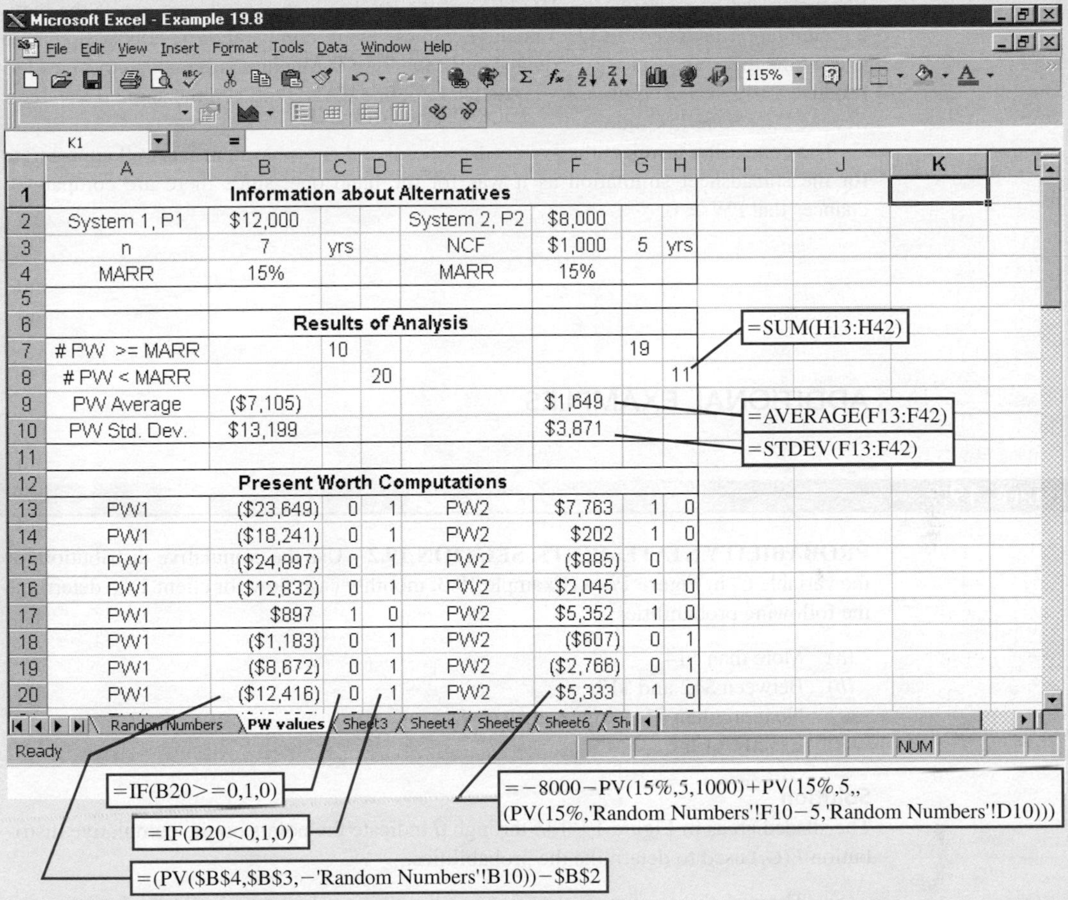

Figure 19–13
Spreadsheet simulation results of 30 PW values, Example 19.8.

of Figure 19–12) was used to calculate PW1 = $897 by Equation [19.19]. Cells in rows 7 and 8 show the number of times in the 30 samples that system 1 and system 2 may return at least the MARR = 15% because the corresponding PW ≥ 0. Sample averages and standard deviations are also indicated.

Comparison between the hand and spreadsheet simulations is presented below.

	System 1 PW			System 2 PW		
	$\bar{X}$, \$	s, \$	No. of PW ≥ 0	$\bar{X}$, \$	s, \$	No. of PW ≥ 0
Hand	−7,729	10,190	8	2,724	4,336	20
Spreadsheet	−7,105	13,199	10	1,649	3,871	19

For the spreadsheet simulation, 10 (33%) of the PW1 values exceed zero, while the manual simulation included 8 (27%) positive values. These comparative results will change every time this spreadsheet is activated since the RAND function is set up (in this case) to produce a new RN each time. It is possible to define RAND to keep the same RN values. See the Excel User's Guide.

The conclusion to reject the system 1 proposal and accept system 2 is still appropriate for the spreadsheet simulation as it was for the hand one, since there are comparable chances that $PW \geq 0$.

ADDITIONAL EXAMPLES

EXAMPLE 19.9

PROBABILITY STATEMENTS, SECTION 19.2   Use the cumulative distribution for the variable C_1 in Figure 19–4 (Example 19.3, monthly cash flow for client 1) to determine the following probabilities:

(a)   More than $14.
(b)   Between $12 and $13.
(c)   No more than $11 or more than $14.
(d)   Exactly $12.

Solution
The shaded areas in Figure 19–14a through d indicate the points on the cumulative distribution $F(C_1)$ used to determine the probabilities.

(a)   The probability of more than $14 per month is easily determined by subtracting the value of $F(C_1)$ at 14 from the value at 15. (Since the probability at a point is zero for

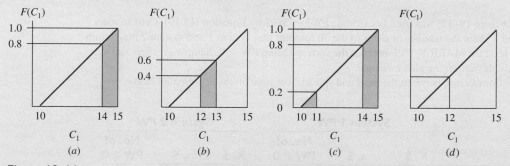

Figure 19–14
Calculation of probabilities from the cumulative distribution for a continuous variable that is distributed uniformly, Example 19.9.

a continuous variable, the equals sign does not change the value of the resulting probability.)

$$P(C_1 > 14) = P(C_1 \leq 15) - P(C_1 \leq 14)$$

$$= F(15) - F(14) = 1.0 - 0.8$$

$$= 0.2$$

(b)　　　　　　$$P(12 \leq C_1 \leq 13) = P(C_1 \leq 13) - P(C_1 \leq 12) = 0.6 - 0.4$$

$$= 0.2$$

(c)　　$$P(C_1 \leq 11) + P(C_1 > 14) = [F(11) - F(10)] + [F(15) - F(14)]$$

$$= (0.2 - 0) + (1.0 - 0.8)$$

$$= 0.2 + 0.2$$

$$= 0.4$$

(d)　　　　　　　$$P(C_1 = 12) = F(12) - F(12) = 0.0$$

There is no area under the cumulative distribution curve at a point for a continuous variable, as mentioned earlier. If two very closely placed points are used, it is possible to obtain a probability, for example, between 12.0 and 12.1 or between 12 and 13, as in part (b).

EXAMPLE 19.10

THE NORMAL DISTRIBUTION, SECTION 19.4　Camilla is the regional safety engineer for a chain of franchise-based gasoline and food stores. The home office has had many complaints and several legal actions from employees and customers about slips and falls due to liquids (water, oil, gas, soda, etc.) on concrete surfaces. Corporate management has authorized each regional engineer to contract locally to apply to all exterior concrete surfaces a newly marketed product that absorbs up to 100 times its own weight in liquid and to charge a home office account for the installation. The authorizing letter to Camilla states that, based upon their simulation and random samples that assume a normal population, the cost of the locally arranged installation should be about $10,000 and almost always is within the range of $8000 to $12,000.

Camilla asks you, TJ, an engineering technology graduate, to write a brief but thorough summary about the normal distribution, explain the $8000 to $12,000 range statement, and explain the phrase "random samples that assume a normal population."

Solution

You kept this book and a basic engineering statistics text when you graduated, and you have developed the following response to Camilla, using them and the letter from the home office.

Camilla,

Here is a brief summary of how the home office appears to be using the normal distribution. As a refresher, I've included a summary of what the normal distribution is all about.

Normal distribution, probabilities, and *random samples*

The normal distribution is also referred to as the bell-shaped curve, the Gaussian distribution, or the error distribution. It is, by far, the most commonly used probability distribution in all applications. It places exactly one-half of the probability on either side of the mean or expected value. It is used for continuous variables over the entire range of numbers. The normal is found to accurately predict many types of outcomes, such as IQ values; manufacturing errors about a specified size, volume, weight, etc.; and the distribution of sales revenues, costs, and many other business parameters around a specified mean, which is why it may apply in this situation.

The normal distribution, identified by the symbol $N(\mu,\sigma^2)$, where μ is the expected value or mean and σ^2 is the variance, or measure of spread, can be described as follows:

- The mean μ locates the probability distribution (Figure 19–15*a*), and the spread of the distribution varies with variance (Figure 19–15*b*), growing wider and flatter for larger variance values.
- When a sample is taken, the estimates are identified as sample mean $\overline{X}$ for μ and sample standard deviation s for σ.
- The normal probability distribution $f(X)$ for a variable X is quite complicated, because its formula is

$$f(X) = \frac{1}{\sigma\sqrt{2\pi}} \exp\left\{-\left[\frac{(X-\mu)^2}{2\sigma^2}\right]\right\}$$

where exp represents the number $e = 2.71828+$ and it is raised to the power of the $-[\ ]$ term. In short, if X is given different values, for a given mean μ and standard deviation σ, a curve looking like those in Figure 19–15*a* and *b* is developed.

Since $f(X)$ is so unwieldy, random samples and probability statements are developed using a transformation, called the *standard normal distribution (SND),* which uses the μ and σ (population) or $\overline{X}$ and s (sample) to compute values of the variable Z.

Population : $$Z = \frac{\text{deviation from mean}}{\text{standard deviation}} = \frac{X-\mu}{\sigma}$$ [19.21]

Sample: $$Z = \frac{X-\overline{X}}{s}$$ [19.22]

The SND for Z (Figure 19–15*c*) is the same as for X, except that it always has a mean of 0 and a standard deviation of 1, and it is identified by the symbol $N(0,1)$. Therefore, the probability values under the SND curve can be stated exactly. It is always possible to transfer back to the original values from sample data by solving Equation [19.21] for X:

$$X = Z\sigma + \mu$$ [19.23]

Several probability statements for Z and X are summarized in the following table and are shown on the distribution curve for Z in Figure 19–15*c*.

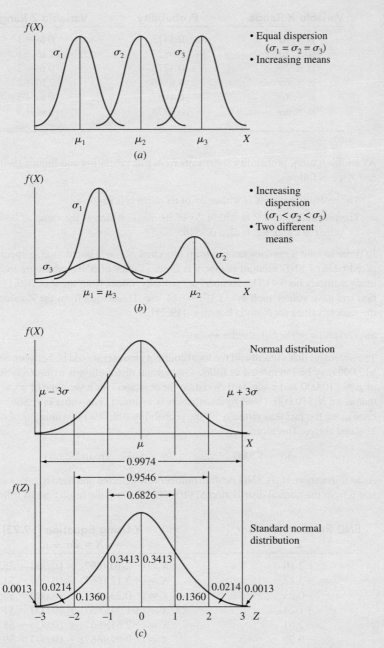

Figure 19–15
Normal distribution showing (a) different mean values μ; (b) different standard deviation values σ; and (c) relation of normal X to standard normal Z.

Variable X Range	Probability	Variable Z Range
$\mu + 1\sigma$	0.3413	0 to +1
$\mu \pm 1\sigma$	0.6826	−1 to +1
$\mu + 2\sigma$	0.4773	0 to +2
$\mu \pm 2\sigma$	0.9546	−2 to +2
$\mu + 3\sigma$	0.4987	0 to +3
$\mu \pm 3\sigma$	0.9974	−3 to +3

As an illustration, probability statements from this tabulation and Figure 19–15c for X and Z are as follows:

The probability that X is within 2σ of its mean is 0.9546.

The probability that Z is within 2σ of its mean, which is the same as between the values −2 and +2, is also 0.9546.

In order to take a random sample from a normal $N(\mu,\sigma^2)$ population, a specially prepared table of SND random numbers is used. (Tables of SND values are available in many statistics books.) The numbers are actually values from the Z or $N(0,1)$ distribution and have values such as −2.10, +1.24, etc. Translation from the Z value back to the sample values for X is via Equation [19.23].

Interpretation of the home office memo

The statement that virtually all the local contract amounts should be between $8000 and $12,000 may be interpreted as follows: A normal distribution is assumed with a mean of $\mu = \$10,000$ and a standard deviation for $\sigma = \$667$, or a variance of $\sigma^2 = (\$667)^2$; that is, an $N[\$10,000, (\$667)^2]$ distribution is assumed. The value $\sigma = \$667$ is calculated using the fact that virtually all the probability (99.74%) is within 3σ of the mean, as stated above. Therefore,

$$3\sigma = \$2000 \quad \text{and} \quad \sigma = \$667 \quad \text{(rounded off)}$$

As an illustration, if six SND random numbers are selected and used to take a sample of size 6 from the normal distribution $N[\$10,000, (\$667)^2]$, the results are as follows:

SND Random Number Z	X Using Equation [19.23] $X = Z\sigma + \mu$
−2.10	$X = (-2.10)(667) + 10,000 = \$8599$
+3.12	$X = (+3.12)(667) + 10,000 = \$12,081$
−0.23	$X = (-0.23)(667) + 10,000 = \$9847$
+1.24	$X = (+1.24)(667) + 10,000 = \$10,827$
−2.61	$X = (-2.61)(667) + 10,000 = \$8259$
−0.99	$X = (-0.99)(667) + 10,000 = \$9340$

If we consider this a sample of six typical concrete surfacing contract amounts for sites in our region, the average is $9825 and five of six values are within the range of $8000 and $12,000, with the sixth being only $81 above the upper limit. So we should have no real problems, but it is important that we keep a close watch on the contract amounts, because the assumption of the normal distribution with a mean of about $10,000 and virtually all contract amounts within ±$2000 of it may not prove to be correct for our region.

If you have any questions about this summary, please contact me.

TJ

CHAPTER SUMMARY

To perform decision making under risk implies that some parameters of an engineering alternative are treated as random variables. Assumptions about the shape of the variable's probability distribution are used to explain how the estimates of parameter values may vary. Additionally, measures such as the expected value and standard deviation describe the characteristic shape of the distribution. In this chapter, we learned several of the simple, but useful, discrete and continuous population distributions used in engineering economy—uniform and triangular—as well as specifying our own distribution or assuming the normal distribution.

Since the population's probability distribution for a parameter is not fully known, a random sample of size *n* is usually taken, and its sample average and standard deviation are determined. The results are used to make probability statements about the parameter, which help make the final decision with risk considered.

The Monte Carlo sampling method is combined with engineering economy relations for a measure of worth such as PW to implement a simulation approach to risk analysis. The results of such an analysis can then be compared with decisions when parameter estimates are made with certainty.

PROBLEMS

Certainty, Risk, and Uncertainty

19.1   For each situation below, determine (1) if the variable(s) is(are) discrete or continuous and (2) if the information involves certainty, risk, and/or uncertainty. Where risk is involved, graph the information in the general form of Figure 19–1.

(a)   A friend in real estate tells you the price per square foot for new houses will go up slowly or rapidly during the next 6 months.

(b)   Your manager informs the staff there is an equal chance that sales will be between 50 and 55 units next month.

(c)   Jane got paid yesterday, and $400 was taken out in income taxes. The amount withheld next month will be larger because of a pay raise between 3% and 5%.

(d) There is a 20% chance of rain and a 30% chance of snow today.

19.2 An engineer learned that production output is between 1000 and 2000 units per week 90% of the time, and it may fall below 1000 or go above 2000. He wants to use $E(output)$ in the decision making process. Identify at least two additional pieces of information that must be obtained or assumed to finalize the output information for this use.

Probability and Distributions

19.3 A survey of households included a question about the number of operating automobiles N currently owned by people living at the residence and the interest rate i on the lowest-rate loan for the cars. The results for 100 households are shown:

Number of Cars N	Households
0	12
1	56
2	26
3	3
≥4	3

Loan Rate i	Households
0.0–2	22
2.01–4	10
4.01–6	12
6.01–8	42
8.01–10	8
10.01–12	6

(a) State whether each variable is discrete or continuous.
(b) Plot the probability distributions and cumulative distributions for N and i.
(c) From the data collected, what is the probability that a household has 1 or 2 cars? Three or more cars?

(d) Use the data for i to estimate the chances that the interest rate is between 7% and 11% per year.

19.4 An officer of the state lottery commission has sampled lottery ticket purchasers over a 1-week period at one location. The amounts distributed back to the purchasers and the associated probabilities for 5000 tickets are as follows:

Distribution, $	0	2	5	10	100
Probability	0.91	0.045	0.025	0.013	0.007

(a) Plot the cumulative distribution of winnings.
(b) Calculate the expected value of the distribution of dollars per ticket.
(c) If tickets cost $2, what is the expected long-term income to the state per ticket, based upon this sample?

19.5 Bob is working on two separate probability-related projects. The first involves a variable N, which is the number of consecutively manufactured parts that weigh in above the weight specification limit. The variable N is described by the formula $(0.5)^N$ because each unit has a 50-50 chance of being below or above the limit. The second involves a battery life L which varies between 2 and 5 months. The probability distribution is triangular with the mode at 5 months, which is the design life. Some batteries fail early, but 2 months is the smallest life experienced thus far. (a) Write out and plot the probability distributions and cumulative distributions for Bob. (b) Determine the probability of N being 1, 2, or 3 consecutive units above the weight limit.

19.6 An alternative to buy and an alternative to lease hydraulic lifting equipment have been formulated. Use the parameter estimates and assumed distribution data shown to plot the probability distributions on one graph for corresponding parameters. Label the parameters carefully.

Purchase Alternative

Parameter	Estimated Value High	Estimated Value Low	Assumed Distribution
First cost, $	25,000	20,000	Uniform; continuous
Salvage value, $	3,000	2,000	Triangular; mode at $2500
Life, years	8	4	Triangular; mode at 6
AOC, $/year	9,000	5,000	Uniform; continuous

Lease Alternative

Parameter	Estimated Value High	Estimated Value Low	Assumed Distribution
Lease first cost, $	2000	1800	Uniform; continuous
AOC, $/year	9000	5000	Triangular; mode at $7000
Lease term, years	2	2	Certainty

19.7 Carla is a statistician with a bank. She has collected debt-to-equity mix data on mature (M) and young (Y) companies. The debt percentages vary from 20% to 80% in her sample. Carla has defined D_M as a variable for the mature companies from 0 to 1, with $D_M = 0$ interpreted as the low of 20% debt and $D_M = 1.0$ as the high of 80% debt. The variable for young corporation debt percentages D_Y is similarly defined. The probability distributions used to describe D_M an D_Y are

$$f(D_M) = 3(1-D_M)^2 \qquad 0 \le D_M \le 1$$
$$f(D_Y) = 2D_Y \qquad 0 \le D_Y \le 1$$

(a) Use different values of the debt percentage between 20% and 80% to calculate values for the probability distributions and then plot them. (b) What can you comment about the probability that a mature company or a young company will have a low debt percentage? A high debt percentage?

19.8 A discrete variable X can take on integer values of 1 to 10. A sample of size 50 results in the following probability estimates:

X_i	1	2	3	6	9	10
$P(X_i)$	0.2	0.2	0.2	0.1	0.2	0.1

(a) Write out and graph the cumulative distribution.
(b) Calculate the following probabilities using the cumulative distribution: X is between 6 and 10, and X has the values 4, 5, or 6.
(c) Use the cumulative distribution to show that $P(X = 7 \text{ or } 8) = 0.0$. Even though this probability is zero, the statement about X is that it can take on integer values of 1 to 10. How do you explain the apparent contradiction in these two statements?

Random Samples

19.9 Use the discrete variable probability distribution in Problem 19.8 to develop a sample of size 25. Estimate the probabilities for each value of X from your sample, and compare them with those of the originating $P(X_i)$ values.

19.10 The percent price increase p on a variety of retail food prices over a 1-year period varied from 5% to 10% in all cases. Because of the distribution of p values, the assumed probability distribution for the next year is

$$f(X) = 2X \qquad 0 \le X \le 1$$

where

$$X = \begin{cases} 0 & \text{when } p = 5\% \\ 1 & \text{when } p = 10\% \end{cases}$$

For a continuous variable the cumulative distribution $F(X)$ is the integral of $f(X)$ over the same range of the variable. In this case

$$F(X) = X^2 \qquad 0 \le X \le 1$$

(a)   Graphically assign RNs to the cumulative distribution, and take a sample of size 30 for the variable. Transform the X values into interest rates.

(b)   Calculate the average p value for the sample.

19.11   Develop a discrete probability distribution of your own for the variable G, the expected grade in this course, where $G = A, B, C, D, F,$ or I (incomplete). Assign random numbers to $F(G)$, and take a sample from it. Now plot the probability values from the sample for each G value.

19.12   Use the RAND or RANDBETWEEN function in Excel (or corresponding random number generator on another spreadsheet) to generate 100 values from a $U(0,1)$ distribution.

(a)   Calculate the average and compare it to 0.5, the expected value for a random sample between 0 and 1.

(b)   For the RAND function sample, cluster the results into cells of 0.1 width, that is 0.0–0.1, 0.1–0.2, etc., where the upper-limit value is excluded from each cell. Determine the probability for each grouping from

the results. Does your sample come close to having approximately 10% in each cell?

Sample Estimates

19.13   Carol sampled the monthly maintenance costs for automated soldering machines a total of 100 times during 1 year. She clustered the costs into $200 cells, for example, $500 to $700, with cell midpoints of $600, $800, $1000, etc. She indicated the number of times (frequency) each cell value was observed. The costs and frequency data are as follows:

Cell Midpoint	Frequency
600	6
800	10
1000	9
1200	15
1400	28
1600	15
1800	7
2000	10

(a)   Estimate the expected value and standard deviation of the maintenance costs the company should anticipate based on Carol's sample.

(b)   What is the best estimate of the percentage of costs that will fall within 2 standard deviations of the mean?

(c)   Develop a probability distribution of the monthly maintenance costs from Carol's sample, and indicate the answers to the previous two questions on it.

19.14   (a) Determine the values of sample average and standard deviation of the data in Problem 19.8. (b) Determine the values 1 and 2 standard deviations from the mean.

Of the 50 sample points, how many fall within these two ranges?

19.15 (a) Use the relations in Section 19.4 for continuous variables to determine the expected value and standard deviation for the distribution of $f(D_Y)$ in Problem 19.7. (b) It is possible to calculate the probability of a continuous variable X between two points (a, b) using the following integral?

$$P(a \leq X \leq b) = \int_a^b f(X) \, dx$$

Determine the probability that D_Y is within 2 standard deviations of the expected value.

19.16 (a) Use the relations in Section 19.4 for continuous variables to determine the expected value and variance for the distribution of D_M in Problem 19.7.

$$f(D_M) = 3(1 - D_M)^2 \qquad 0 \leq D_M \leq 1$$

(b) Determine the probability that D_M is within two standard deviations of the expected value. Use the relation in Problem 19.15.

19.17 Calculate the expected value for the variable N in Problem 19.5.

19.18 A newsstand manager is tracking Y, the number of weekly magazines left on the shelf when the new edition is delivered. Data collected over a 30-week period are summarized by the following probability distribution. Plot the distribution and the estimates for expected value and one standard deviation on either side of $E(Y)$ on the plot.

Y copies	3	7	10	12
P(Y)	1/3	1/4	1/3	1/12

Simulation

19.19 Carl, an engineering colleague, estimated net cash flow after taxes (CFAT) for the project he is working on. The additional CFAT of $2800 in year 10 is the salvage value of capital assets.

Year	CFAT, $
0	−28,800
1–6	5,400
7–10	2,040
10	2,800

The PW value at the current MARR of 7% per year is

$$\text{PW} = -28,800 + 5400(P/A,7\%,6)$$
$$+ 2040(P/A,7\%,4)(P/F,7\%,6)$$
$$+ 2800(P/F,7\%,10)$$
$$= \$2966$$

Carl believes the MARR will vary over a relatively narrow range, as will the CFAT, especially during the out years of 7 through 10. He is willing to accept the other estimates as certain. Use the following probability distribution assumptions for MARR and CFAT to perform a simulation—hand- or computer-based.

MARR. Uniform distribution over the range 6% to 10%.

CFAT, years 7 through 10. Uniform distribution over the range $1600 to $2400 for each year.

Plot the resulting PW distribution. Should the plan be accepted using decision making under certainty? Under risk?

19.20 Repeat Problem 19.19, except use the normal distribution for the CFAT in years 7 through 10 with an expected value of $2040 and a standard deviation of $500.

EXTENDED EXERCISE

USING SIMULATION AND THE EXCEL RNG FOR SENSITIVITY ANALYSIS

Note: This exercise requires you to learn about and use the Random Number Generation (RNG) Data Analysis Tool package of Microsoft Excel. The online help function explains how to initiate and use the RNG to generate random numbers from a variety of probability distributions: normal, uniform (continuous variable), binomial, Poisson, and discrete. The discrete option is used to generate random numbers from a discrete variable distribution that you specify on the worksheet. This option is the one to use below for the discrete uniform distribution.

Reread the situation in Example 18.3 in which three mutually exclusive alternatives are compared. The parameters of salvage value S, annual operating cost (AOC), and life n are varied using the three-estimate approach to sensitivity analysis. Set up a simulation by answering the following questions using the data provided.

Questions

1. Become familiar with the RNG Data Analysis Tool in Excel by clicking on the help button and reading about it, how to install it (if necessary), and apply it.

2. Develop a sample of 10 random numbers from each of the following distributions:
   - Normal with a mean of 100 and a standard deviation of 20.
   - Uniform (continuous) between 5 and 10.
   - Uniform (discrete) between 5 and 10 with a probability of 0.2 for 5 through 7, 0.05 for 8 and 9, and 0.3 for 10.

3. Develop a simulation of 50 sample points of AW values at a MARR of 12% per year for the three alternatives described in Example 18.3. Use the specified probability distributions below. Do the results of your simulation indicate that alternative B is still the clear choice? If not, what is the better choice?

Parameter	Alternative A	B	C
AOC	Normal Mean: $8000 Std. dev.: $1000	Normal Mean: $3000 Std. dev.: $500	Normal Mean: $6000 Std. dev.: $700
S	Uniform 0 to $1000	Uniform $500 to $2000	Fixed at $3000
n	Discrete uniform 3 to 8 years with equal probability	Discrete uniform 3 to 7 years with equal probability	Discrete uniform 5 to 8 years with equal probability

APPENDIX A

USING SPREADSHEETS AND MICROSOFT EXCEL©

This appendix explains the layout of a spreadsheet and the use of Microsoft Excel (hereafter called Excel) functions in engineering economy. Refer to the User's Guide and Excel help system for your particular computer and version of Excel.

A.1 INTRODUCTION TO USING EXCEL

Run Excel on Windows

After booting up the computer, click on the Microsoft Excel icon to start it. If the icon does not show, left-click the Start button located in the lower left corner of the screen. Move the mouse pointer to Programs, and a submenu will appear on the right. Move to the Microsoft Excel icon, and left-click to run.

If the Microsoft Excel icon is not on the Programs submenu, move to the Microsoft Office icon and highlight the Microsoft Excel icon. Left-click to run.

Enter a Formula

Some example computations are detailed below. The = sign is necessary to perform any formula or function computation in a cell.

1. Move the mouse pointer to cell B4 and left-click.
2. Type =4+3, touch <Enter>, and 7 appears in B4.
3. To edit, use the mouse or <arrow keys> to return to B4, touch <F2>, or use the mouse to move to the Formula Bar in the upper section of the spreadsheet.
4. In either location, touch <Backspace> twice to delete +3.
5. Type −3 and touch <Enter>.
6. The answer 1 appears in cell B4.
7. To delete the cell entirely, move to cell B4 and touch the <Delete> key once.
8. To exit, move the mouse pointer to the top left corner and left-click on File in the top bar menu.
9. Move the mouse down the File submenu, highlight Exit, and left-click.
10. When the Save Changes box appears, left-click "No" to exit without saving.
11. If you wish to save your work, left-click "Yes."
12. Type in a file name (e.g., calcs 1) and click on "Save."

The formulas and functions on the worksheet can be displayed by pressing Ctrl and `. The symbol ` is usually in the upper left of the keyboard with the ~ (tilde) symbol. Pressing Ctrl+` a second time hides the formulas and functions.

Use Excel Functions

1. Run Excel.
2. Move to cell C3. (Move the mouse pointer to C3 and left-click.)
3. Type =PV(5%,12,10) and <Enter>. This function will calculate the present value of 12 payments of $10 at a 5% per year interest rate.

Another use: To calculate the future value of 12 payments of $10 at 6% per year interest, do the following:

1. Move to cell B3, and type INTEREST.
2. Move to cell C3, and type 6% or =6/100.
3. Move to cell B4, and type PAYMENT.
4. Move to cell C4, and type 10 (to represent the size of each payment).
5. Move to cell B5, and type NUMBER OF PAYMENTS.
6. Move to cell C5, and type 12 (to represent the number of payments).
7. Move to cell B7, and type FUTURE VALUE.
8. Move to cell C7, and type =FV(C3,C5,C4) and hit <Enter>. The answer will appear in cell C7.

To edit the values in cells (this feature is used repeatedly in sensitivity analysis and breakeven analysis),

1. Move to cell C3 and type =5/100 (the previous value will be replaced).
2. The value in cell C7 will change its answer automatically.

Cell References in Formulas and Functions

If a cell reference is used in lieu of a specific number, it is possible to change the number once and perform sensitivity analysis on any variable (entry) that is referenced by the cell number, such as C5. This approach defines the referenced cell as a *global variable* for the worksheet. There are two types of cell references—relative and absolute.

Relative References   If a cell reference is entered, for example, A1, into a formula or function that is copied or dragged into another cell, the reference is changed relative to the movement of the original cell. If the formula in C5 is =A1, and it is copied into cell C6, the formula is changed to =A2. This feature is used when dragging a function through several cells, and the source entries must change with the column or row.

Absolute References   If adjusting cell references is not desired, place a $ sign in front of the part of the cell reference that is not to be adjusted—the column, row, or both. For example, =$A$1 will retain the formula when it is moved anywhere on the worksheet. Similarly, =$A1 will retain the column A, but the relative reference on 1 will adjust the row number upon movement around the worksheet.

Absolute references are used in engineering economy for sensitivity analysis of parameters such as MARR, first cost, and annual cash flows. In these cases, a change in the absolute-reference cell entry can help determine the sensitivity of a result, such as PW or AW.

Print the Spreadsheet

First define the portion (or all) of the spreadsheet to be printed.

1. Move the mouse pointer to the top left corner of your spreadsheet.
2. Hold down the left click button. (Do not release the left click button.)
3. Drag the mouse to the lower right corner of your spreadsheet or to wherever you want to stop printing.
4. Release the left click button. (It is ready to print.)
5. Left-click the File top bar menu.
6. Move the mouse down to select Print and left-click.
7. In the Print dialog box, left-click the option Selection in the Print What box (or similar command).
8. Left-click the OK button to start printing.

Depending on your computer environment, you may have to select a network printer and queue your printout through a server.

Save the Spreadsheet

You can save your spreadsheet at any time during or after completing your work. It is recommended that you save your work regularly.

1. Left-click the File top bar menu.
2. To save the spreadsheet the first time, left-click the Save As . . . option.
3. Type the file name, e.g., calcs2, and left-click the Save button.

To save the spreadsheet after it has been saved the first time, i.e., a file name has been assigned to it, left-click the File top bar menu, move the mouse pointer down, and left-click on Save.

Create a Column Chart

1. Run Excel.
2. Move to cell A1 and type 1. Move down to cell A2 and type 2. Type 3 in cell A3, 4 in cell A4, and 5 in cell A5.
3. Move to cell B1 and type 4. Type 3.5 in cell B2; 5 in cell B3; 7 in cell B4; and 12 in cell B5.
4. Move the mouse pointer to cell A1, left-click and hold, while dragging the mouse to cell B5. (All the cells with numbers should be highlighted.)

5. Left-click on the Chart Wizard button on the toolbar.
6. Select the Column option in step 1 of 4 and choose the first subtype of column chart.
7. Left-click and hold the Press and Hold to View Sample button to determine you have selected the type and style of chart desired. Click Next.
8. Since the data were highlighted previously, step 2 can be passed. Left-click Next.
9. For step 3 of 4, click the Titles tab and the Chart Title box. Type Sample 1.
10. Left-click Category (X) axis box and type Year, then left-click Value (Y) axis box and type Rate of return. There are other options (gridlines, legend, etc.) on additional tabs. When finished, left-click Next.
11. For step 4 of 4, left-click As Object In; Sheet1 is highlighted.
12. Left-click Finish, and the chart appears on the spreadsheet.
13. To adjust the size of the chart window, left-click anywhere inside the chart to display small dots on the sides and corners. The words Chart Area will appear immediately below the arrow. Move the mouse to a dot, left-click and hold, then drag the dot to change the size of the chart.
14. To move the chart, left-click and hold within the chart frame, but outside of the graphic itself. A small crosshairs indicator will appear as soon as any movement in the mouse takes place. Changing the position of the mouse moves the entire chart to any location on the worksheet.
15. To adjust the size of the plot area (the graphic itself) within the chart frame, left-click within the graphic. The words Plot Area will appear. Left-click and hold any corner or side dot, and move the mouse to change the size of the graphic up to the size of the chart frame.

Other features are available to change the specific characteristics of the chart. Left-click within the chart frame and click the Chart button on the toolbar at the top of the screen. Options are to alter Chart Type, Source Data, and Chart Options. To obtain detailed help on these, see the help function, or experiment with the sample Column Chart.

Create an *xy* (Scatter) Chart

This chart is one of the most commonly used in scientific analysis, including engineering economy. It plots pairs of data and can place multiple series of entries on the Y axis. The xy scatter chart is especially useful for results such as the PW vs. i graph, where i is the X axis and the Y axis displays the results of the NPV function for several alternatives.

1. Run Excel.
2. Enter the following numbers in columns A, B, and C, respectively.
   Column A, cell A1 through A6: Rate $i\%$, 4, 6, 8, 9, 10
   Column B, cell B1 through B6: $ for A, 40, 55, 60, 45, 10
   Column C, cell C1 through C6: $ for B, 100, 70, 65, 50, 30.
3. Move the mouse to A1, left-click, and hold while dragging to cell C6. All cells will be highlighted, including the title cell for each column.

4. If all the columns for the chart are not adjacent to one another, first press and hold the Control key on the keyboard during the entirety of step 3. After dragging over one column of data, momentarily release the left click, then move to the top of the next (nonadjacent) column for the chart. Do not release the Control key until all columns to be plotted have been highlighted.
5. Left-click on the Chart Wizard button on the toolbar.
6. Select the *xy* (scatter) option in step 1 of 4, and choose a subtype of scatter chart.

The rest of the steps (7 and higher) are the same as detailed earlier for the Column chart. The Legend tab in step 3 of 4 of the Chart Wizard process displays the series labels from the highlighted columns. (Only the bottom row of the title can be highlighted.) If titles are not highlighted, the data sets are generically identified as Series 1, Series 2, etc. on the legend.

Obtain Help While Using Excel

1. To get general help information, left-click on the Help top bar menu.
2. Left-click on Microsoft Excel Help Topics.
3. For example, if you want to know more about how to save a file, type the word Save in box 1.
4. Select the appropriate matching words in box 2. You can browse through the selected words in box 2 by left-clicking on suggested words.
5. Observe the listed topics in box 3.
6. If you find a topic listed in box 3 that matches what you are looking for, double-left-click the selected topic in box 3.

A.2   ORGANIZATION (LAYOUT) OF THE SPREADSHEET

A spreadsheet can be used in several ways to obtain answers to numerical questions. The first is as a rapid solution tool, often with the entry of only a few numbers or one predefined function. In the text, this application is identified using the Q-solv icon in the margin.

1. Run Excel.
2. Move the mouse to cell A1 and type =SUM(45,15,−20). The answer of 40 is displayed in the cell.
3. Move the mouse to cell B4 and type =FV(8%,5,−2500). The number $14,666.50 is displayed as the 8% per year future worth at the end of the fifth year of five payments of $2500 each.

Q-Solv

The second application is more formal. The spreadsheet with the results may serve as documentation of what the entries mean; the sheet may be presented to a coworker, a boss, or a professor; or the final sheet may be placed into a report to management. This type of spreadsheet is identified by the e-solve icon in the text. Some fundamental guidelines useful in setting up the spreadsheet follow.

E-Solve

Figure A–1

Sample spreadsheet layout with estimates, results of formulas and functions, and an *xy* scatter chart.

A very simple layout is presented in Figure A–1. As the solutions become more complex, an orderly arrangement of information makes the spreadsheet easier to read and use.

Cluster the data and the answers. It is advisable to organize the given or estimated data in the top left of the spreadsheet. A very brief label should be used to identify the data, for example, MARR = in cell A1 and the value, 12%, in cell B1. Then B1 can be the referenced cell for all entries requiring the MARR. Additionally, it may be worthwhile to cluster the answers into one area and frame it using the Outside Border button on the toolbar. Often, the answers are best placed at the bottom or top of the column of entries used in the formula or predefined function.

Enter titles for columns and rows. Each column or row should be labeled so its entries are clear to the reader. It is very easy to select from the wrong column or row when no brief title is present at the head of the data.

Enter income and cost cash flows separately. When there are both income and cost cash flows involved, it is strongly recommended that the cash flow estimates for revenue (usually positive) and first cost, salvage value, and annual costs (usually negative, with salvage a positive number) be entered into two adjacent columns. Then a formula combining them in a third column displays the net cash flow. There are two immediate advantages to this practice: fewer errors are made when performing the summation and subtraction mentally, changes for sensitivity analysis are more easily made.

Use cell references. The use of absolute and relative cell references is a must when any changes in entries are expected. For example, suppose the MARR is entered in cell B1, and three separate references are made to the MARR in functions on the spreadsheet. The absolute cell reference entry $B$1 in the three functions allows the MARR to be changed one time, not three.

Obtain a final answer through summing and embedding. When the formulas and functions are kept relatively simple, the final answer can be obtained using the SUM function. For example, if the present worth values (PW) of two columns of cash flows are determined separately, then the total PW is the SUM of the subtotals. This practice is especially useful when the cash flow series are complex.

Although the embedding of functions is allowed in Excel, this means more opportunities for entry errors. Separating the computations makes it easier for the reader to understand the entries. A common application in engineering economy of this practice is in the PMT function that finds the annual worth of a cash flow series. The NPV function can be embedded as the present worth (*P*) value into PMT. Alternatively, the NPV function can be applied first, after which the cell with the PW answer can be referenced in the PMT function. (See Section 3.1 for further comments.)

Prepare for a chart. If a chart (graph) will be developed, plan ahead by leaving sufficient room on the right of the data and answers. Charts can be placed on the same worksheet or on a separate worksheet when the Chart Wizard is used, as discussed in Section A.1 on creating charts. Placement on the same worksheet is recommended, especially when the results of sensitivity analysis are plotted.

A.3   EXCEL FUNCTIONS IMPORTANT TO ENGINEERING ECONOMY (alphabetical order)

DB (Declining Balance)

Calculates the depreciation amount for an asset for a specified period n using the declining balance method. The depreciation rate, d, used in the computation is determined from asset values S (salvage value) and B (basis or first cost) as $d = 1 - (S/B)^{1/n}$. This is Equation [16.11]. Three-decimal-place accuracy is used for d.

$$=DB(\text{cost, salvage, life, period, month})$$

cost	First cost or basis of the asset.
salvage	Salvage value.
life	Depreciation life (recovery period).
period	The period, year, for which the depreciation is to be calculated.
month	(optional entry) If this entry is omitted, a full year is assumed for the first year.

Example A new machine costs $100,000 and is expected to last 10 years. At the end of 10 years, the salvage value of the machine is $50,000. What is the depreciation of the machine in the first year and the fifth year?

Depreciation for the first year: =DB(100000,50000,10,1)
Depreciation for the fifth year: =DB(100000,50000,10,5)

DDB (Double Declining Balance)

Calculates the depreciation of an asset for a specified period n using the double declining balance method. A factor can also be entered for some other declining balance depreciation method by specifying a factor in the function.

$$=DDB(cost, salvage, life, period, factor)$$

cost   First cost or basis of the asset.
salvage   Salvage value of the asset.
life   Depreciation life.
period   The period, year, for which the depreciation is to be calculated.
factor   (optional entry) If this entry is omitted, the function will use a double declining method with 2 times the straight line rate. If, for example, the entry is 1.5, the 150% declining balance method will be used.

Example A new machine costs $200,000 and is expected to last 10 years. The salvage value is $10,000. Calculate the depreciation of the machine for the first and the eighth years. Finally, calculate the depreciation for the fifth year using the 175% declining balance method.

Depreciation for the first year: =DDB(200000,10000,10,1)
Depreciation for the eighth year: =DDB(200000,10000,10,8)
Depreciation for the fifth year using 175% DB:
=DDB(200000,10000,10,5,1.75)

FV (Future Value)

Calculates the future value (worth) based on periodic payments at a specific interest rate.

$$=FV(rate, nper, pmt, pv, type)$$

rate   Interest rate per compounding period.
nper   Number of compounding periods.
pmt   Constant payment amount.
pv   The present value amount. If pv is not specified, the function will assume it to be 0.
type   (optional entry) Either 0 or 1. A 0 represents payments made at the end of the period, and 1 represents payments at the beginning of the period. If omitted, 0 is assumed.

Example   Jack wants to start a savings account that can be increased as desired. He will deposit $12,000 to start the account and plans to add $500 to the account at the beginning of each month for the next 24 months. The bank pays 0.25% per month. How much will be in Jack's account at the end of 24 months?

Future value in 24 months: $=$FV(0.25%,24,500,12000,1)

IF (IF Logical Function)

Determines which of two entries is entered into a cell based on the outcome of a logical check on the outcome of another cell. The logical test can be a function or a simple value check, but it must use an equality or inequality sense. If the response is a text string, place it between quote marks (" "). The responses can themselves be IF functions. Up to seven IF functions can be nested for very complex logical tests.

$$= \textbf{IF(logical\_test,value\_if\_true,value\_if\_false)}$$

logical_test   Any worksheet function can be used here,
                    including a mathematical operation.
value_if_true   Result if the logical_test argument is true.
value_if_false   Result if the logical_test argument is false.

Example   The entry in cell B4 should be "selected" if the PW value in cell B3 is greater than or equal to zero and "rejected" if PW $<$ 0.

Entry in cell B4: $=$IF(B3 $>=$0,"selected","rejected")

Example   The entry in cell C5 should be "selected" if the PW value in cell C4 is greater than or equal to zero, "rejected" if PW $<$ 0, and "fantastic" if PW $\geq$ 200.

Entry in cell C5: $=$IF(C4$<$0,"rejected",IF(C4$>=$200,"fantastic",
                        "selected"))

IPMT (Interest Payment)

Calculates the interest accrued for a given period n based on constant periodic payments and interest rate.

$$=\textbf{IPMT(rate, per, nper, pv, fv, type)}$$

rate   Interest rate per compounding period.
per   Period for which interest is to be calculated.
nper   Number of compounding periods.
pv   Present value. If pv is not specified, the function
        will assume it to be 0.
fv   Future value. If fv is omitted, the function will assume it to
        be 0. The fv can also be considered a cash balance after the last
        payment is made.
type   (optional entry) Either 0 or 1. A 0 represents payments made at
        the end of the period, and 1 represents payments made at the
        beginning of the period. If omitted, 0 is assumed.

Example Calculate the interest due in the tenth month for a 48-month, $20,000 loan. The interest rate is 0.25% per month.

Interest due: =IPMT(0.25%,10,48,20000)

IRR (Internal Rate of Return)

Calculates the internal rate of return between −100% and infinity for a series of cash flows at regular periods.

=IRR(values, guess)

values   A set of numbers in a spreadsheet column (or row) for which the rate of return will be calculated. The set of numbers must consist of at least *one* positive and *one* negative number. Negative numbers denote a payment made or cash outflow, and positive numbers denote income or cash inflow.

guess   (optional entry) To reduce the number of iterations, a *guessed rate of return* can be entered. In most cases, a guess is not required, and a 10% rate of return is initially assumed. If the #NUM! error appears, try using different values for guess. Inputting different guess values makes it possible to determine the multiple roots for the rate of return equation of a nonconventional cash flow series.

Example John wants to start a printing business. He will need $25,000 in capital and anticipates that the business will generate the following incomes during the first 5 years. Calculate his rate of return after 3 years and after 5 years.

Year 1	$5,000
Year 2	$7,500
Year 3	$8,000
Year 4	$10,000
Year 5	$15,000

Set up an array in the spreadsheet.

In cell A1, type −25000 (negative for payment).
In cell A2, type 5000 (positive for income).
In cell A3, type 7500.
In cell A4, type 8000.
In cell A5, type 10000.
In cell A6, type 15000.

Therefore, cells A1 through A6 contain the array of cash flows for the first 5 years, including the capital outlay. *Note that any years with a zero cash flow must have a zero entered* to ensure that the year value is correctly maintained for computation purposes.

To calculate the internal rate of return after 3 years, move to cell A7, and type =IRR(A1:A4).

To calculate the internal rate of return after 5 years and specify a guess value of 5%, move to cell A8, and type =IRR(A1:A6,5%).

MIRR (Modified Internal Rate of Return)

Calculates the modified internal rate of return for a series of cash flows and reinvestment of income and interest at a stated rate.

=MIRR(values, finance_rate, reinvest_rate)

values	Refers to an array of cells in the spreadsheet. Negative numbers represent payments, and positive numbers represent income. The series of payments and income must occur at regular periods and must contain at least *one* positive number and *one* negative number.
finance_rate	Interest rate of money used in the cash flows.
reinvest_rate	Interest rate for reinvestment on positive cash flows. (This is not the same reinvestment rate on the net investments when the cash flow series is nonconventional. See Section 7.5 for comments.)

Example   Jane opened a hobby store 4 years ago. When she started the business, Jane borrowed $50,000 from a bank at 12% per year. Since then, the business has yielded $10,000 the first year, $15,000 the second year, $18,000 the third year, and $21,000 the fourth year. Jane reinvests her profits, earning 8% per year. What is the modified rate of return after 3 years and after 4 years?

In cell A1, type −50000.

In cell A2, type 10000.

In cell A3, type 15000.

In cell A4, type 18000.

In cell A5, type 21000.

To calculate the modified rate of return after 3 years, move to cell A6, and type =MIRR(A1:A4,12%,8%).

To calculate the modified rate of return after 4 years, move to cell A7, and type =MIRR(A1:A5,12%,8%).

NPER (Number of Periods)

Calculates the number of periods for the present worth of an investment to equal the future value specified, based on uniform regular payments and a stated interest rate.

=NPER(rate, pmt, pv, fv, type)

rate	Interest rate per compounding period.
pmt	Amount paid during each compounding period.
pv	Present value (lump-sum amount).

| fv | (optional entry) Future value or cash balance after the last payment. If fv is omitted, the function will assume a value of 0. |
| type | (optional entry) Enter 0 if payments are due at the end of the compounding period, and 1 if payments are due at the beginning of the period. If omitted, 0 is assumed. |

Example   Sally plans to open a savings account which pays 0.25% per month. Her initial deposit is $3000, and she plans to deposit $250 at the beginning of every month. How many payments does she have to make to accumulate $15,000 to buy a new car?

Number of payments: =NPER(0.25%,−250,−3000,15000,1)

NPV (Net Present Value)

Calculates the net present value of a series of future cash flows at a stated interest rate.

$$=NPV(rate, series)$$

| rate | Interest rate per compounding period. |
| series | Series of costs and incomes set up in a range of cells in the spreadsheet. |

Example   Mark is considering buying a sports store for $100,000 and expects to receive the following income during the next 6 years of business: $25,000, $40,000, $42,000, $44,000, $48,000, $50,000. The interest rate is 8% per year.

In cell A1, type −100000.
In cell A2, type 25000.
In cell A3, type 40000.
In cell A4, type 42000.
In cell A5, type 44000.
In cell A6, type 48000.
In cell A7, type 50000.
In cell A8, type =NPV(8%,A2:A7)+A1.

The cell A1 value is already a present value. *Any year with a zero cash flow must have a 0 entered* to ensure a correct result.

PMT (Payments)

Calculates equivalent periodic amounts based on present value and/or future value at a constant interest rate.

$$=PMT(rate, nper, pv, fv, type)$$

| rate | Interest rate per compounding period. |
| nper | Total number of periods. |

pv   Present value.

fv   Future value.

type   (optional entry) Enter 0 for payments due at the end of the
            compounding period, and 1 if payment is due at the start
            of the compounding period. If omitted, 0 is assumed.

Example   Jim plans to take a $15,000 loan to buy a new car. The interest rate is 7%. He wants to pay the loan off in 5 years (60 months). What are his monthly payments?

Monthly payments: =PMT(7%/12,60,15000)

PPMT (Principal Payment)

Calculates the payment on the principal based on uniform payments at a specified interest rate.

$$=PPMT(rate, per, nper, pv, fv, type)$$

rate   Interest rate per compounding period.

per   Period for which the payment on the principal is required.

nper   Total number of periods.

pv   Present value.

fv   Future value.

type   (optional entry) Enter 0 for payments that are due at the end
            of the compounding period, and 1 if payments are due at the
            start of the compounding period. If omitted, 0 is assumed.

Example   Jovita is planning to invest $10,000 in equipment which is expected to last 10 years with no salvage value. The interest rate is 5%. What is the principal payment at the end of year 4 and year 8?

At the end of year 4: =PPMT(5%,4,10,−10000)
At the end of year 8: =PPMT(5%,8,10,−10000)

PV (Present Value)

Calculates the present value of a future series of equal cash flows and a single lump sum in the last period at a constant interest rate.

$$=PV(rate, nper, pmt, fv, type)$$

rate   Interest rate per compounding period.

nper   Total number of periods.

pmt   Cash flow at regular intervals. Negative numbers represent
            payments (cash outflows), and positive numbers represent income.

fv   Future value or cash balance at the end of the last period.

type   (optional entry) Enter 0 if payments are due at the end of the
            compounding period, and 1 if payments are due at the start of
            each compounding period. If omitted, 0 is assumed.

There are two primary differences between the PV function and the NPV function: PV allows for end or beginning of period cash flows, and PV requires that all amounts have the same value, whereas they may vary for the NPV function.

Example   Jose is considering leasing a car for $300 a month for 3 years (36 months). After the 36-month lease, he can purchase the car for $12,000. Using an interest rate of 8% per year, find the present value of this option.

Present value: $=PV(8\%/12,36,-300,-12000)$

Note the minus signs on the pmt and fv amounts.

RAND (Random Number)

Returns an evenly distributed number that is (1) ≥ 0 and < 1; (2) ≥ 0 and < 100; or (3) between two specified numbers.

=RAND()	**for range 0 to 1**
=RAND()*100	**for range 0 to 100**
=RAND()*(b−a)+a	**for range a to b**

a = minimum integer to be generated
b = maximum integer to be generated

The Excel function RANDBETWEEN(a,b) may also be used to obtain a random number between two values.

Example   Grace needs random numbers between 5 and 10 with 3 digits after the decimal. What is the Excel function? Here a = 5 and b = 10.

Random number: $=RAND()*5 + 5$

Example   Randi wants to generate random numbers between the limits of -10 and 25. What is the Excel function? The minimum and maximum values are $a = -10$ and $b = 25$, so $b - a = 25 - (-10) = 35$.

Random number: $=RAND()*35 - 10$

RATE (Interest Rate)

Calculates the interest rate per compounding period for a series of payments or incomes.

$$=RATE(nper, pmt, pv, fv, type, guess)$$

nper	Total number of periods.
pmt	Payment amount made each compounding period.
pv	Present value.
fv	Future value (not including the pmt amount).

type   (optional entry) Enter 0 for payments due at the end of the
         compounding period, and 1 if payments are due at the start
         of each compounding period. If omitted, 0 is assumed.

guess   (optional entry) To minimize computing time, include a
         guessed interest rate. If a value of guess is not specified,
         the function will assume a rate of 10%. This function
         usually converges to a solution, if the rate is between 0%
         and 100%.

Example   Mary wants to start a savings account at a bank. She will make an initial deposit of $1000 to open the account and plans to deposit $100 at the beginning of each month. She plans to do this for the next 3 years (36 months). At the
end of 3 years, she wants to have at least $5000. What is the minimum interest
required to achieve this result?

Interest rate: =RATE(36, − 100, − 1000,5000,1)

SLN (Straight Line Depreciation)

Calculates the straight line depreciation of an asset for a given year.

=SLN(cost, salvage, life)

cost   First cost or basis of the asset.
salvage   Salvage value.
life   Depreciation life.

Example   Maria purchased a printing machine for $100,000. The machine has
an allowed depreciation life of 8 years and an estimated salvage value of
$15,000. What is the depreciation each year?

Depreciation: =SLN(100000,15000,8)

SYD (Sum-of-Year-Digits Depreciation)

Calculates the sum-of-year-digits depreciation of an asset for a given year.

=SYD(cost, salvage, life, period)

cost   First cost or basis of the asset.
salvage   Salvage value.
life   Depreciation life.
period   The year for which the depreciation is sought.

Example   Jack bought equipment for $100,000 which has a depreciation life of
10 years. The salvage value is $10,000. What is the depreciation for year 1 and
year 9?

Depreciation for year 1: =SYD(100000,10000,10,1)
Depreciation for year 9: =SYD(100000,10000,10,9)

VDB (Variable Declining Balance)

Calculates the depreciation using the declining balance method with a switch to straight line depreciation in the year in which straight line has a larger depreciation amount. This function automatically implements the switch from DB to SL depreciation, unless specifically instructed to not switch.

=VDB (cost, salvage, life, start_period, end_period, factor, no_switch)

cost	First cost of the asset.
salvage	Salvage value.
life	Depreciation life.
start_period	First period for depreciation to be calculated.
end_period	Last period for depreciation to be calculated.
factor	(optional entry) If omitted, the function will use the double declining rate of $2/n$, or twice the straight line rate. Other entries define the declining balance method, for example, 1.5 for 150% declining balance.
no_switch	(optional entry) If omitted or entered as FALSE, the function will switch from declining balance to straight line depreciation when the latter is greater than DB depreciation. If entered as TRUE, the function will not switch to SL depreciation at any time during the depreciation life.

Example   Newly purchased equipment with a first cost of $300,000 has a depreciable life of 10 years with no salvage value. Calculate the 175% declining balance depreciation for the first year and the ninth year if switching to SL depreciation is acceptable, and if switching is not permitted.

Depreciation for first year, with switching: =VDB(300000,0,10,0,1,1.75)

Depreciation for ninth year, with switching: =VDB(300000,0,10,8,9,1.75)

Depreciation for first year, no switching: =VDB(300000,0,10,0,1,1.75,TRUE)

Depreciation for ninth year, no switching: =VDB(300000,0,10,8,9,1.75,TRUE)

A.4   SOLVER—AN EXCEL TOOL FOR BREAKEVEN AND "WHAT IF?" ANALYSIS

SOLVER is a powerful tool used to change the value in one or more cells based upon a specified (target) cell value. It is especially helpful in performing breakeven and sensitivity analysis to answer "what if" questions. The SOLVER template is shown in Figure A–2.

Set Target Cell box. Enter a cell reference or name. The target cell itself must contain a formula or function. The value in the cell can be maximized (Max), minimized (Min), or restricted to a specified value (Value of).

By Changing Cells box. Enter the cell reference for each cell to be adjusted, using commas between nonadjacent cells. Each cell must be directly or indirectly related to the target cell. SOLVER proposes a value for the changing cell based

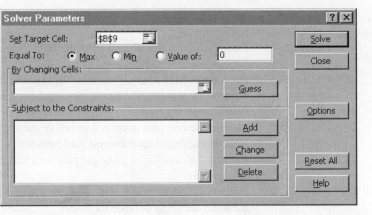

Figure A–2
Excel's SOLVER
template.

on input provided about the target cell. The Guess button will list all possible changing cells related to the target cell.

Subject to the Constraints box. Enter any constraints that may apply, for example, $C$1 < $50,000$. Integer and binary variables are determined in this box.

Options box. Choices here allow the user to specify various parameters of the solution: maximum time and number of iterations allowed, the precision and tolerance of the values determined, and the convergence requirements as the final solution is determined. Also, linear and nonlinear model assumptions can be set here. *If integer or binary variables are involved, the tolerance option must be set to a small number,* say, 0.0001. This is especially important for the binary variables when selecting from independent projects (Chapter 12). If tolerance remains at the default value of 5%, a project may be incorrectly included in the solution set at a very low level.

SOLVER Results box. This appears after Solve is clicked and a solution appears. It is possible, of course, that no solution can be found for the scenario described. It is possible to update the spreadsheet by clicking Keep Solver Solution, or return to the original entries using Restore Original Values.

A.5 LIST OF EXCEL FINANCIAL FUNCTIONS

Here is a listing and brief description of the output of all Excel financial functions. Not all these functions are available on all versions of Microsoft Excel. The Add-ins command can help you determine if the function is available on the system you are using.

ACCRINT	Returns the accrued interest for a security that pays periodic interest.
ACCRINTM	Returns the accrued interest for a security that pays interest at maturity.
AMORDEGRC	Returns the depreciation for each accounting period.
AMORLINC	Returns the depreciation for each accounting period.
COUPDAYBS	Returns the number of days from the beginning of the coupon period to the settlement date.

COUPDAYS	Returns the number of days in the coupon period that contains the settlement date.
COUPDAYSNC	Returns the number of days from the settlement date to the next coupon date.
COUPNCD	Returns the next coupon date after the settlement date.
COUPNUM	Returns the number of coupons payable between the settlement date and maturity date.
COUPPCD	Returns the previous coupon date before the settlement date.
CUMIPMT	Returns the cumulative interest paid between two periods.
CUMPRINC	Returns the cumulative principal paid on a loan between two periods.
DB	Returns the depreciation of an asset for a specified period using the fixed declining balance method.
DDB	Returns the depreciation of an asset for a specified period using the double declining balance method or some other method you specify.
DISC	Returns the discount rate for a security.
DOLLARDE	Converts a dollar price expressed as a fraction to a dollar price expressed as a decimal number.
DOLLARFR	Converts a dollar price expressed as a decimal number to a dollar price expressed as a fraction.
DURATION	Returns the annual duration of a security with periodic interest payments.
EFFECT	Returns the effective annual interest rate.
FV	Returns the future value of an investment.
FVSCHEDULE	Returns the future value of an initial principal after applying a series of compound interest rates.
INTRATE	Returns the interest rate for a fully invested security.
IPMT	Returns the interest payment for an investment for a given period.
IRR	Returns the internal rate of return for a series of cash flows.
ISPMT	Returns the interest paid during a specific period of an investment. (Provides compatibility with Lotus 1-2-3.)
MDURATION	Returns the Macauley modified duration for a security with an assumed par value of $100.
MIRR	Returns the internal rate of return where positive and negative cash flows are financed at different rates.
NOMINAL	Returns the annual nominal interest rate.
NPER	Returns the number of periods for an investment.
NPV	Returns the net present value of an investment based on a series of periodic cash flows and a discount rate.
ODDFPRICE	Returns the price per $100 face value of a security with an odd first period.
ODDFYIELD	Returns the yield of a security with an odd first period.
ODDLPRICE	Returns the price per $100 face value of a security with an odd last period.
ODDLYIELD	Returns the yield of a security with an odd last period.

PMT	Returns the periodic payment for an annuity.
PPMT	Returns the payment on the principal for an investment for a given period.
PRICE	Returns the price per $100 face value of a security that pays periodic interest.
PRICEDISC	Returns the price per $100 face value of a discounted security.
PRICEMAT	Returns the price per $100 face value of a security that pays interest at maturity.
PV	Returns the present value of an investment.
RATE	Returns the interest rate per period of an annuity.
RECEIVED	Returns the amount received at maturity for a fully invested security
SLN	Returns the straight line depreciation of an asset for one period.
SYD	Returns the sum-of-year-digits depreciation of an asset for a specified period.
TBILLEQ	Returns the bond-equivalent yield for a Treasury bill.
TBILLPRICE	Returns the price per $100 face value for a Treasury bill.
TBILLYIELD	Returns the yield for a Treasury bill.
VDB	Returns the depreciation of an asset for a specified or partial period using a declining-balance method with a switch to straight line when it is better.
XIRR	Returns the internal rate of return for a schedule of cash flows that is not necessarily periodic.
XNPV	Returns the net present value for a schedule of cash flows that is not necessarily periodic.
YIELD	Returns the yield on a security that pays periodic interest.
YIELDDISC	Returns the annual yield for a discounted security. For example, a Treasury bill.
YIELDMAT	Returns the annual yield of a security that pays interest at maturity.

There are many more functions available on Excel in other areas: mathematics and trigonometry, statistical, date and time, database, logical, and information.

A.6 ERROR MESSAGES

If Excel is unable to complete a formula or function computation, an error message is displayed. Some of the common messages are as follows:

#DIV/0!	Requires division by zero.
#N/A	Refers to a value that is not available.
#NAME?	Uses a name that Excel doesn't recognize.
#NULL!	Specifies an invalid intersection of two areas.
#NUM!	Uses a number incorrectly.
#REF!	Refers to a cell that is not valid.
#VALUE!	Uses an invalid argument or operand.
#####	Produces a result, or includes a constant numeric value, that is too long to fit in the cell. (Widen the column.)

BASICS OF ACCOUNTING REPORTS AND BUSINESS RATIOS

This appendix provides a fundamental description of financial statements. The documents discussed here will assist in reviewing or understanding basic financial statements and in gathering information useful in an engineering economy study.

LEARNING OBJECTIVES

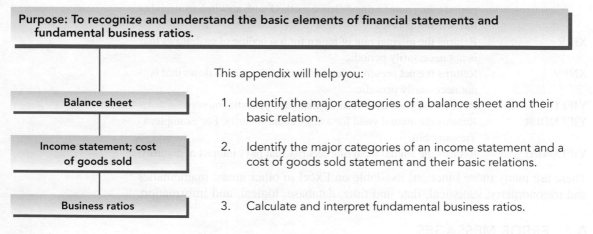

Purpose: To recognize and understand the basic elements of financial statements and fundamental business ratios.

This appendix will help you:

Balance sheet	1. Identify the major categories of a balance sheet and their basic relation.
Income statement; cost of goods sold	2. Identify the major categories of an income statement and a cost of goods sold statement and their basic relations.
Business ratios	3. Calculate and interpret fundamental business ratios.

B.1 THE BALANCE SHEET

The fiscal year and the tax year are defined identically for a corporation or an individual—12 months in length. The fiscal year (FY) is commonly not the calendar year (CY) for a corporation. The U.S. government uses October through September as its FY. For example, October 2005 through September 2006 is FY2006. The fiscal or tax year is always the calendar year for an individual citizen.

At the end of each fiscal year, a company publishes a *balance sheet*. A sample balance sheet for TeamWork Corporation is presented in Table B–1. This is a yearly presentation of the state of the firm at a particular time, for example,

TABLE B–1	Sample Balance Sheet

TEAMWORK CORPORATION
Balance Sheet
December 31, 2006

Assets		Liabilities	
Current			
Cash	$10,500	Accounts payable	$19,700
Accounts receivable	18,700	Dividends payable	7,000
Interest accrued receivable	500	Long-term notes payable	16,000
Inventories	52,000	Bonds payable	20,000
Total current assets	$81,700	Total liabilities	$62,700
		Net Worth	
Fixed			
Land	$25,000	Common stock	$275,000
Building and equipment	438,000	Preferred stock	100,000
Less: Depreciation		Retained earnings	25,000
allowance $82,000	356,000		
Total fixed assets	381,000	Total net worth	400,000
Total assets	$462,700	Total liabilities and net worth	$462,700

December 31, 2006; however, a balance sheet is also usually prepared quarterly and monthly. Note that three main categories are used.

Assets. This section is a summary of all resources owned by or owed to the company. There are two main classes of assets. *Current assets* represent shorter-lived working capital (cash, accounts receivable, etc.), which is more easily converted to cash, usually within 1 year. Longer-lived assets are referred to as *fixed assets* (land, equipment, etc.). Conversion of these holdings to cash in a short period of time would require a major corporate reorientation.

Liabilities. This section is a summary of all financial obligations (debts, mortgages, loans, etc.) of a corporation. Bond indebtedness is included here.

Net worth. Also called *owner's equity,* this section provides a summary of the financial value of ownership, including stocks issued and earnings retained by the corporation.

The balance sheet is constructed using the relation

Assets = liabilities + net worth

In Table B–1 each major category is further divided into standard subcategories. For example, current assets is comprised of cash, accounts receivable, etc. Each subdivision has a specific interpretation, such as accounts receivable, which represents all money owed to the company by its customers.

B.2   INCOME STATEMENT AND COST OF GOODS SOLD STATEMENT

A second important financial statement is the *income statement* (Table B–2). The income statement summarizes the profits or losses of the corporation for a stated period of time. Income statements always accompany balance sheets. The major categories of an income statement are

Revenues. This includes all sales and interest revenue that the company has received in the past accounting period.

Expenses. This is a summary of all expenses for the period. Some expense amounts are itemized in other statements, for example, cost of goods sold and income taxes.

The income statement, published at the same time as the balance sheet, uses the basic equation

$$\text{Revenues} - \text{expenses} = \text{profit (or loss)}$$

The *cost of goods sold* is an important accounting term. It represents the net cost of producing the product marketed by the firm. Cost of goods sold may also be called *factory cost*. A statement of the cost of goods sold, such as that shown in Table B–3, is useful in determining exactly how much it costs to make a particular product over a stated time period, usually a year. Note that the total of the

---

TABLE B–2   Sample Income Statement

TEAMWORK CORPORATION
Income Statement
Year Ended December 31, 2006

Revenues		
Sales	$505,000	
Interest revenue	3,500	
Total revenues		$508,500
Expenses		
Cost of goods sold (from Table B–3)	$290,000	
Selling	28,000	
Administrative	35,000	
Other	12,000	
Total expenses		365,000
Income before taxes		143,500
Taxes for year		64,575
Net profit for year		$ 78,925

TABLE B–3   Sample Cost of Goods Sold Statement

TEAMWORK CORPORATION
Statement of Cost of Goods Sold
Year Ended December 31, 2006

Materials		
Inventory, January 1, 2006	$ 54,000	
Purchases during year	174,500	
Total	$228,500	
Less: Inventory December 31, 2006	50,000	
Cost of materials		$178,500
Direct labor		110,000
Prime cost		288,500
Indirect costs		7,000
Factory cost		295,500
Less: Increase in finished goods inventory during year		5,500
Cost of goods sold (into Table B–2)		$290,000

cost of goods sold statement is entered as an expense item on the income statement. This total is determined using the relations

$$\text{Cost of goods sold} = \text{prime cost} + \text{indirect cost}$$
$$\text{Prime cost} = \text{direct materials} + \text{direct labor}$$

[B.1]

Indirect costs include all indirect and overhead charges made to a product, process, or cost center. Indirect cost allocation methods are discussed in Chapter 15.

B.3 BUSINESS RATIOS

Accountants, financial analysts, and engineering economists frequently utilize business ratio analysis to evaluate the financial health (status) of a company over time and in relation to industry norms. Because the engineering economist must continually communicate with others, she or he should have a basic understanding of several ratios. For comparison purposes, it is necessary to compute the ratios for several companies in the same industry. Industrywide median ratio values are published annually by firms such as Dun and Bradstreet in *Industry Norms and Key Business Ratios*. The ratios are classified according to their role in measuring the corporation.

> **Solvency ratios.** Assess ability to meet short-term and long-term financial obligations.
>
> **Efficiency ratios.** Measure management's ability to use and control assets.
>
> **Profitability ratios.** Evaluate the ability to earn a return for the owners of the corporation.

Numerical data for several important ratios are discussed here and are extracted from the TeamWork balance sheet and income statement, Tables B–1 and B–2.

Secs.
15.5 and 15.6

Indirect costs

Current Ratio   This ratio is utilized to analyze the company's working capital condition. It is defined as

$$\text{Current ratio} = \frac{\text{current assets}}{\text{current liabilities}}$$

Current liabilities include all short-term debts, such as accounts and dividends payable. Note that only balance sheet data are utilized in the current ratio; that is, no association with revenues or expenses is made. For the balance sheet of Table B–1, current liabilities amount to $19,700 + $7000 = $26,700 and

$$\text{Current ratio} = \frac{81,700}{26,700} = 3.06$$

Since current liabilities are those debts payable in the next year, the current ratio value of 3.06 means that the current assets would cover short-term debts approximately 3 times. Current ratio values of 2 to 3 are common.

The current ratio assumes that the working capital invested in inventory can be converted to cash quite rapidly. Often, however, a better idea of a company's *immediate* financial position can be obtained by using the acid test ratio.

Acid Test Ratio (Quick Ratio)   This ratio is

$$\text{Acid-test ratio} = \frac{\text{quick assets}}{\text{current liabilities}}$$

$$= \frac{\text{current assets} - \text{inventories}}{\text{current liabilities}}$$

It is meaningful for the emergency situation when the firm must cover short-term debts using its readily convertible assets. For TeamWork Corporation,

$$\text{Acid test ratio} = \frac{81,700 - 52,000}{26,700} = 1.11$$

Comparison of this and the current ratio shows that approximately 2 times the current debts of the company are invested in inventories. However, an acid test ratio of approximately 1.0 is generally regarded as a strong current position, regardless of the amount of assets in inventories.

Debt Ratio   This ratio is a measure of financial strength since it is defined as

$$\text{Debt ratio} = \frac{\text{total liabilities}}{\text{total assets}}$$

For TeamWork Corporation,

$$\text{Debt ratio} = \frac{62,700}{462,700} = 0.136$$

TeamWork is 13.6% creditor-owned and 86.4% stockholder-owned. A debt ratio in the range of 20% or less usually indicates a sound financial condition, with little fear of forced reorganization because of unpaid liabilities. However, a company with virtually no debts, that is, one with a very low debt ratio, may not have a promising future, because of its inexperience in dealing with short-term and long-term debt financing. The debt-equity (D-E) mix is another measure of financial strength.

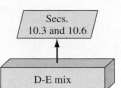

Return on Sales Ratio   This often quoted ratio indicates the profit margin for the company. It is defined as

$$\text{Return on sales} = \frac{\text{net profit}}{\text{net sales}} (100\%)$$

Net profit is the after-tax value from the income statement. This ratio measures profit earned per sales dollar and indicates how well the corporation can sustain adverse conditions over time, such as falling prices, rising costs, and declining sales. For TeamWork Corporation,

$$\text{Return on sales} = \frac{78,925}{505,000} (100\%) = 15.6\%$$

Corporations may point to small return on sales ratios, say, 2.5% to 4.0%, as indications of sagging economic conditions. In truth, for a relatively large-volume, high-turnover business, an income ratio of 3% is quite healthy. Of course, a steadily decreasing ratio indicates rising company expenses, which absorb net profit after taxes.

Return on Assets Ratio   This is the key indicator of profitability since it evaluates the ability of the corporation to transfer assets into operating profit. The definition and value for TeamWork are

$$\text{Return on assets} = \frac{\text{net profit}}{\text{total assets}} (100\%)$$

$$= \frac{78,925}{462,700} (100\%) = 17.1\%$$

Efficient use of assets indicates that the company should earn a high return, while low returns usually accompany lower values of this ratio compared to the industry group ratios.

Inventory Turnover Ratio   Two different ratios are used here. They both indicate the number of times the average inventory value passes through the operations of the company. If turnover of inventory to *net sales* is desired, the formula is

$$\text{Net sales to inventory} = \frac{\text{net sales}}{\text{average inventory}}$$

where average inventory is the figure recorded in the balance sheet. For TeamWork Corporation this ratio is

$$\text{Net sales to inventory} = \frac{505,000}{52,000} = 9.71$$

This means that the average value of the inventory has been sold 9.71 times during the year. Values of this ratio vary greatly from one industry to another.

If inventory turnover is related to *cost of goods sold*, the ratio to use is

$$\text{Cost of goods sold to inventory} = \frac{\text{cost of goods sold}}{\text{average inventory}}$$

Now, average inventory is computed as the average of the beginning and ending inventory values in the statement of cost of goods sold. This ratio is commonly used as a measure of the inventory turnover rate in manufacturing companies. It varies with industries, but management likes to see it remain relatively constant as business increases. For TeamWork, using the values in Table B–3,

$$\text{Cost of goods sold to inventory} = \frac{290,000}{\frac{1}{2}(54,000 + 50,000)} = 5.58$$

There are, of course, many other ratios to use in various circumstances; however, the ones presented here are commonly used by both accountants and economic analysts.

EXAMPLE B.1

Typical values for financial ratios or percentages of four nationally surveyed companies are presented below. Compare the corresponding TeamWork Corporation values with these norms, and comment on differences and similarities.

Ratio or Percentage	Motor Vehicles and Parts Manufacturing 336105*	Air Transportation (Medium-Sized) 481000*	Industrial Machinery Manufacturing 333200*	Home Furnishings 442000*
Current ratio	2.4	0.4	1.7	2.6
Quick ratio	1.6	0.3	0.9	1.2
Debt ratio	59.3%	96.8%	61.5%	52.4%
Return on assets	40.9%	8.1%	6.4%	5.1%

*North American Industry Classification System (NAICS) code for this industry sector.
SOURCE: L. Troy, *Almanac of Business and Industrial Financial Ratios,* 33d annual edition, Prentice-Hall, Paramus, NJ, 2002.

Solution

It is not correct to compare ratios for one company with indexes in different industries, that is, with indexes for different NAICS codes. So, the comparison below is for illustration purposes only. The corresponding values for TeamWork are

$$\text{Current ratio} = 3.06$$
$$\text{Quick ratio} = 1.11$$
$$\text{Debt ratio} = 13.5\%$$
$$\text{Return on assets} = 17.1\%$$

TeamWork has a current ratio larger than all four of these industries, since 3.06 indicates it can cover current liabilities 3 times compared with 2.6 and much less in the case of the "average" air transportation corporation. TeamWork has a significantly lower debt ratio than that of any of the sample industries, so it is likely more financially sound. Return on assets, which is a measure of ability to turn assets into profitability, is not as high at TeamWork as motor vehicles, but TeamWork competes well with the other industry sectors.

To make a fair comparison of TeamWork ratios with other values, it is necessary to have norm values for its industry type as well as ratio values for other corporations in the same NAICS category and about the same size in total assets. Corporate assets are classified in categories by $100,000 units, such as 100 to 250, 1001 to 5000, over 250,000, etc.

PROBLEMS

The following financial data (in thousands of dollars) are for the month of July 20XX for Non-Stop. Use this information in solving Problems B.1 to B.5.

Present Situation, July 31, 20XX

Account	Balance
Accounts payable	$ 35,000
Accounts receivable	29,000
Bonds payable (20-year)	110,000
Buildings (net value)	605,000
Cash on hand	17,000
Dividends payable	8,000
Inventory value (all inventories)	31,000
Land value	450,000
Long-term mortgage payable	450,000
Retained earnings	154,000
Stock value outstanding	375,000

Transactions for July 20XX

Category		Amount
Direct labor		$ 50,000
Expenses		
Insurance	$ 20,000	
Selling	62,000	
Rent and lease	40,000	
Salaries	110,000	
Other	62,000	
Total		294,000
Income taxes		20,000
Increase in finished goods inventory		25,000
Materials inventory, July 1, 20XX		46,000
Materials inventory, July 31, 20XX		25,000
Materials purchases		20,000
Overhead charges		75,000
Revenue from sales		500,000

B.1   Use the account summary information (*a*) to construct a balance sheet for Non-Stop as of July 31, 20XX, and (*b*) to determine the value of each term in the basic equation of the balance sheet.

B.2   What was the net change in materials inventory value during the month?

B.3   Use the summary information to develop (*a*) an income statement for July 20XX and (*b*) the basic equation of the income statement. (*c*) What percentage of revenue is reported as after-tax income?

B.4   (*a*) Compute the value of each business ratio that uses only balance sheet informa-

tion from the statement you constructed in Problem B.1. (*b*) What percentage of the company's current debt is unavailable and in inventory?

B.5   (*a*) Compute the turnover of inventory ratio (based on net sales) for Non-Stop and state its meaning. (*b*) What percentage of each sales dollar can the company rely upon as profit? (*c*) If Non-Stop is an airline, how does its key profitability indicator compare with the median ratio value for its NAICS?

REFERENCE MATERIALS

TEXTBOOKS ON RELATED TOPICS

Bowman, M. S.: *Applied Economic Analysis for Technologists, Engineers, and Managers,* 2d ed., Pearson Prentice-Hall, Upper Saddle River, NJ, 2003.

Bussey, L. E., and T. G. Eschenbach: *The Economic Analysis of Industrial Projects,* 2d ed., Pearson Prentice-Hall, Upper Saddle River, NJ, 1992.

Canada, J. R., W. G. Sullivan, and J. A. White: *Capital Investment Analysis for Engineering and Management,* 2d ed., Pearson Prentice-Hall, Upper Saddle River, NJ, 1996.

Collier, C. A., and C. R. Glagola: *Engineering and Economic Cost Analysis,* 3d ed., Pearson Prentice-Hall, Upper Saddle River, NJ, 1999.

Eschenbach, T. G.: *Engineering Economy: Applying Theory to Practice,* McGraw-Hill, New York, 1995.

Fabrycky, W. J., G. J. Thuesen, and D. Verma: *Economic Decision Analysis,* 3d ed., Pearson Prentice-Hall, Upper Saddle River, NJ, 1998.

Innes, J., F. Mitchell, and T. Yoshikawa: *Activity Costing for Engineers,* John Wiley & Sons, Hoboken, New Jersey, 1994.

Levy, S. M.: *Build, Operate, Transfer: Paving the Way for Tomorrow's Infrastructure,* John Wiley & Sons, Hoboken, New Jersey, 1996.

Newnan, D. G., T. G. Eschenbach, and J. P. Lavelle: *Engineering Economic Analysis,* 9th ed., Oxford University Press, New York, 2004.

Ostwald, P. F.: *Construction Cost Analysis and Estimating,* Pearson Prentice-Hall, Upper Saddle River, NJ, 2001.

Ostwald, P. F., and T. S. McLaren: *Cost Analysis and Estimating for Engineering and Management,* Pearson Prentice-Hall, Upper Saddle River, NJ, 2004.

Park, C. S.: *Contemporary Engineering Economics,* 3d ed., Pearson Prentice-Hall, Upper Saddle River, NJ, 2002.

Park, C. S.: *Fundamentals of Engineering Economics,* Pearson Prentice-Hall, Upper Saddle River, NJ, 2004.

Peurifoy, R. L., and G. D. Oberlender: *Estimating Construction Costs,* 5th ed., McGraw-Hill, New York, 2002.

Stewart, R. D., R. M. Wyskida, and J. D. Johannes: *Cost Estimator's Reference Manual,* 2d ed., John Wiley & Sons, Hoboken, New Jersey, 1995.

Sullivan, W. G., E. Wicks, and J. Luxhoj: *Engineering Economy,* 12th ed., Pearson Prentice-Hall, Upper Saddle River, NJ, 2003.

Thuesen, G. J., and W. J. Fabrycky: *Engineering Economy,* 9th ed., Pearson Prentice-Hall, Upper Saddle River, NJ, 2001.

White, J. A., K. E. Case, D. B. Pratt, and M. H. Agee: *Principles of Engineering Economic Analysis,* 4th ed., John Wiley & Sons, Hoboken, New Jersey, 1997.

Young, D.: *Modern Engineering Economy,* John Wiley & Sons, Hoboken, New Jersey, 1993.

USING EXCEL IN ENGINEERING ECONOMY

Gottfried, B. S.: *Spreadsheet Tools for Engineers Using Excel,* McGraw-Hill, New York, 2003.

WEBSITES

U.S. Internal Revenue Service: www.irs.gov
Revenue Canada: www.ccra-adrc.gc.ca
For this textbook: www.mhhe.com/catalogs
Plant cost estimation index: www.che.com/pindex
Construction cost estimation index: www.construction.com

SELECTED JOURNALS AND PUBLICATIONS

Corporations, Publication 542, Department of the Treasury, Internal Revenue Service, Government Printing Office, Washington, DC, annually.

Engineering News-Record, McGraw-Hill, New York, monthly.

Harvard Business Review, Harvard University Press, Boston, 6 issues per year.

How to Depreciate Property, Publication 946, U.S. Department of the Treasury, Internal Revenue Service, Government Printing Office, Washington, DC, annually.

Journal of Finance, American Finance Association, New York, 5 issues per year.

Sales and Other Dispositions of Assets, Publication 542, Department of the Treasury, Internal Revenue Service, Government Printing Office, Washington, DC, annually.

The Engineering Economist, joint publication of the Engineering Economy Divisions of ASEE and IIE, published by Taylor and Francis, Philadelphia, PA, quarterly.

U.S. Master Tax Guide, Commerce Clearing House, Chicago, annually.

0.25%			TABLE 1	Discrete Cash Flow: Compound Interest Factors				0.25%
	Single Payments		Uniform Series Payments				Arithmetic Gradients	
n	Compound Amount F/P	Present Worth P/F	Sinking Fund A/F	Compound Amount F/A	Capital Recovery A/P	Present Worth P/A	Gradient Present Worth P/G	Gradient Uniform Series A/G
1	1.0025	0.9975	1.00000	1.0000	1.00250	0.9975		
2	1.0050	0.9950	0.49938	2.0025	0.50188	1.9925	0.9950	0.4994
3	1.0075	0.9925	0.33250	3.0075	0.33500	2.9851	2.9801	0.9983
4	1.0100	0.9901	0.24906	4.0150	0.25156	3.9751	5.9503	1.4969
5	1.0126	0.9876	0.19900	5.0251	0.20150	4.9627	9.9007	1.9950
6	1.0151	0.9851	0.16563	6.0376	0.16813	5.9478	14.8263	2.4927
7	1.0176	0.9827	0.14179	7.0527	0.14429	6.9305	20.7223	2.9900
8	1.0202	0.9802	0.12391	8.0704	0.12641	7.9107	27.5839	3.4869
9	1.0227	0.9778	0.11000	9.0905	0.11250	8.8885	35.4061	3.9834
10	1.0253	0.9753	0.09888	10.1133	0.10138	9.8639	44.1842	4.4794
11	1.0278	0.9729	0.08978	11.1385	0.09228	10.8368	53.9133	4.9750
12	1.0304	0.9705	0.08219	12.1664	0.08469	11.8073	64.5886	5.4702
13	1.0330	0.9681	0.07578	13.1968	0.07828	12.7753	76.2053	5.9650
14	1.0356	0.9656	0.07028	14.2298	0.07278	13.7410	88.7587	6.4594
15	1.0382	0.9632	0.06551	15.2654	0.06801	14.7042	102.2441	6.9534
16	1.0408	0.9608	0.06134	16.3035	0.06384	15.6650	116.6567	7.4469
17	1.0434	0.9584	0.05766	17.3443	0.06016	16.6235	131.9917	7.9401
18	1.0460	0.9561	0.05438	18.3876	0.05688	17.5795	148.2446	8.4328
19	1.0486	0.9537	0.05146	19.4336	0.05396	18.5332	165.4106	8.9251
20	1.0512	0.9513	0.04882	20.4822	0.05132	19.4845	183.4851	9.4170
21	1.0538	0.9489	0.04644	21.5334	0.04894	20.4334	202.4634	9.9085
22	1.0565	0.9466	0.04427	22.5872	0.04677	21.3800	222.3410	10.3995
23	1.0591	0.9442	0.04229	23.6437	0.04479	22.3241	243.1131	10.8901
24	1.0618	0.9418	0.04048	24.7028	0.04298	23.2660	264.7753	11.3804
25	1.0644	0.9395	0.03881	25.7646	0.04131	24.2055	287.3230	11.8702
26	1.0671	0.9371	0.03727	26.8290	0.03977	25.1426	310.7516	12.3596
27	1.0697	0.9348	0.03585	27.8961	0.03835	26.0774	335.0566	12.8485
28	1.0724	0.9325	0.03452	28.9658	0.03702	27.0099	360.2334	13.3371
29	1.0751	0.9301	0.03329	30.0382	0.03579	27.9400	386.2776	13.8252
30	1.0778	0.9278	0.03214	31.1133	0.03464	28.8679	413.1847	14.3130
36	1.0941	0.9140	0.02658	37.6206	0.02908	34.3865	592.4988	17.2306
40	1.1050	0.9050	0.02380	42.0132	0.02630	38.0199	728.7399	19.1673
48	1.1273	0.8871	0.01963	50.9312	0.02213	45.1787	1040.06	23.0209
50	1.1330	0.8826	0.01880	53.1887	0.02130	46.9462	1125.78	23.9802
52	1.1386	0.8782	0.01803	55.4575	0.02053	48.7048	1214.59	24.9377
55	1.1472	0.8717	0.01698	58.8819	0.01948	51.3264	1353.53	26.3710
60	1.1616	0.8609	0.01547	64.6467	0.01797	55.6524	1600.08	28.7514
72	1.1969	0.8355	0.01269	78.7794	0.01519	65.8169	2265.56	34.4221
75	1.2059	0.8292	0.01214	82.3792	0.01464	68.3108	2447.61	35.8305
84	1.2334	0.8108	0.01071	93.3419	0.01321	75.6813	3029.76	40.0331
90	1.2520	0.7987	0.00992	100.7885	0.01242	80.5038	3446.87	42.8162
96	1.2709	0.7869	0.00923	108.3474	0.01173	85.2546	3886.28	45.5844
100	1.2836	0.7790	0.00881	113.4500	0.01131	88.3825	4191.24	47.4216
108	1.3095	0.7636	0.00808	123.8093	0.01058	94.5453	4829.01	51.0762
120	1.3494	0.7411	0.00716	139.7414	0.00966	103.5618	5852.11	56.5084
132	1.3904	0.7192	0.00640	156.1582	0.00890	112.3121	6950.01	61.8813
144	1.4327	0.6980	0.00578	173.0743	0.00828	120.8041	8117.41	67.1949
240	1.8208	0.5492	0.00305	328.3020	0.00555	180.3109	19399	107.5863
360	2.4568	0.4070	0.00172	582.7369	0.00422	237.1894	36264	152.8902
480	3.3151	0.3016	0.00108	926.0595	0.00358	279.3418	53821	192.6699

0.5%				TABLE 2	Discrete Cash Flow: Compound Interest Factors			0.5%

	Single Payments		Uniform Series Payments				Arithmetic Gradients	
	Compound Amount	Present Worth	Sinking Fund	Compound Amount	Capital Recovery	Present Worth	Gradient Present Worth	Gradient Uniform Series
n	F/P	P/F	A/F	F/A	A/P	P/A	P/G	A/G
1	1.0050	0.9950	1.00000	1.0000	1.00500	0.9950		
2	1.0100	0.9901	0.49875	2.0050	0.50375	1.9851	0.9901	0.4988
3	1.0151	0.9851	0.33167	3.0150	0.33667	2.9702	2.9604	0.9967
4	1.0202	0.9802	0.24813	4.0301	0.25313	3.9505	5.9011	1.4938
5	1.0253	0.9754	0.19801	5.0503	0.20301	4.9259	9.8026	1.9900
6	1.0304	0.9705	0.16460	6.0755	0.16960	5.8964	14.6552	2.4855
7	1.0355	0.9657	0.14073	7.1059	0.14573	6.8621	20.4493	2.9801
8	1.0407	0.9609	0.12283	8.1414	0.12783	7.8230	27.1755	3.4738
9	1.0459	0.9561	0.10891	9.1821	0.11391	8.7791	34.8244	3.9668
10	1.0511	0.9513	0.09777	10.2280	0.10277	9.7304	43.3865	4.4589
11	1.0564	0.9466	0.08866	11.2792	0.09366	10.6770	52.8526	4.9501
12	1.0617	0.9419	0.08107	12.3356	0.08607	11.6189	63.2136	5.4406
13	1.0670	0.9372	0.07464	13.3972	0.07964	12.5562	74.4602	5.9302
14	1.0723	0.9326	0.06914	14.4642	0.07414	13.4887	86.5835	6.4190
15	1.0777	0.9279	0.06436	15.5365	0.06936	14.4166	99.5743	6.9069
16	1.0831	0.9233	0.06019	16.6142	0.06519	15.3399	113.4238	7.3940
17	1.0885	0.9187	0.05651	17.6973	0.06151	16.2586	128.1231	7.8803
18	1.0939	0.9141	0.05323	18.7858	0.05823	17.1728	143.6634	8.3658
19	1.0994	0.9096	0.05030	19.8797	0.05530	18.0824	160.0360	8.8504
20	1.1049	0.9051	0.04767	20.9791	0.05267	18.9874	177.2322	9.3342
21	1.1104	0.9006	0.04528	22.0840	0.05028	19.8880	195.2434	9.8172
22	1.1160	0.8961	0.04311	23.1944	0.04811	20.7841	214.0611	10.2993
23	1.1216	0.8916	0.04113	24.3104	0.04613	21.6757	233.6768	10.7806
24	1.1272	0.8872	0.03932	25.4320	0.04432	22.5629	254.0820	11.2611
25	1.1328	0.8828	0.03765	26.5591	0.04265	23.4456	275.2686	11.7407
26	1.1385	0.8784	0.03611	27.6919	0.04111	24.3240	297.2281	12.2195
27	1.1442	0.8740	0.03469	28.8304	0.03969	25.1980	319.9523	12.6975
28	1.1499	0.8697	0.03336	29.9745	0.03836	26.0677	343.4332	13.1747
29	1.1556	0.8653	0.03213	31.1244	0.03713	26.9330	367.6625	13.6510
30	1.1614	0.8610	0.03098	32.2800	0.03598	27.7941	392.6324	14.1265
36	1.1967	0.8356	0.02542	39.3361	0.03042	32.8710	557.5598	16.9621
40	1.2208	0.8191	0.02265	44.1588	0.02765	36.1722	681.3347	18.8359
48	1.2705	0.7871	0.01849	54.0978	0.02349	42.5803	959.9188	22.5437
50	1.2832	0.7793	0.01765	56.6452	0.02265	44.1428	1035.70	23.4624
52	1.2961	0.7716	0.01689	59.2180	0.02189	45.6897	1113.82	24.3778
55	1.3156	0.7601	0.01584	63.1258	0.02084	47.9814	1235.27	25.7447
60	1.3489	0.7414	0.01433	69.7700	0.01933	51.7256	1448.65	28.0064
72	1.4320	0.6983	0.01157	86.4089	0.01657	60.3395	2012.35	33.3504
75	1.4536	0.6879	0.01102	90.7265	0.01602	62.4136	2163.75	34.6679
84	1.5204	0.6577	0.00961	104.0739	0.01461	68.4530	2640.66	38.5763
90	1.5666	0.6383	0.00883	113.3109	0.01383	72.3313	2976.08	41.1451
96	1.6141	0.6195	0.00814	122.8285	0.01314	76.0952	3324.18	43.6845
100	1.6467	0.6073	0.00773	129.3337	0.01273	78.5426	3562.79	45.3613
108	1.7137	0.5835	0.00701	142.7399	0.01201	83.2934	4054.37	48.6758
120	1.8194	0.5496	0.00610	163.8793	0.01110	90.0735	4823.51	53.5508
132	1.9316	0.5177	0.00537	186.3226	0.01037	96.4596	5624.59	58.3103
144	2.0508	0.4876	0.00476	210.1502	0.00976	102.4747	6451.31	62.9551
240	3.3102	0.3021	0.00216	462.0409	0.00716	139.5808	13416	96.1131
360	6.0226	0.1660	0.00100	1004.52	0.00600	166.7916	21403	128.3236
480	10.9575	0.0913	0.00050	1991.49	0.00550	181.7476	27588	151.7949

0.75%			TABLE 3	Discrete Cash Flow: Compound Interest Factors			0.75%	
	Single Payments		Uniform Series Payments				Arithmetic Gradients	
	Compound Amount	Present Worth	Sinking Fund	Compound Amount	Capital Recovery	Present Worth	Gradient Present Worth	Gradient Uniform Series
n	F/P	P/F	A/F	F/A	A/P	P/A	P/G	A/G
1	1.0075	0.9926	1.00000	1.0000	1.00750	0.9926		
2	1.0151	0.9852	0.49813	2.0075	0.50563	1.9777	0.9852	0.4981
3	1.0227	0.9778	0.33085	3.0226	0.33835	2.9556	2.9408	0.9950
4	1.0303	0.9706	0.24721	4.0452	0.25471	3.9261	5.8525	1.4907
5	1.0381	0.9633	0.19702	5.0756	0.20452	4.8894	9.7058	1.9851
6	1.0459	0.9562	0.16357	6.1136	0.17107	5.8456	14.4866	2.4782
7	1.0537	0.9490	0.13967	7.1595	0.14717	6.7946	20.1808	2.9701
8	1.0616	0.9420	0.12176	8.2132	0.12926	7.7366	26.7747	3.4608
9	1.0696	0.9350	0.10782	9.2748	0.11532	8.6716	34.2544	3.9502
10	1.0776	0.9280	0.09667	10.3443	0.10417	9.5996	42.6064	4.4384
11	1.0857	0.9211	0.08755	11.4219	0.09505	10.5207	51.8174	4.9253
12	1.0938	0.9142	0.07995	12.5076	0.08745	11.4349	61.8740	5.4110
13	1.1020	0.9074	0.07352	13.6014	0.08102	12.3423	72.7632	5.8954
14	1.1103	0.9007	0.06801	14.7034	0.07551	13.2430	84.4720	6.3786
15	1.1186	0.8940	0.06324	15.8137	0.07074	14.1370	96.9876	6.8606
16	1.1270	0.8873	0.05906	16.9323	0.06656	15.0243	110.2973	7.3413
17	1.1354	0.8807	0.05537	18.0593	0.06287	15.9050	124.3887	7.8207
18	1.1440	0.8742	0.05210	19.1947	0.05960	16.7792	139.2494	8.2989
19	1.1525	0.8676	0.04917	20.3387	0.05667	17.6468	154.8671	8.7759
20	1.1612	0.8612	0.04653	21.4912	0.05403	18.5080	171.2297	9.2516
21	1.1699	0.8548	0.04415	22.6524	0.05165	19.3628	188.3253	9.7261
22	1.1787	0.8484	0.04198	23.8223	0.04948	20.2112	206.1420	10.1994
23	1.1875	0.8421	0.04000	25.0010	0.04750	21.0533	224.6682	10.6714
24	1.1964	0.8358	0.03818	26.1885	0.04568	21.8891	243.8923	11.1422
25	1.2054	0.8296	0.03652	27.3849	0.04402	22.7188	263.8029	11.6117
26	1.2144	0.8234	0.03498	28.5903	0.04248	23.5422	284.3888	12.0800
27	1.2235	0.8173	0.03355	29.8047	0.04105	24.3595	305.6387	12.5470
28	1.2327	0.8112	0.03223	31.0282	0.03973	25.1707	327.5416	13.0128
29	1.2420	0.8052	0.03100	32.2609	0.03850	25.9759	350.0867	13.4774
30	1.2513	0.7992	0.02985	33.5029	0.03735	26.7751	373.2631	13.9407
36	1.3086	0.7641	0.02430	41.1527	0.03180	31.4468	524.9924	16.6946
40	1.3483	0.7416	0.02153	46.4465	0.02903	34.4469	637.4693	18.5058
48	1.4314	0.6986	0.01739	57.5207	0.02489	40.1848	886.8404	22.0691
50	1.4530	0.6883	0.01656	60.3943	0.02406	41.5664	953.8486	22.9476
52	1.4748	0.6780	0.01580	63.3111	0.02330	42.9276	1022.59	23.8211
55	1.5083	0.6630	0.01476	67.7688	0.02226	44.9316	1128.79	25.1223
60	1.5657	0.6387	0.01326	75.4241	0.02076	48.1734	1313.52	27.2665
72	1.7126	0.5839	0.01053	95.0070	0.01803	55.4768	1791.25	32.2882
75	1.7514	0.5710	0.00998	100.1833	0.01748	57.2027	1917.22	33.5163
84	1.8732	0.5338	0.00859	116.4269	0.01609	62.1540	2308.13	37.1357
90	1.9591	0.5104	0.00782	127.8790	0.01532	65.2746	2578.00	39.4946
96	2.0489	0.4881	0.00715	139.8562	0.01465	68.2584	2853.94	41.8107
100	2.1111	0.4737	0.00675	148.1445	0.01425	70.1746	3040.75	43.3311
108	2.2411	0.4462	0.00604	165.4832	0.01354	73.8394	3419.90	46.3154
120	2.4514	0.4079	0.00517	193.5143	0.01267	78.9417	3998.56	50.6521
132	2.6813	0.3730	0.00446	224.1748	0.01196	83.6064	4583.57	54.8232
144	2.9328	0.3410	0.00388	257.7116	0.01138	87.8711	5169.58	58.8314
240	6.0092	0.1664	0.00150	667.8869	0.00900	111.1450	9494.12	85.4210
360	14.7306	0.0679	0.00055	1830.74	0.00805	124.2819	13312	107.1145
480	36.1099	0.0277	0.00021	4681.32	0.00771	129.6409	15513	119.6620

	Single Payments		Uniform Series Payments				Arithmetic Gradients	
	Compound Amount	Present Worth	Sinking Fund	Compound Amount	Capital Recovery	Present Worth	Gradient Present Worth	Gradient Uniform Series
n	F/P	P/F	A/F	F/A	A/P	P/A	P/G	A/G
1	1.0100	0.9901	1.00000	1.0000	1.01000	0.9901		
2	1.0201	0.9803	0.49751	2.0100	0.50751	1.9704	0.9803	0.4975
3	1.0303	0.9706	0.33002	3.0301	0.34002	2.9410	2.9215	0.9934
4	1.0406	0.9610	0.24628	4.0604	0.25628	3.9020	5.8044	1.4876
5	1.0510	0.9515	0.19604	5.1010	0.20604	4.8534	9.6103	1.9801
6	1.0615	0.9420	0.16255	6.1520	0.17255	5.7955	14.3205	2.4710
7	1.0721	0.9327	0.13863	7.2135	0.14863	6.7282	19.9168	2.9602
8	1.0829	0.9235	0.12069	8.2857	0.13069	7.6517	26.3812	3.4478
9	1.0937	0.9143	0.10674	9.3685	0.11674	8.5660	33.6959	3.9337
10	1.1046	0.9053	0.09558	10.4622	0.10558	9.4713	41.8435	4.4179
11	1.1157	0.8963	0.08645	11.5668	0.09645	10.3676	50.8067	4.9005
12	1.1268	0.8874	0.07885	12.6825	0.08885	11.2551	60.5687	5.3815
13	1.1381	0.8787	0.07241	13.8093	0.08241	12.1337	71.1126	5.8607
14	1.1495	0.8700	0.06690	14.9474	0.07690	13.0037	82.4221	6.3384
15	1.1610	0.8613	0.06212	16.0969	0.07212	13.8651	94.4810	6.8143
16	1.1726	0.8528	0.05794	17.2579	0.06794	14.7179	107.2734	7.2886
17	1.1843	0.8444	0.05426	18.4304	0.06426	15.5623	120.7834	7.7613
18	1.1961	0.8360	0.05098	19.6147	0.06098	16.3983	134.9957	8.2323
19	1.2081	0.8277	0.04805	20.8109	0.05805	17.2260	149.8950	8.7017
20	1.2202	0.8195	0.04542	22.0190	0.05542	18.0456	165.4664	9.1694
21	1.2324	0.8114	0.04303	23.2392	0.05303	18.8570	181.6950	9.6354
22	1.2447	0.8034	0.04086	24.4716	0.05086	19.6604	198.5663	10.0998
23	1.2572	0.7954	0.03889	25.7163	0.04889	20.4558	216.0660	10.5626
24	1.2697	0.7876	0.03707	26.9735	0.04707	21.2434	234.1800	11.0237
25	1.2824	0.7798	0.03541	28.2432	0.04541	22.0232	252.8945	11.4831
26	1.2953	0.7720	0.03387	29.5256	0.04387	22.7952	272.1957	11.9409
27	1.3082	0.7644	0.03245	30.8209	0.04245	23.5596	292.0702	12.3971
28	1.3213	0.7568	0.03112	32.1291	0.04112	24.3164	312.5047	12.8516
29	1.3345	0.7493	0.02990	33.4504	0.03990	25.0658	333.4863	13.3044
30	1.3478	0.7419	0.02875	34.7849	0.03875	25.8077	355.0021	13.7557
36	1.4308	0.6989	0.02321	43.0769	0.03321	30.1075	494.6207	16.4285
40	1.4889	0.6717	0.02046	48.8864	0.03046	32.8347	596.8561	18.1776
48	1.6122	0.6203	0.01633	61.2226	0.02633	37.9740	820.1460	21.5976
50	1.6446	0.6080	0.01551	64.4632	0.02551	39.1961	879.4176	22.4363
52	1.6777	0.5961	0.01476	67.7689	0.02476	40.3942	939.9175	23.2686
55	1.7285	0.5785	0.01373	72.8525	0.02373	42.1472	1032.81	24.5049
60	1.8167	0.5504	0.01224	81.6697	0.02224	44.9550	1192.81	26.5333
72	2.0471	0.4885	0.00955	104.7099	0.01955	51.1504	1597.87	31.2386
75	2.1091	0.4741	0.00902	110.9128	0.01902	52.5871	1702.73	32.3793
84	2.3067	0.4335	0.00765	130.6723	0.01765	56.6485	2023.32	35.7170
90	2.4486	0.4084	0.00690	144.8633	0.01690	59.1609	2240.57	37.8724
96	2.5993	0.3847	0.00625	159.9273	0.01625	61.5277	2459.43	39.9727
100	2.7048	0.3697	0.00587	170.4814	0.01587	63.0289	2605.78	41.3426
108	2.9289	0.3414	0.00518	192.8926	0.01518	65.8578	2898.42	44.0103
120	3.3004	0.3030	0.00435	230.0387	0.01435	69.7005	3334.11	47.8349
132	3.7190	0.2689	0.00368	271.8959	0.01368	73.1108	3761.69	51.4520
144	4.1906	0.2386	0.00313	319.0616	0.01313	76.1372	4177.47	54.8676
240	10.8926	0.0918	0.00101	989.2554	0.01101	90.8194	6878.60	75.7393
360	35.9496	0.0278	0.00029	3494.96	0.01029	97.2183	8720.43	89.6995
480	118.6477	0.0084	0.00008	11765	0.01008	99.1572	9511.16	95.9200

1%　　**TABLE 4**　Discrete Cash Flow: Compound Interest Factors　　**1%**

1.25%			**TABLE 5**	Discrete Cash Flow: Compound Interest Factors				**1.25%**

	Single Payments		Uniform Series Payments				Arithmetic Gradients	
	Compound Amount	Present Worth	Sinking Fund	Compound Amount	Capital Recovery	Present Worth	Gradient Present Worth	Gradient Uniform Series
n	F/P	P/F	A/F	F/A	A/P	P/A	P/G	A/G
1	1.0125	0.9877	1.00000	1.0000	1.01250	0.9877		
2	1.0252	0.9755	0.49680	2.0125	0.50939	1.9631	0.9755	0.4969
3	1.0380	0.9634	0.32920	3.0377	0.34170	2.9265	2.9023	0.9917
4	1.0509	0.9515	0.24536	4.0756	0.25786	3.8781	5.7569	1.4845
5	1.0641	0.9398	0.19506	5.1266	0.20756	4.8178	9.5160	1.9752
6	1.0774	0.9282	0.16153	6.1907	0.17403	5.7460	14.1569	2.4638
7	1.0909	0.9167	0.13759	7.2680	0.15009	6.6627	19.6571	2.9503
8	1.1045	0.9054	0.11963	8.3589	0.13213	7.5681	25.9949	3.4348
9	1.1183	0.8942	0.10567	9.4634	0.11817	8.4623	33.1487	3.9172
10	1.1323	0.8832	0.09450	10.5817	0.10700	9.3455	41.0973	4.3975
11	1.1464	0.8723	0.08537	11.7139	0.09787	10.2178	49.8201	4.8758
12	1.1608	0.8615	0.07776	12.8604	0.09026	11.0793	59.2967	5.3520
13	1.1753	0.8509	0.07132	14.0211	0.08382	11.9302	69.5072	5.8262
14	1.1900	0.8404	0.06581	15.1964	0.07831	12.7706	80.4320	6.2982
15	1.2048	0.8300	0.06103	16.3863	0.07353	13.6005	92.0519	6.7682
16	1.2199	0.8197	0.05685	17.5912	0.06935	14.4203	104.3481	7.2362
17	1.2351	0.8096	0.05316	18.8111	0.06566	15.2299	117.3021	7.7021
18	1.2506	0.7996	0.04988	20.0462	0.06238	16.0295	130.8958	8.1659
19	1.2662	0.7898	0.04696	21.2968	0.05946	16.8193	145.1115	8.6277
20	1.2820	0.7800	0.04432	22.5630	0.05682	17.5993	159.9316	9.0874
21	1.2981	0.7704	0.04194	23.8450	0.05444	18.3697	175.3392	9.5450
22	1.3143	0.7609	0.03977	25.1431	0.05227	19.1306	191.3174	10.0006
23	1.3307	0.7515	0.03780	26.4574	0.05030	19.8820	207.8499	10.4542
24	1.3474	0.7422	0.03599	27.7881	0.04849	20.6242	224.9204	10.9056
25	1.3642	0.7330	0.03432	29.1354	0.04682	21.3573	242.5132	11.3551
26	1.3812	0.7240	0.03279	30.4996	0.04529	22.0813	260.6128	11.8024
27	1.3985	0.7150	0.03137	31.8809	0.04387	22.7963	279.2040	12.2478
28	1.4160	0.7062	0.03005	33.2794	0.04255	23.5025	298.2719	12.6911
29	1.4337	0.6975	0.02882	34.6954	0.04132	24.2000	317.8019	13.1323
30	1.4516	0.6889	0.02768	36.1291	0.04018	24.8889	337.7797	13.5715
36	1.5639	0.6394	0.02217	45.1155	0.03467	28.8473	466.2830	16.1639
40	1.6436	0.6084	0.01942	51.4896	0.03192	31.3269	559.2320	17.8515
48	1.8154	0.5509	0.01533	65.2284	0.02783	35.9315	759.2296	21.1299
50	1.8610	0.5373	0.01452	68.8818	0.02702	37.0129	811.6738	21.9295
52	1.9078	0.5242	0.01377	72.6271	0.02627	38.0677	864.9409	22.7211
55	1.9803	0.5050	0.01275	78.4225	0.02525	39.6017	946.2277	23.8936
60	2.1072	0.4746	0.01129	88.5745	0.02379	42.0346	1084.84	25.8083
72	2.4459	0.4088	0.00865	115.6736	0.02115	47.2925	1428.46	30.2047
75	2.5388	0.3939	0.00812	123.1035	0.02062	48.4890	1515.79	31.2605
84	2.8391	0.3522	0.00680	147.1290	0.01930	51.8222	1778.84	34.3258
90	3.0588	0.3269	0.00607	164.7050	0.01857	53.8461	1953.83	36.2855
96	3.2955	0.3034	0.00545	183.6411	0.01795	55.7246	2127.52	38.1793
100	3.4634	0.2887	0.00507	197.0723	0.01757	56.9013	2242.24	39.4058
108	3.8253	0.2614	0.00442	226.0226	0.01692	59.0865	2468.26	41.7737
120	4.4402	0.2252	0.00363	275.2171	0.01613	61.9828	2796.57	45.1184
132	5.1540	0.1940	0.00301	332.3198	0.01551	64.4781	3109.35	48.2234
144	5.9825	0.1672	0.00251	398.6021	0.01501	66.6277	3404.61	51.0990
240	19.7155	0.0507	0.00067	1497.24	0.01317	75.9423	5101.53	67.1764
360	87.5410	0.0114	0.00014	6923.28	0.01264	79.0861	5997.90	75.8401
480	388.7007	0.0026	0.00003	31016	0.01253	79.7942	6284.74	78.7619

1.5%			TABLE 6	Discrete Cash Flow: Compound Interest Factors			1.5%	
	Single Payments		Uniform Series Payments				Arithmetic Gradients	
n	Compound Amount F/P	Present Worth P/F	Sinking Fund A/F	Compound Amount F/A	Capital Recovery A/P	Present Worth P/A	Gradient Present Worth P/G	Gradient Uniform Series A/G
1	1.0150	0.9852	1.00000	1.0000	1.01500	0.9852		
2	1.0302	0.9707	0.49628	2.0150	0.51128	1.9559	0.9707	0.4963
3	1.0457	0.9563	0.32838	3.0452	0.34338	2.9122	2.8833	0.9901
4	1.0614	0.9422	0.24444	4.0909	0.25944	3.8544	5.7098	1.4814
5	1.0773	0.9283	0.19409	5.1523	0.20909	4.7826	9.4229	1.9702
6	1.0934	0.9145	0.16053	6.2296	0.17553	5.6972	13.9956	2.4566
7	1.1098	0.9010	0.13656	7.3230	0.15156	6.5982	19.4018	2.9405
8	1.1265	0.8877	0.11858	8.4328	0.13358	7.4859	25.6157	3.4219
9	1.1434	0.8746	0.10461	9.5593	0.11961	8.3605	32.6125	3.9008
10	1.1605	0.8617	0.09343	10.7027	0.10843	9.2222	40.3675	4.3772
11	1.1779	0.8489	0.08429	11.8633	0.09929	10.0711	48.8568	4.8512
12	1.1956	0.8364	0.07668	13.0412	0.09168	10.9075	58.0571	5.3227
13	1.2136	0.8240	0.07024	14.2368	0.08524	11.7315	67.9454	5.7917
14	1.2318	0.8118	0.06472	15.4504	0.07972	12.5434	78.4994	6.2582
15	1.2502	0.7999	0.05994	16.6821	0.07494	13.3432	89.6974	6.7223
16	1.2690	0.7880	0.05577	17.9324	0.07077	14.1313	101.5178	7.1839
17	1.2880	0.7764	0.05208	19.2014	0.06708	14.9076	113.9400	7.6431
18	1.3073	0.7649	0.04881	20.4894	0.06381	15.6726	126.9435	8.0997
19	1.3270	0.7536	0.04588	21.7967	0.06088	16.4262	140.5084	8.5539
20	1.3469	0.7425	0.04325	23.1237	0.05825	17.1686	154.6154	9.0057
21	1.3671	0.7315	0.04087	24.4705	0.05587	17.9001	169.2453	9.4550
22	1.3876	0.7207	0.03870	25.8376	0.05370	18.6208	184.3798	9.9018
23	1.4084	0.7100	0.03673	27.2251	0.05173	19.3309	200.0006	10.3462
24	1.4295	0.6995	0.03492	28.6335	0.04992	20.0304	216.0901	10.7881
25	1.4509	0.6892	0.03326	30.0630	0.04826	20.7196	232.6310	11.2276
26	1.4727	0.6790	0.03173	31.5140	0.04673	21.3986	249.6065	11.6646
27	1.4948	0.6690	0.03032	32.9867	0.04532	22.0676	267.0002	12.0992
28	1.5172	0.6591	0.02900	34.4815	0.04400	22.7267	284.7958	12.5313
29	1.5400	0.6494	0.02778	35.9987	0.04278	23.3761	302.9779	12.9610
30	1.5631	0.6398	0.02664	37.5387	0.04164	24.0158	321.5310	13.3883
36	1.7091	0.5851	0.02115	47.2760	0.03615	27.6607	439.8303	15.9009
40	1.8140	0.5513	0.01843	54.2679	0.03343	29.9158	524.3568	17.5277
48	2.0435	0.4894	0.01437	69.5652	0.02937	34.0426	703.5462	20.6667
50	2.1052	0.4750	0.01357	73.6828	0.02857	34.9997	749.9636	21.4277
52	2.1689	0.4611	0.01283	77.9249	0.02783	35.9287	796.8774	22.1794
55	2.2679	0.4409	0.01183	84.5296	0.02683	37.2715	868.0285	23.2894
60	2.4432	0.4093	0.01039	96.2147	0.02539	39.3803	988.1674	25.0930
72	2.9212	0.3423	0.00781	128.0772	0.02281	43.8447	1279.79	29.1893
75	3.0546	0.3274	0.00730	136.9728	0.02230	44.8416	1352.56	30.1631
84	3.4926	0.2863	0.00602	166.1726	0.02102	47.5786	1568.51	32.9668
90	3.8189	0.2619	0.00532	187.9299	0.02032	49.2099	1709.54	34.7399
96	4.1758	0.2395	0.00472	211.7202	0.01972	50.7017	1847.47	36.4381
100	4.4320	0.2256	0.00437	228.8030	0.01937	51.6247	1937.45	37.5295
108	4.9927	0.2003	0.00376	266.1778	0.01876	53.3137	2112.13	39.6171
120	5.9693	0.1675	0.00302	331.2882	0.01802	55.4985	2359.71	42.5185
132	7.1370	0.1401	0.00244	409.1354	0.01744	57.3257	2588.71	45.1579
144	8.5332	0.1172	0.00199	502.2109	0.01699	58.8540	2798.58	47.5512
240	35.6328	0.0281	0.00043	2308.85	0.01543	64.7957	3870.69	59.7368
360	212.7038	0.0047	0.00007	14114	0.01507	66.3532	4310.72	64.9662
480	1269.70	0.0008	0.00001	84580	0.01501	66.6142	4415.74	66.2883

2%				TABLE 7	Discrete Cash Flow: Compound Interest Factors			2%
	Single Payments		Uniform Series Payments				Arithmetic Gradients	
	Compound Amount	Present Worth	Sinking Fund	Compound Amount	Capital Recovery	Present Worth	Gradient Present Worth	Gradient Uniform Series
n	F/P	P/F	A/F	F/A	A/P	P/A	P/G	A/G
1	1.0200	0.9804	1.00000	1.0000	1.02000	0.9804		
2	1.0404	0.9612	0.49505	2.0200	0.51505	1.9416	0.9612	0.4950
3	1.0612	0.9423	0.32675	3.0604	0.34675	2.8839	2.8458	0.9868
4	1.0824	0.9238	0.24262	4.1216	0.26262	3.8077	5.6173	1.4752
5	1.1041	0.9057	0.19216	5.2040	0.21216	4.7135	9.2403	1.9604
6	1.1262	0.8880	0.15853	6.3081	0.17853	5.6014	13.6801	2.4423
7	1.1487	0.8706	0.13451	7.4343	0.15451	6.4720	18.9035	2.9208
8	1.1717	0.8535	0.11651	8.5830	0.13651	7.3255	24.8779	3.3961
9	1.1951	0.8368	0.10252	9.7546	0.12252	8.1622	31.5720	3.8681
10	1.2190	0.8203	0.09133	10.9497	0.11133	8.9826	38.9551	4.3367
11	1.2434	0.8043	0.08218	12.1687	0.10218	9.7868	46.9977	4.8021
12	1.2682	0.7885	0.07456	13.4121	0.09456	10.5753	55.6712	5.2642
13	1.2936	0.7730	0.06812	14.6803	0.08812	11.3484	64.9475	5.7231
14	1.3195	0.7579	0.06260	15.9739	0.08260	12.1062	74.7999	6.1786
15	1.3459	0.7430	0.05783	17.2934	0.07783	12.8493	85.2021	6.6309
16	1.3728	0.7284	0.05365	18.6393	0.07365	13.5777	96.1288	7.0799
17	1.4002	0.7142	0.04997	20.0121	0.06997	14.2919	107.5554	7.5256
18	1.4282	0.7002	0.04670	21.4123	0.06670	14.9920	119.4581	7.9681
19	1.4568	0.6864	0.04378	22.8406	0.06378	15.6785	131.8139	8.4073
20	1.4859	0.6730	0.04116	24.2974	0.06116	16.3514	144.6003	8.8433
21	1.5157	0.6598	0.03878	25.7833	0.05878	17.0112	157.7959	9.2760
22	1.5460	0.6468	0.03663	27.2990	0.05663	17.6580	171.3795	9.7055
23	1.5769	0.6342	0.03467	28.8450	0.05467	18.2922	185.3309	10.1317
24	1.6084	0.6217	0.03287	30.4219	0.05287	18.9139	199.6305	10.5547
25	1.6406	0.6095	0.03122	32.0303	0.05122	19.5235	214.2592	10.9745
26	1.6734	0.5976	0.02970	33.6709	0.04970	20.1210	229.1987	11.3910
27	1.7069	0.5859	0.02829	35.3443	0.04829	20.7069	244.4311	11.8043
28	1.7410	0.5744	0.02699	37.0512	0.04699	21.2813	259.9392	12.2145
29	1.7758	0.5631	0.02578	38.7922	0.04578	21.8444	275.7064	12.6214
30	1.8114	0.5521	0.02465	40.5681	0.04465	22.3965	291.7164	13.0251
36	2.0399	0.4902	0.01923	51.9944	0.03923	25.4888	392.0405	15.3809
40	2.2080	0.4529	0.01656	60.4020	0.03656	27.3555	461.9931	16.8885
48	2.5871	0.3865	0.01260	79.3535	0.03260	30.6731	605.9657	19.7556
50	2.6916	0.3715	0.01182	84.5794	0.03182	31.4236	642.3606	20.4420
52	2.8003	0.3571	0.01111	90.0164	0.03111	32.1449	678.7849	21.1164
55	2.9717	0.3365	0.01014	98.5865	0.03014	33.1748	733.3527	22.1057
60	3.2810	0.3048	0.00877	114.0515	0.02877	34.7609	823.6975	23.6961
72	4.1611	0.2403	0.00633	158.0570	0.02633	37.9841	1034.06	27.2234
75	4.4158	0.2265	0.00586	170.7918	0.02586	38.6771	1084.64	28.0434
84	5.2773	0.1895	0.00468	213.8666	0.02468	40.5255	1230.42	30.3616
90	5.9431	0.1683	0.00405	247.1567	0.02405	41.5869	1322.17	31.7929
96	6.6929	0.1494	0.00351	284.6467	0.02351	42.5294	1409.30	33.1370
100	7.2446	0.1380	0.00320	312.2323	0.02320	43.0984	1464.75	33.9863
108	8.4883	0.1178	0.00267	374.4129	0.02267	44.1095	1569.30	35.5774
120	10.7652	0.0929	0.00205	488.2582	0.02205	45.3554	1710.42	37.7114
132	13.6528	0.0732	0.00158	632.6415	0.02158	46.3378	1833.47	39.5676
144	17.3151	0.0578	0.00123	815.7545	0.02123	47.1123	1939.79	41.1738
240	115.8887	0.0086	0.00017	5744.44	0.02017	49.5686	2374.88	47.9110
360	1247.56	0.0008	0.00002	62328	0.02002	49.9599	2482.57	49.7112
480	13430	0.0001			0.02000	49.9963	2498.03	49.9643

TABLE 8 Discrete Cash Flow: Compound Interest Factors

	Single Payments		Uniform Series Payments				Arithmetic Gradients	
n	Compound Amount F/P	Present Worth P/F	Sinking Fund A/F	Compound Amount F/A	Capital Recovery A/P	Present Worth P/A	Gradient Present Worth P/G	Gradient Uniform Series A/G
1	1.0300	0.9709	1.00000	1.0000	1.03000	0.9709		
2	1.0609	0.9426	0.49261	2.0300	0.52261	1.9135	0.9426	0.4926
3	1.0927	0.9151	0.32353	3.0909	0.35353	2.8286	2.7729	0.9803
4	1.1255	0.8885	0.23903	4.1836	0.26903	3.7171	5.4383	1.4631
5	1.1593	0.8626	0.18835	5.3091	0.21835	4.5797	8.8888	1.9409
6	1.1941	0.8375	0.15460	6.4684	0.18460	5.4172	13.0762	2.4138
7	1.2299	0.8131	0.13051	7.6625	0.16051	6.2303	17.9547	2.8819
8	1.2668	0.7894	0.11246	8.8923	0.14246	7.0197	23.4806	3.3450
9	1.3048	0.7664	0.09843	10.1591	0.12843	7.7861	29.6119	3.8032
10	1.3439	0.7441	0.08723	11.4639	0.11723	8.5302	36.3088	4.2565
11	1.3842	0.7224	0.07808	12.8078	0.10808	9.2526	43.5330	4.7049
12	1.4258	0.7014	0.07046	14.1920	0.10046	9.9540	51.2482	5.1485
13	1.4685	0.6810	0.06403	15.6178	0.09403	10.6350	59.4196	5.5872
14	1.5126	0.6611	0.05853	17.0863	0.08853	11.2961	68.0141	6.0210
15	1.5580	0.6419	0.05377	18.5989	0.08377	11.9379	77.0002	6.4500
16	1.6047	0.6232	0.04961	20.1569	0.07961	12.5611	86.3477	6.8742
17	1.6528	0.6050	0.04595	21.7616	0.07595	13.1661	96.0280	7.2936
18	1.7024	0.5874	0.04271	23.4144	0.07271	13.7535	106.0137	7.7081
19	1.7535	0.5703	0.03981	25.1169	0.06981	14.3238	116.2788	8.1179
20	1.8061	0.5537	0.03722	26.8704	0.06722	14.8775	126.7987	8.5229
21	1.8603	0.5375	0.03487	28.6765	0.06487	15.4150	137.5496	8.9231
22	1.9161	0.5219	0.03275	30.5368	0.06275	15.9369	148.5094	9.3186
23	1.9736	0.5067	0.03081	32.4529	0.06081	16.4436	159.6566	9.7093
24	2.0328	0.4919	0.02905	34.4265	0.05905	16.9355	170.9711	10.0954
25	2.0938	0.4776	0.02743	36.4593	0.05743	17.4131	182.4336	10.4768
26	2.1566	0.4637	0.02594	38.5530	0.05594	17.8768	194.0260	10.8535
27	2.2213	0.4502	0.02456	40.7096	0.05456	18.3270	205.7309	11.2255
28	2.2879	0.4371	0.02329	42.9309	0.05329	18.7641	217.5320	11.5930
29	2.3566	0.4243	0.02211	45.2189	0.05211	19.1885	229.4137	11.9558
30	2.4273	0.4120	0.02102	47.5754	0.05102	19.6004	241.3613	12.3141
31	2.5001	0.4000	0.02000	50.0027	0.05000	20.0004	253.3609	12.6678
32	2.5751	0.3883	0.01905	52.5028	0.04905	20.3888	265.3993	13.0169
33	2.6523	0.3770	0.01816	55.0778	0.04816	20.7658	277.4642	13.3616
34	2.7319	0.3660	0.01732	57.7302	0.04732	21.1318	289.5437	13.7018
35	2.8139	0.3554	0.01654	60.4621	0.04654	21.4872	301.6267	14.0375
40	3.2620	0.3066	0.01326	75.4013	0.04326	23.1148	361.7499	15.6502
45	3.7816	0.2644	0.01079	92.7199	0.04079	24.5187	420.6325	17.1556
50	4.3839	0.2281	0.00887	112.7969	0.03887	25.7298	477.4803	18.5575
55	5.0821	0.1968	0.00735	136.0716	0.03735	26.7744	531.7411	19.8600
60	5.8916	0.1697	0.00613	163.0534	0.03613	27.6756	583.0526	21.0674
65	6.8300	0.1464	0.00515	194.3328	0.03515	28.4529	631.2010	22.1841
70	7.9178	0.1263	0.00434	230.5941	0.03434	29.1234	676.0869	23.2145
75	9.1789	0.1089	0.00367	272.6309	0.03367	29.7018	717.6978	24.1634
80	10.6409	0.0940	0.00311	321.3630	0.03311	30.2008	756.0865	25.0353
84	11.9764	0.0835	0.00273	365.8805	0.03273	30.5501	784.5434	25.6806
85	12.3357	0.0811	0.00265	377.8570	0.03265	30.6312	791.3529	25.8349
90	14.3005	0.0699	0.00226	443.3489	0.03226	31.0024	823.6302	26.5667
96	17.0755	0.0586	0.00187	535.8502	0.03187	31.3812	858.6377	27.3615
108	24.3456	0.0411	0.00129	778.1863	0.03129	31.9642	917.6013	28.7072
120	34.7110	0.0288	0.00089	1123.70	0.03089	32.3730	963.8635	29.7737

4%			TABLE 9	Discrete Cash Flow: Compound Interest Factors				4%
	Single Payments		Uniform Series Payments				Arithmetic Gradients	
	Compound Amount F/P	Present Worth P/F	Sinking Fund A/F	Compound Amount F/A	Capital Recovery A/P	Present Worth P/A	Gradient Present Worth P/G	Gradient Uniform Series A/G
n								
1	1.0400	0.9615	1.00000	1.0000	1.04000	0.9615		
2	1.0816	0.9246	0.49020	2.0400	0.53020	1.8861	0.9246	0.4902
3	1.1249	0.8890	0.32035	3.1216	0.36035	2.7751	2.7025	0.9739
4	1.1699	0.8548	0.23549	4.2465	0.27549	3.6299	5.2670	1.4510
5	1.2167	0.8219	0.18463	5.4163	0.22463	4.4518	8.5547	1.9216
6	1.2653	0.7903	0.15076	6.6330	0.19076	5.2421	12.5062	2.3857
7	1.3159	0.7599	0.12661	7.8983	0.16661	6.0021	17.0657	2.8433
8	1.3686	0.7307	0.10853	9.2142	0.14853	6.7327	22.1806	3.2944
9	1.4233	0.7026	0.09449	10.5828	0.13449	7.4353	27.8013	3.7391
10	1.4802	0.6756	0.08329	12.0061	0.12329	8.1109	33.8814	4.1773
11	1.5395	0.6496	0.07415	13.4864	0.11415	8.7605	40.3772	4.6090
12	1.6010	0.6246	0.06655	15.0258	0.10655	9.3851	47.2477	5.0343
13	1.6651	0.6006	0.06014	16.6268	0.10014	9.9856	54.4546	5.4533
14	1.7317	0.5775	0.05467	18.2919	0.09467	10.5631	61.9618	5.8659
15	1.8009	0.5553	0.04994	20.0236	0.08994	11.1184	69.7355	6.2721
16	1.8730	0.5339	0.04582	21.8245	0.08582	11.6523	77.7441	6.6720
17	1.9479	0.5134	0.04220	23.6975	0.08220	12.1657	85.9581	7.0656
18	2.0258	0.4936	0.03899	25.6454	0.07899	12.6593	94.3498	7.4530
19	2.1068	0.4746	0.03614	27.6712	0.07614	13.1339	102.8933	7.8342
20	2.1911	0.4564	0.03358	29.7781	0.07358	13.5903	111.5647	8.2091
21	2.2788	0.4388	0.03128	31.9692	0.07128	14.0292	120.3414	8.5779
22	2.3699	0.4220	0.02920	34.2480	0.06920	14.4511	129.2024	8.9407
23	2.4647	0.4057	0.02731	36.6179	0.06731	14.8568	138.1284	9.2973
24	2.5633	0.3901	0.02559	39.0826	0.06559	15.2470	147.1012	9.6479
25	2.6658	0.3751	0.02401	41.6459	0.06401	15.6221	156.1040	9.9925
26	2.7725	0.3607	0.02257	44.3117	0.06257	15.9828	165.1212	10.3312
27	2.8834	0.3468	0.02124	47.0842	0.06124	16.3296	174.1385	10.6640
28	2.9987	0.3335	0.02001	49.9676	0.06001	16.6631	183.1424	10.9909
29	3.1187	0.3207	0.01888	52.9663	0.05888	16.9837	192.1206	11.3120
30	3.2434	0.3083	0.01783	56.0849	0.05783	17.2920	201.0618	11.6274
31	3.3731	0.2965	0.01686	59.3283	0.05686	17.5885	209.9556	11.9371
32	3.5081	0.2851	0.01595	62.7015	0.05595	17.8736	218.7924	12.2411
33	3.6484	0.2741	0.01510	66.2095	0.05510	18.1476	227.5634	12.5396
34	3.7943	0.2636	0.01431	69.8579	0.05431	18.4112	236.2607	12.8324
35	3.9461	0.2534	0.01358	73.6522	0.05358	18.6646	244.8768	13.1198
40	4.8010	0.2083	0.01052	95.0255	0.05052	19.7928	286.5303	14.4765
45	5.8412	0.1712	0.00826	121.0294	0.04826	20.7200	325.4028	15.7047
50	7.1067	0.1407	0.00655	152.6671	0.04655	21.4822	361.1638	16.8122
55	8.6464	0.1157	0.00523	191.1592	0.04523	22.1086	393.6890	17.8070
60	10.5196	0.0951	0.00420	237.9907	0.04420	22.6235	422.9966	18.6972
65	12.7987	0.0781	0.00339	294.9684	0.04339	23.0467	449.2014	19.4909
70	15.5716	0.0642	0.00275	364.2905	0.04275	23.3945	472.4789	20.1961
75	18.9453	0.0528	0.00223	448.6314	0.04223	23.6804	493.0408	20.8206
80	23.0498	0.0434	0.00181	551.2450	0.04181	23.9154	511.1161	21.3718
85	28.0436	0.0357	0.00148	676.0901	0.04148	24.1085	526.9384	21.8569
90	34.1193	0.0293	0.00121	827.9833	0.04121	24.2673	540.7369	22.2826
96	43.1718	0.0232	0.00095	1054.30	0.04095	24.4209	554.9312	22.7236
108	69.1195	0.0145	0.00059	1702.99	0.04059	24.6383	576.8949	23.4146
120	110.6626	0.0090	0.00036	2741.56	0.04036	24.7741	592.2428	23.9057
144	283.6618	0.0035	0.00014	7066.55	0.04014	24.9119	610.1055	24.4906

5%   **TABLE 10**   Discrete Cash Flow: Compound Interest Factors   **5%**

	Single Payments		Uniform Series Payments				Arithmetic Gradients	
	Compound Amount	Present Worth	Sinking Fund	Compound Amount	Capital Recovery	Present Worth	Gradient Present Worth	Gradient Uniform Series
n	F/P	P/F	A/F	F/A	A/P	P/A	P/G	A/G
1	1.0500	0.9524	1.00000	1.0000	1.05000	0.9524		
2	1.1025	0.9070	0.48780	2.0500	0.53780	1.8594	0.9070	0.4878
3	1.1576	0.8638	0.31721	3.1525	0.36721	2.7232	2.6347	0.9675
4	1.2155	0.8227	0.23201	4.3101	0.28201	3.5460	5.1028	1.4391
5	1.2763	0.7835	0.18097	5.5256	0.23097	4.3295	8.2369	1.9025
6	1.3401	0.7462	0.14702	6.8019	0.19702	5.0757	11.9680	2.3579
7	1.4071	0.7107	0.12282	8.1420	0.17282	5.7864	16.2321	2.8052
8	1.4775	0.6768	0.10472	9.5491	0.15472	6.4632	20.9700	3.2445
9	1.5513	0.6446	0.09069	11.0266	0.14069	7.1078	26.1268	3.6758
10	1.6289	0.6139	0.07950	12.5779	0.12950	7.7217	31.6520	4.0991
11	1.7103	0.5847	0.07039	14.2068	0.12039	8.3064	37.4988	4.5144
12	1.7959	0.5568	0.06283	15.9171	0.11283	8.8633	43.6241	4.9219
13	1.8856	0.5303	0.05646	17.7130	0.10646	9.3936	49.9879	5.3215
14	1.9799	0.5051	0.05102	19.5986	0.10102	9.8986	56.5538	5.7133
15	2.0789	0.4810	0.04634	21.5786	0.09634	10.3797	63.2880	6.0973
16	2.1829	0.4581	0.04227	23.6575	0.09227	10.8378	70.1597	6.4736
17	2.2920	0.4363	0.03870	25.8404	0.08870	11.2741	77.1405	6.8423
18	2.4066	0.4155	0.03555	28.1324	0.08555	11.6896	84.2043	7.2034
19	2.5270	0.3957	0.03275	30.5390	0.08275	12.0853	91.3275	7.5569
20	2.6533	0.3769	0.03024	33.0660	0.08024	12.4622	98.4884	7.9030
21	2.7860	0.3589	0.02800	35.7193	0.07800	12.8212	105.6673	8.2416
22	2.9253	0.3418	0.02597	38.5052	0.07597	13.1630	112.8461	8.5730
23	3.0715	0.3256	0.02414	41.4305	0.07414	13.4886	120.0087	8.8971
24	3.2251	0.3101	0.02247	44.5020	0.07247	13.7986	127.1402	9.2140
25	3.3864	0.2953	0.02095	47.7271	0.07095	14.0939	134.2275	9.5238
26	3.5557	0.2812	0.01956	51.1135	0.06956	14.3752	141.2585	9.8266
27	3.7335	0.2678	0.01829	54.6691	0.06829	14.6430	148.2226	10.1224
28	3.9201	0.2551	0.01712	58.4026	0.06712	14.8981	155.1101	10.4114
29	4.1161	0.2429	0.01605	62.3227	0.06605	15.1411	161.9126	10.6936
30	4.3219	0.2314	0.01505	66.4388	0.06505	15.3725	168.6226	10.9691
31	4.5380	0.2204	0.01413	70.7608	0.06413	15.5928	175.2333	11.2381
32	4.7649	0.2099	0.01328	75.2988	0.06328	15.8027	181.7392	11.5005
33	5.0032	0.1999	0.01249	80.0638	0.06249	16.0025	188.1351	11.7566
34	5.2533	0.1904	0.01176	85.0670	0.06176	16.1929	194.4168	12.0063
35	5.5160	0.1813	0.01107	90.3203	0.06107	16.3742	200.5807	12.2498
40	7.0400	0.1420	0.00828	120.7998	0.05828	17.1591	229.5452	13.3775
45	8.9850	0.1113	0.00626	159.7002	0.05626	17.7741	255.3145	14.3644
50	11.4674	0.0872	0.00478	209.3480	0.05478	18.2559	277.9148	15.2233
55	14.6356	0.0683	0.00367	272.7126	0.05367	18.6335	297.5104	15.9664
60	18.6792	0.0535	0.00283	353.5837	0.05283	18.9293	314.3432	16.6062
65	23.8399	0.0419	0.00219	456.7980	0.05219	19.1611	328.6910	17.1541
70	30.4264	0.0329	0.00170	588.5285	0.05170	19.3427	340.8409	17.6212
75	38.8327	0.0258	0.00132	756.6537	0.05132	19.4850	351.0721	18.0176
80	49.5614	0.0202	0.00103	971.2288	0.05103	19.5965	359.6460	18.3526
85	63.2544	0.0158	0.00080	1245.09	0.05080	19.6838	366.8007	18.6346
90	80.7304	0.0124	0.00063	1594.61	0.05063	19.7523	372.7488	18.8712
95	103.0347	0.0097	0.00049	2040.69	0.05049	19.8059	377.6774	19.0689
96	108.1864	0.0092	0.00047	2143.73	0.05047	19.8151	378.5555	19.1044
98	119.2755	0.0084	0.00042	2365.51	0.05042	19.8323	380.2139	19.1714
100	131.5013	0.0076	0.00038	2610.03	0.05038	19.8479	381.7492	19.2337

6%			TABLE 11	Discrete Cash Flow: Compound Interest Factors				6%
	Single Payments		**Uniform Series Payments**				**Arithmetic Gradients**	
	Compound Amount F/P	Present Worth P/F	Sinking Fund A/F	Compound Amount F/A	Capital Recovery A/P	Present Worth P/A	Gradient Present Worth P/G	Gradient Uniform Series A/G
n								
1	1.0600	0.9434	1.00000	1.0000	1.06000	0.9434		
2	1.1236	0.8900	0.48544	2.0600	0.54544	1.8334	0.8900	0.4854
3	1.1910	0.8396	0.31411	3.1836	0.37411	2.6730	2.5692	0.9612
4	1.2625	0.7921	0.22859	4.3746	0.28859	3.4651	4.9455	1.4272
5	1.3382	0.7473	0.17740	5.6371	0.23740	4.2124	7.9345	1.8836
6	1.4185	0.7050	0.14336	6.9753	0.20336	4.9173	11.4594	2.3304
7	1.5036	0.6651	0.11914	8.3938	0.17914	5.5824	15.4497	2.7676
8	1.5938	0.6274	0.10104	9.8975	0.16104	6.2098	19.8416	3.1952
9	1.6895	0.5919	0.08702	11.4913	0.14702	6.8017	24.5768	3.6133
10	1.7908	0.5584	0.07587	13.1808	0.13587	7.3601	29.6023	4.0220
11	1.8983	0.5268	0.06679	14.9716	0.12679	7.8869	34.8702	4.4213
12	2.0122	0.4970	0.05928	16.8699	0.11928	8.3838	40.3369	4.8113
13	2.1329	0.4688	0.05296	18.8821	0.11296	8.8527	45.9629	5.1920
14	2.2609	0.4423	0.04758	21.0151	0.10758	9.2950	51.7128	5.5635
15	2.3966	0.4173	0.04296	23.2760	0.10296	9.7122	57.5546	5.9260
16	2.5404	0.3936	0.03895	25.6725	0.09895	10.1059	63.4592	6.2794
17	2.6928	0.3714	0.03544	28.2129	0.09544	10.4773	69.4011	6.6240
18	2.8543	0.3503	0.03236	30.9057	0.09236	10.8276	75.3569	6.9597
19	3.0256	0.3305	0.02962	33.7600	0.08962	11.1581	81.3062	7.2867
20	3.2071	0.3118	0.02718	36.7856	0.08718	11.4699	87.2304	7.6051
21	3.3996	0.2942	0.02500	39.9927	0.08500	11.7641	93.1136	7.9151
22	3.6035	0.2775	0.02305	43.3923	0.08305	12.0416	98.9412	8.2166
23	3.8197	0.2618	0.02128	46.9958	0.08128	12.3034	104.7007	8.5099
24	4.0489	0.2470	0.01968	50.8156	0.07968	12.5504	110.3812	8.7951
25	4.2919	0.2330	0.01823	54.8645	0.07823	12.7834	115.9732	9.0722
26	4.5494	0.2198	0.01690	59.1564	0.07690	13.0032	121.4684	9.3414
27	4.8223	0.2074	0.01570	63.7058	0.07570	13.2105	126.8600	9.6029
28	5.1117	0.1956	0.01459	68.5281	0.07459	13.4062	132.1420	9.8568
29	5.4184	0.1846	0.01358	73.6398	0.07358	13.5907	137.3096	10.1032
30	5.7435	0.1741	0.01265	79.0582	0.07265	13.7648	142.3588	10.3422
31	6.0881	0.1643	0.01179	84.8017	0.07179	13.9291	147.2864	10.5740
32	6.4534	0.1550	0.01100	90.8898	0.07100	14.0840	152.0901	10.7988
33	6.8406	0.1462	0.01027	97.3432	0.07027	14.2302	156.7681	11.0166
34	7.2510	0.1379	0.00960	104.1838	0.06960	14.3681	161.3192	11.2276
35	7.6861	0.1301	0.00897	111.4348	0.06897	14.4982	165.7427	11.4319
40	10.2857	0.0972	0.00646	154.7620	0.06646	15.0463	185.9568	12.3590
45	13.7646	0.0727	0.00470	212.7435	0.06470	15.4558	203.1096	13.1413
50	18.4202	0.0543	0.00344	290.3359	0.06344	15.7619	217.4574	13.7964
55	24.6503	0.0406	0.00254	394.1720	0.06254	15.9905	229.3222	14.3411
60	32.9877	0.0303	0.00188	533.1282	0.06188	16.1614	239.0428	14.7909
65	44.1450	0.0227	0.00139	719.0829	0.06139	16.2891	246.9450	15.1601
70	59.0759	0.0169	0.00103	967.9322	0.06103	16.3845	253.3271	15.4613
75	79.0569	0.0126	0.00077	1300.95	0.06077	16.4558	258.4527	15.7058
80	105.7960	0.0095	0.00057	1746.60	0.06057	16.5091	262.5493	15.9033
85	141.5789	0.0071	0.00043	2342.98	0.06043	16.5489	265.8096	16.0620
90	189.4645	0.0053	0.00032	3141.08	0.06032	16.5787	268.3946	16.1891
95	253.5463	0.0039	0.00024	4209.10	0.06024	16.6009	270.4375	16.2905
96	268.7590	0.0037	0.00022	4462.65	0.06022	16.6047	270.7909	16.3081
98	301.9776	0.0033	0.00020	5016.29	0.06020	16.6115	271.4491	16.3411
100	339.3021	0.0029	0.00018	5638.37	0.06018	16.6175	272.0471	16.3711

7%			TABLE 12	Discrete Cash Flow: Compound Interest Factors				7%
	Single Payments		Uniform Series Payments				Arithmetic Gradients	
n	Compound Amount F/P	Present Worth P/F	Sinking Fund A/F	Compound Amount F/A	Capital Recovery A/P	Present Worth P/A	Gradient Present Worth P/G	Gradient Uniform Series A/G
1	1.0700	0.9346	1.00000	1.0000	1.07000	0.9346		
2	1.1449	0.8734	0.48309	2.0700	0.55309	1.8080	0.8734	0.4831
3	1.2250	0.8163	0.31105	3.2149	0.38105	2.6243	2.5060	0.9549
4	1.3108	0.7629	0.22523	4.4399	0.29523	3.3872	4.7947	1.4155
5	1.4026	0.7130	0.17389	5.7507	0.24389	4.1002	7.6467	1.8650
6	1.5007	0.6663	0.13980	7.1533	0.20980	4.7665	10.9784	2.3032
7	1.6058	0.6227	0.11555	8.6540	0.18555	5.3893	14.7149	2.7304
8	1.7182	0.5820	0.09747	10.2598	0.16747	5.9713	18.7889	3.1465
9	1.8385	0.5439	0.08349	11.9780	0.15349	6.5152	23.1404	3.5517
10	1.9672	0.5083	0.07238	13.8164	0.14238	7.0236	27.7156	3.9461
11	2.1049	0.4751	0.06336	15.7836	0.13336	7.4987	32.4665	4.3296
12	2.2522	0.4440	0.05590	17.8885	0.12590	7.9427	37.3506	4.7025
13	2.4098	0.4150	0.04965	20.1406	0.11965	8.3577	42.3302	5.0648
14	2.5785	0.3878	0.04434	22.5505	0.11434	8.7455	47.3718	5.4167
15	2.7590	0.3624	0.03979	25.1290	0.10979	9.1079	52.4461	5.7583
16	2.9522	0.3387	0.03586	27.8881	0.10586	9.4466	57.5271	6.0897
17	3.1588	0.3166	0.03243	30.8402	0.10243	9.7632	62.5923	6.4110
18	3.3799	0.2959	0.02941	33.9990	0.09941	10.0591	67.6219	6.7225
19	3.6165	0.2765	0.02675	37.3790	0.09675	10.3356	72.5991	7.0242
20	3.8697	0.2584	0.02439	40.9955	0.09439	10.5940	77.5091	7.3163
21	4.1406	0.2415	0.02229	44.8652	0.09229	10.8355	82.3393	7.5990
22	4.4304	0.2257	0.02041	49.0057	0.09041	11.0612	87.0793	7.8725
23	4.7405	0.2109	0.01871	53.4361	0.08871	11.2722	91.7201	8.1369
24	5.0724	0.1971	0.01719	58.1767	0.08719	11.4693	96.2545	8.3923
25	5.4274	0.1842	0.01581	63.2490	0.08581	11.6536	100.6765	8.6391
26	5.8074	0.1722	0.01456	68.6765	0.08456	11.8258	104.9814	8.8773
27	6.2139	0.1609	0.01343	74.4838	0.08343	11.9867	109.1656	9.1072
28	6.6488	0.1504	0.01239	80.6977	0.08239	12.1371	113.2264	9.3289
29	7.1143	0.1406	0.01145	87.3465	0.08145	12.2777	117.1622	9.5427
30	7.6123	0.1314	0.01059	94.4608	0.08059	12.4090	120.9718	9.7487
31	8.1451	0.1228	0.00980	102.0730	0.07980	12.5318	124.6550	9.9471
32	8.7153	0.1147	0.00907	110.2182	0.07907	12.6466	128.2120	10.1381
33	9.3253	0.1072	0.00841	118.9334	0.07841	12.7538	131.6435	10.3219
34	9.9781	0.1002	0.00780	128.2588	0.07780	12.8540	134.9507	10.4987
35	10.6766	0.0937	0.00723	138.2369	0.07723	12.9477	138.1353	10.6687
40	14.9745	0.0668	0.00501	199.6351	0.07501	13.3317	152.2928	11.4233
45	21.0025	0.0476	0.00350	285.7493	0.07350	13.6055	163.7559	12.0360
50	29.4570	0.0339	0.00246	406.5289	0.07246	13.8007	172.9051	12.5287
55	41.3150	0.0242	0.00174	575.9286	0.07174	13.9399	180.1243	12.9215
60	57.9464	0.0173	0.00123	813.5204	0.07123	14.0392	185.7677	13.2321
65	81.2729	0.0123	0.00087	1146.76	0.07087	14.1099	190.1452	13.4760
70	113.9894	0.0088	0.00062	1614.13	0.07062	14.1604	193.5185	13.6662
75	159.8760	0.0063	0.00044	2269.66	0.07044	14.1964	196.1035	13.8136
80	224.2344	0.0045	0.00031	3189.06	0.07031	14.2220	198.0748	13.9273
85	314.5003	0.0032	0.00022	4478.58	0.07022	14.2403	199.5717	14.0146
90	441.1030	0.0023	0.00016	6287.19	0.07016	14.2533	200.7042	14.0812
95	618.6697	0.0016	0.00011	8823.85	0.07011	14.2626	201.5581	14.1319
96	661.9766	0.0015	0.00011	9442.52	0.07011	14.2641	201.7016	14.1405
98	757.8970	0.0013	0.00009	10813	0.07009	14.2669	201.9651	14.1562
100	867.7163	0.0012	0.00008	12382	0.07008	14.2693	202.2001	14.1703

| 8% | | | TABLE 13 | Discrete Cash Flow: Compound Interest Factors | | | | 8% |

	Single Payments		Uniform Series Payments				Arithmetic Gradients	
	Compound Amount	Present Worth	Sinking Fund	Compound Amount	Capital Recovery	Present Worth	Gradient Present Worth	Gradient Uniform Series
n	F/P	P/F	A/F	F/A	A/P	P/A	P/G	A/G
1	1.0800	0.9259	1.00000	1.0000	1.08000	0.9259		
2	1.1664	0.8573	0.48077	2.0800	0.56077	1.7833	0.8573	0.4808
3	1.2597	0.7938	0.30803	3.2464	0.38803	2.5771	2.4450	0.9487
4	1.3605	0.7350	0.22192	4.5061	0.30192	3.3121	4.6501	1.4040
5	1.4693	0.6806	0.17046	5.8666	0.25046	3.9927	7.3724	1.8465
6	1.5869	0.6302	0.13632	7.3359	0.21632	4.6229	10.5233	2.2763
7	1.7138	0.5835	0.11207	8.9228	0.19207	5.2064	14.0242	2.6937
8	1.8509	0.5403	0.09401	10.6366	0.17401	5.7466	17.8061	3.0985
9	1.9990	0.5002	0.08008	12.4876	0.16008	6.2469	21.8081	3.4910
10	2.1589	0.4632	0.06903	14.4866	0.14903	6.7101	25.9768	3.8713
11	2.3316	0.4289	0.06008	16.6455	0.14008	7.1390	30.2657	4.2395
12	2.5182	0.3971	0.05270	18.9771	0.13270	7.5361	34.6339	4.5957
13	2.7196	0.3677	0.04652	21.4953	0.12652	7.9038	39.0463	4.9402
14	2.9372	0.3405	0.04130	24.2149	0.12130	8.2442	43.4723	5.2731
15	3.1722	0.3152	0.03683	27.1521	0.11683	8.5595	47.8857	5.5945
16	3.4259	0.2919	0.03298	30.3243	0.11298	8.8514	52.2640	5.9046
17	3.7000	0.2703	0.02963	33.7502	0.10963	9.1216	56.5883	6.2037
18	3.9960	0.2502	0.02670	37.4502	0.10670	9.3719	60.8426	6.4920
19	4.3157	0.2317	0.02413	41.4463	0.10413	9.6036	65.0134	6.7697
20	4.6610	0.2145	0.02185	45.7620	0.10185	9.8181	69.0898	7.0369
21	5.0338	0.1987	0.01983	50.4229	0.09983	10.0168	73.0629	7.2940
22	5.4365	0.1839	0.01803	55.4568	0.09803	10.2007	76.9257	7.5412
23	5.8715	0.1703	0.01642	60.8933	0.09642	10.3711	80.6726	7.7786
24	6.3412	0.1577	0.01498	66.7648	0.09498	10.5288	84.2997	8.0066
25	6.8485	0.1460	0.01368	73.1059	0.09368	10.6748	87.8041	8.2254
26	7.3964	0.1352	0.01251	79.9544	0.09251	10.8100	91.1842	8.4352
27	7.9881	0.1252	0.01145	87.3508	0.09145	10.9352	94.4390	8.6363
28	8.6271	0.1159	0.01049	95.3388	0.09049	11.0511	97.5687	8.8289
29	9.3173	0.1073	0.00962	103.9659	0.08962	11.1584	100.5738	9.0133
30	10.0627	0.0994	0.00883	113.2832	0.08883	11.2578	103.4558	9.1897
31	10.8677	0.0920	0.00811	123.3459	0.08811	11.3498	106.2163	9.3584
32	11.7371	0.0852	0.00745	134.2135	0.08745	11.4350	108.8575	9.5197
33	12.6760	0.0789	0.00685	145.9506	0.08685	11.5139	111.3819	9.6737
34	13.6901	0.0730	0.00630	158.6267	0.08630	11.5869	113.7924	9.8208
35	14.7853	0.0676	0.00580	172.3168	0.08580	11.6546	116.0920	9.9611
40	21.7245	0.0460	0.00386	259.0565	0.08386	11.9246	126.0422	10.5699
45	31.9204	0.0313	0.00259	386.5056	0.08259	12.1084	133.7331	11.0447
50	46.9016	0.0213	0.00174	573.7702	0.08174	12.2335	139.5928	11.4107
55	68.9139	0.0145	0.00118	848.9232	0.08118	12.3186	144.0065	11.6902
60	101.2571	0.0099	0.00080	1253.21	0.08080	12.3766	147.3000	11.9015
65	148.7798	0.0067	0.00054	1847.25	0.08054	12.4160	149.7387	12.0602
70	218.6064	0.0046	0.00037	2720.08	0.08037	12.4428	151.5326	12.1783
75	321.2045	0.0031	0.00025	4002.56	0.08025	12.4611	152.8448	12.2658
80	471.9548	0.0021	0.00017	5886.94	0.08017	12.4735	153.8001	12.3301
85	693.4565	0.0014	0.00012	8655.71	0.08012	12.4820	154.4925	12.3772
90	1018.92	0.0010	0.00008	12724	0.08008	12.4877	154.9925	12.4116
95	1497.12	0.0007	0.00005	18702	0.08005	12.4917	155.3524	12.4365
96	1616.89	0.0006	0.00005	20199	0.08005	12.4923	155.4112	12.4406
98	1885.94	0.0005	0.00004	23562	0.08004	12.4934	155.5176	12.4480
100	2199.76	0.0005	0.00004	27485	0.08004	12.4943	155.6107	12.4545

| 9% | | | | | | | TABLE 14 Discrete Cash Flow: Compound Interest Factors | | | | | | 9% |

	Single Payments		Uniform Series Payments				Arithmetic Gradients	
	Compound Amount	Present Worth	Sinking Fund	Compound Amount	Capital Recovery	Present Worth	Gradient Present Worth	Gradient Uniform Series
n	F/P	P/F	A/F	F/A	A/P	P/A	P/G	A/G
1	1.0900	0.9174	1.00000	1.0000	1.09000	0.9174		
2	1.1881	0.8417	0.47847	2.0900	0.56847	1.7591	0.8417	0.4785
3	1.2950	0.7722	0.30505	3.2781	0.39505	2.5313	2.3860	0.9426
4	1.4116	0.7084	0.21867	4.5731	0.30867	3.2397	4.5113	1.3925
5	1.5386	0.6499	0.16709	5.9847	0.25709	3.8897	7.1110	1.8282
6	1.6771	0.5963	0.13292	7.5233	0.22292	4.4859	10.0924	2.2498
7	1.8280	0.5470	0.10869	9.2004	0.19869	5.0330	13.3746	2.6574
8	1.9926	0.5019	0.09067	11.0285	0.18067	5.5348	16.8877	3.0512
9	2.1719	0.4604	0.07680	13.0210	0.16680	5.9952	20.5711	3.4312
10	2.3674	0.4224	0.06582	15.1929	0.15582	6.4177	24.3728	3.7978
11	2.5804	0.3875	0.05695	17.5603	0.14695	6.8052	28.2481	4.1510
12	2.8127	0.3555	0.04965	20.1407	0.13965	7.1607	32.1590	4.4910
13	3.0658	0.3262	0.04357	22.9534	0.13357	7.4869	36.0731	4.8182
14	3.3417	0.2992	0.03843	26.0192	0.12843	7.7862	39.9633	5.1326
15	3.6425	0.2745	0.03406	29.3609	0.12406	8.0607	43.8069	5.4346
16	3.9703	0.2519	0.03030	33.0034	0.12030	8.3126	47.5849	5.7245
17	4.3276	0.2311	0.02705	36.9737	0.11705	8.5436	51.2821	6.0024
18	4.7171	0.2120	0.02421	41.3013	0.11421	8.7556	54.8860	6.2687
19	5.1417	0.1945	0.02173	46.0185	0.11173	8.9501	58.3868	6.5236
20	5.6044	0.1784	0.01955	51.1601	0.10955	9.1285	61.7770	6.7674
21	6.1088	0.1637	0.01762	56.7645	0.10762	9.2922	65.0509	7.0006
22	6.6586	0.1502	0.01590	62.8733	0.10590	9.4424	68.2048	7.2232
23	7.2579	0.1378	0.01438	69.5319	0.10438	9.5802	71.2359	7.4357
24	7.9111	0.1264	0.01302	76.7898	0.10302	9.7066	74.1433	7.6384
25	8.6231	0.1160	0.01181	84.7009	0.10181	9.8226	76.9265	7.8316
26	9.3992	0.1064	0.01072	93.3240	0.10072	9.9290	79.5863	8.0156
27	10.2451	0.0976	0.00973	102.7231	0.09973	10.0266	82.1241	8.1906
28	11.1671	0.0895	0.00885	112.9682	0.09885	10.1161	84.5419	8.3571
29	12.1722	0.0822	0.00806	124.1354	0.09806	10.1983	86.8422	8.5154
30	13.2677	0.0754	0.00734	136.3075	0.09734	10.2737	89.0280	8.6657
31	14.4618	0.0691	0.00669	149.5752	0.09669	10.3428	91.1024	8.8083
32	15.7633	0.0634	0.00610	164.0370	0.09610	10.4062	93.0690	8.9436
33	17.1820	0.0582	0.00556	179.8003	0.09556	10.4644	94.9314	9.0718
34	18.7284	0.0534	0.00508	196.9823	0.09508	10.5178	96.6935	9.1933
35	20.4140	0.0490	0.00464	215.7108	0.09464	10.5668	98.3590	9.3083
40	31.4094	0.0318	0.00296	337.8824	0.09296	10.7574	105.3762	9.7957
45	48.3273	0.0207	0.00190	525.8587	0.09190	10.8812	110.5561	10.1603
50	74.3575	0.0134	0.00123	815.0836	0.09123	10.9617	114.3251	10.4295
55	114.4083	0.0087	0.00079	1260.09	0.09079	11.0140	117.0362	10.6261
60	176.0313	0.0057	0.00051	1944.79	0.09051	11.0480	118.9683	10.7683
65	270.8460	0.0037	0.00033	2998.29	0.09033	11.0701	120.3344	10.8702
70	416.7301	0.0024	0.00022	4619.22	0.09022	11.0844	121.2942	10.9427
75	641.1909	0.0016	0.00014	7113.23	0.09014	11.0938	121.9646	10.9940
80	986.5517	0.0010	0.00009	10951	0.09009	11.0998	122.4306	11.0299
85	1517.93	0.0007	0.00006	16855	0.09006	11.1038	122.7533	11.0551
90	2335.53	0.0004	0.00004	25939	0.09004	11.1064	122.9758	11.0726
95	3593.50	0.0003	0.00003	39917	0.09003	11.1080	123.1287	11.0847
96	3916.91	0.0003	0.00002	43510	0.09002	11.1083	123.1529	11.0866
98	4653.68	0.0002	0.00002	51696	0.09002	11.1087	123.1963	11.0900
100	5529.04	0.0002	0.00002	61423	0.09002	11.1091	123.2335	11.0930

10%				TABLE 15	Discrete Cash Flow: Compound Interest Factors			10%

	Single Payments		Uniform Series Payments				Arithmetic Gradients	
	Compound Amount	Present Worth	Sinking Fund	Compound Amount	Capital Recovery	Present Worth	Gradient Present Worth	Gradient Uniform Series
n	F/P	P/F	A/F	F/A	A/P	P/A	P/G	A/G
1	1.1000	0.9091	1.00000	1.0000	1.10000	0.9091		
2	1.2100	0.8264	0.47619	2.1000	0.57619	1.7355	0.8264	0.4762
3	1.3310	0.7513	0.30211	3.3100	0.40211	2.4869	2.3291	0.9366
4	1.4641	0.6830	0.21547	4.6410	0.31547	3.1699	4.3781	1.3812
5	1.6105	0.6209	0.16380	6.1051	0.26380	3.7908	6.8618	1.8101
6	1.7716	0.5645	0.12961	7.7156	0.22961	4.3553	9.6842	2.2236
7	1.9487	0.5132	0.10541	9.4872	0.20541	4.8684	12.7631	2.6216
8	2.1436	0.4665	0.08744	11.4359	0.18744	5.3349	16.0287	3.0045
9	2.3579	0.4241	0.07364	13.5795	0.17364	5.7590	19.4215	3.3724
10	2.5937	0.3855	0.06275	15.9374	0.16275	6.1446	22.8913	3.7255
11	2.8531	0.3505	0.05396	18.5312	0.15396	6.4951	26.3963	4.0641
12	3.1384	0.3186	0.04676	21.3843	0.14676	6.8137	29.9012	4.3884
13	3.4523	0.2897	0.04078	24.5227	0.14078	7.1034	33.3772	4.6988
14	3.7975	0.2633	0.03575	27.9750	0.13575	7.3667	36.8005	4.9955
15	4.1772	0.2394	0.03147	31.7725	0.13147	7.6061	40.1520	5.2789
16	4.5950	0.2176	0.02782	35.9497	0.12782	7.8237	43.4164	5.5493
17	5.0545	0.1978	0.02466	40.5447	0.12466	8.0216	46.5819	5.8071
18	5.5599	0.1799	0.02193	45.5992	0.12193	8.2014	49.6395	6.0526
19	6.1159	0.1635	0.01955	51.1591	0.11955	8.3649	52.5827	6.2861
20	6.7275	0.1486	0.01746	57.2750	0.11746	8.5136	55.4069	6.5081
21	7.4002	0.1351	0.01562	64.0025	0.11562	8.6487	58.1095	6.7189
22	8.1403	0.1228	0.01401	71.4027	0.11401	8.7715	60.6893	6.9189
23	8.9543	0.1117	0.01257	79.5430	0.11257	8.8832	63.1462	7.1085
24	9.8497	0.1015	0.01130	88.4973	0.11130	8.9847	65.4813	7.2881
25	10.8347	0.0923	0.01017	98.3471	0.11017	9.0770	67.6964	7.4580
26	11.9182	0.0839	0.00916	109.1818	0.10916	9.1609	69.7940	7.6186
27	13.1100	0.0763	0.00826	121.0999	0.10826	9.2372	71.7773	7.7704
28	14.4210	0.0693	0.00745	134.2099	0.10745	9.3066	73.6495	7.9137
29	15.8631	0.0630	0.00673	148.6309	0.10673	9.3696	75.4146	8.0489
30	17.4494	0.0573	0.00608	164.4940	0.10608	9.4269	77.0766	8.1762
31	19.1943	0.0521	0.00550	181.9434	0.10550	9.4790	78.6395	8.2962
32	21.1138	0.0474	0.00497	201.1378	0.10497	9.5264	80.1078	8.4091
33	23.2252	0.0431	0.00450	222.2515	0.10450	9.5694	81.4856	8.5152
34	25.5477	0.0391	0.00407	245.4767	0.10407	9.6086	82.7773	8.6149
35	28.1024	0.0356	0.00369	271.0244	0.10369	9.6442	83.9872	8.7086
40	45.2593	0.0221	0.00226	442.5926	0.10226	9.7791	88.9525	9.0962
45	72.8905	0.0137	0.00139	718.9048	0.10139	9.8628	92.4544	9.3740
50	117.3909	0.0085	0.00086	1163.91	0.10086	9.9148	94.8889	9.5704
55	189.0591	0.0053	0.00053	1880.59	0.10053	9.9471	96.5619	9.7075
60	304.4816	0.0033	0.00033	3034.82	0.10033	9.9672	97.7010	9.8023
65	490.3707	0.0020	0.00020	4893.71	0.10020	9.9796	98.4705	9.8672
70	789.7470	0.0013	0.00013	7887.47	0.10013	9.9873	98.9870	9.9113
75	1271.90	0.0008	0.00008	12709	0.10008	9.9921	99.3317	9.9410
80	2048.40	0.0005	0.00005	20474	0.10005	9.9951	99.5606	9.9609
85	3298.97	0.0003	0.00003	32980	0.10003	9.9970	99.7120	9.9742
90	5313.02	0.0002	0.00002	53120	0.10002	9.9981	99.8118	9.9831
95	8556.68	0.0001	0.00001	85557	0.10001	9.9988	99.8773	9.9889
96	9412.34	0.0001	0.00001	94113	0.10001	9.9989	99.8874	9.9898
98	11389	0.0001	0.00001		0.10001	9.9991	99.9052	9.9914
100	13781	0.0001	0.00001		0.10001	9.9993	99.9202	9.9927

11%			TABLE 16	Discrete Cash Flow: Compound Interest Factors				11%
	Single Payments			Uniform Series Payments			Arithmetic Gradients	
n	Compound Amount F/P	Present Worth P/F	Sinking Fund A/F	Compound Amount F/A	Capital Recovery A/P	Present Worth P/A	Gradient Present Worth P/G	Gradient Uniform Series A/G
1	1.1100	0.9009	1.00000	1.0000	1.11000	0.9009		
2	1.2321	0.8116	0.47393	2.1100	0.58393	1.7125	0.8116	0.4739
3	1.3676	0.7312	0.29921	3.3421	0.40921	2.4437	2.2740	0.9306
4	1.5181	0.6587	0.21233	4.7097	0.32233	3.1024	4.2502	1.3700
5	1.6851	0.5935	0.16057	6.2278	0.27057	3.6959	6.6240	1.7923
6	1.8704	0.5346	0.12638	7.9129	0.23638	4.2305	9.2972	2.1976
7	2.0762	0.4817	0.10222	9.7833	0.21222	4.7122	12.1872	2.5863
8	2.3045	0.4339	0.08432	11.8594	0.19432	5.1461	15.2246	2.9585
9	2.5580	0.3909	0.07060	14.1640	0.18060	5.5370	18.3520	3.3144
10	2.8394	0.3522	0.05980	16.7220	0.16980	5.8892	21.5217	3.6544
11	3.1518	0.3173	0.05112	19.5614	0.16112	6.2065	24.6945	3.9788
12	3.4985	0.2858	0.04403	22.7132	0.15403	6.4924	27.8388	4.2879
13	3.8833	0.2575	0.03815	26.2116	0.14815	6.7499	30.9290	4.5822
14	4.3104	0.2320	0.03323	30.0949	0.14323	6.9819	33.9449	4.8619
15	4.7846	0.2090	0.02907	34.4054	0.13907	7.1909	36.8709	5.1275
16	5.3109	0.1883	0.02552	39.1899	0.13552	7.3792	39.6953	5.3794
17	5.8951	0.1696	0.02247	44.5008	0.13247	7.5488	42.4095	5.6180
18	6.5436	0.1528	0.01984	50.3959	0.12984	7.7016	45.0074	5.8439
19	7.2633	0.1377	0.01756	56.9395	0.12756	7.8393	47.4856	6.0574
20	8.0623	0.1240	0.01558	64.2028	0.12558	7.9633	49.8423	6.2590
21	8.9492	0.1117	0.01384	72.2651	0.12384	8.0751	52.0771	6.4491
22	9.9336	0.1007	0.01231	81.2143	0.12231	8.1757	54.1912	6.6283
23	11.0263	0.0907	0.01097	91.1479	0.12097	8.2664	56.1864	6.7969
24	12.2392	0.0817	0.00979	102.1742	0.11979	8.3481	58.0656	6.9555
25	13.5855	0.0736	0.00874	114.4133	0.11874	8.4217	59.8322	7.1045
26	15.0799	0.0663	0.00781	127.9988	0.11781	8.4881	61.4900	7.2443
27	16.7386	0.0597	0.00699	143.0786	0.11699	8.5478	63.0433	7.3754
28	18.5799	0.0538	0.00626	159.8173	0.11626	8.6016	64.4965	7.4982
29	20.6237	0.0485	0.00561	178.3972	0.11561	8.6501	65.8542	7.6131
30	22.8923	0.0437	0.00502	199.0209	0.11502	8.6938	67.1210	7.7206
31	25.4104	0.0394	0.00451	221.9132	0.11451	8.7331	68.3016	7.8210
32	28.2056	0.0355	0.00404	247.3236	0.11404	8.7686	69.4007	7.9147
33	31.3082	0.0319	0.00363	275.5292	0.11363	8.8005	70.4228	8.0021
34	34.7521	0.0288	0.00326	306.8374	0.11326	8.8293	71.3724	8.0836
35	38.5749	0.0259	0.00293	341.5896	0.11293	8.8552	72.2538	8.1594
40	65.0009	0.0154	0.00172	581.8261	0.11172	8.9511	75.7789	8.4659
45	109.5302	0.0091	0.00101	986.6386	0.11101	9.0079	78.1551	8.6763
50	184.5648	0.0054	0.00060	1668.77	0.11060	9.0417	79.7341	8.8185
55	311.0025	0.0032	0.00035	2818.20	0.11035	9.0617	80.7712	8.9135
60	524.0572	0.0019	0.00021	4755.07	0.11021	9.0736	81.4461	8.9762
65	883.0669	0.0011	0.00012	8018.79	0.11012	9.0806	81.8819	9.0172
70	1488.02	0.0007	0.00007	13518	0.11007	9.0848	82.1614	9.0438
75	2507.40	0.0004	0.00004	22785	0.11004	9.0873	82.3397	9.0610
80	4225.11	0.0002	0.00003	38401	0.11003	9.0888	82.4529	9.0720
85	7119.56	0.0001	0.00002	64714	0.11002	9.0896	82.5245	9.0790

12%			TABLE **17**	Discrete Cash Flow: Compound Interest Factors				12%
	Single Payments		Uniform Series Payments				Arithmetic Gradients	
	Compound Amount F/P	Present Worth P/F	Sinking Fund A/F	Compound Amount F/A	Capital Recovery A/P	Present Worth P/A	Gradient Present Worth P/G	Gradient Uniform Series A/G
n								
1	1.1200	0.8929	1.00000	1.0000	1.12000	0.8929		
2	1.2544	0.7972	0.47170	2.1200	0.59170	1.6901	0.7972	0.4717
3	1.4049	0.7118	0.29635	3.3744	0.41635	2.4018	2.2208	0.9246
4	1.5735	0.6355	0.20923	4.7793	0.32923	3.0373	4.1273	1.3589
5	1.7623	0.5674	0.15741	6.3528	0.27741	3.6048	6.3970	1.7746
6	1.9738	0.5066	0.12323	8.1152	0.24323	4.1114	8.9302	2.1720
7	2.2107	0.4523	0.09912	10.0890	0.21912	4.5638	11.6443	2.5512
8	2.4760	0.4039	0.08130	12.2997	0.20130	4.9676	14.4714	2.9131
9	2.7731	0.3606	0.06768	14.7757	0.18768	5.3282	17.3563	3.2574
10	3.1058	0.3220	0.05698	17.5487	0.17698	5.6502	20.2541	3.5847
11	3.4785	0.2875	0.04842	20.6546	0.16842	5.9377	23.1288	3.8953
12	3.8960	0.2567	0.04144	24.1331	0.16144	6.1944	25.9523	4.1897
13	4.3635	0.2292	0.03568	28.0291	0.15568	6.4235	28.7024	4.4683
14	4.8871	0.2046	0.03087	32.3926	0.15087	6.6282	31.3624	4.7317
15	5.4736	0.1827	0.02682	37.2797	0.14682	6.8109	33.9202	4.9803
16	6.1304	0.1631	0.02339	42.7533	0.14339	6.9740	36.3670	5.2147
17	6.8660	0.1456	0.02046	48.8837	0.14046	7.1196	38.6973	5.4353
18	7.6900	0.1300	0.01794	55.7497	0.13794	7.2497	40.9080	5.6427
19	8.6128	0.1161	0.01576	63.4397	0.13576	7.3658	42.9979	5.8375
20	9.6463	0.1037	0.01388	72.0524	0.13388	7.4694	44.9676	6.0202
21	10.8038	0.0926	0.01224	81.6987	0.13224	7.5620	46.8188	6.1913
22	12.1003	0.0826	0.01081	92.5026	0.13081	7.6446	48.5543	6.3514
23	13.5523	0.0738	0.00956	104.6029	0.12956	7.7184	50.1776	6.5010
24	15.1786	0.0659	0.00846	118.1552	0.12846	7.7843	51.6929	6.6406
25	17.0001	0.0588	0.00750	133.3339	0.12750	7.8431	53.1046	6.7708
26	19.0401	0.0525	0.00665	150.3339	0.12665	7.8957	54.4177	6.8921
27	21.3249	0.0469	0.00590	169.3740	0.12590	7.9426	55.6369	7.0049
28	23.8839	0.0419	0.00524	190.6989	0.12524	7.9844	56.7674	7.1098
29	26.7499	0.0374	0.00466	214.5828	0.12466	8.0218	57.8141	7.2071
30	29.9599	0.0334	0.00414	241.3327	0.12414	8.0552	58.7821	7.2974
31	33.5551	0.0298	0.00369	271.2926	0.12369	8.0850	59.6761	7.3811
32	37.5817	0.0266	0.00328	304.8477	0.12328	8.1116	60.5010	7.4586
33	42.0915	0.0238	0.00292	342.4294	0.12292	8.1354	61.2612	7.5302
34	47.1425	0.0212	0.00260	384.5210	0.12260	8.1566	61.9612	7.5965
35	52.7996	0.0189	0.00232	431.6635	0.12232	8.1755	62.6052	7.6577
40	93.0510	0.0107	0.00130	767.0914	0.12130	8.2438	65.1159	7.8988
45	163.9876	0.0061	0.0074	1358.23	0.12074	8.2825	66.7342	8.0572
50	289.0022	0.0035	0.00042	2400.02	0.12042	8.3045	67.7624	8.1597
55	509.3206	0.0020	0.00024	4236.01	0.12024	8.3170	68.4082	8.2251
60	897.5969	0.0011	0.00013	7471.64	0.12013	8.3240	68.8100	8.2664
65	1581.87	0.0006	0.00008	13174	0.12008	8.3281	69.0581	8.2922
70	2787.80	0.0004	0.00004	23223	0.12004	8.3303	69.2103	8.3082
75	4913.06	0.0002	0.00002	40934	0.12002	8.3316	69.3031	8.3181
80	8658.48	0.0001	0.00001	72146	0.12001	8.3324	69.3594	8.3241
85	15259	0.0001	0.00001		0.12001	8.3328	69.3935	8.3278

TABLE 18 Discrete Cash Flow: Compound Interest Factors

	Single Payments		Uniform Series Payments				Arithmetic Gradients	
	Compound Amount	Present Worth	Sinking Fund	Compound Amount	Capital Recovery	Present Worth	Gradient Present Worth	Gradient Uniform Series
n	F/P	P/F	A/F	F/A	A/P	P/A	P/G	A/G
1	1.1400	0.8772	1.00000	1.0000	1.14000	0.8772		
2	1.2996	0.7695	0.46729	2.1400	0.60729	1.6467	0.7695	0.4673
3	1.4815	0.6750	0.29073	3.4396	0.43073	2.3216	2.1194	0.9129
4	1.6890	0.5921	0.20320	4.9211	0.34320	2.9137	3.8957	1.3370
5	1.9254	0.5194	0.15128	6.6101	0.29128	3.4331	5.9731	1.7399
6	2.1950	0.4556	0.11716	8.5355	0.25716	3.8887	8.2511	2.1218
7	2.5023	0.3996	0.09319	10.7305	0.23319	4.2883	10.6489	2.4832
8	2.8526	0.3506	0.07557	13.2328	0.21557	4.6389	13.1028	2.8246
9	3.2519	0.3075	0.06217	16.0853	0.20217	4.9464	15.5629	3.1463
10	3.7072	0.2697	0.05171	19.3373	0.19171	5.2161	17.9906	3.4490
11	4.2262	0.2366	0.04339	23.0445	0.18339	5.4527	20.3567	3.7333
12	4.8179	0.2076	0.03667	27.2707	0.17667	5.6603	22.6399	3.9998
13	5.4924	0.1821	0.03116	32.0887	0.17116	5.8424	24.8247	4.2491
14	6.2613	0.1597	0.02661	37.5811	0.16661	6.0021	26.9009	4.4819
15	7.1379	0.1401	0.02281	43.8424	0.16281	6.1422	28.8623	4.6990
16	8.1372	0.1229	0.01962	50.9804	0.15962	6.2651	30.7057	4.9011
17	9.2765	0.1078	0.01692	59.1176	0.15692	6.3729	32.4305	5.0888
18	10.5752	0.0946	0.01462	68.3941	0.15462	6.4674	34.0380	5.2630
19	12.0557	0.0829	0.01266	78.9692	0.15266	6.5504	35.5311	5.4243
20	13.7435	0.0728	0.01099	91.0249	0.15099	6.6231	36.9135	5.5734
21	15.6676	0.0638	0.00954	104.7684	0.14954	6.6870	38.1901	5.7111
22	17.8610	0.0560	0.00830	120.4360	0.14830	6.7429	39.3658	5.8381
23	20.3616	0.0491	0.00723	138.2970	0.14723	6.7921	40.4463	5.9549
24	23.2122	0.0431	0.00630	158.6586	0.14630	6.8351	41.4371	6.0624
25	26.4619	0.0378	0.00550	181.8708	0.14550	6.8729	42.3441	6.1610
26	30.1666	0.0331	0.00480	208.3327	0.14480	6.9061	43.1728	6.2514
27	34.3899	0.0291	0.00419	238.4993	0.14419	6.9352	43.9289	6.3342
28	39.2045	0.0255	0.00366	272.8892	0.14366	6.9607	44.6176	6.4100
29	44.6931	0.0224	0.00320	312.0937	0.14320	6.9830	45.2441	6.4791
30	50.9502	0.0196	0.00280	356.7868	0.14280	7.0027	45.8132	6.5423
31	58.0832	0.0172	0.00245	407.7370	0.14245	7.0199	46.3297	6.5998
32	66.2148	0.0151	0.00215	465.8202	0.14215	7.0350	46.7979	6.6522
33	75.4849	0.0132	0.00188	532.0350	0.14188	7.0482	47.2218	6.6998
34	86.0528	0.0116	0.00165	607.5199	0.14165	7.0599	47.6053	6.7431
35	98.1002	0.0102	0.00144	693.5727	0.14144	7.0700	47.9519	6.7824
40	188.8835	0.0053	0.00075	1342.03	0.14075	7.1050	49.2376	6.9300
45	363.6791	0.0027	0.00039	2590.56	0.14039	7.1232	49.9963	7.0188
50	700.2330	0.0014	0.00020	4994.52	0.14020	7.1327	50.4375	7.0714
55	1348.24	0.0007	0.00010	9623.13	0.14010	7.1376	50.6912	7.1020
60	2595.92	0.0004	0.00005	18535	0.14005	7.1401	50.8357	7.1197
65	4998.22	0.0002	0.00003	35694	0.14003	7.1414	50.9173	7.1298
70	9623.64	0.0001	0.00001	68733	0.14001	7.1421	50.9632	7.1356
75	18530	0.0001	0.00001		0.14001	7.1425	50.9887	7.1388
80	35677				0.14000	7.1427	51.0030	7.1406
85	68693				0.14000	7.1428	51.0108	7.1416

15%			TABLE 19	Discrete Cash Flow: Compound Interest Factors				15%
	Single Payments			Uniform Series Payments			Arithmetic Gradients	
	Compound Amount	Present Worth	Sinking Fund	Compound Amount	Capital Recovery	Present Worth	Gradient Present Worth	Gradient Uniform Series
n	F/P	P/F	A/F	F/A	A/P	P/A	P/G	A/G
1	1.1500	0.8696	1.00000	1.0000	1.15000	0.8696		
2	1.3225	0.7561	0.46512	2.1500	0.61512	1.6257	0.7561	0.4651
3	1.5209	0.6575	0.28798	3.4725	0.43798	2.2832	2.0712	0.9071
4	1.7490	0.5718	0.20027	4.9934	0.35027	2.8550	3.7864	1.3263
5	2.0114	0.4972	0.14832	6.7424	0.29832	3.3522	5.7751	1.7228
6	2.3131	0.4323	0.11424	8.7537	0.26424	3.7845	7.9368	2.0972
7	2.6600	0.3759	0.09036	11.0668	0.24036	4.1604	10.1924	2.4498
8	3.0590	0.3269	0.07285	13.7268	0.22285	4.4873	12.4807	2.7813
9	3.5179	0.2843	0.05957	16.7858	0.20957	4.7716	14.7548	3.0922
10	4.0456	0.2472	0.04925	20.3037	0.19925	5.0188	16.9795	3.3832
11	4.6524	0.2149	0.04107	24.3493	0.19107	5.2337	19.1289	3.6549
12	5.3503	0.1869	0.03448	29.0017	0.18448	5.4206	21.1849	3.9082
13	6.1528	0.1625	0.02911	34.3519	0.17911	5.5831	23.1352	4.1438
14	7.0757	0.1413	0.02469	40.5047	0.17469	5.7245	24.9725	4.3624
15	8.1371	0.1229	0.02102	47.5804	0.17102	5.8474	26.6930	4.5650
16	9.3576	0.1069	0.01795	55.7175	0.16795	5.9542	28.2960	4.7522
17	10.7613	0.0929	0.01537	65.0751	0.16537	6.0472	29.7828	4.9251
18	12.3755	0.0808	0.01319	75.8364	0.16319	6.1280	31.1565	5.0843
19	14.2318	0.0703	0.01134	88.2118	0.16134	6.1982	32.4213	5.2307
20	16.3665	0.0611	0.00976	102.4436	0.15976	6.2593	33.5822	5.3651
21	18.8215	0.0531	0.00842	118.8101	0.15842	6.3125	34.6448	5.4883
22	21.6447	0.0462	0.00727	137.6316	0.15727	6.3587	35.6150	5.6010
23	24.8915	0.0402	0.00628	159.2764	0.15628	6.3988	36.4988	5.7040
24	28.6252	0.0349	0.00543	184.1678	0.15543	6.4338	37.3023	5.7979
25	32.9190	0.0304	0.00470	212.7930	0.15470	6.4641	38.0314	5.8834
26	37.8568	0.0264	0.00407	245.7120	0.15407	6.4906	38.6918	5.9612
27	43.5353	0.0230	0.00353	283.5688	0.15353	6.5135	39.2890	6.0319
28	50.0656	0.0200	0.00306	327.1041	0.15306	6.5335	39.8283	6.0960
29	57.5755	0.0174	0.00265	377.1697	0.15265	6.5509	40.3146	6.1541
30	66.2118	0.0151	0.00230	434.7451	0.15230	6.5660	40.7526	6.2066
31	76.1435	0.0131	0.00200	500.9569	0.15200	6.5791	41.1466	6.2541
32	87.5651	0.0114	0.00173	577.1005	0.15173	6.5905	41.5006	6.2970
33	100.6998	0.0099	0.00150	664.6655	0.15150	6.6005	41.8184	6.3357
34	115.8048	0.0086	0.00131	765.3654	0.15131	6.6091	42.1033	6.3705
35	133.1755	0.0075	0.00113	881.1702	0.15113	6.6166	42.3586	6.4019
40	267.8635	0.0037	0.00056	1779.09	0.15056	6.6418	43.2830	6.5168
45	538.7693	0.0019	0.00028	3585.13	0.15028	6.6543	43.8051	6.5830
50	1083.66	0.0009	0.00014	7217.72	0.15014	6.6605	44.0958	6.6205
55	2179.62	0.0005	0.00007	14524	0.15007	6.6636	44.2558	6.6414
60	4384.00	0.0002	0.00003	29220	0.15003	6.6651	44.3431	6.6530
65	8817.79	0.0001	0.00002	58779	0.15002	6.6659	44.3903	6.6593
70	17736	0.0001	0.00001		0.15001	6.6663	44.4156	6.6627
75	35673				0.15000	6.6665	44.4292	6.6646
80	71751				0.15000	6.6666	44.4364	6.6656
85					0.15000	6.6666	44.4402	6.6661

16%				TABLE 20 Discrete Cash Flow: Compound Interest Factors				16%
	Single Payments		Uniform Series Payments				Arithmetic Gradients	
n	Compound Amount F/P	Present Worth P/F	Sinking Fund A/F	Compound Amount F/A	Capital Recovery A/P	Present Worth P/A	Gradient Present Worth P/G	Gradient Uniform Series A/G
1	1.1600	0.8621	1.00000	1.0000	1.16000	0.8621		
2	1.3456	0.7432	0.46296	2.1600	0.62296	1.6052	0.7432	0.4630
3	1.5609	0.6407	0.28526	3.5056	0.44526	2.2459	2.0245	0.9014
4	1.8106	0.5523	0.19738	5.0665	0.35738	2.7982	3.6814	1.3156
5	2.1003	0.4761	0.14541	6.8771	0.30541	3.2743	5.5858	1.7060
6	2.4364	0.4104	0.11139	8.9775	0.27139	3.6847	7.6380	2.0729
7	2.8262	0.3538	0.08761	11.4139	0.24761	4.0386	9.7610	2.4169
8	3.2784	0.3050	0.07022	14.2401	0.23022	4.3436	11.8962	2.7388
9	3.8030	0.2630	0.05708	17.5185	0.21708	4.6065	13.9998	3.0391
10	4.4114	0.2267	0.04690	21.3215	0.20690	4.8332	16.0399	3.3187
11	5.1173	0.1954	0.03886	25.7329	0.19886	5.0286	17.9941	3.5783
12	5.9360	0.1685	0.03241	30.8502	0.19241	5.1971	19.8472	3.8189
13	6.8858	0.1452	0.02718	36.7862	0.18718	5.3423	21.5899	4.0413
14	7.9875	0.1252	0.02290	43.6720	0.18290	5.4675	23.2175	4.2464
15	9.2655	0.1079	0.01936	51.6595	0.17936	5.5755	24.7284	4.4352
16	10.7480	0.0930	0.01641	60.9250	0.17641	5.6685	26.1241	4.6086
17	12.4677	0.0802	0.01395	71.6730	0.17395	5.7487	27.4074	4.7676
18	14.4625	0.0691	0.01188	84.1407	0.17188	5.8178	28.5828	4.9130
19	16.7765	0.0596	0.01014	98.6032	0.17014	5.8775	29.6557	5.0457
20	19.4608	0.0514	0.00867	115.3797	0.16867	5.9288	30.6321	5.1666
22	26.1864	0.0382	0.00635	157.4150	0.16635	6.0113	32.3200	5.3765
24	35.2364	0.0284	0.00467	213.9776	0.16467	6.0726	33.6970	5.5490
26	47.4141	0.0211	0.00345	290.0883	0.16345	6.1182	34.8114	5.6898
28	63.8004	0.0157	0.00255	392.5028	0.16255	6.1520	35.7073	5.8041
30	85.8499	0.0116	0.00189	530.3117	0.16189	6.1772	36.4234	5.8964
32	115.5196	0.0087	0.00140	715.7475	0.16140	6.1959	36.9930	5.9706
34	155.4432	0.0064	0.00104	965.2698	0.16104	6.2098	37.4441	6.0299
35	180.3141	0.0055	0.00089	1120.71	0.16089	6.2153	37.6327	6.0548
36	209.1643	0.0048	0.00077	1301.03	0.16077	6.2201	37.8000	6.0771
38	281.4515	0.0036	0.00057	1752.82	0.16057	6.2278	38.0799	6.1145
40	378.7212	0.0026	0.00042	2360.76	0.16042	6.2335	38.2992	6.1441
45	795.4438	0.0013	0.00020	4965.27	0.16020	6.2421	38.6598	6.1934
50	1670.70	0.0006	0.00010	10436	0.16010	6.2463	38.8521	6.2201
55	3509.05	0.0003	0.00005	21925	0.16005	6.2482	38.9534	6.2343
60	7370.20	0.0001	0.00002	46058	0.16002	6.2492	39.0063	6.2419

18%			TABLE 21	Discrete Cash Flow: Compound Interest Factors				18%
	Single Payments			Uniform Series Payments			Arithmetic Gradients	
n	Compound Amount F/P	Present Worth P/F	Sinking Fund A/F	Compound Amount F/A	Capital Recovery A/P	Present Worth P/A	Gradient Present Worth P/G	Gradient Uniform Series A/G
1	1.1800	0.8475	1.00000	1.0000	1.18000	0.8475		
2	1.3924	0.7182	0.45872	2.1800	0.63872	1.5656	0.7182	0.4587
3	1.6430	0.6086	0.27992	3.5724	0.45992	2.1743	1.9354	0.8902
4	1.9388	0.5158	0.19174	5.2154	0.37174	2.6901	3.4828	1.2947
5	2.2878	0.4371	0.13978	7.1542	0.31978	3.1272	5.2312	1.6728
6	2.6996	0.3704	0.10591	9.4420	0.28591	3.4976	7.0834	2.0252
7	3.1855	0.3139	0.08236	12.1415	0.26236	3.8115	8.9670	2.3526
8	3.7589	0.2660	0.06524	15.3270	0.24524	4.0776	10.8292	2.6558
9	4.4355	0.2255	0.05239	19.0859	0.23239	4.3030	12.6329	2.9358
10	5.2338	0.1911	0.04251	23.5213	0.22251	4.4941	14.3525	3.1936
11	6.1759	0.1619	0.03478	28.7551	0.21478	4.6560	15.9716	3.4303
12	7.2876	0.1372	0.02863	34.9311	0.20863	4.7932	17.4811	3.6470
13	8.5994	0.1163	0.02369	42.2187	0.20369	4.9095	18.8765	3.8449
14	10.1472	0.0985	0.01968	50.8180	0.19968	5.0081	20.1576	4.0250
15	11.9737	0.0835	0.01640	60.9653	0.19640	5.0916	21.3269	4.1887
16	14.1290	0.0708	0.01371	72.9390	0.19371	5.1624	22.3885	4.3369
17	16.6722	0.0600	0.01149	87.0680	0.19149	5.2223	23.3482	4.4708
18	19.6733	0.0508	0.00964	103.7403	0.18964	5.2732	24.2123	4.5916
19	23.2144	0.0431	0.00810	123.4135	0.18810	5.3162	24.9877	4.7003
20	27.3930	0.0365	0.00682	146.6280	0.18682	5.3527	25.6813	4.7978
22	38.1421	0.0262	0.00485	206.3448	0.18485	5.4099	26.8506	4.9632
24	53.1090	0.0188	0.00345	289.4945	0.18345	5.4509	27.7725	5.0950
26	73.9490	0.0135	0.00247	405.2721	0.18247	5.4804	28.4935	5.1991
28	102.9666	0.0097	0.00177	566.4809	0.18177	5.5016	29.0537	5.2810
30	143.3706	0.0070	0.00126	790.9480	0.18126	5.5168	29.4864	5.3448
32	199.6293	0.0050	0.00091	1103.50	0.18091	5.5277	29.8191	5.3945
34	277.9638	0.0036	0.00065	1538.69	0.18065	5.5356	30.0736	5.4328
35	327.9973	0.0030	0.00055	1816.65	0.18055	5.5386	30.1773	5.4485
36	387.0368	0.0026	0.00047	2144.65	0.18047	5.5412	30.2677	5.4623
38	538.9100	0.0019	0.00033	2988.39	0.18033	5.5452	30.4152	5.4849
40	750.3783	0.0013	0.00024	4163.21	0.18024	5.5482	30.5269	5.5022
45	1716.68	0.0006	0.00010	9531.58	0.18010	5.5523	30.7006	5.5293
50	3927.36	0.0003	0.00005	21813	0.18005	5.5541	30.7856	5.5428
55	8984.84	0.0001	0.00002	49910	0.18002	5.5549	30.8268	5.5494
60	20555			114190	0.18001	5.5553	30.8465	5.5526

20%			TABLE 22	Discrete Cash Flow: Compound Interest Factors				20%
	Single Payments		Uniform Series Payments				Arithmetic Gradients	
n	Compound Amount F/P	Present Worth P/F	Sinking Fund A/F	Compound Amount F/A	Capital Recovery A/P	Present Worth P/A	Gradient Present Worth P/G	Gradient Uniform Series A/G
1	1.2000	0.8333	1.00000	1.0000	1.20000	0.8333		
2	1.4400	0.6944	0.45455	2.2000	0.65455	1.5278	0.6944	0.4545
3	1.7280	0.5787	0.27473	3.6400	0.47473	2.1065	1.8519	0.8791
4	2.0736	0.4823	0.18629	5.3680	0.38629	2.5887	3.2986	1.2742
5	2.4883	0.4019	0.13438	7.4416	0.33438	2.9906	4.9061	1.6405
6	2.9860	0.3349	0.10071	9.9299	0.30071	3.3255	6.5806	1.9788
7	3.5832	0.2791	0.07742	12.9159	0.27742	3.6046	8.2551	2.2902
8	4.2998	0.2326	0.06061	16.4991	0.26061	3.8372	9.8831	2.5756
9	5.1598	0.1938	0.04808	20.7989	0.24808	4.0310	11.4335	2.8364
10	6.1917	0.1615	0.03852	25.9587	0.23852	4.1925	12.8871	3.0739
11	7.4301	0.1346	0.03110	32.1504	0.23110	4.3271	14.2330	3.2893
12	8.9161	0.1122	0.02526	39.5805	0.22526	4.4392	15.4667	3.4841
13	10.6993	0.0935	0.02062	48.4966	0.22062	4.5327	16.5883	3.6597
14	12.8392	0.0779	0.01689	59.1959	0.21689	4.6106	17.6008	3.8175
15	15.4070	0.0649	0.01388	72.0351	0.21388	4.6755	18.5095	3.9588
16	18.4884	0.0541	0.01144	87.4421	0.21144	4.7296	19.3208	4.0851
17	22.1861	0.0451	0.00944	105.9306	0.20944	4.7746	20.0419	4.1976
18	26.6233	0.0376	0.00781	128.1167	0.20781	4.8122	20.6805	4.2975
19	31.9480	0.0313	0.00646	154.7400	0.20646	4.8435	21.2439	4.3861
20	38.3376	0.0261	0.00536	186.6880	0.20536	4.8696	21.7395	4.4643
22	55.2061	0.0181	0.00369	271.0307	0.20369	4.9094	22.5546	4.5941
24	79.4968	0.0126	0.00255	392.4842	0.20255	4.9371	23.1760	4.6943
26	114.4755	0.0087	0.00176	567.3773	0.20176	4.9563	23.6460	4.7709
28	164.8447	0.0061	0.00122	819.2233	0.20122	4.9697	23.9991	4.8291
30	237.3763	0.0042	0.00085	1181.88	0.20085	4.9789	24.2628	4.8731
32	341.8219	0.0029	0.00059	1704.11	0.20059	4.9854	24.4588	4.9061
34	492.2235	0.0020	0.00041	2456.12	0.20041	4.9898	24.6038	4.9308
35	590.6682	0.0017	0.00034	2948.34	0.20034	4.9915	24.6614	4.9406
36	708.8019	0.0014	0.00028	3539.01	0.20028	4.9929	24.7108	4.9491
38	1020.67	0.0010	0.00020	5098.37	0.20020	4.9951	24.7894	4.9627
40	1469.77	0.0007	0.00014	7343.86	0.20014	4.9966	24.8469	4.9728
45	3657.26	0.0003	0.00005	18281	0.20005	4.9986	24.9316	4.9877
50	9100.44	0.0001	0.00002	45497	0.20002	4.9995	24.9698	4.9945
55	22645		0.00001		0.20001	4.9998	24.9868	4.9976

22%			TABLE 23	Discrete Cash Flow: Compound Interest Factors				22%
	Single Payments		Uniform Series Payments				Arithmetic Gradients	
n	Compound Amount F/P	Present Worth P/F	Sinking Fund A/F	Compound Amount F/A	Capital Recovery A/P	Present Worth P/A	Gradient Present Worth P/G	Gradient Uniform Series A/G
1	1.2200	0.8197	1.00000	1.0000	1.22000	0.8197		
2	1.4884	0.6719	0.45045	2.2200	0.67045	1.4915	0.6719	0.4505
3	1.8158	0.5507	0.26966	3.7084	0.48966	2.0422	1.7733	0.8683
4	2.2153	0.4514	0.18102	5.5242	0.40102	2.4936	3.1275	1.2542
5	2.7027	0.3700	0.12921	7.7396	0.34921	2.8636	4.6075	1.6090
6	3.2973	0.3033	0.09576	10.4423	0.31576	3.1669	6.1239	1.9337
7	4.0227	0.2486	0.07278	13.7396	0.29278	3.4155	7.6154	2.2297
8	4.9077	0.2038	0.05630	17.7623	0.27630	3.6193	9.0417	2.4982
9	5.9874	0.1670	0.04411	22.6700	0.26411	3.7863	10.3779	2.7409
10	7.3046	0.1369	0.03489	28.6574	0.25489	3.9232	11.6100	2.9593
11	8.9117	0.1122	0.02781	35.9620	0.24781	4.0354	12.7321	3.1551
12	10.8722	0.0920	0.02228	44.8737	0.24228	4.1274	13.7438	3.3299
13	13.2641	0.0754	0.01794	55.7459	0.23794	4.2028	14.6485	3.4855
14	16.1822	0.0618	0.01449	69.0100	0.23449	4.2646	15.4519	3.6233
15	19.7423	0.0507	0.01174	85.1922	0.23174	4.3152	16.1610	3.7451
16	24.0856	0.0415	0.00953	104.9345	0.22953	4.3567	16.7838	3.8524
17	29.3844	0.0340	0.00775	129.0201	0.22775	4.3908	17.3283	3.9465
18	35.8490	0.0279	0.00631	158.4045	0.22631	4.4187	17.8025	4.0289
19	43.7358	0.0229	0.00515	194.2535	0.22515	4.4415	18.2141	4.1009
20	53.3576	0.0187	0.00420	237.9893	0.22420	4.4603	18.5702	4.1635
22	79.4175	0.0126	0.00281	356.4432	0.22281	4.4882	19.1418	4.2649
24	118.2050	0.0085	0.00188	532.7501	0.22188	4.5070	19.5635	4.3407
26	175.9364	0.0057	0.00126	795.1653	0.22126	4.5196	19.8720	4.3968
28	261.8637	0.0038	0.00084	1185.74	0.22084	4.5281	20.0962	4.4381
30	389.7579	0.0026	0.00057	1767.08	0.22057	4.5338	20.2583	4.4683
32	580.1156	0.0017	0.00038	2632.34	0.22038	4.5376	20.3748	4.4902
34	863.4441	0.0012	0.00026	3920.20	0.22026	4.5402	20.4582	4.5060
35	1053.40	0.0009	0.00021	4783.64	0.22021	4.5411	20.4905	4.5122
36	1285.15	0.0008	0.00017	5837.05	0.22017	4.5419	20.5178	4.5174
38	1912.82	0.0005	0.00012	8690.08	0.22012	4.5431	20.5601	4.5256
40	2847.04	0.0004	0.00008	12937	0.22008	4.5439	20.5900	4.5314
45	7694.71	0.0001	0.00003	34971	0.22003	4.5449	20.6319	4.5396
50	20797		0.00001	94525	0.22001	4.5452	20.6492	4.5431
55	56207				0.22000	4.5454	20.6563	4.5445

24%			TABLE 24	Discrete Cash Flow: Compound Interest Factors			24%	
	Single Payments		Uniform Series Payments				Arithmetic Gradients	
n	Compound Amount F/P	Present Worth P/F	Sinking Fund A/F	Compound Amount F/A	Capital Recovery A/P	Present Worth P/A	Gradient Present Worth P/G	Gradient Uniform Series A/G
1	1.2400	0.8065	1.00000	1.0000	1.24000	0.8065		
2	1.5376	0.6504	0.44643	2.2400	0.68643	1.4568	0.6504	0.4464
3	1.9066	0.5245	0.26472	3.7776	0.50472	1.9813	1.6993	0.8577
4	2.3642	0.4230	0.17593	5.6842	0.41593	2.4043	2.9683	1.2346
5	2.9316	0.3411	0.12425	8.0484	0.36425	2.7454	4.3327	1.5782
6	3.6352	0.2751	0.09107	10.9801	0.33107	3.0205	5.7081	1.8898
7	4.5077	0.2218	0.06842	14.6153	0.30842	3.2423	7.0392	2.1710
8	5.5895	0.1789	0.05229	19.1229	0.29229	3.4212	8.2915	2.4236
9	6.9310	0.1443	0.04047	24.7125	0.28047	3.5655	9.4458	2.6492
10	8.5944	0.1164	0.03160	31.6434	0.27160	3.6819	10.4930	2.8499
11	10.6571	0.0938	0.02485	40.2379	0.26485	3.7757	11.4313	3.0276
12	13.2148	0.0757	0.01965	50.8950	0.25965	3.8514	12.2637	3.1843
13	16.3863	0.0610	0.01560	64.1097	0.25560	3.9124	12.9960	3.3218
14	20.3191	0.0492	0.01242	80.4961	0.25242	3.9616	13.6358	3.4420
15	25.1956	0.0397	0.00992	100.8151	0.24992	4.0013	14.1915	3.5467
16	31.2426	0.0320	0.00794	126.0108	0.24794	4.0333	14.6716	3.6376
17	38.7408	0.0258	0.00636	157.2534	0.24636	4.0591	15.0846	3.7162
18	48.0386	0.0208	0.00510	195.9942	0.24510	4.0799	15.4385	3.7840
19	59.5679	0.0168	0.00410	244.0328	0.24410	4.0967	15.7406	3.8423
20	73.8641	0.0135	0.00329	303.6006	0.24329	4.1103	15.9979	3.8922
22	113.5735	0.0088	0.00213	469.0563	0.24213	4.1300	16.4011	3.9712
24	174.6306	0.0057	0.00138	723.4610	0.24138	4.1428	16.6891	4.0284
26	268.5121	0.0037	0.00090	1114.63	0.24090	4.1511	16.8930	4.0695
28	412.8642	0.0024	0.00058	1716.10	0.24058	4.1566	17.0365	4.0987
30	634.8199	0.0016	0.00038	2640.92	0.24038	4.1601	17.1369	4.1193
32	976.0991	0.0010	0.00025	4062.91	0.24025	4.1624	17.2067	4.1338
34	1500.85	0.0007	0.00016	6249.38	0.24016	4.1639	17.2552	4.1440
35	1861.05	0.0005	0.00013	7750.23	0.24013	4.1664	17.2734	4.1479
36	2307.71	0.0004	0.00010	9611.28	0.24010	4.1649	17.2886	4.1511
38	3548.33	0.0003	0.00007	14781	0.24007	4.1655	17.3116	4.1560
40	5455.91	0.0002	0.00004	22729	0.24004	4.1659	17.3274	4.1593
45	15995	0.0001	0.00002	66640	0.24002	4.1664	17.3483	4.1639
50	46890		0.00001		0.24001	4.1666	17.3563	4.1653
55					0.24000	4.1666	17.3593	4.1663

25%		TABLE 25	Discrete Cash Flow: Compound Interest Factors					25%
	Single Payments		Uniform Series Payments				Arithmetic Gradients	
	Compound Amount	Present Worth	Sinking Fund	Compound Amount	Capital Recovery	Present Worth	Gradient Present Worth	Gradient Uniform Series
n	F/P	P/F	A/F	F/A	A/P	P/A	P/G	A/G
1	1.2500	0.8000	1.00000	1.0000	1.25000	0.8000		
2	1.5625	0.6400	0.44444	2.2500	0.69444	1.4400	0.6400	0.4444
3	1.9531	0.5120	0.26230	3.8125	0.51230	1.9520	1.6640	0.8525
4	2.4414	0.4096	0.17344	5.7656	0.42344	2.3616	2.8928	1.2249
5	3.0518	0.3277	0.12185	8.2070	0.37185	2.6893	4.2035	1.5631
6	3.8147	0.2621	0.08882	11.2588	0.33882	2.9514	5.5142	1.8683
7	4.7684	0.2097	0.06634	15.0735	0.31634	3.1611	6.7725	2.1424
8	5.9605	0.1678	0.05040	19.8419	0.30040	3.3289	7.9469	2.3872
9	7.4506	0.1342	0.03876	25.8023	0.28876	3.4631	9.0207	2.6048
10	9.3132	0.1074	0.03007	33.2529	0.28007	3.5705	9.9870	2.7971
11	11.6415	0.0859	0.02349	42.5661	0.27349	3.6564	10.8460	2.9663
12	14.5519	0.0687	0.01845	54.2077	0.26845	3.7251	11.6020	· 3.1145
13	18.1899	0.0550	0.01454	68.7596	0.26454	3.7801	12.2617	3.2437
14	22.7374	0.0440	0.01150	86.9495	0.26150	3.8241	12.8334	3.3559
15	28.4217	0.0352	0.00912	109.6868	0.25912	3.8593	13.3260	3.4530
16	35.5271	0.0281	0.00724	138.1085	0.25724	3.8874	13.7482	3.5366
17	44.4089	0.0225	0.00576	173.6357	0.25576	3.9099	14.1085	3.6084
18	55.5112	0.0180	0.00459	218.0446	0.25459	3.9279	14.4147	3.6698
19	69.3889	0.0144	0.00366	273.5558	0.25366	3.9424	14.6741	3.7222
20	86.7362	0.0115	0.00292	342.9447	0.25292	3.9539	14.8932	3.7667
22	135.5253	0.0074	0.00186	538.1011	0.25186	3.9705	15.2326	3.8365
24	211.7582	0.0047	0.00119	843.0329	0.25119	3.9811	15.4711	3.8861
26	330.8722	0.0030	0.00076	1319.49	0.25076	3.9879	15.6373	3.9212
28	516.9879	0.0019	0.00048	2063.95	0.25048	3.9923	15.7524	3.9457
30	807.7936	0.0012	0.00031	3227.17	0.25031	3.9950	15.8316	3.9628
32	1262.18	0.0008	0.00020	5044.71	0.25020	3.9968	15.8859	3.9746
34	1972.15	0.0005	0.00013	7884.61	0.25013	3.9980	15.9229	3.9828
35	2465.19	0.0004	0.00010	9856.76	.025010	3.9984	15.9367	3.9858
36	3081.49	0.0003	0.00008	12322	0.25008	3.9987	15.9481	3.9883
38	4814.82	0.0002	0.00005	19255	0.25005	3.9992	15.9651	3.9921
40	7523.16	0.0001	0.00003	30089	0.25003	3.9995	15.9766	3.9947
45	22959		0.00001	91831	0.25001	3.9998	15.9915	3.9980
50	70065				0.25000	3.9999	15.9969	3.9993
55					0.25000	4.0000	15.9989	3.9997

30%			TABLE 26 Discrete Cash Flow: Compound Interest Factors				30%	
	Single Payments		Uniform Series Payments				Arithmetic Gradients	
n	Compound Amount F/P	Present Worth P/F	Sinking Fund A/F	Compound Amount F/A	Capital Recovery A/P	Present Worth P/A	Gradient Present Worth P/G	Gradient Uniform Series A/G
1	1.3000	0.7692	1.00000	1.0000	1.30000	0.7692		
2	1.6900	0.5917	0.43478	2.3000	0.73478	1.3609	0.5917	0.4348
3	2.1970	0.4552	0.25063	3.9900	0.55063	1.8161	1.5020	0.8271
4	2.8561	0.3501	0.16163	6.1870	0.46163	2.1662	2.5524	1.1783
5	3.7129	0.2693	0.11058	9.0431	0.41058	2.4356	3.6297	1.4903
6	4.8268	0.2072	0.07839	12.7560	0.37839	2.6427	4.6656	1.7654
7	6.2749	0.1594	0.05687	17.5828	0.35687	2.8021	5.6218	2.0063
8	8.1573	0.1226	0.04192	23.8577	0.34192	2.9247	6.4800	2.2156
9	10.6045	0.0943	0.03124	32.0150	0.33124	3.0190	7.2343	2.3963
10	13.7858	0.0725	0.02346	42.6195	0.32346	3.0915	7.8872	2.5512
11	17.9216	0.0558	0.01773	56.4053	0.31773	3.1473	8.4452	2.6833
12	23.2981	0.0429	0.01345	74.3270	0.31345	3.1903	8.9173	2.7952
13	30.2875	0.0330	0.01024	97.6250	0.31024	3.2233	9.3135	2.8895
14	39.3738	0.0254	0.00782	127.9125	0.30782	3.2487	9.6437	2.9685
15	51.1859	0.0195	0.00598	167.2863	0.30598	3.2682	9.9172	3.0344
16	66.5417	0.0150	0.00458	218.4722	0.30458	3.2832	10.1426	3.0892
17	86.5042	0.0116	0.00351	285.0139	0.30351	3.2948	10.3276	3.1345
18	112.4554	0.0089	0.00269	371.5180	0.30269	3.3037	10.4788	3.1718
19	146.1920	0.0068	0.00207	483.9734	0.30207	3.3105	10.6019	3.2025
20	190.0496	0.0053	0.00159	630.1655	0.30159	3.3158	10.7019	3.2275
22	321.1839	0.0031	0.00094	1067.28	0.30094	3.3230	10.8482	3.2646
24	542.8008	0.0018	0.00055	1806.00	0.30055	3.3272	10.9433	3.2890
25	705.6410	0.0014	0.00043	2348.80	0.30043	3.3286	10.9773	3.2979
26	917.3333	0.0011	0.00033	3054.44	0.30033	3.3297	11.0045	3.3050
28	1550.29	0.0006	0.00019	5164.31	0.30019	3.3312	11.0437	3.3153
30	2620.00	0.0004	0.00011	8729.99	0.30011	3.3321	11.0687	3.3219
32	4427.79	0.0002	0.00007	14756	0.30007	3.3326	11.0845	3.3261
34	7482.97	0.0001	0.00004	24940	0.30004	3.3329	11.0945	3.3288
35	9727.86	0.0001	0.00003	32423	0.30003	3.3330	11.0980	3.3297

| 35% | TABLE 27 | | Discrete Cash Flow: Compound Interest Factors | | | | | 35% |

	Single Payments		Uniform Series Payments				Arithmetic Gradients	
	Compound Amount	Present Worth	Sinking Fund	Compound Amount	Capital Recovery	Present Worth	Gradient Present Worth	Gradient Uniform Series
n	F/P	P/F	A/F	F/A	A/P	P/A	P/G	A/G
1	1.3500	0.7407	1.00000	1.0000	1.35000	0.7407		
2	1.8225	0.5487	0.42553	2.3500	0.77553	1.2894	0.5487	0.4255
3	2.4604	0.4064	0.23966	4.1725	0.58966	1.6959	1.3616	0.8029
4	3.3215	0.3011	0.15076	6.6329	0.50076	1.9969	2.2648	1.1341
5	4.4840	0.2230	0.10046	9.9544	0.45046	2.2200	3.1568	1.4220
6	6.0534	0.1652	0.06926	14.4384	0.41926	2.3852	3.9828	1.6698
7	8.1722	0.1224	0.04880	20.4919	0.39880	2.5075	4.7170	1.8811
8	11.0324	0.0906	0.03489	28.6640	0.38489	2.5982	5.3515	2.0597
9	14.8937	0.0671	0.02519	39.6964	0.37519	2.6653	5.8886	2.2094
10	20.1066	0.0497	0.01832	54.5902	0.36832	2.7150	6.3363	2.3338
11	27.1439	0.0368	0.01339	74.6967	0.36339	2.7519	6.7047	2.4364
12	36.6442	0.0273	0.00982	101.8406	0.35982	2.7792	7.0049	2.5205
13	49.4697	0.0202	0.00722	138.4848	0.35722	2.7994	7.2474	2.5889
14	66.7841	0.0150	0.00532	187.9544	0.35532	2.8144	7.4421	2.6443
15	90.1585	0.0111	0.00393	254.7385	0.35393	2.8255	7.5974	2.6889
16	121.7139	0.0082	0.00290	344.8970	0.35290	2.8337	7.7206	2.7246
17	164.3138	0.0061	0.00214	466.6109	0.35214	2.8398	7.8180	2.7530
18	221.8236	0.0045	0.00158	630.9247	0.35158	2.8443	7.8946	2.7756
19	299.4619	0.0033	0.00117	852.7483	0.35117	2.8476	7.9547	2.7935
20	404.2736	0.0025	0.00087	1152.21	0.35087	2.8501	8.0017	2.8075
22	736.7886	0.0014	0.00048	2102.25	0.35048	2.8533	8.0669	2.8272
24	1342.80	0.0007	0.00026	3833.71	0.35026	2.8550	8.1061	2.8393
25	1812.78	0.0006	0.00019	5176.50	0.35019	2.8556	8.1194	2.8433
26	2447.25	0.0004	0.00014	6989.28	0.35014	2.8560	8.1296	2.8465
28	4460.11	0.0002	0.00008	12740	0.35008	2.8565	8.1435	2.8509
30	8128.55	0.0001	0.00004	23222	0.35004	2.8568	8.1517	2.8535
32	14814	0.0001	0.00002	42324	0.35002	2.8569	8.1565	2.8550
34	26999		0.00001	77137	0.35001	2.8570	8.1594	2.8559
35	36449		0.00001		0.35001	2.8571	8.1603	2.8562

40%			TABLE 28	Discrete Cash Flow: Compound Interest Factors				40%
	Single Payments			Uniform Series Payments			Arithmetic Gradients	
	Compound Amount	Present Worth	Sinking Fund	Compound Amount	Capital Recovery	Present Worth	Gradient Present Worth	Gradient Uniform Series
n	F/P	P/F	A/F	F/A	A/P	P/A	P/G	A/G
1	1.4000	0.7143	1.00000	1.0000	1.40000	0.7143		
2	1.9600	0.5102	0.41667	2.4000	0.81667	1.2245	0.5102	0.4167
3	2.7440	0.3644	0.22936	4.3600	0.62936	1.5889	1.2391	0.7798
4	3.8416	0.2603	0.14077	7.1040	0.54077	1.8492	2.0200	1.0923
5	5.3782	0.1859	0.09136	10.9456	0.49136	2.0352	2.7637	1.3580
6	7.5295	0.1328	0.06126	16.3238	0.46126	2.1680	3.4278	1.5811
7	10.5414	0.0949	0.04192	23.8534	0.44192	2.2628	3.9970	1.7664
8	14.7579	0.0678	0.02907	34.3947	0.42907	2.3306	4.4713	1.9185
9	20.6610	0.0484	0.02034	49.1526	0.42034	2.3790	4.8585	2.0422
10	28.9255	0.0346	0.01432	69.8137	0.41432	2.4136	5.1696	2.1419
11	40.4957	0.0247	0.01013	98.7391	0.41013	2.4383	5.4166	2.2215
12	56.6939	0.0176	0.00718	139.2348	0.40718	2.4559	5.6106	2.2845
13	79.3715	0.0126	0.00510	195.9287	0.40510	2.4685	5.7618	2.3341
14	111.1201	0.0090	0.00363	275.3002	0.40363	2.4775	5.8788	2.3729
15	155.5681	0.0064	0.00259	386.4202	0.40259	2.4839	5.9688	2.4030
16	217.7953	0.0046	0.00185	541.9883	0.40185	2.4885	6.0376	2.4262
17	304.9135	0.0033	0.00132	759.7837	0.40132	2.4918	6.0901	2.4441
18	426.8789	0.0023	0.00094	1064.70	0.40094	2.4941	6.1299	2.4577
19	597.6304	0.0017	0.00067	1491.58	0.40067	2.4958	6.1601	2.4682
20	836.6826	0.0012	0.00048	2089.21	0.40048	2.4970	6.1828	2.4761
22	1639.90	0.0006	0.00024	4097.24	0.40024	2.4985	6.2127	2.4866
24	3214.20	0.0003	0.00012	8033.00	0.40012	2.4992	6.2294	2.4925
25	4499.88	0.0002	0.00009	11247	0.40009	2.4994	6.2347	2.4944
26	6299.83	0.0002	0.00006	15747	0.40006	2.4996	6.2387	2.4959
28	12348	0.0001	0.00003	30867	0.40003	2.4998	6.2438	2.4977
30	24201		0.00002	60501	0.40002	2.4999	6.2466	2.4988
32	47435		0.00001		0.40001	2.4999	6.2482	2.4993
34	92972				0.40000	2.5000	6.2490	2.4996
35					0.40000	2.5000	6.2493	2.4997

50%		TABLE 29	Discrete Cash Flow: Compound Interest Factors				50%
	Single Payments		Uniform Series Payments				Arithmetic Gradients	
	Compound Amount F/P	Present Worth P/F	Sinking Fund A/F	Compound Amount F/A	Capital Recovery A/P	Present Worth P/A	Gradient Present Worth P/G	Gradient Uniform Series A/G
n								
1	1.5000	0.6667	1.00000	1.0000	1.50000	0.6667		
2	2.2500	0.4444	0.40000	2.5000	0.90000	1.1111	0.4444	0.4000
3	3.3750	0.2963	0.21053	4.7500	0.71053	1.4074	1.0370	0.7368
4	5.0625	0.1975	0.12308	8.1250	0.62308	1.6049	1.6296	1.0154
5	7.5938	0.1317	0.07583	13.1875	0.57583	1.7366	2.1564	1.2417
6	11.3906	0.0878	0.04812	20.7813	0.54812	1.8244	2.5953	1.4226
7	17.0859	0.0585	0.03108	32.1719	0.53108	1.8829	2.9465	1.5648
8	25.6289	0.0390	0.02030	49.2578	0.52030	1.9220	3.2196	1.6752
9	38.4434	0.0260	0.01335	74.8867	0.51335	1.9480	3.4277	1.7596
10	57.6650	0.0173	0.00882	113.3301	0.50882	1.9653	3.5838	1.8235
11	86.4976	0.0116	0.00585	170.9951	0.50585	1.9769	3.6994	1.8713
12	129.7463	0.0077	0.00388	257.4927	0.50388	1.9846	3.7842	1.9068
13	194.6195	0.0051	0.00258	387.2390	0.50258	1.9897	3.8459	1.9329
14	291.9293	0.0034	0.00172	581.8585	0.50172	1.9931	3.8904	1.9519
15	437.8939	0.0023	0.00114	873.7878	0.50114	1.9954	3.9224	1.9657
16	656.8408	0.0015	0.00076	1311.68	0.50076	1.9970	3.9452	1.9756
17	985.2613	0.0010	0.00051	1968.52	0.50051	1.9980	3.9614	1.9827
18	1477.89	0.0007	0.00034	2953.78	0.50034	1.9986	3.9729	1.9878
19	2216.84	0.0005	0.00023	4431.68	0.50023	1.9991	3.9811	1.9914
20	3325.26	0.0003	0.00015	6648.51	0.50015	1.9994	3.9868	1.9940
22	7481.83	0.0001	0.00007	14962	0.50007	1.9997	3.9936	1.9971
24	16834	0.0001	0.00003	33666	0.50003	1.9999	3.9969	1.9986
25	25251		0.00002	50500	0.50002	1.9999	3.9979	1.9990
26	37877		0.00001	75752	0.50001	1.9999	3.9985	1.9993
28	85223		0.00001		0.50001	2.0000	3.9993	1.9997
30					0.50000	2.0000	3.9997	1.9998
32					0.50000	2.0000	3.9998	1.9999
34					0.50000	2.0000	3.9999	2.0000
35					0.50000	2.0000	3.9999	2.0000

INDEX

A

A, 23, 26
Absolute cell referencing, 247
Accelerated Cost Recovery System (ACRS), 534
Accelerated write-off, 532, 578
Accounting
  ratios, 719–23
  statements, 718–19
Acid-test ratio, 720
Acquisition phase, 190
Activity based costing (ABC), 512–16
A/F factor, 60
After-tax
  and alternative selection, 11, 603
  and annual worth, 586–88
  cash flow, 575–77, 586–92, 606
  debt *versus* equity financing, 617–19
  and depreciation, 532
  international, 603–5
  and MARR, 353, 586
  rate or return, 588–91
  and WACC, 356
After-tax replacement analysis, 595–99
A/G factor, 68. *See also* Gradients, arithmetic
Alternative depreciation system (ADS), 545
Alternatives
  attribute-based rating, 368–69
  and breakeven analysis, 287
  cash flow estimate types, 220–21
  defined, 9–10
  and EVA™, 600–601
  and incremental rate-of-return, 588–90
  independent, 170–72, 283, 330 (*see also* Capital budgeting)
  infinite life, 228–31, 330
  mutually exclusive, 170, 223–28, 348–51
  and payback analysis, 186
  revenue, 172
  selection, 11
  service, 172
  in simulation, 678, 679
Amortization. *See* Depreciation
Annual income, estimated, 10
Annual interest rate
  effective, 130–36
  nominal, 130, 131

Annual operating costs (AOC), 10, 220, 391–92, 451
  and estimation, 496–97
  in Excel® spreadsheet, 702–3
Annual Percentage Rate (APR), 127–28
Annual Percentage Yield (APY), 127–28
Annual worth
  advantages, 218, 254
  after-tax analysis, 586–88
  of annual operating costs, 391
  and B/C analysis, 321–23, 326, 333
  and breakeven analysis, 451
  and capital-recovery-plus-interest, 221–23
  components, 221–23
  computer solutions, 225, 227–28, 230–31
  equivalent uniform, 221, 229
  evaluation by, 218–31
  and EVA™, 600–603
  and future worth, 218
  and incremental rate of return, 588–91
  of infinite-life projects, 228–31
  and inflation, 218, 485–86
  and present worth, 218
  and rate of return, 248, 291–92
  and replacement analysis, 388, 390, 392–96, 404–10
  and sensitivity analysis, 623
  when to use, 350 (table)
AOC. *See* Annual operating costs
A/P factor, 58–59, 221
APR, 127–28
APY, 127–28
Arithmetic gradients. *See* Gradients, arithmetic
Assets. *See also* Book value; Depletion; Depreciation; Life; Salvage value
  in balance sheet, 717
  capital recovery, 221
  replacement studies, 220 (*see also* Replacement analysis)
  return on, 721
  sunk cost, 389
Attributes
  evaluating multiple, 369–71
  identifying, 365–66
  weighting, 366–69
Average. *See* Expected value
Average cost per unit, 448
Average tax rate, 571–72

B

Balance sheet
  basic equation, 717
  and business ratios, 719
  categories, 717
Base amount
  defined, 69
  and shifted gradients, 103, 108
Basis, unadjusted, 532
B/C. *See* Benefit/cost ratio
Before-tax rate of return
  and after-tax, 354, 589
  calculation, 240–49
Bell-shaped curve. *See* Normal distribution
Benefit and cost difference, 319–21
Benefit/cost ratio
  calculation, 319–24
  conventional, 319–20
  incremental analysis, 324–27
  modified, 320
  for three or more alternatives, 327–33
  for two alternatives, 324–27
  when to use, 350 (table)
Benefits
  direct *versus* implied, 327
  in public projects, 315, 320, 327, 333
β, 360, 361
Bonds
  and debt financing, 352, 357–59, 363–64
  and inflation, 492–93
  interest computation, 194
  payment periods, 194
  present worth, 194–95
  for public sector projects, 315
  rate of return, 261–63
  types, 194–95, 315
Book depreciation, 532, 534, 600
Book value
  by declining-balance method, 539
  defined, 532
  by double declining balance method, 539
  and EVA™, 600
  by MACRS, 541, 543
  versus market value, 533
  straight line method, 535
  sum of year digits method, 555
Borrowed money. *See* Debt capital
Bottom-up approach, 497–99
Breakeven analysis. *See also* PW *vs. i* graph; Sensitivity analysis; SOLVER

  and annual worth, 220
  average cost per unit, 448
  description, 442
  fixed costs, 444–48
  and make-buy decisions, 453
  and payback, 185, 448–50
  and rate or return, 287, 291, 591
  versus sensitivity analysis, 442, 622–23
  single project, 444–50, 459
  spreadsheet application, 455–58
  three or more alternatives, 434
  two alternatives, 451–54, 459
  variable, 451
Breakeven point, 444, 628–29
Budgeting. *See* Capital budgeting
Bundles, 424, 426

C

c. See External rate of return
Canada, depreciation and taxes, 603–4
Capital
  cost of (*see* Cost of capital)
  cost of invested, 600–601
  debt *versus* equity, 29, 352
  limited, 353, 424–36
  unrecovered, 389
  working, 720
Capital asset pricing model (CAPM), 360–61
Capital budgeting
  description, 422, 424–26
  equal life projects, 426–28
  linear programming, 432–36
  mutually exclusive bundles, 426
  and net cash flow, 436
  present-worth use, 426–32
  reinvestment assumption, 425–27
  spreadsheet solution, 433–36
  unequal life projects, 428–32
Capital cost
  in alternative evaluation, 179–85
  and present worth, 179
  and public projects, 314–15, 331
Capital expense deduction, 533–34
Capital financing. *See also* Cost of capital
  debt, 29, 352
  debt *versus* equity, 382–83, 617–19
  equity, 29–30, 352, 617–19
  mixed (debt and equity), 355–57, 362–64

Capital gains
    defined, 581
    short-term and long-term, 582–83
    taxes for, 583–85
Capital investment, and alternative evaluation, 172
Capital losses
    defined, 582
    taxes for, 582–83
Capital recovery, 220–23. *See also* A/P factor;
        Depreciation
    decreasing cost of, 391–92
    defined, 221
    and EVA™, 603
    and inflation, 485–86
    and replacement analysis, 395, 403, 407
Capital recovery factor
    and equivalent annual worth, 221
    and random single amounts, 98
Capital recovery for assets. *See* Depreciation
CAPM. *See* Capital asset pricing model
Carry-back and carry-forward, 575
Case studies
    alternative description, 46–47
    annual worth, 236–37
    breakeven analysis, 464–67
    compound interest, 46–47, 90–91
    cost estimation, 525–29
    debt *versus* equity financing, 382–83, 617–19
    house financing, 162–64
    independent project selection, 440–41
    multiple interest rates, 310–11
    payback analysis, 213–15
    public project, 343–45
    rates of return, 273–75
    replacement analysis, 420–21
    sale of business, 309–10
    sensitivity analysis, 525–27
Cash flow. *See also* Discrete cash flows; Gradients,
        arithmetic; Payment period
    actual *versus* incremental, 591
    after tax, 575–77
        and EVA™, 600–603
    beyond study period, 175
    continuous, 151
    as continuous variable, 664–66
    conventional series, 248, 249
    defined, 11
    diagram, conventional gradient, 70
    diagramming, 32–34, 39–41, 70
    discounted, 168, 632
    estimating, 11, 30–34, 220–21, 315–16

future, 705–6
incremental, 279–82, 291–92
    after tax, 590–94
inflow and outflow, 30–31, 244
net, 31, 98, 574–75
    in Excel®, 702–3
    and payback period, 186–87
nonconventional, 249–55
and NPV function, 198
periodic, 180–82
positive net, and ROR, 255
and public sector projects, 315–16
recurring and nonrecurring, 180
and replacement analysis, 403, 595–99
revenue *versus* service, 172
series factors, 71–73
series with single amounts, 98–103
before tax, 575–77
zero, 111, 198, 700, 708
Cash flow after taxes (CFAT), 575–77, 586–92, 606
    and EVA™, 600–603
Cash flow before taxes (CFBT), 575–77, 606,
        639–40
Cash-flow diagrams, 32–34, 39–41
    partitioned, 70
Cell references, 703
    absolute, 247
    sign, 247, 705–6
Certainty, 656, 658, 667
CFAT. *See* Cash flow after taxes
CFBT. *See* Cash flow before taxes
Challenger
    in multiple-alternative evaluation. 293, 327
    in replacement analysis, 388, 389, 391, 404, 410
Charts in Excel®, 701–3
Class life, 545
Common stock, 352, 359
Comparison types, selection of, 348–51
Composite rate of return (CRR), 255–61, 283
Compound amount factors
    single payment (F/P), 50
    uniform series (F/A), 60
Compound interest, 18–22, 38–39, 699–715.
        See also Compounding
Compounding
    annual, 130–36, 151
    continuous, 149–51
    doubling time, 18–22, 35
    frequency, 128, 135, 136
    interperiod, 147–49
    and simple interest, 17–22

Compounding period
  continuous, 149–51
  defined, 128
  and effective annual rate, 132 (table)
  monthly, 141 (table)
  number per year, 130
  and payment period, 139–49
Computers, uses, 26–28, 36–39. *See also* Spreadsheets
Construction costs, change in, 71
Construction stage, 191
Contingent projects, 424
Continuous compounding, 149–51
Contracts, types, 318
Conventional benefit/cost ratio, 319–20
Conventional cash flow series, 249, 261
Conventional gradient, 66–70
Convertible bonds, 195
Corporations
  financial worth, 600–601
  leveraged, 363–64
Cost, first, 10
Cost, life-cycle, 190–93
Cost-capacity equations, 503–5
Cost centers, 508–10
Cost components, 496–98
Cost depletion. *See* Depletion
Cost drivers, 512–16
Cost estimation
  approaches, 497–99
  cost-capacity method, 503–5
  and cost indexes, 499–503
  factor method, 505–7
  and inflation, 14–15
  unit method, 499
Cost indexes, 499–503
Cost of capital
  and debt-equity mix, 355–57, 362–64
  for debt financing, 357–59
  defined, 28–30, 351–52
  for equity financing, 359–62
  versus MARR, 28, 351–54
  weighted average, 29–30, 355–57
Cost-of-goods-sold, 508, 510, 718–19
  statement, 718–19
Cost of invested capital, 600–601
Cost pool, 512–16
Costs. *See also* Capital cost; Incremental costs;
        Opportunity cost; Total cost relation
  and alternative evaluation, 172
  and annual worth, 221

  and apparent savings, 193
  of asset ownership, 221
  direct, 496–99, 505–6, 507
  estimating, 496–99
  EUAW (*see* Annual worth, equivalent
        uniform)
  fixed, 444–45, 450
  indirect, 505–7, 508–16, 719
  of invested capital, 351–54, 580
  life-cycle, 190–93
  marginal, 394
  operating, 71–73
  periodic, 180
  in public projects, 315
  sign convention, 235
  sunk, 389
  variable, 444–45, 451, 454
Coupon rate, 194
CRR. *See* Composite rate of return
Cumulative cash flow sign test, 250
Cumulative distribution, 661–66
Current assets, 717
Current liabilities, 720
Current ratio, 720

D

DB function, 537, 703–4
DDB function, 537–40, 704
Debenture bonds, 195
Debt capital, 352, 357–59
Debt-equity mix, 355–57, 362–64
Debt financing, 29, 352
  on balance sheet, 717
  costs of, 357–59
  and inflation, 479–80
  leveraging, 363–64
Debt ratio, 720–21
Decision making
  attributes, 365–69
  under certainty, 658, 667
  engineering economy role, 7–9
  guideline, 351
  under risk, 658–59, 660–66
  under uncertainty, 659
Decision trees, 635–36
Declining balance depreciation, 533, 536–37
  in Excel®, 703–4, 711–12
Decreasing gradients, 69, 108–10

Defender
    in multiple alternative evaluation, 293, 327
    in replacement analysis, 388–89, 391, 595–99
Deflation, 473–74
Delphi method, 365–66
Dependent projects, 424
Depletion
    cost, 546–47, 548
    percentage, 547–48
Depreciation. *See also* Rate of depreciation; Recaptured
            depreciation; Replacement analysis
    accelerated, 532
    ACRS, 534
    alternative system, 545
    book, 532, 534, 600
    class life, 547–48
    declining balance, 533, 536–37, 711–12
    defined, 532
    double declining balance, 537–40
    and EVA, 606
    GDS, 545
    half-year convention, 533, 542
    and income taxes, 530, 534–35, 575–76
    MACRS, 534, 541–45, 564–67
        present worth, 560
        recovery period for, 541, 545–46, 549, 562
    not used, 534
    rate of, 533
    recovery rate, 533
    straight line, 533, 535–36, 549
        Excel® functions, 710–13
    straight line alternative, 541–42
    sum-of-year digits, 533, 555–57, 711
    switching methods, 557–62, 711–12
    tax, 532
Descartes' rule, 249–50, 251, 253
Design stages, preliminary and detailed, 190, 193
Design-to-cost approach, 497–99
Different-life alternatives, 174–77, 197–99
Direct benefits, 327–28
Direct costs, 496–99, 505–7
Disbenefits, 315, 319–20
Disbursements, 31, 244
Discount rate, 316, 333
Discounted-cash-flow, 168, 633
Discounted payback analysis, 185–89
Discrete cash flows
    compound interest factors (tables), 727–55
    discrete compounding, 147, 150
Disposal stage, 191

Distribution
    normal, 688–91
    standard normal, 688–91
    triangular, 663, 665–66
    uniform, 663, 664, 670
Distribution estimates, 31
Dividends, 194, 359–60
Dollars, today *versus* future, 472–73
Do-nothing alternative, 11, 170, 173
    and B/C analysis, 325
    and independent projects, 425, 426
    and rate of return, 282–83, 292–93, 295
Double declining balance, 537, 543–45
    in Excel®, 540, 704
    in switching, 557–62
    and taxes, 578
Doubling time, 35

E

Economic equivalence. *See* Equivalence
Economic service life (ESL), 388, 391–97, 402
Economic value added, 220, 600–603, 606
Effective interest rate
    annual, 130–35
    for any time period, 136–38
    of bonds, 196
    and compounding periods, 132 (table), 136–38
    for continuous compounding, 149–51
    defined, 127–28, 153
    flow chart, 165
    and nominal rate, 571, 595
Effective tax rate, 571, 595
Efficiency ratios, 719
End-of-period convention, 31–32
Engineering economy
    defined, 6
    role in decision-making, 7–9
    study approach, 9–11
    terminology and symbols, 23
    uses, 6–9
Equal-service alternatives, 172–74, 282, 297–300, 592
Equity financing, 29–30, 352
    cost of, 359–62
Equivalence, 15–17, 20–22, 243
    compounding period greater than payment
            period, 147–49
    compounding period less than payment
            period, 139–46

Equivalent annual cost. *See* Annual worth

Equivalent annual worth. *See* Annual worth

Equivalent uniform annual cost, 388. *See also* Annual worth

Equivalent uniform annual worth. *See* Annual worth

Error distribution. *See* Normal distribution

ESL. *See* Economic service life

E-solve, 28

Estimation

    and alternatives, 10

    of cash flow, 11, 30–34, 315–16

    of doubling time, 35

    factor method, 505–7

    of interest rates, 35

    and sensitivity analysis, 622, 629–31

    before tax MARR, 589

    and uncertainty, 7 (*see also* Uncertainty)

EUAC. *See* Annual worth

EUAW. *See* Annual worth, equivalent uniform

Evaluation criteria, 11, 348–51

Evaluation method, 348–51

EVA. *See* Economic value added

Excel®. *See also* Spreadsheets; *specific functions*

    after-tax AW and PW, 587–88

    basics, 697–701

    and breakeven analysis, 455–58

    charts, 703

    and depreciation, 533, 535, 537–40, 542, 544–45, 558, 560–61

    displaying entries, 39

    embedding functions, 703

    error messages, 715

    functions listing, 713–15

    introduction, 26–29

    and linear programming, 433–36

    random number generation, 696

    and rate of return, 254, 296

    and replacement value, 402–3

    and simulation, 683–86

    spreadsheet layout, 701–3

Expected value

    computation, 632

    and decisions under risk, 659

    and decision trees, 638–40

    defined, 671

    in simulation, 678, 684, 685

Expenses, 570, 718. *See also* Cost estimation; Costs

External rate of return, 256. *See also* Rate of return

Extra investment, 282–83, 292–95

F

F, 23, 26

F/A factor, 60. *See also* Uniform series, compound amount factor

F/G factor, 69. *See also* Gradients, arithmetic

Face value, of bonds, 194

Factor method estimation, 505–7

Factors. *See also* Present-worth factors

    capital recovery (*see* Capital recovery factor)

    continuous compound interest, 151

    derivations, 50–74

    discrete compound interest, 147, 150

    gradient

        arithmetic, 65–71

        geometric, 71–74

    intangible, 6

    multiple, 92

    notation, 51, 58, 61

    single payment, 50–56

    sinking fund, 60–63

    tables, 51–52, 727–55

    uniform-series, 56–58, 60–63, 80–81

Factory cost, 510, 718

Financial worth of corporations, 600–601

Financing. *See* Debt financing; Equity financing

Finite-life alternatives, 183

First cost, 10, 388–89, 451. *See also* Initial investment

    and depreciation, 532, 541, 581

    and estimation, 496–98

    in Excel® spreadsheet, 702

    and sensitivity analysis, 629

Fiscal year, 716

Fixed assets, 717

Fixed costs, 444–45, 451

Fixed-income investment, 492–93. *See also* Bonds

Fixed percentage method. *See* Declining balance depreciation

F/P factor, 50. *See also* Single payment factors

Future worth. *See also* Sensitivity analysis

    from annual worth, 218

    calculation, 69

    and effective interest rate, 130

    evaluation by, 177–79

    in Excel®, 704–5

    and inflation, 480–84, 486

    of shifted series, 94–96, 98

    when to use, 177, 350 (table)

FV function, 704–5
  and random single amounts, 102–3
  and shifted uniform series, 111
  and single payment factors, 52–53

G

Gains and losses. *See* Capital gains;
      Capital losses
Gaussian distribution. *See* Normal distribution
General depreciation system (GDS) value, 545
Geometric gradients, 71–73
  factors, 71–73
  and inflation, 476
  shifted, 106
Geothermal deposits. *See* Depletion
Government projects. *See* Public sector projects
Gradients, arithmetic
  amount, 65
  base amount, 69, 103
  conventional, 66, 70, 103
  decreasing, 69, 108–10
  defined, 65
  derivation of factors for, 67–68
  equivalent uniform annual series, 60
  increasing, 69
  present worth, 67
  shifted, 103–7
  spreadsheet use, 69
  uniform, 68
Graduated tax rates, 571
Gross income, 570

H

Half-year convention, 533, 542, 545
Highly leveraged corporations, 363–64
Hurdle rate. *See* Minimum attractive rate of return
Hyperinflation, 473, 484

I

i, 23, 26. *See also* Effective interest rate; Interest rate;
      Internal rate of return
i'. *See* Composite rate of return
i*, 243–48. *See also* MARR; Rate of return
IF function, in Excel®, 705
Implementation stage, 191

Income
  estimated annual, 10
  in Excel® spreadsheet, 702
  gross and net, 570
  taxable, 570, 583
Income statement
  basic equation, 718
  categories, 718
  ratios, 721
Income tax
  and annual worth, 220
  average tax rate, 571–72
  and capital gains and losses, 581–83
  and cash flow, 575–77, 586–92, 606
  corporate, 571 (table)
  defined, 570
  and depreciation, 530, 534–35, 575–76
      recaptured, 577, 582
  effective rates, 571, 595
  and individual taxpayers, 573–74
  international, 603–5
  negative, 575–76
  present worth of, 578–81
  and rate of return, 588–95
  rates, 570–72
  and replacement studies, 595–99
  tax savings, 575–76
Incremental benefit/cost analysis
  for three or more alternatives, 327–33
  for two alternatives, 324–27
Incremental cash flow, 279–82, 291–92, 588–91
Incremental costs
  and benefit/cost analysis, 324–33
  definition, 282
  and rate-of-return, 283–97
Incremental rate of return
  for multiple alternatives, 292–97, 588–91
  for two alternatives, 283–92, 588–91
  unequal lives, 349, 432
Independent alternatives, 170–72, 283, 331. *See also*
      Capital budgeting
Indexing, income taxes, 571
Indirect costs
  and activity-based costing, 512–16
  allocation variance, 510–11
  charge, 510
  in cost of goods sold statement, 719
  and factor method, 505–7
  rates, 508–10
Infinite life, 179, 228–31, 314, 331

Inflation
    assumption in PW and AW, 175, 218
    and capital recovery, 485–86
    definition, 14, 473
    and future worth, 480–84
    high, 484, 486
    impact, 14, 472–73
    and interest rates, 473
    and MARR, 473, 482–84
    and present worth, 473–80
    and sensitivity analysis, 622
Initial investment. *See also* First cost
    defined, 220
    larger, 279, 281, 287, 293
    lower, 295–97
    in replacement analysis, 388–89, 403–4
    in spreadsheet analysis, 197
Installation costs, 389
Installment financing, 241–42
Intangible factors, 6. *See also* Multiple attribute evaluation;
        Noneconomic factors
Integer linear programming, 432–36
Interest
    accrued, in Excel® (*see* IPMT function)
    from bonds, 194
    compound, 18–22, 38–39, 46–47
    continuous compounding, 149–51
    defined, 12
    interperiod, 147–49
    rate (*see* Interest rate(s))
    simple, 17–18, 35, 36–37
Interest period, 12, 14
Interest rate(s). *See also* Effective interest rate
    and breakeven analysis, 451
    definition, 12–13
    estimation, 35
    in Excel®, 710–11
    expressions, 129
    inflation-adjusted, 473
    inflation free (real), 473, 480–84, 486
    interpolation, 64–65
    market, 473
    multiple, 310–11
    nominal *versus* effective, 126–29
    for public sector, 316
    and risk, 360–61
    and sensitivity analysis, 622
    unknown, 74–77
    on unrecovered balance (ROR), 240–42
    varying over time, 151–53

Interest tables
    discrete compounding, 727–55
    interpolation, 63–65, 133
Internal rate of return, 255. *See also*
        Rate of return
International aspects
    after taxes, 603–5
    contracts, 318
    cost estimation, 498
    deflation, 473–74
    depreciation, 532, 534–36, 603–5
    design and manufacturing, 6
    hyperinflation, 484
Interperiod interest, 147–49
Interpolation, in interest rate tables, 63–65, 133
Inventory turnover ratio, 721–22
Invested capital, cost of, 600–601
Investment(s). *See also* Initial investment
    extra, 282–83, 292–95
    fixed-income, 492–93
    net, 256–60
    permanent, 228–31
    safe, 194–95
Investment opportunity, 353
IPMT function, 705–6
IRR function, 75–77, 245, 706–7
    incremental after-tax, 592–93

L

Land, 533
Lang factors, 505
Least common multiple, 174–75, 177
    and annual worth, 218, 226
    in evaluation methods, 349–50
    and future worth, 177
    and incremental cash flow, 279–81, 291
    and incremental rate of return, 283, 291–92,
        590–91
    and independent projects, 432
    in spreadsheet analysis, 197
Leveraging, 363–64
Liabilities, 717
Life
    finite, 183
    infinite or very long, 179, 228, 314, 331
    minimum cost, 391
    recovery (tax), 535
    in simulation, 696

unknown, 77–78
useful, 10, 545
Life cycle, and annual worth, 218
Life-cycle costs, 190–93
Likert scale, 368–69
Linear programming, 432–36
Lives
    common multiple, 174–75
    equal, 172–74, 279–80, 426–28
    and independent projects, 425
    perpetual, 179, 225, 314
    and rate of return, 279–81
    unequal, 280–81, 297, 333, 349, 403, 451
Loan payment, 20–22

M

m. See Compounding period, number per year
MACRS (Modified Accelerated Cost
        Recovery System), 541–45
    in CFAT example, 576–77
    computer use, 544–45
    depreciation rates, 541–45
    PW of, depreciation, 557
    recovery period, 541, 545–46, 562
    straight line alternative (ADS), 545,
        546 (table)
    switching, 557–62
    U.S., required, 534, 544
Maintenance and operating (M&O) costs, 320, 570.
        See also Annual operating costs;
        Life-cycle costs
Make-or-buy decisions, 220, 453. See also
        Breakeven analysis
Marginal costs, 395–97
Marginal tax rates, 571
Market interest rate. See Interest rate(s)
Market value
    and depreciation, 533
    in ESL analysis, 394
    estimating, 395–97
    and PW, different life alternatives, 175
    in replacement analysis, 388–90
    as salvage value, 220, 389
MARR. See Minimum attractive
        rate of return
Materials. See Direct costs
Mean. See Expected value
Mean squared deviation, 674

Measure of worth, 9, 11, 678
Median, 672–73
Mexico, depreciation and taxes, 604
Minimum attractive rate of return
    after-tax, 353, 586, 591, 603
    in alternative evaluation, 172–73, 174–75,
        177–79, 223–24
    and bonds, 195–96
    and capital budgeting, 425–32
    definition, 28
    establishing, 28–30, 242–43, 351–54
    as hurdle rate, 28
    and independent projects, 425–32
    inflation-adjusted, 473, 482–84
    and rate of return, 282–83, 284, 287, 291, 293
    in sensitivity analysis, 623
    before tax, 589
    and WACC, 29–30, 361–62
Minimum cost life of asset. See Economic
        service life
MIRR function, 261, 707
M&O costs. See Annual operating costs;
        Maintenance and operating costs
Mode, 664, 672–73
Modified benefit/cost ratio, 320
Money
    financial units, 11
    and inflation, 14–15, 472–73
    time value of, 9
Monte Carlo simulation, 677–86
Mortgage bonds, 194
Most likely estimate, 629–31
Multiple alternatives
    benefit/cost analysis for, 324–33
    incremental rate of return, 283–97
    independent, 170–72, 283, 331 (see also Capital
        budgeting)
    mutually exclusive, 170, 223, 348–51
Multiple attribute evaluation, 365–72
Multiple rate of return. See also Net investment
        procedure
    definition, 248
    determining, 250–52
    presence of, 249–50
    removing, 255
Municipal bonds, 194, 315, 492–93
Mutually exclusive alternatives, 170
    and annual worth, 223–28
    evaluation method selection, 348–51
    and present worth, 172–77

N

n, 23, 26, 103–5. *See also* Interpolation in interest rate
     tables; Least common multiple; Payback
     analysis; Study period
Natural resources. *See* Depletion
Net cash flow, 31, 436, 574–75
Net income (NI), 570
Net investment procedure, 256–61
Net operating profit after taxes (NOPAT), 570
Net present value. *See* NPV function; Present worth
Net profit after taxes (NPAT), 570, 602–6
Net worth, 717
Nominal interest rate, 126, 128
    annual, 130, 131
    any payment period, 130, 131
    of bonds, 196–97
    and effective rates, 133 (table), 138
Nonconventional cash flow series, 249–55
Noneconomic factors, 10
Nonrecurring cash flows, 180
Nonsimple cash flow series, 249–55
No-return (simple) payback, 186–87
Normal distribution, 663, 687–91
Norstrom's criterion, 250, 593
Notation for factors, 51, 58, 61, 67, 72
NPER function, 707–8
    and annual worth, 230–31
    and unknown *n,* 77–78
NPV function, 708
    after-tax, 587, 588
    for arithmetic gradients, 69
    embedding in PMT, 227, 394, 703
    geometric gradients, 73
    independent projects, 430
    and present worth, 197–200
    in PV *vs. i* graphs, 245, 247
    sensitivity analysis, 626, 628
    and shifting, 96, 98, 107, 110–11

O

Obsolescence, 388
One-additional-year replacement studies, 398
Operating costs. *See* Annual operating costs;
        Cost estimation
Operations phase, 191, 192–93
Opportunity cost, 30, 352
    and replacement analysis, 391, 405

Optimistic estimate, 629–31
Overhead rates. *See* Indirect costs
Owner's equity, 352, 717

P

P, 23, 26
P/A factor, 56–57, 71–73. *See also* Geometric gradients;
        Uniform series, present worth factors
Parameter variation, 622–29
Payback analysis
    calculation, 185–89, 200–201
    definition, 185
    limitations, 187
    spreadsheet analysis, 200–202
    uses, 186, 188–89
Payment period
    of bonds, 194
    defined, 136
    equals compounding period, 142 (table)
    Excel® function, 708–9
    longer than compounding period, 139–46
    single amount, 139–42, 147–49
Payout period. *See* Payback analysis
Percentage depletion. *See* Depletion
Periodic cash flows, 180–82
Permanent investments, 228–31. *See also* Capital cost
Perpetual investment. *See* Capital cost
Personal property, 533, 541, 545–46
Perspective
    and public sector, 316–19
    for replacement analysis, 389–90
Pessimistic estimate, 629–31
P/F factor, 50. *See also* Single-payment present
        worth factor
P/G factor, 67. *See also* Gradients, arithmetic
Phaseout stage, 191
Phases of systems, 190–91
Planning horizon. *See* Study period
Plant costs, 505–7
PMT function, 708–9
    and after-tax analysis, 587, 588
    and annual worth, 225, 349
    and arithmetic gradients, 69
    and B/C analysis, 323
    and capital recovery, 223
    and capitalized cost, 184–85
    and economic service life, 393
    and embedded NPV, 703
    and geometric gradients, 73

and random single amounts, 102–3
and replacement analysis, 392–93
and shifted series, 96–98, 106
and sinking fund factor, 62
and uniform series present worth, 59
Point estimates, 31, 632, 660
Power law and sizing model, 503–4
PPMT function, 709
Preferred stock, 352, 359
Present value net. *See* NPV function; Present worth
Present worth, 170–202
    after-tax analysis, 586–88
    in alternative evaluation, 168
    and annual worth, 218
    assumptions, 175
    of bonds, 194–97
    and breakeven analysis, 451
    and capital budgeting, 432–36
    of depreciation, 560
    for equal lives, 172–74
    evaluation method, 350 (table)
    geometric gradient series, 71–74
    income taxes, 586
    and independent projects, 428–34
    and inflation, 473–80
    and life-cycle cost, 190
    and multiple interest rates, 249–54, 310–11
    and rate of return, 244–47, 283–91
    and sensitivity analysis, 623–26
    in shifted series, 94–96
    in simulation, 678, 679
    single-payment factor, 50–52
    for unequal lives, 174–76
Present-worth factors
    gradient, 65–71
    single-payment factor, 50–52
    uniform-series, 56–58
Probability
    in decision trees, 637–40
    defined, 632, 661
    and expected value, 634–35, 671–72
Probability distribution, 661–77
    of continuous variables, 661, 677
    of discrete variables, 662–63
    and Excel®, 696
    properties, 673, 677
    and samples, 667–73
    in simulation, 678
Productive hour rate, 510
Profitability index, 238. *See also* Rate of return

Profitability ratios, 719
Profit-and-loss statement. *See* Income statement
Project net-investment, 256–61
Property class, 545–46
Property of independent random variables, 677
Public sector projects, 314–19
    and annual worth, 228
    benefit/cost analysis, 319–24
    BOT contracts, 319
    capitalized cost, 179–85
    estimation, 315, 333, 499
    joint ventures, 318
Purchasing power, 480–81, 482–83
PV function, 26, 59, 709–10
    and bonds, 197
    and NPV function, 708
    and present worth, 349
    and shifted uniform series, 111
    and single payment, 52, 55
    and uniform series present worth, 59
PW *vs. i* graph, 245, 287–91, 595

Q

Q-Solv, 28
Quick ratio, 720

R

r. See Nominal interest rate
RAND function, 683–86, 710
Random numbers, 668 (table)
    generation, 710
Random samples, 666–71, 678
Random variables
    continuous, 660, 661, 664–66, 670–71, 676–77, 679–80, 696
    cumulative distribution, 661–66
    discrete, 660–63, 667–70, 673–76, 679–80, 696
    expected value, 671–73
    probability distribution of, 661, 662–63, 667, 671
    standard deviation, 673–73
Range, 676
Rank and rate technique, 369
Ranking inconsistency, 291
RATE function, 76–77, 245, 710–11

Rate of depreciation
    declining balance, 537
    MACRS, 562–65
    straight line, 533
    sum-of-year digits, 555–56
Rate of return. *See also* Incremental rate of return
    after-tax, 588–94
    and annual worth, 248, 291–92
    on bonds, 261–63
    breakeven, 287, 591
    in capital budgeting, 432–36
    cautions, 248–49
    composite, 255–61, 283
    computer solutions, 245, 247, 248–49, 252, 253, 263
    on debt capital, 357–59
    defined, 13, 240
    determining, 74–76, 242–49
    on equity capital, 359–62
    evaluation method, 350 (table)
    in Excel®, 26, 706, 710–11
    external, 256
    on extra investment, 282–83
    incremental, 278–82
    and independent projects, 283
    and inflation, 14–15, 473, 482–83
    installment financing, 241–42
    internal, 75–76, 255, 706–7
    minimum attractive (*see* Minimum attractive rate
        of return)
    multiple, 248, 249–61, 273–75
    and mutually exclusive alternatives, 278
    and present worth, 242–47, 283–91
    ranking inconsistency, 291
    and reinvestment rate, 256
    and sensitivity analysis, 623
Ratios, accounting, 719–23
Real interest rate, 473, 476, 478, 483
Real property, 533, 541–42
Reasonable estimate, 629–31
Recaptured depreciation
    definition, 582
    in replacement studies, 595–99
    and taxes, 582, 586–87, 606
Recovery period
    defined, 533
    effect on taxes, 580–82
    MACRS, 541, 544, 545–46, 561, 570
    straight line option, 549
Recovery rate. *See* Rate of depreciation
Recurring cash flows, 180
Reinvestment, assumption in capital budgeting, 425–27

Reinvestment rate, 256
Repayment of loans, 20–22
Replacement analysis
    after-tax, 595–99
    annual worth, 390, 392–96, 404–10
    before-tax, 386–410
    cash flow approach, 403
    depreciation recapture, 595
    economic service life, 388, 391–97
    first costs, 388–89
    gains and losses, 595
    market value, 388–90, 392–93
    need for, 388
    one-additional year, 398
    opportunity cost approach, 389, 403
    overview, 398
    and study periods, 404–10
    sunk costs, 389
    unequal-life assets, 403, 404, 410
Replacement life. *See* Economic service life
Replacement value, 403–5
Retained earnings, 29, 352, 360
Retirement life. *See* Economic service life
Return on assets ratio, 721
Return on invested capital, 256
Return on investment (ROI), 14, 238. *See also* Rate
        of return
Return on sales ratio, 721
Revenue alternatives, 172, 279, 293, 591
RIC. *See* Return on invested capital
Risk
    and debt-equity mix, 362–64
    and decision making, 656–59, 660–66
    description, 656
    and MARR, 353, 362
    and payback analysis, 186, 188–89
    and random sampling, 667
RNG function, 696
ROI. *See* Return on investment
Root mean square deviation, 674
ROR. *See* Rate of return
Round-off errors, 53, 103
Rule of 72, 35, 179
Rule of 100, 35
Rule of signs, 249–50, 251, 253, 299

S

s. See Standard deviation
Safe investment, 28-29, 194–95, 360

Sales, return on, 721
Salvage value. *See also* Trade-in value
    and capital recovery, 221
    defined, 10, 220
    and depreciation, 533, 537, 539, 543, 557
    and market value, 388, 392, 397
    and public projects, 319
    and PW over LCM, 175
    in replacement analysis, 386, 388, 391–92
    updating for study period, 349
Sampling, 666–71
    Monte Carlo, 677–84, 685–88
Savings, tax, 576, 595
Scatter charts. *See* xy Excel® charts
Screening projects, 186, 188–89
Section 179 property, 533–34
Section 1231 transactions, 583
Securities, defined, 361
Selection of evaluation method, 348–51
Sensitivity analysis. *See also* Breakeven analysis;
        SOLVER
    approach, 7–8, 623
    and Excel® cell referencing, 247, 698–99
    of one parameter, 622–29
    with three estimates, 629–31
    two alternatives, 626–29
Service alternatives, 172, 279
    and incremental rate of return, 591
Service life, 174–75. *See also* Economic
        service life
Shifted gradients. *See* Gradients, arithmetic
Shifted series, 94–103
Sign changes, number of, 249–54
Simple cash flow series, 249
Simple interest, 17–18, 36–37
Simulation, 659, 678–86
Single-payment compound amount factor, 50
Single payment factors, 50–56
Single-payment present worth factor, 51–52
Sinking fund (A/F) factor, 60–63
SLN function, 535, 711
Social discount rate, 316, 333
Solvency ratios, 719
SOLVER, 434–36, 455–59, 712–13
Spreadsheet. *See also* Excel®
    absolute cell referencing, 247, 703
    annual worth, 225, 227–28
    and B/C analysis, 323, 330–31
    and breakeven analysis, 451–52
    and CFAT with depreciation, 578–79
    and EVA™, 601–3

and independent projects, 430, 434–36
and inflation, 480
and present worth, 197–202
and rate of return, 75–77, 247, 252, 263, 287, 289,
        297–300
replacement analysis, 395–97, 402–5
    after-tax, 598–99
and sensitivity analysis, 625–26, 628–29
and shifted uniform series, 110–11
for simple and compound interest, 36–39
    incremental, after tax, 592–95
Standard deviation
    for continuous variable, 676–77
    definition, 673–74
    for discrete variable, 673–76
Standard normal distribution, 688–89
Stocks
    CAPM model, 360–61
    common, 352, 359
    in equity financing, 352, 359
    preferred, 352, 359
Straight line alternative, in MACRS,
        541–42, 549
Straight line depreciation, 535–36, 580–86.
        See also Alternatives
Straight line rate, 533, 535
Study period
    and annual worth, 218, 226–28
    and capital budgeting, 425
    equal service, 173, 427
    and FW analysis, 177
    and PW evaluation, 174–77
    and replacement analysis, 390, 400, 404–10
    and salvage value, 220
    spreadsheet example, 197–200
SUM function, 703
Sum-of-year digits depreciation, 533,
        535–57, 711
Sunk costs, 389
SYD function, 556, 711
Symbols, 23
System, phases of, 190–91

T

Tables
    compound interest factors, 727–55
    effective interest rates, 133
    interpolation, 63–65
Tax depreciation, 532

Taxable income, 570–71, 583
  and CFAT, 577
  and depreciation, 572
  negative, 577–78
  and taxes, 572
Taxes. *See also* After-tax; Income tax; Taxable income
  and debt capital, 357–59
  and depreciation, 530
  and equity capital, 359–60
  and MARR, 353–54
  rates, 573, 575
Time
  interest rates over, 151–53
  units of, 23
Time placement of dollars, 31–32, 472–73
Time value of money, 9, 127. *See also* Inflation
  and capitalized cost, 179–80
  and equivalence, 15–17
  and no-return payback, 187
Total cost relation, 445. *See also* Breakeven analysis
Trade-in value, 389, 533. *See also* Market value;
      Salvage value
Treasury securities, 194–95, 315
Triangular distribution, 663, 665–66

U

Unadjusted basis, 532
Uncertainty, 656, 659
Uniform distribution, 663, 664, 670
Uniform gradient. *See* Gradient, arithmetic
Uniform percentage method. *See* Declining balance
      depreciation
Uniform series
  compound amount factor, 60–63
  compounding period greater than payment period,
      147–49
  compounding period less than payment period,
      142–46
  description, 23
  present worth factors, 56–58, 80–81, 114
  shifted, 94, 95, 110–14
Unit method, 498–99
Unknown interest rate, 74–76
Unknown years (life), 77–78
Unrecovered balance, 240–41, 255–56

Unrecovered capital. *See* Sunk costs
Usage stage, 191

V

Value, resale, 10. *See also* Salvage value; Trade-in value
Value added analysis, after tax, 599–603. *See also*
      Economic value added
Variable. *See* Random variables
Variable costs, 444–48
Variance
  in cost allocation, 509–11
  definition, 673
  formula for, 674, 677
  and normal distribution, 688
VDB function, 558, 561–62, 712

W

WACC. *See* Weighted average cost of capital
Websites, IRS, 568
  Blank and Tarquin, xviii
Weighted attribute method, 371–72
Weighted average cost of capital, 29–30, 355–57
Working capital, 720
Worth, measures of, 9, 678

X

xy Excel® charts, 247, 700–701

Y

Year(s). *See also* Half-year convention;
      One-additional-year replacement studies;
      Sum-of-year digits depreciation
  and end-of-period convention, 31–32
  fiscal *versus* calendar, 716
  symbols, 23
  unknown, 77–78

Z

Zero cash flow in Excel® functions, 111, 198, 706, 708

Format for Commonly Used Spreadsheet Functions on Excel©

Present worth:

    PV(i%,n,A,F)   for constant A series

    NPV(i%,second_cell:last_cell) + first_cell   for varying cash flow series

Future worth:

    FV(i%,n,A,P)   for constant A series

Annual worth:

    PMT(i%,n,P,F)   for single amounts with no A series

Number of periods (years):

    NPER(i%,A,P,F)   for constant A series

(Note: The PV, FV, and PMT functions change the sense of the sign. Place a minus in front of the function to retain the same sign. The NPV and IRR functions take the sign from the tabulated cash flows.)

Rate of return:

    RATE(n,A,P,F)   for constant A series

    IRR(first_cell:last_cell)   for varying cash flow series

Depreciation:

    SLN(P,S,n)   Straight line depreciation for each period

    DDB(P,S,n,t,d)   Double declining balance depreciation for period t at rate d (optional)

    DB(P,S,n,t)   Declining balance, rate determined by the function

    SYD(P,S,n,t)   Sum-of-year-digits depreciation for period t

Logical IF function:

    IF (logical_test,value_if_true,value_if_false)   For logical two-branch operations

One function may be imbedded into another function.

All functions must be preceded by an = sign.

Relations for Discrete Cash Flows with End-of-Period Compounding

Type	Find/Given	Factor Notation and Formula	Relation	Sample Cash Flow Diagram
Single Amount	F/P Compound amount	$(F/P,i,n) = (1 + i)^n$	$F = P(F/P,i,n)$	
	P/F Present worth	$(P/F,i,n) = \dfrac{1}{(1 + i)^n}$	$P = F(P/F,i,n)$ (Sec. 2.1)	
Uniform Series	P/A Present worth	$(P/A,i,n) = \dfrac{(1 + i)^n - 1}{i(1 + i)^n}$	$P = A(P/A,i,n)$	
	A/P Capital recovery	$(A/P,i,n) = \dfrac{i(1 + i)^n}{(1 + i)^n - 1}$	$A = P(A/P,i,n)$ (Sec. 2.2)	
	F/A Compound amount	$(F/A,i,n) = \dfrac{(1 + i)^n - 1}{i}$	$F = A(F/A,i,n)$	
	A/F Sinking fund	$(A/F,i,n) = \dfrac{i}{(1 + i)^n - 1}$	$A = F(A/F,i,n)$ (Sec. 2.3)	
Arithmetic Gradient	P_G/G Present worth	$(P/G,i,n) = \dfrac{(1 + i)^n - in - 1}{i^2(1 + i)^n}$	$P_G = G(P/G,i,n)$	
	A_G/G Uniform series	$(A/G,i,n) = \dfrac{1}{i} - \dfrac{n}{(1 + i)^n - 1}$	$A_G = G(A/G,i,n)$ (Sec. 2.5)	
Geometric Gradient	P_g/A_1 and g Present worth	$P_g = \begin{cases} \dfrac{A_1\left[1 - \left(\dfrac{1 + g}{1 + i}\right)^n\right]}{i - g} & g \neq i \\[4mm] A_1\dfrac{n}{1 + i} & g = i \end{cases}$	(Sec. 2.6)	

Comparison of Mutually Exclusive Alternatives Using Different Evaluation Methods

Evaluation Method	Equivalence Relation	Lives of Alternatives	Time Period for Analysis	Evaluate This Series	Interest Rate	Select* Alternative With	Section Reference
Present Worth	PW	Equal	Lives	Cash flows	MARR	Best PW	5.2
	PW	Unequal	LCM	Cash flows	MARR	Best PW	5.3
	PW	Study period	Study period	Updated cash flows	MARR	Best PW	5.3
	Capitalized cost (CC)	Long to infinite	Infinity	Cash flows	MARR	Best CC	5.5
Future Worth	FW		Same as present worth for equal lives, unequal lives, and study period				5.4
Annual Worth	AW	Equal or unequal	Lives	Cash flows	MARR	Best AW	6.3
	AW	Study period	Study period	Updated cash flows	MARR	Best AW	6.3
	AW	Long to infinite	Infinity	Cash flows	MARR	Best AW	6.4
Rate of Return	PW or AW	Equal	Lives	Incremental cash flows	Find Δi^*	Last $\Delta i^* >$ MARR	8.4
	PW or AW	Unequal	LCM of pair	Incremental cash flows	Find Δi^*	Last $\Delta i^* >$ MARR	8.4
	AW	Unequal	Lives	Cash flows	Find Δi^*	Last $\Delta i^* >$ MARR	8.5
	PW or AW	Study period	Study period	Updated incremental cash flows	Find Δi^*	Last $\Delta i^* >$ MARR	8.4
Benefit/ Cost	AW	Equal or unequal	Lives	Incremental cash flows	Discount rate	Last ΔB/C > 1.0	9.3
	AW or PW	Long to infinite	Infinity	Incremental cash flows	Discount rate	Last ΔB/C > 1.0	9.3
	PW	Equal or unequal	LCM of pairs	Incremental cash flows	Discount rate	Last ΔB/C > 1.0	9.3

* Alternative with the numerically largest value has the lowest equivalent cost or highest equivalent income.